T0137925

Grundlehren der mathematischen Wissenschaften   349

A Series of Comprehensive Studies in Mathematics

For further volumes:
www.springer.com/series/138

Peter Bürgisser · Felipe Cucker

Condition

The Geometry of Numerical Algorithms

 Springer

Peter Bürgisser
Institut für Mathematik
Technische Universität Berlin
Berlin, Germany

Felipe Cucker
Department of Mathematics
City University of Hong Kong
Hong Kong, Hong Kong SAR

ISSN 0072-7830   Grundlehren der mathematischen Wissenschaften
ISBN 978-3-642-44012-0   ISBN 978-3-642-38896-5 (eBook)
DOI 10.1007/978-3-642-38896-5
Springer Heidelberg New York Dordrecht London

Mathematics Subject Classification (2010): 15A12, 52A22, 60D05, 65-02, 65F22, 65F35, 65G50, 65H04, 65H10, 65H20, 90-02, 90C05, 90C31, 90C51, 90C60, 68Q25, 68W40, 68Q87

Printed on acid-free paper

Springer is part of Springer Science+Business Media (www.springer.com)

Dedicated to the memory of
Walter Bürgisser and Gritta Bürgisser-Glogau
and of
Federico Cucker and Rosemary Farkas
in love and gratitude

Preface

Motivation   A combined search at `Mathscinet` and `Zentralblatt` shows more than 800 articles with the expression "condition number" in their title. It is reasonable to assume that the number of articles dealing with conditioning, in one way or another, is a substantial multiple of this quantity. This is not surprising. The occurrence of condition numbers in the accuracy analysis of numerical algorithms is pervasive, and its origins are tied to those of the digital computer. Indeed, the expression "condition number" itself was first introduced in 1948, in a paper by Alan M. Turing in which he studied the propagation of errors for linear equation solving with the then nascent computing machinery [221]. The same subject occupied John von Neumann and Herman H. Goldstine, who independently found results similar to those of Turing [226]. Ever since then, condition numbers have played a leading role in the study of both accuracy and complexity of numerical algorithms.

To the best of our knowledge, and in stark contrast to this prominence, there is no book on the subject of conditioning. Admittedly, most books on numerical analysis have a section or chapter devoted to it. But their emphasis is on algorithms, and the links between these algorithms and the condition of their data are not pursued beyond some basic level (for instance, they contain almost no instances of probabilistic analysis of algorithms via such analysis for the relevant condition numbers).

Our goal in writing this book has been to fill this gap. We have attempted to provide a unified view of conditioning by making condition numbers the primary object of study and by emphasizing the many aspects of condition numbers in their relation to numerical algorithms.

Structure   The book is divided into three parts, which approximately correspond to themes of conditioning in linear algebra, linear programming, and polynomial equation solving, respectively. The increase in technical requirements for these subjects is reflected in the different paces for their expositions. Part I proceeds leisurely and can be used for a semester course at the undergraduate level. The tempo increases in Part II and reaches its peak in Part III with the exposition of the recent advances in and partial solutions to the 17th of the problems proposed by Steve Smale for the mathematicians of the 21st century, a set of results in which conditioning plays a paramount role [27, 28, 46].

As in a symphonic poem, these changes in cadence underlie a narration in which, as mentioned above, condition numbers are the main character. We introduce them, along with the cast of secondary characters making up the *dramatis personae* of this narration, in the *Overture* preceding Part I.

We mentioned above that Part I can be used for a semester course at the undergraduate level. Part II (with some minimal background from Part I) can be used as an undergraduate course as well (though a notch more advanced). Briefly stated, it is a "condition-based" exposition of linear programming that, unlike more elementary accounts based on the simplex algorithm, sets the grounds for similar expositions of convex programming. Part III is also a course on its own, now on computation with polynomial systems, but it is rather at the graduate level.

Overlapping with the primary division of the book into its three parts there is another taxonomy. Most of the results in this book deal with condition numbers of specific problems. Yet there are also a few discussions and general results applying either to condition numbers in general or to large classes of them. These discussions are in most of the *Overture*, the two *Intermezzi* between parts, Sects. 6.1, 6.8, 9.5, and 14.3, and Chaps. 20 and 21. Even though few, these pages draft a general theory of condition, and most of the remainder of the book can be seen as worked examples and applications of this theory.

The last structural attribute we want to mention derives from the technical characteristics of our subject, which prominently features probability estimates and, in Part III, demands some nonelementary geometry. A possible course of action in our writing could have been to act like Plato and deny access to our edifice to all those not familiar with geometry (and, in our case, probabilistic analysis). We proceeded differently. Most of the involved work in probability takes the form of estimates—of either distributions' tails or expectations—for random variables in a very specific context. We therefore included within the book a *Crash Course on Probability* providing a description of this context and the tools we use to compute these estimates. It goes without saying that probability theory is vast, and alternative choices in its toolkit could have been used as well. A penchant for brevity, however, prevented us to include these alternatives. The course is supplied in installments, six in total, and contains the proofs of most of its results. Geometry requirements are of a more heterogeneous nature, and consequently, we have dealt with them differently. Some subjects, such as Euclidean and spherical convexity, and the basic properties of projective spaces, are described in detail within the text. But we could not do so with the basic notions of algebraic, differential, and integral geometry. We therefore collected these notions in an appendix, providing only a few proofs.

Paderborn, Germany   Peter Bürgisser
Hong Kong, Hong Kong SAR   Felipe Cucker
May 2013

Acknowledgements

A substantial part of the material in this book formed the core of several graduate courses taught by PB at the University of Paderborn. Part of the material was also used in a graduate course at the Fields Institute held in the fall of 2009. We thank all the participants of these courses for valuable feedback. In particular, Dennis Amelunxen, Christian Ikenmeyer, Stefan Mengel, Thomas Rothvoss, Peter Scheiblechner, Sebastian Schrage, and Martin Ziegler, who attended the courses in Paderborn, had no compassion in pointing to the lecturer the various forms of typos, redundancies, inaccuracies, and plain mathematical mistakes that kept popping up in the early drafts of this book used as the course's main source. We thank Dennis Amelunxen for producing a first LATEX version of the lectures in Paderborn, which formed the initial basis of the book. In addition, Dennis was invaluable in producing the TikZ files for the figures occurring in the book.

Also, Diego Armentano, Dennis Cheung, Martin Lotz, and Javier Peña read various chapters and have been pivotal in shaping the current form of these chapters. We have pointed out in the Notes the places where their input is most notable.

Finally, we want to emphasize that our viewpoint about conditioning and its central role in the foundations of numerical analysis evolved from hours of conversations and exchange of ideas with a large group of friends working in similar topics. Among them it is impossible not to mention Carlos Beltrán, Lenore Blum, Irenée Briquel, Jean-Pierre Dedieu, Alan Edelman, Raphael Hauser, Gregorio Malajovich, Luis Miguel Pardo, Jim Renegar, Vera Roshchina, Michael Shub, Steve Smale, Henryk Woźniakowski, and Mario Wschebor. We are greatly indebted to all of them.

The financial support of the German Research Foundation (individual grants BU 1371/2-1 and 1371/2-2) and the GRF (grant CityU 100810) is gratefully acknowledged. We also thank the Fields Institute in Toronto for hospitality and financial support during the thematic program on the Foundations of Computational Mathematics in the fall of 2009, where a larger part of this monograph took definite form.

We thank the staff at Springer-Verlag in Basel and Heidelberg for their help and David Kramer for the outstanding editing work he did on our manuscript.

Finally, we are grateful to our families for their support, patience, and understanding of the commitment necessary to carry out such a project while working on different continents.

Contents

Overture: On the Condition of Numerical Problems

O.1 The Size of Errors

> *Since none of the numbers we take out from logarithmic or trigonometric tables admit of absolute precision, but are all to a certain extent approximate only, the results of all calculations performed by the aid of these numbers can only be approximately true. [...] It may happen, that in special cases the effect of the errors of the tables is so augmented that we may be obliged to reject a method, otherwise the best, and substitute another in its place.*
>
> Carl Friedrich Gauss, *Theoria Motus*

The heroes of numerical mathematics (Euler, Gauss, Lagrange, ...) developed a good number of the algorithmic procedures which constitute the essence of numerical analysis. At the core of these advances was the invention of calculus. And underlying the latter, the field of real numbers.

The dawn of the digital computer, in the decade of the 1940s, allowed the execution of these procedures on increasingly large data, an advance that, however, made even more patent the fact that real numbers cannot be encoded with a finite number of bits and therefore that computers had to work with approximations only. With the increased length of computations, the systematic rounding of all occurring quantities could now accumulate to a greater extent. Occasionally, as already remarked by Gauss, the errors affecting the outcome of a computation were so big as to make it irrelevant.

Expressions like "the error is big" lead to the question, how does one measure an error? To approach this question, let us first assume that the object whose error we are considering is a single number x encoding a quantity that may take values on an open real interval. An error of magnitude 1 may yield another real number $\tilde{x}$ with value either $x - 1$ or $x + 1$. Intuitively, this will be harmless or devastating depending on the magnitude of x itself. Thus, for $x = 10^6$, the error above is hardly noticeable, but for $x = 10^{-3}$, it certainly is (and may even change basic features of

x such as being positive). A relative measure of the error appears to convey more meaning. We therefore define[1]

$$\mathsf{RelError}(x) = \frac{|\tilde{x} - x|}{|x|}.$$

Note that this expression is well defined only when $x \neq 0$.

How does this measure extend to elements $x \in \mathbb{R}^m$? We want to consider relative errors as well, but how does one relativize? There are essentially two ways:

Componentwise: Here we look at the relative error in each component, taking as error for x the maximum of them. That is, for $x \in \mathbb{R}^m$ such that $x_i \neq 0$ for $i = 1, \ldots, m$, we define

$$\mathsf{RelError}(x) = \max_{i \leq m} \mathsf{RelError}(x_i).$$

Normwise: Endowing $\mathbb{R}^m$ with a norm allows one to mimic, for $x \neq 0$, the definition for the scalar case. We obtain

$$\mathsf{RelError}(x) = \frac{\|\tilde{x} - x\|}{\|x\|}.$$

Needless to say, the normwise measure depends on the choice of the norm.

O.2 The Cost of Erring

How do round-off errors affect computations? The answer to this question depends on a number of factors: the problem being solved, the data at hand, the algorithm used, the machine precision (as well as other features of the computer's arithmetic). While it is possible to consider all these factors together, a number of idealizations leading to the consideration of simpler versions of our question appears as a reasonable—if not necessary—course of action. The notion of condition is the result of some of these idealizations. More specifically, assume that the problem being solved can be described by a function

$$\varphi : \mathcal{D} \subseteq \mathbb{R}^m \to \mathbb{R}^q,$$

where $\mathcal{D}$ is an open subset of $\mathbb{R}^m$. Assume as well that the computation of φ is performed by an algorithm with infinite precision (that is, there are no round-off errors during the execution of this algorithm). All errors in the computed value arise as a consequence of possible errors in reading the input (which we will call *perturbations*). Our question above then takes the following form:

How large is the output error with respect to the input perturbation?

---

[1] To be completely precise, we should write $\mathsf{RelError}(x, \tilde{x})$. In all what follows, however, to simplify notation, we will omit the perturbation $\tilde{x}$ and write simply $\mathsf{RelError}(x)$.

The *condition number* of input $a \in \mathcal{D}$ (with respect to problem φ) is, roughly speaking, the worst possible magnification of the output error with respect to a small input perturbation. More formally,

$$\mathrm{cond}^{\varphi}(a) = \lim_{\delta \to 0} \sup_{\mathrm{RelError}(a) \leq \delta} \frac{\mathrm{RelError}(\varphi(a))}{\mathrm{RelError}(a)}. \qquad (O.1)$$

This expression defines the condition number as a limit. For small values of δ we can consider the approximation

$$\mathrm{cond}^{\varphi}(a) \approx \sup_{\mathrm{RelError}(a) \leq \delta} \frac{\mathrm{RelError}(\varphi(a))}{\mathrm{RelError}(a)}$$

and, for practical purposes, the approximate bound

$$\mathrm{RelError}(\varphi(a)) \lesssim \mathrm{cond}^{\varphi}(a)\mathrm{RelError}(a), \qquad (O.2)$$

or yet, using "little oh" notation[2] for $\mathrm{RelError}(a) \to 0$,

$$\mathrm{RelError}(\varphi(a)) \leq \mathrm{cond}^{\varphi}(a)\mathrm{RelError}(a) + o(\mathrm{RelError}(a)). \qquad (O.3)$$

Expression (O.1) defines a family of condition numbers for the pair (φ, a). Errors can be measured either componentwise or normwise, and in the latter case, there is a good number of norms to choose from. The choice of normwise or componentwise measures for the errors has given rise to three kinds of condition numbers (condition numbers for normwise perturbations and componentwise output errors are not considered in the literature).

		PERTURBATION	
		normwise	componentwise
OUTPUT	normwise	*normwise*	*mixed*
ERROR	componentwise		*componentwise*

We will generically denote normwise condition numbers by $\mathrm{cond}^{\varphi}(a)$, mixed condition numbers by $M^{\varphi}(a)$, and componentwise condition numbers by $Cw^{\varphi}(a)$. We may skip the superscript φ if it is clear from the context. In the case of componentwise condition numbers one may be interested in considering the relative error for each of the output components separately. Thus, for $j \leq q$ one defines

$$Cw_{j}^{\varphi}(a) = \lim_{\delta \to 0} \sup_{\mathrm{RelError}(a) \leq \delta} \frac{\mathrm{RelError}(\varphi(a)_{j})}{\mathrm{RelError}(a)},$$

and one has $Cw^{\varphi}(a) = \max_{j \leq q} Cw_{j}^{\varphi}(a)$.

---

[2] A short description of the little oh and other asymptotic notations is in the Appendix, Sect. A.1.

The consideration of a normwise, mixed, or componentwise condition number will be determined by the characteristics of the situation at hand. To illustrate this, let's look at data perturbation. The two main reasons to consider such perturbations are inaccurate data reading and backward-error analysis.

In the first case the idea is simple. We are given data that we know to be inaccurate. This may be because we obtained it by measurements with finite precision (e.g., when an object is weighed, the weight is displayed with a few digits only) or because our data are the result of an inaccurate computation.

The idea of backward error analysis is less simple (but very elegant). For a problem φ we may have many algorithms that solve it. While all of them ideally compute φ when endowed with infinite precision, under the presence of errors they will compute only approximations of this function. At times, for a problem φ and a finite-precision algorithm $\mathcal{A}^\varphi$ solving it, it is possible to show that for all $a \in \mathcal{D}$ there exists $e \in \mathbb{R}^m$ with $a + e \in \mathcal{D}$ satisfying

$$(*) \quad \mathcal{A}^\varphi(a) = \varphi(a + e), \quad \text{and}$$

$$(**) \quad e \text{ is small with respect to } a.$$

In this situation—to which we refer by saying that $\mathcal{A}^\varphi$ is *backward-stable*—information on how small exactly e is (i.e., how large $\mathsf{RelError}(a)$ is) together with the condition number of a directly yields bounds on the error of the computed quantity $\mathcal{A}^\varphi(a)$. For instance, if $(**)$ above takes the form

$$\|e\| \leq m^3 10^{-6} \|a\|,$$

we will deduce, using (O.2), that

$$\left\| \mathcal{A}^\varphi(a) - \varphi(a) \right\| \lesssim \mathsf{cond}^\varphi(a) m^3 10^{-6} \left\| \varphi(a) \right\|. \tag{O.4}$$

No matter whether due to inaccurate data reading or because of a backward-error analysis, we will measure the perturbation of a in accordance with the situation at hand. If, for instance, we are reading data in a way that each component a_i satisfies $\mathsf{RelError}(a_i) \leq 5 \times 10^{-8}$, we will measure perturbations in a componentwise manner. If, in contrast, a backward-error analysis yields an e satisfying $\|e\| \leq m^3 \|a\| 10^{-6}$, we will have to measure perturbations in a normwise manner.

While we may have more freedom in the way we measure the output error, there are situations in which a given choice seems to impose itself. Such a situation could arise when the outcome of the computation at hand is going to be the data of another computation. If perturbations of the latter are measured, say, componentwise, we will be interested in doing the same with the output error of the former. A striking example in which error analysis can be only appropriately explained using componentwise conditioning is the solution of triangular systems of equations. We will return to this issue in Chap. 3.

At this point it is perhaps convenient to emphasize a distinction between *condition* and (backward) *stability*. Given a problem φ, the former is a property of the input only. That is, it is independent on the possible algorithms used to compute φ.

In contrast, backward stability, at least in the sense defined above, is a property of an algorithm $\mathcal{A}^\varphi$ computing φ that holds for all data $a \in \mathcal{D}$ (and is therefore independent of particular data instances).

Expressions like (O.4) are known as forward-error analyses, and algorithms $\mathcal{A}^\varphi$ yielding a small value of $\frac{\|\mathcal{A}^\varphi(a)-\varphi(a)\|}{\|\varphi(a)\|}$ are said to be *forward-stable*. It is important to mention that while backward-error analyses immediately yield forward-error bounds, some problems do not admit backward-error analysis, and therefore, their error analysis must be carried forward.

It is time to have a closer look at the way errors are produced in a computer.

O.3 Finite-Precision Arithmetic and Loss of Precision

O.3.1 Precision ...

Although the details of computer arithmetic may vary with computers and software implementations, the basic idea was agreed upon shortly after the dawn of digital computers. It consisted in fixing positive integers $\beta \geq 2$ (the *basis* of the representation), t (its *precision*), and e_0, and approximating nonzero real numbers by rational numbers of the form

$$z = \pm \frac{m}{\beta^t} \beta^e$$

with $m \in \{1, \ldots, \beta^t\}$ and $e \in \{-e_0, \ldots, e_0\}$. The fraction $\frac{m}{\beta^t}$ is called the *mantissa* of z and the integer e its *exponent*. The condition $|e| \leq e_0$ sets limits on how big (and how small) z may be. Although these limits may give rise to situations in which (the absolute value of) the number to be represented is too large (*overflow*) or too small (*underflow*) for the possible values of z, the value of e_0 in most implementations is large enough to make these phenomena rare in practice. Idealizing a bit, we may assume $e_0 = \infty$.

As an example, taking $\beta = 10$ and $t = 12$, we can approximate

$$\pi^8 \approx 0.948853101607 \times 10^4.$$

The relative error in this approximation is bounded by 1.1×10^{-12}. Note that t is the number of correct digits of the approximation. Actually, for any real number x, by appropriately rounding and truncating an expansion of x we can obtain a number $\tilde{x}$ as above satisfying $\tilde{x} = x(1 + \delta)$ with $|\delta| \leq \frac{\beta^{-t+1}}{2}$. That is,

$$\text{RelError}(x) \leq \frac{\beta^{-t+1}}{2}.$$

More generally, whenever a real number x is approximated by $\tilde{x}$ satisfying an inequality like the one above, we say that $\tilde{x}$ *approximates* x *with* t *correct digits.*[3]

Leaving aside the details such as the choice of basis and the particular way a real number is truncated to obtain a number as described above, we may summarize the main features of computer arithmetic (recall that we assume $e_0 = \infty$) by stating the existence of a subset $\mathbb{F} \subset \mathbb{R}$ containing 0 (the *floating-point numbers*), a *rounding map* round : $\mathbb{R} \to \mathbb{F}$, and a *round-off unit* (also called *machine epsilon*) $0 < \epsilon_{\mathrm{mach}} < 1$, satisfying the following properties:

(a) For any $x \in \mathbb{F}$, $\mathrm{round}(x) = x$. In particular $\mathrm{round}(0) = 0$.
(b) For any $x \in \mathbb{R}$, $\mathrm{round}(x) = x(1 + \delta)$ with $|\delta| \le \epsilon_{\mathrm{mach}}$.

Furthermore, one can take $\epsilon_{\mathrm{mach}} = \frac{\beta^{-t+1}}{2}$ and therefore $|\log_\beta \epsilon_{\mathrm{mach}}| = t - \log_\beta \frac{\beta}{2}$.

Arithmetic operations on $\mathbb{F}$ are defined following the scheme

$$x \, \tilde{\circ} \, y = \mathrm{round}(x \circ y)$$

for any $x, y \in \mathbb{F}$ and $\circ \in \{+, -, \times, /\}$ and therefore

$$\tilde{\circ} : \mathbb{F} \times \mathbb{F} \to \mathbb{F}.$$

It follows from (b) above that for any $x, y \in \mathbb{F}$ we have

$$x \, \tilde{\circ} \, y = (x \circ y)(1 + \delta), \quad |\delta| \le \epsilon_{\mathrm{mach}}.$$

Other operations may also be considered. Thus, a floating-point version $\tilde{\sqrt{\ }}$ of the square root would similarly satisfy

$$\tilde{\sqrt{x}} = \sqrt{x}(1 + \delta), \quad |\delta| \le \epsilon_{\mathrm{mach}}.$$

When combining many operations in floating-point arithmetic, expressions such as $(1 + \delta)$ above naturally appear. To simplify round-off analyses it is useful to consider the quantities, for $k \ge 1$ and $k\epsilon_{\mathrm{mach}} < 1$,

$$\gamma_k := \frac{k\epsilon_{\mathrm{mach}}}{1 - k\epsilon_{\mathrm{mach}}} \tag{O.5}$$

and to denote by θ_k *any* number satisfying $|\theta_k| \le \gamma_k$. In this sense, θ_k represents a set of numbers, and different occurrences of θ_k in a proof may denote different numbers. Note that

$$\gamma_k \le (k + 1)\epsilon_{\mathrm{mach}} \quad \text{if } k(k+1) \le \epsilon_{\mathrm{mach}}^{-1}. \tag{O.6}$$

The proof of the following proposition can be found in Chap. 3 of [121].

Proposition O.1 *The following relations hold (assuming all quantities are well defined):*

---

[3]This notion reflects the intuitive idea of significant figures modulo carry differences. The number 0.9999 approximates 1 with a precision $t = 10^{-4}$. Yet their first significant digits are different.

(a) $(1 + \theta_k)(1 + \theta_j) = 1 + \theta_{k+j}$,

(b)

$$\frac{1 + \theta_k}{1 + \theta_j} = \begin{cases} 1 + \theta_{k+j} & \text{if } j \leq k, \\ 1 + \theta_{k+2j} & \text{if } j > k, \end{cases}$$

(c) $\gamma_k \gamma_j \leq \gamma_{\min\{k,j\}}$ if $\max\{k\epsilon_{\mathrm{mach}}, j\epsilon_{\mathrm{mach}}\} \leq 1/2$,

(d) $i\gamma_k \leq \gamma_{ik}$,

(e) $\gamma_k + \epsilon_{\mathrm{mach}} \leq \gamma_{k+1}$,

(f) $\gamma_k + \gamma_j + \gamma_k\gamma_j \leq \gamma_{k+j}$. □

O.3.2 ... and the Way We Lose It

In computing an arithmetic expression q with a round-off algorithm, errors will accumulate, and we will obtain another quantity, which we denote by $\mathrm{fl}(q)$. We will also write $\mathrm{Error}(q) = |q - \mathrm{fl}(q)|$, so that $\mathrm{RelError}(q) = \frac{\mathrm{Error}(q)}{|q|}$.

Assume now that q is computed with a real-number algorithm $\mathcal{A}$ executed using floating-point arithmetic from data a (a formal model for real-number algorithms was given in [37]). No matter how precise the representation we are given of the entries of a, these entries will be rounded to t digits. Hence t (or, being roughly the same, $|\log_\beta \epsilon_{\mathrm{mach}}|$) is the precision of our data. On the other hand, the number of correct digits in $\mathrm{fl}(q)$ is approximately $-\log_\beta \mathrm{RelError}(q)$. Therefore, the value

$$\mathrm{LoP}(q) := \log_\beta \frac{\mathrm{RelError}(q)}{\epsilon_{\mathrm{mach}}} = |\log_\beta \epsilon_{\mathrm{mach}}| - |\log_\beta \mathrm{RelError}(q)|$$

quantifies the *loss of precision* in the computation of q. To extend this notion to the computation of vectors $v = (v_1, \ldots, v_q) \in \mathbb{R}^q$, we need to fix a measure for the precision of the computed $\mathrm{fl}(e) = (\mathrm{fl}(v_1), \ldots, \mathrm{fl}(v_q))$: componentwise or normwise.

In the componentwise case, we have

$$-\log_\beta \mathrm{RelError}(e) = -\log_\beta \max_{i \leq q} \frac{|\mathrm{fl}(v_i) - v_i|}{|v_i|} = \min_{i \leq q} \left(-\log_\beta \frac{|\mathrm{fl}(v_i) - v_i|}{|v_i|} \right),$$

so that the precision of v is the smallest of the precisions of its components.

For the normwise measure, we take the precision of v to be

$$-\log_\beta \mathrm{RelError}(e) = -\log_\beta \frac{\|\mathrm{fl}(e) - v\|}{\|v\|}.$$

This choice has both the pros and cons of viewing v as a whole and not as the aggregation of its components.

For both the componentwise and the normwise measures we can consider ϵ_{mach} as a measure of the worst possible relative error $\mathrm{RelError}(a)$ when we read data a with round-off unit ϵ_{mach}, since in both cases

$$\max_{|\tilde{a}_i - a_i| \leq \epsilon_{\mathrm{mach}}|a_i|} \mathrm{RelError}(a) = \epsilon_{\mathrm{mach}}.$$

Hence, $|\log_\beta \epsilon_{\text{mach}}|$ represents in both cases the precision of the data. We therefore define the loss of precision in the computation of $\varphi(a)$ to be

$$\text{LoP}(\varphi(a)) := \log_\beta \frac{\text{RelError}(\varphi(a))}{\epsilon_{\text{mach}}} = |\log_\beta \epsilon_{\text{mach}}| + \log_\beta \text{RelError}(\varphi(a)). \quad (O.7)$$

Remark O.2 By associating $\text{RelError}(a) \approx \epsilon_{\text{mach}}$, we may view the logarithm of a condition number $\log_\beta \text{cond}^\varphi(a)$ as a measure of the worst possible loss of precision in a computation of $\varphi(a)$ in which the only error occurs in reading the data.

To close this section we prove a result putting together—and making precise—a number of issues dealt with so far. For data $a \in \mathcal{D} \subseteq \mathbb{R}^m$ we call m the *size* of a and we write $\text{size}(a) = m$. Occasionally, this size is a function of a few integers, the *dimensions* of a, the set of which we denote by $\text{dims}(a)$. For instance, a $p \times q$ matrix has dimensions p and q and size pq.

Theorem O.3 *Let $\mathcal{A}^\varphi$ be a finite-precision algorithm with round-off unit ϵ_{mach} computing a function $\varphi : \mathcal{D} \subseteq \mathbb{R}^m \to \mathbb{R}^q$. Assume $\mathcal{A}^\varphi$ satisfies the following backward bound: for all $a \in \mathcal{D}$ there exists $\tilde{a} \in \mathcal{D}$ such that*

$$\mathcal{A}^\varphi(a) = \varphi(\tilde{a})$$

and

$$\text{RelError}(a) \leq f(\text{dims}(a))\epsilon_{\text{mach}} + o(\epsilon_{\text{mach}})$$

for some positive function f, and where the "little oh" is for $\epsilon_{\text{mach}} \to 0$. Then the computed $\mathcal{A}^\varphi(a)$ satisfies the forward bound

$$\text{RelError}(\varphi(a)) \leq f(\text{dims}(a))\text{cond}^\varphi(a)\epsilon_{\text{mach}} + o(\epsilon_{\text{mach}}),$$

and the loss of precision in the computation (in base β) is bounded as

$$\text{LoP}(\varphi(a)) \leq \log_\beta f(\text{dims}(a)) + \log_\beta \text{cond}^\varphi(a) + o(1).$$

Here cond^φ refers to the condition number defined in (O.1) with the same measures (normwise or componentwise) for $\text{RelError}(a)$ and $\text{RelError}(\varphi(a))$ as those in the backward and forward bounds above, respectively.

Proof The forward bound immediately follows from the backward bound and (O.3). For the loss of precision we have

$$\log_\beta \text{RelError}(\varphi(a)) \leq \log_\beta f(\text{dims}(a))\text{cond}^\varphi(a)\epsilon_{\text{mach}}(1 + o(1))$$

$$\leq \log_\beta f(\text{dims}(a)) + \log_\beta \text{cond}^\varphi(a) - |\log_\beta \epsilon_{\text{mach}}| + o(1),$$

from which the statement follows. $\square$

O.4 An Example: Matrix–Vector Multiplication

It is perhaps time to illustrate the notions introduced so far by analyzing a simple problem, namely, matrix–vector multiplication. We begin with a (componentwise) backward stability analysis.

Proposition O.4 *There is a finite-precision algorithm $\mathcal{A}$ that with input $A \in \mathbb{R}^{m \times n}$ and $x \in \mathbb{R}^n$, computes the product Ax. If $\epsilon_{\mathrm{mach}}(\lceil \log_2 n \rceil + 2)^2 < 1$, then the computed vector $\mathrm{fl}(Ax)$ satisfies $\mathrm{fl}(Ax) = \tilde{A}x$ with*

$$|\tilde{a}_{ij} - a_{ij}| \le (\lceil \log_2 n \rceil + 2)\epsilon_{\mathrm{mach}}|a_{ij}|.$$

Proof Let $b = Ax$. For $i = 1, \ldots, m$ we have

$$b_i = a_{i1}x_1 + a_{i2}x_2 + \cdots + a_{in}x_n.$$

For the first product on the right-hand side we have $\mathrm{fl}(a_{i1}x_1) = a_{i1}x_1(1 + \delta)$ with $|\delta| \le \epsilon_{\mathrm{mach}} \le \frac{\epsilon_{\mathrm{mach}}}{1 - \epsilon_{\mathrm{mach}}} = \gamma_1$. That is, $\mathrm{fl}(a_{i1}x_1) = a_{i1}x_1(1 + \theta_1)$ and similarly $\mathrm{fl}(a_{i2}x_2) = a_{i2}x_2(1 + \theta_1)$. Note that the two occurrences of θ_1 here denote two different quantities. Hence, using Proposition O.1,

$$\mathrm{fl}(a_{i1}x_1 + a_{i2}x_2) = \big(a_{i1}x_1(1 + \theta_1) + a_{i2}x_2(1 + \theta_1)\big)(1 + \theta_1)$$
$$= a_{i1}x_1(1 + \theta_2) + a_{i2}x_2(1 + \theta_2).$$

By the same reasoning, $\mathrm{fl}(a_{i3}x_3 + a_{i4}x_4) = a_{i3}x_3(1 + \theta_2) + a_{i4}x_4(1 + \theta_2)$, and therefore

$$\mathrm{fl}(a_{i1}x_1 + a_{i2}x_2 + a_{i3}x_3 + a_{i4}x_4)$$
$$= \big(a_{i1}x_1(1 + \theta_2) + a_{i2}x_2(1 + \theta_2) + a_{i3}x_3(1 + \theta_2) + a_{i4}x_4(1 + \theta_2)\big)(1 + \theta_1)$$
$$= a_{i1}x_1(1 + \theta_3) + a_{i2}x_2(1 + \theta_3) + a_{i3}x_3(1 + \theta_3) + a_{i4}x_4(1 + \theta_3).$$

Continuing in this way, we obtain

$$\mathrm{fl}(b_i) = \tilde{a}_{i1}x_1 + \tilde{a}_{i2}x_2 + \cdots + \tilde{a}_{in}x_n$$

with $\tilde{a}_{ij} = a_{ij}(1 + \theta_{\lceil \log_2 n \rceil + 1})$. The result follows from the estimate (O.6), setting $k = \lceil \log_2 n \rceil + 1$. $\qquad \square$

Remark O.5 Note that the algorithm computing Ax is implicitly given in the proof of Proposition O.4. This algorithm uses a balanced treelike structure for the sums. The order of the sums cannot be arbitrarily altered: the operations $\tilde{+}$ and $\tilde{\cdot}$ are nonassociative.

We next estimate the componentwise condition number of matrix–vector multiplication. In doing so, we note that in the backward analysis of Proposition O.4, only the entries of A are perturbed. Those of x are not. This feature allows one to consider the condition of data (A, x) for perturbations of A only. Such a situation is common and also arises when data are structured (e.g., unit upper-triangular matrices have zeros below the diagonal and ones on the diagonal) or contain entries that are known to be integers.

Proposition O.6 *The componentwise condition numbers* $\mathsf{Cw}_i(A, x)$ *of matrix–vector multiplication, for perturbations of A only, satisfy*

$$\mathsf{Cw}_i(A, x) \le \left| \sec(a_i, x) \right|,$$

where a_i denotes the ith row of A and $\sec(a_i, x) = \frac{1}{\cos(a_i, x)}$ denotes the secant of the angle it makes with x (we assume $a_i, x \ne 0$).

Proof Let $\tilde{A} = A + E$ be a perturbation of A with $E = (e_{ij})$. By definition, $|e_{ij}| \le \mathsf{RelError}(A)|a_{ij}|$ for all i, j, whence $\|e_i\| \le \mathsf{RelError}(A)\|a_i\|$ for all i (here $\| \ \|$ denotes the Euclidean norm in $\mathbb{R}^n$). We obtain

$$\mathsf{RelError}\big((Ax)_i\big) = \frac{|e_i^{\mathsf{T}} x|}{|a_i^{\mathsf{T}} x|} \le \frac{\|e_i\|\,\|x\|}{|a_i^{\mathsf{T}} x|} \le \mathsf{RelError}(A) \frac{\|a_i\|\,\|x\|}{|a_i^{\mathsf{T}} x|}.$$

This implies that

$$\mathsf{Cw}_i(A, x) = \lim_{\delta \to 0} \ \sup_{\mathsf{RelError}(A) \le \delta} \frac{\mathsf{RelError}((Ax)_i)}{\mathsf{RelError}(A)}$$

$$\le \frac{\|a_i\|\,\|x\|}{|a_i^{\mathsf{T}} x|} = \frac{1}{|\cos(a_i, x)|} = \left| \sec(a_i, x) \right|. \qquad \square$$

A bound for the loss of precision in the componentwise context follows.

Corollary O.7 *In the componentwise setting, for all i such that $b_i = (Ax)_i \ne 0$,*

$$\mathsf{RelError}(b_i) \le \left| \sec(a_i, x) \right| \big(\lceil \log_2 n \rceil + 2\big) \epsilon_{\mathsf{mach}} + o(\epsilon_{\mathsf{mach}}),$$

$$\mathsf{LoP}(b_i) \le \log_\beta \left| \sec(a_i, x) \right| + \log_\beta \big(\lceil \log_2 n \rceil + 2\big) + o(1),$$

provided $\log_2 n \le \epsilon_{\mathsf{mach}}^{-1/2} + 3$.

Proof Immediate from Propositions O.4 and O.6 and Theorem O.3. $\qquad \square$

The corollary above states that if we are working with $|\log_\beta \epsilon_{\mathsf{mach}}|$ bits of precision, we compute a vector $\mathsf{fl}(Ax)$ whose nonzero entries have, approximately, at least

$$|\log_\beta \epsilon_{\mathsf{mach}}| - \log_\beta \left| \sec(a_i, x) \right| - \log_\beta \log_2 n$$

bits of precision. (The required bound on n is extremely weak and will be satisfied in all cases of interest.) This is a satisfying result. One may, nevertheless, wonder about the (absolute) error for the zero components of Ax. In this case, a normwise analysis may be more appropriate.

To proceed with a normwise analysis we first need to choose a norm in the space of $m \times n$ matrices. For simplicity, we choose

$$\|A\|_\infty = \max_{\|x\|_\infty = 1} \|Ax\|_\infty.$$

It is well known that

$$\|A\|_\infty = \max_{i \leq n} \|a_i\|_1. \tag{O.8}$$

Now note that it follows from Proposition O.4 that the perturbation $\tilde{A}$ in its statement satisfies, for n not too large,

$$\|\tilde{A} - A\|_\infty \leq \left(\lceil \log_2 n \rceil + 2\right) \epsilon_{\text{mach}}. \tag{O.9}$$

Therefore, we do have a normwise backward-error analysis. In addition, a normwise version of Proposition O.6 can be easily obtained.

Proposition O.8 *The normwise condition number* $\text{cond}(A, x)$ *of matrix–vector multiplication, for perturbations on A only, satisfies, for $Ax \neq 0$,*

$$\text{cond}(A, x) = \frac{\|A\|_\infty \|x\|_\infty}{\|Ax\|_\infty}.$$

Proof We have

$$\text{cond}(A, x) = \lim_{\delta \to 0} \sup_{\text{RelError}(A) \leq \delta} \frac{\text{RelError}(Ax)}{\text{RelError}(A)}$$

$$= \lim_{\delta \to 0} \sup_{\|\tilde{A} - A\|_\infty \leq \delta \|A\|_\infty} \frac{\|\tilde{A}x - Ax\|_\infty}{\|Ax\|_\infty} \frac{\|A\|_\infty}{\|\tilde{A} - A\|_\infty}$$

$$\leq \frac{\|A\|_\infty \|x\|_\infty}{\|Ax\|_\infty}.$$

Actually, equality holds. In order to see this, assume, without loss of generality, that $\|x\|_\infty = |x_1|$. Set $\tilde{A} = A + E$, where $e_{11} = \delta$ and $e_{ij} = 0$ otherwise. Then we have $\|\tilde{A}x - Ax\|_\infty = \|Ex\|_\infty = \delta |x_1| = \|E\|_\infty \|x\|_\infty = \|\tilde{A} - A\|_\infty \|x\|_\infty$. □

Again, a bound for the loss of precision immediately follows.

Corollary O.9 *In the normwise setting, when $Ax \neq 0$,*

$$\text{LoP}(Ax) \leq \log_\beta \left(\frac{\|A\|_\infty \|x\|_\infty}{\|Ax\|_\infty}\right) + \log_\beta \left(\lceil \log_2 n \rceil + 2\right) + o(1),$$

provided $\log_2 n \leq \epsilon_{\text{mach}}^{-1/2} + 3$.

Proof It is an immediate consequence of (O.9), Proposition O.8, and Theorem O.3. □

Remark O.10 If $m = n$ and A is invertible, it is possible to give a bound on the normwise condition that is independent of x. Using that $x = A^{-1}Ax$, we deduce $\|x\|_\infty \leq \|A^{-1}\|_\infty \|Ax\|_\infty$ and therefore, by Proposition O.8, $\text{cond}(A, x) \leq \|A^{-1}\|_\infty \|A\|_\infty$. A number of readers may find this expression familiar.

O.5 The Many Faces of Condition

The previous sections attempted to introduce condition numbers by retracing the way these numbers were introduced: as a way of measuring the effect of data perturbations. The expression "condition number" was first used by Turing [221] to denote a condition number for linear equation solving, independently introduced by him and by von Neumann and Goldstine [226] in the late 1940s. Expressions like "ill-conditioned set [of equations]" to denote systems with a large condition number were also introduced in [221].

Conditioning, however, was eventually related to issues in computation other than error-propagation analysis and this fact—together with the original role of conditioning in error analysis—triggered research on different aspects of the subject. We briefly describe some of them in what follows.

O.5.1 Condition and Complexity

In contrast with direct methods (such as Gaussian elimination), the number of times that a certain basic procedure is repeated in iterative methods is not data-independent. In the analysis of this dependence on the data at hand it was early realized that, quite often, one could express it using its condition number. That is, the number of iterations the algorithm $\mathcal{A}^\varphi$ would perform with data $a \in \mathbb{R}^m$ could be bounded by a function of m, $\text{cond}^\varphi(a)$, and—in the case of an algorithm computing an ε-approximation of the desired solution—the accuracy ε. A very satisfying bound for the number of iterations # iterations$(\mathcal{A}^\varphi(a))$ of algorithm $\mathcal{A}^\varphi$ would have the form

$$\text{\# iterations}\big(\mathcal{A}^\varphi(a)\big) \leq \left(m + \log \text{cond}^\varphi(a) + \log\left(\frac{1}{\varepsilon}\right) \right)^{\mathcal{O}(a)}, \qquad (O.10)$$

and a less satisfying (but often still acceptable) bound would have $\log \text{cond}^\varphi(a)$ replaced by $\text{cond}^\varphi(a)$ and/or $\log(\frac{1}{\varepsilon})$ replaced by $\frac{1}{\varepsilon}$. We will encounter several instances of this *condition-based complexity analysis* in the coming chapters.

O.5.2 Computing Condition Numbers

Irrespective of whether relative errors are measured normwise or componentwise, the expression (O.1) defining the condition number of a (for the problem φ) is hardly usable. Not surprisingly then, one of the main lines of research regarding condition numbers has focused on finding equivalent expressions for $\text{cond}^\varphi(a)$ that would be directly computable or, if this appears to be out of reach, tight enough bounds with this property. We have done so for the problem of matrix–vector multiplication in Propositions O.6 and O.8 (for the componentwise and normwise cases, respectively). In fact, in many examples the condition number can be succinctly expressed in terms of the norm of a derivative, which facilitates its analysis (cf. Sect. 14.1).

O.5.3 Condition of Random Data

How many iterations does an iterative algorithm need to perform to compute $\varphi(a)$? To answer this question we need $\text{cond}^\varphi(a)$. And to compute $\text{cond}^\varphi(a)$ we would like a simple expression like those in Propositions O.6 and O.8. A second look at these expressions, however, shows that they seem to require $\varphi(a)$, the quantity in which we were interested in the first place. For in the componentwise case, we need to compute $\sec(a_i, x)$—and hence $a_i^T x$—for $i = 1, \ldots, n$, and in the normwise case the expression $\|Ax\|_\infty$ speaks for itself. Worst of all, this is not an isolated situation. We will see that the condition number of a matrix A with respect to matrix inversion is expressed in terms of A^{-1} (or some norm of this inverse) and that a similar phenomenon occurs for each of the problems we consider. So, even though we do not formalize this situation as a mathematical statement, we can informally describe it by saying that the computation of a condition number $\text{cond}^\varphi(a)$ is never easier than the computation of $\varphi(a)$. The most elaborate reasoning around this issue was done by Renegar [164].

A similar problem appears with perturbation considerations. If we are given only a perturbation $\tilde{a}$ of data a, how can we know how accurate $\varphi(\tilde{a})$ is? Even assuming that we can compute cond^φ accurately and fast, the most we could do is to compute $\text{cond}^\varphi(\tilde{a})$, not $\text{cond}^\varphi(a)$.

There are a number of ways in which this seemingly circular situation can be broken. Instead of attempting to make a list of them (an exercise that can only result in boredom), we next describe a way out pioneered by John von Neumann (e.g., in [108]) and strongly advocated by Steve Smale in [201]. It consists in randomizing the data (i.e., in assuming a probabilistic distribution $\mathcal{D}$ in $\mathbb{R}^m$) and considering the tail

$$\operatorname*{Prob}_{a \sim \mathcal{D}} \{ \text{cond}^\varphi(a) \geq t \}$$

or the expected value (for $q \geq 1$)

$$\operatorname*{\mathbb{E}}_{a \sim \mathcal{D}} \left(\log^q \text{cond}^\varphi(a) \right).$$

The former, together with a bound as in (O.10), would allow one to bound the probability that $\mathcal{A}^\varphi$ needs more than a given number of iterations. The latter, taking q to be the constant in the $\mathcal{O}(a)$ notation, would make it possible to estimate the expected number of iterations. Furthermore, the latter again, now with $q = 1$, can be used to obtain an estimate of the average loss of precision for a problem φ (together with a backward stable algorithm $\mathcal{A}^\varphi$ if we are working with finite-precision arithmetic).

For instance, for the example that formed the substance of Sect. O.4, we will prove for a matrix $A \in \mathbb{R}^{m \times n}$ with standard Gaussian entries that

$$\mathbb{E}\big(\log_\beta \mathsf{Cw}_i(A)\big) \leq \frac{1}{2} \log_\beta n + 2.$$

In light of Corollary O.7, this bound implies that the expected loss of precision in the computation of $(Ax)_i$ is at most $\frac{1}{2} \log_\beta n + \log_\beta \log_2 n + \mathcal{O}(1)$.

The probabilistic analysis proposed by von Neumann and Smale relies on the assumption of "evenly spread random data." A different approach was recently proposed that relies instead on the assumption of "nonrandom data affected by random noise." We will develop both approaches in this book.

O.5.4 Ill-posedness and Condition

Let us return once more to the example of matrix–vector multiplication. If A and x are such that $Ax = 0$, then the denominator in $\frac{\|A\|_\infty \|x\|_\infty}{\|Ax\|_\infty}$ is zero, and we can define $\mathsf{cond}(A, x) = \infty$. This reflects the fact that no matter how small the absolute error in computing Ax, the relative error will be infinite. The quest for any relative precision is, in this case, a battle lost in advance. It is only fair to refer to instances like this with a name that betrays this hopelessness. We say that a is *ill-posed for* φ when $\mathsf{cond}^\varphi(a) = \infty$. Again, one omits the reference to φ when the problem is clear from the context, but it goes without saying that the notion of ill-posedness, like that of condition, is with respect to a problem. It also depends on the way we measure errors. For instance, in our example, $\mathsf{Cw}(A, x) = \infty$ if and only if there exists $i \leq n$ such that $a_i^\mathsf{T} x = 0$, while for $\mathsf{cond}(A, x)$ to be infinity, it is necessary (and sufficient) that $Ax = 0$.

The subset of $\mathbb{R}^m$ of ill-posed inputs is denoted by Σ^φ (or simply by Σ), and it has played a distinguished role in many developments in conditioning. To see why, let us return (yes, once again) to matrix–vector multiplication, say in the componentwise setting. Recall that we are considering x as fixed (i.e., not subject to perturbations). In this situation we take $\Sigma \subset \mathbb{R}^{n \times m}$ to be the set of matrices A such that $\mathsf{Cw}(A, x) = \infty$. We have $\Sigma = \bigcup_{i \leq n} \Sigma_i$ with

$$\Sigma_i = \big\{ A \in \mathbb{R}^{n \times m} \mid \mathsf{Cw}_i(A, x) = \infty \big\} = \big\{ A \in \mathbb{R}^{n \times m} \mid a_i^\mathsf{T} x = 0 \big\}.$$

Now recall $\mathrm{Cw}_i(A, x) \leq \frac{1}{|\cos(a_i, x)|}$. If we denote by $\bar{a}_i$ the orthogonal projection of a_i on the space $x^\perp = \{y \in \mathbb{R}^m \mid y^\mathsf{T} x = 0\}$, then

$$\frac{1}{|\cos(a_i, x)|} = \frac{\|a_i\|}{\|a_i - \bar{a}_i\|},$$

and it follows that

$$\mathrm{Cw}_i(A, x) \leq \frac{\|a_i\|}{\mathrm{dist}(A, \Sigma_i)}. \tag{O.11}$$

That is, componentwise, the condition number of (A, x) is bounded by the inverse of the relativized distance from A to ill-posedness.

This is not an isolated phenomenon. On the contrary, it is a common occurrence that condition numbers can be expressed as, or at least bounded by, the inverse of a relativized distance to ill-posedness. We will actually meet this theme repeatedly in this book.

Part I
Condition in Linear Algebra
(*Adagio*)

Chapter 1
Normwise Condition of Linear Equation Solving

Every invertible matrix $A \in \mathbb{R}^{n \times n}$ can be uniquely factored as $A = QR$, where Q is an orthogonal matrix and R is upper triangular with positive diagonal entries. This is called the *QR factorization* of A, and in numerical linear algebra, different ways for computing it are studied. From the QR factorization one obtains the solution of the system $Ax = b$ by $y = Q^{\mathrm{T}}b$ and $x = R^{-1}y$, where the latter is easily computed by back substitution.

The *Householder QR factorization method* is an algorithm for computing the QR-decomposition of a given matrix (compare Sect. 4.1.2). It is one of the main engines in numerical linear algebra. The following result states a backward analysis for this algorithm.

Theorem 1.1 *Let $A \in \mathbb{R}^{n \times n}$ be invertible and $b \in \mathbb{R}^n$. If the system $Ax = b$ is solved using the Householder QR factorization method, then the computed solution $\tilde{x}$ satisfies*

$$\tilde{A}\tilde{x} = \tilde{b},$$

where $\tilde{A}$ and $\tilde{b}$ satisfy the relative error bounds

$$\|\tilde{A} - A\|_F \leq n\gamma_{cn}\|A\|_F \quad and \quad \|\tilde{b} - b\| \leq n\gamma_{cn}\|b\|$$

for a small constant c and with γ_{cn} as defined in (O.5). $\qquad\square$

This yields $\|\tilde{A} - A\| \leq n^{3/2}\gamma_{cn}\|A\|$ when the Frobenius norm is replaced by the spectral norm. It follows from this backward stability result, (O.6), and Theorem O.3 that the relative error for the computed solution $\tilde{x}$ satisfies

$$\frac{\|\tilde{x} - x\|}{\|x\|} \leq cn^{5/2}\epsilon_{\mathsf{mach}}\mathsf{cond}(A, b) + o(\epsilon_{\mathsf{mach}}), \tag{1.1}$$

and the loss of precision is bounded by

$$\mathsf{LoP}(A^{-1}b) \leq \frac{5}{2}\log_\beta n + \log_\beta \mathsf{cond}(A, b) + \log_\beta c + o(1). \tag{1.2}$$

P. Bürgisser, F. Cucker, *Condition*,
Grundlehren der mathematischen Wissenschaften 349,
DOI 10.1007/978-3-642-38896-5_1, © Springer-Verlag Berlin Heidelberg 2013

Table 1.1 Equivalence of
vector norms

	1	2	∞
1	=	$\sqrt{n}$	n
2	1	=	$\sqrt{n}$
∞	1	1	=

Here $\mathrm{cond}(A, b)$ is the normwise condition number for linear equation solving,

$$\mathrm{cond}(A, b) = \lim_{\delta \to 0} \; \sup_{\max\{\mathrm{RelError}(A), \mathrm{RelError}(b)\} \leq \delta} \frac{\mathrm{RelError}(A^{-1}b)}{\max\{\mathrm{RelError}(A), \mathrm{RelError}(b)\}},$$

where $\mathrm{RelError}(A)$ is defined with respect to the spectral norm and $\mathrm{RelError}(b)$ with respect to the Euclidean norm. Inequality (1.1) calls for a deeper understanding of what $\mathrm{cond}(A, b)$ is than the equality above. The pursuit of this understanding is the goal of this chapter.

1.1 Vector and Matrix Norms

The condition number $\mathrm{cond}(A, b)$ in the introduction is a normwise one. For this reason, we begin by providing a brief review of norms.

The three most useful norms in error analysis on the real vector space $\mathbb{R}^n$ are the following:

$$\|x\|_1 := \sum_{i=1}^{n} |x_i|, \quad \|x\|_2 := \left(\sum_{i=1}^{n} |x_i|^2 \right)^{1/2}, \quad \|x\|_\infty := \max_{1 \leq i \leq n} |x_i|.$$

Any two of them are equivalent, and the equivalence constants are given in Table 1.1, whose (i, j)th entry shows the smallest constant k for which $\| \; \|_i \leq k \| \; \|_j$.

These norms are special cases of the Hölder r-norm

$$\|x\|_r := \left(\sum_{i=1}^{n} |x_i|^r \right)^{1/r}$$

defined for a real number $r \geq 1$. Even though we will need only the cases $r \in \{1, 2, \infty\}$, stating the results for general Hölder norms avoids case distinctions and thus saves space.

For a given $r \geq 1$ there is exactly one $r^* \geq 1$ such that $1/r + 1/r^* = 1$. The well-known Hölder inequality states that for $x, z \in \mathbb{R}^n$, we have

$$\left| x^{\mathsf{T}} z \right| \leq \|x\|_r \|z\|_{r^*}.$$

Moreover, equality holds if $(|x_i|^r)$ and $(|z_i|^{r^*})$ are linearly dependent. This easily implies that for any $x \in \mathbb{R}^n$,

$$\max_{\|z\|_{r^*}=1} x^{\mathrm{T}} z = \|x\|_r. \tag{1.3}$$

For this reason, one calls $\| \ \|_{r^*}$ the *dual norm* of $\| \ \|_r$. In particular, for each $x \in \mathbb{R}^n$ with $\|x\|_r = 1$ there exists $z \in \mathbb{R}^n$ such that $\|z\|_{r^*} = 1$ and $z^{\mathrm{T}} x = 1$.

We will adopt the notational convention $\| \ \| := \| \ \|_2$ for the Euclidean vector norm. Note that this norm is dual to itself. Note as well that $\| \ \|_1$ and $\| \ \|_\infty$ are dual to each other.

To the vector norms $\| \ \|_r$ on a domain space $\mathbb{R}^n$ and $\| \ \|_s$ on a range space $\mathbb{R}^m$, one associates the *subordinate matrix norm* $\| \ \|_{rs}$ on the vector space of linear operators $A: \mathbb{R}^n \to \mathbb{R}^m$ defined by

$$\|A\|_{rs} := \sup_{\substack{x \in \mathbb{R}^p \\ x \neq 0}} \frac{\|Ax\|_s}{\|x\|_r} = \sup_{\|x\|_r=1} \|Ax\|_s. \tag{1.4}$$

By compactness of the unit sphere, the supremum is a minimum. In case $r = s$, we write $\| \ \|_r$ instead of $\| \ \|_{rr}$. (We recall that we already met $\| \ \|_\infty$ in Sect. O.4.) Furthermore, when $r = 2$, $\| \ \|_2$ is called the *spectral norm*, and it is written simply as $\| \ \|$.

We note that the following submultiplicativity property of matrix norms holds: for $r, s, t \geq 1$ and matrices A, B we have

$$\|AB\|_{rs} \leq \|A\|_{ts} \|B\|_{rt}, \tag{1.5}$$

provided the matrix product is defined.

Most of what we will need about operator norms is stated in the following simple lemma.

Lemma 1.2

(a) *For $y \in \mathbb{R}^m$ and $v \in \mathbb{R}^n$ we have $\|yv^{\mathrm{T}}\|_{rs} = \|y\|_s \|v\|_{r^*}$.*

(b) *Suppose that $x \in \mathbb{R}^n$ and $y \in \mathbb{R}^m$ satisfy $\|x\|_r = \|y\|_s = 1$. Then there exists $B \in \mathbb{R}^{m \times n}$ such that $\|B\|_{rs} = 1$ and $Bx = y$.*

(c) $\|A^{\mathrm{T}}\|_{rs} = \|A\|_{s^* r^*}$.

Proof (a) We have

$$\left\|yv^{\mathrm{T}}\right\|_{rs} = \max_{\|x\|_r=1} \left\|yv^{\mathrm{T}}x\right\|_s = \|y\|_s \max_{\|x\|_r=1} \left|v^{\mathrm{T}}x\right| = \|y\|_s \|v\|_{r^*},$$

where the last equality holds due to (1.3).

(b) By (1.3) there exists $z \in \mathbb{R}^n$ such that $\|z\|_{r^*} = 1$ and $z^{\mathrm{T}} x = 1$. For $B := yz^{\mathrm{T}}$ we have $Bx = y$, and by part (a) $\|B\|_{rs} = \|y\|_s \|z\|_{r^*} = 1$.

(c) We have

$$\left\| A^{\mathrm{T}} \right\|_{rs} = \max_{\|x\|_r = 1} \left\| A^{\mathrm{T}} x \right\|_s \overset{(1.3)}{=} \max_{\|x\|_r = 1} \max_{\|z\|_{s*} = 1} x^{\mathrm{T}} A z$$

$$\overset{(1.3)}{=} \max_{\|z\|_{s*} = 1} \|A z\|_{r*} = \|A\|_{s*r*}. \qquad \square$$

Lemma 1.2 allows one to provide friendly characterizations of some operator norms.

Corollary 1.3 *For all* r, $\|A\|_{r\infty} = \max_{i \leq m} \|a_i\|_{r*}$, *where* a_i *is the* i*th row of* A. *In particular:*

(a) $\|A\|_{1\infty} = \max_{i \leq m, j \leq n} |a_{ij}|$,
(b) $\|A\|_{2\infty} = \max_{i \leq m} \|a_i\|$,
(c) $\|A\|_\infty = \max_{i \leq m} \sum_{j \leq n} |a_{ij}|$,
(d) $\|A\|_1 = \max_{j \leq n} \sum_{i \leq m} |a_{ij}|$,
(e) $\|A\|_{12} = \max_{j \leq n} \|a_{.j}\|$ ($a_{.j}$ *denoting the* j*th column of* A).

Proof Using (1.3) we obtain

$$\|A\|_{r\infty} = \max_{\|x\|_r = 1} \|A x\|_\infty = \max_{i \leq m} \max_{\|x\|_r = 1} \left| a_i^{\mathrm{T}} x \right| = \max_{i \leq m} \|a_i\|_{r*}.$$

The particular cases follow from the definition of vector norms $\| \ \|_1$, $\| \ \|_2$, and $\| \ \|_\infty$ and the use of Lemma 1.2(c). $\qquad \square$

Considering a matrix $A = (a_{ij}) \in \mathbb{R}^{m \times n}$ as an element in $\mathbb{R}^{mn}$ yields at least two more matrix norms (corresponding to the 1-norm and 2-norm in this space). Of them, the most frequently used is the *Frobenius norm*,

$$\|A\|_F := \left(\sum_{i=1}^m \sum_{j=1}^n a_{ij}^2 \right)^{1/2},$$

which corresponds to the Euclidean norm of A as an element of $\mathbb{R}^{mn}$. The advantage of the Frobenius norm is that it is induced by an inner product on $\mathbb{R}^{m \times n}$.

Just like the vector norms, all matrix norms are equivalent. A table showing equivalence constants for the matrix norms we have described above is shown next as Table 1.2. Most of these bounds follow from those in Table 1.1, while a few will be shown below (Proposition 1.15(h)).

1.2 Turing's Condition Number

We now proceed to exhibit a characterization of the normwise condition number for linear equation solving, pursuing the theme described in Sect. O.5.2.

Let $m = n$ and fix norms $\| \ \|_r$ and $\| \ \|_s$ on $\mathbb{R}^n$. Also, let

$$\Sigma := \left\{ A \in \mathbb{R}^{n \times n} \mid \det(A) = 0 \right\}$$

Table 1.2 Equivalence of matrix norms

	1	2	∞	12	1∞	21	2∞	∞1	∞2	F
1	=	$\sqrt{m}$	m	$\sqrt{m}$	m	1	m	1	$\sqrt{m}$	$\sqrt{m}$
2	$\sqrt{n}$	=	$\sqrt{m}$	$\sqrt{n}$	$\sqrt{mn}$	1	$\sqrt{m}$	1	1	1
∞	n	$\sqrt{n}$	=	n	n	$\sqrt{n}$	$\sqrt{n}$	1	1	$\sqrt{n}$
12	1	1	$\sqrt{m}$	=	$\sqrt{m}$	1	$\sqrt{m}$	1	1	1
1∞	1	1	1	1	=	1	1	1	1	1
21	$\sqrt{n}$	$\sqrt{m}$	m	$\sqrt{mn}$	$m\sqrt{n}$	=	m	1	$\sqrt{m}$	$\sqrt{m}$
2∞	$\sqrt{n}$	1	1	$\sqrt{n}$	$\sqrt{n}$	1	=	1	1	1
∞1	n	$\sqrt{mn}$	m	$n\sqrt{m}$	mn	$\sqrt{n}$	$m\sqrt{n}$	=	$\sqrt{m}$	$\sqrt{mn}$
∞2	n	$\sqrt{n}$	$\sqrt{m}$	n	$n\sqrt{m}$	$\sqrt{n}$	$\sqrt{mn}$	1	=	$\sqrt{n}$
F	$\sqrt{n}$	$\sqrt{\operatorname{rank}(A)}$	$\sqrt{m}$	$\sqrt{n}$	$\sqrt{mn}$	$\sqrt{\operatorname{rank}(A)}$	$\sqrt{m}$	1	$\sqrt{\operatorname{rank}(A)}$	=

denote the *set of ill-posed matrices* and put $\mathcal{D} := \mathbb{R}^{n \times n} \setminus \Sigma$. We define the map $\kappa_{rs} : \mathcal{D} \to \mathbb{R}$ by

$$\kappa_{rs}(A) := \|A\|_{rs} \|A^{-1}\|_{sr}.$$

Note that $\kappa_{rs}(A) \geq 1$, since $1 = \|I\|_r \leq \|A\|_{rs} \|A^{-1}\|_{sr} = \kappa_{rs}(A)$.

Theorem 1.4 *Let $\varphi : \mathcal{D} \times \mathbb{R}^n \to \mathbb{R}^n$ be given by $\varphi(A, b) = A^{-1}b$. We measure the relative error in $\mathcal{D} \times \mathbb{R}^n$ by*

$$\mathsf{RelError}(A, b) = \max\left\{ \frac{\|\tilde{A} - A\|_{rs}}{\|A\|_{rs}}, \frac{\|\tilde{b} - b\|_s}{\|b\|_s} \right\},$$

and we measure the relative error in the solution space normwise with respect to $\| \ \|_r$. Then

$$\mathsf{cond}^\varphi(A, b) = \kappa_{rs}(A) + \frac{\|A^{-1}\|_{sr} \|b\|_s}{\|A^{-1}b\|_r}.$$

In particular, we have

$$\kappa_{rs}(A) \leq \mathsf{cond}^\varphi(A, b) \leq 2\kappa_{rs}(A).$$

Proof Let $\tilde{A} = A - E$ and $\tilde{b} = b + f$. By definition, $\|E\|_{rs} \leq \mathsf{R}\|A\|_{rs}$ and $\|f\|_s \leq \mathsf{R}\|b\|_s$, where for simplicity, $\mathsf{R} = \mathsf{RelError}(A, b)$. We have, for $\mathsf{R} \to 0$,

$$(A - E)^{-1} = A^{-1}(I - EA^{-1})^{-1} = A^{-1}(I + EA^{-1} + o(\mathsf{R}))$$
$$= A^{-1} + A^{-1}EA^{-1} + o(\mathsf{R}).$$

This implies, writing $x := A^{-1}b$ and $\tilde{x} := \tilde{A}^{-1}\tilde{b}$,

$$\tilde{x} - x = (A - E)^{-1}(b + f) - x = A^{-1}Ex + A^{-1}f + o(\mathsf{R}). \qquad (1.6)$$

Taking norms and using (1.5), we conclude that

$$\|\tilde{x} - x\|_r \le \|A^{-1}\|_{sr} \|E\|_{rs} \|x\|_r + \|A^{-1}\|_{sr} \|f\|_s + o(\mathsf{R})$$

$$\le \|A^{-1}\|_{sr} \|A\|_{rs} \|x\|_r \mathsf{R} + \|A^{-1}\|_{sr} \|b\|_s \mathsf{R} + o(\mathsf{R}),$$

and hence

$$\frac{\|\tilde{x} - x\|_r}{\mathsf{R}\|x\|_r} \le \kappa_{rs}(A) + \frac{\|A^{-1}\|_{sr} \|b\|_s}{\|x\|_r},$$

which shows the upper bound in the claimed equality.

For the corresponding lower bound we choose $y \in \mathbb{R}^n$ such that $\|y\|_s = 1$ and $\|A^{-1}y\|_r = \|A^{-1}\|_{sr}$. Further, we choose $v \in \mathbb{R}^n$ such that $\|v\|_{r*} = 1$ and $v^{\mathsf{T}}x = \|x\|_r$, which is possible by (1.3). Now we put

$$E := \mathsf{R}\|A\|_{rs} yv^{\mathsf{T}}, \qquad f := \pm\mathsf{R}\|b\|_s y. \tag{1.7}$$

We note that

$$\|E\|_{rs} = \mathsf{R}\|A\|_{rs}, \qquad \|f\|_s = \mathsf{R}\|b\|_s,$$

the first equality holding since by Lemma 1.2(a), $\|yv^{\mathsf{T}}\|_{rs} = \|y\|_s \|v\|_{r*} = 1$. We have

$$A^{-1}Ex = \mathsf{R}\|A\|_{rs} A^{-1}y v^{\mathsf{T}}x = \mathsf{R}\|A\|_{rs}\|x\|_r A^{-1}y$$

and hence $\|A^{-1}Ex\|_r = \kappa_{rs}(A)\|x\|_r\mathsf{R}$. Similarly, $A^{-1}f = \pm R\|b\|_s A^{-1}y$ and $\|A^{-1}f\|_r = \|A^{-1}\|_{sr}\|b\|_s\mathsf{R}$. Since $A^{-1}Ex$ and $A^{-1}f$ are both proportional to $A^{-1}y$, we obtain from (1.6),

$$\|\tilde{x} - x\|_r = \kappa_{rs}(A)\|x\|_r\mathsf{R} + \|A^{-1}\|_{sr}\|b\|_s\mathsf{R},$$

if we choose the sign for f in (1.7) appropriately. This proves the claimed lower bound. $\qquad\square$

The next result shows that κ_{rs} actually coincides with the condition number for the problem of matrix inversion.

Theorem 1.5 *Let* $\psi: \mathcal{D} \to \mathbb{R}^{n \times n}$ *be given by* $\psi(A) = A^{-1}$. *We measure the relative error on the data space and solution space with respect to* $\| \; \|_{rs}$ *and* $\| \; \|_{sr}$, *respectively. Then we have*

$$\mathrm{cond}^{\psi}(A) = \kappa_{rs}(A).$$

Proof Let $E \in \mathbb{R}^{n \times n}$ be such that $\tilde{A} = A - E$. Then $\mathsf{RelError}(A) = \frac{\|E\|_{rs}}{\|A\|_{rs}}$. As in the proof of Theorem 1.4, we have for $\|E\| \to 0$,

$$\|\tilde{A}^{-1} - A^{-1}\|_{sr} = \|A^{-1}EA^{-1}\|_{sr} + o(\|E\|). \tag{1.8}$$

Hence, $\|A^{-1}EA^{-1}\|_{sr} \leq \|A^{-1}\|_{sr}\|E\|_{rs}A^{-1}\|_{sr}$. Consequently, we obtain

$$\mathsf{RelError}\left(A^{-1}\right) = \frac{\|\tilde{A}^{-1} - A^{-1}\|_{sr}}{\|A^{-1}\|_{sr}} \leq \|A^{-1}\|_{sr}\|E\|_{rs} + o\left(\|E\|\right).$$

We conclude that

$$\frac{\mathsf{RelError}(A^{-1})}{\mathsf{RelError}(A)} \leq \|A\|_{rs}\|A^{-1}\|_{sr} + o(1)$$

and hence $\mathrm{cond}^{\psi}(A) \leq \kappa_{rs}(A)$.

To prove the reverse inequality it is enough to find arbitrarily small matrices E such that $\|A^{-1}EA^{-1}\|_{sr} = \|A^{-1}\|_{sr}^2\|E\|_{rs}$, since then we can proceed from (1.8) as we did in Theorem 1.4 from (1.6).

To do so, let $y \in \mathbb{R}^n$ be such that $\|y\|_s = 1$ and $\|A^{-1}y\|_r = \|A^{-1}\|_{sr}$. Define $x := \frac{1}{\|A^{-1}\|_{sr}}A^{-1}y$, so that $A^{-1}y = \|A^{-1}\|_{sr}x$ and $\|x\|_r = \|y\|_s = 1$. For any $B \in \mathbb{R}^{n \times n}$ we have

$$\|A^{-1}BA^{-1}\|_{sr} \geq \|A^{-1}BA^{-1}y\|_r = \|A^{-1}\|_{sr} \cdot \|A^{-1}Bx\|_r.$$

By Lemma 1.2(b) there exists $B \in \mathbb{R}^{n \times n}$ such that $Bx = y$ and $\|B\|_{rs} = 1$. Therefore,

$$\|A^{-1}BA^{-1}\|_{sr} \geq \|A^{-1}\|_{sr} \cdot \|A^{-1}y\|_r = \|A^{-1}\|_{sr}^2.$$

Taking $E = \delta B$ with arbitrarily small δ finishes the proof. $\square$

The most often considered case is $r = s = 2$, that is, when the error in both the input and the output space is measured with the Euclidean norm. The resulting condition number $\kappa(A) := \kappa_{22}(A)$ is so pervasive in numerical linear algebra that it is commonly referred to as "the condition number of A"—without mention of the function of A whose condition we want to measure. We remark that $\kappa(A)$ was originally introduced by Turing [221] and by von Neumann and Goldstine [226] (Turing actually considered norms other than the spectral).

Theorem 1.4—together with (1.2)—immediately yields a bound for the loss of precision in linear equation solving.

Corollary 1.6 *Let $A \in \mathbb{R}^{n \times n}$ be invertible and $b \in \mathbb{R}^n$. If the system $Ax = b$ is solved using the Householder QR factorization method, then the computed solution $\tilde{x}$ satisfies, for a small constant c,*

$$\mathsf{LoP}(A^{-1}b) \leq 2\log_\beta n + \log_\beta \kappa(A) + \log_\beta c + o(1),$$

where $o(1)$ is for $\epsilon_{\mathsf{mach}} \to 0$. $\square$

1.3 Condition and Distance to Ill-posedness

A goal of this section, now revisiting the discussion in Sect. O.5.4, is to show that the condition number $\kappa_{rs}(A)$ can be expressed as the relativized inverse of the distance from the square matrix A to the set Σ of singular matrices: a large $\kappa_{rs}(A)$ means that A is close to a singular matrix. In order to make this precise, we introduce the distance of $A \in \mathbb{R}^{n \times n}$ to the set Σ of singular matrices,

$$d_{rs}(A, \Sigma) := \min\{\|A - B\|_{rs} \mid B \in \Sigma\}, \tag{1.9}$$

defined with respect to the norm $\|\ \|_{rs}$. For the spectral norm we just write $d(A, \Sigma) := d_{22}(A, \Sigma)$.

Theorem 1.7 *Let $A \in \mathbb{R}^{n \times n}$ be nonsingular. Then*

$$d_{rs}(A, \Sigma) = \frac{1}{\|A^{-1}\|_{sr}}.$$

Proof Let A be nonsingular and let $A + E$ be singular. Then there exists an $x \in \mathbb{R}^n \setminus \{0\}$ such that $(A + E)x = 0$. This means that $x = -A^{-1}Ex$ and hence

$$\|x\|_r \leq \|A^{-1}E\|_{rr} \cdot \|x\|_r \leq \|A^{-1}\|_{sr} \cdot \|E\|_{rs} \cdot \|x\|_r,$$

which implies $\|E\|_{rs} \geq \|A^{-1}\|_{sr}^{-1}$. Therefore $d_{rs}(A, \Sigma) \geq \|A^{-1}\|_{sr}^{-1}$.

To prove the other inequality, it suffices to find a singular matrix $\tilde{A}$ with $d_{rs}(A, \tilde{A}) \leq \|A^{-1}\|_{sr}^{-1}$. Let $y \in \mathbb{R}^n$ be such that $\|A^{-1}\|_{sr} = \|A^{-1}y\|_r$ and $\|y\|_s = 1$. Writing $x := A^{-1}y$, we have $\|x\|_r = \|A^{-1}\|_{sr}$, in particular $x \neq 0$. By Lemma 1.2(b), there exists $B \in \mathbb{R}^{n \times n}$ such that $\|B\|_{rs} = 1$ and

$$B\frac{x}{\|x\|_r} = -y.$$

Hence $E := \|x\|_r^{-1}B$ satisfies $Ex = -y$, and hence $(A + E)x = 0$. So the matrix $\tilde{A} := A + E$ must be singular. In addition, we have

$$d_{rs}(A, \tilde{A}) = \|E\|_{rs} = \|x\|_r^{-1}\|B\|_{rs} = \|A^{-1}\|_{sr}^{-1} \cdot \|B\|_{rs} = \|A^{-1}\|_{sr}^{-1},$$

which finishes the proof.   □

Defining $\kappa_{rs}(A) := \infty$ for a singular matrix, we immediately obtain the following result, which is known as the "condition number theorem."

Corollary 1.8 *For nonzero $A \in \mathbb{R}^{n \times n}$ we have*

$$\kappa_{rs}(A) = \frac{\|A\|_{rs}}{d_{rs}(A, \Sigma)}.$$

                                                                                □

Thus the condition number $\kappa_{rs}(A)$ can be seen as the inverse of a normalized distance of A to the set of ill-posed inputs Σ.

Notation 1.9 *In this book we will consider matrices given by their columns or by their rows. In order to emphasize this distinction and avoid ambiguities, given vectors $a_1, \ldots, a_n \in \mathbb{R}^m$, we write $(a_1, \ldots, a_n)$ for the matrix in $\mathbb{R}^{n \times m}$ whose rows are $a_1, \ldots, a_n$, and $[a_1, \ldots, a_n]$ for the matrix in $\mathbb{R}^{m \times n}$ whose columns are these vectors. Note that this notation relieves us from having to transpose $(x_1, \ldots, x_n)$ when we want to emphasize that this is a column vector.*

For a matrix $A \in \mathbb{R}^{n \times m}$, a vector $c \in \mathbb{R}^n$, and an index $j \in [m]$, we denote by $A(j : c)$ the matrix obtained by replacing the jth row of A by c. The meaning of $A[j : c]$ is defined similarly.

We draw now a consequence of Theorem 1.7 that will be used in several variations throughout the book.

Proposition 1.10 *For $A \in \mathbb{R}^{n \times n}$ and $r, s \geq 1$ there exist $j \in [n]$ and $c \in \mathbb{R}^n$ such that $A[j : c] \in \Sigma$ and $\|a_j - c\|_s \leq n^{1/r} d_{rs}(A, \Sigma)$.*

Proof Theorem 1.7 states that $\|A^{-1}\|_{sr} = \varepsilon^{-1}$, where $\varepsilon := d_{rs}(A, \Sigma)$. There exists $b \in \mathbb{R}^n$ such that $\|b\|_s = 1$ and $\|A^{-1}b\|_r = \|A^{-1}\|_{sr}$. So if we put $v := A^{-1}b$, then $\|v\|_r \geq \varepsilon^{-1}$. This implies $\|v\|_\infty \geq n^{-1/r}\|v\|_r \geq n^{-1/r}\varepsilon^{-1}$. Without loss of generality we may assume that $|v_n| = \|v\|_\infty$.

Since $Av = b$, we can express v_n by Cramer's rule as follows:

$$v_n = \frac{\det([a_1, \ldots, a_{n-1}, b])}{\det(A)}.$$

This implies

$$0 = \det(A) - v_n^{-1} \det\big([a_1, \ldots, a_{n-1}, b]\big) = \det\big([a_1, \ldots, a_{n-1}, a_n - v_n^{-1}b]\big).$$

Thus if we put $c := a_n - v_n^{-1}b$, we have $A[i : c] \in \Sigma$ and

$$\|a_n - c\|_s = |v_n|^{-1}\|b\|_s = |v_n|^{-1} \leq n^{1/r}\varepsilon. \qquad \square$$

1.4   An Alternative Characterization of Condition

Theorem 1.7 characterizes $\|A^{-1}\|_{sr}$—and hence $\kappa_{rs}(A)$—as the inverse of the distance from A to Σ. The underlying geometry is on the space $\mathbb{R}^{n \times n}$ of matrices. The following result characterizes $\|A^{-1}\|_{sr}$ in different terms, with underlying geometry on $\mathbb{R}^n$. Even though its proof is very simple, the idea behind this alternative characterization can (and will) be useful in more complex settings.

For $a \in \mathbb{R}^n$ and $\delta > 0$ denote by $B_r(a, \delta)$ the closed ball with center a and radius δ in $\mathbb{R}^n$ with the norm $\|\ \|_r$.

Proposition 1.11 *For $A \in \mathbb{R}^{n \times n} \setminus \Sigma$,*

$$\left\| A^{-1} \right\|_{sr}^{-1} = \sup\{\delta \mid B_s(0, \delta) \subseteq A(B_r(0, 1))\}.$$

Proof It is immediate from the fact that

$$B_s(0, \delta) \subseteq A(B_r(0, 1)) \quad \Longleftrightarrow \quad A^{-1}(B_s(0, 1)) \subseteq B_r\left(0, \frac{1}{\delta}\right). \qquad \square$$

1.5 The Singular Value Decomposition

The singular value decomposition of a matrix is the numerically appropriate way to discuss matrix rank. It also leads to a natural generalization of Theorem 1.7.

In this section we mainly work with the spectral and the Frobenius norms. Both of them are invariant under orthogonal transformations.

Lemma 1.12 *For $A \in \mathbb{R}^{m \times n}$ and orthogonal matrices $U \in \mathbb{R}^{m \times m}$ and $V \in \mathbb{R}^{n \times n}$ we have $\|UAV\|_F = \|A\|_F$ and $\|UAV\| = \|A\|$.*

Proof For the first assertion, let $s_1, \ldots, s_n$ denote the columns of A. Then $U s_i$ is the ith column of UA. Since U is orthogonal, we have $\|U s_i\| = \|s_i\|$ and therefore

$$\|UA\|_F^2 = \sum_{i \leq n} \|U s_i\|^2 = \sum_{i \leq n} \|s_i\|^2 = \|A\|_F^2.$$

In the same way, one shows that $\|AV\|_F = \|A\|_F$. The second assertion is proved as follows:

$$\begin{aligned}
\|UAV\| &= \sup_{\|x\|=1} \|UAVx\| = \sup_{\|x\|=1} \left\| U(AVx) \right\| \\
&= \sup_{\|x\|=1} \|AVx\| = \sup_{\|x\|=1} \left\| A(Vx) \right\| \\
&= \sup_{\|x'\|=1} \left\| Ax' \right\| = \|A\|.
\end{aligned} \qquad \square$$

For conveniently stating the singular value decomposition, we extend the usual notation for diagonal matrices from square to rectangular $m \times n$ matrices. We put $p := \min\{n, m\}$ and define, for $a_1, \ldots, a_p \in \mathbb{R}$,

$$\text{diag}_{m,n}(a_1, \ldots, a_p) := (b_{ij}) \in \mathbb{R}^{m \times n} \quad \text{with } b_{ij} := \begin{cases} a_i & \text{if } i = j, \\ 0 & \text{otherwise.} \end{cases}$$

For notational convenience we usually drop the index, the format being clear from the context.

The next result is known as the "singular value decomposition theorem" (or, for short, the "SVD theorem").

Theorem 1.13 *For $A \in \mathbb{R}^{m \times n}$ there exist orthogonal matrices $U \in \mathbb{R}^{m \times m}$ and $V \in \mathbb{R}^{n \times n}$ such that*

$$U^T A V = \mathrm{diag}(\sigma_1, \ldots, \sigma_p),$$

with $p = \min\{m, n\}$ and $\sigma_1 \geq \sigma_2 \geq \cdots \geq \sigma_p \geq 0$.

Proof Let $x \in \mathbb{R}^n$, $\|x\| = 1$ be such that $\sigma := \|A\| = \|Ax\|$ and define $y := \sigma^{-1} Ax \in \mathbb{R}^m$, so that $\|y\| = 1$ and $Ax = \sigma y$. There exist matrices $V_2 \in \mathbb{R}^{n \times (n-1)}$ and $U_2 \in \mathbb{R}^{m \times (m-1)}$ such that $V := [x, V_2]$ and $U := [y, U_2]$ are orthogonal.

We have for some $w \in \mathbb{R}^{n-1}$ and $B \in \mathbb{R}^{(m-1) \times (n-1)}$ that

$$U^T A V = \begin{bmatrix} y^T \\ U_2^T \end{bmatrix} A[x, V_2] = \begin{bmatrix} y^T \\ U_2^T \end{bmatrix} [\sigma y, A V_2] = \begin{bmatrix} \sigma & w^T \\ 0 & B \end{bmatrix} =: A_1.$$

Note that $\|A_1\| = \|A\|$ by the orthogonal invariance of the spectral norm. Moreover, we have for $v \in \mathbb{R}^{n-1}$,

$$\|Bv\| = \left\| \begin{bmatrix} \sigma & w^T \\ 0 & B \end{bmatrix} \begin{bmatrix} 0 \\ v \end{bmatrix} \right\| = \left\| U^T A V \begin{bmatrix} 0 \\ v \end{bmatrix} \right\| \leq \|U^T A V\| \, \|v\| \leq \|A\| \, \|v\|,$$

whence $\|B\| \leq \|A\|$.

We claim that $w = 0$. To see this, note that

$$A_1 \begin{bmatrix} \sigma \\ w \end{bmatrix} = \begin{bmatrix} \sigma^2 + w^T w \\ * \end{bmatrix}$$

and therefore

$$\left\| A_1 \begin{bmatrix} \sigma \\ w \end{bmatrix} \right\| \geq \sigma^2 + \|w\|^2.$$

On the other hand,

$$\left\| A_1 \begin{bmatrix} \sigma \\ w \end{bmatrix} \right\| \leq \|A\| (\sigma^2 + \|w\|^2)^{1/2} = \sigma (\sigma^2 + \|w\|^2)^{1/2}.$$

It follows that $w = 0$. The argument can now be completed by induction. $\qquad\square$

The nonnegative numbers σ_i in Theorem 1.13 are called the *singular values* of A and are sometimes written $\sigma_i(A)$. We will see soon enough (Corollary 1.18) that they are uniquely determined by A. Sometimes one writes $\sigma_{\max}$ and $\sigma_{\min}$ for σ_1 and σ_p, respectively. The ith columns u_i and v_i of U and V in Theorem 1.13 are called ith *left singular vector* and ith *right singular vector* of A, respectively (in general, those are not uniquely determined).

Remark 1.14 If $A \in \mathbb{R}^{n \times n}$ is symmetric, then there exists $V \in \mathbb{R}^{n \times n}$ orthogonal such that $V^T A V = \mathrm{diag}(\lambda_1, \ldots, \lambda_n)$, where $\lambda_1, \ldots, \lambda_n \in \mathbb{R}$ are the eigenvalues of A (spectral theorem). Hence $|\lambda_1|, \ldots, |\lambda_n|$ are the singular values of A.

The following result summarizes the main properties of the singular value decomposition.

Proposition 1.15 *Suppose that $\sigma_1 \geq \sigma_2 \geq \cdots \geq \sigma_r > \sigma_{r+1} = \cdots = \sigma_p = 0$ are the singular values of $A \in \mathbb{R}^{m \times n}$ and u_i, v_i are left and right singular vectors of A. Then:*

(a) $A = \sum_{i=1}^{r} \sigma_i u_i v_i^{\mathrm{T}}$ *(singular value decomposition of A)*,
(b) $\mathrm{rank}(A) = r$,
(c) $\ker(A) = \mathrm{span}\{v_{r+1}, \ldots, v_n\}$, $\mathrm{Im}(A) = \mathrm{span}\{u_1, \ldots, u_r\}$,
(d) $\|A\| = \sigma_1$, $\|A\|_F^2 = \sigma_1^2 + \cdots + \sigma_p^2$,
(e) $\min_{\|x\|=1} \|Ax\| = \sigma_n$ *if $m \geq n$*,
(f) $\kappa(A) = \sigma_1/\sigma_n$ *if $m = n$, $A \neq 0$*,
(g) *A and A^{T} have the same singular values; in particular, $\|A\| = \|A^{\mathrm{T}}\|$,*
(h) $\|A\| \leq \|A\|_F \leq \sqrt{\mathrm{rank}(A)}\|A\|$.

Proof In the case $p = m \leq n$, we have

$$A = U \cdot \mathrm{diag}_{m,n}(\sigma_1, \ldots, \sigma_m) \cdot V^{\mathrm{T}} = [u_1 \quad \cdots \quad u_m] \begin{bmatrix} \sigma_1 v_1^{\mathrm{T}} \\ \vdots \\ \sigma_m v_m^{\mathrm{T}} \end{bmatrix} = \sum_{i=1}^{m} \sigma_i u_i v_i^{\mathrm{T}}.$$

$$(1.10)$$

The case $n > m$ is treated similarly, which proves the first assertion. The second assertion is immediate from the diagonal form of $U^{\mathrm{T}} A V$.

For showing (c), note that

$$(Av_1, \ldots, Av_n) = AV = U \mathrm{diag}(\sigma_1, \ldots, \sigma_r, 0, \ldots, 0)$$

$$= (\sigma_1 u_1, \ldots, \sigma_r u_r, 0, \ldots, 0)$$

implies the inclusions $\mathrm{span}\{v_{r+1}, \ldots, v_n\} \subseteq \ker(A)$ and $\mathrm{span}\{u_1, \ldots, u_r\} \subseteq \mathrm{Im}(A)$. Equality follows by comparing the dimensions.

Assertion (d) is an immediate consequence of the orthogonal invariance of the spectral norm and the Frobenius norm; cf. Lemma 1.12. For (e) note that

$$\min_{\|x\|=1} \|Ax\| = \min_{\|x\|=1} \|\mathrm{diag}_{m,n}(\sigma_1, \ldots, \sigma_p)x\| = \begin{cases} \sigma_n & \text{if } m \geq n, \\ 0 & \text{otherwise.} \end{cases}$$

For proving (f), suppose $m = n$ and $A \in \mathbb{R}^{n \times n}$ invertible. Then

$$V^{\mathrm{T}} A^{-1} U = \mathrm{diag}(\sigma_1^{-1}, \ldots, \sigma_n^{-1}).$$

Hence $\sigma_n^{-1} \geq \sigma_{n-1}^{-1} \geq \cdots \geq \sigma_1^{-1}$ are the singular values of A^{-1}. Assertion (d) implies $\|A^{-1}\| = \sigma_n^{-1}$. Hence

$$\kappa(A) = \|A\| \cdot \|A^{-1}\| = \frac{\sigma_1}{\sigma_n}.$$

The first part of assertion (g) is trivial; the second easily follows from (d). Finally, assertion (h) follows from (d) by noting that $\sigma_1^2 + \cdots + \sigma_r^2 \leq r\sigma_1^2$. □

We draw now some conclusions from the singular value decomposition. For a square matrix we always have $\kappa(A) \geq 1$. So the best condition one can hope for is $\kappa(A) = 1$. Orthogonal matrices A satisfy this property, since $\|A\| = 1$ (and A^{-1} is orthogonal as well). Of course, any nonzero multiple λA of an orthogonal matrix A also satisfies

$$\kappa(A) = \|\lambda A\| \cdot \|\lambda^{-1} A^{-1}\| = \lambda \lambda^{-1} \|A\| = 1.$$

Proposition 1.15(f) implies that these are all matrices with $\kappa(A) = 1$.

Corollary 1.16 *If $\kappa(A) = 1$, then $\sigma_1 = \cdots = \sigma_n$. This implies that $U^{\mathrm{T}} A V = \sigma_1 I$, and hence $\sigma_1^{-1} A$ is orthogonal.* □

The following results extend Theorem 1.7 in the case of spectral norms.

Theorem 1.17 *Let $A = \sum_{i=1}^{r} \sigma_i u_i v_i^{\mathrm{T}}$ be a singular value decomposition of $A \in \mathbb{R}^{m \times n}$ and $0 \leq k < r = \mathrm{rank}(A)$. Then we have*

$$\min_{\mathrm{rank}(B) \leq k} \|A - B\| = \|A - A_k\| = \sigma_{k+1},$$

where $A_k := \sum_{i=1}^{k} \sigma_i u_i v_i^{\mathrm{T}}$.

Proof As in (1.10) we get $U^{\mathrm{T}} A_k V = \mathrm{diag}(\sigma_1, \ldots, \sigma_k, 0, \ldots, 0)$, which implies that $\mathrm{rank}(A_k) = k$. Moreover, $U^{\mathrm{T}}(A - A_k) V = \mathrm{diag}(0, \ldots, 0, \sigma_{k+1}, \ldots, \sigma_p)$, which implies that $\|A - A_k\| = \sigma_{k+1}$.

Let now $B \in \mathbb{R}^{m \times n}$ with $\mathrm{rank}(B) \leq k$. Then $\dim(\ker B) \geq n - k$ and therefore $\mathrm{span}\{v_1, \ldots, v_{k+1}\} \cap \ker B \neq 0$. Let z be an element of this intersection with $\|z\| = 1$. Then

$$Az = \sum_{i=1}^{r} \sigma_i u_i v_i^{\mathrm{T}} z = \sum_{i=1}^{r} \sigma_i \langle v_i, z \rangle u_i,$$

and hence

$$\|Az\|^2 = \sum_{i=1}^{r} \sigma_i^2 \langle v_i, z \rangle^2 \geq \sum_{i=1}^{k+1} \sigma_i^2 \langle v_i, z \rangle^2 \geq \sigma_{k+1}^2 \sum_{i=1}^{k+1} \langle v_i, z \rangle^2 = \sigma_{k+1}^2.$$

Therefore,

$$\|A - B\|^2 \geq \|(A - B)z\|^2 = \|Az\|^2 \geq \sigma_{k+1}^2,$$

completing the proof of the theorem. □

Corollary 1.18 *The singular values σ_i of A are uniquely determined.* □

We can now extend some of the discussion in Sect. 1.2 from square to rectangular matrices. Put $p := \min\{m, n\}$ and consider the *set of ill-posed matrices*

$$\Sigma := \{A \in \mathbb{R}^{m \times n} \mid \mathrm{rank}(A) < p\}.$$

We may measure the distance to ill-posedness from a matrix $A \in \mathbb{R}^{m \times n}$, similarly as in (1.9), by the spectral norm, resulting in $d(A, \Sigma)$. Alternatively, we may also measure the distance from A to Σ with respect to the Frobenius norm and define

$$d_F(A, \Sigma) := \min\{\|A - B\|_F \mid B \in \Sigma\}.$$

It turns out that this gives the same distance as that given by the spectral norm.

Corollary 1.19 *For $A \in \mathbb{R}^{m \times n}$ we have $d(A, \Sigma) = d_F(A, \Sigma) = \sigma_{\min}(A)$.*

Proof It is sufficient to show that $d_F(A, \Sigma) \leq d(A, \Sigma)$, since the other inequality is obvious. Theorem 1.7 with $k = p - 1$ tells us that $d(A, \Sigma)$ equals the smallest singular value σ_p of A. Let now $A = \sum_{i=1}^{p} \sigma_i u_i v_i^{\mathrm{T}}$ be a singular value decomposition of A. Then $B = \sum_{i=1}^{p-1} \sigma_i u_i v_i^{\mathrm{T}}$ lies in Σ, and $A - B = \sigma_n u_n v_n^{\mathrm{T}}$ has Frobenius norm σ_n. Therefore $d_F(A, \Sigma) \leq \sigma_p$, completing the proof. □

Remark 1.20 The singular value decomposition has a natural extension to complex matrices, and so have all the results in this and the previous sections. More specifically, for $A \in \mathbb{C}^{m \times n}$ there exist unitary matrices $U \in \mathbb{C}^{m \times m}$, $V \in \mathbb{C}^{n \times n}$ as well as real numbers $\sigma_1 \geq \sigma_2 \geq \cdots \geq \sigma_p \geq 0$ such that $U^* A V = \mathrm{diag}_{m,n}(\sigma_1, \ldots, \sigma_p)$. Here $A^* = \bar{A}^{\mathrm{T}}$ is the conjugate transpose of A, and $p = \min\{m, n\}$.

We finish this section with two results that will be needed in Chap. 9. Recall that $\sigma_{\min}(A)$ denotes the smallest singular value of A.

Lemma 1.21 *Let $A \in \mathbb{R}^{m \times n}$ with $n \geq m$ and $\sigma_{\min}(A) > 0$. Denote by B_m and B_n the closed unit balls in $\mathbb{R}^m$ and $\mathbb{R}^n$, respectively. Then we have*

$$\sigma_{\min}(A) = \sup\{\lambda > 0 \mid \lambda B_m \subseteq A(B_n)\}.$$

Proof By Theorem 1.13 we assume without loss of generality that $A = \mathrm{diag}(\sigma_1, \ldots, \sigma_m)$. It follows that

$$A(B_n) = \left\{ y \in \mathbb{R}^m \;\middle|\; \frac{y_1^2}{\sigma_1^2} + \cdots + \frac{y_m^2}{\sigma_m^2} \leq 1 \right\},$$

which is a hyperellipsoid with semiaxes σ_i. This proves the assertion (see Fig. 1.1). □

Remark 1.22 It is sometimes useful to visualize the singular values of A as the lengths of the semiaxes of the hyperellipsoid $\{Ax \mid \|x\| = 1\}$.

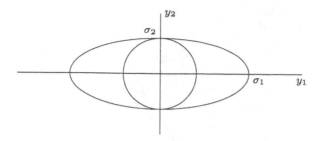

Fig. 1.1 Ball of maximal radius σ_2 contained in an ellipse

We will also need the following perturbation result.

Lemma 1.23 *For $A, B \in \mathbb{R}^{m \times n}$ we have*

$$\left| \sigma_{\min}(A + B) - \sigma_{\min}(A) \right| \le \|B\|.$$

Proof Since A and A^{T} have the same singular values, we assume without loss of generality that $n \ge m$. According to the characterization of $\sigma_{\min}$ in Proposition 1.15, there exists $x \in \mathbb{R}^n$ with $\|x\| = 1$ such that $\|Ax\| = \sigma_{\min}(A)$. Then

$$\sigma_{\min}(A + B) \le \left\| (A + B)x \right\| \le \|Ax\| + \|Bx\| \le \sigma_{\min}(A) + \|B\|.$$

Since A, B were arbitrary, we also get

$$\sigma_{\min}(A) = \sigma_{\min}\big((A + B) + (-B)\big) \le \sigma_{\min}(A + B) + \|B\|.$$

This proves the assertion. $\square$

1.6 Least Squares and the Moore–Penrose Inverse

In Sect. 1.2 we studied the condition of solving a square system of linear equations. If instead, there are more equations than variables (overdetermined case) or fewer equations than variables (underdetermined case), the Moore–Penrose inverse and its condition naturally enter the game.

Let $A \in \mathbb{R}^{m \times n}$ be of maximal rank $p = \min\{m, n\}$ with a singular value decomposition

$$U^{\mathrm{T}} A V = \mathrm{diag}_{m,n}(\sigma_1, \ldots, \sigma_p),$$

where $\sigma_1 \ge \sigma_2 \ge \cdots \ge \sigma_p > 0$. We define the *Moore–Penrose inverse* of A to be the matrix

$$A^{\dagger} = V \mathrm{diag}_{n,m}\big(\sigma_1^{-1}, \ldots, \sigma_p^{-1}\big) U^{\mathrm{T}}.$$

From the geometric description of $A^{\dagger}$ given below, it follows that $A^{\dagger}$ is in fact independent of the choice of the orthogonal matrices U and V.

Fig. 1.2 The spaces $\mathsf{Im}(A)$, $\mathsf{Im}(A)^\perp$ and the points b and c in $\mathbb{R}^m$

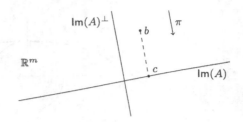

Lemma 1.24

(a) *Suppose that $m \geq n$ and $A \in \mathbb{R}^{m \times n}$ has rank n. Then the matrix A defines a linear isomorphism A_1 of $\mathbb{R}^n$ onto $\mathsf{Im}(A)$, and we have $A^\dagger = A_1^{-1} \circ \pi$, where $\pi : \mathbb{R}^m \to \mathsf{Im}(A)$ denotes the orthogonal projection. In particular, $A^\dagger A = \mathrm{I}$. Moreover, $A^\mathrm{T} A$ is invertible and $A^\dagger = (A^\mathrm{T} A)^{-1} A^\mathrm{T}$.*

(b) *Suppose that $n \geq m$ and $A \in \mathbb{R}^{m \times n}$ has rank m. Then the matrix A defines an isomorphism $A_2 : (\ker A)^\perp \to \mathbb{R}^m$, and we have $A^\dagger = \iota \circ A_2^{-1}$, where $\iota : (\ker A)^\perp \to \mathbb{R}^n$ denotes the embedding. In particular, $AA^\dagger = \mathrm{I}$. Moreover, AA^T is invertible and $A^\dagger = A^\mathrm{T}(AA^\mathrm{T})^{-1}$.*

Proof The claims are obvious for the diagonal matrix $A = \mathsf{diag}_{m,n}(\sigma_1, \ldots, \sigma_p)$ and easily extend to the general case by orthogonal invariance. □

The following is obvious from the definition of $A^\dagger$.

Corollary 1.25 *We have $\|A^\dagger\| = \frac{1}{\sigma_{\min}(A)}$.* □

Suppose we are given a matrix $A \in \mathbb{R}^{m \times n}$, with $m > n$ and $\mathrm{rank}(A) = n$, as well as $b \in \mathbb{R}^m$. Since A, as a linear map, is not surjective, the system $Ax = b$ may have no solutions. We might therefore attempt to find the point $x \in \mathbb{R}^n$ with Ax closest to b, that is, to solve the *linear least squares* problem

$$\min_{x \in \mathbb{R}^n} \|Ax - b\|^2. \tag{1.11}$$

Since A is injective, there is a unique minimizer x for (1.11), namely the preimage of the projection c of b onto $\mathsf{Im}(A)$. From Lemma 1.24(a) it follows immediately that the minimizer can be expressed as $x = A^\dagger b$ (see Fig. 1.2).

For the case of underdetermined systems, we consider instead the case $m < n$ and $\mathrm{rank}(A) = m$. For each $b \in \mathbb{R}^m$, the set of solutions of $Ax = b$ is an affine subspace of $\mathbb{R}^n$ of dimension $n - m$ and therefore contains a unique point of minimal norm. We want to find this point, i.e., to solve

$$\min_{x \mid Ax = b} \|x\|^2. \tag{1.12}$$

Lemma 1.24(b) implies that the solution of (1.12) again satisfies $x = A^\dagger b$.

So the Moore–Penrose inverse naturally yields the solution of linear least squares problems and of underdetermined systems. What is the condition of computing the Moore–Penrose inverse? Theorem 1.5 has a natural extension showing that the quantity

$$\kappa_{rs}(A) := \|A\|_{rs} \|A^\dagger\|_{sr}$$

equals the normwise condition for the computation of the Moore–Penrose inverse.

Theorem 1.26 *Consider*

$$\psi : \left\{ A \in \mathbb{R}^{m \times n} \mid \mathrm{rank}(A) = \min\{m, n\} \right\} \to \mathbb{R}^{m \times n}, \quad A \mapsto A^\dagger.$$

Then we have $\mathrm{cond}^\psi(A) = \kappa_{rs}(A)$ *when errors are measured on the data space with respect to* $\|\ \|_{rs}$ *and on the solution space with respect to* $\|\ \|_{sr}$.

Proof Let $\tilde{A} = A - E$. We claim that for $\|E\| \to 0$, we have

$$\tilde{A}^\dagger - A^\dagger = A^\dagger E A^\dagger + o(\|E\|).$$

For proving this we may assume without loss of generality that $m \geq n$, hence $A^\dagger = (A^T A)^{-1} A^T$, and perform a computation similar to that in the proof of Theorem 1.5. We leave the straightforward details to the reader. The remaining arguments then follow in exactly the same way as in the proof of Theorem 1.5, just by replacing A^{-1} by $A^\dagger$.   □

We note that the solution of linear least squares problems and underdetermined systems has, in contrast to Moore–Penrose inversion, a normwise condition that is only loosely approximated by $\kappa(A)$. Indeed, in 1973, P.-Å. Wedin gave tight upper bounds for the normwise condition numbers $\mathrm{cond}^{\mathsf{LLS}}$ and $\mathrm{cond}^{\mathsf{ULS}}$ for these problems. It follows from these bounds that

$$\Omega\big(\kappa(A)\big) \leq \mathrm{cond}^{\mathsf{LLS}}, \mathrm{cond}^{\mathsf{ULS}} \leq \mathcal{O}\big(\kappa(A)^2\big). \tag{1.13}$$

Interestingly, in contrast to Theorem 1.4, the normwise condition for solving $\min \|Ax - b\|$ depends on b as well as on A.

We finally note that Theorem 1.7 has a natural extension: $\kappa(A)$ is again the relativized inverse of the distance to ill-posedness, where the latter now amounts to rank-deficiency. The following is an immediate consequence of Corollary 1.19.

Corollary 1.27 *For* $A \in \mathbb{R}^{m \times n}$ *we have*

$$\kappa(A) = \frac{\|A\|}{d(A, \Sigma)} = \frac{\|A\|}{d_F(A, \Sigma)},$$

where $\Sigma = \{A \in \mathbb{R}^{m \times n} \mid \mathrm{rank}(A) < \min\{m, n\}\}$.   □

Remark 1.28 The extension of Corollary 1.27 to more general norms as in Corollary 1.8 is false in general.

Chapter 2
Probabilistic Analysis

Recall Corollary 1.6. It tells us that the loss of precision in linear equation solving (via QR Householder factorization) is bounded as

$$\mathsf{LoP}\big(A^{-1}b\big) \leq (2 + C_{rs}) \log_\beta n + \log_\beta \kappa_{rs}(A) + \log_\beta c + o(1),$$

where c, C_{rs} are small constants. While the terms $(2 + C_{rs}) \log_\beta n + \log_\beta c$ point to a loss of approximately $(2 + C_{rs}) \log_\beta n$ figures of precision independently of the data (A, b), the quantity $\log_\beta \kappa_{rs}(A)$, i.e., $\log_\beta \|A\|_{rs} + \log_\beta \|A^{-1}\|_{sr}$, depends on A and does not appear to be a priori estimable.

We already discussed this problem in Sect. O.5.3, where we pointed to a way out consisting in randomizing the data and analyzing the effects of such randomization on the condition number at hand (which now becomes a random variable). In this chapter we become more explicit and actually perform such an analysis for $\kappa_{rs}(A)$.

A cursory look at the current literature shows two different ideas of randomization for the underlying data. In the first one, which lacking a better name we will call *classical* or *average*, data are supposed to be drawn from "evenly spread" distributions. If the space M where data live is compact, a uniform measure is usually assumed. If instead, data are taken from $\mathbb{R}^n$, the most common choice is the multivariate isotropic Gaussian centered at the origin. In the case of condition numbers (which are almost invariably scale-invariant), this choice is essentially equivalent to the uniform measure on the sphere $\mathbb{S}^{n-1}$ of dimension $n - 1$. We will make this precise in Sect. 2.2. Data randomly drawn from these evenly spread distributions are meant to be "average" (whence the name), and the analysis performed for such a randomization is meant to describe the behavior of the analyzed quantity for such an "average Joe" inhabitant of M.

The second idea for randomization, known as *smoothed analysis*, replaces this average data by a small random perturbation of worst-case data. That is, it considers an arbitrary element $\bar{x}$ in M (and thus, in particular, the instance at hand) and assumes that $\bar{x}$ is affected by random noise. The distribution for this perturbed input is usually taken to be centered and isotropic around $\bar{x}$, and with a small variance.

An immediate advantage of smoothed analysis is its robustness with respect to the distribution governing the random noise (see Sect. 2.2 below). This is in con-

P. Bürgisser, F. Cucker, *Condition*,
Grundlehren der mathematischen Wissenschaften 349,
DOI 10.1007/978-3-642-38896-5_2, © Springer-Verlag Berlin Heidelberg 2013

trast to the most common critique of average-case analysis: "A bound on the performance of an algorithm under one distribution says little about its performance under another distribution, and may say little about the inputs that occur in practice" [207].

The main results of this chapter show bounds for both the classical and smoothed analysis of $\log_\beta \kappa_{rs}(A)$, for all choices of $r, s \in \{1, 2, \ldots, \infty\}$. In the first case we obtain $\mathbb{E}(\log_\beta \kappa_{rs}(A)) = \mathcal{O}(\log_\beta n)$. In the second, that for all $\overline{A} \in \mathbb{R}^{n \times n}$, $\mathbb{E}(\log_\beta \kappa_{rs}(A)) = \mathcal{O}(\log_\beta n) + \log_\beta \frac{1}{\sigma}$, where A is randomly drawn from a distribution centered at $\overline{A}$ with dispersion σ (we will be more explicit in Sect. 2.4.3). Therefore, the first result implies that for random data (A, b) we have

$$\mathbb{E}\big(\mathsf{LoP}(A^{-1}b)\big) = \mathcal{O}(\log_\beta n),$$

and the second that for all data $(\overline{A}, \overline{b})$ and random perturbations (A, b) of it,

$$\mathbb{E}\big(\mathsf{LoP}(A^{-1}b)\big) = \mathcal{O}(\log_\beta n) + \log_\beta \frac{1}{\sigma}.$$

2.1 A Crash Course on Integration

Our use of probabilities in the first two parts in this book is limited to the following situation. We endow a space (of data) with a probability distribution and consider a certain real-valued function $g(x)$ of a point x in this space (the running time of a given algorithm with input x, a condition number of $x, \ldots$) as a random variable. The goal is to estimate some quantities (the probability that $g(x)$ is at least K for a given K, the expected value of $g, \ldots$) that provide some picture of the behavior of g (what is the probability that the algorithm just mentioned will need more than a certain number of iterations, which condition should we expect for a random input $x, \ldots$).

Data Spaces and Measures A first step towards a formal setting for this background is the description of our spaces of data. For a time to come we will confine these spaces to being of a precise form, which, lacking some established name, we will call by one to suit our development.

We give the name *data space* to any open subset M in a product of Euclidean spaces and spheres. That is, there exist $m, n_1, n_2, \ldots, n_k \in \mathbb{N}$ such that

$$M \subseteq \mathbb{R}^m \times \mathbb{S}^{n_1-1} \times \mathbb{S}^{n_2-1} \times \cdots \times \mathbb{S}^{n_k-1}$$

is an open subset. In a second step we will endow the data space M with a probability measure describing the law governing data sampling from M.

Before doing so, we briefly recall some basic concepts of integration, tailored to our purposes. It is not our goal to dwell on the subtleties of measure theory. Rather, we intend to collect here in a coherent (and correct) way the basic facts of integration that are needed in the later developments of the book.

Before defining measures on our data space M, we need to introduce the abstract notion of measurability. By a *measurable* (or *Borel-measurable*) set A in a data space M we understand a subset $A \subseteq M$ that can be obtained from open and closed subsets of M by countably many operations of taking unions and intersections. In particular, open and closed subsets of M are measurable.

Let N be a further data space. A function $f : M \to N$ is called *measurable* if $f^{-1}(B)$ is measurable for all measurable sets B in N. In particular, continuous functions are measurable. The *indicator function* $\mathbb{1}_A$ of a subset $A \subseteq M$ is defined by

$$\mathbb{1}_A(x) = \begin{cases} 1 & \text{if } x \in A, \\ 0 & \text{otherwise.} \end{cases}$$

Clearly, $\mathbb{1}_A$ is measurable iff A is measurable.

A *measure* on the data space M is a function μ assigning a value $\mu(A) \in [0, \infty]$ to each measurable set A of M such that $\mu(\emptyset) = 0$ and countable additivity holds, that is,

$$\mu\left(\bigcup_{i=0}^{\infty} A_i \right) = \sum_{i=0}^{\infty} \mu(A_i)$$

for each sequence A_i of pairwise disjoint measurable sets in M.

On the data space M we have a natural measure vol_M that can be interpreted as the volume in a higher-dimensional sense. In the case $M = \mathbb{R}^1$, the measure vol_M is characterized by giving the length of intervals $[a, b]$, that is, by the requirement $\mathsf{vol}_M([a, b]) = b - a$ for all $a \leq b$. In the case $M = \mathbb{S}^1$, vol_M is similarly characterized as measuring angles.

Products of Data Spaces, Fubini's Theorem   One can build up vol_M from simpler components by the product measure construction. Assume that μ_i is a measure on a data space M_i for $i = 1, 2$. It can be shown that there exists a uniquely determined measure μ on $M_1 \times M_2$, called the *product measure*, with the property that

$$\mu(A_1 \times A_2) = \mu(A_1) \cdot \mu(A_2) \tag{2.1}$$

for all measurable sets A_i in M_i. One can formally define $\mathsf{vol}_{\mathbb{R}^m}$ as the m-fold product of the measures $\mathsf{vol}_{\mathbb{R}}$. The measure $\mathsf{vol}_{\mathbb{S}^{n-1}}$ on the sphere can be defined by setting $\mathsf{vol}_{\mathbb{S}^{n-1}}(A) := n \, \mathsf{vol}_{\mathbb{R}^n}(B_A)$, where $B_A := \{tx \mid x \in A, \ 0 \leq t \leq 1\}$. (In the case $n = 2$ this gives the angle, as mentioned above.) Altogether, by the product construction, we have a well-defined measure vol_M on M. We say that a *property* of elements of M holds *almost everywhere* if it holds for all elements except those in a set of measure zero (with respect to vol_M).

We turn now to the topic of integration. One can assign to any measurable function $f : M \to [0, \infty]$, in a unique way, a value

$$\int_M f = \int_{x \in M} f(x)\, dx,$$

the integral of f over M, by some limit process along with the basic requirement that

$$\int_M \sum_{i=1}^n c_i \mathbb{1}_{A_i} = \sum_{i=1}^n c_i \operatorname*{vol}_M(A_i)$$

for measurable sets A_i and $c_i \geq 0$. The function f is called *integrable* if $\int_M f$ is finite. One can show that in this case, $\{x \in M \mid f(x) = \infty\}$ has measure zero and thus is irrelevant for the integration. A measurable function $f : M \to \mathbb{R}$ is called *integrable* if it can be written as the difference of two integrable functions with non-negative values. The map $f \mapsto \int_M f$ can be shown to be linear and continuous with respect to the L_1-norm given by $\|f\|_1 := \int_M |f|$. We note that changing the value of a function f on a set of measure zero does not alter the value of the integral. We will therefore write $\int_M f$ even when f is only defined almost everywhere. We also write $\int_A f := \int_M f \mathbb{1}_A$ if the set $A \subseteq M$ and the function f on M are measurable.

How can one possibly compute such integrals? An important tool is *Fubini's theorem*, which allows one to reduce the computation of integrals over a product $M \times N$ to integrals over the factor spaces M, N. Suppose that $A \subseteq M \times N$ is measurable. For $x \in M$ we define the set $A_x := \{y \in N \mid (x, y) \in A\}$ (which can be shown to be measurable). Then Fubini's theorem states that for an integrable function $f : M \times N \to \mathbb{R}$, the map $x \mapsto \int_{y \in A_x} f(x, y)\, dy$ is integrable for almost all $x \in M$, and we have

$$\int_{(x,y) \in A} f(x, y)\, d(x, y) = \int_{x \in M} \left(\int_{y \in A_x} f(x, y)\, dy \right) dx. \qquad (2.2)$$

By this theorem one can in principle reduce the computation of integrals over $\mathbb{R}^m$ to integrals over the real line $\mathbb{R}$, a good number of which are known to us from elementary calculus. *Tonelli's theorem* is a subtle variant of Fubini's theorem. It says that (2.2) holds for any *nonnegative* measurable function $f : M \times N \to \mathbb{R}$ (without the assumption of f being integrable).

The Transformation Formula   Another important tool to compute integrals is the *change of variables* or *transformation formula*. Suppose we have a linear isomorphism $\Lambda : \mathbb{R}^m \to \mathbb{R}^m$. It is a well-known fact that the determinant of Λ is the volume-stretching factor of the map Λ. More precisely, we have $\operatorname{vol}_{\mathbb{R}^m}(\Lambda(A)) = |\det \Lambda| \cdot \operatorname{vol}_{\mathbb{R}^m}(A)$ for a measurable subset A of M.

The transformation formula extends this finding to the nonlinear case. Suppose that M and N are open subsets of $\mathbb{R}^m$, and $\psi : M \to N$ is a *diffeomorphism* (here and in what follows we will assume that diffeomorphisms have continuous derivatives). This means that ψ is bijective and both ψ and its inverse ψ^{-1} are differentiable. The derivative of ψ at $x \in M$ is the linearization of ψ given by the *Jacobian matrix* $D\psi(x) = [\partial_{x_j} \psi_i(x)] \in \mathbb{R}^{m \times m}$. The absolute value of its determinant,

$$\mathrm{J}\psi(x) := |\det D\psi(x)|,$$

is called the *Jacobian of ψ at x*. The transformation formula for $\mathbb{R}^m$ states that for any integrable function $f : N \to \mathbb{R}$ we have

$$\int_{y \in N} f(y)\, dy = \int_{x \in M} f(\psi(x))\, \mathrm{J}\psi(x)\, dx. \tag{2.3}$$

Data spaces are more general than open subsets of Euclidean spaces. Fortunately, formula (2.3) carries over to this more general situation. The only thing we need to clarify is the notion of the Jacobian in this more general setting.

Suppose first that $M = \mathbb{S}^{n-1}$ and $p \in M$. The orthogonal projection

$$\gamma : \mathbb{S}^{n-1} \to \mathbb{R}p^{\perp}, \quad x \mapsto x - \langle x, p \rangle p, \tag{2.4}$$

defines a bijection of the hemisphere given by $\|x\| = 1$, $\langle x, p \rangle > 0$, to the open unit ball in the orthogonal complement $\mathbb{R}p^{\perp} \simeq \mathbb{R}^{n-1}$ of $\mathbb{R}p$. We call the map γ the *standard chart* of $\mathbb{S}^{n-1}$ at p. Note that in the special case $p = e_n = (0, \ldots, 0, 1)$, writing $B = \{x \in \mathbb{R}^{n-1} \mid \sum_i x_i^2 < 1\}$, the inverse of this map reads as

$$B \to \mathbb{S}^{n-1}, \quad (x_1, \ldots, x_{n-1}) \mapsto \left(x_1, \ldots, x_{n-1}, \sqrt{1 - x_1^2 - \cdots - x_{n-1}^2}\right). \tag{2.5}$$

The standard chart of $\mathbb{R}^m$ is defined to be the identity map on $\mathbb{R}^m$. By taking products, we arrive at a notion of a *standard chart at a point p of a data space M*, which is a bijection $\gamma_{M,p}$ of a certain open neighborhood of p in M to an open subset of $\mathbb{R}^n$, where n is the dimension of M.

Suppose now we have a bijection $\psi : M \to N$ between data spaces such that $n = \dim M = \dim N$. For any $p \in M$ we can form the composition $\tilde{\psi}_p := \gamma_{N,\psi(p)} \circ \psi \circ \gamma_{M,p}^{-1}$ of ψ with the standard charts. Then, $\tilde{\psi}_p : U \subseteq \mathbb{R}^n \to \mathbb{R}^n$ for some open subset U. We say that ψ is differentiable at p if $\tilde{\psi}_p$ is. In this case, we define the *Jacobian of ψ at p* by $\mathrm{J}\psi(p) := \mathrm{J}\tilde{\psi}_p(p')$, where $\gamma_{M,p}(p') = p$.

Theorem 2.1 (Transformation formula) *Let $\psi : M \to N$ be a diffeomorphism between data spaces and let $f : N \to \mathbb{R}$ be an integrable function. Then we have*

$$\int_{y \in N} f(y)\, dy = \int_{x \in M} f(\psi(x)) \cdot \mathrm{J}\psi(x)\, dx. \qquad \square$$

An important application of this formula is *integration in polar coordinates* (in Euclidean space). Consider the diffeomorphism

$$\psi_{pc} : \mathbb{S}^{n-1} \times (0, \infty) \to \mathbb{R}^n \setminus \{0\}, \quad (u, r) \mapsto ru, \tag{2.6}$$

describing polar coordinates u, r on $\mathbb{R}^n$. The next result shows the usefulness of polar coordinates in integration. Before stating it, let us point out a remarkable symmetry property of this map.

Let $\mathrm{GL}_n(\mathbb{R})$ denote the *general linear group* over $\mathbb{R}$, i.e., the group of invertible $n \times n$ real matrices. Also, let $\mathscr{O}(n) := \{g \in \mathrm{GL}_n(\mathbb{R}) \mid gg^{\mathrm{T}} = \mathrm{I}_n\}$ denote the *orthogonal group* (i.e., the group of orthogonal linear endomorphisms of $\mathbb{R}^n$). This group acts on $\mathbb{R}^n$ via $g \cdot x := gx$, and the induced action on the sphere $\mathbb{S}^{n-1}$ is transitive. We may also let $\mathscr{O}(n)$ act on $\mathbb{S}^{n-1} \times (0, \infty)$ by setting $g \cdot (u, r) := (gu, r)$. Then the map ψ_{po} is $\mathscr{O}(n)$-equivariant, that is, $\psi_{pc}(g \cdot (u, r)) = g \cdot \psi_{pc}(u, r)$ for all $g \in \mathscr{O}(n)$ and $(u, r) \in \mathbb{S}^{n-1} \times (0, \infty)$. From this property, it is straightforward to derive that the Jacobian of ψ_{pc} is *invariant under the action of $\mathscr{O}(n)$*, that is, $\mathrm{J}\psi_{pc}(g \cdot (u, r)) = \mathrm{J}\psi_{pc}(u, r)$. This observation often allows us to simplify the writing of proofs considerably. In fact, the use of orthogonal (or some other group) invariance will be pervasive in this book.

Corollary 2.2 *For any integrable function $f : \mathbb{R}^n \to \mathbb{R}$ we have*

$$
\int_{y \in \mathbb{R}^n} f(y)\, dy = \int_{(u, r) \in \mathbb{S}^{n-1} \times (0, \infty)} f\big(\psi_{pc}(u, r)\big) \cdot r^{n-1}\, du\, dr.
$$

Proof By Theorem 2.1 it is sufficient to show that $\mathrm{J}\psi_{pc}(u, r) = r^{n-1}$. By orthogonal invariance it suffices to prove this at $u = e_n = (0, 0, \dots, 0, 1)$. From (2.5) it follows that the inverse of the standard chart of $\mathbb{S}^{n-1} \times (0, \infty)$ at (e_n, r) is the map $B(0, 1) \times (0, \infty) \to \mathbb{S}^{n-1} \times (0, \infty)$ given by

$$
(u_1, \dots, u_{n-1}, r) \mapsto \left(u_1, \dots, u_{n-1}, \sqrt{1 - u_1^2 - \cdots - u_{n-1}^2}, r\right).
$$

By composing with ψ_{pc} this gives the map $\tilde{\psi}_{pc} : B(0, 1) \times (0, \infty) \to \mathbb{R}^n$,

$$
(u_1, \dots, u_{n-1}, r) \mapsto \left(ru_1, \dots, ru_{n-1}, r\sqrt{1 - u_1^2 - \cdots - u_{n-1}^2}\right).
$$

It is clear that $D\tilde{\psi}_{pc}(0, r) = \mathrm{diag}(r, \dots, r, 1)$. Hence we obtain $\mathrm{J}\psi_{pc}(e_n, r) = \mathrm{J}\tilde{\psi}_{pc}(0, r) = r^{n-1}$ as claimed. $\qquad\square$

A second application of the transformation formula is *integration in polar coordinates on a sphere*. Let $p = e_{n+1}$ be the "north pole" of $\mathbb{S}^n$ and consider the diffeomorphism

$$
\psi_{pcs} : \mathbb{S}^{n-1} \times (0, \pi) \to \mathbb{S}^n \setminus \{\pm e_{n+1}\}, \quad (u, \theta) \mapsto \big((\sin\theta)u, \cos\theta\big). \tag{2.7}
$$

Note that we may interpret u, θ as polar coordinates on the sphere $\mathbb{S}^n$ with respect to (the center) e_{n+1}.

Corollary 2.3 *For any integrable function* $f : \mathbb{S}^n \to \mathbb{R}$ *we have*

$$\int_{y \in \mathbb{S}^n} f(y) \, dy = \int_{(u,\theta) \in \mathbb{S}^{n-1} \times (0,\pi)} f\big(\psi_{pcs}(u,\theta)\big) \cdot (\sin\theta)^{n-1} \, du \, d\theta.$$

Proof By Theorem 2.1 it is sufficient to show that $J\psi_{pcs}(u,\theta) = (\sin\theta)^{n-1}$. By orthogonal invariance it suffices to prove this at $u = e_n$. We fix $\overline{\theta}$ and put

$$\overline{y} := \psi_{pcs}(e_n, \overline{\theta}) = (0, \ldots, 0, \sin\overline{\theta}, \cos\overline{\theta}).$$

Equation (2.4) and a short calculation imply that the standard chart of $\mathbb{S}^n$ at $\overline{y}$ is given by

$$\mathbb{S}^n \to \mathbb{R}\overline{y}^\perp, \quad y \mapsto y - \langle y, \overline{y}\rangle \overline{y} = \big(y_1, \ldots, y_{n-1}, y_n \cos^2\overline{\theta} - y_{n+1} \cos\overline{\theta} \sin\overline{\theta}\big).$$

To get coordinates for $\mathbb{R}\overline{y}^\perp$, we use the orthogonal map

$$\mathbb{R}\overline{x}^\perp \to \mathbb{R}^n,$$

$$(y_1, \ldots, y_{n-1}, y_n, y_{n+1}) \mapsto (y_1, \ldots, y_{n-1}, y_n / \cos\overline{\theta}).$$

This gives the standard chart

$$\mathbb{S}^n \to \mathbb{R}^{n-1}, \quad y \mapsto (y_1, \ldots, y_{n-1}, y_n \cos\overline{\theta} - y_{n+1} \sin\overline{\theta}),$$

of $\mathbb{S}^n$ at $\overline{y}$. Recall that the inverse of the standard chart of $\mathbb{S}^{n-1}$ at e_n is given by (2.5). By composing ψ_{pcs} with these standard charts we obtain the map $\tilde{\psi}_{pcs}$ given by

$$(u_1, \ldots, u_{n-1}, \theta,)$$

$$\mapsto \left(u_1 \sin\theta, \ldots, u_{n-1} \sin\theta, \left(1 - \sum_{i=1}^{n-1} u_i^2\right)^{1/2} \sin\theta \cos\overline{\theta} - \cos\theta \sin\overline{\theta}\right).$$

A calculation shows that $D\tilde{\psi}_{pcs}(e_n, \overline{\theta}) = \mathrm{diag}(\sin\overline{\theta}, \ldots, \sin\overline{\theta}, 1)$, which implies that $J\psi_{pcs}(e_n, \overline{\theta}) = (\sin\overline{\theta})^{n-1}$ and completes the proof. $\qquad\square$

2.2 A Crash Course on Probability: I

We develop here some basics of probability theory and show how to apply them in our cases of interest, which are mainly Gaussian distributions in Euclidean spaces, uniform distributions on spheres, and their products on data spaces.

2.2.1 Basic Facts

Densities and Probabilities By a *probability measure on a data space M* one understands a measure μ on M such that $\mu(M) = 1$. All the measures we are interested in can be defined in terms of a probability density, defined as follows.

Definition 2.4 A *(probability) density* on a data space M is a measurable function $f : M \to [0, \infty]$ such that $\int_M f = 1$.

A density f on M defines a probability measure μ on M by

$$\mu(A) := \int_M \mathbb{1}_A.$$

The additivity properties of the integral readily imply that μ is indeed a probability measure. Up to changes on a set of measure zero, the density f is uniquely determined by μ.

Example 2.5 Let M be a data space of finite volume. Then the constant function on M with value $\operatorname{vol}_M(M)^{-1}$ is a density on M. The corresponding probability measure is called the *uniform distribution* $U(M)$ on M. More generally, let A be a measurable subset of a data space M such that $\operatorname{vol}_M(A)$ is finite. Then $\operatorname{vol}_M(A)^{-1}\mathbb{1}_A$ is a density on M, and one calls the corresponding probability measure the *uniform distribution on A*.

It is common to say that via the density f, we endow M with a *probability distribution*, or simply a *distribution*. Even though we will sometimes use interchangeably the terms "probability distribution" and "probability measure" induced by f, we tend to denote them differently (and use expressions such as $U(M)$ to denote a uniform distribution on M and $N(0, I_n)$ to denote the standard normal distribution on $\mathbb{R}^n$; see below). In this context, it is also common to call any measurable subset of M an *event*.

Let M and N be two data spaces and let $f : M \times N \to [0, \infty]$ be a density on the product $M \times N$. We can associate with f its *marginal densities* on M and N defined as follows:

$$f_M(x) := \int_{y \in N} f(x, y)\, dy \quad \text{and} \quad f_N(y) = \int_{x \in M} f(x, y)\, dx. \tag{2.8}$$

It follows from Fubini's theorem (2.2) that these are indeed probability densities.

One says that *M and N are independent* if $f(x, y) = f_M(x) f_N(y)$ for all $x \in M$ and $y \in N$. We note that in this case, we have $\mu(A \times B) = \mu_M(A) \cdot \mu_N(B)$, where μ, μ_M, and μ_N denote the measures associated with the densities f, f_M, and f_N, respectively, and $A \subseteq M$ and $B \subseteq N$ are measurable sets. In other words, μ is the product measure of μ_M and μ_N. We also note that if we start with any densities

f_M and f_N on M and N, respectively, and endow $M \times N$ with the product density $f_M(x) f_N(y)$, then M and N become independent by construction.

In the situation where M and N are not independent with respect to the density f on $M \times N$, it is convenient to introduce conditional densities. The *conditional density of x, given y*, is defined as

$$f(x \mid y) := \frac{f(x, y)}{f_N(y)} \tag{2.9}$$

(we assume here $f_N(y) \neq 0$). It is clear from the definition that $M \to \mathbb{R}, x \mapsto f(x \mid y)$ is actually a density on M. Note that $f(x \mid y) = f_M(x)$ if M and N are independent. When we happen to know (or we can bound) the conditional density, then we can derive information on f by means of the equation

$$f(x, y) = f(x \mid y) \cdot f_N(y), \tag{2.10}$$

which is just a rewriting of (2.9).

Remark 2.6 Equation (2.10) can be interpreted in the following operational way in terms of random sampling. First we sample $y \in N$ according to the marginal density f_N. In a second step, we sample $x \in M$ according to the conditional density $f(\cdot \mid y)$. Then the obtained pair (x, y) is random according to the density f.

Random Variables   One of the most fundamental notions in probability is that of a random variable.

Definition 2.7 Let M be endowed with the probability density f. A *random variable* defined on M is a measurable function $Z: M \to \mathbb{R}$ (defined almost everywhere). The *expected value* or *expectation* of Z is defined by

$$\mathbb{E}(Z) := \int_{x \in M} Z(x) f(x) \, dx$$

if the integral is well defined. The *variance* of Z is defined as

$$\mathsf{Var}(Z) := \mathbb{E}\big((Z - \mathbb{E}(Z))^2\big).$$

Example 2.5 (continued)   Let Z be the map $(0, 1) \hookrightarrow \mathbb{R}, x \mapsto x$. The expected value of Z for the uniform distribution on $(0, 1)$ is $\frac{1}{2}$, and its variance is $\frac{1}{12}$. The expected value of the function $Z(x) = e^x$ is $\int_0^1 e^x \, dx = e - 1$.

A few words on notation. If $\mathcal{D}$ denotes a probability distribution on M associated with the probability measure μ, and R is a (measurable) predicate on M, we will write

$$\underset{x \sim \mathcal{D}}{\mathsf{Prob}}\{R(x)\} := \mu\big(\{x \in M \mid R(x)\}\big).$$

Also, for a random variable Z on M, we define the measure μ_Z on $\mathbb{R}$ given by $\mu_Z(A) := \mu\{Z^{-1}(A)\}$. This is a probability measure on the data space $\mathbb{R}$. It is common to use the shorthand notation $\mathsf{Prob}\{Z \geq t\} := \mu_Z\{[t, \infty)\}$.

In case the random variable takes only nonnegative values, we can express its expectation differently.

Proposition 2.8 *Let $Z\colon M \to [0, \infty)$ be integrable. Then*

$$\mathbb{E}(Z) = \int_0^\infty \mathsf{Prob}\{Z \geq t\}\, dt$$

Proof We apply Fubini's theorem (2.2) to the set $A := \{(x, t) \in M \times [0, \infty) \mid Z(x) \geq t\}$, obtaining

$$\int_{(x,t)\in A} f(x)\, d(x, t) = \int_0^\infty \left(\int_{x \in A_t} f(x)\, dx \right) dt = \int_0^\infty \mathsf{Prob}\{Z \geq t\}\, dt.$$

Applying Fubini again (and thus interchanging the order of integration) yields

$$\int_{(x,t)\in A} f(x)\, d(x, t) = \int_{x \in M} \left(\int_{t \in A_x} f(x)\, dt \right) dx = \int_{x \in M} f(x) Z(x) = \mathbb{E}(Z).$$

$\square$

The following simple corollary is at the core of numerous probability tail estimates.

Corollary 2.9 (Markov's inequality) *Let $Z\colon M \to [0, \infty)$ be integrable. Then for all $t > 0$, we have*

$$\mathsf{Prob}\{Z \geq t\} \leq \frac{1}{t}\mathbb{E}(Z).$$

Proof Proposition 2.8 implies that

$$\mathbb{E}(Z) \geq \int_0^t \mathsf{Prob}\{Z \geq \tau\}\, d\tau \leq t\,\mathsf{Prob}\{Z \geq t\}.$$

$\square$

Here is a general result relating expectation and variance of a random variable.

Proposition 2.10 *Let Z be a random variable on a data space M endowed with a density f. Then $\mathbb{E}(Z) \leq \sqrt{\mathbb{E}(Z^2)}$.*

Proof The functional $(Y, Z) \mapsto \int_{x \in M} Y(x) Z(x) f(x)\, dx$ defines an inner product on the linear space of random variables X on M satisfying $\mathbb{E}(X^2) < \infty$. When $Y = 1$, the Cauchy–Schwarz inequality yields

$$\mathbb{E}(Z) = \int_M Z(x) f(x)\, dx \leq \sqrt{\int_M Z(x)^2 f(x)\, dx} \sqrt{\int_M f(x)\, dx} = \sqrt{\mathbb{E}(Z^2)}. \qquad \square$$

Pushforward Measures Suppose that $\psi : M \to N$ is any measurable map between data spaces. In general, a probability measure μ_M on M induces a probability measure on N via ψ defined by

$$\mu_N(B) := \mu_M\big(\psi^{-1}(B)\big)$$

for measurable sets $B \subseteq N$. One calls μ_N the *pushforward measure* of μ_M.

For instance, a random variable $Z : M \to \mathbb{R}$ has an associated probability distribution on $\mathbb{R}$, sometimes called the *distribution of Z*, which is nothing but the pushforward of μ_M with respect to Z. We already met this distribution when we were introducing the notation $\mathsf{Prob}\{Z \geq t\} := \mu(\{x \in M \mid Z(x) \geq t\})$.

In our situations of interest, μ_M is given by a density f_M. If ψ happens to be a diffeomorphism, then the pushforward of μ_M has a density as well that can be explicitly calculated with the transformation formula.

Proposition 2.11 *Let* $\psi : M \to N$ *be a diffeomorphism of data spaces and let* μ_M *be a probability measure on* M *with the density* f_M. *Then the pushforward measure* μ_N *of* μ_M *has the density*

$$f_N(y) = \frac{f_M(x)}{\mathsf{J}\psi(x)}, \qquad \text{where } x = \psi^{-1}(y).$$

Moreover, for any random variable $Z : N \to \mathbb{R}$, *we have*

$$\mathbb{E}_{\mu_N}(Z) = \mathbb{E}_{\mu_M}(Z \circ \psi), \qquad (2.11)$$

where the expectations refer to μ_N *and* μ_M *respectively.*

Proof Let $B \subseteq N$ be measurable and set $A := \psi^{-1}(B)$. Applying Theorem 2.1 to the function $f_N \mathbb{1}_B$ implies

$$\int_{y \in B} f_N(y)\, dy = \int_{x \in A} f_M(x)\, dx.$$

Hence μ_N has the density f_N. The second assertion follows by applying Theorem 2.1 to $Z f_N$. $\qquad \square$

Remark 2.12 Equation (2.11) also holds when ψ is not a diffeomorphism, but we will not prove this here. Instead, we will see a general result extending both Propositions 2.14 and 2.11 in Sect. 17.3.

Independence Suppose that Z and W are random variables on the data space M, endowed with the density f. Let us denote by μ_Z and μ_W their probability measures on $\mathbb{R}$. The map $M \to \mathbb{R}^2$, $x \mapsto (Z(x), W(x))$, induces a pushforward $\mu_{Z,W}$ on $\mathbb{R}^2$. One calls the random variables Z and W *independent* if $\mu_{Z,W}$ is the product measure of μ_Z and μ_W; compare (2.1). This means that for all $s, t \in \mathbb{R}$ we have

$$\mathrm{Prob}\{Z \geq s \text{ and } W \geq t\} = \mathrm{Prob}\{Z \geq s\} \cdot \mathrm{Prob}\{W \geq t\}.$$

Proposition 2.13 *Suppose that Z and W are independent random variables on M. Then $\mathbb{E}(ZW) = \mathbb{E}(Z)\mathbb{E}(W)$, provided Z and W are integrable.*

Sketch of proof Suppose first that $Z = \mathbb{1}_A$ and $W = \mathbb{1}_B$ are indicator functions of $A, B \subseteq M$. The independence of Z and W means that $\mathrm{Prob}(A \cap B) = \mathrm{Prob}(A)\,\mathrm{Prob}(B)$. Since $ZW = \mathbb{1}_{A \cap B}$, we have

$$\mathbb{E}(ZW) = \mathrm{Prob}(A \cap B) = \mathrm{Prob}(A)\,\mathrm{Prob}(B) = \mathbb{E}(Z)\mathbb{E}(W).$$

By the linearity of expectation this immediately extends to random variables that are a finite linear combination of indicator functions. Finally, the assertion follows by a limit argument (compare the definition of the integral in Sect. 2.1). $\square$

Conditional Expectations Let M, N be data spaces and let f be a density in $M \times N$. Let f_M and f_N be the marginal densities on M and N, respectively. Finally, let $\varphi : M \times N \to \mathbb{R}$ be a random variable.

An element $x_0 \in M$ determines a random variable $\varphi_{x_0} : N \to \mathbb{R}$ given by $\varphi_{x_0}(y) := \varphi(x_0, y)$. The *conditional expectation* of φ (with respect to the event $x = x_0$), which we denote by $\mathbb{E}_{N_{x_0}} \varphi_{x_0}$ or sometimes by $\mathbb{E}_{y \in N_{x_0}}(\varphi(x, y) \mid x = x_0)$, is the expectation of φ_{x_0} with respect to the conditional density $f(y \mid x_0)$, i.e.,

$$\mathop{\mathbb{E}}_{N_{x_0}} \varphi_{x_0} := \int_{y \in N} \varphi_{x_0}(y) f(y \mid x_0)\, dy,$$

with the left-hand side defined only if both $f(y \mid x_0)$ and the integral on the right-hand side exist. Here we wrote N_{x_0} at the left to emphasize that the distribution on N is the one given by the conditional density $f(y \mid x_0)$. By construction, this is a random variable on M,

$$x_0 \mapsto \mathop{\mathbb{E}}_{N_{x_0}} \varphi_{x_0}.$$

The following result ties the expectation of this random variable to that of φ. It will be helpful in many computations in which conditional expectations are easier to estimate than unrestricted ones.

Proposition 2.14 *For all integrable $\varphi : M \times N \to \mathbb{R}$ we have*

$$\mathop{\mathbb{E}}_{M \times N} \varphi = \mathop{\mathbb{E}}_{x_0 \in M} \mathop{\mathbb{E}}_{y \in N_{x_0}} \big(\varphi(x, y) \mid x = x_0\big).$$

In particular, if $\mathbb{E}_{N_{x_0}} \varphi_{x_0} \leq K$ for almost all $x_0 \in M$, then $\mathbb{E}_{M \times N} \varphi \leq K$, and the same is true for lower bounds.

Proof We have

$$\underset{M \times N}{\mathbb{E}} \varphi = \int_{M \times N} \varphi(x, y) f(x, y) \, dy \, dx$$

$$= \int_{M \times N} \varphi(x, y) f(y \mid x) f_M(x) \, dy \, dx$$

$$\overset{(2.2)}{=} \int_M \left(\int_N \varphi_x(y) f(y \mid x) \, dy \right) f_M(x) \, dx$$

$$= \underset{x_0 \in M}{\mathbb{E}} \underset{N_{x_0}}{\mathbb{E}} \varphi_{x_0}. \qquad \square$$

We next describe the basic probability distributions occurring in this book.

2.2.2 Gaussian Distributions

Take $M = \mathbb{R}^n$ for some n. The most important example of a density in this context is the *isotropic multivariate Gaussian*. For a point $a \in \mathbb{R}^n$ and $\sigma > 0$, we consider the density $\varphi_n^{a,\sigma} : \mathbb{R}^n \to (0, \infty)$ given by

$$\varphi_n^{a,\sigma}(x) = \frac{1}{(2\pi\sigma^2)^{\frac{n}{2}}} e^{-\frac{\|x-a\|^2}{2\sigma^2}}.$$

Using Fubini's theorem and the well-known fact that

$$\frac{1}{\sqrt{2\pi\sigma^2}} \int_{-\infty}^{\infty} e^{-\frac{x^2}{2\sigma^2}} \, dx = 1,$$

one sees that $\varphi_n^{a,\sigma}$ is indeed a density on $\mathbb{R}^n$.

We denote the distribution associated with $\varphi_n^{a,\sigma}$ by $N(a, \sigma^2 I_n)$, and its induced measure on $\mathbb{R}^n$ by $\gamma_n^{a,\sigma}$. When $a = 0$ and $\sigma = 1$, it is commonly referred to as the *standard normal* (or *standard Gaussian*) distribution in $\mathbb{R}^n$, and its density and measure are denoted by φ_n and γ_n.

Lemma 2.15 *We have $\mathbb{E}(x_i) = a_i$ and $\mathbb{E}((x_i - a_i)(x_j - a_j)) = \sigma^2 \delta_{ij}$.*

Proof Since the density $\varphi_n^{a,\sigma}$ is invariant under the map $x \mapsto 2a - x$ (reflection at a), we have $\mathbb{E}(x_i) = 2a_i - \mathbb{E}(x_i)$ and hence $\mathbb{E}(x_i) = a_i$. For the second property we may assume without loss of generality that $a = 0$ and $\sigma = 1$. A direct calculation shows that

$$\frac{1}{\sqrt{2\pi}} \int_{-\infty}^{\infty} t^2 e^{-\frac{t^2}{2}} \, dt = 1,$$

which implies $\mathbb{E}(x_i^2) = 1$. Suppose now $i \neq j$. Since the density $\varphi_n^{0,1}$ is invariant under $x_i \mapsto -x_i$, leaving the other coordinates of x fixed, we have $\mathbb{E}(x_i x_j) = \mathbb{E}(-x_i x_j)$ and hence $\mathbb{E}(x_i x_j) = 0$. $\square$

We may interpret $x = (x_1, \ldots, x_n)$ as a *random vector* on $\mathbb{R}^n$. The expectation $\mathbb{E}(x)$ of x is componentwise defined and thus equals a, which is called the *center* of the Gaussian distribution. The distribution is called *centered* if $a = 0$.

Furthermore, one calls $\mathrm{Cov}(x) := [\mathbb{E}((x_i - a_i)(x_j - a_j))]_{1 \leq i, j \leq n}$ the *covariance matrix* of the random vector x. Thus we have $\mathrm{Cov}(x) = \sigma^2 I_n$ in our situation. We call σ^2 the *variance* of the isotropic Gaussian distribution. Note also that if x has the distribution $N(a, \sigma^2 I_n)$, then $x - v$ has the distribution $N(a - v, \sigma^2 I_n)$, for any $v \in \mathbb{R}^n$.

It is common to denote the distribution function of the one-dimensional standard Gaussian distribution γ_1 by

$$\Phi(t) := \frac{1}{\sqrt{2\pi}} \int_{-\infty}^t e^{-\frac{t^2}{2}} dt$$

and to set $\Psi(t) := 1 - \Phi(t)$ for $t \in \mathbb{R}$. This function can be bounded as follows.

Lemma 2.16 *We have* $\Psi(t) \leq \frac{1}{t\sqrt{2\pi}} e^{-\frac{t^2}{2}}$ *for* $t > 0$.

Proof Making the substitution $x = \tau^2/2$, we obtain

$$\Psi(t) = \frac{1}{\sqrt{2\pi}} \int_t^\infty e^{-\frac{\tau^2}{2}} d\tau = \frac{1}{\sqrt{2\pi}} \int_{t^2/2}^\infty \frac{1}{\tau} e^{-x} dx \leq \frac{1}{t\sqrt{2\pi}} e^{-\frac{t^2}{2}}.$$ $\square$

The Gaussian distribution has several properties that together with its common occurrence in practice and its role in the central limit theorem, explain why it is so frequently used.

A first such property is *orthogonal invariance* when $a = 0$. If $g \in \mathcal{O}(n)$ is an orthogonal linear map of $\mathbb{R}^n$, then $\varphi_n^{0,\sigma}(x) = \varphi_n^{0,\sigma}(gx)$. This is obvious from the fact that $\varphi_n^{0,\sigma}(x)$ depends on $\|x\|$ only. Using Theorem 2.1, it follows that for all measurable subsets $B \subseteq \mathbb{R}^n$, $\gamma_n^{0,\sigma}(g(B)) = \gamma_n^{0,\sigma}(B)$.

A second such property is that the isotropic Gaussian density decomposes as a product of lower-dimensional standard Gaussians in the following sense. Take n_1, n_2 such that $n = n_1 + n_2$ and consider the decomposition $\mathbb{R}^n = \mathbb{R}^{n_1} \times \mathbb{R}^{n_2}$. For a point $x \in \mathbb{R}^n$ we thus write $x = (x_1, x_2)$. Then

$$\varphi_n^{a,\sigma}(x) = \frac{1}{(2\pi\sigma^2)^{\frac{n}{2}}} e^{-\frac{\|x-a\|^2}{2\sigma^2}} = \frac{1}{(2\pi\sigma^2)^{\frac{n_1}{2}}} e^{-\frac{\|x_1 - a_1\|^2}{2\sigma^2}} \frac{1}{(2\pi\sigma^2)^{\frac{n_2}{2}}} e^{-\frac{\|x_2 - a_2\|^2}{2\sigma^2}}$$

$$= \varphi_{n_1}^{a_1,\sigma}(x_1) \, \varphi_{n_2}^{a_2,\sigma}(x_2), \tag{2.12}$$

and it is clear that $\varphi_{n_1}^{a_1,\sigma}$ and $\varphi_{n_2}^{a_2,\sigma}$ are the marginals of $\varphi_n^{a,\sigma}$. Hence the distributions induced by $\varphi_n^{a,\sigma}$ on $\mathbb{R}^{n_1}$ and $\mathbb{R}^{n_2}$ are also isotropic Gaussians, and x_1 and x_2 are independent (compare Sect. 2.2.1).

A third property of Gaussians is that they are preserved by linear combinations in the following sense.

Proposition 2.17 *Suppose that $x \in \mathbb{R}^n$ and $y \in \mathbb{R}^n$ are independent isotropic Gaussian vectors with centers $a \in \mathbb{R}^n$ and $b \in \mathbb{R}^n$ and variance σ^2 and τ^2, respectively. Then the distribution of $\alpha x + \beta y$ is isotropic Gaussian with center $\alpha a + \beta b$ and variance $\alpha^2 \sigma^2 + \beta^2 \tau^2$.*

Proof Without loss of generality we assume that $a = b = 0$. We first consider the case $n = 1$. When we write $x = \sigma x'$ and $y = \tau y'$, then (x', y') is standard Gaussian distributed. We put

$$r := \sqrt{\alpha^2 \sigma^2 + \beta^2 \tau^2}, \quad c := \alpha \sigma / r, \quad d := \beta \tau / r.$$

Then $z := \alpha x + \beta y = r(cx' + dy')$ and $c^2 + d^2 = 1$. We look at the distribution of

$$\begin{pmatrix} z' \\ w' \end{pmatrix} = \begin{pmatrix} c & d \\ -d & c \end{pmatrix} \begin{pmatrix} x' \\ y' \end{pmatrix}.$$

Since this is a transformation with an orthogonal matrix of the standard Gaussian (x', y'), the resulting (z', w') is standard Gaussian as well. Hence the marginal distribution of z' is standard Gaussian. It follows that $z = rz'$ has the distribution $N(0, r^2)$, which was to be shown.

This shows that in the general case, $n \geq 1$, $z_i = \alpha x_i + \beta y_i$ has the distribution $N(0, r^2)$, for all i. Since the z_i are independent, the assertion follows. □

A last property is that standard Gaussians are preserved by pushforwards under norm-preserving diffeomorphisms with Jacobian identically one.

Corollary 2.18 *Let $\psi : \mathbb{R}^n \to \mathbb{R}^n$ be a diffeomorphism satisfying $\|\psi(x)\| = \|x\|$ and $J\psi(x) = 1$ for all x. Then the pushforward of the standard Gaussian distribution under the map ψ is again the standard Gaussian distribution.*

Proof This is an immediate consequence of Proposition 2.11. □

2.2.3 The χ^2 Distribution

Suppose that $x \in \mathbb{R}^n$ is standard Gaussian distributed. The induced distribution of $q := \|x\|^2 := x_1^2 + \cdots + x_n^2$ is called the χ^2 *distribution with n degrees of freedom.* It is also denoted χ_n^2. We note that

$$\mathbb{E}(q) = \mathbb{E}(\|x\|^2) = \mathbb{E}(x_1^2) + \cdots + \mathbb{E}(x_n^2) = n,$$

so that n equals the expectation of a χ^2-distributed random variable with n degrees of freedom.

In the following we are going to derive a formula for the density of q as well as for the volume $\mathcal{O}_{n-1}$ of the sphere $\mathbb{S}^{n-1}$ in terms of the gamma function, which is defined as

$$\Gamma(x) := \int_0^\infty t^{x-1} e^{-t} \, dt \quad \text{for } x > 0. \tag{2.13}$$

This is an extension of the factorial in the sense that it satisfies $\Gamma(x+1) = x\Gamma(x)$ for all $x > 0$. In particular, we have $\Gamma(n+1) = n!$ for $n \in \mathbb{N}$. It can be tightly approximated by the well-known Stirling bounds

$$\sqrt{2\pi} x^{x+\frac{1}{2}} e^{-x} < \Gamma(x+1) < \sqrt{2\pi} x^{x+\frac{1}{2}} e^{-x+\frac{1}{12x}} \quad \text{for all } x > 0. \tag{2.14}$$

Proposition 2.19

(a) *The volume of the sphere $\mathbb{S}^{n-1}$ is given by the formula*

$$\mathcal{O}_{n-1} = \operatorname{vol} \mathbb{S}^{n-1} = \frac{2\pi^{n/2}}{\Gamma(\frac{n}{2})}.$$

(b) *The χ^2-distribution with n degrees of freedom has the density, for $q \geq 0$,*

$$\rho(q) = \frac{1}{2^{\frac{n}{2}} \Gamma(\frac{n}{2})} q^{\frac{n}{2}-1} e^{-\frac{q}{2}}.$$

(c) *The pushforward density of the standard Gaussian distribution on $\mathbb{R}^n$ with respect to the map*

$$\Psi : \mathbb{R}^n \setminus \{0\} \to \mathbb{S}^{n-1} \times (0, \infty), \qquad x \mapsto (u, q) := (x/\|x\|, \|x\|^2),$$

has the density $\rho(u, q) = \frac{1}{\mathcal{O}_{n-1}} \cdot \rho(q)$. In particular, u and q are independent.

Proof Recall the diffeomorphism ψ_{pc} from (2.6) introducing polar coordinates in $\mathbb{R}^n$. The inverse of this map is given by

$$\psi_{pc}^{-1} : \mathbb{R}^n \setminus \{0\} \to \mathbb{S}^{n-1} \times (0, \infty), \qquad x \mapsto (u, r) := (x/\|x\|, \|x\|).$$

Making the further change of variable $q = r^2$, we arrive at the diffeomorphism Ψ defined above. By Corollary 2.2 we know that $J\psi_{pc}(u, r) = r^{n-1}$. It follows that

$$J\Psi(x) = \frac{1}{r^{n-1}} \cdot 2r = \frac{2}{2r^{n-2}} = \frac{2}{q^{n/2-1}}.$$

Hence, by Proposition 2.11, the pushforward density ρ on $\mathbb{S}^{n-1} \times (0, \infty)$ of the standard Gaussian induced via Ψ equals

$$\rho(u, q) = \frac{1}{2(2\pi)^{n/2}} q^{n/2-1} e^{-q/2}.$$

Integrating over $\mathbb{S}^{n-1} \times (0, \infty)$ yields, using Fubini,

$$1 = \int_{\mathbb{S}^{n-1} \times (0,\infty)} \rho = \frac{\mathcal{O}_{n-1}}{2(2\pi)^{n/2}} \cdot \int_0^\infty q^{n/2-1} e^{-q/2} \, dq = \frac{\mathcal{O}_{n-1}}{2\pi^{n/2}} \Gamma\left(\frac{n}{2}\right),$$

where we have used the definition (2.13) for the last equality (and made the change of variable $t = q/2$). The stated formula for the volume $\mathcal{O}_{n-1}$ of the sphere is an immediate consequence.

Using this formula for $\mathcal{O}_{n-1}$, we can rewrite the density ρ as

$$\rho(u, q) = \frac{1}{\mathcal{O}_{n-1}} \frac{\mathcal{O}_{n-1}}{2(2\pi)^{n/2}} q^{n/2-1} e^{-q/2} = \frac{1}{\mathcal{O}_{n-1}} \frac{1}{2^{n/2} \Gamma(n/2)} q^{n/2-1} e^{-q/2}$$

and arrive at the second assertion of the proposition. The third assertion is now obvious. □

Corollary 2.20 *The n-dimensional unit ball $B_n := \{x \in \mathbb{R}^n \mid |x| = 1\}$ has the volume* $\mathrm{vol}\, B_n = \mathcal{O}_{n-1}/n$.

Proof The diffeomorphism ψ_{pc} from (2.6) maps $\mathbb{S}^{n-1} \times [0, 1]$ to B_n. Using polar coordinates and Fubini, we obtain

$$\mathrm{vol}\, B_n = \int_{B_n} 1 = \int_{(u,r) \in \mathbb{S}^{n-1} \times (0,1)} r^{n-1} \, d(u, r) = \mathcal{O}_{n-1} \int_0^1 r^{n-1} \, dr = \frac{\mathcal{O}_{n-1}}{n}. \quad \square$$

The following result will be needed later on.

Proposition 2.21 *For all $n \geq 2$, $\mathbb{E}(\ln \chi_n^2) \geq 0$.*

Proof It is enough to prove the statement for $n = 2$. In this case we have

$$\mathbb{E}\left(\ln \chi_n^2\right) = \frac{1}{2\pi} \int_{\mathbb{R}^2} \ln\left(x^2 + y^2\right) e^{-\frac{x^2+y^2}{2}} \, dx \, dy$$

$$= \frac{1}{2\pi} \int_0^{2\pi} \int_0^\infty \ln r^2 e^{-\frac{r^2}{2}} r \, dr \, d\theta$$

$$= \int_0^\infty \ln r^2 e^{-\frac{r^2}{2}} r \, dr = -\gamma + \ln 2 \approx 0.115932,$$

where the last equality is obtained using software for symbolic integration and γ is the Euler–Mascheroni constant, which is approximately 0.577. □

In Part III of this book we will need the following fact (which we state without proof; see the notes for a reference to a proof). A *median* median(X) of a random variable X is any value $m \in \mathbb{R}$ for which

$$\mathrm{Prob}\{X \leq m\} \geq \frac{1}{2} \quad \text{and} \quad \mathrm{Prob}\{X \geq m\} \leq \frac{1}{2}.$$

Gaussian random variables have a unique median, which equals its expectation. For the χ_n^2 distribution the following inequality holds.

Proposition 2.22 *For all* $n \geq 1$, $\mathrm{median}(\chi_n^2) \leq \mathbb{E}(\chi_n^2) = n$. $\square$

2.2.4 Uniform Distributions on Spheres

Take now $M = \mathbb{S}^{n-1} := \{x \in \mathbb{R}^n \mid \|x\| = 1\}$ for some n. The simplest (and again, the most important) example of a probability distribution in this case is the *uniform distribution*, which we denote by $U(\mathbb{S}^{n-1})$. Its density is given by $1/\mathcal{O}_{n-1}$.

A function $g : \mathbb{R}^n \to \mathbb{R}$ is *scale-invariant* when for all $a \in \mathbb{R}^n$ and all $\lambda > 0$, $g(\lambda a) = g(a)$. We noted in the introduction to this chapter that the behavior of such a function over random points from a standard Gaussian distribution is essentially equivalent to the behavior of its restriction to $\mathbb{S}^{n-1}$ when points are drawn from the uniform distribution on this sphere. This was formally proved in Sect. 2.2.3, where we showed that $U(\mathbb{S}^{n-1})$ arises as the pushforward distribution of the standard Gaussian on $\mathbb{R}^n$ under the map $\mathbb{R}^n \setminus \{0\} \to \mathbb{S}^{n-1}, x \mapsto x/\|x\|$.

Proposition 2.11 immediately implies the following.

Corollary 2.23 *Let* $g \colon \mathbb{R}^n \to \mathbb{R}$ *be a scale-invariant, integrable function and denote by* $g_{|\mathbb{S}^{n-1}}$ *its restriction to* $\mathbb{S}^{n-1}$. *Then we have, for all* $t \in \mathbb{R}$,

$$\Pr_{N(0,I_n)}\{g \geq t\} = \Pr_{U(\mathbb{S}^{n-1})}\{g_{|\mathbb{S}^{n-1}} \geq t\}$$

and

$$\mathbb{E}_{N(0,I_n)}(g) = \mathbb{E}_{U(\mathbb{S}^{n-1})}(g_{|\mathbb{S}^{n-1}}).$$ $\square$

Remark 2.24 A function $g : \mathbb{R}^{n_1} \times \cdots \times \mathbb{R}^{n_k} \to \mathbb{R}$ is *scale-invariant by blocks* when $g(\lambda_1 a_1, \ldots, \lambda_k a_k) = g(a_1, \ldots, a_k)$ for all $\lambda_1, \ldots, \lambda_k > 0$.

An extension of Corollary 2.23 to such functions is immediate. More precisely, one can prove that for all $t \in \mathbb{R}$,

$$\Pr_{N(0,I_n)}\{g \geq t\} = \Pr_{U(\mathbb{S}^{n_1-1}) \times \cdots \times U(\mathbb{S}^{n_k-1})}\{g_{|\mathbb{S}^{n_1-1} \times \cdots \times \mathbb{S}^{n_k-1}} \geq t\}$$

and

$$\mathbb{E}_{N(0,I_n)} g = \mathbb{E}_{U(\mathbb{S}^{n_1-1}) \times \cdots \times U(\mathbb{S}^{n_k-1})} g_{|\mathbb{S}^{n_1-1} \times \cdots \times \mathbb{S}^{n_k-1}}.$$

Here $n = n_1 + \cdots + n_k$.

We close this subsection with some useful bounds for quotients $\frac{\mathcal{O}_{n-1}}{\mathcal{O}_n}$ of volumes of spheres. To do so, we first prove a general result on expectations.

Lemma 2.25 *Suppose that $Z \in \mathbb{R}^n$ is standard normal distributed. Then*

$$\frac{n}{\sqrt{2\pi(n+1)}} \leq \frac{\mathcal{O}_{n-1}}{\mathcal{O}_n} = \frac{1}{\sqrt{\pi}} \frac{\Gamma(\frac{n+1}{2})}{\Gamma(\frac{n}{2})} = \frac{1}{\sqrt{2\pi}} \mathbb{E}(\|Z\|) \leq \sqrt{\frac{n}{2\pi}}.$$

Proof The left-hand equality follows immediately from the formula for $\mathcal{O}_{n-1}$ given in Proposition 2.19. Further, using polar coordinates and the variable transformation $u = \rho^2/2$, we get

$$\mathbb{E}(\|Z\|) = \frac{\mathcal{O}_{n-1}}{(2\pi)^{\frac{n}{2}}} \int_0^\infty \rho^n e^{-\frac{\rho^2}{2}} d\rho = \frac{\mathcal{O}_{n-1}}{(2\pi)^{\frac{n}{2}}} 2^{\frac{n-1}{2}} \int_0^\infty u^{\frac{n-1}{2}} e^{-u} du$$

$$= \frac{\mathcal{O}_{n-1}}{(2\pi)^{\frac{n}{2}}} 2^{\frac{n-1}{2}} \Gamma\left(\frac{n+1}{2}\right) = \sqrt{2} \frac{\Gamma(\frac{n+1}{2})}{\Gamma(\frac{n}{2})},$$

where we used the definition of the gamma function for the second-to-last equality and again the formula for $\mathcal{O}_{n-1}$ for the last equality. This gives the right-hand equality in the statement.

To obtain the right-hand inequality we use Proposition 2.10 with $X = \|Z\|$ and note that $\mathbb{E}(\|Z\|^2) = n$, since $\|Z\|^2$ is χ^2-distributed with n degrees of freedom.

To obtain the left-hand inequality we use the formula for $\mathcal{O}_n$ and the recurrence $\Gamma(x+1) = x\Gamma(x)$ to get

$$\frac{\mathcal{O}_{n-1}}{\mathcal{O}_n} = \frac{1}{\sqrt{\pi}} \frac{\Gamma(\frac{n+1}{2})}{\Gamma(\frac{n}{2})} = \frac{1}{\sqrt{\pi}} \frac{\Gamma(\frac{n+2}{2})}{\Gamma(\frac{n}{2})} \frac{\Gamma(\frac{n+1}{2})}{\Gamma(\frac{n+2}{2})} = \frac{n}{2\sqrt{\pi}} \frac{\Gamma(\frac{n+1}{2})}{\Gamma(\frac{n+2}{2})}.$$

The assertion follows now from the estimate

$$\frac{\Gamma(\frac{n+1}{2})}{\Gamma(\frac{n+2}{2})} \geq \sqrt{\frac{2}{n+1}},$$

which we have just proved. $\square$

2.2.5 Expectations of Nonnegative Random Variables

The following result allows one to quickly derive bounds for the expectation of the logarithm of a random variable X, provided certain bounds on the tail $\text{Prob}\{X \geq t\}$ are known.

Proposition 2.26 *Let X be a random variable taking values in $[1, \infty]$ such that*

$$\forall t \geq t_0: \quad \text{Prob}\{X \geq t\} \leq Kt^{-\alpha},$$

where K, α, and $t_0 \geq 1$ are positive real numbers. Then we have ($\beta \geq 2$)

$$\mathbb{E}(\log_\beta X) \leq \log_\beta t_0 + \frac{K}{\alpha} t_0^\alpha \log_\beta e.$$

Consequently, if $t_0^\alpha \geq K$, then $\mathbb{E}(\log_\beta X) \leq \log_\beta t_0 + \frac{1}{\alpha} \log_\beta e$.

Proof We put $Z := \log_\beta X$ and $s_0 := \log_\beta t_0$. Then $\text{Prob}\{Z \geq s\} \leq K\beta^{-\alpha s}$ for all $s \geq s_0$. Therefore, using Proposition 2.8,

$$\mathbb{E}(Z) = \int_0^{+\infty} \text{Prob}\{Z \geq s\}\, ds \leq s_0 + \int_{s_0}^{+\infty} K\beta^{-\alpha s}\, ds$$

$$= s_0 + \frac{K}{\alpha \ln \beta} \beta^{-\alpha s} \Big|_{+\infty}^{s_0} = s_0 + \frac{K}{\alpha t_0^\alpha \ln \beta}. \qquad \square$$

Sometimes, we want to infer bounds on moments of a random variable from tail bounds. For this, the following result is useful.

Proposition 2.27 Let X be a random variable taking values in $[0, \infty]$ and $K, \alpha, B > 0$ such that for all $t \geq B$,

$$\text{Prob}\{X \geq t\} \leq K t^{-\alpha}.$$

Then for all $k < \alpha$, we have

$$\mathbb{E}(X^k) \leq B + \frac{K}{\frac{\alpha}{k} - 1} B^{1 - \frac{\alpha}{k}}.$$

If $B \leq K^{\frac{k}{\alpha}}$, we actually have

$$\mathbb{E}(X^k) \leq \frac{\alpha}{\alpha - k} K^{\frac{k}{\alpha}}.$$

Proof We have $\text{Prob}\{Z^k \geq t\} = \text{Prob}\{Z \geq t^{\frac{1}{k}}\} \leq K t^{-\frac{\alpha}{k}}$ for all $t \geq 0$. Then, using Proposition 2.8,

$$\mathbb{E}(Z^k) = \int_0^{+\infty} \text{Prob}\{Z^k \geq t\}\, dt \leq B + \int_B^{+\infty} K t^{-\frac{\alpha}{k}}\, dt$$

$$= B + K \frac{t^{1 - \frac{\alpha}{k}}}{1 - \frac{\alpha}{k}} \Big|_B^{+\infty} = B + \frac{K}{\frac{\alpha}{k} - 1} B^{1 - \frac{\alpha}{k}}.$$

If $B \leq K^{\frac{k}{\alpha}}$, then the reasoning above, splitting the integral at $K^{\frac{k}{\alpha}}$ instead of at B, shows that

$$\mathbb{E}(Z^k) = K^{\frac{k}{\alpha}} + \frac{K}{\frac{\alpha}{k} - 1} K^{\frac{k}{\alpha}(1 - \frac{\alpha}{k})} = K^{\frac{k}{\alpha}} \left(1 + \frac{1}{\frac{\alpha}{k} - 1}\right) = \frac{\alpha}{\alpha - k} K^{\frac{k}{\alpha}}. \qquad \square$$

We finish this subsection with a classical result. Recall that a function $\phi: I \to \mathbb{R}$ defined on an interval I is called *concave* if

$$t\phi(x) + (1 - t)\phi(y) \leq \phi\big(tx + (1 - t)y\big)$$

for all $x, y \in I$ and $t \in [0, 1]$. This easily implies that $\sum_{i=1}^{N} t_i \phi(x_i) \leq \phi(\sum_{i=1}^{N} t_i x_i)$ for $x_1, \ldots, x_N \in I$ and $t_i \geq 0$ such that $\sum_{i=1}^{N} t_i = 1$. For instance, the logarithm functions are concave.

Proposition 2.28 (Jensen's inequality) *Let Z be an integrable random variable on the data space M taking values in the interval $I \subseteq \mathbb{R}$ and assume that $\phi: I \to \mathbb{R}$ is concave. Then*

$$\mathbb{E}(\phi \circ Z) \leq \phi\big(\mathbb{E}(Z)\big).$$

In particular, for any positive random variable Z we have $\mathbb{E}(\log Z) \leq \log \mathbb{E}(Z)$.

Sketch of proof Suppose that $\{A_1, \ldots, A_N\}$ is a finite partition of the data space M and $Z = \sum_{i=1}^{N} c_i \mathbb{1}_{A_i}$ for $c_i \in \mathbb{R}$. Then $\sum_i \mathsf{vol}(A_i) = 1$. We have $\mathbb{E}(Z) = \sum_i c_i \, \mathsf{vol}(A_i)$ and $\mathbb{E}(\phi \circ Z) = \sum_i \phi(c_i) \, \mathsf{vol}(A_i)$. The concavity of ϕ implies that $\mathbb{E}(\phi \circ Z) \leq \phi(\mathbb{E}(Z))$. The general case follows from this by a limit argument. $\quad\square$

Note that if the function ϕ is convex, then the reverse inequality $\phi(\mathbb{E}(Z)) \leq \mathbb{E}(\phi \circ Z)$ holds (just replace ϕ by $-\phi$). Taking $\phi = \exp$, we obtain the following useful result.

Corollary 2.29 *For an integrable random variable Z on the data space M we have* $e^{\mathbb{E}(Z)} \leq \mathbb{E}(e^Z)$. $\quad\square$

In the case of centered isotropic Gaussians we can be more precise.

Lemma 2.30 *If Z is a standard Gaussian distributed random variable, then we have $\mathbb{E}(e^{aZ}) = e^{\frac{a^2}{2}}$ for all $a \in \mathbb{R}$.*

Proof

$$\frac{1}{\sqrt{2\pi}} \int_{-\infty}^{\infty} e^{at - \frac{t^2}{2}} \, dt = \frac{1}{\sqrt{2\pi}} \int_{-\infty}^{\infty} e^{-\frac{(t-a)^2}{2}} e^{\frac{a^2}{2}} \, dt = e^{\frac{a^2}{2}}. \qquad\square$$

2.2.6 Caps and Tubes in Spheres

When we are working with uniform distributions on spheres, a number of objects and notions repeatedly occur. We collect some of them in what follows. We begin with spherical caps.

The *spherical cap* in $\mathbb{S}^{n-1}$ with center $p \in \mathbb{S}^{n-1}$ and radius $\alpha \in [0, \pi]$ is defined as

$$\mathrm{cap}(p, \alpha) := \{x \in \mathbb{S}^{n-1} \mid \langle x, p \rangle \geq \cos \alpha\}.$$

The *uniform distribution on* $\mathrm{cap}(p, \alpha)$ has as density the function that equals $1/\mathrm{vol}(\mathrm{cap}(p, \alpha))$ at points in $\mathrm{cap}(p, \alpha)$ and zero elsewhere.

Lemma 2.31 *The volume of* $\mathrm{cap}(p, \alpha)$ *in* $\mathbb{S}^{n-1}$ *satisfies*

$$\mathrm{vol}\,\mathrm{cap}(p, \alpha) = \mathcal{O}_{n-2} \cdot \int_0^\alpha (\sin \theta)^{n-2}\, d\theta.$$

Proof We may assume without loss of generality that $p = e_n$. The spherical cap $\mathrm{cap}(p, \alpha)$ is the image of $\mathbb{S}^{n-2} \times [0, \alpha]$ under the map ψ_{pcs} defined in Eq. (2.7) (with n replaced by $n - 1$), which gives polar coordinates on $\mathbb{S}^{n-1}$ with respect to the center e_n. Corollary 2.3 implies that

$$\mathrm{vol}\,\mathrm{cap}(p, \alpha) = \int_{(u,\theta) \in \mathbb{S}^{n-2} \times (0,\alpha)} (\sin \theta)^{n-2}\, d(u, \theta) = \mathcal{O}_{n-2} \int_0^\alpha (\sin \theta)^{n-2}\, d\theta.$$

$\square$

The sphere $\mathbb{S}^{n-1}$ has a natural metric given by the angular distance between points. The following derived distance allows for a number of short and elegant statements related to condition numbers.

Definition 2.32 Let $a, b \in \mathbb{S}^{n-1}$. The *sine distance* of a and b is defined as

$$d_{\sin}(a, b) := \sin \theta \in [0, 1],$$

where $\theta \in [0, \pi]$ is the angle between a and b, i.e., $\langle a, b \rangle = \cos \theta$.

Remark 2.33 One can easily verify the triangle inequality: $d_{\sin}(a, c) \leq d_{\sin}(a, b) + d_{\sin}(b, c)$. Note, however, that $d_{\sin}$ is not a metric on $\mathbb{S}^{n-1}$, since $d_{\sin}(a, -a) = 0$. Nevertheless, we observe that $d_{\sin}$ defines a metric on the real projective space. Note as well that a and b have the maximal possible distance, namely 1, if and only if they are orthogonal.

The sine distance gives an alternative way to describe small caps. For $\sigma \in [0, 1]$ it will be convenient to use the notation

$$B(p, \sigma) := \{x \in \mathbb{S}^{n-1} \mid d_{\sin}(x, p) \leq \sigma \text{ and } \langle x, p \rangle \geq 0\}.$$

Note that $\{x \in \mathbb{S}^{n-1} \mid d_{\sin}(x, p) \leq \sigma\} = \mathrm{cap}(p, \sigma) \cup \mathrm{cap}(-p, \sigma)$. It is immediate that we have $\mathrm{cap}(p, \alpha) = B(p, \sin \alpha)$ for $\alpha \in [0, \frac{\pi}{2}]$.

Lemma 2.34 *The volume of $B(p, \sigma)$ is bounded as*

$$\frac{1}{\sqrt{2\pi n}}\sigma^{n-1} \leq \frac{1}{n-1}\frac{\mathcal{O}_{n-2}}{\mathcal{O}_{n-1}}\sigma^{n-1} \leq \frac{\text{vol } B(p, \sigma)}{\mathcal{O}_{n-1}} \leq \frac{1}{2}\sigma^{n-1},$$

where $n \geq 2$ and $0 \leq \sigma \leq 1$.

Proof The left-hand inequality follows from Lemma 2.25. The asserted lower bound on vol $B(p, \sigma)$ follows from Lemma 2.31 together with

$$\int_0^\alpha (\sin\theta)^{n-2}\, d\theta \geq \int_0^\alpha (\sin\theta)^{n-2}\cos\theta\, d\theta = \frac{(\sin\alpha)^{n-1}}{n-1} = \frac{\sigma^{n-1}}{n-1},$$

where we have put $\sigma := \arcsin\alpha$. For the upper bound one easily checks that

$$\left[0, \frac{\pi}{2}\right] \to \mathbb{R}, \qquad \alpha \mapsto \frac{1}{(\sin\alpha)^{n-1}}\int_0^\alpha (\sin\theta)^{n-2}\, d\theta,$$

is monotonically increasing by computing the derivative of this function. Hence

$$\frac{1}{(\sin\alpha)^{n-1}}\int_0^\alpha (\sin\theta)^{n-2}\, d\theta \leq \int_0^{\pi/2} (\sin\theta)^{n-2}\, d\theta. \qquad (2.15)$$

On the other hand,

$$\frac{1}{2}\mathcal{O}_{n-1} = \text{vol cap}\left(p, \frac{\pi}{2}\right) = \mathcal{O}_{n-2}\int_0^{\pi/2} (\sin\theta)^{n-2}\, d\theta. \qquad (2.16)$$

Inequalities (2.15) and (2.16) together with Lemma 2.31 yield the desired bound. $\square$

The following estimate tells us how likely it is that a random point on $\mathbb{S}^{m-1}$ will lie in a fixed spherical cap.

Lemma 2.35 *Let $u \in \mathbb{S}^{m-1}$ be fixed, $m \geq 2$. Then, for all $\xi \in [0, 1]$,*

$$\Prob_{v \sim U(\mathbb{S}^{m-1})}\left\{\left|u^T v\right| \geq \xi\right\} \geq \sqrt{\frac{2}{\pi m}}(1-\xi^2)^{\frac{m-1}{2}}.$$

Proof We put $\theta = \arccos\xi$ and let $\text{cap}(u, \theta)$ denote the spherical cap in $\mathbb{S}^{m-1}$ with center u and angular radius θ. Using the left-hand bound in Lemma 2.34, we get

$$\Prob_{v \sim U(\mathbb{S}^{m-1})}\left\{\left|u^T v\right| \geq \xi\right\} = \frac{2\,\text{vol cap}(u, \theta)}{\text{vol }\mathbb{S}^{m-1}} \geq \frac{2\mathcal{O}_{m-2}}{\mathcal{O}_{m-1}}\frac{(1-\xi^2)^{\frac{m-1}{2}}}{(m-1)}.$$

The result now follows from Lemma 2.25. $\square$

Fig. 2.1 A tube (and a
neighborhood) around
$U \subseteq \mathbb{S}^{n-2}$

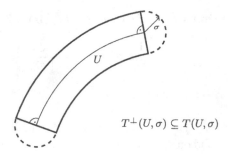

$$T^{\perp}(U, \sigma) \subseteq T(U, \sigma)$$

For the last point of this section let us consider a fixed embedding of $\mathbb{S}^{n-2}$ in $\mathbb{S}^{n-1}$. For example, we can choose $\{x = (x_1, \ldots, x_n) \in \mathbb{S}^{n-1} \mid x_n = 0\}$. The ε-*neighborhood* of $\mathbb{S}^{n-2}$ in $\mathbb{S}^{n-1}$ is defined as

$$T(\mathbb{S}^{n-2}, \varepsilon) := \{x \in \mathbb{S}^{n-1} \mid d_{\sin}(x, \mathbb{S}^{n-2}) \leq \varepsilon\}.$$

Hereby, $d_{\sin}(x, \mathbb{S}^{m-2}) := \min\{d_{\sin}(x, y) \mid y \in \mathbb{S}^{m-2}\}$.

More generally, let $U \subseteq \mathbb{S}^{n-2}$ be a closed subset. The ε-*tube* $T^{\perp}(U, \varepsilon)$ around U in $\mathbb{S}^{n-1}$ is the set of those $x \in \mathbb{S}^{n-1}$ such that there is a segment of a great circle in $\mathbb{S}^{n-1}$ from x to a point $y \in U$, of length at most $\arcsin \varepsilon$, that intersects $\mathbb{S}^{n-2}$ orthogonally at y. Figure 2.1 attempts to convey the difference between the ε-tube (whose boundary is an unbroken line) and the ε-neighborhood (which adds to the tube the two extremities with dashed boundaries).

In other words, $T^{\perp}(U, \varepsilon)$ is the image of $U \times [\pi/2 - \alpha, \pi/2 + \alpha]$ under the map

$$\psi_{pcs} \colon \mathbb{S}^{n-2} \times (0, \pi) \to \mathbb{S}^{n-1} \setminus \{\pm e_n\}, \quad (u, \theta) \mapsto (u \sin \theta, \cos \theta),$$

defining polar coordinates on $\mathbb{S}^{n-1}$ with respect to the center e_n; compare (2.7).

The next lemma gives a formula, as well as a useful upper bound, for the volume of $T^{\perp}(U, \sigma)$.

Lemma 2.36 *We suppose that $n > 2$.*

(a) *For a closed subset $U \subseteq \mathbb{S}^{n-2}$, $0 \leq \alpha \leq \frac{\pi}{2}$, and $\varepsilon = \sin \alpha$, we have*

$$\operatorname{vol} T^{\perp}(U, \varepsilon) = \operatorname{vol} U \int_{-\alpha}^{\alpha} (\cos \rho)^{n-2} \, d\rho.$$

(b) *We have $\operatorname{vol} T^{\perp}(U, \varepsilon) \leq 2\varepsilon \operatorname{vol} U$. In particular, $\operatorname{vol} T(\mathbb{S}^{n-2}, \varepsilon) \leq 2\varepsilon \mathcal{O}_{n-2}$.*

Proof (a) Use Corollary 2.3 and substitute $\rho := \pi/2 - \theta$.

For (b) note that if $n > 2$, then $\int_0^\alpha (\cos \rho)^{n-2} \, d\rho \leq \int_0^\alpha \cos \rho \, d\rho = \sin \alpha = \varepsilon$. This proves the first inequality. The second one follows by taking $U = \mathbb{S}^{n-2}$ and noting that $T(\mathbb{S}^{n-2}, \varepsilon) = T^{\perp}(\mathbb{S}^{n-2}, \varepsilon)$. $\square$

The following lemma will be essential in various smoothed analysis results.

Fig. 2.2 An illustration of
the quantities in the proof of
Lemma 2.37

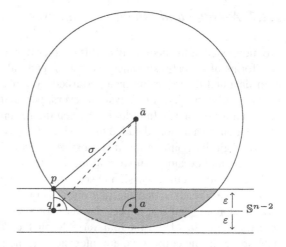

Lemma 2.37 *Let $0 < \varepsilon, \sigma \leq 1$, $n > 2$, and $\bar{a} \in \mathbb{S}^{n-1}$. Fix also an embedding $\mathbb{S}^{n-2} \hookrightarrow \mathbb{S}^{n-1}$. Then*

$$\frac{\text{vol}(T(\mathbb{S}^{n-2}, \varepsilon) \cap B(\bar{a}, \sigma))}{\text{vol } B(\bar{a}, \sigma)} \leq 2(n-1)\left(1 + \frac{\varepsilon}{\sigma}\right)^{n-2} \frac{\varepsilon}{\sigma}.$$

Proof Let $a \in \mathbb{S}^{n-2}$ satisfy $d_{\sin}(\bar{a}, a) = d_{\sin}(\bar{a}, \mathbb{S}^{n-2})$. We consider the $(\sigma + \varepsilon)$-neighborhood of a

$$U := B_{\mathbb{S}^{n-2}}(a, \sigma + \varepsilon) \cup B_{\mathbb{S}^{n-2}}(-a, \sigma + \varepsilon)$$

with respect to the sine distance. We claim that

$$T\left(\mathbb{S}^{n-2}, \varepsilon\right) \cap B(\bar{a}, \sigma) \subseteq T^{\perp}(U, \varepsilon).$$

For proving this, take any $p \in T(\mathbb{S}^{n-2}, \varepsilon) \cap B(\bar{a}, \sigma)$ and let $q \in \mathbb{S}^{n-2}$ be such that $d_{\sin}(p, q) = d_{\sin}(p, \mathbb{S}^{n-2})$. By the triangle inequality we have

$$d_{\sin}(q, \bar{a}) \leq d_{\sin}(q, p) + d_{\sin}(p, \bar{a}) \leq \varepsilon + \sigma.$$

An elementary geometric argument shows that $d_{\sin}(q, a) \leq d_{\sin}(q, \bar{a})$. Hence $q \in U$, and the claim follows (cf. Fig. 2.2).

We have $\text{vol } T^{\perp}(U, \varepsilon) \leq 2\varepsilon \text{ vol } U$ by Lemma 2.36. Moreover, Lemma 2.34 yields the estimates

$$\text{vol}(U) = 2 \text{ vol } B_{\mathbb{S}^{n-2}}(a, \sigma + \varepsilon) \leq \mathcal{O}_{n-2}(\sigma + \varepsilon)^{n-2},$$

$$\text{vol } B(\bar{a}, \sigma) \geq \frac{1}{n-1}\mathcal{O}_{n-2}\sigma^{n-1}.$$

The assertion follows by combining these observations.   □

2.2.7 Average and Smoothed Analyses

We mentioned at the beginning of this chapter that a glimpse at the literature shows two forms of probabilistic analysis that, in general terms, model the ideas of random data and deterministic data perturbed by noise, respectively. We will exhibit instances of both kinds of analysis at several points in this book (beginning with the section after the next). In order to sharpen the meaning of these analyses, we close this section with a more detailed discussion of the two approaches.

We begin by pointing out that most of the condition numbers occurring in the literature (and certainly all we will meet in this book) are scale-invariant. We can therefore confine our discussion to scale-invariant random variables.

We have already noted that the underlying distribution in average analysis is "evenly spread" and that expected values for such distributions are meant to capture the behavior of the random variable on the "average Joe" inhabitant of the data space. In the context we are interested in, the probabilistic analysis of condition numbers, this data space is usually a Euclidean space $\mathbb{R}^n$. So the question that poses itself is which distribution should we endow $\mathbb{R}^n$ with for the analysis. Assume we have a "natural" system of coordinates in $\mathbb{R}^n$. A first step then towards an answer consists in noting that such a distribution should be rotationally invariant. The undistinguished character of "average Joe" cannot favor any particular direction. Furthermore, scale invariance allows us to give a second step. Indeed, for such a function on $\mathbb{R}^n$ the value of the density along a half-line with origin at 0 is not relevant. It follows that we can take *any* rotationally invariant distribution on $\mathbb{R}^n$, and the collection of features of the standard Gaussian we listed in Sect. 2.2.2 make this distribution the obvious choice.

Scale invariance also suggests the choice of an underlying distribution that is evenly spread by definition, namely, the uniform distribution on the unit sphere $\mathbb{S}^{n-1}$. This requires us to consider data on the sphere only, a requirement easily achieved by means of the map

$$\mathbb{R}^n \setminus \{0\} \to \mathbb{S}^{n-1}, \qquad a \mapsto \frac{a}{\|a\|}.$$

Proposition 2.23 shows that this choice is equivalent to the standard Gaussian on $\mathbb{R}^n$.

All the above, however, is subject to a "natural" system of coordinates in $\mathbb{R}^n$. And while some situations may suggest such a system (we will argue for one in Sect. 16.1), its choice remains a bone of contention for average-case analysis. The most common objection to average-case analysis is that its underlying probability measures may not accurately reflect the distributions occurring in practice, in particular, that they may be "optimistic" in the sense that they may put more probability mass on the instances for which the values of the function ψ under consideration are small. Such an optimism would produce an expectation $\mathbb{E}\psi$ smaller than the true one.

Smoothed analysis was introduced mainly to overcome this objection. Its underlying idea, we recall, was to look at the behavior of a function for small perturbations of arbitrary data. In the case of a condition number, this amounts to understanding the condition of slight perturbations of ill-posed data. When compared with the average analysis, it replaces the goal of showing that

for a random a, it is unlikely that $\mathsf{cond}(a)$ will be large

by the following one:

for all $\overline{a}$, it is unlikely that a slight random perturbation $a + \Delta a$ will have $\mathsf{cond}(\overline{a} + \Delta a)$ large.

To perform a smoothed analysis, a family of distributions (parameterized by a parameter r controlling the size of the perturbation) is considered with the following characteristics:

(a) the density of an element a depends only on the distance $\|a - \overline{a}\|$.
(b) the value of r is closely related to the variance of $\|a - \overline{a}\|$.

A first possible choice for this family of distributions is the set of Gaussians $N(\overline{a}, \sigma^2 I_n)$. The role of r is in this case played by $\sigma > 0$.

Because of scale invariance, one usually assumes that data live in $\mathbb{S}^{n-1}$. In this way, the value of the parameter r controlling the size of the perturbations is directly comparable with the size of $\overline{a}$. Note that in this case, a Gaussian $N(\overline{a}, \sigma^2 I_n)$ induces on the sphere a distribution different from the uniform, the density being higher when close to $\overline{a}$.

A different choice of distributions consists in taking, for each $\alpha \in (0, \pi]$, the uniform measure on the spherical cap $\mathsf{cap}(\overline{a}, \alpha)$ or even on $B(\overline{a}, \sigma)$ for each $\sigma \in (0, 1]$.

The following table shows a schematic comparison of the quantities computed in worst-case, average-case, and smoothed analyses for a scale-invariant function $\psi : \mathbb{R}^n \to \mathbb{R}$ in the uniform case (the Gaussian case is obtained in the obvious manner).

worst-case	average-case	smoothed
$\displaystyle\sup_{a \in \mathbb{S}^{n-1}} \psi(a)$	$\displaystyle\mathbb{E}_{a \sim U(\mathbb{S}^{n-1})} \psi(a)$	$\displaystyle\sup_{\overline{a} \in \mathbb{S}^{n-1}} \mathbb{E}_{a \sim U(\mathsf{cap}(\overline{a}, \alpha))} \psi(a)$

Usually, the quantities estimated in the first two columns are functions of n. For the estimate in a smoothed analysis there is, in addition, a dependence on α. This dependence appears to interpolate between worst-case and average-case. Indeed, when α approaches 0, the value of

$$\sup_{\overline{a} \in \mathbb{S}^{n-1}} \mathbb{E}_{a \sim U(\mathsf{cap}(\overline{a}, \alpha))} \psi(a)$$

approaches $\sup_{a \in \mathbb{R}^n} \psi(a)$, while when $\alpha = \pi$ this value coincides with $\mathbb{E}_{a \sim U(\mathbb{S}^{n-1})} \psi(a)$ (since $\mathsf{cap}(\overline{a}, \pi) = \mathbb{S}^{n-1}$ for all $\overline{a}$). In case $\psi(-a) = \psi(a)$ for all $a \in \mathbb{S}^{n-1}$, a common occurrence when ψ is a condition number, it is immediate to see that

$$\mathop{\mathbb{E}}_{a\sim U(\mathbb{S}^{n-1})} \psi(a) = \mathop{\mathbb{E}}_{a\sim U(B(\overline{a},1))} \psi(a)$$

for all $\overline{a} \in \mathbb{S}^{n-1}$. Therefore, in this case,

$$\sup_{\overline{a}\in\mathbb{S}^{n-1}} \mathop{\mathbb{E}}_{a\sim U(B(\overline{a},\sigma))} \psi(a)$$

interpolates between the worst-case and average-case analyses of ψ as σ varies between 0 and 1.

The local nature of randomization in smoothed analysis, coupled with its worst case dependence on the input data, removes from smoothed analysis the objection to average-case analysis mentioned above. A satisfying result in this context (usually a low-degree polynomial bound in the input size n and in the inverse of the dispersion parameter r) is consequently considered a much more reliable indication that one may expect low values for the function ψ in practice. In addition, there is an emerging impression that smoothed analysis is robust in the sense that its dependence on the chosen family of measures is low. This tenet is supported in Sect. 21.8, where the uniform measure is replaced by an adversarial measure (one having a pole at $\overline{a}$) without a significant loss in the estimated averages.

2.3 Probabilistic Analysis of $\mathrm{Cw}_i(A, x)$

As a first illustration we perform an average-case analysis of the componentwise condition numbers $\mathrm{Cw}_i(A, x)$ of matrix–vector multiplication that were introduced in Sect. O.4. For the average analysis we shall suppose that $A \in \mathbb{R}^{m\times n}$ and $x \in \mathbb{R}^n$ are both standard Gaussian distributed and that they are independent.

As will be the case often in this book, the starting point of the probabilistic analysis is a "condition number theorem," which in the situation at hand is expressed by

$$\mathrm{Cw}_i(A, x) \leq \frac{1}{d_{\mathsf{sin}}(\breve{a}_i, \Sigma_i(x))}, \tag{2.17}$$

where $\Sigma_i(x) := \{b \in \mathbb{S}^{n-1} \mid b^\mathrm{T} x = 0\}$. This bound is an easy consequence of (O.11) seen in Sect. O.5.4.

Let a_i denote the ith row of A. By the rotational invariance of the standard Gaussian distribution, the normalized ith row $\breve{a}_i := a_i/\|a_i\|$ of A is then uniformly distributed in the sphere $\mathbb{S}^{n-1}$. We note that by its definition, $\mathrm{Cw}_i(A, x)$ depends only on $a_i/\|a_i\|$ and $x/\|x\|$.

The average-case analysis of Cw_i is summarized in the following result.

Theorem 2.38 *Let $n > 2$. For $A \in \mathbb{R}^{m\times n}$ and $x \in \mathbb{R}^n$ standard Gaussian distributed and independent, we have, for all $0 < \varepsilon \leq 1$ and all $i \in [m]$,*

$$\mathsf{Prob}\{\mathrm{Cw}_i(A, x) \geq \varepsilon^{-1}\} \leq \sqrt{\frac{2n}{\pi}}\varepsilon.$$

Moreover, for $\beta \geq 2$,

$$\mathbb{E}\big(\log_\beta Cw_i(A, x)\big) \leq \frac{1}{2} \log_\beta n + 2.$$

Proof We fix i throughout the proof. By Proposition 2.19, $\breve{a}_i$ is uniformly distributed in $\mathbb{S}^{n-1}$. Therefore, we obtain from (2.17) for fixed $x \in \mathbb{R}^n$ that

$$\Prob_{A \sim N(0, I_{m \times n})} \big\{ Cw_i(A, x) \geq \varepsilon^{-1} \big\} \leq \Prob_{\breve{a}_i \sim U(\mathbb{S}^{n-1})} \big\{ d_{\sin}(\breve{a}_i, \Sigma) \leq \varepsilon \big\} = \frac{\mathrm{vol}\, T(\Sigma, \varepsilon)}{\mathcal{O}_{n-1}},$$

where we have written $\Sigma := \Sigma_i(x)$ to simplify notation. Since Σ is isometric to the subsphere $\mathbb{S}^{n-2}$ of $\mathbb{S}^{n-1}$, Lemma 2.36(b) implies that $\mathrm{vol}\, T(\Sigma, \varepsilon) \leq 2\mathcal{O}_{n-2}\,\varepsilon$ (here we use $n > 2$). From Lemma 2.25 we get the tail bound

$$\Prob_A \big\{ Cw_i(A, x) \geq \varepsilon^{-1} \big\} \leq \frac{2\mathcal{O}_{n-2}\,\varepsilon}{\mathcal{O}_{n-1}} \leq \sqrt{\frac{2(n-1)}{\pi}}\,\varepsilon \leq \sqrt{\frac{2n}{\pi}}\,\varepsilon.$$

Since this bound is independent of x, we conclude, with φ_m denoting the density of the standard normal distribution on $\mathbb{R}^m$, that

$$\Prob_{A, x} \big\{ Cw_i(A, x) \geq \varepsilon^{-1} \big\} = \int_{x \in \mathbb{R}^n} \int_{A \in \mathbb{R}^{m \times n}} \mathbb{1}_{Cw_i(A,x) \geq \varepsilon^{-1}} \varphi_{n^2}(A)\, dA\, \varphi_n(x)\, dx$$

$$\leq \int_{\mathbb{R}^n} \sqrt{\frac{2n}{\pi}}\,\varepsilon\, \varphi_n(x)\, dx = \sqrt{\frac{2n}{\pi}}\,\varepsilon.$$

Furthermore, applying Proposition 2.26 to the random variable $Cw_i(A, x) \geq 1$ (with $\alpha = 1$ and $K = t_0 = \sqrt{2n/\pi}$), we obtain

$$\mathbb{E}\big(\log_\beta Cw_i(A, x)\big) \leq \frac{1}{2} \log_\beta \frac{2n}{\pi} + \log_\beta e < \frac{1}{2} \log_\beta n + 2,$$

as claimed.   □

We turn now to the smoothed analysis of $Cw_i(A, x)$. Fix any matrix $\overline{A} \in (\mathbb{S}^{n-1})^m$ and any vector $\overline{x} \in \mathbb{S}^{n-1}$. Let $0 < \sigma \leq 1$ and suppose that A is a random matrix such that the ith row a_i of A is chosen uniformly at random in the σ-ball $B(\overline{a}_i, \sigma)$ of $\overline{a}_i$ and x is a random vector uniformly chosen in $B(\overline{x}_i, \sigma)$ (note that $Cw_i(A, x) = Cw_i(-A, x)$, so that the discussion at the end of Sect. 2.2.7 applies here).

Theorem 2.39 *Let $n > 2$, $\overline{A} \in (\mathbb{S}^{n-1})^m$ and $\overline{x} \in \mathbb{S}^{n-1}$. Then, for all $i \in [m]$, all $\sigma \in (0, 1]$, and all $0 < \varepsilon \leq \frac{\sigma}{n}$,*

$$\Prob_{\substack{a_i \in B(\overline{a}_i, \sigma) \\ x \in B(\overline{x}, \sigma)}} \big\{ Cw_i(A, x) \geq \varepsilon^{-1} \big\} \leq 2en\,\frac{\varepsilon}{\sigma}.$$

Moreover, we have for $\beta \geq 2$ and all i,

$$\mathbb{E}\left(\log_\beta \mathrm{Cw}_i(A, x)\right) \leq \log_\beta n + \log_\beta \frac{1}{\sigma} + 4.$$

Proof Again let i be fixed throughout the proof. By (2.17) we have for fixed $x \in B(\overline{x}, \sigma)$,

$$\Prob_{a_i \in B(\overline{a}_i, \sigma)} \left\{ \mathrm{Cw}_i(A, x) \geq \varepsilon^{-1} \right\} \leq \frac{\mathrm{vol}(T(\Sigma_i(x), \varepsilon) \cap B(\overline{a}_i, \sigma))}{\mathrm{vol}\, B(\overline{a}_i, \sigma)}.$$

Since $\Sigma_i(x)$ is isometric to $\mathbb{S}^{n-2}$, Lemma 2.37 implies that

$$\Prob_{a_i \in B(\overline{a}_i, \sigma)} \left\{ \mathrm{Cw}_i(A, x) \geq \varepsilon^{-1} \right\} \leq 2(n-1)\left(1 + \frac{\varepsilon}{\sigma}\right)^{n-2} \frac{\varepsilon}{\sigma}.$$

We have $(1 + \frac{\varepsilon}{\sigma})^{n-2} \leq (1 + \frac{1}{n-2})^{n-2} \leq e$ if $\varepsilon^{-1} \geq \frac{n-2}{\sigma}$. Hence, under this assumption, we have

$$\Prob_{a_i \in B(\overline{a}_i, \sigma)} \left\{ \mathrm{Cw}_i(A, x) \geq \varepsilon^{-1} \right\} \leq 2e(n-1)\frac{\varepsilon}{\sigma} \leq 2en\frac{\varepsilon}{\sigma},$$

proving the tail bound for fixed $x \in B(\overline{x}, \sigma)$. The desired tail bound (for both A and x random) follows as in Theorem 2.38.

For the expectation, Proposition 2.26 (with $\alpha = 1$ and $t_0 = K = \frac{2en}{\sigma}$) implies that

$$\mathbb{E}\left(\log_\beta \mathrm{Cw}_i(A, x)\right) \leq \log_\beta n + \log_\beta \frac{1}{\sigma} + \log_\beta\left(2e^2\right) \leq \log_\beta n + \log_\beta \frac{1}{\sigma} + 4,$$

as claimed.   □

Remark 2.40 As we noted in Sect. 2.2.7, we can obtain average-case bounds from Theorem 2.39 by taking $\sigma = 1$. A comparison with Theorem 2.38 shows that the bounds thus obtained are slightly worse than those obtained from a standard average-case analysis. This is due to the nonoptimal estimation in Lemma 2.34.

2.4 Probabilistic Analysis of $\kappa_{rs}(A)$

Recall the set $\Sigma = \{A \in \mathbb{R}^{n \times n} \mid \det A = 0\}$ of ill-posed matrices. The starting point of our probabilistic analysis is the condition number theorem, Corollary 1.8, stating that

$$\kappa_{rs}(A) = \frac{\|A\|_{rs}}{d_{rs}(A, \Sigma)}. \tag{2.18}$$

This characterization of the condition number is the key to the geometric way of reasoning below.

2.4.1 Preconditioning

Corollary 1.3 shows that the operator norm $\|\ \|_{2\infty}$ defined in (1.4) admits a simple, easily computable, characterization. As before, let $a_1, \ldots, a_n$ denote the rows of A.

Think now of a matrix A poorly conditioned with respect to $\kappa_{2\infty}$. Because of Eq. (2.18) and Corollary 1.3, this means that for a certain matrix $S \in \Sigma$ with rows s_i, we have that

$$\kappa_{2\infty}(A) = \frac{\max_i \|a_i\|}{\max_j \|a_j - s_j\|} \tag{2.19}$$

is large. A possible reason for this poor condition may be a bad scaling of A—that is, the existence of $i \neq j$ such that the numerator in (2.19) is maximized at i, its denominator is maximized at j, and $\frac{\|a_i\|}{\|a_j - s_j\|}$ is large because $\|a_i\|$ is large compared with $\|a_j\|$.

Since the solution of $Ax = b$ is not changed when we divide the rows of $[A, b]$ by nonzero scalars, a way to avoid poor condition due to a possible bad scaling is to solve instead the system $\check{A}x = \check{b}$, where

$$\check{A} = \left(\frac{a_1}{\|a_1\|}, \ldots, \frac{a_n}{\|a_n\|} \right) \quad \text{and} \quad \check{b} = \left(\frac{b_1}{\|a_1\|}, \ldots, \frac{b_n}{\|a_n\|} \right).$$

The following result justifies doing so.

Proposition 2.41 *We have $\kappa_{2\infty}(\check{A}) \leq \kappa_{2\infty}(A)$ whenever the left-hand side is defined.*

Proof For any $S \in \Sigma$ we have by (2.19),

$$\frac{1}{\kappa_{2\infty}(A)} \leq \frac{\max_j \|a_j - s_j\|}{\max_i \|a_i\|} = \max_j \frac{\|a_j\|}{\max_i \|a_i\|} \left\| \check{a}_j - \frac{s_j}{\|a_j\|} \right\|$$

$$\leq \max_j \left\| \check{a}_j - \frac{s_j}{\|a_j\|} \right\| = d_{2\infty}(\check{A}, \check{S}),$$

where $\check{S}$ is the matrix with columns $s_j / \|s_j\|$. Hence $\kappa_{2\infty}(A)^{-1} \leq d_{2\infty}(\check{A}, \Sigma)$, and we get

$$\kappa_{2\infty}(A) \geq \frac{1}{d_{2\infty}(\check{A}, \Sigma)} = \frac{\|\check{A}\|_{2\infty}}{d_{2\infty}(\check{A}, \Sigma)} = \kappa_{2\infty}(\check{A}). \qquad \square$$

The passage from A to $\check{A}$ is called *preconditioning* in the literature. In general (and this is the case in our example), such a process has a negligible computational cost and can only improve the condition of the data at hand. Furthermore, the loss of precision in the computation of the pair $(\check{A}, \check{b})$ is negligible as well. Hence, the quantity controlling the loss of precision to compute $A^{-1}b$ from this pair is

$$\overline{\kappa}_{2\infty}(A) := \kappa_{2\infty}(\check{A}).$$

Note that by construction, $\overline{\kappa}_{2\infty}(A)$ is scale-invariant in each row of A. We are therefore in the situation of a block scale-invariant random variable described in Remark 2.24. To study its behavior when $A \sim N(0, I_{n^2})$ amounts to studying the behavior of $\breve{A} \sim U(S)$ with $S = (\mathbb{S}^{n-1})^n$. We next proceed to do so.

In the sequel we will interpret a tuple $A = (a_1, \ldots, a_n) \in (\mathbb{S}^{n-1})^n$ as the matrix in $\mathbb{R}^{n \times n}$ with rows $a_1, \ldots, a_n$. Because $\overline{\kappa}_{2\infty}(A)$ is block scale-invariant, it is possible to formulate the condition number theorem (Corollary 1.8) in terms of an extension of $d_{\sin}$ to a product of spheres. In order to do so we define for $A = (a_1, \ldots, a_n) \in (\mathbb{S}^{n-1})^n$ and $B = (b_1, \ldots, b_n) \in (\mathbb{S}^{n-1})^n$ the distance

$$d_{\sin}(A, B) := \max_{i \leq n} d_{\sin}(a_i, b_i).$$

By Remark 2.33, $d_{\sin}$ satisfies the triangle inequality (but recall that it is not a distance on $(\mathbb{S}^{n-1})^n$, since $d_{\sin}(A, -A) = 0$). We put

$$\Sigma_{\mathbb{S}} := \Sigma \cap (\mathbb{S}^{n-1})^n = \left\{ A \in (\mathbb{S}^{n-1})^n \mid a_1, \ldots, a_n \text{ are linearly dependent} \right\}$$

and set

$$d_{\sin}(A, \Sigma_{\mathbb{S}}) := \min \left\{ d_{\sin}(A, B) \mid B \in \Sigma_{\mathbb{S}} \right\}.$$

Theorem 1.7 for $r = 2$, $s = \infty$ combined with Lemma 2.43 stated below immediately yields the following intrinsic characterization of $\overline{\kappa}_{2\infty}(A)$.

Corollary 2.42 *For all $A \notin \Sigma$ with nonzero rows we have*

$$\overline{\kappa}_{2\infty}(A) = \frac{1}{d_{\sin}(\breve{A}, \Sigma_{\mathbb{S}})}.$$

Lemma 2.43 *We have $d_{2\infty}(A, \Sigma) = d_{\sin}(A, \Sigma_{\mathbb{S}})$ for $A \in (\mathbb{S}^{n-1})^n$.*

Proof For $a, s \in \mathbb{S}^{n-1}$ let s^* denote the orthogonal projection of a onto $\mathbb{R}s$. By definition we have $d_{\sin}(a, s) = \|a - s^*\|$. Moreover, $d_{\sin}(a, s) \leq \|a - s\|$. From these observations it immediately follows that $d_{2\infty}(A, \Sigma^*) = d_{\sin}(A, \Sigma_{\mathbb{S}})$, where $\Sigma^* := \{B \in \Sigma \mid b_i \neq 0 \text{ for } i \in [n]\}$. Since Σ^* is a dense subset of Σ, we have $d_{2\infty}(A, \Sigma) = d_{2\infty}(A, \Sigma^*)$. Hence the assertion follows.   □

Recall that for a matrix $A \in \mathbb{R}^{m \times n}$, a vector $c \in \mathbb{R}^n$, and an index $i \in [m]$, we denote by $A(i : c)$ the matrix obtained by replacing the ith row of A by c.

The following result will be used twice in this chapter.

Proposition 2.44 *For $A \in (\mathbb{S}^{n-1})^n$ there exist $i \in [n]$ and $\bar{c} \in \mathbb{S}^{n-1}$ such that we have $A(i : \bar{c}) \in \Sigma_{\mathbb{S}}$ and $d_{\sin}(a_i, \bar{c}) \leq n d_{\sin}(A, \Sigma_{\mathbb{S}})$.*

Proof We first note that $d(A, \Sigma) \leq \sqrt{n} \, d_{2\infty}(a, \Sigma)$ due to the norm inequalities $n^{-1/2} \|x\| \leq \|x\|_\infty \leq \|x\|$ holding for $x \in \mathbb{R}^n$. Moreover, by Lemma 2.43 we have $d_{2\infty}(A, \Sigma) = d_{\sin}(A, \Sigma_{\mathbb{S}})$.

We apply Proposition 1.10 to the matrix A^{T} (having the columns a_i) and the spectral norms ($r = s = 2$). This shows the existence of $i \in [n]$ and $c \in \mathbb{R}^n$ such that $A(i : c) \in \Sigma$ and $\|a_i - c\| \leq \sqrt{n}\, d(A, \Sigma) \leq n\, d_{\sin}(a, \Sigma_{\mathbb{S}})$.

If $c = 0$, then $1 = \|a_i\| \leq n\, d_{\sin}(a, \Sigma_{\mathbb{S}})$, in which case the assertion is trivial (note that $d_{\sin} \leq 1$).

Hence we may suppose that $c \neq 0$ and put $\bar{c} := c/\|c\|$. Then we have $d_{\sin}(a_i, \bar{c}) \leq \|a_i - c\|$ and therefore $d_{\sin}(a_i, \bar{c}) \leq n\, d_{\sin}(a, \Sigma_{\mathbb{S}})$ as claimed. $\square$

2.4.2 Average Analysis

We assume now that A is randomly chosen in $(\mathbb{S}^{n-1})^n$ according to the uniform distribution and investigate the distribution of the random variable $\overline{\kappa}_{2\infty}(A)$.

Theorem 2.45 *We have, for any $n > 2$ and $0 < \varepsilon \leq 1$,*

$$\mathrm{Prob}\{\kappa_{2\infty}(A) \geq \varepsilon^{-1}\} \leq \sqrt{\frac{2}{\pi}}\, n^{5/2}\, \varepsilon,$$

if A is chosen randomly in $(\mathbb{S}^{n-1})^n$ according to the uniform distribution. Moreover, for $\beta \geq 2$,

$$\mathbb{E}\big(\log_\beta \kappa_{2\infty}(A)\big) \leq \frac{5}{2}\log_\beta n + 2.$$

Proof Let $T(\Sigma_{\mathbb{S}}, \varepsilon)$ denote the ε-neighborhood of $\Sigma_{\mathbb{S}}$ in $(\mathbb{S}^{n-1})^n$, i.e.,

$$T(\Sigma_{\mathbb{S}}, \varepsilon) := \big\{A \in (\mathbb{S}^{n-1})^n \mid d_{\sin}(A, \Sigma_{\mathbb{S}}) \leq \varepsilon\big\}.$$

By Proposition 2.42 we know that

$$T(\Sigma_{\mathbb{S}}, \varepsilon) = \big\{A \in (\mathbb{S}^{n-1})^n \mid \kappa_{2\infty}(A) \geq \varepsilon^{-1}\big\},$$

and hence

$$\mathrm{Prob}\{\kappa_{2\infty}(A) \geq \varepsilon^{-1}\} = \frac{\mathrm{vol}\, T(\Sigma_{\mathbb{S}}, \varepsilon)}{\mathrm{vol}(\mathbb{S}^{n-1})^n}.$$

We thus need to bound the volume of $T(\Sigma_{\mathbb{S}}, \varepsilon)$. Proposition 2.44 implies that

$$T(\Sigma_{\mathbb{S}}, \varepsilon) \subseteq \bigcup_{i=1}^{n} W_i,$$

where

$$W_i := \big\{A \in (\mathbb{S}^{n-1})^n \mid \exists c \in \mathbb{S}^{n-1} : d_{\sin}(a_i, c) \leq n\varepsilon \text{ and } A(i : c) \in \Sigma_{\mathbb{S}}\big\}.$$

By symmetry, $\mathsf{vol}\, W_i = \mathsf{vol}\, W_n$, whence

$$\mathsf{Prob}\{\kappa_{2\infty}(A) \geq \varepsilon^{-1}\} \leq n\frac{\mathsf{vol}\, W_n}{\mathsf{vol}(\mathbb{S}^{n-1})^n}.$$

Consider the projection

$$\pi : W_n \to (\mathbb{S}^{n-1})^{n-1}, \quad (a_1, \ldots, a_n) \mapsto (a_1, \ldots, a_{n-1}).$$

Without loss of generality we may assume that $u_1, \ldots, u_{n-1} \in \mathbb{S}^{n-1}$ are linearly independent. Then the set

$$H := \mathsf{span}\{a_1, \ldots, a_{n-1}\} \cap \mathbb{S}^{n-1}$$

is isometric to $\mathbb{S}^{n-2}$ and $A(i : c) \in \Sigma_{\mathbb{S}} \iff c \in H$. We therefore have

$$\pi^{-1}(a_1, \ldots, a_{n-1}) \simeq T(H, n\varepsilon) := \{a_n \in \mathbb{S}^{n-1} \mid d_{\sin}(a_n, H) \leq n\varepsilon\}.$$

Clearly, $\mathsf{vol}\, T(H, n\varepsilon) = \mathsf{vol}\, T(\mathbb{S}^{n-2}, n\varepsilon)$. Then, by Lemma 2.36(b),

$$\mathsf{vol}\, \pi^{-1}(a_1, \ldots, a_{n-1}) = \mathsf{vol}\, T(\mathbb{S}^{n-2}, n\varepsilon) \leq 2n\varepsilon \mathcal{O}_{n-2}.$$

Using Fubini's theorem, we obtain

$$\mathsf{vol}\, W_n = \int_{(\mathbb{S}^{n-1})^{n-1}} \mathsf{vol}\, \pi^{-1}(a_1, \ldots, a_{n-1})\, d(\mathbb{S}^{n-1})^{n-1} \leq 2n\varepsilon \mathcal{O}_{n-2} \mathcal{O}_{n-1}^{n-1}.$$

Now using Lemma 2.25, we get

$$\frac{\mathsf{vol}\, W_n}{\mathsf{vol}(\mathbb{S}^{n-1})^n} \leq \frac{\mathcal{O}_{n-2}}{\mathcal{O}_{n-1}} 2n\varepsilon \leq \sqrt{\frac{2}{\pi}} n^{3/2} \varepsilon.$$

Altogether,

$$\mathsf{Prob}\{\kappa_{2\infty}(A) \geq \varepsilon^{-1}\} \leq \sqrt{\frac{2}{\pi}} n^{5/2} \varepsilon,$$

which is the desired tail bound.

In order to prove the bound on the expectation we apply Proposition 2.26 (with $\alpha = 1$ and $K = t_0 = \sqrt{\frac{2}{\pi}} n^{5/2}$) to the random variable $\kappa_{2\infty}(A) \geq 1$ to obtain

$$\mathbb{E}(\log_\beta \kappa_{2\infty}(A)) \leq \log_\beta \left(\sqrt{\frac{2}{\pi}} n^{5/2}\right) + \log_\beta e \leq \frac{5}{2} \log_\beta n + 2,$$

as claimed. $\square$

2.4.3 Uniform Smoothed Analysis

Let $\overline{A} = (\overline{a}_1, \ldots, \overline{a}_n) \in (\mathbb{S}^{n-1})^n$ and $0 \leq \sigma \leq 1$. We define the σ-*neighborhood* of $\overline{A}$ in $(\mathbb{S}^{m-1})^n$ as

$$B(\overline{A}, \sigma) := B(\overline{a}_1, \sigma) \times \cdots \times B(\overline{a}_n, \sigma).$$

Our smoothed analysis result is the following.

Theorem 2.46 *For any $n > 2$, any $0 < \sigma, \varepsilon \leq 1$, and any $\overline{A} \in (\mathbb{S}^{n-1})^n$ we have*

$$\Prob_{A \in B(\overline{A}, \sigma)} \left\{ \kappa_{2\infty}(A) \geq \varepsilon^{-1} \right\} \leq 2n^3 \left(1 + \frac{n\varepsilon}{\sigma} \right)^{n-2} \frac{\varepsilon}{\sigma}.$$

Moreover, for $\beta \geq 2$,

$$\mathbb{E}_{A \in B(\overline{A}, \sigma)} \left(\log_\beta \kappa_{2\infty}(A) \right) \leq 3 \log_\beta n + \log_\beta \frac{1}{\sigma} + 4.$$

Proof We proceed as in the proof of Theorem 2.45. Fix $\overline{A} \in (\mathbb{S}^{n-1})^n$ and $0 < \varepsilon, \sigma \leq 1$. We have

$$\Prob_{A \in B(\overline{A}, \sigma)} \left\{ \kappa_{2\infty}(A) \geq \varepsilon^{-1} \right\} = \frac{\mathrm{vol}(T(\Sigma_\mathbb{S}, \varepsilon) \cap B(\overline{A}, \sigma))}{\mathrm{vol}\, B(\overline{A}, \sigma)}.$$

By Proposition 2.44 we have $T(\Sigma_\mathbb{S}, \varepsilon) \subseteq \bigcup_{i=1}^{n} W_i$, where

$$W_i := \left\{ A \in \left(\mathbb{S}^{n-1} \right)^n \mid \exists c \in \mathbb{S}^{n-1} : d_{\sin}(a_i, c) \leq n\varepsilon \text{ and } A(i : c) \in \Sigma_\mathbb{S} \right\}.$$

Fix $i = n$ and consider the projection

$$\pi : W_n \cap B(\overline{A}, \sigma) \to \prod_{i=1}^{n-1} B(\overline{a}_i, \sigma), \quad (a_1, \ldots, a_n) \mapsto (a_1, \ldots, a_{n-1}).$$

Without loss of generality we assume that the components a_i of $(a_1, \ldots, a_{n-1})$ in $\prod_{i=1}^{n-1} B(\overline{a}_i, \sigma)$ are linearly independent. Then its fiber can be described by

$$\pi^{-1}(a_1, \ldots, a_{n-1}) \simeq \left\{ a_n \in B(\overline{a}_n, \sigma) \mid d_{\sin}(a_n, H) \leq n\varepsilon \right\}$$
$$= T(H, n\varepsilon) \cap B(\overline{a}_n, \sigma),$$

where $H := \mathrm{span}\{a_1, \ldots, a_{n-1}\} \cap \mathbb{S}^{n-1}$. Lemma 2.37 implies that

$$\frac{\mathrm{vol}(T(H, n\varepsilon) \cap B(\mathrm{vol}\, a_n, \sigma))}{\mathrm{vol}\, B(\overline{a}_n, \sigma)} \leq 2(n-1) \left(1 + \frac{n\varepsilon}{\sigma} \right)^{n-2} \frac{n\varepsilon}{\sigma}. \tag{2.20}$$

Using Fubini's theorem we get

$$\text{vol}\big(W_n \cap B(\overline{A},\sigma)\big) = \int_{\prod_{i=1}^{n-1} B(\overline{a}_i,\sigma)} \text{vol}\,\pi^{-1}(a_1,\dots,a_{n-1})\,d\big(\mathbb{S}^{n-1}\big)^{n-1},$$

and hence, by Eq. (2.20),

$$\frac{\text{vol}(W_n \cap B(\overline{A},\sigma))}{\text{vol}\,B(\overline{A},\sigma)} \le 2n^2\Big(1+\frac{n\varepsilon}{\sigma}\Big)^{n-2}\frac{\varepsilon}{\sigma}. \tag{2.21}$$

The same bound holds for W_i for $i=1,\dots,n$. Noting that

$$\frac{\text{vol}(T(\Sigma_{\mathbb{S}},\varepsilon) \cap B(\overline{A},\sigma))}{\text{vol}\,B(\overline{A},\sigma)} \le \sum_{i=1}^{n} \frac{\text{vol}(W_i \cap B(\overline{A},\sigma))}{\text{vol}\,B(\overline{A},\sigma)},$$

the desired tail estimate follows.

For the bound on the expectation, note that $(1+\frac{n\varepsilon}{\sigma})^n \le (1+\frac{1}{n})^n \le e$ if $\varepsilon^{-1} \ge \frac{n^2}{\sigma}$. Hence, for $t \ge \frac{n^2}{\sigma}$,

$$\text{Prob}\big\{\kappa_{2\infty}(A) \ge t\big\} \le 2\frac{en^3}{\sigma}t^{-1}.$$

Proposition 2.26 (with $\alpha=1$ and $t_0 = K = 2\frac{en^3}{\sigma}$) implies that

$$\mathbb{E}\big(\log_\beta \kappa_{2\infty}(A)\big) \le \log_\beta\Big(\frac{2en^3}{\sigma}\Big) + \log_\beta e \le 3\log_\beta n + \log_\beta \frac{1}{\sigma} + 4. \qquad \square$$

2.5 Additional Considerations

2.5.1 Probabilistic Analysis for Other Norms

The analyses in the previous sections took advantage of the block scale invariance of $\overline{\kappa}_{2\infty}$ and therefore, ultimately, of the properties of $\|\ \|_{2\infty}$. With some additional effort, such analyses can be performed for other matrix norms by working in $\mathbb{S}^{n^2-1}$ instead of $(\mathbb{S}^{n-1})^n$. But we can also obtain similar results by preconditioning and using the equivalence between different matrix norms in Table 1.2. We show how for the case of the spectral norm.

Lemma 2.47 *We have* $\frac{1}{\sqrt{n}}\kappa(A) \le \kappa_{2\infty}(A) \le \sqrt{n}\kappa(A)$ *for* $A \in \mathbb{R}^{n\times n} \setminus \{0\}$.

Proof The norm inequality $n^{-1/2}\|y\|_2 \le \|y\|_\infty \le \|y\|_2$ for $y \in \mathbb{R}^n$ implies $n^{-1/2}\|A\| \le \|A\|_{2\infty} \le \|A\|$ and hence $n^{-1/2}d(A,\Sigma) \le d_{2\infty}(A,\Sigma) \le d(A,\Sigma)$.

Theorem 1.7 implies

$$\kappa(A) = \frac{\|A\|}{d(A, \Sigma)} \quad \text{and} \quad \kappa_{2\infty}(A) = \frac{\|A\|_{2\infty}}{d_{2\infty}(A, \Sigma)},$$

from which the statement immediately follows.   □

We may consider the preconditioning $A \mapsto \check{A}$ and define $\bar{\kappa}(A) := \kappa(\check{A})$. In contrast with the case of $\| \ \|_{2\infty}$, this procedure may increase the condition, but it is easy to see that it is not by much.

Proposition 2.48 *For all* $A \notin \Sigma$ *we have* $\kappa(\check{A}) \leq n\kappa(A)$. *Moreover,* $\bar{\kappa}(A) \leq \sqrt{n}\,\bar{\kappa}_{2\infty}(A)$.

Proof Using Lemma 2.47 twice and Proposition 2.41, we obtain

$$\kappa(\check{A}) \leq \sqrt{n}\,\kappa_{2\infty}(\check{A}) \leq \sqrt{n}\,\kappa_{2\infty}(A) \leq n\,\kappa(A).$$

For the second statement we use Lemma 2.47 to get

$$\bar{\kappa}(A) = \kappa(\check{A}) \leq \sqrt{n}\,\kappa_{2\infty}(\check{A}) = \sqrt{n}\,\bar{\kappa}_{2\infty}(A).$$ □

The following analyses are an immediate consequence of Theorem 2.45 and Theorem 2.46.

Proposition 2.49 *For* A *chosen randomly in* $(\mathbb{S}^{n-1})^n$ *from the uniform distribution, we have*

$$\mathbb{E}\big(\log_\beta \kappa(A)\big) \leq 3\log_\beta n + 1,$$

and for $n > 2$ *and any* $\overline{A} \in (\mathbb{S}^{n-1})^n$,

$$\mathop{\mathbb{E}}_{A \in B(\overline{A}, \sigma)} \big(\log_\beta \kappa(A)\big) \leq \frac{7}{2}\log_\beta n + \log_\beta \frac{1}{\sigma} + 5.$$

                                                                                      □

2.5.2   Probabilistic Analysis for Gaussian Distributions

Throughout this chapter, for investigating the condition of a random matrix $A \in \mathbb{R}^{n \times n}$, we assumed the rows a_i of A to be normalized to Euclidean norm 1 and viewed A as the point $(a_1, \ldots, a_n)$ in the product $(\mathbb{S}^{n-1})^n$ of spheres. We then assumed A to be chosen uniformly at random in $(\mathbb{S}^{n-1})^n$ (or in a product of spherical caps thereof). We have chosen this probabilistic model for two reasons. On the one hand, the average and smoothed analyses in this model are the most elementary instances of such analyses we are aware of. On the other hand, our considerations are a natural preparation for similar, but more involved, studies regarding the GCC condition number of linear programming treated later in this book (see Chap. 13).

However, this chapter would be incomplete without mentioning what is known in the Gaussian model, in particular for smoothed analysis, since in contrast to average analysis (where the use of the uniform measure on $\mathbb{S}^{n-1}$ is equivalent to the use of a standard Gaussian in $\mathbb{R}^n$ due to Corollary 2.23), the smoothed analyses for both measures are not equivalent. The following result is due to Mario Wschebor.

Theorem 2.50 *There is a universal constant $K \geq 1$ such that for all $\overline{A} \in \mathbb{R}^{n \times n}$ with $\|\overline{A}\| \leq 1$, all $0 < \sigma \leq 1$, and all $t \geq 1$, we have*

$$\Prob_{A \sim N(\overline{A}, \sigma^2 I)} \{\kappa(A) \geq t\} \leq K n \frac{1}{\sigma t}. \qquad \square$$

We will not prove this result here. Techniques to study the condition of matrices in the Gaussian model will be developed in Chap. 4, where we shall derive bounds that are even sharper for rectangular matrices.

Combining Theorem 2.50 with Proposition 2.26 immediately implies the following.

Corollary 2.51 *There exists $K \geq 1$ such that for all $\sigma \in (0, 1]$, all $\beta \geq 2$, and all $\overline{A} \in \mathbb{R}^{n \times n}$ with $\|\overline{A}\| \leq 1$, we have*

$$\mathop{\mathbb{E}}_{A \sim N(\overline{A}, \sigma^2 I)} \left(\log_\beta \kappa(A)\right) \leq \log_\beta n + \log_\beta \frac{1}{\sigma} + 2 + \log_\beta K.$$

In particular, taking $\overline{A} = 0$ and $\sigma = 1$,

$$\mathop{\mathbb{E}}_{A \sim N(0, I)} \left(\log_\beta \kappa(A)\right) \leq \log_\beta n + 2 + \log_\beta K. \qquad \square$$

Note that the second bound in the corollary above is better than the one (obtained for the same quantity but with humbler means) in Proposition 2.49.

Chapter 3
Error Analysis of Triangular Linear Systems

The vast majority of the occurrences of condition numbers in the study of linear systems of equations involve the normwise condition number $\kappa(A)$. Almost invariably, the use of $\kappa(A)$ is enough to provide a satisfying explanation of the phenomena observed in practice.

The case of triangular systems of linear equations provides, in contrast, an example in which $\kappa(A)$ turns out to be inadequate. Practitioners long observed that triangular systems of equations are generally solved to high accuracy in spite of being, in general, ill-conditioned. Thus, for instance, J.H. Wilkinson in [235, p. 105]: "In practice one almost invariably finds that if L is ill-conditioned, so that $\|L\|\|L^{-1}\| \gg 1$, then the computed solution of $Lx = b$ (or the computed inverse) is far more accurate than [what forward stability analysis] would suggest."

A first goal in this chapter is to give a precise meaning to the feeling that triangular matrices are, in general, ill-conditioned. We prove that if $L \in \mathbb{R}^{n \times n}$ is a lower-triangular matrix whose entries are independent standard Gaussian random variables (i.e., L is drawn from $N(0, I_{\frac{n(n+1)}{2}})$), then $\mathbb{E}(\log_\beta \kappa(L)) = \Omega(n)$. Corollary 1.6 then yields an expected loss of precision satisfying

$$\mathbb{E}\big(\mathsf{LoP}\big(L^{-1}b\big)\big) = \mathcal{O}(n).$$

Were the loss of precision in the solution of triangular systems to conform to this bound, we would not be able to accurately find these solutions. The reason we actually do find them can be briefly stated. The error analysis of triangular systems reveals that we may use a componentwise condition number $\mathsf{Cw}(L, b)$ instead of the normwise condition number. The second goal of this chapter is to prove that when L is drawn from $N(0, I_{\frac{n(n+1)}{2}})$ and $b \in \mathbb{R}^n$ is drawn from $N(0, I_n)$, then we have $\mathbb{E}(\log \mathsf{Cw}(L, b)) = \mathcal{O}(\log n)$. This bound, together with some backward error analysis, yields bounds for $\mathbb{E}(\mathsf{LoP}(L^{-1}b))$ that are much smaller than the one above, as well as closer to the loss of precision observed in practice.

P. Bürgisser, F. Cucker, *Condition*,
Grundlehren der mathematischen Wissenschaften 349,
DOI 10.1007/978-3-642-38896-5_3, © Springer-Verlag Berlin Heidelberg 2013

3.1 Random Triangular Matrices Are Ill-conditioned

The main result of this section states that random lower-triangular matrices are ill-conditioned with respect to the normwise (classical) condition number.

Theorem 3.1 *Let $L = (\ell_{ij}) \in \mathbb{R}^{n \times n}$ be a random lower-triangular matrix with independent standard Gaussian random entries ℓ_{ij} for $i \geq j$. Then we have*

$$\mathbb{E}\big(\ln \kappa(L)\big) \geq (\ln 2)n - \ln n - 1.$$

As a warm-up, we first show a related result—with a very simple proof—that already indicates that on average, $\kappa(L)$ grows exponentially in n. For this we focus on unit lower-triangular matrices L, that is, we additionally assume that $\ell_{ii} = 1$.

Proposition 3.2 *Let $L = (\ell_{ij})$ denote a random unit lower-triangular matrix with $\ell_{ii} = 1$ and with independent standard Gaussian random entries ℓ_{ij} for $i > j$. Then we have*

$$\mathbb{E}\big(\|L^{-1}\|_F^2\big) = 2^n - 1.$$

In particular, $\mathbb{E}(\|L\|_F^2 \|L^{-1}\|_F^2) \geq n(2^n - 1)$; hence $\mathbb{E}(\kappa(L)^2)$ grows exponentially in n.

Proof The first column $(s_1, \ldots, s_n)$ of L^{-1} is characterized by $s_1 = 1$ and the recurrence relation

$$s_i = -\sum_{j=1}^{i-1} \ell_{ij} s_j \quad \text{for } i = 2, \ldots, n.$$

This implies that s_i is a function of the first i rows of L. Hence the random variable s_i is independent of the entries of L in the rows with index larger than i. By squaring we obtain for $i \geq 2$,

$$s_i^2 = \sum_{\substack{j \neq k \\ j,k < i}} \ell_{ij} \ell_{ik} s_j s_k + \sum_{j < i} \ell_{ij}^2 s_j^2.$$

By the preceding observation, $s_j s_k$ is independent of $\ell_{ij} \ell_{ik}$ for $j, k < i$. If additionally $j \neq k$, we get

$$\mathbb{E}(\ell_{ij} \ell_{ik} s_j s_k) = \mathbb{E}(\ell_{ij} \ell_{ik}) \mathbb{E}(s_j s_k) = \mathbb{E}(\ell_{ij}) \mathbb{E}(\ell_{ik}) \mathbb{E}(s_j s_k) = 0,$$

since ℓ_{ij} and ℓ_{ik} are independent and centered. So the expectations of the mixed terms vanish, and we obtain, using $\mathbb{E}(\ell_{ij}^2) = 1$, that

$$\mathbb{E}(s_i^2) = \sum_{j=1}^{i-1} \mathbb{E}(s_j^2) \quad \text{for } i \geq 2.$$

Solving this recurrence with $\mathbb{E}(s_1^2) = 1$ yields

$$\mathbb{E}(s_i^2) = 2^{i-2} \quad \text{for } i \geq 2.$$

Therefore, the first column v_1 of L^{-1} satisfies

$$\mathbb{E}(\|v_1\|^2) = \mathbb{E}\left(\sum_{i=1}^{n} s_i^2\right) = 2^{n-1}.$$

By an analogous argument one shows that

$$\mathbb{E}(\|v_k\|^2) = 2^{n-k}$$

for the kth column v_k of L^{-1}. Altogether, we obtain

$$\mathbb{E}(\|L^{-1}\|_F^2) = \mathbb{E}\left(\sum_{k=1}^{n} \|v_k\|^2\right) = \sum_{k=1}^{n} \mathbb{E}(\|v_k\|^2) = 2^n - 1.$$

Finally, we note that $\|L\|_F^2 \geq n$, since $\ell_{ii} = 1$. Hence,

$$\mathbb{E}(\|L\|_F^2 \|L^{-1}\|_F^2) \geq n \, \mathbb{E}(\|L^{-1}\|_F^2) \geq n(2^n - 1).$$

The last assertion follows from $\kappa(L) \geq \frac{1}{n} \|L\|_F \|L^{-1}\|_F$. $\square$

We turn now to the general situation. Consider a lower-triangular matrix $L = (\ell_{ij}) \in \mathbb{R}^{n \times n}$ that is invertible, i.e., $\det L = \ell_{11} \cdots \ell_{nn} \neq 0$. We denote by $t_1, \ldots, t_n$ the entries of the first column of L^{-1}. These entries can be recursively computed as follows:

$$t_1 = \ell_{11}^{-1},$$

$$t_2 = -\ell_{22}^{-1} \ell_{21} t_1,$$

$$t_3 = -\ell_{33}^{-1} (\ell_{31} t_1 + \ell_{32} t_2),$$

$$\vdots$$

$$t_n = -\ell_{nn}^{-1} (\ell_{n1} t_1 + \cdots + \ell_{n,n-1} t_{n-1}).$$

We suppose that the ℓ_{ij} are independent standard Gaussian random variables. The next lemma provides a recurrence formula for the joint probability density function f_k of $(t_1, \ldots, t_k)$. We introduce the notation $T_k := \sqrt{t_1^2 + \cdots + t_k^2}$.

Lemma 3.3 *The joint probability density function $f_k(t_1, \ldots, t_k)$ satisfies the following recurrence:*

$$f_1 = \frac{1}{\sqrt{2\pi t_1^2}} e^{-\frac{1}{2t_1^2}}, \qquad f_k = \frac{1}{\pi} \frac{T_{k-1}}{T_k^2} f_{k-1} \quad \text{for } k > 1.$$

Proof We have $t_1 = 1/x$, where $x = \ell_{11}$ is standard Gaussian with density $\varphi(x) = (2\pi)^{-1/2} e^{-\frac{1}{2}x^2}$. Therefore, by Proposition 2.11 (with $n = 1$, $\psi(x) = 1/x$, and $f_M = \varphi$), the density ρ of the random variable t_1 satisfies

$$\rho(t_1) = \left|\frac{dt_1}{dx}\right|^{-1} \varphi(x) = x^2 \varphi(x) = \frac{1}{\sqrt{2\pi} t_1^2} e^{-\frac{1}{2t_1^2}},$$

as claimed.

To obtain the recurrence expression for f_k, we consider the random variable

$$\tau_k := \ell_{k1} t_1 + \cdots + \ell_{k,k-1} t_k.$$

For fixed values of $t_1, \ldots, t_{k-1}$, the conditional distribution of τ_k is Gaussian with mean 0 and variance T_{k-1}^2. Therefore, the joint probability density of $(t_1, \ldots, t_{k-1}, \tau_k)$ is given by

$$f_{k-1} \cdot \frac{1}{\sqrt{2\pi} T_{k-1}} e^{-\frac{\tau_k^2}{2T_{k-1}^2}}.$$

The variable t_k is obtained as $t_k = \tau_k/\ell$, where $\ell = -\ell_{kk}$ is an independent standard Gaussian random variable. Note that the joint probability density of $(t_1, \ldots, t_{k-1}, \tau_k, \ell)$ is given by

$$f_{k-1} \cdot \frac{1}{\sqrt{2\pi} T_{k-1}} e^{-\frac{\tau_k^2}{2T_{k-1}^2}} \frac{1}{\sqrt{2\pi}} e^{-\frac{\ell^2}{2}}.$$

We make now the change of variables $(t_1, \ldots, t_{k-1}, \tau_k, \ell) \overset{\psi}{\mapsto} (t_1, \ldots, t_{k-1}, t_k, \ell)$, which satisfies $\det D\Psi(t_1, \ldots, t_{k-1}, t_k, \ell) = \ell^{-1}$. Proposition 2.11 implies that the density g of $(t_1, \ldots, t_{k-1}, t_k, \ell)$ satisfies

$$g = f_{k-1} \cdot \frac{1}{\sqrt{2\pi} T_{k-1}} e^{-\frac{\ell^2 t_k^2}{2T_{k-1}^2}} \frac{1}{\sqrt{2\pi}} e^{-\frac{\ell^2}{2}} \cdot |\ell|.$$

A straightforward calculation, making the change of variables $b = \ell^2/2$, shows that

$$f_k(t_1, \ldots, t_k) = \int_{-\infty}^{\infty} g(t_1, \ldots, t_k, \ell) \, d\ell = \frac{f_{k-1}}{2\pi T_{k-1}} 2 \int_0^{\infty} e^{-\frac{\ell^2}{2}\left(\frac{t_k^2}{T_{k-1}^2}+1\right)} \ell \, d\ell$$

$$= \frac{f_{k-1}}{\pi T_{k-1}} \frac{1}{\frac{t_k^2}{T_{k-1}^2}+1} = \frac{f_{k-1}}{\pi T_{k-1}} \frac{T_{k-1}^2}{T_k^2} = \frac{f_{k-1}}{\pi} \frac{T_{k-1}}{T_k^2},$$

which proves the desired recurrence. $\square$

The recursive description of the joint probability density functions f_k in Lemma 3.3 yields the following recurrence for $\mathbb{E}(\ln T_k^2)$.

Lemma 3.4 *We have* $\mathbb{E}(\ln T_k^2) = \mathbb{E}(\ln T_{k-1}^2) + 2\ln 2$ *for* $k > 1$.

Proof By Lemma 3.3 we have, omitting the arguments t_i to avoid cluttering the notation,

$$\mathbb{E}\left(\ln T_k^2\right) = \int_{\mathbb{R}^k} f_k \ln T_k^2 \, dt_1 \cdots dt_k = \int_{\mathbb{R}^{k-1}} \frac{f_{k-1}T_{k-1}}{\pi} \int_{\mathbb{R}} \frac{\ln T_k^2}{T_k^2} \, dt_k \, dt_1 \cdots dt_{k-1}.$$

We fix $t_1, \ldots, t_{k-1}$ and rewrite the inner integral by making the change of variable $y = t_k/T_{k-1}$. Hence $T_k^2 = T_{k-1}^2(1 + y^2)$, and we get

$$\frac{1}{\pi} \int_{\mathbb{R}} \frac{\ln T_k^2}{T_k^2} \, dt_k = \frac{1}{T_{k-1}} \frac{1}{\pi} \int_{\mathbb{R}} \frac{\ln T_{k-1}^2 + \ln(1+y^2)}{1+y^2} \, dy.$$

The function $y \mapsto 1/(\pi(1+y^2))$ is a probability density on $\mathbb{R}$, and a straightforward calculation shows that

$$\frac{1}{\pi} \int_{\mathbb{R}} \frac{\ln(1+y^2)}{1+y^2} \, dy = 2\ln 2.$$

Hence we obtain for the inner integral

$$\frac{1}{\pi} \int_{\mathbb{R}} \frac{\ln T_k^2}{T_k^2} \, dt_k = \frac{1}{T_{k-1}} \left(\ln T_{k-1}^2 + 2\ln 2\right).$$

Plugging this expression into the integral above, we obtain the stated recurrence

$$\mathbb{E}\left(\ln T_k^2\right) = \mathbb{E}\left(\ln T_{k-1}^2\right) + 2\ln 2. \qquad \square$$

Proof of Theorem 3.1 Using the expression for the density function f_1 provided by Lemma 3.3, we obtain, using software for symbolic integration,

$$\mathbb{E}\left(\ln T_1^2\right) = \frac{1}{\sqrt{2\pi}} \int_{\mathbb{R}} \frac{1}{t_1^2} e^{-\frac{1}{2t_1^2}} \ln t_1^2 \, dt_1 = \ln 2 + \gamma,$$

where $\gamma \approx 0.577$ denotes the Euler–Mascheroni constant. Combining this with the recursive expression of Lemma 3.4, we get

$$\mathbb{E}\left(\ln T_n^2\right) = (2\ln 2)(n-1) + \ln 2 + \gamma \geq (2\ln 2)n - 0.12.$$

Recalling that T_n equals the Euclidean norm of the first column of L^{-1}, this implies

$$\mathbb{E}\left(\ln \|L^{-1}\|_F\right) \geq \mathbb{E}(\ln T_n) \geq (\ln 2)n - 0.06.$$

Since $\|L\|_F^2$ is chi-square distributed with $n(n+1)/2$ degrees of freedom, we have, by Proposition 2.21, that $\mathbb{E}(\ln \|L\|_F) \geq 0$ if $n > 1$. Therefore

$$\mathbb{E}\big(\ln\big(\|L\|_F\big\|L^{-1}\big\|_F\big)\big) \geq \mathbb{E}(\ln T_n) \geq (\ln 2)n - 0.06.$$

Using that $\|L\|\|L^{-1}\| \geq \frac{1}{n}\|L\|_F\|L^{-1}\|_F$, the assertion follows. $\square$

3.2 Backward Analysis of Triangular Linear Systems

Let $L = (\ell_{ij}) \in \mathbb{R}^{n \times n}$ be a nonsingular lower-triangular matrix and $b \in \mathbb{R}^n$. We are interested in solving the system $Lx = b$. Algorithmically, this is very simple, and the components $x_1, \ldots, x_n$ of the solution x are sequentially obtained by forward substitution as in Algorithm 3.1.

Algorithm 3.1 FS

Input: $L \in \mathbb{R}^{n \times n}, b \in \mathbb{R}^n$

Preconditions: L is lower-triangular, nonsingular

```
x₁ := b₁/ℓ₁₁
for  i = 2...n  do
        compute  w := ℓᵢ₁x₁ + ··· + ℓᵢ,ᵢ₋₁xᵢ₋₁
        compute  xᵢ := (bᵢ−w)/ℓᵢᵢ
return  x = (x₁,...,xₙ)
```

Output: $x \in \mathbb{R}^n$

Postconditions: $Lx = b$

It is straightforward to obtain a backward error analysis for Algorithm 3.1 from the results we proved in the Overture. We use notation introduced in Sect. O.3.

Proposition 3.5 *There is a round-off implementation of algorithm* FS *that with input* $L \in \mathbb{R}^{n \times n}$ *lower-triangular and* $b \in \mathbb{R}^n$ *computes the solution* x *of* $Lx = b$. *If* $\epsilon_{\mathrm{mach}}(\lceil \log_2 n \rceil + 1) < 1$, *then the computed value* $\mathrm{fl}(x)$ *satisfies* $(L + E)\mathrm{fl}(x) = b$ *with* $|e_{ij}| \leq \gamma_{\lceil \log_2 i \rceil + 1}|\ell_{ij}|$.

Proof By induction on n. If $n = 1$, then

$$\mathrm{fl}(x_1) = \frac{b_1}{\ell_{11}}(1 + \theta_1) = \frac{b_1}{(1 + \theta_1)\ell_{11}},$$

and the statement follows since $|\theta_1| \leq \gamma_1$.

Now assume $n > 1$ and let $\bar{x} = (x_1, \ldots, x_{n-1})$, $\bar{b} = (b_1, \ldots, b_{n-1})$, and $\bar{L} \in \mathbb{R}^{(n-1) \times (n-1)}$ be the matrix obtained by removing the nth row and the nth column of L. Then, $\bar{L}$ is lower-triangular, nonsingular, and $\bar{L}\bar{x} = \bar{b}$. By the induction hypothesis the point $\overline{\mathrm{fl}(x)} = (\mathrm{fl}(x_1), \ldots, \mathrm{fl}(x_{n-1}))$ computed at the first $(n - 2)$ iterations of FS satisfies $(\bar{L} + \bar{E})\overline{\mathrm{fl}(x)} = \bar{b}$ with $|\bar{e}_{ij}| \leq \gamma_{\lceil \log_2 i \rceil + 1}|\ell_{ij}|$.

We now use Proposition O.4 to perform the $(n-1)$th iteration (which computes x_n) with $A = (\ell_{n1}, \ldots, \ell_{n,n-1}) \in \mathbb{R}^{1 \times (n-1)}$. By this proposition, we compute the product $A \overline{\mathrm{fl}(x)} = \ell_{n1}\mathrm{fl}(x_1) + \cdots + \ell_{n,n-1}\mathrm{fl}(x_{n-1})$ and obtain $\mathrm{fl}(w)$ satisfying

$$\mathrm{fl}(w) = (\ell_{n1} + e_{n1})\mathrm{fl}(x_1) + \cdots + (\ell_{n,n-1} + e_{n,n-1})\mathrm{fl}(x_{n-1})$$

with $|e_{nj}| \leq \gamma_{\lceil \log_2(n-1) \rceil + 1}|\ell_{nj}|$ for $j \leq n-1$. We then compute x_n, and we obtain

$$\mathrm{fl}(x_n) = \mathrm{fl}\left(\frac{b_n - \mathrm{fl}(w)}{\ell_{nn}}\right) = \left(\frac{(b_n - \mathrm{fl}(w))(1 + \theta_1)}{\ell_{nn}}\right)(1 + \theta_1)$$

$$= \frac{b_n - (\ell_{n1} + e_{n1})\mathrm{fl}(x_1) + \cdots + (\ell_{n,n-1} + e_{n,n-1})\mathrm{fl}(x_{n-1})}{\ell_{nn}(1 + \theta_2)},$$

and the result follows by taking $e_{nn} = \ell_{nn}\theta_2$ and E the matrix obtained by putting $\bar{E}$ in its upper-left $(n-1) \times (n-1)$ corner, appending $(e_{n1}, \ldots, e_{nn})$ as the nth row, and filling the remainder of the nth column with zeros. $\square$

3.3   Componentwise Condition of Random Sparse Matrices

Proposition 3.5 justifies the componentwise measure of relative errors and, as a consequence, the use of componentwise condition numbers in the error analysis. The goal of this section is to give a (classical) probabilistic analysis for these condition numbers.

We will work in the more general context of *sparse matrices* (which, in this section, are matrices with a fixed pattern of zeros).[1] Therefore, the following results apply not only to triangular matrices but to other classes of sparse matrices such as, for instance, tridiagonal matrices. Also, in the process of proving our main result we will estimate as well the average componentwise condition for the computation of the determinant and for matrix inversion.

3.3.1   Componentwise Condition Numbers

Recall that for a function $\varphi : \mathcal{D} \subseteq \mathbb{R}^m \to \mathbb{R}^q$ and a point $a \in \mathcal{D}$ with $a_i \neq 0$ and $\varphi_j(a) \neq 0$ for all $i \leq m$ and $j \leq q$, we defined in (O.1) the componentwise condition number

$$\mathrm{Cw}^\varphi(a) = \lim_{\delta \to 0} \sup_{\mathrm{RelError}(a) \leq \delta} \frac{\mathrm{RelError}(\varphi(a))}{\mathrm{RelError}(a)},$$

---

[1]The word "sparse" is also used to denote matrices with a large number of zeros, not necessarily in fixed positions.

where both RelError(a) and RelError($\varphi(a)$) are measured componentwise and we follow the convention that $\frac{0}{0} = 1$. That is,

$$\mathsf{RelError}(a) = \max_{i \leq m} \frac{|\tilde{a}_i - a_i|}{|a_i|},$$

and similarly for $\varphi(a)$. In fact, we have $\mathsf{Cw}^\varphi(a) = \max_{j \leq q} \mathsf{Cw}_j^\varphi(a)$, where for $j \leq q$,

$$\mathsf{Cw}_j^\varphi(a) = \lim_{\delta \to 0} \sup_{\mathsf{RelError}(a) \leq \delta} \frac{\mathsf{RelError}(\varphi(a)_j)}{\mathsf{RelError}(a)};$$

compare Sect. O.2.

Componentwise condition numbers behave nicely with respect to multiplication and division.

Lemma 3.6 *Let $\varphi, \psi : \mathcal{D} \subseteq \mathbb{R}^m \to \mathbb{R}$ be functions and $a \in \mathcal{D}$ such that $a_i \neq 0$ for all i and $\varphi(a)\psi(a) \neq 0$. Then we have*

$$\mathsf{Cw}^{\varphi\psi}(a) \leq \mathsf{Cw}^\varphi(a) + \mathsf{Cw}^\psi(a), \qquad \mathsf{Cw}^{\varphi/\psi}(a) \leq \mathsf{Cw}^\varphi(a) + \mathsf{Cw}^\psi(a).$$

Proof The first statement follows from the identity

$$\frac{\tilde{x}\tilde{y} - xy}{xy} = \frac{\tilde{x} - x}{x} + \frac{\tilde{y} - y}{y} + \frac{\tilde{x} - x}{x}\frac{\tilde{y} - y}{y}.$$

For the second statement, we use instead

$$\frac{\tilde{x}/\tilde{y} - x/y}{x/y} = \frac{\frac{\tilde{x}-x}{x} - \frac{\tilde{y}-y}{y}}{1 + \frac{\tilde{y}-y}{y}}. \qquad \square$$

Example 3.7 The multiplication $\varphi : \mathbb{R}^m \to \mathbb{R}, (a_1, \ldots, a_m) \mapsto a_1 \cdots a_m$ satisfies $\mathsf{Cw}^\varphi(a) = m$ if $a_i \neq 0$ for all i. In fact, $\mathsf{Cw}^\varphi(a) \leq m$ follows immediately from Lemma 3.6 by taking into account that the componentwise condition number of a projection $a \mapsto a_i$ equals one. It is easy to see that equality holds.

Componentwise condition numbers also behave nicely with respect to addition and subtraction. We leave the straightforward proof to the reader.

Lemma 3.8 *Let $\varphi, \psi : \mathcal{D} \subseteq \mathbb{R}^m \to \mathbb{R}$ be functions and $a \in \mathcal{D}$ such that $a_i \neq 0$ for all i and $\varphi(a)\psi(a) \neq 0$. Then we have*

$$\mathsf{Cw}^{\varphi \pm \psi}(a) \leq \max\{\mathsf{Cw}^\varphi(a), \mathsf{Cw}^\psi(a)\},$$

provided the left-hand side is defined (i.e., $\varphi(a) \pm \psi(a) \neq 0$). $\qquad \square$

In all that follows, for $n \in \mathbb{N}$, we denote the set $\{1, \ldots, n\}$ by $[n]$ and write, as usual, $[n]^2 = [n] \times [n]$.

Definition 3.9 We denote by $\mathcal{M}$ the set of $n \times n$ real matrices and by Σ its subset of singular matrices. Also, for a subset $S \subseteq [n]^2$ we define

$$\mathcal{M}_S := \left\{ A \in \mathcal{M} \mid \text{if } (i, j) \notin S \text{ then } a_{ij} = 0 \right\}$$

and write $|S|$ for the cardinality of S. We denote by $\mathcal{R}_S$ the space of random $n \times n$ matrices obtained by setting $a_{ij} = 0$ if $(i, j) \notin S$ and drawing all other entries independently from the standard Gaussian $N(0, 1)$. As above, if $S = [n]^2$, we write simply $\mathcal{R}$.

Lemma 3.10 *If $\mathcal{M}_S \nsubseteq \Sigma$, then $\text{Prob}_{A \in \mathcal{R}_S}(A \text{ is singular}) = 0$.*

Proof The set of singular matrices in $\mathcal{M}_S$ is the zero set of the restriction of the determinant to $\mathcal{M}_S$. This restriction is a polynomial in $\mathbb{R}^{|S|}$ whose zero set, if different from $\mathbb{R}^{|S|}$, has dimension smaller than $|S|$. $\qquad\square$

3.3.2 Determinant Computation

We consider here the problem of computing the determinant of a sparse matrix $A \in \mathcal{M}_S \simeq \mathbb{R}^{|S|}$ and its componentwise condition number $\text{Cw}^{\text{det}}(A)$, which is defined by taking $\varphi \colon \mathcal{M}_S \to \mathbb{R}$, $A \mapsto \det A$. We shall suppose that $\mathcal{M}_S \nsubseteq \Sigma$. Then $\text{Cw}^{\text{det}}(A)$ is almost surely defined by Lemma 3.10, since $\det A \prod_{ij} a_{ij} \neq 0$ holds almost surely.

Our goal is to derive probability tail estimates for $\text{Cw}^{\text{det}}(A)$. We begin with a simple observation on $\text{Cw}^{\text{det}}(A)$ for triangular matrices, which is in stark contrast to our findings in Sect. 3.1 on the normwise condition number of such matrices.

Proposition 3.11 *Let S be such that $\mathcal{M}_S$ equals the set of upper-triangular $n \times n$ matrices. Then we have $\text{Cw}^{\text{det}}(A) = n$, provided $\det A \neq 0$.*

Proof This is an immediate consequence of Example 3.7. $\qquad\square$

Our main result for $\text{Cw}^{\text{det}}(A)$ is the following.

Theorem 3.12 *Suppose $S \subseteq [n]^2$ such that $\mathcal{M}_S \nsubseteq \Sigma$. Then, for $t \geq 2|S|$,*

$$\Prob_{A \in \mathcal{R}_S} \left\{ \text{Cw}^{\text{det}}(A) \geq t \right\} \leq |S|^2 \frac{1}{t}.$$

We may use this result to estimate the average componentwise condition number for the computation of the determinant.

Corollary 3.13 *For a base $\beta \geq 2$ and a set $S \subseteq [n]^2$ with $|S| \geq 2$, we have $\mathbb{E}(\log_\beta \text{Cw}^{\text{det}}(A)) \leq 2\log_\beta |S| + \log_\beta e$, where $\mathbb{E}$ denotes expectation over $A \in \mathcal{R}_S$.*

Proof Use Proposition 2.26, taking $X = \mathrm{Cw}^{\mathrm{det}}(A)$, $\alpha = 1$, and $t_0 = K = |S|^2$ (note that $|S|^2 \geq 2|S|$, since $|S| \geq 2$), together with Theorem 3.12.   □

We move now to the proof of Theorem 3.12. First we give a closed formula for $\mathrm{Cw}^{\mathrm{det}}(A)$. We shall denote by A_{ij} the submatrix of A obtained by deleting its ith row and its jth column.

Lemma 3.14 *For $A \in \mathcal{M}_S \setminus \Sigma$, we have*

$$\mathrm{Cw}^{\mathrm{det}}(A) = \sum_{(i,j) \in S} \left| \frac{a_{ij} \det A_{ij}}{\det A} \right|.$$

In particular, $\mathrm{Cw}^{\mathrm{det}}(A)$ does not depend on S.

Proof For any $i \in [n]$, expanding by the ith row, we have

$$\det A = \sum_{j} (-1)^{i+j} a_{ij} \det A_{ij}.$$

Hence, for all $i, j \in [n]$, we get

$$\frac{\partial \det A}{\partial a_{ij}} = (-1)^{i+j} \det A_{ij}.$$

Fix $A \in \mathcal{M}_S \setminus \Sigma$ and $\delta > 0$. Let $\tilde{A} \in \mathcal{M}_S$ be such that $\mathrm{RelError}(A) = \delta$. Then $|\tilde{a}_{ij} - a_{ij}| \leq \delta |a_{ij}|$ for all $(i, j) \in S$. Using Taylor's expansion and the equalities above, we obtain for $\delta \to 0$,

$$\det \tilde{A} = \det A + \sum_{i,j} (-1)^{i+j} (\tilde{a}_{ij} - a_{ij}) \det A_{ij} + \mathcal{O}(\delta^2).$$

It follows that

$$\frac{|\det \tilde{A} - \det A|}{\delta |\det A|} \leq \sum_{i,j} \frac{|a_{ij} \det A_{ij}|}{|\det A|} + \mathcal{O}(\delta).$$

Hence, by the definition (O.1), we obtain

$$\mathrm{Cw}^{\mathrm{det}}(A) \leq \sum_{i,j} \frac{|a_{ij} \det A_{ij}|}{|\det A|}.$$

To see that equality holds we choose $\tilde{A}$ by taking $\tilde{a}_{ij} = a_{ij}(1 \pm \delta)$, where we take the plus sign if $(-1)^{i+j} \det A_{ij} \geq 0$ and the minus sign otherwise. Then the terms $(-1)^{i+j}(\tilde{a}_{ij} - a_{ij}) \det A_{ij}$ have the same sign for all $i, j \in [n]$.   □

We proceed with a general tail estimate for Gaussian random variables.

Lemma 3.15 *Let p, q be two fixed vectors in $\mathbb{R}^n$ such that $\|p\| \leq \|q\|$, $q \neq 0$. If $x \sim N(0, I_n)$, then for all $t \geq 2$,*

$$\text{Prob}\left\{\left|\frac{x^\mathsf{T} p}{x^\mathsf{T} q}\right| \geq t\right\} \leq \frac{1}{t}.$$

Proof Let $v = \|q\|$. By the rotational invariance of $N(0, I_n)$ we may assume that $q = (v, 0, \ldots, 0)$. Also, by appropriately scaling, we may assume that $v = 1$. Note that then $\|p\| \leq 1$. We therefore have

$$\text{Prob}\left\{\left|\frac{x^\mathsf{T} p}{x^\mathsf{T} q}\right| \geq t\right\} = \text{Prob}\left\{\left|p_1 + \sum_{i=2}^{n} \frac{x_i p_i}{x_1}\right| \geq t\right\}$$

$$= \text{Prob}\left\{\left|p_1 + \frac{1}{x_1}\alpha Z\right| \geq t\right\}$$

$$= \text{Prob}\left\{\frac{Z}{x_1} \geq \frac{t - p_1}{\alpha}\right\} + \text{Prob}\left\{\frac{Z}{x_1} \leq \frac{-t - p_1}{\alpha}\right\}, \quad (3.1)$$

where $Z = N(0, 1)$ is independent of x_1 and $\alpha = \sqrt{p_2^2 + \cdots + p_n^2} \leq 1$. Here we used that a sum of independent centered Gaussians is a centered Gaussian whose variance is the sum of the terms' variances (cf. Sect. 2.2.2). Note that in case $\alpha = 0$, the statement of the lemma is trivially true.

The random variables x_1 and Z are independent $N(0, 1)$. It therefore follows from Proposition 2.19 that the angle $\theta = \arctan(Z/x_1)$ is uniformly distributed in $[-\pi/2, \pi/2]$. Hence, for $\gamma \in [0, \infty)$,

$$\text{Prob}\left\{\frac{Z}{x_1} \geq \gamma\right\} = \text{Prob}\{\theta \geq \arctan \gamma\} = \frac{1}{\pi}\left(\frac{\pi}{2} - \arctan \gamma\right)$$

$$= \frac{1}{\pi}\int_\gamma^\infty \frac{1}{1 + t^2}\, dt \leq \frac{1}{\pi}\int_\gamma^\infty \frac{1}{t^2}\, dt = \frac{1}{\pi \gamma}.$$

Similarly, one shows for $\sigma \in (-\infty, 0]$, that

$$\text{Prob}\left\{\frac{Z}{x_1} \leq \sigma\right\} \leq \frac{1}{\pi(-\sigma)}.$$

Using these bounds in (3.1) with $\gamma = \frac{t - p_1}{\alpha}$ and $\sigma = \frac{-t - p_1}{\alpha}$, we obtain

$$\text{Prob}\left\{\left|\frac{x^\mathsf{T} p}{x^\mathsf{T} q}\right| \geq t\right\} \leq \frac{1}{\pi}\left(\frac{\alpha}{t - p_1} + \frac{\alpha}{t + p_1}\right) = \frac{\alpha}{\pi}\frac{2t}{t^2 - p_1^2} \leq \frac{2}{\pi}\frac{t}{t^2 - 1} \leq \frac{1}{t},$$

the last since $t \geq 2$.   □

Proof of Theorem 3.12 From Lemma 3.14 we obtain

$$\text{Prob}\{\mathsf{Cw}^{\text{det}}(A) \geq t\} = \text{Prob}\left\{ \sum_{(i,j)\in S} \left| \frac{a_{ij} \det A_{ij}}{\det A} \right| \geq t \right\}$$

$$\leq \sum_{(i,j)\in S} \text{Prob}\left\{ \left| \frac{a_{ij} \det A_{ij}}{\det A} \right| \geq \frac{t}{|S|} \right\}. \tag{3.2}$$

It is therefore enough to prove that for all $(i, j) \in S$ and all $z \geq 2$,

$$\text{Prob}\left\{ \left| \frac{a_{ij} \det A_{ij}}{\det A} \right| \geq z \right\} \leq \frac{1}{z}. \tag{3.3}$$

Without loss of generality, take $(i, j) = (1, 1)$. Let $x = a_1$ be the first column of A. Also, let $I = \{i \in [n] \mid (i, 1) \in S\}$ and let x_I be the vector obtained by removing entries x_i with $i \notin I$. Then,

$$x_I \sim N(0, \mathrm{I}_{|I|}). \tag{3.4}$$

For $i \in [n]$ write $q_i = (-1)^{i+1} \det A_{i1}$. Let $q = (q_1, \ldots, q_n)$ and let q_I be the vector obtained by removing entries q_i with $i \notin I$. Clearly, q_I is independent of x_I. Using this notation, the expansion by the first column yields

$$\det A = \sum_{i\in[n]} (-1)^{i+1} a_{i1} \det A_{i1} = x_I^{\mathsf{T}} q_I.$$

In addition, $a_{11} \det A_{11} = x_I^{\mathsf{T}}(q_1 e_1)$, where e_1 is the vector with the first entry equal to 1 and all others equal to 0. Hence,

$$\frac{a_{11} \det A_{11}}{\det A} = \frac{x_I^{\mathsf{T}}(q_1 e_1)}{x_I^{\mathsf{T}} q_I}.$$

Let ρ be the density of the random vector q_I. Then, for $z \geq 2$,

$$\text{Prob}\left\{ \left| \frac{a_{11} \det A_{11}}{\det A} \right| \geq z \right\}$$

$$= \text{Prob}\left\{ \left| \frac{x_I^{\mathsf{T}}(q_1 e_1)}{x_I^{\mathsf{T}} q_I} \right| \geq z \right\}$$

$$= \int_{u\in\mathbb{R}^{|I|}} \text{Prob}\left\{ \left| \frac{x_I^{\mathsf{T}}(q_1 e_1)}{x_I^{\mathsf{T}} q_I} \right| \geq z \,\middle|\, q_I = u \right\} \rho(u)\,du$$

$$\leq \int_{u\in\mathbb{R}^{|I|}} \frac{1}{z} \rho(u)\,du = \frac{1}{z}.$$

Here the middle line is Proposition 2.14, and the inequality follows since x_I is independent of q_1 and q_I, and therefore we can use (3.4) and Lemma 3.15 (with $p = u_1 e_1$ and $q = u$). This proves (3.3) and hence the lemma. $\square$

3.3.3   Matrix Inversion

We now focus on the problem of inverting a matrix A. The (i, j)th entry γ_{ij} of the inverse A^{-1} is given by Cramer's rule: $\gamma_{ij} = (-1)^{i+j} \det A_{ji} / \det A$.

Fix $S \subseteq [n]^2$ such that $\mathcal{M}_S$ is not contained in Σ. Let J_S denote the set of all $(i, j) \in [n^2]$ such that there exists an invertible $A \in \mathcal{M}_S$ with $\det A_{ji} \neq 0$. Note that the entries of A^{-1} vanish at the positions outside J_S for $A \in \mathcal{M}_S \setminus \Sigma$ and are thus uninteresting. For instance, if $S = \{(i, i) \mid i \in [n]\}$, then $J_S = S$. As for Lemma 3.10, we can show that $\gamma_{ij} \neq 0$ with probability one for $A \in \mathcal{R}_S$.

We define the componentwise condition number $\mathrm{Cw}^{\dagger}(A)$ as in (O.1) for the map $\varphi: \mathcal{M} \setminus \Sigma \to \mathbb{R}^{J_S}, A \mapsto A^{-1}$. By the previous reasoning, $\mathrm{Cw}^{\dagger}(A)$ is almost surely defined.

Theorem 3.16 *Let $S \subseteq [n]^2$ be such that $\mathcal{M}_S \nsubseteq \Sigma$. Then, for all $t \geq 4|S|$,*

$$\mathop{\mathrm{Prob}}_{A \in \mathcal{R}_S} \left\{ \mathrm{Cw}^{\dagger}(A) \geq t \right\} \leq 4|S|^2 n^2 \frac{1}{t}.$$

Proof By the definition of $\mathrm{Cw}^{\dagger}(A)$ we have

$$\mathrm{Prob}\{\mathrm{Cw}^{\dagger}(A) \geq t\} = \mathrm{Prob}\left\{ \max_{i, j \in [n]} \mathrm{Cw}^{\dagger}_{ij}(A) \geq t \right\} \leq \sum_{i, j \in [n]} \mathrm{Prob}\{\mathrm{Cw}^{\dagger}_{ij}(A) \geq t\}.$$

Cramer's rule

$$\gamma_{ij} = (-1)^{i+j} \det A_{ji} / \det A$$

combined with Lemma 3.6 yields

$$\mathrm{Cw}^{\dagger}_{ij}(A) \leq \mathrm{Cw}^{\mathrm{det}}(A) + \mathrm{Cw}^{\mathrm{det}}(A_{ji}).$$

We conclude that

$$\mathrm{Prob}\{\mathrm{Cw}^{\dagger}_{ij}(A) \geq t\} \leq \mathrm{Prob}\left\{\mathrm{Cw}^{\mathrm{det}}(A) \geq \frac{t}{2}\right\} + \mathrm{Prob}\left\{\mathrm{Cw}^{\mathrm{det}}(A_{ji}) \geq \frac{t}{2}\right\}$$

$$\leq 4|S|^2 \frac{1}{t},$$

obtaining the last inequality by applying Theorem 3.12 to A and A_{ji}. The statement now follows. $\square$

Combining Theorem 3.16 with Proposition 2.26, we obtain the following corollary.

Corollary 3.17 *Let $S \subseteq [n]^2$ be such that $\mathcal{M}_S \nsubseteq \Sigma$. Then,*

$$\mathbb{E}\left(\log_{\beta}\left(\mathrm{Cw}^{\dagger}(A)\right)\right) \leq 2\log_{\beta} n + 2\log_{\beta} |S| + \log_{\beta} 4e,$$

where $\mathbb{E}$ denotes expectation over $A \in \mathcal{R}_S$.

3.3.4 Solving Linear Equations

We finally deal with the problem of solving linear systems of equations. That is, we consider a matrix $A \in \mathscr{M}_S$ and a vector $b \in \mathbb{R}^n$, and we want to solve $Ax = b$. We denote by $\mathsf{Cw}(A, b)$ the corresponding componentwise condition number obtained from the definition (O.1) by taking $\varphi : (\mathscr{M}_S \setminus \Sigma) \times \mathbb{R}^n \to \mathbb{R}^n$ given by $\varphi(A, b) = A^{-1}b$. We note that $\mathsf{Cw}(A, b)$ is almost surely defined.

Theorem 3.18 *Let* $S \subset [n]^2$ *be such that* $\mathscr{M}_S \nsubseteq \Sigma$. *Then, for all* $t \geq 4(|S| + n)$,

$$\mathsf{Prob}\{\mathsf{Cw}(A, b) \geq t\} \leq 10|S|^2 n \frac{1}{t},$$

where Prob *denotes the probability over* (A, b) *with respect to the product measure* $\mathcal{R}_S \times N(0, I_n)$.

Proof Cramer's rule states that

$$x_i = \frac{\det A[i : b]}{\det A},$$

where $A[i : b]$ denotes the matrix obtained by replacing the ith column of A by b. Hence, Lemma 3.6 implies that

$$\mathsf{Cw}_i(A, b) \leq \mathsf{Cw}^{\det}(A) + \mathsf{Cw}^{\det}(A[i : b]). \tag{3.5}$$

As in the proof of Theorem 3.16, we have

$$\mathsf{Prob}\{\mathsf{Cw}(A, b) \geq t\} \leq \sum_{i \in [n]} \mathsf{Prob}\{\mathsf{Cw}_i(A, b) \geq t\}.$$

Hence, applying the estimate (3.5) and Theorem 3.12 (using $\frac{t}{2} \geq 2|S|$), we get

$$\mathsf{Prob}\{\mathsf{Cw}_i(A, b) \geq t\} \leq \mathsf{Prob}\left\{\mathsf{Cw}^{\det}(A) \geq \frac{t}{2}\right\} + \mathsf{Prob}\left\{\mathsf{Cw}^{\det}(A[i : b]) \geq \frac{t}{2}\right\}$$

$$\leq 2|S|^2 \frac{1}{t} + 2(|S| + n)^2 \frac{1}{t} \leq 10|S|^2 \frac{1}{t}.$$

For the second inequality we used the fact that since $\mathscr{M}_S \nsubseteq \Sigma$, we have $|S| \geq n$. The statement now follows.   □

Theorem 3.18, again combined with Proposition 2.26, yields the following.

Corollary 3.19 *Let* $S \subseteq [n]^2$ *be such that* $\mathscr{M}_S \nsubseteq \Sigma$. *Then,*

$$\mathbb{E}\big(\log_\beta(\mathsf{Cw}(A, b))\big) \leq \log_\beta n + 2 \log_\beta |S| + \log_\beta 10e.$$

3.4 Error Bounds for Triangular Linear Systems

We may now use the results in the preceding sections to estimate the expected loss of precision in the solution of a triangular system $Lx = b$.

Theorem 3.20 *Assume that we solve $Lx = b$ using Algorithm* FS. *Then, for standard Gaussian L and b we have*

$$\mathbb{E}\big(\mathsf{LoP}(L^{-1}b)\big) \leq 5\log_\beta n + \log_\beta\big(\lceil\log_2 n\rceil + 1\big) + \log_\beta 10e + o(1).$$

Proof By Proposition 3.5 and Theorem O.3 (where we take $f(\mathrm{dims}(L, b)) = \lceil\log_2 n\rceil + 1$) we have

$$\mathsf{LoP}(L^{-1}b) \leq \log_\beta\big(\lceil\log_2 n\rceil + 1\big) + \log_\beta \mathsf{Cw}(L, b) + o(1).$$

Therefore, using Corollary 3.19 with $|S| = \frac{n^2+n}{2}$,

$$\mathbb{E}\big(\mathsf{LoP}(L^{-1}b)\big) \leq \log_\beta\big(\lceil\log_2 n\rceil + 1\big) + \mathbb{E}\big(\log_\beta \mathsf{Cw}(L, b)\big) + o(1)$$
$$\leq \log_\beta\big(\lceil\log_2 n\rceil + 1\big) + 5\log_\beta n + \log_\beta 10e + o(1). \qquad \square$$

If $\mathsf{fl}(x) = (\mathsf{fl}(x_1), \ldots, \mathsf{fl}(x_n))$ is the solution of $Lx = b$ computed by FS with ϵ_{mach} sufficiently small, the number of correct significant figures of its ith component is

$$\left\lfloor \log_\beta \frac{|\mathsf{fl}(x_i) - x_i|}{|x_i|} \right\rfloor.$$

We can rephrase Theorem 3.20 by stating that for standard Gaussian L and b,

$$\mathbb{E}\left(\min_{i\leq n}\left|\log_\beta \frac{|\mathsf{fl}(x_i) - x_i|}{|x_i|}\right|\right)$$
$$\geq t - \big(5\log_\beta n + \log_\beta\big(\lceil\log_2 n\rceil + 1\big) + \log_\beta 10e + o(1)\big),$$

where $t = |\log_\beta \epsilon_{\mathrm{mach}}|$ is the number of significant figures the machine works with (compare Sect. O.3.2).

3.5 Additional Considerations

3.5.1 On Norms and Mixed Condition Numbers

A norm $\|\ \|$ on $\mathbb{R}^q$ is said to be *monotonic* if whenever $|u_i| \leq |v_i|$ for $i = 1, \ldots, q$, we have $\|u\| \leq \|v\|$. It is well known that a norm is monotonic if and only if $\|(u_1, \ldots, u_q)\| = \|(|u_1|, \ldots, |u_q|)\|$, for all $u \in \mathbb{R}^q$. All norms we deal with in this book are monotonic.

For $a \in \mathbb{R}^q$ and $\delta > 0$ define

$$\mathcal{S}(a,\delta) = \left\{ a' \in \mathbb{R}^d \mid |a'_i - a_i| \leq \delta |a_i|,\ i = 1, \ldots, q \right\}.$$

Proposition 3.21 *For all* $a \in \mathcal{D}$ *and any monotonic norm in* $\mathbb{R}^q$, $\mathsf{M}^\varphi(a) \leq \mathsf{Cw}^\varphi(a)$.

Proof For all $x \in \mathcal{S}(a,\delta)$ and all $i \leq q$, $|\varphi(x)_i - \varphi(a)_i| \leq d(\varphi(x), \varphi(a))|\varphi(a)_i|$. Since $\| \ \|$ is monotonic, this implies $\|\varphi(x) - \varphi(a)\| \leq d(\varphi(x), \varphi(a))\|\varphi(a)\|$ and hence the statement.   □

Using a reasoning similar to that in Sect. 3.3.3, for a norm $\| \ \|$ on $\mathcal{M}$, we have

$$\mathsf{M}^\dagger(A) = \lim_{\delta \to 0}\ \sup_{A' \in \mathcal{S}(A,\delta)} \frac{\|(A')^{-1} - A^{-1}\|}{\delta \|A^{-1}\|},$$

and for a norm $\| \ \|$ in $\mathbb{R}^n$, we have

$$\mathsf{M}(A,b) = \lim_{\delta \to 0}\ \sup_{(A',b') \in \mathcal{S}((A,b),\delta)} \frac{\|x' - x\|}{\delta \|x\|},$$

where $x = A^{-1}b$ and $x' = (A')^{-1}b'$.

For all monotonic norms on $\mathcal{M}_S$, the bounds for $\mathsf{Cw}^{\mathrm{det}}(A), \mathsf{Cw}^\dagger(A)$, and $\mathsf{Cw}(A,b)$ hold as well for $\mathsf{M}^{\mathrm{det}}(A), \mathsf{M}^\dagger(A)$, and $\mathsf{M}(A,b)$ by Proposition 3.21.

3.5.2 *On the Underlying Probability Measure*

The main result in Sect. 3.4 gives a possible explanation of why triangular systems are solved with great accuracy that steers clear of the statement "random triangular systems are poorly normwise conditioned." The truth of this statement, however, should be taken with a grain of salt.

The reason is that the triangular matrices L occurring in the solution of systems $Lx = b$ are usually the result of a process applied to a matrix A that is almost invariably not triangular. The two such processes that are consistently used are the *LU factorization* (underlying Gaussian elimination) and the QR factorization. We already mentioned the latter in the introduction to Chap. 1 along with the fact that one solves the system $Ax = b$ by decomposing $A = QR$ and then, using that R is upper-triangular, solves $Rx = Q^\mathsf{T}b$ by back substitution. We mention now that the version of this decomposition producing a lower-triangular matrix (which we consider only for consistency with the rest of this chapter) is known as *QL factorization*.

If A is invertible and we require that the diagonal elements of L be positive, which we can do without loss of generality, then both Q and L are unique. Hence, the QL factorization defines a map

$$\psi : \mathrm{GL}_n(\mathbb{R}) \to \mathrm{GL}_n(\mathsf{Triang}),$$

where we have written GL_n(Triang) for the subgroup of $GL_n(\mathbb{R})$ of invertible lower-triangular matrices. A reasonable choice for randomness in the former is obtained by endowing the latter with the standard Gaussian measure and then pushing forward this measure (note that there is no loss of generality in considering triangular matrices with positive diagonal entries only). Let $\mathscr{P}$ be the measure thus obtained on GL_n(Triang). The next result shows that under $\mathscr{P}$, the normwise condition number has a completely different behavior from the one shown in Theorem 3.1.

Proposition 3.22

$$\underset{L\sim\mathscr{P}}{\mathbb{E}} \log\kappa(L) = \log n + \mathcal{O}(1).$$

Proof Let $A \in GL_n(\mathbb{R})$ and $A = QL$ with Q orthogonal and L lower-triangular, so that $\psi(A) = L$. Let the SVD (recall Theorem 1.13) of L be given by $L = UDV^{\mathrm{T}}$ with $D = \mathrm{diag}(\sigma_1,\ldots,\sigma_n)$. Then $A = QUDV^{\mathrm{T}}$, and since both QU and V are orthogonal, it follows that A has the same singular values as L. In particular, $\kappa(L) = \kappa(A) = \frac{\sigma_1}{\sigma_n}$. The statement now follows from (2.11) and Corollary 2.51. $\square$

Chapter 4
Probabilistic Analysis of Rectangular Matrices

We started Chap. 1 by stating a backward analysis for linear equation solving that was a particular case of Theorem 19.3 of [121]. We may now quote this result in full.

Theorem 4.1 *Let $A \in \mathbb{R}^{q \times n}$ have full rank, $q \geq n$, $b \in \mathbb{R}^q$, and suppose the least-squares problem $\min_x \|b - Ax\|$ is solved using the Householder QR factorization method. The computed solution $\tilde{x}$ is the exact solution to*

$$\min_{x \in \mathbb{R}^n} \|\tilde{b} - \tilde{A}x\|,$$

where $\tilde{A}$ and $\tilde{b}$ satisfy the relative error bounds

$$\|\tilde{A} - A\|_F \leq n\gamma_{cq}\|A\|_F \quad and \quad \|\tilde{b} - b\| \leq n\gamma_{cq}\|b\|$$

for a small constant c and with γ_{cq} as defined in (O.5). ☐

Replacing the Frobenius norm by the spectral norm, which yields

$$\|\tilde{A} - A\| \leq n^{3/2}\gamma_{cq}\|A\|,$$

it follows from this backward stability result, (O.6), and Theorem O.3 that the relative error for the computed solution $\tilde{x}$ satisfies

$$\frac{\|\tilde{x} - x\|}{\|x\|} \leq cn^{3/2}q\,\epsilon_{\text{mach}}\text{cond}(A, b) + o(\epsilon_{\text{mach}})$$

and the loss of precision is bounded by

$$\text{LoP}(A^{\dagger}b) \leq \log_\beta n^{3/2}q + \log_\beta \text{cond}(A, b) + \log_\beta c + o(1), \tag{4.1}$$

P. Bürgisser, F. Cucker, *Condition*,
Grundlehren der mathematischen Wissenschaften 349,
DOI 10.1007/978-3-642-38896-5_4, © Springer-Verlag Berlin Heidelberg 2013

where $\mathrm{cond}(A, b)$ is the normwise condition number for linear least squares (with respect to the spectral norm), which is defined as

$$\mathrm{cond}(A, b) = \lim_{\delta \to 0} \sup_{\max\{\mathsf{RelError}(A), \mathsf{RelError}(b)\} \leq \delta} \frac{\mathsf{RelError}(A^\dagger b)}{\delta}.$$

We mentioned in Sect. 1.6 that this condition number, even though not tightly approximated by $\kappa(A)$, is bounded by a constant times $\kappa(A)^2$. Consequently, to obtain expected bounds (or a smoothed analysis) for the loss of precision $\mathsf{LoP}(A^\dagger b)$ from Eq. (4.1) it is enough to perform the corresponding analysis for $\log_\beta \kappa(A)$.

The goal of this chapter is to do so. For consistency with other chapters in this book, we will consider matrices $A \in \mathbb{R}^{m \times n}$ with $m \leq n$ and study $\kappa(A)$, which, we note, coincides with $\kappa(A^\mathsf{T})$. One of the main results we prove is the following theorem.

Theorem 4.2 *For all $\lambda_0 \in (0, 1)$ there exists n_0 such that for all $1 \leq m \leq n$ with $\lambda = \frac{m-1}{n} \leq \lambda_0$ and $n \geq n_0$, we have for all σ with $\frac{1}{\sqrt{m}} \leq \sigma \leq 1$ and all $\overline{A} \in \mathbb{R}^{m \times n}$ with $\|\overline{A}\| \leq 1$, that*

$$\mathbb{E}_{A \sim N(\overline{A}, \sigma^2 \mathrm{I})}\big(\kappa(A)\big) \leq \frac{20.1}{1 - \lambda}.$$

Jensen's inequality (Proposition 2.28) immediately yields the following consequence.

Corollary 4.3 *Under the hypothesis of Theorem 4.2,*

$$\sup_{\|\overline{A}\| \leq 1} \mathbb{E}_{A \sim N(\overline{A}, \sigma^2 \mathrm{I})}\big(\log_\beta \kappa(A)\big) \leq \log_\beta\left(\frac{20.1}{1 - \lambda}\right). \qquad \square$$

It is worth noting that the bounds in Theorem 4.2 and Corollary 4.3 are independent of n and depend only on the bound λ_0 on the elongation. Furthermore, surprisingly, they are also independent of σ. In fact, Corollary 4.3 indicates that for large reasonably elongated matrices, one may expect the loss of precision in the solution of least-squares problems to derive mostly from the backward error bounds of the algorithm used.

We also mention here that the bounds obtained in this chapter are sharper than those derived in Sect. 2.4. The methods used to prove them are, in exchange, more involved.

4.1 A Crash Course on Probability: II

We continue our crash course on probability with some results of a more advanced nature.

4.1.1 Large Deviations

Let $f : \mathbb{R}^n \to \mathbb{R}$ be a Lipschitz continuous function with Lipschitz constant L. This means that $|f(x) - f(y)| \leq L\|x - y\|$ for all $x, y \in \mathbb{R}^n$, where $\| \ \|$ denotes the Euclidean norm. We claim that if f is differentiable, then

$$\|\operatorname{grad} f(x)\| \leq L.$$

Indeed, for given x there exists a unit-length vector v such that

$$\frac{d}{ds} f(x + sv)|_{s=0} = \|\operatorname{grad} f(x)\|.$$

Using that $\frac{1}{s}|f(x + sv) - f(x)| \leq L$, the claim follows.

We shall now prove a powerful and general large-deviation result.

Theorem 4.4 *Let $f : \mathbb{R}^n \to \mathbb{R}$ be an almost everywhere differentiable and Lipschitz continuous function with Lipschitz constant L. Then we have, for all $t \geq 0$ and $x \in \mathbb{R}^n$ drawn from the standard Gaussian distribution γ_n, that*

$$\operatorname*{Prob}_{x \sim N(0, I_n)} \left\{ f(x) \geq \mathbb{E}(f) + t \right\} \leq e^{-\frac{2}{\pi^2 L^2} t^2}.$$

Proof Note first that the integrability of f follows from the Lipschitz property. Without loss of generality we may assume that $\mathbb{E}(f) = 0$. We recall that φ_n denotes the density of γ_n. By Markov's inequality (Corollary 2.9) we have, for any $\lambda > 0$ (to be chosen later),

$$\operatorname{Prob}\{f(x) \geq t\} = \operatorname{Prob}\{e^{\lambda f(x)} \geq e^{\lambda t}\} \leq e^{-\lambda t} \, \mathbb{E}\big(e^{\lambda f}\big).$$

By Corollary 2.29 we have $1 = e^{\mathbb{E}(-\lambda f)} \leq \mathbb{E}(e^{-\lambda f})$. This implies, using Fubini,

$$\mathbb{E}\big(e^{\lambda f}\big) \leq \mathbb{E}\big(e^{\lambda f}\big) \cdot \mathbb{E}\big(e^{-\lambda f}\big) = \int_{\mathbb{R}^n \times \mathbb{R}^n} e^{\lambda(f(x) - f(y))} \, \varphi_n(x) \varphi_n(y) \, dx \, dy. \qquad (4.2)$$

Now we set for $\theta \in [0, \pi/2]$,

$$x(\theta) := x \sin\theta + y \cos\theta, \qquad x'(\theta) := x \cos\theta - y \sin\theta.$$

(Note that $x'(\theta)$ is the derivative of $x(\theta)$ with respect to θ.) It is a consequence of the orthogonal invariance of the standard Gaussian distribution that if (x, y) is standard Gaussian distributed on $\mathbb{R}^n \times \mathbb{R}^n$, then so is the induced random vector $(x(\theta), x'(\theta))$, for fixed θ.

We have, for all $x, y \in \mathbb{R}^n$,

$$f(x) - f(y) = \int_0^{\pi/2} \frac{d}{d\theta} f\big(x(\theta)\big) \, d\theta = \int_0^{\pi/2} \big\langle \operatorname{grad} f\big(x(\theta)\big), x'(\theta) \big\rangle \, d\theta.$$

This implies, applying Corollary 2.29 to the uniform distribution on $[0, \pi/2]$,

$$e^{\lambda(f(x)-f(y))} = e^{\frac{2}{\pi} \int_0^{\pi/2} \frac{\pi\lambda}{2} \langle \operatorname{grad} f(x(\theta)), x'(\theta) \rangle \, d\theta} \leq \frac{2}{\pi} \int_0^{\pi/2} e^{\frac{\pi\lambda}{2} \langle \operatorname{grad} f(x(\theta)), x'(\theta) \rangle} \, d\theta.$$

Interchanging integrals, we get from (4.2),

$$\mathbb{E}(e^{\lambda f}) \leq \frac{2}{\pi} \int_0^{\pi/2} \int_{\mathbb{R}^n \times \mathbb{R}^n} e^{\frac{\pi\lambda}{2} \langle \operatorname{grad} f(x(\theta)), x'(\theta) \rangle} \varphi_n(x) \varphi_n(y) \, dx \, dy \, d\theta.$$

Since for fixed θ, $(x(\theta), x'(\theta))$ is standard Gaussian distributed in $\mathbb{R}^n \times \mathbb{R}^n$, the integral on the right-hand side simplifies and we obtain

$$\mathbb{E}(e^{\lambda f}) \leq \int_{\mathbb{R}^n \times \mathbb{R}^n} e^{\frac{\pi\lambda}{2} \langle \operatorname{grad} f(x), y \rangle} \varphi_n(x) \varphi_n(y) \, dx \, dy$$

$$= \int_{\mathbb{R}^n} \left(\int_{\mathbb{R}^n} e^{\frac{\pi\lambda}{2} \langle \operatorname{grad} f(x), y \rangle} \varphi_n(y) \, dy \right) \varphi_n(x) \, dx.$$

By Fubini, the inner integral on the right-hand side equals

$$\prod_{k=1}^n \int_{\mathbb{R}} e^{\frac{\pi\lambda}{2} \partial_{x_k} f(x) y_k} \varphi_1(y_k) \, dy_k = \prod_{k=1}^n e^{\frac{1}{2}(\frac{\pi\lambda}{2} \partial_{x_k} f(x))^2} = e^{\frac{\pi^2 \lambda^2}{8} \|\operatorname{grad} f(x)\|^2},$$

with the second equality due to Lemma 2.30. Since the last expression is bounded by $e^{\frac{\pi^2 \lambda^2 L^2}{8}}$, we conclude that

$$\mathbb{E}(e^{\lambda f}) \leq e^{\frac{\pi^2 \lambda^2 L^2}{8}}.$$

So we have shown that for any positive λ we have

$$\mathsf{Prob}\{f(x) \geq t\} \leq e^{-\lambda t} \mathbb{E}(e^{\lambda f}) \leq e^{-\lambda t + \frac{\pi^2 L^2 \lambda^2}{8}}.$$

Choosing $\lambda = \frac{4t}{\pi^2 L^2}$ minimizes the right-hand side, and we obtain

$$\mathsf{Prob}\{f(x) \geq t\} \leq e^{-\frac{2t^2}{\pi^2 L^2}},$$

as claimed.   □

Remark 4.5 Theorem 4.4 applied to f and $-f$ implies the *concentration inequalities*

$$\mathop{\mathsf{Prob}}_{x \sim N(0, I_n)} \{|f(x) - \mathbb{E}(f)| \geq t\} \leq 2e^{-\frac{2t^2}{\pi^2 L^2}},$$

valid for $t \geq 0$. With some additional work [136], this inequality can be improved to

$$\Prob_{x \sim N(0, \mathrm{I}_n)} \left\{ |f(x) - \mathbb{E}(f)| \geq t \right\} \leq e^{-\frac{t^2}{2L^2}}.$$

Here is a first and important application of Theorem 4.4, which will be needed several times.

Corollary 4.6 *If $x \in \mathbb{R}^n$ is chosen from the standard Gaussian distribution, then for $t > 0$,*

$$\Prob_{x \sim N(0, \mathrm{I}_n)} \left\{ \|x\| \geq \sqrt{n} + t \right\} \leq e^{-\frac{t^2}{2}}.$$

Proof The norm function $f(x) = \|x\|$ is Lipschitz continuous with Lipschitz constant $L = 1$. Noting that $\mathbb{E}(\|x\|^2) = \sum_{i=1}^n \mathbb{E}(x_i^2) = n$ and using Proposition 2.10, we get $\mathbb{E}(\|x\|) \leq \sqrt{\mathbb{E}(\|x\|^2)} = \sqrt{n}$. The assertion follows now from Theorem 4.4, where the better exponent is due to Remark 4.5. $\square$

Remark 4.7 Let us illustrate the power of Theorem 4.4 with a simple example. Suppose that $x_1, x_2, \ldots$ are independent standard Gaussian random variables. Put $f_n(x_1, \ldots, x_n) := (x_1 + \cdots + x_n)/\sqrt{n}$. The *central limit theorem* states that

$$\lim_{n \to \infty} \Prob_{x \sim N(0, \mathrm{I}_n)} \left\{ f_n(x) \geq t \right\} = \Psi(t) = \frac{1}{\sqrt{2\pi}} \int_t^\infty e^{-\frac{x^2}{2}} \, dx \leq \frac{1}{t\sqrt{2\pi}} e^{-\frac{t^2}{2}}$$

(the inequality is due to Lemma 2.16). Theorem 4.4 immediately implies a corresponding nonasymptotic result. Namely, note that $f_n \colon \mathbb{R}^n \to \mathbb{R}$ is a Lipschitz continuous function with Lipschitz constant $L = 1$. Hence, for all $t \geq 0$,

$$\Prob_{x \sim N(0, \mathrm{I}_n)} \left\{ f_n(x) \geq t \right\} \leq e^{-\frac{2t^2}{\pi^2}}.$$

4.1.2 Random Gaussian Matrices

We begin by recalling some facts about Householder matrices. Assume that $v \in \mathbb{R}^m$ is nonzero. One checks immediately that the reflection H_v at the hyperplane orthogonal to v is given by the linear map

$$H_v x = x - \frac{2}{v^\mathsf{T} v} v v^\mathsf{T} x. \tag{4.3}$$

The matrix corresponding to H_v is called the *Householder matrix* associated with the vector v. It is clear that H_v is orthogonal.

It is geometrically evident that for given $w \in \mathbb{R}^m$, there exists a reflection H_v that maps w to a multiple of the first standard basis vector e_1. The following lemma tells us how to compute v.

Lemma 4.8 *We have $H_v w = \|w\| e_1$ for $v = e_1 - w/\|w\|$ if $w \neq 0$.*

Proof We have $v^\mathsf{T} w = w_1 - \|w\|$ and $v^\mathsf{T} v = 2(1 - w_1/\|w\|)$, hence $\frac{2v^\mathsf{T} w}{v^\mathsf{T} v} = -\|w\|$. It follows that

$$H_v w = w - \frac{2v^\mathsf{T} w}{v^\mathsf{T} v} v = w + \|w\| v = w + \|w\| \left(e_1 - \frac{w}{\|w\|} \right) = \|w\| e_1. \qquad \square$$

Lemma 4.8 can be used to transform a given matrix $A = [a_1, \ldots, a_n] \in \mathbb{R}^{m \times n}$ into an orthogonally equivalent one with few nonzero entries. For the first step of this transformation we assume $a_1 \neq 0$, put $v = e_1 - a_1/\|a_1\|$, and form the transformed matrix $H_v A = [H_v a_1, H_v a_2, \ldots, H_v a_n]$. The first column $H_v a_1$ of $H_v A$ equals $\|a_1\| e_1$; hence all of its entries, except the first one, are zero. Note that if the given matrix A is standard Gaussian distributed, then $\|a_1\|^2$ is χ^2-distributed with m degrees of freedom. Moreover, the next lemma guarantees that the remaining matrix $[H_v a_2, \ldots, H_v a_n]$ is standard Gaussian distributed and independent of a_1, which will allow an inductive continuation of the argument.

Lemma 4.9 *If $[a_1, \ldots, a_n] \in \mathbb{R}^{m \times n}$ is standard Gaussian distributed, then $[a_1, H_v a_2, \ldots, H_v a_n]$ is standard Gaussian distributed as well. Here, v is defined in terms of a_1 by $v = e_1 - a_1/\|a_1\|$.*

Proof According to Corollary 2.18 it suffices to show that the diffeomorphism

$$\psi : [a_1, a_2, \ldots, a_n] \mapsto [a_1, H_v a_2, \ldots, H_v a_n]$$

preserves the Euclidean norm and has Jacobian identically one. The first property is obvious, since H_v is orthogonal. For the latter, using that v depends on a_1 only, one sees that the derivative of ψ has a block lower-triangular form with entries $I_m, H_v, \ldots, H_v$ on the diagonal. Hence $\mathrm{J}\psi(A) = 1$ for all A. $\qquad \square$

We show now that every $X \in \mathbb{R}^{m \times n}$ can be transformed to a bidiagonal matrix by performing Householder transformations on the left- and right-hand sides of X.

To begin, we apply the transformation of Lemma 4.8 to X^T in order to find a Householder matrix H_1 such that $X H_1^\mathsf{T} = (\|x_1'\| e_1, A)$ with $A \in \mathbb{R}^{(m-1) \times n}$. Here, x_1' denotes the first row of X. We then apply a similar transformation to A in order to find a Householder matrix H_2 such that $H_2 A = [\|a_1\| e_1, B]$ with $B \in \mathbb{R}^{(m-1) \times (n-1)}$. Continuing in this way, we construct orthogonal matrices $g \in \mathcal{O}(m)$ and $h \in \mathcal{O}(n)$ (products of Householder matrices) such that gXh has the following bidiagonal form:

$$
gXh = \begin{bmatrix}
v_n & & & 0 & \cdots & 0 \\
w_{m-1} & v_{n-1} & & \vdots & & \vdots \\
& \ddots & \ddots & \vdots & & \vdots \\
& & w_1 & v_{n-m+1} & 0 & \cdots & 0
\end{bmatrix}. \tag{4.4}
$$

The following proposition is an immediate consequence of Lemma 4.9.

Proposition 4.10 *If $X \in \mathbb{R}^{m \times n}$ is standard Gaussian, then the nonzero entries v_i, w_i of the bidiagonal matrix in (4.4) resulting from the above described procedure are independent random variables. Moreover, v_i^2 and w_i^2 are χ^2-distributed with i degrees of freedom.* □

If $X \in \mathbb{R}^{m \times n}$ is standard Gaussian, then the distribution of the matrix XX^T is called the *Wishart distribution* $W(m, n)$. As an application of Proposition 4.10, we determine the distribution of $\det(XX^T)$.

Corollary 4.11 *If $X \in \mathbb{R}^{m \times n}$ is standard Gaussian, $m \leq n$, then $\det(XX^T)$ has the same distribution as $v_n^2 v_{n-1}^2 \cdots v_{n-m+1}^2$, where $v_{n-m+1}^2, \ldots, v_n^2$ are independent random variables and v_i^2 is χ^2-distributed with i degrees of freedom. In particular, $\mathbb{E} \det(XX^T) = n!/(n - m)!$.*

Proof Proposition 4.10 implies that $\det(XX^T)$ has the same distribution as $\det(YY^T)$, where Y denotes the bidiagonal matrix in (4.4).

In the case $m = n$ we have $\det Y = v_n \cdots v_1$, and hence $\det(YY^T) = (\det Y)^2 = v_n^2 \cdots v_1^2$ is as claimed. More generally, $\det(YY^T)$ can be interpreted as the square of the m-dimensional volume of the parallelepiped spanned by the rows of Y. It has the same volume as the parallelepiped spanned by the orthogonal vectors $v_n e_1, v_{n-1} e_2, \ldots, v_{n-m+1} e_m$, where e_i denotes the ith standard basis vector in $\mathbb{R}^n$. It follows that $\det(YY^T) = v_n^2 \cdots v_{n-m+1}^2$. □

The previous result easily extends to complex matrices. We call a random variable $z \in \mathbb{C}$ standard Gaussian if it is standard Gaussian when we identify $\mathbb{C}$ with $\mathbb{R}^2$. Moreover, we call $X \in \mathbb{C}^{m \times n}$ standard Gaussian when its entries are independent standard Gaussian distributions in $\mathbb{C}$. The following result will be needed in Part III of the book.

Lemma 4.12 *Let $N(0, I)$ denote the standard normal distribution on the set of $n \times n$ complex matrices. Then*

$$
\mathbb{E}_{A \sim N(0,I)} \det(XX^*) = 2^n n!.
$$

Proof It is immediate to see that the proof of Proposition 4.10 holds for complex Gaussian matrices as well. The proof of Corollary 4.11 carries over to show that

$c = \mathbb{E}v_n^2 v_{n-1}^2 \cdots v_1^2$, where v_i is a χ^2-distributed random variable with $2i$ degrees of freedom. Since the expectation of v_i^2 equals $2i$, the result follows. $\square$

4.1.3 A Bound on the Expected Spectral Norm

We make the general assumption that $1 \le m \le n$. For a standard Gaussian $X \in \mathbb{R}^{m \times n}$ we put

$$Q(m,n) := \frac{1}{\sqrt{n}}\mathbb{E}\big(\|X\|\big). \tag{4.5}$$

The function $\mathbb{R}^{m \times n} \to \mathbb{R}$ mapping a matrix X to its spectral norm $\|X\|$ is Lipschitz continuous with Lipschitz constant 1, since $\|X - Y\| \le \|X - Y\|_F$. Hence Theorem 4.4 implies that for $t > 0$,

$$\mathsf{Prob}\big\{\|X\| \ge Q(m,n)\sqrt{n} + t\big\} \le e^{-\frac{2t^2}{\pi^2}}. \tag{4.6}$$

This tail bound easily implies the following large-deviation result.

Proposition 4.13 *Let* $\overline{A} \in \mathbb{R}^{m \times n}$, $\|\overline{A}\| \le 1$, *and* $\sigma \in (0, 1]$. *If* $A \in \mathbb{R}^{m \times n}$ *follows the law* $N(\overline{A}, \sigma^2 I)$, *then for* $t > 0$,

$$\mathop{\mathsf{Prob}}_{A \sim N(\overline{A}, \sigma^2 I)} \big\{\|A\| \ge Q(m,n)\sigma\sqrt{n} + t + 1\big\} \le e^{-\frac{2t^2}{\pi^2}}.$$

Proof We note that $\|A\| \ge Q(m,n)\sigma\sqrt{n} + t + 1$ implies that $\|A - \overline{A}\| \ge \|A\| - \|\overline{A}\| \ge Q(m,n)\sqrt{n} + t$. Moreover, if $A \in \mathbb{R}^{m \times n}$ follows the law $N(\overline{A}, \sigma^2 I)$, then $X := \frac{A - \overline{A}}{\sigma}$ is standard Gaussian in $\mathbb{R}^{m \times n}$. The assertion follows from (4.6). $\square$

We derive now an upper bound on $Q(m,n)$.

Lemma 4.14 *For* $n > 2$ *we have*

$$\sqrt{\frac{n}{n+1}} \le Q(m,n) \le 2\left(1 + \sqrt{\frac{2\ln(2m-1)}{n}} + \frac{2}{\sqrt{n}}\right) \le 6.$$

In particular, for standard Gaussian matrices $A \in \mathbb{R}^{m \times n}$, *we have*

$$\frac{n}{\sqrt{n+1}} \le \mathbb{E}\|A\| \le 6\sqrt{n}.$$

The proof relies on the following lemma.

Lemma 4.15 *Let $r_1, \ldots, r_n$ be independent random variables with nonnegative values such that r_i^2 is χ^2-distributed with f_i degrees of freedom. Then,*

$$\mathbb{E}\left(\max_{1 \le i \le n} r_i\right) \le \max_{1 \le i \le n} \sqrt{f_i} + \sqrt{2 \ln n} + 1.$$

Proof Suppose that $r_1, \ldots, r_n$ are independent random variables with nonnegative values such that r_i^2 is χ^2-distributed with f_i degrees of freedom. Put $f := \max_i f_i$. Corollary 4.6 tells us that for all i and all $t > 0$,

$$\mathrm{Prob}\{r_i \ge \sqrt{f} + t\} \le e^{-\frac{t^2}{2}},$$

and hence, by the union bound,

$$\mathrm{Prob}\left\{\max_{1 \le i \le n} r_i \ge \sqrt{f} + t\right\} \le n e^{-\frac{t^2}{2}}.$$

For a fixed parameter $b \ge 1$ (to be determined later), this implies

$$\mathbb{E}\left(\max_{1 \le i \le n} r_i\right) \le \sqrt{f} + b + \int_{\sqrt{f}+b}^{\infty} \mathrm{Prob}\left\{\max_{1 \le i \le n} r_i \ge T\right\} dT$$

$$= \sqrt{f} + b + \int_{b}^{\infty} \mathrm{Prob}\left\{\max_{1 \le i \le n} r_i \ge \sqrt{f} + t\right\} dt$$

$$\le \sqrt{f} + b + n \int_{b}^{\infty} e^{-\frac{t^2}{2}} dt.$$

Using Lemma 2.16 we get, for $b \ge \pi/2$,

$$\frac{1}{\sqrt{2\pi}} \int_{b}^{\infty} e^{-\frac{t^2}{2}} dt \le \frac{1}{b\sqrt{2\pi}} e^{-\frac{b^2}{2}} \le \frac{1}{\sqrt{2\pi}} e^{-\frac{b^2}{2}}.$$

Hence we obtain

$$\mathbb{E}\left(\max_{1 \le i \le n} r_i\right) \le \sqrt{f} + b + n e^{-\frac{b^2}{2}}.$$

Finally, choosing $b := \sqrt{2 \ln n}$, we get

$$\mathbb{E}\left(\max_{1 \le i \le n} r_i\right) \le \sqrt{f} + \sqrt{2 \ln n} + 1,$$

as claimed.   □

Proof of Lemma 4.14 According to Proposition 4.10, the spectral norm $\|X\|$ of a standard Gaussian matrix $X \in \mathbb{R}^{m \times n}$ has the same distribution as the spectral norm of the random bidiagonal matrix Y defined in (4.4). The occurring entries v_i^2 and w_i^2 are χ^2-distributed with i degrees of freedom.

The spectral norm of Y is bounded by $\max_i v_i + \max_j w_j \leq 2r$, where r denotes the maximum of the values v_i and w_j. Lemma 4.15 implies that for $n > 2$,

$$\mathbb{E}(r) \leq \sqrt{n} + \sqrt{2\ln(2m-1)} + 1 \leq 3\sqrt{n}.$$

This proves the claimed upper bound on $Q(m, n)$. For the lower bound we note that $\|Y\| \geq |v_n|$, which gives $\mathbb{E}(\|Y\|) \geq \mathbb{E}(|v_n|)$. The claimed lower bound now follows from Lemma 2.25, which states that $\mathbb{E}(|v_n|) \geq \sqrt{\frac{n}{n+1}}$. $\square$

4.2 Tail Bounds for $\kappa(A)$

Prior to proving Theorem 4.2, we want to prove tail bounds for $\kappa(A)$ under local Gaussian perturbations. To state this result we need to introduce some notation. We still assume $1 \leq m \leq n$.

We define for $\lambda \in (0, 1)$ the quantity

$$c(\lambda) := \sqrt{\frac{1+\lambda}{2(1-\lambda)}}. \tag{4.7}$$

Note that $\lim_{\lambda \to 0} c(\lambda) = \frac{1}{\sqrt{2}}$, $\lim_{\lambda \to 1} c(\lambda) = \infty$, and $c(\lambda)$ is monotonically increasing. Further, for $1 \leq m \leq n$ and $0 < \sigma \leq 1$, we define the *elongation* $\lambda := \frac{m-1}{n}$ and introduce the quantity

$$\zeta_\sigma(m, n) := \left(Q(m, n) + \frac{1}{\sigma\sqrt{n}} \right) c(\lambda)^{\frac{1}{n-m+1}}. \tag{4.8}$$

Let $\overline{A} \in \mathbb{R}^{m \times n}$ and $\sigma > 0$. Since there is no risk of confusion, we will denote the density of the Gaussian $N(\overline{A}, \sigma^2 \mathrm{I})$ with center $\overline{A}$ and covariance matrix $\sigma^2 \mathrm{I}$ by $\varphi^{\overline{A},\sigma}$ (instead of $\varphi_{m \times n}^{\overline{A},\sigma}$). We recall that

$$\varphi^{\overline{A},\sigma}(A) := \frac{1}{(2\pi)^{\frac{mn}{2}}} e^{-\frac{\|A-\overline{A}\|_F^2}{2\sigma^2}}.$$

Theorem 4.16 *Suppose that $\overline{A} \in \mathbb{R}^{m \times n}$ satisfies $\|\overline{A}\| \leq 1$ and let $0 < \sigma \leq 1$. Put $\lambda := \frac{m-1}{n}$. Then for $z \geq \zeta_\sigma(m, n)$, we have*

$$\Prob_{A \sim N(\overline{A}, \sigma^2 \mathrm{I})} \left\{ \kappa(A) \geq \frac{ez}{1-\lambda} \right\}$$

$$\leq 2c(\lambda) \left[\left(Q(m, n) + \sqrt{\frac{\pi^2}{2}\ln(2z)} + \frac{1}{\sigma\sqrt{n}} \right) \frac{1}{z} \right]^{n-m+1}.$$

Remark 4.17 When $\sigma = 1$ and $\overline{A} = 0$, Theorem 4.16 yields tail bounds for the usual average case. In the notes at the end of the book there is a comparison of these bounds with bounds derived ad hoc for the average case.

Lemma 4.18 *For $\lambda \in (0, 1)$ we have $\lambda^{-\frac{\lambda}{1-\lambda}} \le e$.*

Proof Writing $u = 1/\lambda$, the assertion is equivalent to $u^{\frac{1}{u-1}} \le e$, or $u \le e^{u-1}$, which is certainly true for $u \ge 1$. □

4.2.1 Tail Bounds for $\|A^\dagger\|$

The main work in proving Theorem 4.16 is the following tail bound on $\|A^\dagger\|$.

Proposition 4.19 *Let $\overline{A} \in \mathbb{R}^{m \times n}$, $\sigma > 0$, and put $\lambda := \frac{m-1}{n}$. For random $A \sim N(\overline{A}, \sigma^2 I)$ we have, for any $t > 0$,*

$$\underset{A \sim N(\overline{A}, \sigma^2 I)}{\text{Prob}} \left\{ \|A^\dagger\| \ge \frac{t}{1-\lambda} \right\} \le c(\lambda) \left(\frac{e}{\sigma \sqrt{n} t} \right)^{(1-\lambda)n}.$$

Before proving Proposition 4.19, we note a consequence of it for square matrices.

Corollary 4.20 *Let $\overline{A} \in \mathbb{R}^{n \times n}$ and $\sigma > 0$. For any $t > 0$,*

$$\underset{A \sim N(\overline{A}, \sigma^2 I)}{\text{Prob}} \left\{ \|A^{-1}\| \ge t \right\} \le \frac{ne}{\sigma t}$$

and

$$\underset{A \sim N(\overline{A}, \sigma^2 I)}{\mathbb{E}} \log \|A^{-1}\| \le \log \frac{n}{\sigma} + \log e.$$

Proof The tail estimate follows from Proposition 4.19 by noting that $\lambda = \frac{n-1}{n} = 1 - \frac{1}{n}$ and $c(\lambda) = \sqrt{\frac{2n-1}{2}} \le \sqrt{n}$. The expectation then follows from Proposition 2.26 with $t_0 = K = \frac{ne}{\sigma}$. □

We next prove Proposition 4.19, starting with the following result.

Proposition 4.21 *For all $v \in \mathbb{S}^{m-1}$, $\overline{A} \in \mathbb{R}^{m \times n}$, $\sigma > 0$, and $\xi > 0$ we have*

$$\underset{A \sim N(\overline{A}, \sigma^2 I)}{\text{Prob}} \left\{ \|A^\dagger v\| \ge \xi \right\} \le \frac{1}{(\sqrt{2\pi})^{n-m+1}} \frac{\mathcal{O}_{n-m}}{n-m+1} \left(\frac{1}{\sigma \xi} \right)^{n-m+1}.$$

Proof We first claim that because of orthogonal invariance, we may assume that $v = e_m := (0, \ldots, 0, 1)$. To see this, take $\Psi \in \mathcal{O}(m)$ such that $v = \Psi e_m$. Consider the isometric map $A \mapsto B = \Psi^{-1} A$, which transforms the density $\varphi^{\overline{A}, \sigma}(A)$ into a density of the same form, namely $\varphi_{\Psi^{-1} \overline{A}, \sigma}(B)$. Thus the assertion for e_m and random B implies the assertion for v and A, noting that $A^\dagger v = B^\dagger e_m$. This proves the claim.

We are going to characterize the norm of $w := A^\dagger e_m$ in a geometric way. Let a_i denote the ith row of A. Almost surely, the rows $a_1, \ldots, a_m$ are linearly independent; hence, we assume so in what follows. Let

$$R := \mathsf{span}\{a_1, \ldots, a_m\}, \qquad S := \mathsf{span}\{a_1, \ldots, a_{m-1}\}.$$

Let $S^\perp$ denote the orthogonal complement of S in $\mathbb{R}^n$. We decompose $a_m = a_m^\perp + a_m^S$, where $a_m^\perp$ denotes the orthogonal projection of a_m onto $S^\perp$ and $a_m^S \in S$. Then $a_m^\perp \in R$, since both a_m and a_m^S are in R. It follows that $a_m^\perp \in R \cap S^\perp$.

We claim that $w \in R \cap S^\perp$ as well. Indeed, note that R equals the orthogonal complement of the kernel of A in $\mathbb{R}^n$. Therefore, by definition of the Moore–Penrose inverse, $w = A^\dagger e_m$ lies in R. Moreover, since $A A^\dagger = I$, we have $\langle w, a_i \rangle = 0$ for $i = 1, \ldots, m - 1$ and hence $w \in S^\perp$ as well.

It is immediate to see that $\dim R \cap S^\perp = 1$. It then follows that $R \cap S^\perp = \mathbb{R} w = \mathbb{R} a_m^\perp$. Since $\langle w, a_m \rangle = 1$, we get $1 = \langle w, a_m \rangle = \langle w, a_m^\perp \rangle = \|w\| \, \|a_m^\perp\|$ and therefore

$$\left\| A^\dagger e_m \right\| = \frac{1}{\|a_m^\perp\|}. \tag{4.9}$$

Let $A_m \in \mathbb{R}^{(m-1) \times n}$ denote the matrix obtained from A by omitting a_m. The density $\varphi^{\overline{A}, \sigma}$ factors as $\varphi^{\overline{A}, \sigma}(A) = \varphi_1(A_n) \varphi_2(a_n)$, where φ_1 and φ_2 denote the density functions of $N(\overline{A}_m, \sigma^2 I)$ and $N(\overline{a}_m, \sigma^2 I)$, respectively (the meaning of $\overline{A}_m$ and $\overline{a}_m$ being clear). Fubini's theorem combined with (4.9) yields, for $\xi > 0$,

$$\mathop{\mathrm{Prob}}_{N(\overline{A}, \sigma^2 I)} \left\{ \left\| A^\dagger e_m \right\| \geq \xi \right\} = \int_{\|A^\dagger e_m\| \geq \xi} \varphi_{\overline{A}, \sigma^2 I}(A) \, dA$$

$$= \int_{A_m \in \mathbb{R}^{(m-1) \times n}} \varphi_1(A_m) \cdot \left(\int_{\|a_m^\perp\| \leq 1/\xi} \varphi_2(a_m) \, da_m \right) dA_m. \tag{4.10}$$

To complete the proof it is sufficient to prove the bound

$$\int_{\|a_m^\perp\| \leq \frac{1}{\xi}} \varphi_2(a_m) \, da_m \leq \frac{1}{(\sqrt{2\pi})^{n-m+1}} \frac{\mathcal{O}_{n-m}}{n-m+1} \left(\frac{1}{\sigma \xi} \right)^{n-m+1} \tag{4.11}$$

for fixed, linearly independent $a_1, \ldots, a_{m-1}$ and $\xi > 0$.

To prove (4.11), note that $a_m^\perp \sim N(\overline{a}_m^\perp, \sigma^2 I)$ in $S^\perp \simeq \mathbb{R}^{n-m+1}$, where $\overline{a}_m^\perp$ is the orthogonal projection of $\overline{a}_m$ onto $S^\perp$. Let B_r denote the ball of radius r in $\mathbb{R}^p$

centered at the origin. It is easy to see that $\mathrm{vol}\,B_r = \mathcal{O}_{p-1}r^p/p$. For any $\bar{x} \in \mathbb{R}^p$ and any $\sigma > 0$ we have

$$
\begin{aligned}
\mathop{\mathrm{Prob}}_{x\sim N(\bar{x},\sigma^2 I)} \{\|x\| \leq \varepsilon\} &\leq \mathop{\mathrm{Prob}}_{x\sim N(0,\sigma^2 I)} \{\|x\| \leq \varepsilon\} = \frac{1}{(\sigma\sqrt{2\pi})^p} \int_{\|x\|\leq\varepsilon} e^{-\frac{\|x\|^2}{2\sigma^2}} \, dx \\
&\overset{x=\sigma z}{=} \frac{1}{(\sqrt{2\pi})^p} \int_{\|z\|\leq\frac{\varepsilon}{\sigma}} e^{-\frac{\|z\|^2}{2}} \, dz \\
&\leq \frac{1}{(\sqrt{2\pi})^p} \mathrm{vol}\,B_{\frac{\varepsilon}{\sigma}} = \frac{1}{(\sqrt{2\pi})^p} \left(\frac{\varepsilon}{\sigma}\right)^p \mathrm{vol}\,B_1 \\
&= \frac{1}{(\sqrt{2\pi})^p} \left(\frac{\varepsilon}{\sigma}\right)^p \frac{\mathcal{O}_{p-1}}{p}.
\end{aligned}
$$

Taking $\bar{x} = \bar{a}_m^{\perp}$, $\varepsilon = \frac{1}{\xi}$, and $p = n - m + 1$, the claim (4.11) follows. $\qquad\square$

Proof of Proposition 4.19 For $A \in \mathbb{R}^{m\times n}$ there exists $u_A \in \mathbb{S}^{m-1}$ such that $\|A^{\dagger}\| = \|A^{\dagger}u_A\|$. Moreover, for almost all A, the vector u_A is uniquely determined up to sign. Using the singular value decomposition, it is easy to show that for all $v \in \mathbb{S}^{m-1}$,

$$
\|A^{\dagger}v\| \geq \|A^{\dagger}\| \cdot |u_A^{\mathrm{T}}v|. \tag{4.12}
$$

Now take $A \sim N(\overline{A}, \sigma^2 I)$ and $v \sim U(\mathbb{S}^{m-1})$ independently. Then for any $s \in (0,1)$ and $t > 0$ we have

$$
\begin{aligned}
\mathop{\mathrm{Prob}}_{A,v} &\{\|A^{\dagger}v\| \geq t\sqrt{1-s^2}\} \\
&\geq \mathop{\mathrm{Prob}}_{A,v}\{\|A^{\dagger}\| \geq t \ \& \ |u_A^{\mathrm{T}}v| \geq \sqrt{1-s^2}\} \\
&= \mathop{\mathrm{Prob}}_{A}\{\|A^{\dagger}\| \geq t\} \cdot \mathop{\mathrm{Prob}}_{A,v}\{|u_A^{\mathrm{T}}v| \geq \sqrt{1-s^2} \mid \|A^{\dagger}\| \geq t\} \\
&\geq \mathop{\mathrm{Prob}}_{A}\{\|A^{\dagger}\| \geq t\} \cdot \sqrt{\frac{2}{\pi m}}\, s^{m-1}, \tag{4.13}
\end{aligned}
$$

the last line by Proposition 2.14 and Lemma 2.35 with $\xi = \sqrt{1-s^2}$. Now we use Proposition 4.21 with $\xi = t\sqrt{1-s^2}$ to deduce that

$$
\begin{aligned}
\mathop{\mathrm{Prob}}_{A}\{\|A^{\dagger}\| \geq t\} &\leq \sqrt{\frac{\pi m}{2}}\, \frac{1}{s^{m-1}} \mathop{\mathrm{Prob}}_{A,v}\{\|A^{\dagger}v\| \geq t\sqrt{1-s^2}\} \\
&\leq \frac{\sqrt{m}}{2s^{m-1}} \frac{1}{(\sqrt{2\pi})^{n-m}} \frac{\mathcal{O}_{n-m}}{n-m+1} \left(\frac{1}{\sigma t\sqrt{1-s^2}}\right)^{n-m+1}.
\end{aligned}
$$

$$\tag{4.14}$$

We next choose $s \in (0, 1)$ to minimize the bound above. To do so amounts to maximizing $(1 - x)^{\frac{n-m+1}{2}} x^{\frac{m-1}{2}}$, where $x = s^2 \in (0, 1)$, or equivalently to maximizing

$$g(x) = \left((1 - x)^{\frac{n-m+1}{2}} x^{\frac{m-1}{2}}\right)^{\frac{2}{n}} = (1 - x)^{\frac{n-m+1}{n}} x^{\frac{m-1}{n}} = (1 - x)^{1-\lambda} x^{\lambda}.$$

We have $\frac{d}{dx} \ln g(x) = \frac{\lambda}{x} - \frac{1-\lambda}{1-x}$ with the only zero attained at $x^* = \lambda$.

Replacing s^2 by λ in (4.14), we obtain the bound

$$\operatorname*{Prob}_{A}\{\|A^{\dagger}\| \geq t\} \leq \frac{\sqrt{\lambda n + 1}}{2\lambda^{\frac{\lambda n}{2}}} \frac{1}{(\sqrt{2\pi})^{n-m}} \frac{\mathcal{O}_{n-m}}{(1-\lambda)n} \left(\frac{1}{\sigma t \sqrt{1-\lambda}}\right)^{(1-\lambda)n}.$$

Lemma 4.18 implies

$$\lambda^{-\frac{\lambda n}{2}} = \left(\lambda^{-\frac{\lambda}{2(1-\lambda)}}\right)^{(1-\lambda)n} \leq e^{\frac{(1-\lambda)n}{2}}.$$

So we get

$$\operatorname*{Prob}_{A}\{\|A^{\dagger}\| \geq t\}$$

$$\leq \frac{\sqrt{\lambda n + 1}}{2} \frac{1}{(\sqrt{2\pi})^{n-m}} \frac{\mathcal{O}_{n-m}}{(1-\lambda)n} \left(\frac{\sqrt{e}}{\sigma t \sqrt{1-\lambda}}\right)^{(1-\lambda)n}$$

$$= \frac{\sqrt{\lambda n + 1}}{2} \left(\frac{e}{1-\lambda}\right)^{\frac{(1-\lambda)n}{2}} \frac{1}{(\sqrt{2\pi})^{n-m}} \frac{\mathcal{O}_{n-m}}{(1-\lambda)n} \left(\frac{1}{\sigma t}\right)^{(1-\lambda)n}$$

$$= \frac{1}{2(1-\lambda)} \sqrt{\lambda + \frac{1}{n}} \frac{1}{\sqrt{n}} \left(\frac{e}{1-\lambda}\right)^{\frac{(1-\lambda)n}{2}} \frac{\mathcal{O}_{n-m}}{(\sqrt{2\pi})^{n-m}} \left(\frac{1}{\sigma t}\right)^{(1-\lambda)n}$$

$$\leq \frac{\sqrt{\lambda + 1}}{2(1-\lambda)} \frac{1}{\sqrt{n}} \left(\frac{e}{1-\lambda}\right)^{\frac{(1-\lambda)n}{2}} \frac{2\pi^{\frac{n-m+1}{2}}}{\Gamma(\frac{n-m+1}{2})(\sqrt{2\pi})^{n-m}} \left(\frac{1}{\sigma t}\right)^{(1-\lambda)n}$$

$$= \frac{\sqrt{1+\lambda}}{1-\lambda} \frac{1}{\sqrt{n}} \left(\frac{e}{1-\lambda}\right)^{\frac{(1-\lambda)n}{2}} \frac{\sqrt{2\pi}}{\Gamma(\frac{n(1-\lambda)}{2})2^{\frac{(1-\lambda)n}{2}}} \left(\frac{1}{\sigma t}\right)^{(1-\lambda)n}.$$

We next estimate $\Gamma(\frac{(1-\lambda)n}{2})$. To do so, recall Stirling's bound (2.14), which yields, using $\Gamma(x + 1) = x\Gamma(x)$, the bound $\Gamma(x) > \sqrt{2\pi/x}\,(x/e)^x$. We use this with $x = \frac{(1-\lambda)n}{2}$ to obtain

$$\Gamma\left(\frac{(1-\lambda)n}{2}\right) \geq \sqrt{\frac{4\pi}{(1-\lambda)n}} \left(\frac{(1-\lambda)n}{2e}\right)^{\frac{(1-\lambda)n}{2}}.$$

Plugging this into the above, we obtain (observe the crucial cancellation of $\sqrt{n}$)

$$\text{Prob}_{A}\{\|A^{\dagger}\| \geq t\}$$

$$\leq \sqrt{\frac{1+\lambda}{(1-\lambda)^2 n}}\left(\frac{e}{1-\lambda}\right)^{\frac{(1-\lambda)n}{2}}\sqrt{2\pi}\sqrt{\frac{(1-\lambda)n}{4\pi}}\left(\frac{e}{(1-\lambda)n}\right)^{\frac{(1-\lambda)n}{2}}\left(\frac{1}{\sigma t}\right)^{(1-\lambda)n}$$

$$= c(\lambda)\left(\frac{e}{1-\lambda}\right)^{(1-\lambda)n}\left(\frac{1}{n}\right)^{\frac{(1-\lambda)n}{2}}\left(\frac{1}{\sigma t}\right)^{(1-\lambda)n} = c(\lambda)\left(\frac{e}{\sigma\sqrt{n}(1-\lambda)t}\right)^{(1-\lambda)n},$$

which completes the proof of the proposition.   □

4.2.2 Proof of Theorem 4.16

To simplify notation we write $c := c(\lambda)$ and $Q := Q(m, n)$. Proposition 4.19 implies that for any $\varepsilon > 0$ we have

$$\text{Prob}_{A\sim N(\overline{A},\sigma^2 I)}\left\{\|A^{\dagger}\| \geq \frac{e}{1-\lambda}\frac{1}{\sigma\sqrt{n}}\left(\frac{c}{\varepsilon}\right)^{\frac{1}{(1-\lambda)n}}\right\} \leq \varepsilon. \qquad (4.15)$$

Similarly, letting $\varepsilon = e^{-\frac{2t^2}{\pi^2\sigma^2}}$ in Proposition 4.13 and solving for t, we deduce that for any $\varepsilon \in (0, 1]$,

$$\text{Prob}\left\{\|A\| \geq Q\sigma\sqrt{n} + \sigma\sqrt{\frac{\pi^2}{2}\ln\frac{1}{\varepsilon}} + 1\right\} \leq \varepsilon. \qquad (4.16)$$

It is a trivial observation that for nonnegative random variables X, Y and positive α, β we have

$$\text{Prob}\{XY \geq \alpha\beta\} \leq \text{Prob}\{X \geq \alpha\} + \text{Prob}\{Y \geq \beta\}. \qquad (4.17)$$

Using this, we conclude that

$$\text{Prob}_{A\sim N(\overline{A},\sigma^2 I)}\left\{\kappa(A) \geq \frac{ez(\varepsilon)}{1-\lambda}\right\} \leq 2\varepsilon, \qquad (4.18)$$

where we have set, for $\varepsilon \in (0, 1]$,

$$z(\varepsilon) := \left(Q + \sqrt{\frac{\pi^2}{2n}\ln\frac{1}{\varepsilon}} + \frac{1}{\sigma\sqrt{n}}\right)\left(\frac{c}{\varepsilon}\right)^{\frac{1}{(1-\lambda)n}}. \qquad (4.19)$$

We note that $z(a) = \zeta := \zeta_\sigma(m, n)$; cf. Eq. (4.8). Moreover, $\lim_{\varepsilon\to 0} z(\varepsilon) = \infty$ and z is decreasing in the interval $(0, 1]$. Hence, for $z \geq \zeta$, there exists $\varepsilon = \varepsilon(z) \in (0, 1]$ such that $z = z(\varepsilon)$.

We need to bound $\varepsilon(z)$ from above as a function of z. To do so, we start with a weak lower bound on $\varepsilon(z)$ and claim that for $z \geq \zeta$,

$$\frac{1}{n} \ln \frac{1}{\varepsilon} \leq \ln\big(2z(\varepsilon)\big). \tag{4.20}$$

To prove this, note first that $Q \geq \sqrt{\frac{n}{n+1}} \geq \frac{1}{\sqrt{2}}$ due to Lemma 4.14. Hence $\zeta \geq Q \geq \frac{1}{\sqrt{2}}$, and it follows that $\sqrt{2}z \leq 1$ for $z \geq \zeta$. Equation (4.19) implies that

$$z(\varepsilon) \geq \frac{1}{\sqrt{2}} \left(\frac{c}{\varepsilon}\right)^{\frac{1}{(1-\lambda)n}}.$$

Using $c \geq \frac{1}{\sqrt{2}}$, we get

$$\left(\sqrt{2}z\right)^n \geq \left(\sqrt{2}z\right)^{(1-\lambda)n} \geq \frac{c}{\varepsilon} \geq \frac{1}{\sqrt{2}\,\varepsilon}.$$

Hence

$$(2z)^n \geq \frac{1}{\varepsilon},$$

which proves the claimed inequality (4.20).

Using the bound (4.20) in Eq. (4.19), we get, again writing $z = z(\varepsilon)$, that for all $z \geq \zeta$,

$$z \leq \left(Q + \sqrt{\frac{\pi^2}{2} \ln(2z)} + \frac{1}{\sigma\sqrt{n}}\right)\left(\frac{c}{\varepsilon}\right)^{\frac{1}{(1-\lambda)n}},$$

which means that

$$\varepsilon \leq c\left[\left(Q + \sqrt{\frac{\pi^2}{2} \ln(2z)} + \frac{1}{\sigma\sqrt{n}}\right)\frac{1}{z}\right]^{(1-\lambda)n}.$$

By (4.18) this completes the proof. $\square$

4.3 Expectations: Proof of Theorem 4.2

Fix $\lambda_0 \in (0, 1)$ and put $c := c(\lambda_0)$. Suppose that $m \leq n$ satisfy $\lambda = (m-1)/n \leq \lambda_0$. Then $n - m + 1 = (1-\lambda)n \geq (1-\lambda_0)n$, and in order to have $n - m$ sufficiently large, it suffices to require that n be sufficiently large. Thus, $c^{\frac{1}{n-m+1}} \leq 1.1$ if n is sufficiently large. Similarly, because of Lemma 4.14, $Q(m, n) \leq 2.1$ for large enough n. This implies that for $\frac{1}{\sqrt{m}} \leq \sigma \leq 1$, we have

$$Q(m, n) + \frac{1}{\sigma\sqrt{n}} \leq 2.1 + \frac{1}{\sigma\sqrt{n}} \leq 2.1 + \sqrt{\frac{m}{n}} \leq 2.1 + \sqrt{\lambda_0 + \frac{1}{n}} < 3.1,$$

provided n is large enough. Then $\zeta_\sigma(m, n) \leq 3.1 \cdot 1.1 = 3.41$.

By Theorem 4.16, the random variable $Z := (1 - \lambda)\kappa(A)/e$ satisfies, for any $\overline{A}$ with $\|\overline{A}\| \leq 1$ and any $z \geq 3.41$,

$$\operatorname*{Prob}_{A \sim N(\overline{A}, \sigma^2 I)} \{Z \geq z\} \leq 2c \left[\left(Q(m,n) + \sqrt{\frac{\pi^2}{2} \ln(2z)} + \frac{1}{\sigma \sqrt{n}} \right) \frac{1}{z} \right]^{n-m+1}$$

$$\leq 2c \left[\left(3.1 + \sqrt{\frac{\pi^2}{2} \ln(2z)} \right) \frac{1}{z} \right]^{n-m+1},$$

for large enough n. Since

$$3.1 + \sqrt{\frac{\pi^2}{2} \ln(2z)} \leq e\sqrt{z}$$

for all $z \geq e^2$, we deduce that for all such z,

$$\operatorname*{Prob}_{A \sim N(\overline{A}, \sigma^2 I)} \{Z \geq z\} \leq 2c \left(\frac{e}{\sqrt{z}} \right)^{n-m+1}.$$

Using this tail bound to compute $\mathbb{E}(Z)$, we get

$$\mathbb{E}(Z) = \int_0^\infty \operatorname{Prob}\{Z \geq z\}\, dz \leq e^2 + 2c \int_{e^2}^\infty \left(\frac{e^2}{z} \right)^{\frac{n-m+1}{2}} dz$$

$$\overset{z=e^2 y}{=} e^2 + 2c \int_1^\infty \left(\frac{1}{y} \right)^{\frac{n-m+1}{2}} e^2\, dy = e^2 + \frac{4ce^2}{n-m-1}.$$

We can now conclude, since

$$\mathbb{E}\big((1-\lambda)\kappa(A) \big) = \mathbb{E}(eZ) = e\mathbb{E}(Z) \leq e^3 + \frac{4ce^3}{n-m-1} \leq 20.1,$$

where the inequality again follows by taking n large enough. $\qquad\square$

4.4 Complex Matrices

In this and the preceding chapters we have assumed data to be given by real numbers. For a number of problems in scientific computation, however, data is better assumed to be complex. All of the results we have shown can be given, without major modifications, a complex version.

A difference stands out, nonetheless, and it is the fact that—in contrast to the situation over the real numbers—condition numbers for complex Gaussian data have in general a finite expectation. The reasons for this general phenomenon will become clear in Chap. 20. In this section we compute bounds for the probability analysis of some complex condition numbers. They give a first taste of the difference with the real case and will, in addition, be crucial to some of our arguments in Part III.

In the following we fix $\overline{A} \in \mathbb{C}^{n \times n}$, $\sigma > 0$ and denote by $\varphi^{\overline{A},\sigma}$ the Gaussian density of $N(\overline{A}, \sigma^2 I)$ on $\mathbb{C}^{n \times n}$. Moreover, we consider the related density

$$\rho^{\overline{A},\sigma}(A) = c_{\overline{A},\sigma}^{-1} |\det A|^2 \varphi^{\overline{A},\sigma}(A), \tag{4.21}$$

where

$$c_{\overline{A},\sigma} := \mathop{\mathbb{E}}_{A \sim N(\overline{A}, \sigma^2 I)} \left(|\det A|^2 \right).$$

The following result is akin to a smoothed analysis of the matrix condition number $\kappa(A) = \|A\| \cdot \|A^{-1}\|$, with respect to the probability densities $\rho^{\overline{A},\sigma}$ that are not Gaussian, but closely related to Gaussians.

Proposition 4.22 *For $\overline{A} \in \mathbb{C}^{n \times n}$ and $\sigma > 0$ we have*

$$\mathbb{E}_{A \sim \rho^{\overline{A},\sigma}} \left(\left\| A^{-1} \right\|^2 \right) \le \frac{e(n+1)}{2\sigma^2}.$$

Before embarking on the proof, we note that in the centered case $\overline{A} = 0$, the constant in (4.21) evaluates to

$$c_{0,\sigma} = \mathop{\mathbb{E}}_{A \sim N(0,I)} \left| \det(A) \right|^2 = 2^n n!$$

by Lemma 4.12. In this case, Proposition 4.22 implies the following result.

Corollary 4.23 *Let $N(0, I)$ denote the standard Gaussian on $\mathbb{C}^{n \times n}$. Then,*

$$\mathop{\mathbb{E}}_{A \sim N(0,I)} \left(\left\| A^{-1} \right\|^2 |\det A|^2 \right) \le 2^n n! \frac{e(n+1)}{2}. \qquad \square$$

We turn now to the proof of Proposition 4.22. Actually, we will prove tail bounds from which the stated bound on the expectation easily follows.

Let us denote by $\mathbb{S}(\mathbb{C}^n) := \{ \zeta \in \mathbb{C}^n \mid \|\zeta\| = 1 \}$ the unit sphere in $\mathbb{C}^n$. Also, let $\mathscr{U}(n)$ be the *unitary group*, which is defined as

$$\mathscr{U}(n) := \left\{ u \in GL_n(\mathbb{C}) \mid uu^* = I_n \right\},$$

where u^* denotes the *adjoint* of u, i.e., $(u^*)_{ij} = \bar{u}_{ji}$.

Lemma 4.24 *Let $\overline{A} \in \mathbb{C}^{n \times n}$ and $\sigma > 0$. For any $v \in \mathbb{S}(\mathbb{C}^n)$ and any $t > 0$, we have*

$$\mathop{\mathrm{Prob}}_{A \sim \rho^{\overline{A},\sigma}} \left\{ \left\| A^{-1} v \right\| \ge t \right\} \le \frac{1}{4\sigma^4 t^4}.$$

Proof We argue similarly as for Proposition 4.21. We first claim that because of unitary invariance, we may assume that $v = e_n := (0, \ldots, 0, 1)$. To see this, take $u \in U(n)$ such that $v = u e_n$. Consider the isometric map $A \mapsto B = u^{-1}A$ that transforms the density $\rho^{\bar{A},\sigma}(A)$ to a density of the same form, namely

$$\rho^{\bar{B},\sigma}(B) = \rho^{\bar{A},\sigma}(A) = c^{-1} |\det A|^2 \varphi^{\bar{A},\sigma}(A) = c^{-1} |\det B|^2 \varphi^{\bar{B},\sigma}(B),$$

where $\bar{B} := u^{-1}\bar{A}$ and $c = \mathbb{E}_{\varphi^{\bar{A},\sigma}}(|\det A|^2) = \mathbb{E}_{\varphi^{\bar{B},\sigma}}(|\det B|^2)$. Thus the assertion for e_n and random B (chosen from any isotropic Gaussian distribution) implies the assertion for v and A, noting that $A^{-1}v = B^{-1}e_n$. This proves the claim.

Let a_i denote the ith row of A. Almost surely, the rows $a_1, \ldots, a_{n-1}$ are linearly independent. We are going to characterize $\|A^{-1}e_n\|$ in a geometric way. Let $S_n := \mathrm{span}\{a_1, \ldots, a_{n-1}\}$ and denote by $a_n^\perp$ the orthogonal projection of a_n onto $S_n^\perp$. Consider $w := A^{-1}e_n$, which is the nth column of A^{-1}. Since $AA^{-1} = I$, we have $\langle w, a_i \rangle = 0$ for $i = 1, \ldots, n-1$ and hence $w \in S_n^\perp$. Moreover, $\langle w, a_n \rangle = 1$, so $\|w\| \|a_n^\perp\| = 1$, and we arrive at

$$\|A^{-1}e_n\| = \frac{1}{\|a_n^\perp\|}. \tag{4.22}$$

Let $A_n \in \mathbb{C}^{(n-1)\times n}$ denote the matrix obtained from A by omitting a_n. We shall write $\mathrm{vol}(A_n) = \det(AA^*)^{1/2}$ for the $(n-1)$-dimensional volume of the parallelepiped spanned by the rows of A_n. Similarly, $|\det A|$ can be interpreted as the n-dimensional volume of the parallelepiped spanned by the rows of A.

Now we write $\varphi(A) := \varphi^{\bar{A},\sigma}(A) = \varphi_1(A_n)\varphi_2(a_n)$, where φ_1 and φ_2 are the density functions of $N(\bar{A}_n, \sigma^2 I)$ and $N(\bar{a}_n, \sigma^2 I)$, respectively (the meaning of $\bar{A}_n$ and $\bar{a}_n$ being clear). Moreover, note that

$$\mathrm{vol}(A)^2 = \mathrm{vol}(A_n)^2 \|a_n^\perp\|^2.$$

Fubini's theorem combined with (4.22) yields for $t > 0$,

$$\int_{\|A^{-1}e_n\| \geq t} \mathrm{vol}(A)^2 \varphi(A)\, dA = \int_{A_n \in \mathbb{C}^{(n-1)\times n}} \mathrm{vol}(A_n)^2 \varphi_1(A_n)$$
$$\cdot \left(\int_{\|a_n^\perp\| \leq 1/t} \|a_n^\perp\|^2 \varphi_2(a_n)\, da_n \right) dA_n. \tag{4.23}$$

We next show that for fixed, linearly independent $a_1, \ldots, a_{n-1}$ and $\lambda > 0$,

$$\int_{\|a_n^\perp\| \leq \lambda} \|a_n^\perp\|^2 \varphi_2(a_n)\, da_n \leq \frac{\lambda^4}{2\sigma^2}. \tag{4.24}$$

For this, note that $a_n^\perp \sim N(\bar{a}_n^\perp, \sigma^2 I)$ in $S_n^\perp \simeq \mathbb{C}$, where $\bar{a}_n^\perp$ is the orthogonal projection of $\bar{a}_n$ onto $S_n^\perp$. Thus, proving (4.24) amounts to showing that

$$\int_{|z| \leq \lambda} |z|^2 \varphi_{\bar{z},\sigma}(z)\, dz \leq \frac{\lambda^4}{2\sigma^2}$$

for the Gaussian density $\varphi^{\bar{z},\sigma}(z) = \frac{1}{2\pi\sigma^2} e^{-\frac{1}{2\sigma^2}|z-\bar{z}|^2}$ of $z \in \mathbb{C}$, where $\bar{z} \in \mathbb{C}$. Clearly, it is enough to show that

$$\int_{|z|\leq\lambda} \varphi^{\bar{z},\sigma}(z)\,dz \leq \frac{\lambda^2}{2\sigma^2}.$$

Without loss of generality we may assume that $\bar{z} = 0$, since the integral on the left-hand side is maximized at this value of $\bar{z}$. The substitution $z = \sigma w$ yields $dz = \sigma^2 dw$ (dz denotes the Lebesgue measure on $\mathbb{R}^2$), and we get

$$\int_{|z|\leq\lambda} \varphi_{0,\sigma}(z)\,dz = \int_{|w|\leq\frac{\lambda}{\sigma}} \frac{1}{2\pi} e^{-\frac{1}{2}|w|^2}\,dw = \int_0^{\frac{\lambda}{\sigma}} \frac{1}{2\pi} e^{-\frac{1}{2}r^2} 2\pi r\,dr$$

$$= -e^{-\frac{1}{2}r^2}\Big|_0^{\frac{\lambda}{\sigma}} = 1 - e^{-\frac{\lambda^2}{2\sigma^2}} \leq \frac{\lambda^2}{2\sigma^2},$$

which proves inequality (4.24).

A similar argument shows that

$$2\sigma^2 \leq \int |z|^2 \varphi^{\bar{z},\sigma}(z)\,dz = \int \|a_n^\perp\|^2 \varphi_2(a_n)\,da_n. \tag{4.25}$$

Plugging in this inequality into (4.23) (with $t = 0$), we conclude that

$$2\sigma^2\, \mathbb{E}_{\varphi_1}\big(\mathrm{vol}(A_n)^2\big) \leq \mathbb{E}_\varphi\big(\mathrm{vol}(A)^2\big). \tag{4.26}$$

On the other hand, plugging in (4.11) with $\lambda = \frac{1}{t}$ into (4.23), we obtain

$$\int_{\|A^{-1}e_n\|\geq t} \mathrm{vol}(A)^2 \varphi(A)\,dA \leq \frac{1}{2\sigma^2 t^4}\, \mathbb{E}_{\varphi_1}\big(\mathrm{vol}(A_n)^2\big).$$

Combined with (4.26), this yields

$$\int_{\|A^{-1}e_n\|\geq t} \mathrm{vol}(A)^2 \varphi(A)\,dA \leq \frac{1}{4\sigma^4 t^4}\, \mathbb{E}_\varphi\big(\mathrm{vol}(A)^2\big).$$

By the definition of the density $\rho^{\bar{A},\sigma}$, this means that

$$\operatorname*{Prob}_{A\sim\rho}\{\|A^{-1}e_n\| \geq t\} \leq \frac{1}{4\sigma^4 t^4},$$

which was to be shown.   □

Lemma 4.25 *For fixed $u \in \mathbb{S}(\mathbb{C}^n)$, $0 \leq s \leq 1$, and random v uniformly chosen in $\mathbb{S}(\mathbb{C}^n)$ we have*

$$\operatorname*{Prob}_v\{|u^\mathsf{T} v| \geq s\} = (1 - s^2)^{n-1}.$$

Proof By unitary invariance we may assume without loss of generality that $u = (1, 0, \ldots, 0)$. Also, we may assume that $s < 1$. Note that if $v \in \mathbb{C}^n$ is standard Gaussian, then $\|v\|^{-1} v$ is uniformly distributed in the sphere $\mathbb{S}(\mathbb{C}^n)$; see Proposition 2.19. Therefore, we need to prove that the probability

$$p := \underset{v \in \mathbb{S}(\mathbb{C}^n)}{\mathrm{Prob}} \{|v_1| \geq s\} = \underset{v \sim N(0;I)}{\mathrm{Prob}} \{|v_1| \geq s\|v\|\}$$

equals $(1 - s^2)^{n-1}$. For this, it is convenient to identify $\mathbb{C}^n$ with $\mathbb{R}^{2n}$ and to write $v = (x, y)$, where $x \in \mathbb{R}^2$ and $y \in \mathbb{R}^{2n-2}$. So we have $v_1 = x_1 + ix_2$. Note that $\|x\|^2 \geq s^2(\|x\|^2 + \|y\|^2)$ iff $\|x\| \geq \lambda\|y\|$, where $\lambda := \frac{s}{\sqrt{1-s^2}}$. Therefore, we can write

$$p = \frac{1}{(2\pi)^{n-1}} \int_{\mathbb{R}^{2n-2}} e^{-\frac{1}{2}\|y\|^2} \left(\frac{1}{2\pi} \int_{\|x\| \geq \lambda\|y\|} e^{-\frac{1}{2}\|x\|^2} dx \right) dy.$$

Integrating in polar coordinates (cf. Corollary 2.2), we obtain for $r \geq 0$,

$$\frac{1}{2\pi} \int_{\|x\| \geq r} e^{-\frac{1}{2}\|x\|^2} dx = \frac{1}{2\pi} \int_{\rho \geq r} e^{-\frac{1}{2}\rho^2} 2\pi\rho \, d\rho = e^{-\frac{1}{2}r^2}.$$

This implies

$$p = \frac{1}{(2\pi)^{n-1}} \int_{\mathbb{R}^{2n-2}} e^{-\frac{1}{2}\|y\|^2} e^{-\frac{s^2\|y\|^2}{2(1-s^2)}} dy = \frac{1}{(2\pi)^{n-1}} \int_{\mathbb{R}^{2n-2}} e^{-\frac{\|y\|^2}{2(1-s^2)}} dy.$$

Making the substitution $\eta := (1 - s^2)^{-1/2}\|y\|$, we get

$$p = \frac{1}{(2\pi)^{n-1}} \int_{\mathbb{R}^{2n-2}} e^{-\frac{1}{2}\|\eta\|^2} \left(1 - s^2\right)^{\frac{2n-2}{2}} d\eta = \left(1 - s^2\right)^{n-1}. \qquad \square$$

Lemma 4.26 *Let* $\overline{A} \in \mathbb{C}^{n \times n}$ *and* $\sigma > 0$. *For any* $t > 0$ *we have*

$$\underset{A \sim \rho_{\overline{A},\sigma}}{\mathrm{Prob}} \left\{ \|A^{-1}\| \geq t \right\} \leq \frac{e^2(n+1)^2}{16\sigma^4} \frac{1}{t^4}.$$

Proof We proceed similarly as for Proposition 4.19. For any invertible $A \in \mathbb{C}^{n \times n}$ there exists $u \in \mathbb{S}(\mathbb{C}^n)$ such that $\|A^{-1}u\| = \|A^{-1}\|$. For almost all A, the vector u is uniquely determined up to a scaling factor θ of modulus 1. We shall denote by u_A a representative of such u.

The following is an easy consequence of the singular value decomposition of $\|A^{-1}\|$: for any $v \in \mathbb{S}(\mathbb{C}^n)$ we have

$$\|A^{-1}v\| \geq \|A^{-1}\| \cdot |u_A^T v|. \tag{4.27}$$

We choose now a random pair (A, v) with A following the law ρ and, independently, $v \in \mathbb{S}(\mathbb{C}^n)$ from the uniform distribution. Lemma 4.24 implies that

$$\Prob_{A,v}\left\{\|A^{-1}v\| \geq t\sqrt{\frac{2}{n+1}}\right\} \leq \frac{(n+1)^2}{16\sigma^4 t^4}.$$

On the other hand, we have by (4.27)

$$\Prob_{A,v}\{\|A^{-1}v\| \geq t\sqrt{2/(n+1)}\}$$

$$\geq \Prob_{A,v}\{\|A^{-1}\| \geq t \text{ and } |u_A^{\mathsf{T}}v| \geq \sqrt{2/(n+1)}\}$$

$$\geq \Prob_{A}\{\|A^{-1}\| \geq t\}\,\Prob_{A,v}\{|u_A^{\mathsf{T}}v| \geq \sqrt{2/(n+1)}\ \big|\ \|A^{-1}\| \geq t\}.$$

Lemma 4.25 tells us that for any fixed $u \in \mathbb{S}(\mathbb{C}^n)$ we have

$$\Prob_{v}\{|u^{\mathsf{T}}v| \geq \sqrt{2/(n+1)}\} = \left(1 - 2/(n+1)\right)^{n-1} \geq e^{-2},$$

the last inequality following from $(\frac{n+1}{n-1})^{n-1} = (1 + \frac{2}{n-1})^{n-1} \leq e^2$. We thus obtain

$$\Prob_{A}\{\|A^{-1}\| \geq t\} \leq e^2 \Prob_{A,v}\left\{\|A^{-1}v\| \geq t\sqrt{\frac{2}{n+1}}\right\} \leq \frac{e^2(n+1)^2}{16\sigma^4 t^4},$$

as claimed.   □

Proof of Proposition 4.22 By Lemma 4.26 we obtain, for any $T_0 > 0$,

$$\mathbb{E}(\|A^{-1}\|^2) = \int_0^\infty \Prob\{\|A^{-1}\|^2 \geq T\}\,dT$$

$$\leq T_0 + \int_{T_0}^\infty \Prob\{\|A^{-1}\|^2 \geq T\}\,dT \leq T_0 + \frac{e^2(n+1)^2}{16\sigma^4}\frac{1}{T_0},$$

using $\int_{T_0}^\infty T^{-2}\,dT = T_0^{-1}$. Now choose $T_0 = \frac{e(n+1)}{4\sigma^2}$.   □

We have already mentioned that all of the probabilistic analyses for random real matrices in this chapter extend, without major modifications, to a complex version. We refrain from stating these obvious extensions and only record here the following variant of Proposition 4.19 for the particular case of complex $m \times (m+1)$ matrices (and average analysis). This result will be needed in Sect. 17.8.

Proposition 4.27 *For a standard Gaussian $A \in \mathbb{C}^{m \times (m+1)}$ and for any $t > 0$,*

$$\Prob_{A \sim N(0,\mathrm{I})}\{\|A^\dagger\| \geq t\} \leq \frac{m^2}{8e}\frac{1}{t^4}.$$

Moreover, $\mathbb{E}\|A^\dagger\|^2 \le 1 + \frac{m^2}{8e}$ is finite.

Remark 4.28

(a) The fact that the expectation of $\|A^\dagger\|^2$ is finite is a key property used in Chap. 17 for analyzing the running time of a certain algorithm LV for computing a zero of a system of complex polynomial equations.

(b) In Chap. 20 we will see that the exponent 4 in the tail bound t^{-4} comes in naturally as twice the complex codimension of the projective variety in $\mathbb{P}(\mathbb{C}^{m \times (m+1)})$ corresponding to the rank-deficient matrices $A \in \mathbb{C}^{m \times (m+1)}$.

For the proof of Proposition 4.27 we need the following lemma.

Lemma 4.29 *For fixed $v \in \mathbb{S}(\mathbb{C}^m)$ and a standard Gaussian matrix $A \in \mathbb{C}^{m \times (m+1)}$ we have for all $\varepsilon > 0$,*

$$\Prob_{A \sim N(0,I)} \left\{ \|A^\dagger v\| \ge \varepsilon^{-1} \right\} \le \frac{1}{8} \varepsilon^4.$$

Proof This is very similar to the proof of Proposition 4.21, so that it is sufficient to point out the few modifications needed. We adopt the notation from there. So we assume $v = e_m$ and note that (4.10) holds. To complete the proof, it is sufficient to establish the bound

$$\int_{\|a_m^\perp\| \le \varepsilon^{-1}} \varphi_2(a_m) \, da_m \le \frac{1}{8} \varepsilon^4$$

for fixed, linearly independent $a_1, \ldots, a_{m-1} \in \mathbb{C}^{m+1}$ and $\varepsilon > 0$. Note that the orthogonal projection $a_m^\perp$ of a_m onto the span S of $a_1, \ldots, a_{m-1}$ is standard normal distributed in $S^\perp \sim \mathbb{C}^2 \simeq \mathbb{R}^4$. It is therefore sufficient to verify that

$$\Prob_{\substack{x \in \mathbb{R}^4 \\ x \sim N(0,I)}} \left\{ \|x\| \le \varepsilon \right\} \le \left(\frac{1}{\sqrt{2\pi}} \right)^4 \operatorname{vol} B_\varepsilon = \left(\frac{1}{\sqrt{2\pi}} \right)^4 \frac{O_3}{4} \varepsilon^4 = \frac{1}{8} \varepsilon^4. \qquad \square$$

Proof of Proposition 4.27 We proceed similarly as for Proposition 4.19 and adopt the notation from there. Similarly as for (4.13) we have for $s \in (0, 1)$ and $t > 0$,

$$\Prob_{A,v} \left\{ \|A^\dagger v\| \ge t\sqrt{1-s^2} \right\}$$

$$\ge \Prob_{A,v} \left\{ \|A^\dagger\| \ge t \ \& \ |u_A^T v| \ge \sqrt{1-s^2} \right\}$$

$$= \Prob_A \left\{ \|A^\dagger\| \ge t \right\} \cdot \Prob_{A,v} \left\{ |u_A^T v| \ge \sqrt{1-s^2} \mid \|A^\dagger\| \ge t \right\}$$

$$\ge \Prob_A \left\{ \|A^\dagger\| \ge t \right\} \cdot s^{2(m-1)},$$

the last line by Lemma 4.25 (replacing s by $\sqrt{1 - s^2}$). Using Lemma 4.29, we obtain

$$\operatorname*{Prob}_{A}\{\|A^{\dagger}\| \geq t\} \leq \frac{1}{s^{2m-2}} \operatorname*{Prob}_{v} \operatorname*{Prob}_{A}\{\|A^{\dagger}v\| \geq t\sqrt{1 - s^2}\}$$

$$\leq \frac{1}{8t^4} \frac{1}{s^{2m-2}(1 - s^2)^2}.$$

We choose now $s_* := \sqrt{1 - \frac{1}{m}}$ to minimize the right-hand side. This gives

$$s_*^{2m-2}(1 - s_*^2)^2 = \left(1 - \frac{1}{m}\right)^{m-1} \frac{1}{m^2} \geq \frac{e}{m^2}.$$

Hence the tail estimate $\operatorname*{Prob}\{\|A^{\dagger}\| \geq t\} \leq \frac{m^2}{8e} \frac{1}{t^4}$ follows.

The expectation can be bounded as usual:

$$\mathbb{E}\|A^{\dagger}\|^2 \leq 1 + \int_1^{\infty} \operatorname*{Prob}\{\|A^{\dagger}\|^2 \geq s\}\, ds \leq 1 + \frac{m^2}{8e} \int_1^{\infty} \frac{ds}{s^2} = 1 + \frac{m^2}{8e}. \qquad \square$$

Remark 4.30 A similar argument for a real standard Gaussian matrix $A \in \mathbb{R}^{m \times (m+1)}$ reveals that $\operatorname*{Prob}\{\|A^{\dagger}\|^2 \geq t\}$ decays with t^{-2}. From this one can deduce that $\mathbb{E}\|A^{\dagger}\|^2 = \infty$. This difference between this and the complex case is responsible for the fact that a version of an adaptive homotopy algorithm for solving polynomial equations must fail over the reals (on average).

Chapter 5
Condition Numbers and Iterative Algorithms

Consider a full-rank rectangular matrix $R \in \mathbb{R}^{q \times n}$ with $q > n$, a vector $c \in \mathbb{R}^q$, and the least-squares problem

$$\min_{v \in \mathbb{R}^n} \| Rv - c \|.$$

We saw in Sect. 1.6 that the solution $x \in \mathbb{R}^n$ of this problem is given by

$$x = R^\dagger c = \left(R^{\mathrm{T}} R \right)^{-1} R^{\mathrm{T}} c.$$

It follows that we can find x as the solution of the system $Ax = b$ with $A := R^{\mathrm{T}} R$, $A \in \mathbb{R}^{n \times n}$, and $b := R^{\mathrm{T}} c$.

A key remark at this stage is that by construction, A is symmetric and positive definite. One may therefore consider algorithms exploiting symmetry and positive definiteness. We do so in this chapter.

The algorithms we describe, steepest descent and conjugate gradient, will serve to deepen our understanding of the only facet of conditioning—among those described in Sect. O.5—that we have not dealt with up to now: the relationship between condition and complexity. To better focus on this issue, we will disregard all issues concerning finite precision and assume, instead, infinite precision in all computations. Remarkably, the condition number $\kappa(A)$ of A will naturally occur in the analysis of the running time for these algorithms. And this occurrence leads us to the last issue we discuss in this introduction.

Complexity bounds in terms of $\kappa(A)$ are not directly applicable, since $\kappa(A)$ is not known a priori. We have already argued that one can remove $\kappa(A)$ from these bounds by trading worst-case for, say, average-case complexity. This passes through an average analysis of $\kappa(A)$, and in turn, such an analysis assumes that the set of matrices A is endowed with a probability distribution. When A is arbitrary in $\mathbb{R}^{n \times n}$, we endow this space with a standard Gaussian. In our case, when A is positive definite, this choice is no longer available. A look at our original computational problem may, however, shed some light. Matrix A is obtained as $A = R^{\mathrm{T}} R$. It then makes sense to consider R as our primary random data—and for R we can assume Gaussianity—and endow A with the distribution inherited from that of R.

P. Bürgisser, F. Cucker, *Condition*,
Grundlehren der mathematischen Wissenschaften 349,
DOI 10.1007/978-3-642-38896-5_5, © Springer-Verlag Berlin Heidelberg 2013

Furthermore, as we will see, one has $\kappa(A) = \kappa^2(R)$. Therefore, the analysis of $\kappa(A)$ for this inherited distribution reduces to the analysis of $\kappa(R)$ when R is Gaussian.

5.1 The Cost of Computing: A Primer in Complexity

Before stepping into the description and analysis of algorithms, it is convenient to agree on some basic notions of complexity.

Since our interest in this book is limited to the analysis of specific algorithms, we do not need to formally describe machine models.[1] We will instead consider algorithms $\mathscr{A}$ described in a high-level language (such as Algorithm FS in Sect. 3.2) and define, for a given input $a \in \mathcal{D} \subseteq \mathbb{R}^m$, the *cost* or *running time* of $\mathscr{A}$ on input a to be the number $\mathsf{cost}^{\mathscr{A}}(a)$ of arithmetic operations (and square roots if the algorithm performs any) and comparisons performed by $\mathscr{A}$ during the execution with input a. A simple counting argument shows that with input $L \in \mathbb{R}^{n \times n}$ lower-triangular and $b \in \mathbb{R}^n$, Algorithm FS performs n^2 arithmetic operations.

The object of interest is the growth rate of the running time with respect to the input size. For a given $a \in \mathcal{D} \subseteq \mathbb{R}^m$ we say that m is the *size* of a, and we write $\mathsf{size}(a)$ for the latter. This is the number of reals (i.e., floating-point numbers) we feed the algorithm with. In our example, $\mathsf{size}(L, b) = \frac{n(n+3)}{2}$ (we represent L by its $\frac{n(n+1)}{2}$ lower entries). Hence, the running time of Algorithm FS on input (L, b) is about (actually less than) twice $\mathsf{size}(L, b)$: a linear growth rate.

Another example of this idea is given by Gaussian elimination (we omit describing the algorithm, since the reader certainly knows it). It is easily checked that for a given pair (A, b), where $A \in \mathbb{R}^{n \times n}$ and $b \in \mathbb{R}^n$, the cost of producing the triangular system $Lx = c$ whose solution is $A^{-1}b$ is $\frac{2}{3}(n-1)n(n+1) + n(n-1)$. To solve the system, we need to add the $n^2 - 1$ operations required by Algorithm FS. In this case, $\mathsf{size}(A, b) = n(n+1)$, and hence the cost of solving $Ax = b$ using Gaussian elimination (plus backward substitution) is of order $\frac{2}{3}\mathsf{size}(A, b)^{3/2}$.

Backward substitution and Gaussian elimination are said to be *direct methods*. One has an a priori bound on $\mathsf{cost}^{\mathscr{A}}(a)$ depending on $\mathsf{size}(a)$ (or on $\mathsf{dims}(a)$ if this bound communicates better). In contrast to this kind of algorithms, *iterative methods* may not possess such a bound. These algorithms iterate a basic procedure until a certain condition is met, for instance that an approximation of $\varphi(a)$ has been found with $\mathsf{RelError}(\varphi(a)) \leq \varepsilon$. In this case, the cost on a given input a will depend on ε. As we will see in this chapter (as well as in some others), it often depends as well on the condition of a.

We close this section with a word of caution. Most of the algorithms considered in this book are numerical algorithms: the data they handle are floating-point numbers, and the basic operations performed on these data are floating-point arithmetic operations, whence the overall justice of defining data size as the number

---

[1] These theoretical constructions are a must, however, if one wants to prove lower bounds for the complexity of a computational problem.

of floating-point numbers in the description of the data and cost as the number of such operations performed. Alongside numerical algorithms there is the vast class of discrete algorithms, which handle discrete data (rational numbers, combinatorial structures such as graphs and lists, etc.). In this case, the size of a certain datum (say a positive integer number ℓ) is the number of bits necessary to describe it (the number of bits in the binary expansion of ℓ, which is roughly $\lceil \log_2 \ell \rceil$). Also, the basic operations are elementary bit operations (read a bit, change the value of a bit, write a bit), and the cost of any procedure on given data is the number of such elementary operations performed during the execution of the procedure. For example, the way we learned to multiply integers (modified to work in base 2) performs $\mathcal{O}(\text{size}(\ell)\text{size}(q))$ bit operations to compute the product ℓq.

It is a basic principle that one should analyze discrete problems with a discrete model for cost such as the one just described—we call it *bit cost*—and numerical problems with a numerical one—the one above, usually known as *algebraic cost*. In particular, the restriction of a given numerical problem (e.g., matrix multiplication) to discrete data (e.g., matrices with integer coefficients) entails a change of cost model in the analysis of a given algorithm solving the problem. We will see a striking example of this issue in Chap. 7.

5.2   The Method of Steepest Descent

The method of steepest descent, also called the *gradient method*, is one of the oldest and most widely known algorithms for minimizing a function.

Let $A \in \mathbb{R}^{n \times n}$ be positive definite, $b \in \mathbb{R}^n$, and consider the quadratic function

$$f(x) := \frac{1}{2} x^{\mathsf{T}} A x - b^{\mathsf{T}} x.$$

Its gradient at x is $\text{grad } f(x) = Ax - b$. Let $\bar{x} = A^{-1} b$ be the solution of the linear system of equations

$$Ax = b. \tag{5.1}$$

Then $\bar{x}$ is the unique minimum of f, since f is strictly convex. The idea is to (approximately) compute $\bar{x}$ by minimizing f. This turns out to be faster for large sparse matrices A (see Remark 5.4 below).

The method works as follows. Let $x_0 \in \mathbb{R}^n$ be a starting point. We iteratively compute a sequence of approximations $x_0, x_1, x_2, \ldots$ by taking

$$x_{k+1} = x_k - \alpha_k \text{ grad } f(x_k), \tag{5.2}$$

where α_k is found by minimizing the quadratic univariate function

$$\mathbb{R} \to \mathbb{R}, \quad \alpha \mapsto f\big(x_k - \alpha \text{ grad } f(x_k)\big).$$

We call $\epsilon_k := x_k - \bar{x}$ the kth *error* and $r_k = b - Ax_k$ the kth *residual*. Note that $-r_k$ is the gradient of f at x_k:

$$r_k = -\operatorname{grad} f(x_k) = -A\epsilon_k. \tag{5.3}$$

Lemma 5.1 *We have*

$$\alpha_k = \frac{r_k^T r_k}{r_k^T A r_k} = \frac{\|r_k\|^2}{r_k^T A r_k}.$$

Proof Put

$$g(\alpha) := f(x_k + \alpha r_k) = \frac{1}{2}(x_k^T + \alpha r_k^T)A(x_k + \alpha r_k) - b^T(x_k + \alpha r_k).$$

Hence

$$g'(\alpha) = \frac{1}{2}r_k^T A x_k + \frac{1}{2}x_k^T A r_k + \alpha r_k^T A r_k - b^T r_k.$$

Writing $Ax_k = b - r_k$, this yields

$$g'(\alpha) = -\frac{1}{2}r_k^T r_k + \frac{1}{2}r_k^T b + \frac{1}{2}b^T r_k - \frac{1}{2}r_k^T r_k + \alpha r_k^T A r_k - b^T r_k$$

$$= \alpha r_k^T A r_k - r_k^T r_k.$$

Solving for $g'(\alpha) = 0$ yields the assertion. $\square$

For the analysis it is useful to define the A-norm

$$\|x\|_A := (x^T A x)^{1/2}$$

coming from the scalar product

$$\langle x, y \rangle_A := x^T A y.$$

We claim that

$$f(x) = \frac{1}{2}\|x - \bar{x}\|_A^2 + f(\bar{x}).$$

Indeed, recall that $\bar{x} = A^{-1}b$. Then,

$$f(\bar{x}) - f(x) + \frac{1}{2}\|x - \bar{x}\|_A^2$$

$$= \frac{1}{2}\bar{x}^T A \bar{x} - b^T \bar{x} - \frac{1}{2}x^T A x + b^T x + \frac{1}{2}(x - \bar{x})^T A(x - \bar{x})$$

$$= \frac{1}{2}(\bar{x}^T A \bar{x} - x^T A x + x^T A x - \bar{x}^T A x - x^T A \bar{x} + \bar{x}^T A \bar{x}) - b^T \bar{x} + b^T x$$

$$\bar{x} \stackrel{=A^{-1}b}{=} \frac{1}{2}\left((A^{-1}b)^{\mathrm{T}}b - (A^{-1}b)^{\mathrm{T}}Ax - x^{\mathrm{T}}b + (A^{-1}b)^{\mathrm{T}}b\right) - b^{\mathrm{T}}A^{-1}b + b^{\mathrm{T}}x$$

$$= \frac{1}{2}\left(x^{\mathrm{T}}b - (A^{-1}b)^{\mathrm{T}}Ax\right)$$

$$\stackrel{A^{\mathrm{T}}=A}{=} 0.$$

Note that with this notation, Lemma 5.1 can be written as

$$\alpha_k = \left(\frac{\|r_k\|_2}{\|r_k\|_A}\right)^2.$$

Our goal is to prove the following result.

Theorem 5.2 (Convergence of steepest descent) *For any starting point $x_0 \in \mathbb{R}^n$, the sequence x_k defined by (5.2) converges to the unique minimum $\bar{x}$ and*

$$\|\epsilon_k\|_A \le \left(\frac{\kappa - 1}{\kappa + 1}\right)^k \|\epsilon_0\|_A,$$

where $\kappa = \kappa(A)$ is the condition number of A.

A bound for the number of iterations needed to decrease the A-norm of the error by a given factor immediately follows.

Corollary 5.3 *For all $\varepsilon > 0$, we have $\|\epsilon_t\|_A \le \varepsilon\|\epsilon_0\|_A$ whenever*

$$t \ge \log\left(1 + \frac{2}{\kappa(A) - 1}\right)\log\left(\frac{1}{\varepsilon}\right) \approx \frac{1}{2}\kappa(A)\log\left(\frac{1}{\varepsilon}\right).$$

Remark 5.4 Suppose A has s nonzero entries with $s \ll n^2$ (A is "sparse"). Then one iteration of the method $(x_k, r_k) \mapsto (x_{k+1}, r_{k+1})$,

$$\alpha_k = \frac{r_k^{\mathrm{T}}r_k}{r_k^{\mathrm{T}}Ar_k}, \quad x_{k+1} = x_k + \alpha_k r_k, \quad r_{k+1} = b - Ax_{k+1},$$

costs $\mathcal{O}(s + n)$ arithmetic operations. If, in addition, we are satisfied with an approximate solution for which the bound in Corollary 5.3 is $\mathcal{O}(n)$, then the total complexity—i.e., the total number of arithmetic operations performed to compute this solution—is $\mathcal{O}(n(n + s))$. In this case we might want to use steepest descent instead of Gaussian elimination, which we recall from Sect. 5.1, has a complexity of $\mathcal{O}(n^3)$. In the next two sections we will describe an improvement of steepest

Fig. 5.1 The method of
steepest descent for
$A = \text{diag}(1, 9)$, $b = (0, 0)$,
and $x_0 = (18, 2)$

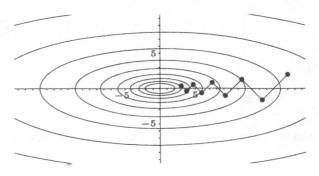

descent, called conjugate gradient, and additional reasons to replace, in a number of
situations, the use of Gaussian elimination with that of conjugate gradient.

Example 5.5 Let $A = \text{diag}(1, 9)$, $b = (0, 0)$, and $x_0 = (18, 2)$. Figure 5.1 shows
the level curves of the function $f(x) = \frac{1}{2}x^{\mathrm{T}}Ax$ for $c \in \{2^k \mid k = -2, -1, \ldots, 6\}$.
Moreover, it depicts the first iterates $x_0, x_1, \ldots, x_8$.

We next prove Theorem 5.2. We begin with a simple lemma.

Lemma 5.6 *We have*

$$\|\epsilon_{k+1}\|_A^2 = \left(1 - \frac{\|r_k\|^4}{\|r_k\|_A^2 \cdot \|r_k\|_{A^{-1}}^2}\right)\|\epsilon_k\|_A^2.$$

Proof By definition,

$$\epsilon_{k+1} = x_{k+1} - \bar{x} = \epsilon_k + \alpha_k r_k,$$

and therefore

$$\begin{aligned}
\|\epsilon_{k+1}\|_A^2 &= (\epsilon_k + \alpha_k r_k)^{\mathrm{T}} A(\epsilon_k + \alpha_k r_k) \\
&= \epsilon_k^{\mathrm{T}} A\epsilon_k + \alpha_k r_k^{\mathrm{T}} A\epsilon_k + \alpha_k \epsilon_k^{\mathrm{T}} Ar_k + \alpha_k^2 r_k^{\mathrm{T}} Ar_k \\
&= \|\epsilon_k\|_A^2 + \alpha_k^2 r_k^{\mathrm{T}} Ar_k + 2\alpha_k r_k^{\mathrm{T}} A\epsilon_k \\
&= \|\epsilon_k\|_A^2 + \alpha_k^2 r_k^{\mathrm{T}} Ar_k - 2\alpha_k r_k^{\mathrm{T}} r_k,
\end{aligned}$$

the last by (5.3). Plugging in the formula for α_k, Lemma 5.1 yields

$$\|\epsilon_{k+1}\|_A^2 - \|\epsilon_k\|_A^2 = \frac{(r_k^{\mathrm{T}} r_k)^2}{r_k^{\mathrm{T}} Ar_k} - \frac{2(r_k^{\mathrm{T}} r_k)^2}{r_k^{\mathrm{T}} Ar_k} = -\frac{(r_k^{\mathrm{T}} r_k)^2}{r_k^{\mathrm{T}} Ar_k} = \frac{\|r_k\|^4}{\|r_k\|_A^2}.$$

The claim follows, using (5.3) again, by noting that

$$\|\epsilon_k\|_A^2 = \epsilon_k^{\mathrm{T}} A\epsilon_k = \left(A^{-1}r_k\right)^{\mathrm{T}} r_k = r_k^{\mathrm{T}} A^{-1} r_k = \|r_k\|_{A^{-1}}^2. \qquad \square$$

Proposition 5.7 (Kantorovich's inequality) *For a positive definite matrix $A \in \mathbb{R}^{n \times n}$ with largest eigenvalue λ_1 and smallest eigenvalue λ_n, we have for any $x \in \mathbb{R}^n$,*

$$\|x\|_A \cdot \|x\|_{A^{-1}} \leq \frac{\lambda_1 + \lambda_n}{2\sqrt{\lambda_1 \lambda_n}} \|x\|^2. \tag{5.4}$$

Proof Without loss of generality, we can assume that $A = \text{diag}(\lambda_1, \ldots, \lambda_n)$ with $\lambda_1 \geq \cdots \geq \lambda_n > 0$ (by an orthogonal transformation) as well as $\|x\| = 1$ (by homogeneity). Then

$$\frac{\|x\|^4}{\|x\|_A^2 \cdot \|x\|_{A^{-1}}^2} = \frac{1}{(\sum_i \lambda_i x_i^2)(\sum_i \lambda_i^{-1} x_i^2)} = \frac{\phi(\sum_i w_i \lambda_i)}{\sum_i w_i \phi(\lambda_i)},$$

where $\phi(t) := \frac{1}{t}$, $w_i := x_i^2$, $\sum_i w_i = 1$. The linear function $L(t) = -\frac{1}{\lambda_1 \lambda_n} t + \frac{\lambda_1 + \lambda_n}{\lambda_1 \lambda_n}$ satisfies $L(\lambda_1) = \frac{1}{\lambda_1}$ and $L(\lambda_n) = \frac{1}{\lambda_n}$. Furthermore, for $\tilde{t} := \sum_i w_i \lambda_i$ we have $\tilde{t} \in [\lambda_n, \lambda_1]$ and $L(\tilde{t}) = \sum_i w_i \phi(\lambda_i)$. Therefore we have

$$\frac{\phi(\sum_i w_i \lambda_i)}{\sum_i w_i \phi(\lambda_i)} = \frac{\phi(\tilde{t})}{L(\tilde{t})} \geq \min_{\lambda_1 \geq t \geq \lambda_n} \frac{\phi(t)}{L(t)}.$$

The minimum is achieved at $t = \frac{\lambda_1 + \lambda_n}{2}$, and has the value $\frac{4\lambda_1 \lambda_n}{(\lambda_1 + \lambda_n)^2}$. So we get

$$\|x\|_A^2 \cdot \|x\|_{A^{-1}}^2 = \left(\frac{\phi(\sum_i w_i \lambda_i)}{\sum_i w_i \phi(\lambda_i)} \right)^{-1} \leq \frac{(\lambda_1 + \lambda_n)^2 / 4}{\lambda_1 \lambda_n}. \qquad \square$$

Proof of Theorem 5.2 Combining inequality (5.4) with Lemma 5.6 and using that $\kappa(A) = \frac{\lambda_1}{\lambda_n}$, we get

$$\|\epsilon_{k+1}\|_A^2 \leq \left(1 - \frac{\|r_k\|^4}{\|r_k\|_A^2 \cdot \|r_k\|_{A^{-1}}^2} \right) \|\epsilon_k\|_A^2 \leq \left(1 - \frac{4\lambda_1 \lambda_n}{(\lambda_1 + \lambda_n)^2} \right) \|\epsilon_k\|_A^2$$

$$= \frac{(\lambda_1 - \lambda_n)^2}{(\lambda_1 + \lambda_n)^2} \|\epsilon_k\|_A^2 = \left(\frac{\kappa - 1}{\kappa + 1} \right)^2 \|\epsilon_k\|_A^2,$$

which implies the theorem. $\qquad \square$

5.3   The Method of Conjugate Gradients

The method of conjugate gradients can be seen as an improvement of the method of steepest descent in the sense that the convergence is much faster, with the number of arithmetic operations per iteration step being roughly the same. As in the previous section, $A \in \mathbb{R}^{n \times n}$ is positive definite. The function f, the error e, the residual r, and the A-inner product $\langle \, , \, \rangle_A$ are similarly defined.

We say that vectors x, y are A-*orthogonal* (or *conjugated*) if $\langle x, y \rangle_A = 0$. Let $(d_0, \ldots, d_{n-1})$ be an A-orthogonal basis of $\mathbb{R}^n$, i.e., $\langle d_i, d_j \rangle_A = 0$ for $i \neq j$. Moreover, let $\bar{x}$ be the solution of $Ax = b$. Let $x_0 \in \mathbb{R}^n$ be any starting point. Expand

$$\bar{x} - x_0 = \sum_{i=0}^{n-1} \alpha_i d_i,$$

with $\alpha_i \in \mathbb{R}$. Then by A-orthogonality,

$$\alpha_k = \frac{\langle \bar{x} - x_0, d_k \rangle_A}{\|d_k\|_A^2}.$$

Define

$$x_k := \sum_{i=0}^{k-1} \alpha_i d_i + x_0.$$

Then we have $\langle x_k - x_0, d_\ell \rangle_A = 0$, for all $\ell = k, \ldots, n-1$. In particular, taking $\ell = k$,

$$\langle \bar{x} - x_0, d_k \rangle_A = \langle \bar{x} - x_k, d_k \rangle_A = d_k^{\mathrm{T}} A(\bar{x} - x_k) = d_k^{\mathrm{T}} r_k,$$

with $r_k := b - Ax_k = A\bar{x} - Ax_k$. We obtain that

$$\alpha_k = \frac{\langle d_k, r_k \rangle}{\|d_k\|_A^2}.$$

Note that α_k depends only on d_k and r_k. We have proved the following.

Lemma 5.8 *Let* $(d_0, \ldots, d_{n-1})$ *be an* A-*orthogonal basis and* $x_0 \in \mathbb{R}^n$. *Define* $x_1, \ldots, x_n \in \mathbb{R}^n$ *by*

$$x_{k+1} = x_k + \alpha_k d_k$$

with

$$\alpha_k = \frac{\langle d_k, r_k \rangle}{\|d_k\|_A^2}, \quad r_k = b - Ax_k.$$

Then $x_n = \bar{x}$. $\square$

The following insight is crucial. Recall that

$$f(x) = \frac{1}{2} x^{\mathrm{T}} Ax - b^{\mathrm{T}} x = \frac{1}{2} \|x - \bar{x}\|_A^2 + f(\bar{x}).$$

Proposition 5.9

(a) *For $k \geq 1$, x_k minimizes the function $x \mapsto \|x - \bar{x}\|_A$ on the line $x_{k-1} + \mathbb{R}d_{k-1}$ as well as on the affine space $x_0 + \mathsf{span}\{d_0, \ldots, d_{k-1}\}$.*
(b) *We have $\langle r_k, d_i \rangle = 0$ for $i < k$.*

Proof For part (a) note that the point $x_k - x_0$ is the A-orthogonal projection of $\bar{x} - x_0$ onto $\mathsf{span}\{d_0, \ldots, d_{k-1}\}$. Therefore, $x_k - x_0$ minimizes the A-distance to $\bar{x}$,

$$x \mapsto \|x - \bar{x}\|_A = \|(x - x_0) - (\bar{x} - x_0)\|_A,$$

on $x_0 + \mathsf{span}\{d_0, \ldots, d_{k-1}\}$.

By part (a), x_k minimizes f on $x_0 + \mathsf{span}\{d_0, \ldots, d_{k-1}\}$. Hence, $\mathsf{grad}\, f(x_k)$ is orthogonal to $\mathsf{span}\{d_0, \ldots, d_{k-1}\}$. But $\mathsf{grad}\, f(x_k) = -r_k$ by (5.3). This proves (b). $\square$

So far we have assumed that we are already given an A-orthogonal basis (d_i). We next show how one actually computes such a basis. The idea is to sequentially choose the directions d_k as conjugates of the gradients $-r_k$ as the method progresses. It turns out that this can be achieved with little cost. The following example should illustrate this idea.

Example 5.10 Let us start with $d_0 := r_0 := b - Ax_0 \neq 0$. Then we get $\alpha_0 := \frac{\langle d_0, r_0 \rangle}{\|d_0\|_A^2} = \frac{\|d_0\|^2}{\|d_0\|_A^2}$. Setting $x_1 := x_0 + \alpha_0 d_0$ and $r_1 := b - Ax_1 \neq 0$ (otherwise $x_1 = \bar{x}$ and we are done), we get from Proposition 5.9(b) $\langle r_1, d_0 \rangle = 0$. Now take $d_1 := r_1 + \beta_0 d_0$. The requirement $0 = \langle d_1, d_0 \rangle_A = \langle r_1, d_0 \rangle_A + \beta_0 \langle d, d_0 \rangle_A$ implies $\beta_0 = -\frac{\langle r_1, d_0 \rangle_A}{\|d_0\|_A^2}$, which can be used as a definition for β_0. In this way we get the second basis vector d_1.

The extension of this example gives us the (full) conjugate gradient algorithm (Algorithm 5.1 below).

Remark 5.11 Before proceeding with the analysis of Algorithm 5.1 (mostly, with the analysis of the number of iterations needed to reach a given accuracy, see Theorem 5.13 and Corollary 5.14 below) we can have a look at the cost of each iteration of the algorithm.

Note that the cost of computing an inner product, such as $\langle r_k, d_k \rangle$, is $2n - 1$. Consequently, the cost of a matrix–vector multiplication, such as Ax_k, is $2n^2 - n$. It follows that computing an A-inner product costs $2n^2 + n - 1$. At each iteration of Conj_Grad the computation of each of x_{k+1} and d_{k+1} thus takes $\mathcal{O}(n)$ arithmetic operations, and those of α_k, r_{k+1}, and β_k take $2n^2 + \mathcal{O}(n)$ each (note that for the latter we use the already computed $\|d_k\|_A^2$). That is, the cost of an iteration of Conj_Grad is $6n^2 + \mathcal{O}(n)$.

Algorithm 5.1 Conj_Grad

Input: $A \in \mathbb{R}^{n \times n}, b \in \mathbb{R}^n, x_0 \in \mathbb{R}^n$

Preconditions: A is positive definite

```
r_0 := d_0 := b − Ax_0
k := 0
while d_k ≠ 0 do
```
$$\alpha_k := \frac{\langle r_k, d_k \rangle}{\|d_k\|_A^2}$$
$$x_{k+1} := x_k + \alpha_k d_k$$
$$r_{k+1} := b − Ax_{k+1}$$
$$\beta_k := -\frac{\langle r_{k+1}, d_k \rangle_A}{\|d_k\|_A^2}$$
$$d_{k+1} := r_{k+1} + \beta_k d_k$$
$$k := k + 1$$
```
end while
return x_k
```

Output: $x \in \mathbb{R}^n$

Postconditions: $Ax = b$

Theorem 5.12 *Let $\bar{k}$ be the last k such that $d_k \neq 0$. Then, for all $k = 0, \dots, \bar{k}$:*

(a) *(Krylov spaces)*

$$\mathsf{span}\{d_0, \dots, d_k\} = \mathsf{span}\{r_0, \dots, r_k\} = \mathsf{span}\{r_0, Ar_0, \dots, A^k r_0\}.$$

(b) *Algorithm* Conj_Grad *produces a sequence $(d_0, d_1, \dots, d_k)$ of A-orthogonal vectors.*

Proof The proof goes by induction on k. The start $k = 0$ is clear. We go from k to $k + 1$. Define

$$D_k := \mathsf{span}\{d_0, \dots, d_k\}, \qquad R_k := \mathsf{span}\{r_0, \dots, r_k\},$$

$$S_k := \mathsf{span}\{r_0, Ar_0, \dots, A^k r_0\}.$$

Then, by the induction hypothesis, $D_k = R_k = S_k$. The equality $D_{k+1} = R_{k+1}$ is trivial. To see that $R_{k+1} = S_{k+1}$ we note that

$$r_{k+1} = -Ax_{k+1} + b = -A(x_k + \alpha_k d_k) + b = r_k - \alpha_k A d_k;$$

hence $r_{k+1} \in R_k + A(D_k) = S_k + A(S_k) \subseteq S_{k+1}$.

For the reverse inclusion suppose $r_{k+1} \neq 0$. According to Proposition 5.9(b), r_{k+1} is orthogonal to S_k. Hence $r_{k+1} \notin S_k$. We obtain

$$\mathsf{span}\{r_0, \dots, r_k, r_{k+1}\} = S_{k+1}.$$

This proves part (a).

For part (b) it remains to prove that d_{k+1} is A-orthogonal to $d_1, \ldots, d_k$. We have $\langle d_{k+1}, d_k \rangle_A$ by the choice of β_k. Furthermore, for $i < k$,

$$\langle d_{k+1}, d_i \rangle_A = \langle r_{k+1}, d_i \rangle_A + \beta_k \langle d_k, d_i \rangle_A = \langle r_{k+1}, d_i \rangle_A = r_{k+1}^T A d_i.$$

Now $A d_i \in A(S_i) \subseteq S_{i+1} \subseteq S_k$. Therefore

$$r_{k+1}^T A d_i = \langle r_{k+1}, A d_i \rangle = 0,$$

since r_{k+1} is orthogonal to S_k by Proposition 5.9(b).   □

We turn now to the analysis of convergence. The main result in this section is the following.

Theorem 5.13 *The error at the kth step of the conjugate gradient method satisfies*

$$\|\epsilon_k\|_A \le 2 \left(\frac{\sqrt{\kappa(A)} - 1}{\sqrt{\kappa(A)} + 1} \right)^k \cdot \|\epsilon_0\|_A.$$

Corollary 5.14 *For all $\delta > 0$ we have $\|\epsilon_k\|_A \le \delta \|\epsilon_0\|_A$ whenever*

$$k \ge \ln \left(\frac{\sqrt{\kappa(A)} + 1}{\sqrt{\kappa(A)} - 1} \right)^{-1} \ln \left(\frac{2}{\delta} \right) \approx \frac{1}{2} \sqrt{\kappa(A)} \ln \left(\frac{1}{\delta} \right).$$

Each iteration step takes $\mathcal{O}(n + s)$ arithmetic operations if A has s nonzero entries.   □

Remark 5.15 The $6n^2 + \mathcal{O}(n)$ cost of each iteration of Algorithm Conj_Grad together with the convergence rate in Corollary 5.14 suggests that for reasonable $\delta > 0$ and large n, computing an approximation of the solution x of $Ax = b$ using Algorithm Conj_Grad may be faster than computing x with, say, Gaussian elimination. We will return to this question in Sect. 5.4 below.

Towards the proof of Theorem 5.13 we introduce some notation. We denote by $\mathcal{P}_k$ the linear space of all real polynomials in one variable X with degree at most k. We also write $\mathcal{Q}_k$ for the subset of $\mathcal{P}_k$ of polynomials with constant coefficient 1.

Theorem 5.16 *The error ϵ_k at the kth step of the conjugate gradient method satisfies*

$$\|\epsilon_k\|_A^2 \le \min_{q \in \mathcal{Q}_k} \max_{j \le n} q(\lambda_j)^2 \cdot \|\epsilon_0\|_A^2,$$

where $\lambda_1, \ldots, \lambda_n$ are the eigenvalues of A.

Proof By Proposition 5.9(a) we know that x_k minimizes the A-distance of $\bar{x}$ to the affine space

$$x_0 + S_{k-1} = x_0 + \mathsf{span}\{d_0, \ldots, d_{k-1}\} = x_0 + \mathsf{span}\{r_0, A r_0, \ldots, A^{k-1} r_0\}.$$

An element x of $x_0 + S_{k-1}$ can therefore be written as

$$x = x_0 - p(A)r_0,$$

with $p \in \mathcal{P}_{k-1}$, and conversely, for any such polynomial we obtain $x \in x_0 + S_{k-1}$. Using $r_0 = b - Ax_0 = A(\bar{x} - x_0)$ we get

$$x - \bar{x} = x_0 - \bar{x} - p(A)A(\bar{x} - x_0) = x_0 - \bar{x} + Ap(A)(x_0 - \bar{x})$$

$$= \big(I + Ap(A)\big)(x_0 - \bar{x}).$$

It follows that the error $\epsilon_k = x_k - \bar{x}$ at the kth step of the conjugate gradient method satisfies

$$\|\epsilon_k\|_A = \min_{x \in x_0 + S_{k-1}} \|x - \bar{x}\|_A = \min_{p \in \mathcal{P}_{k-1}} \big\|\big(I + Ap(A)\big)\epsilon_0\big\|_A. \tag{5.5}$$

Suppose that $v_1, \ldots, v_n$ is an orthonormal basis of eigenvectors of A corresponding to the eigenvalues $\lambda_1 \geq \cdots \geq \lambda_n > 0$. Write

$$\epsilon_0 = x_0 - \bar{x} = \sum_{j=1}^{n} \xi_j v_j$$

for some $\xi_1, \ldots, \xi_n \in \mathbb{R}$. Then

$$\|\epsilon_0\|_A^2 = \sum_{j=1}^{n} \sum_{k=1}^{n} \xi_j \xi_k \, v_j^T A v_k = \sum_{j=1}^{n} \lambda_j \xi_j^2.$$

Moreover, for any polynomial $p \in \mathcal{P}_{k-1}$,

$$\big(I + Ap(A)\big)\epsilon_0 = \sum_{j=1}^{n} \big(1 + \lambda_j p(\lambda_j)\big)\xi_j v_j.$$

Therefore

$$\big\|\big(I + Ap(A)\big)\epsilon_0\big\|_A^2 = \sum_{j=1}^{n}\big(1 + \lambda_j p(\lambda_j)\big)^2 \lambda_j \xi_j^2 \leq \max_{j \leq n}\big(1 + \lambda_j p(\lambda_j)\big)^2 \sum_{j=1}^{n} \lambda_j \xi_j^2,$$

and using (5.5),

$$\|\epsilon_k\|_A^2 \leq \min_{p \in \mathcal{P}_{k-1}} \max_{j \leq n}\big(1 + \lambda_j p(\lambda_j)\big)^2 \sum_{j=1}^{n} \lambda_j \xi_j^2.$$

The result now follows by observing that $\mathcal{Q}_k = 1 + X\mathcal{P}_k$.   □

Theorem 5.16 is hard to apply in concrete situations. It depends on all the eigenvalues of A and it optimizes a function of them over the space $\mathcal{Q}_k$. It is nevertheless a building block in the proof of Theorem 5.13. We proceed to see why.

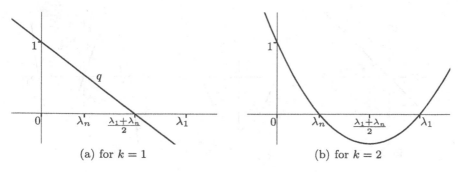

(a) for $k = 1$ (b) for $k = 2$

Fig. 5.2 The optimal choice of q

Example 5.17 For $k = 1$ it is optimal to choose $q(\lambda) = 1 - \frac{2}{\lambda_1 + \lambda_n}\lambda$ (see Fig. 5.2) and hence

$$\|\epsilon_1\|_A^2 \le \left(\frac{\kappa(A) - 1}{\kappa(A) + 1}\right)^2 \|\epsilon_0\|_A^2.$$

The first step of Algorithm 5.1 is just one step of steepest descent. Thus we have re-proved Theorem 5.2. See Fig. 5.2 for an optimal choice of q for $k = 2$.

Remark 5.18 Suppose that A has only $m < n$ distinct eigenvalues. Then there exists $q \in \mathcal{Q}_m$ such that q vanishes on all the eigenvalues. Hence $\epsilon_m = 0$ and the method terminates after m steps. (This can also be easily seen directly.) So multiple eigenvalues decrease the number of steps of Algorithm Conj_Grad.

Suppose that $\lambda_n = a$ and $\lambda_1 = b$, for $0 < a < b$. What are optimal polynomials $q \in \mathcal{Q}_k$ if nothing is known about the location of the eigenvalues λ except that $\lambda \in [a, b]$? In this case we have to minimize the quantity

$$\max_{a \le \lambda \le b} q(\lambda)^2$$

over all real polynomials $q \in \mathcal{Q}_k$. This minimization problem can be considered as well for arbitrary $a, b \in \mathbb{R}$ with $a < b$. In the particular case $a = -1$, $b = 1$ it turns out that its solution is given by the *Chebyshev polynomials* T_k defined by

$$\cos(k\phi) = T_k\big(\cos(\phi)\big).$$

For instance

$$T_0(X) = 1, \qquad T_1(X) = X, \qquad T_2(X) = 2X^2 - 1, \qquad T_3(X) = 4X^3 - 3X,$$

and more generally, for $i \ge 2$,

$$T_i(X) = 2X\, T_{i-1}(X) - T_{i-2}(X).$$

See Fig. 5.3 for a display of some of these polynomials.

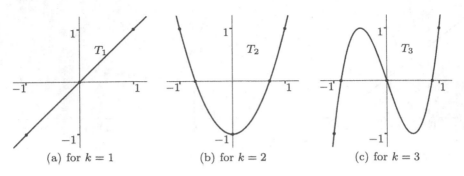

(a) for $k = 1$ (b) for $k = 2$ (c) for $k = 3$

Fig. 5.3 The Chebyshev polynomial T_k

The next proposition lists some important properties of Chebyshev polynomials T_k on $[-1, 1]$. Its easy proof is left to the reader.

Proposition 5.19

(a) $|T_k(x)| \leq 1$ *for* $x \in [-1, 1]$,

(b) T_k *has degree* k,

(c) T_k *has exactly* k *roots in* $[-1, 1]$, *namely* $\cos((j + \frac{1}{2})\frac{\pi}{k})$ *for* $j \in [k]$. $\square$

Proposition 5.20 *For all* $0 < a < b$ *we have*

$$\min_{\substack{\deg q \leq k \\ q(0)=1}} \max_{a \leq \lambda \leq b} q(\lambda)^2 = \frac{1}{T_k(\frac{b+a}{b-a})^2}.$$

Proof The polynomial

$$p(\lambda) := \frac{T_k(\frac{b+a-2\lambda}{b-a})}{T_k(\frac{b+a}{b-a})}$$

has degree k, satisfies $p(0) = 1$, and

$$\max_{a \leq \lambda \leq b} p(\lambda)^2 = \frac{1}{T_k(\frac{b+a}{b-a})^2} =: c^2.$$

(Note that $\lambda \mapsto \frac{b+a-2\lambda}{b-a}$ maps $[a, b]$ onto $[-1, 1]$ and T_k has maximal value 1 on this interval.)

In order to show the optimality of p, suppose that for some polynomial $q \in \mathcal{Q}_k$,

$$\max_{a \leq \lambda \leq b} |q(\lambda)| < \max_{a \leq \lambda \leq b} |p(\lambda)|.$$

Since p alternately takes the values $c, -c, c, -c, \ldots$ exactly $k+1$ times in the interval $[a, b]$, we conclude that the polynomial $p - q$ has at least k roots in this interval (intermediate value theorem). Moreover, $p(0) - q(0) = 1 - 1 = 0$. Hence $p - q$ has degree at most k and $k + 1$ distinct roots. It follows that $p = q$. $\square$

Proof of Theorem 5.13 To estimate the optimal value in Proposition 5.20, note that for $z \in \mathbb{C}$, $z = \cos(\phi) + i \sin(\phi) = x + i\sqrt{1 - x^2}$, we have

$$T_k(x) = \mathrm{Re}(z^k) = \frac{1}{2}(z^k + \bar{z}^k) = \frac{1}{2}\left((x + \sqrt{x^2 - 1})^k + (x - \sqrt{x^2 - 1})^k\right).$$

Now take again $a = \lambda_n$, $b = \lambda_1$, the smallest and largest eigenvalues of A. Then $\kappa = \kappa(A) = \frac{b}{a}$. For $x = \frac{b+a}{b-a} = \frac{\kappa+1}{\kappa-1}$ we get

$$x^2 - 1 = \frac{(\kappa + 1)^2 - (\kappa - 1)^2}{(\kappa - 1)^2} = \frac{4\kappa}{(\kappa - 1)^2}$$

and therefore

$$x + \sqrt{x^2 - 1} = \frac{\kappa + 1 + 2\sqrt{\kappa}}{\kappa - 1} = \frac{(\sqrt{\kappa} + 1)^2}{\kappa - 1} = \frac{\sqrt{\kappa} + 1}{\sqrt{\kappa} - 1}.$$

Hence

$$T_k(x) \geq \frac{1}{2}\left(\frac{\sqrt{\kappa} + 1}{\sqrt{\kappa} - 1}\right)^k.$$ $\square$

We finish this section by noting that the estimate in Theorem 5.13 may be pessimistic in certain situations. Suppose that the matrix A has only m large eigenvalues, while the remaining ones are relatively close to the smallest. The following consequence of Theorem 5.16 shows that one can avoid the bad effect of the m largest eigenvalues by performing only m steps of the conjugate gradient method.

Proposition 5.21 *Suppose the positive definite matrix $A \in \mathbb{R}^{n \times n}$ has $n - m + 1$ eigenvalues in the interval $[a, b']$ and the remaining $m - 1$ eigenvalues are greater than b'. Let $x_0 \in \mathbb{R}^n$ be any starting point. Then*

$$\|\epsilon_m\|_A \leq \frac{b' - a}{b' + a}\|\epsilon_0\|_A.$$

Proof Let $\lambda_1 \geq \lambda_2 \geq \cdots \geq \lambda_{m-1}$ be the $m - 1$ largest eigenvalues of A. Let q be of degree m such that

$$q(\lambda_1) = \cdots = q(\lambda_{m-1}) = 0, \qquad q(0) = 1, \qquad q\left(\frac{b' + a}{2}\right) = 0.$$

Then by Theorem 5.16,

$$\|\epsilon_m\|_A \leq \max_{a \leq \lambda \leq b'} |q(\lambda)| \cdot \|\epsilon_0\|_A.$$

It is not hard to see that

$$q(\lambda) \begin{cases} \leq 1 - \frac{2\lambda}{a+b'} & \text{if } 0 \leq \lambda \leq \frac{a+b'}{2}, \\ \geq 1 - \frac{2\lambda}{a+b'} & \text{if } \frac{a+b'}{2} \leq \lambda \leq b'. \end{cases}$$

Hence, for $\lambda \in [a, b']$, we have

$$\left| q(\lambda) \right| \leq \left| 1 - \frac{2\lambda}{a+b'} \right| \leq \frac{b'-a}{b'+a},$$

which proves the assertion.   □

5.4 Conjugate Gradient on Random Data

We close this chapter by returning to one of our recurring themes: the randomization of data as a way to obtain complexity bounds independent of the condition number.

We have noted in the introduction of this chapter that in many situations, the positive definite matrix $A \in \mathbb{R}^{n \times n}$ given as input to Algorithm 5.1 (Conj_Grad) is obtained as $A = R^{\mathrm{T}}R$ with $R \in \mathbb{R}^{q \times n}$, $q > n$. By Theorem 1.13, there exist orthogonal matrices U and V and positive reals $\sigma_1 \geq \sigma_2 \geq \cdots \geq \sigma_n > 0$ such that $R = U \operatorname{diag}(\sigma_1, \sigma_2, \ldots, \sigma_n)V^{\mathrm{T}}$. Then

$$A = R^{\mathrm{T}}R = V \operatorname{diag}(\sigma_1^2, \sigma_2^2, \ldots, \sigma_n^2)V^{\mathrm{T}}.$$

It follows from this equality that $\kappa(A) = \kappa^2(R)$. Therefore, the analysis of $\kappa(A)$ for this inherited distribution reduces to the analysis of $\kappa(R)$ when R is Gaussian.

In the particular case that R is standard Gaussian, this inherited distribution of A is known as the *Wishart distribution with q degrees of freedom*. It extends the chi-square distribution, since it coincides with a χ_q^2 when $n = 1$.

Corollary 5.14 shows that the number k of iterations that are needed to decrease $\|x_0 - \bar{x}\|_A$ by a factor of ε is proportional to $\sqrt{\kappa(A)}$, that is, proportional to $\kappa(R)$. We are therefore interested in this quantity for Gaussian rectangular matrices R, a theme we have dealt with in Chap. 4. Indeed, in Theorem 4.2 we showed that for all $\lambda_0 \in (0, 1)$ and all $0 < \sigma \leq 1$ there exists q_0 such that for all $1 \leq n \leq q$ we have

$$\sup_{\|\overline{R}\| \leq 1} \mathop{\mathbb{E}}_{R \sim N(\overline{R}, \sigma^2 I)} \big(\kappa(R) \big) \leq \frac{20.1}{1 - \lambda},$$

provided $\lambda = \frac{n-1}{q} \leq \lambda_0$ and $q \geq q_0$.

It follows that if A is obtained as $R^{\mathrm{T}}R$ for a large elongated rectangular matrix R, then we should expect to effect the decrease mentioned above with about $\frac{1}{2}\frac{20.1}{1-\lambda}\ln(\frac{1}{\varepsilon})$ iterations (where $\lambda = \frac{q}{n}$ is the elongation). Since each iteration takes $6n^2 + \mathcal{O}(n)$

arithmetic operations (cf. Remark 5.11), the expected cost is

$$3n^2 \frac{20.1}{1-\lambda} \ln \frac{1}{\varepsilon} + \mathcal{O}(n) = \frac{60.3n^2}{1-\lambda} \ln \frac{1}{\varepsilon} + \mathcal{O}(n).$$

The leading term in this expression is smaller than the $\frac{2}{3}n^3$ operations performed by Gaussian elimination as long as

$$\varepsilon \geq e^{-\frac{n(1-\lambda)}{91}}.$$

For large n (and λ not too close to 1) this bound produces very small values of ε, and therefore, Conj_Grad yields, on average (both for a Gaussian distribution of data and for Gaussian perturbations of arbitrary data), remarkably good approximations of $\bar{x}$.

Intermezzo I: Condition of Structured Data

The themes of Chaps. 3 and 5 introduced, *sotto voce*, the issue of structured data. In both cases we had a general set of data, the space $\mathbb{R}^{n \times n}$ of $n \times n$ real matrices, and a subset S whose elements are the valid inputs of a given algorithm: triangular matrices for FS and symmetric positive definite matrices for CGA.

It is apparent that the analysis pattern we have developed till now—an analysis of the relevant measure of performance for the considered algorithm (loss of precision or running time) in terms of a condition number, followed by a probabilistic analysis of the latter—needs to be adjusted. For the probabilistic analysis, the underlying measure will have to be chosen with support in S. We have already done so in Chap. 3, by drawing from $N(0, 1)$ only the matrix entries that are not fixed to be zero, as well as in Chap. 5, where the more elaborated family of Wishart distributions was imposed on the set of symmetric positive definite matrices.

As for the object of analysis itself, the condition number, its actual shape will have to depend on the situation at hand. Yet, even though there is no standard way to "structure" a condition number, a couple of ways occur frequently enough to be described in detail.

(a) **Structured perturbations**. When the analysis is based on data perturbations (e.g., in accuracy analyses), it is often the case that the only admissible perturbations are those respecting the structure of the data a, that is, those for which $\tilde{a} \in S$ as well. This naturally leads to the following "structuring" of (O.1):

$$\operatorname{cond}_S^\varphi(a) := \lim_{\delta \to 0} \sup_{\substack{\operatorname{RelError}(a) \leq \delta \\ \tilde{a} \in S}} \frac{\operatorname{RelError}(\varphi(a))}{\operatorname{RelError}(a)}. \tag{I.1}$$

In the case of triangular linear systems, the backward analysis of algorithm FS in Sect. 3.2 produced componentwise perturbation bounds that automatically force the perturbed matrix $\tilde{L}$ to be lower triangular as well. But this need not be the case.

(b) **Distance to structured ill-posedness**. We will soon see (in Chap. 6, after this intermezzo) that for a large class of problems (those having a discrete set of

P. Bürgisser, F. Cucker, *Condition*,
Grundlehren der mathematischen Wissenschaften 349,
DOI 10.1007/978-3-642-38896-5, © Springer-Verlag Berlin Heidelberg 2013

values, notably the decisional problems), the notion of condition given by (O.1) is inadequate and that a common, appropriate replacement is given by taking

$$\mathscr{Q}(a) := \frac{\|a\|}{d(a, \Sigma)}$$

for the condition number of a. Here Σ is a natural set of ill-posed data. It is therefore not surprising that in many of the situations in which such a condition number is considered and data are restricted to some subset $\mathcal{S}$, the useful way to structure $\mathscr{Q}(a)$ is by taking

$$\mathscr{Q}_{\mathcal{S}}(a) := \frac{\|a\|}{d(a, \Sigma \cap \mathcal{S})}. \tag{I.2}$$

The difference between $\mathscr{Q}$ and $\mathscr{Q}_{\mathcal{S}}$ can be large. A case at hand is that of triangular matrices. For any such matrix L, the condition number theorem (Theorem 1.7) shows that $d(L, \Sigma) = \|L^{-1}\|^{-1}$ and therefore $\mathscr{Q}(L) = \kappa(L)$. Theorem 3.1 then shows that $\mathbb{E} \log \mathscr{Q}(L) = \Omega(n)$. In contrast, we will see in Sect. 21.7 that $\mathbb{E} \log \mathscr{Q}_{\mathsf{Triang}}(L) = \mathcal{O}(\log n)$.

Occasionally, there is no need for a structuring of the condition number. This was the case, for instance, in the complexity analysis of the conjugate gradient method in Chap. 5. This analysis revealed a dependence of the number of iterations of Conj_Grad on the standard condition number $\kappa(A)$ of the input matrix A; the only influence of this matrix being symmetric positive definite was on the underlying distribution in the probabilistic analysis.

Part II
Condition in Linear Optimization
(*Andante*)

Chapter 6
A Condition Number for Polyhedral Conic Systems

The second part of this book is essentially a self-contained course on linear programming. Unlike the vast majority of expositions of this subject, our account is "condition-based." It emphasizes the numerical aspects of linear programming and derives probabilistic (average and smoothed) analyses of the relevant algorithms by reducing the object of these analyses from the algorithm to the condition number of the underlying problem.

In contrast to the exposition of the previous chapters, in this second part of the book we will use conditioning mainly for complexity analyses. It won't be until Sect. 9.5 that we discuss finite-precision analysis.

In this chapter we begin the development of our course. We do so based on a particular problem, the feasibility of polyhedral conic systems. Briefly stated, the feasibility problem we consider is whether a polyhedral cone given by homogeneous linear inequalities is nontrivial (i.e., has a point other than the coordinate origin). A goal of Sect. 6.1 below is to see that for this problem (as well as for numerous others), the notion of conditioning as defined in the Overture does not help in any analysis. An idea pioneered by Renegar is, in these situations, to define conditioning in terms of distance to ill-posedness. The main character in this chapter, the condition number $\mathscr{C}(A)$—here A is the matrix stipulating the linear inequalities—is defined in these terms. As the chapter evolves, we will see that it can, in addition, be characterized in a number of different ways. The last section of the chapter shows that $\mathscr{C}(A)$ is a natural parameter in the analysis of some classical simple algorithms to find points in feasible cones. In subsequent chapters, it will feature in the analysis of more sophisticated algorithms. The characterizations we just mentioned will turn out to be helpful in these analyses.

6.1 Condition and Continuity

Consider the problem φ that maps any pair (b, c) of real numbers to the number of real roots of the polynomial $f = X^2 + bX + c$. Since the possible values for this

P. Bürgisser, F. Cucker, *Condition*,
Grundlehren der mathematischen Wissenschaften 349,
DOI 10.1007/978-3-642-38896-5_6, © Springer-Verlag Berlin Heidelberg 2013

problem are the elements in $\{0, 1, 2\}$, the set of inputs is partitioned as $\mathcal{D}_0 \cup \mathcal{D}_1 \cup \mathcal{D}_2$ with $\mathcal{D}_i = \{(b, c) \in \mathbb{R}^2 \mid \varphi(b, c) = i\}$. We know that

$$\mathcal{D}_2 = \{(b, c) \in \mathbb{R}^2 \mid b^2 > 4c\},$$

$$\mathcal{D}_1 = \{(b, c) \in \mathbb{R}^2 \mid b^2 = 4c\},$$

$$\mathcal{D}_0 = \{(b, c) \in \mathbb{R}^2 \mid b^2 < 4c\},$$

so that $\dim(\mathcal{D}_2) = \dim(\mathcal{D}_0) = 2$ and $\dim(\mathcal{D}_1) = 1$. Actually, the boundaries $\partial \mathcal{D}_2$ and $\partial \mathcal{D}_0$ are the same and coincide with the parabola $\mathcal{V}_1$.

What is the, say normwise, condition number for this problem? If $(b, c) \in \mathcal{D}_2$, then all sufficiently small perturbations $(\tilde{b}, \tilde{c})$ of (b, c) will also be in $\mathcal{D}_2$. Hence, for these perturbations $\mathsf{RelError}(\varphi(b, c)) = 0$, and therefore we have $\mathsf{cond}(b, c) = 0$. A similar argument yields the same equality when $(b, c) \in \mathcal{D}_0$. In contrast, when $(b, c) \in \mathcal{D}_1$, one can find arbitrarily small perturbations $(\tilde{b}, \tilde{c})$ in $\mathcal{D}_2$ as well as arbitrarily small perturbations in $\mathcal{D}_0$. Therefore, for these perturbations, the quotient $\frac{\mathsf{RelError}(\varphi(b, c))}{\mathsf{RelError}(b, c)}$ can be arbitrarily large, and it follows that $\mathsf{cond}(b, c) = \infty$ when $(b, c) \in \mathcal{D}_1$. In summary,

$$\mathsf{cond}(b, c) = \begin{cases} 0 & \text{if} (b, c) \in \mathcal{D}_0 \cup \mathcal{D}_2, \\ \infty & \text{if} (b, c) \in \mathcal{D}_1. \end{cases}$$

No matter whether for complexity or for finite-precision analysis, it is apparent that $\mathsf{cond}(b, c)$ cannot be of any relevance.

The problem considered above has no computational mysteries. We have chosen it simply for illustration purposes. The discussion above will nevertheless carry over to any *discrete-valued problem* (one with values in a discrete set) and, with the appropriate modifications, to any *decision problem* (one with values in $\{\mathsf{Yes}, \mathsf{No}\}$). For these problems a different development is needed.

Firstly, a different format for finite-precision analysis appears to be a must, the one discussed in the Overture making no sense in this context. The relevant question is no longer how many correct significant figures are lost in the computation but rather how many we need to start with (i.e., how small should ϵ_{mach} be) to ensure a correct output.

Secondly, a different way of measuring condition, appropriate for the goal just described, should be devised. One also expects such a measure to be of use in complexity analyses.

It won't be until Sect. 9.5 that we will deal with the first issue above. We can, in contrast, briefly tackle the second one now. To do so, assume we have a decision problem. At the boundary where the output of the problem changes—i.e., the boundary between the sets of data with output Yes and No—the usual condition is infinity: arbitrarily small perturbations may change this output from any of these values to the other. This boundary is therefore the set Σ of data that are ill-posed for the problem (recall Sect. O.5.4), and Renegar's idea is to define the condition of *a* as the (normalized, if appropriate) inverse of the distance $d(a, \Sigma)$ to ill-posedness, that is, in the normalized case, to take as condition number $C(a)$ of a the following:

$$C(a) := \frac{\|a\|}{d(a, \Sigma)}.$$

In other words, we do not prove a condition number theorem for $C(A)$: we impose it.

This idea extends straightforwardly to discrete-valued problems and will appear systematically in this second part and, more sporadically, in the third and last part of this book.

6.2 Basic Facts on Convexity

We explain here the basic notions related to convexity and recall some of the main fundamental results in this context.

6.2.1 Convex Sets

A subset $K \subseteq \mathbb{R}^m$ is called *convex* when

$$\forall x, y \in K \; \forall t \in [0, 1], \quad tx + (1-t)y \in K.$$

That is, K contains the line segment with endpoints x, y for all $x, y \in K$. The *convex hull* of a set of points $a_1, \dots, a_n \in \mathbb{R}^m$ is defined as

$$\mathrm{conv}\{a_1, \dots, a_n\} := \left\{ \sum_{i=1}^{n} t_i a_i \; \middle| \; t_1, \dots, t_n \geq 0, \sum_{i=1}^{n} t_i = 1 \right\}.$$

This is easily seen to be closed and the smallest convex set containing $a_1, \dots, a_n$. The *affine hull* of $a_1, \dots, a_n$ is defined as

$$\mathrm{aff}\{a_1, \dots, a_n\} := \left\{ \sum_{i=1}^{n} t_i a_i \; \middle| \; t_1, \dots, t_n \in \mathbb{R}, \sum_{i=1}^{n} t_i = 1 \right\}.$$

This is the smallest affine subspace of $\mathbb{R}^m$ containing $a_1, \dots, a_n$. We define the convex hull $\mathrm{conv}(M)$ of a subset $M \subseteq \mathbb{R}^m$ as the union of all $\mathrm{conv}\{a_1, \dots, a_n\}$, where $\{a_1, \dots, a_n\}$ runs over all finite subsets of M. Similarly, we define $\mathrm{aff}(M)$ as the union of all $\mathrm{aff}\{a_1, \dots, a_n\}$, where $a_1, \dots, a_n \in M$. The *dimension* of a convex set K is defined as the dimension of its affine hull.

The *separating hyperplane theorem* is a fundamental result in convexity theory. Throughout we denote by $\langle x, y \rangle := x^\mathsf{T} y = \sum_i x_i y_i$ the standard inner product on $\mathbb{R}^m$.

Theorem 6.1 *Let $K \subseteq \mathbb{R}^m$ be closed and convex. For $p \notin K$ there exist $y \in \mathbb{R}^m \setminus \{0\}$ and $\lambda \in \mathbb{R}$ such that*

$$\forall x \in K \quad \langle x, y \rangle < \lambda < \langle p, y \rangle \quad (strict\ separation).$$

If $p \in \partial K$, there exists $y \in \mathbb{R}^m \setminus \{0\}$ such that

$$\forall x \in K \quad \langle x, y \rangle \leq \langle p, y \rangle \quad (supporting\ hyperplane). \qquad \square$$

A *closed half-space* $H \subseteq \mathbb{R}^m$ is a set $H = \{z \in \mathbb{R}^m \mid \langle z, y \rangle \leq 0\}$ for some $y \in \mathbb{R}^m \setminus \{0\}$. Similarly, we say that $H^\cup = \{z \in \mathbb{R}^m \mid \langle z, y \rangle < 0\}$ is an *open half-space*.

A *convex cone* in $\mathbb{R}^m$ is a subset that is closed under addition and multiplication by nonnegative scalars. We denote by $\mathsf{cone}(M)$ the convex cone generated by a subset $M \subseteq \mathbb{R}^m$. More specifically, the convex cone generated by points $a_1, \dots, a_k \in \mathbb{R}^m$ is given by

$$\mathsf{cone}\{a_1, \dots, a_k\} := \left\{ x \in \mathbb{R}^m \mid \exists \lambda_1 \geq 0, \dots, \lambda_k \geq 0 \ x = \sum_{i=1}^{k} \lambda_i a_i \right\}.$$

This is easily seen to be a closed set.

Definition 6.2 *The* dual cone $\check{C}$ *of a convex cone $C \subseteq \mathbb{R}^m$ is defined as*

$$\check{C} := \left\{ y \in \mathbb{R}^m \mid \forall x \in C \ \langle y, x \rangle \leq 0 \right\}.$$

It is clear that $\check{C}$ is a closed convex cone. Moreover, $C_1 \subseteq C_2$ implies $\check{C}_1 \supseteq \check{C}_2$.

Proposition 6.3 *Let $C \subseteq \mathbb{R}^m$ be a closed convex cone. Then the dual cone of $\check{C}$ equals C.*

Proof It is clear that C is contained in the dual cone of $\check{C}$. Conversely, suppose that $p \notin C$. Theorem 6.1 implies that there exist $y \in \mathbb{R}^m \setminus \{0\}$ and $\lambda \in \mathbb{R}$ such that $\langle x, y \rangle < \lambda < \langle p, y \rangle$ for all $x \in C$. Setting $x = 0$ yields $0 < \lambda$.

If we had $\langle x, y \rangle > 0$ for some $x \in C$, then $\langle kx, y \rangle \geq \lambda$ for some $k > 0$, which is a contradiction to $kx \in C$. Therefore, we must have $y \in \check{C}$. Finally, $0 < \lambda < \langle y, p \rangle$; hence p is not in the dual of $\check{C}$. $\qquad \square$

Here is an important consequence of the previous duality result.

Lemma 6.4 (Farkas's lemma)   *Let $A \in \mathbb{R}^{m \times n}$ and $b \in \mathbb{R}^m$. There exists $x \in \mathbb{R}^m$, $x \geq 0$, such that $Ax = b$ if and only if for each $y \in \mathbb{R}^m$ satisfying $A^\mathsf{T} y \leq 0$ one has $b^\mathsf{T} y \leq 0$.*

Proof Suppose $x \in \mathbb{R}^m$ satisfies $Ax = b$ and $x \geq 0$ and let $y \in \mathbb{R}^m$ be such that $A^\mathsf{T} y \leq 0$. Then we have $b^\mathsf{T} y = x^\mathsf{T} A^\mathsf{T} y \leq 0$. This proves one direction of the assertion.

To prove the other direction consider the cone $C := \text{cone}\{a_1, \ldots, a_n\}$ generated by the columns a_i of A. Note that the condition

$$\exists x \in \mathbb{R}^m, \ x \geq 0, \quad Ax = b,$$

in geometric terms, just means that $b \in C$.

Assume now $b \notin C$. Proposition 6.3 implies that b does not lie in the dual cone of $\check{C}$. This means that there exists $y_0 \in \check{C}$ such that $b^T y_0 > 0$. But $A^T y_0 \leq 0$, since $y_0 \in \check{C}$. $\square$

We also state without proof the following result due to Carathéodory.

Theorem 6.5 *Let $a_1, \ldots, a_n \in \mathbb{R}^m$ and $x \in \text{cone}\{a_1, \ldots, a_n\}$. Then there exists $I \subseteq [n]$ with $|I| \leq m$ such that $x \in \text{cone}\{a_i \mid i \in I\}$.* $\square$

An affine version of Carathéodory's result follows easily.

Corollary 6.6 *Let $a_1, \ldots, a_n \in \mathbb{R}^m$ with d-dimensional affine hull. Then for any $x \in \text{conv}\{a_1, \ldots, a_n\}$ there exists $I \subseteq [n]$ with $|I| \leq d+1$ such that $x \in \text{conv}\{a_i \mid i \in I\}$.*

Proof By replacing $\mathbb{R}^m$ with the affine hull of $a_1, \ldots, a_n$ we may assume without loss of generality that $d = m$. Let $x = \sum_i \lambda_i a_i$ with $\lambda_i \geq 0$ and $\sum_i \lambda_i = 1$. Define the following elements of $\mathbb{R}^{m+1}$: $\tilde{a}_i := (a_i, 1)$ and $\tilde{x} := (x, 1)$. Then $\tilde{x} = \sum_i \lambda_i \tilde{a}_i$. The assertion follows by applying Theorem 6.5 to these points. $\square$

Corollary 6.7 *Assume that I is as in Corollary 6.6 with minimal cardinality. Then the affine hull of $\{a_i \mid i \in I\}$ must have dimension $k = |I| - 1$, that is, $(a_i)_{i \in I}$ are affinely independent.*

Proof If we had $k < |I| - 1$, then Corollary 6.6 applied to the subset $\{a_i \mid i \in I\}$ would yield the existence of $J \subseteq I$ with $x \in \text{conv}\{a_j \mid j \in J\}$ and $|J| \leq k+1 < |I|$, which contradicts the minimality of I. $\square$

We define the *relative interior* of $K = \text{conv}\{a_1, \ldots, a_n\}$ by

$$\text{relint}(\text{conv}\{a_1, \ldots, a_n\}) := \left\{ \sum_{i=1}^n t_i a_i \ \middle| \ t_1, \ldots, t_n > 0, \sum_{i=1}^n t_i = 1 \right\}.$$

One can show that this set can be intrinsically characterized by

$$\text{relint}(K) = \left\{ a \mid \exists \varepsilon > 0 \, \forall a' \in \text{aff}(K) : \|a' - a\| < \varepsilon \Rightarrow a' \in K \right\}.$$

This also provides the definition of $\text{relint}(K)$ for an arbitrary convex set K. We define the *relative boundary* ∂K of K as $\partial K := \overline{K} \setminus \text{relint}(K)$. Here $\overline{K}$ is the topological closure of K in $\mathbb{R}^m$.

For later use in Chap. 13 we also state without proof *Helly's theorem*, which is another basic result in convex geometry.

Theorem 6.8 (Helly's theorem) *Let $K_1, \ldots, K_t \subseteq \mathbb{R}^m$ be a family of convex subsets such that any $n+1$ of them have a nonempty intersection. Then $K_1 \cap \cdots \cap K_t \neq \emptyset$.* $\square$

6.2.2 Polyhedra

Let $a_1, \ldots, a_n \in \mathbb{R}^m$ and $b_1, \ldots, b_n \in \mathbb{R}$. The set $P = \{x \in \mathbb{R}^m \mid a_i^{\mathsf{T}} x \leq b_i, i = 1, \ldots, n\}$ is called a *polyhedron*. Since a polyhedron is an intersection of convex sets, polyhedra are convex as well. It is easy to prove that there exists a subset $I \subseteq [n]$ such that

$$\mathsf{aff}(P) = \{x \in \mathbb{R}^m \mid \forall i \notin I \ a_i^{\mathsf{T}} x = b_i\},$$

$$\partial P = \{x \in P \mid \exists i \in I \ a_i^{\mathsf{T}} x = b_i\},$$

where we recall that ∂P denotes the relative boundary of P. We say that a subset $F \subseteq P$ is a *face* of P when there exists $J \subseteq [n]$ such that

$$F = \{x \in P \mid \forall i \in J \ a_i^{\mathsf{T}} x = b_i\}. \tag{6.1}$$

A face of P is called *proper* when it is strictly included in P. We note that ∂P is the union of the proper faces of P.

Clearly, faces of a polyhedron are themselves polyhedra. In particular, they are convex. Hence, a zero-dimensional face consists of a single point. These faces are called *vertices* of P. We note the following important fact, whose easy proof is left to the reader.

Lemma 6.9 *A face F of a polyhedron P given as in (6.1) is a vertex of P if and only if the corresponding matrix A_J, whose columns are the vectors a_i with $i \in J$, satisfies* $\mathsf{rank}\, A_J = m$. $\square$

The faces of a polyhedron are not arbitrarily placed in space. The following result gives a restriction.

Lemma 6.10 *Let F, F' be faces of a polyhedron P such that neither $F \subseteq F'$ nor $F' \subseteq F$. Then $\dim \mathsf{aff}(F \cup F') > \max\{\dim F, \dim F'\}$.*

Proof Without loss of generality, assume $\max\{\dim F, \dim F'\} = \dim F$. Let $I \subseteq [n]$ be such that $\mathsf{aff}(F) = \{x \in \mathbb{R}^m \mid \forall i \notin I \ a_i^{\mathsf{T}} x = b_i\}$. Then

$$F = \{x \in \mathbb{R}^m \mid \forall i \notin I \ a_i^{\mathsf{T}} x = b_i \text{ and } \forall i \in I \ a_i^{\mathsf{T}} x \leq b_i\}.$$

Since $F' \not\subseteq F$, there exists $x_0 \in F'$ such that $x_0 \notin F$. Since $x_0 \in F'$, we have $a_i^{\mathsf{T}} x_0 \leq b_i$ for all $i \leq n$. Therefore, since $x_0 \notin F$, there exists $j \notin I$ such that $a_j^{\mathsf{T}} x_0 < b_j$. This implies that $x_0 \notin \mathsf{aff}(F)$ and hence that

$$\dim \mathsf{conv}(F \cup F') \geq \dim \mathsf{conv}(F \cup \{x_0\}) = \dim \mathsf{aff}(F \cup \{x_0\}) > \dim F. \qquad \square$$

Lemma 6.11 *If a line ℓ is contained in a polyhedron P given by $a_i^{\mathrm{T}} x \leq b_i$, $i = 1, \ldots, n$, then the matrix A with columns $a_1, \ldots, a_n$ satisfies* rank $A < m$. *In particular, P has no vertices.*

Proof Let $v \neq 0$ be a direction vector for ℓ. Since $\ell \subseteq P$, we have $\ell \subseteq \{x \mid a_i^{\mathrm{T}} x = b_i\}$ for all i. This implies $a_i^{\mathrm{T}} v = 0$ for all i, that is, $v \in \ker A$. Hence rank $A < m$. The second assertion follows now from Lemma 6.9.   □

Lemma 6.12 *If F is a face of a polyhedron of minimal dimension (among nonempty faces), then* aff$(F) = F$. *In particular, if P has no vertices, then it contains a line.*

Proof Let $I \subseteq [n]$ be such that aff$(F) = \{x \in \mathbb{R}^m \mid \forall i \in I \ a_i^{\mathrm{T}} x = b_i\}$.

Assume that $F \neq$ aff(F). Then, there exists a point $x_N \in$ aff$(F) \setminus F$. In particular, there exists $j \notin I$ such that $a_j^{\mathrm{T}} x_N > b_j$. Let x_F be any point in F and let

$$\left\{ x_t := t x_N + (1-t) x_F \mid t \in [0,1] \right\}$$

be the segment with extremities x_F and x_N. Clearly, this segment is contained in aff(F). Let

$$\bar{t} = \inf\left\{ t \in [0,1] \mid \exists j \notin I \text{ s.t. } a_j^{\mathrm{T}} x_t > b_j \right\}.$$

Then there exists $\bar{j} \notin I$ such that $x_{\bar{t}} \in F$, $a_{\bar{j}}^{\mathrm{T}} x_{\bar{t}} = b_{\bar{j}}$, but for all $\varepsilon > 0$, $a_{\bar{j}}^{\mathrm{T}} x_{\bar{t}+\varepsilon} > b_{\bar{j}}$. This shows that the face defined by the set $\bar{I} := I \cup \{\bar{j}\}$ is nonempty and has dimension smaller than dim F, a contradiction.

The second statement is a trivial consequence of the first.   □

The following result immediately follows from Lemmas 6.11 and 6.12.

Corollary 6.13 *A polyhedron possesses vertices if and only if it does not contain lines.*   □

6.3   The Polyhedral Cone Feasibility Problem

For $A \in \mathbb{R}^{m \times n}$, consider the *primal feasibility problem*

$$\exists x \in \mathbb{R}^n \setminus \{0\}, \quad Ax = 0, \quad x \geq 0, \tag{PF}$$

and the *dual feasibility problem*

$$\exists y \in \mathbb{R}^m \setminus \{0\}, \quad A^{\mathrm{T}} y \leq 0. \tag{DF}$$

We say that A is *primal feasible* or *dual feasible* when (PF), or (DF), respectively, is satisfied. In both cases we talk about *strict* feasibility when the satisfied inequality is strict. The following result shows that strict primal feasibility and strict dual

Fig. 6.1 A partition of $\mathbb{R}^{m \times n}$
with respect to feasibility

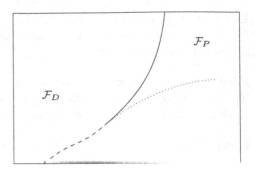

feasibility are incompatible. To simplify its statement we introduce some notation. Let $\mathcal{F}_P$ and $\mathcal{F}_D$ denote the sets of matrices A where (PF) and (DF) are satisfied, respectively. Moreover, let

$$\mathcal{F}_P^\circ = \left\{ A \in \mathbb{R}^{m \times n} \mid \exists x \in \mathbb{R}^n\, Ax = 0, x > 0 \right\},$$
$$\mathcal{F}_D^\circ = \left\{ A \in \mathbb{R}^{m \times n} \mid \exists y \in \mathbb{R}^m\, A^\mathsf{T} y < 0 \right\}$$

be the sets of strictly primal and strictly dual feasible matrices. Finally, let $\mathcal{R} := \{ A \in \mathbb{R}^{m \times n} \mid \operatorname{rank} A = m \}$ and

$$\Sigma := \mathcal{F}_P \cap \mathcal{F}_D.$$

Denote by $\operatorname{int}(M)$, $\overline{M}$, and $\partial M = \overline{M} \setminus \operatorname{int}(M)$, the interior, closure, and boundary of a subset M of Euclidean space.

One can easily show that if $n \leq m$, then $\mathcal{F}_D = \mathbb{R}^{m \times n}$. The situation of interest is therefore the case $n > m$, and in what follows we will assume this inequality.

Theorem 6.14 *Both $\mathcal{F}_P$ and $\mathcal{F}_D$ are closed subsets of $\mathbb{R}^{m \times n}$. In addition, this space is partitioned as*

$$\mathbb{R}^{m \times n} = \operatorname{int}\!\left(\mathcal{F}_P^\circ\right) \cup \operatorname{int}\!\left(\mathcal{F}_D^\circ\right) \cup \Sigma,$$

and we have

$$\Sigma = \partial \mathcal{F}_P = \partial \mathcal{F}_D.$$

Furthermore, $\mathcal{F}_P^\circ \supseteq \operatorname{int}(\mathcal{F}_P)$, $\mathcal{F}_P^\circ \setminus \operatorname{int}(\mathcal{F}_P) = \mathcal{F}_P^\circ \cap \Sigma = \{ A \in \Sigma \mid \operatorname{rank} A < m \}$, and $\mathcal{F}_D^\circ = \operatorname{int}(\mathcal{F}_D)$.

For this case, Fig. 6.1 provides a schematic picture derived from Theorem 6.14. On it, the 2-dimensional space corresponds to the set of all matrices. The curve corresponds to the set Σ, which is divided into three parts. All matrices in Σ are in $\mathcal{F}_D \setminus \mathcal{F}_D^\circ$: those on the full part of the curve correspond to full-rank matrices that are also in $\mathcal{F}_P \setminus \mathcal{F}_P^\circ$, those on the dashed part to rank-deficient matrices with this property, and those on the dotted part to rank-deficient matrices that are in $\mathcal{F}_P^\circ$.

We observe that there are rank-deficient matrices that are not in Σ, all of them being in $\mathcal{F}_D^\circ$.

The set Σ, just as in the picture, is of dimension smaller than mn.

We see that for matrices in Σ, arbitrarily small perturbations can lead to a change with respect to feasibility. In contrast, in the set $\mathcal{D} = \mathbb{R}^{m \times n} \setminus \Sigma$ the following problem is well-defined:

> *Given $A \in \mathcal{D}$ decide whether $A \in \mathcal{F}_P^\circ$ or $A \in \mathcal{F}_D^\circ$.*

We call this the *polyhedral cone feasibility problem* (and we denote it by PCFP). For all $A \in \Sigma$ the problem is ill-posed.

The polyhedral cone feasibility problem fits the situation described in the introduction of this chapter. The approach to condition described in the Overture cannot be applied here (note that even the values of this problem—the tags "strictly primal feasible" and "strictly dual feasible"—are not elements in a Euclidean space). We need a different measure of condition. We will define this measure in the next section. Before doing so, however, we will prove Theorem 6.14 and get some understanding about the partition depicted in Fig. 6.1.

We begin with a simple result (and recall Notation 1.9 for its statement).

Proposition 6.15 *Both $\mathcal{F}_P$ and $\mathcal{F}_D$ are closed subsets of $\mathbb{R}^{m \times n}$ and closed under multiplication by scalars $\lambda_i \geq 0$. That is, if $[a_1, \ldots, a_n] \in \mathcal{F}_P$, then $[\lambda_1 a_1, \ldots, \lambda_n a_n]$ $\in \mathcal{F}_P$, and similarly for $\mathcal{F}_D$.*

Proof Let $\mathbb{S}^{m-1} := \{y \in \mathbb{R}^m \mid \|y\| = 1\}$ denote the $(m-1)$-dimensional unit sphere. The compactness of $\mathbb{S}^{m-1}$ easily implies that

$$\mathcal{F}_D = \big\{ A \mid \exists y \in \mathbb{S}^{m-1} \, \langle a_1, y \rangle \leq 0, \ldots, \langle a_n, y \rangle \leq 0 \big\}$$

is closed. Similarly, one shows that $\mathcal{F}_P$ is closed. The second statement is trivial. $\square$

Let $A \in \mathbb{R}^{m \times n}$ and denote by $a_1, \ldots, a_n \in \mathbb{R}^m$ its columns. We have the following geometric characterizations:

$$
\begin{aligned}
A \in \mathcal{F}_P &\quad\Leftrightarrow\quad 0 \in \mathsf{conv}\{a_1, \ldots, a_n\}, \\
A \in \mathcal{F}_P^\circ &\quad\Leftrightarrow\quad 0 \in \mathsf{relint}\big(\mathsf{conv}\{a_1, \ldots, a_n\}\big).
\end{aligned}
\tag{6.2}
$$

Also, by definition, we have

$$
\begin{aligned}
A \in \mathcal{F}_D &\quad\Leftrightarrow\quad \exists H \text{ closed half-space such that } \mathsf{conv}\{a_1, \ldots, a_n\} \subseteq H, \\
A \in \mathcal{F}_D^\circ &\quad\Leftrightarrow\quad \exists H^\circ \text{ open half-space such that } \mathsf{conv}\{a_1, \ldots, a_n\} \subseteq H^\circ.
\end{aligned}
$$

From the definition of Σ and the first equivalence in (6.2) we obtain the following characterization:

$$A \in \Sigma \quad\Leftrightarrow\quad A \in \mathcal{F}_D \text{ and } 0 \in \mathsf{conv}\{a_1, \ldots, a_n\}. \tag{6.3}$$

Lemma 6.16 *For* $A \in \mathbb{R}^{m \times n}$ *we have*

(a) $A \notin \mathcal{F}_D^\circ \Leftrightarrow A \in \mathcal{F}_P$.
(b) $A \notin \mathcal{F}_P^\circ \Rightarrow A \in \mathcal{F}_D$. *The converse is true if* rank $A = m$.

Proof (a) We prove the contrapositive. Suppose $A \in \mathcal{F}_D^\circ$. Then there exists $y \in \mathbb{R}^m \setminus \{0\}$ such that $\langle a_i, y \rangle < 0$ for all i. If we had $\sum_i x_i a_i = 0$ for some $x_i \geq 0$ with $\sum_i x_i = 1$, then $\sum_i x_i \langle a_i, y \rangle = \langle \sum_i x_i a_i, y \rangle = 0$. Hence $x_i = 0$ for all i, which is a contradiction.

Conversely, suppose that $A \notin \mathcal{F}_P$, that is, $0 \notin \mathrm{conv}\{a_1, \ldots, a_n\}$. Theorem 6.1 (strict separation) implies that $A \in \mathcal{F}_D^\circ$.

(b) Suppose $A \notin \mathcal{F}_P^\circ$. Then $0 \notin \mathrm{relint}(\mathrm{conv}\{a_1, \ldots, a_n\})$, and therefore $0 \notin \mathrm{int}(\mathrm{conv}\{a_1, \ldots, a_n\})$. Theorem 6.1 implies $A \in \mathcal{F}_D$. For the other direction assume that $A \in \mathcal{F}_D$, say $\langle a_i, y \rangle \leq 0$ for all i and some $y \neq 0$. If we had $A \in \mathcal{F}_P^\circ$, then $\sum_i x_i a_i = 0$ for some $x_i > 0$. Therefore $\sum_i x_i \langle a_i, y \rangle = 0$, whence $\langle a_i, y \rangle = 0$ for all i. This implies $\mathrm{rank}(A) \leq m - 1$. $\square$

Remark 6.17 For the converse of part (b) of Lemma 6.16 we indeed need the rank assumption. To see this, take, for example, $a_1, \ldots, a_n \in \mathbb{R}^{m-1}$ such that $0 \in \mathrm{relint}(\mathrm{conv}\{a_1, \ldots, a_n\})$. Then $A \in \mathcal{F}_D \cap \mathcal{F}_P^\circ$.

Lemma 6.16 implies that $\mathcal{F}_P^\circ$ and $\mathcal{F}_D^\circ$ are disjoint,

$$\mathcal{F}_D \setminus \mathcal{F}_D^\circ = \Sigma, \qquad \mathcal{F}_P \setminus \mathcal{F}_P^\circ \subseteq \Sigma,$$

and the right-hand inclusion becomes an equality when the matrices are restricted to being of rank m. Moreover, again using Lemma 6.16,

$$\mathbb{R}^{m \times n} = \mathcal{F}_P \cup \mathcal{F}_D = \mathcal{F}_P^\circ \cup \mathcal{F}_D^\circ \cup \Sigma. \tag{6.4}$$

Furthermore, since Σ is closed, $\mathcal{F}_D^\circ$ is open. It is somewhat confusing that $\mathcal{F}_P^\circ$ is not open. To see this, consider again $a_1, \ldots, a_n \in \mathbb{R}^{m-1}$ such that $0 \in \mathrm{relint}(\mathrm{conv}\{a_1, \ldots, a_n\})$. Then $A \in \mathcal{F}_P^\circ$, but there are arbitrarily small perturbations of A that lie in $\mathcal{F}_D^\circ$.

Lemma 6.18

(a) $\mathcal{F}_D \subseteq \overline{\mathcal{F}_D^\circ}$.
(b) $\mathcal{F}_P \subseteq \overline{\mathcal{F}_P^\circ \cap \mathcal{R}}$.

Proof (a) Let $A = [a_1, \ldots, a_n] \in \mathcal{F}_D$. Hence there exists $y \in \mathbb{S}^{m-1}$ such that $\langle a_i, y \rangle \leq 0$ for all i. For $\varepsilon > 0$ put $a_i(\varepsilon) := a_i - \varepsilon y$. Then $\langle a_i(\varepsilon), y \rangle = \langle a_i, y \rangle - \varepsilon \leq -\varepsilon$; hence $A(\varepsilon) = [a_1(\varepsilon), \ldots, a_n(\varepsilon)] \in \mathcal{F}_D^\circ$. Moreover, $\lim_{\varepsilon \to 0} A(\varepsilon) = A$.

(b) Let $A = [a_1, \ldots, a_n] \in \mathcal{F}_P$. Put $W := \mathrm{span}\{a_1, \ldots, a_n\}$ and $d := \dim W$. The first equivalence in (6.2) implies that $0 \in \mathrm{conv}\{a_1, \ldots, a_n\}$. Note that the affine hull of $\{a_1, \ldots, a_n\}$ equals W. By Carathéodory's Corollary 6.6, we may assume without loss of generality that $0 = x_1 a_1 + \cdots + x_k a_k$ with $x_i > 0$, $\sum_{i=1}^{k} x_i = 1$, and $k \leq d + 1$. Moreover, by Corollary 6.7, we may assume that the affine hull of $a_1, \ldots, a_k$ has dimension $k - 1$. Without loss of generality we may assume that $a_1, \ldots, a_{k-1}$ are linearly independent and that $a_1, \ldots, a_{k-1}, a_{k+1}, \ldots, a_{d+1}$ is a basis of W. Let $b_{d+2}, \ldots, b_{m+1}$ be a basis of the orthogonal complement $W^\perp$. We define now

$$v(\varepsilon) := a_{k+1} + \cdots + a_{d+1} + (a_{d+2} + \varepsilon b_{d+2}) + \cdots + (a_{m+1} + \varepsilon b_{m+1})$$
$$+ a_{m+2} + \cdots + a_n.$$

(Here we used the assumption $n \geq m + 1$.) Moreover, we put

$$a_i(\varepsilon) := \begin{cases} a_i - \varepsilon v(\varepsilon) & \text{for } 1 \leq i \leq k, \\ a_i & \text{for } k + 1 \leq i \leq d + 1, \\ a_i + \varepsilon b_i & \text{for } d + 2 \leq i \leq m + 1, \\ a_i & \text{for } m + 2 \leq i \leq n. \end{cases}$$

Note that $v(\varepsilon) = \sum_{i=k+1}^{n} a_i(\varepsilon)$. It is clear that $A(\varepsilon) := [a_1(\varepsilon), \ldots, a_n(\varepsilon)]$ converges to A for $\varepsilon \to 0$. Also, $W = \text{span}\{a_1, \ldots, a_{k-1}, a_{k+1}, \ldots, a_{d+1}\}$, and using this fact, it follows that $\text{span}\{a_1(\varepsilon), \ldots, a_n(\varepsilon)\} = \mathbb{R}^m$, i.e., that $\text{rank}(A(\varepsilon)) = m$. Finally, we have

$$0 = \sum_{i=1}^{k} x_i a_i = \sum_{i=1}^{k} x_i a_i(\varepsilon) + \varepsilon v(\varepsilon) = \sum_{i=1}^{k} x_i a_i(\varepsilon) + \sum_{j=k+1}^{n} \varepsilon a_j(\varepsilon).$$

Hence $A(\varepsilon) \in \mathcal{F}_P^\circ$. $\square$

Corollary 6.19 *Suppose $n > m$. Then*

(a) $\Sigma = \partial \mathcal{F}_D$, $\text{int}(\mathcal{F}_D) = \mathcal{F}_D^\circ$,
(b) $\Sigma = \partial \mathcal{F}_P$, $\text{int}(\mathcal{F}_P) \subseteq \mathcal{F}_P^\circ$.

Proof (a) We have $\mathcal{F}_D^\circ \subseteq \text{int}(\mathcal{F}_D)$, since $\mathcal{F}_D^\circ$ is open. Hence $\partial \mathcal{F}_D = \mathcal{F}_D \setminus \text{int}(\mathcal{F}_D) \subseteq \mathcal{F}_D \setminus \mathcal{F}_D^\circ = \Sigma$. Suppose $A \in \Sigma$. By Lemma 6.18 there is a sequence $A_k \to A$ such that $\text{rank} A_k = m$ and $A_k \in \mathcal{F}_P^\circ$. Lemma 6.16 shows that $A_k \notin \mathcal{F}_D$. Hence $A \in \partial \mathcal{F}_D$. It follows that $\partial \mathcal{F}_D = \Sigma$ and $\text{int}(\mathcal{F}_D) = \mathcal{F}_D^\circ$.

(b) Let $A \in \Sigma$. By Lemma 6.18 there is a sequence $A_k \to A$ such that $A_k \in \mathcal{F}_D^\circ$; hence $A_k \notin \mathcal{F}_P$. Therefore $A \in \partial \mathcal{F}_P$. It follows that $\Sigma \subseteq \partial \mathcal{F}_P$. On the other hand,

$$\partial \mathcal{F}_P \subseteq \overline{\mathbb{R}^{m \times n} \setminus \mathcal{F}_P} = \overline{\mathcal{F}_D^\circ} \subseteq \mathcal{F}_D,$$

and hence $\partial \mathcal{F}_P \subseteq \mathcal{F}_P \cap \mathcal{F}_D = \Sigma$. It follows that $\Sigma = \partial \mathcal{F}_P$. Finally,

$$\text{int}(\mathcal{F}_P) = \mathcal{F}_P \setminus \partial \mathcal{F}_P = \mathcal{F}_P \setminus \Sigma \subseteq \mathcal{F}_P^\circ.$$ $\square$

It may seem disturbing that int($\mathcal{F}_P$) is properly contained in $\mathcal{F}_P^\circ$. However, the difference $\mathcal{F}_P^\circ \setminus \text{int}(\mathcal{F}_P)$ lies in Σ and thus has measure zero, so that this will not harm us (see Fig. 6.1).

Proof of Theorem 6.14 It immediately follows from the results in this section.   □

6.4   The GCC Condition Number and Distance to Ill-posedness

We want to define a condition number for PCFP. A way of doing so relies on the condition number theorem (Corollary 1.8). This result characterized the condition number of linear equation solving, or matrix inversion, as the inverse of the relativized distance from the matrix at hand to the set of ill-posed matrices. Instead of defining condition in terms of perturbations (which we have seen is now useless), we can take the characterization of the condition number theorem as definition. We have shown in the previous section that for PCFP, the set of ill-posed instances is the boundary between feasible and infeasible instances. This motivates the following definition.

Definition 6.20 Let $A \in \mathbb{R}^{m \times n}$ be given with nonzero columns a_i. Suppose $A \notin \Sigma$ and $A \in \mathcal{F}_S^\circ$ for $S \in \{P, D\}$. We define

$$\Delta(A) := \sup \left\{ \delta > 0 \;\middle|\; \forall A' \in \mathbb{R}^{m \times n} \left(\max_{i \leq n} \frac{\|a_i' - a_i\|}{\|a_i\|} < \delta \Rightarrow A' \in \mathcal{F}_S^\circ \right) \right\},$$

where a_i' stands for the ith column of A'. The *GCC condition number* of A is defined as

$$\mathscr{C}(A) := \frac{1}{\Delta(A)}.$$

If $A \in \Sigma$, we set $\Delta(A) = 0$ and $\mathscr{C}(A) = \infty$.

We note that the suprema are over nonempty bounded sets and hence welldefined, since $\mathcal{F}_S^\circ \setminus \Sigma = \text{int}(\mathcal{F}_S)$ for $S \in \{P, D\}$ due to Corollary 6.19.

We have written the definition in such a way that it becomes clear that we measure the *relative* size of the perturbation for each row a_i, where the relativization is with respect to the norm of a_i. Also, it is clear from the definition that $\Delta(A)$ is scale-invariant in the sense that

$$\Delta([\lambda_1 a_1, \ldots, \lambda_n a_n]) = \Delta([a_1, \ldots, a_n]) \quad \text{for } \lambda_i > 0.$$

For the analysis of Δ we may therefore assume, without loss of generality, that $\|a_i\| = 1$ for all i. Hence we can see the matrix A with columns $a_1, \ldots, a_n$ as an element in the product $(\mathbb{S}^{m-1})^n$ of spheres. The scale invariance of $\mathscr{C}(A)$, together with the characterization of $\| \; \|_{12}$ in Corollary 1.3, yields immediately the following result.

Proposition 6.21 $\mathscr{C}(A) = \dfrac{\|A\|_{12}}{d_{12}(A, \Sigma)}.$ $\square$

We now want to rewrite Definition 6.20 in a way that follows the ideas of Sect. 2.4. Let $d_{\mathbb{S}}(a, b) \in [0, \pi]$ denote the angular distance

$$d_{\mathbb{S}}(a, b) := \arccos(\langle a, b \rangle).$$

It is clear that this defines a metric on $\mathbb{S}^{m-1}$. We extend this metric to $(\mathbb{S}^{m-1})^n$ by taking

$$d_{\mathbb{S}}(A, B) := \max_{1 \le i \le n} d_{\mathbb{S}}(a_i, b_i).$$

Further, for a nonempty subset $M \subseteq (\mathbb{S}^{m-1})^n$ we write

$$d_{\mathbb{S}}(A, M) := \inf\{d_{\mathbb{S}}(A, B) \mid B \in M\}.$$

For simplicity of notation, we shall denote $\mathcal{F}_P \cap (\mathbb{S}^{m-1})^n$ also by the symbol $\mathcal{F}_P$ and similarly for $\mathcal{F}_P^\circ, \mathcal{F}_D, \mathcal{F}_D^\circ$, and Σ. This should not lead to any confusion.

The fact that $\Sigma = \partial \mathcal{F}_P = \partial \mathcal{F}_D$ (cf. Corollary 6.19) immediately tells us that

$$\begin{aligned} d_{\mathbb{S}}\big(A, (\mathbb{S}^{m-1})^n \setminus \mathcal{F}_P^\circ\big) &= d_{\mathbb{S}}(A, \Sigma) \quad \text{for } A \in \mathcal{F}_P^\circ, \\ d_{\mathbb{S}}\big(A, (\mathbb{S}^{m-1})^n \setminus \mathcal{F}_D^\circ\big) &= d_{\mathbb{S}}(A, \Sigma) \quad \text{for } A \in \mathcal{F}_D^\circ. \end{aligned} \tag{6.5}$$

We postpone the proof of the following result (compare Theorem 6.27).

Lemma 6.22 *For $A \in (\mathbb{S}^{m-1})^n$ we have $d_{\mathbb{S}}(A, \Sigma) \le \frac{\pi}{2}$. Moreover, $d_{\mathbb{S}}(A, \Sigma) = \frac{\pi}{2}$ iff $A = (a, a, \ldots, a)$ for some $a \in \mathbb{S}^{m-1}$.*

We can now give a geometric characterization of the GCC condition number. Recall the definition of $d_{\sin}$ in a product of spheres (Sect. 2.4.1).

Proposition 6.23 *For $A \in (\mathbb{S}^{m-1})^n$ we have $\Delta(A) = d_{\sin}(A, \Sigma)$. Hence*

$$\mathscr{C}(A) = \frac{1}{d_{\sin}(A, \Sigma)}.$$

Proof Without loss of generality $A \notin \Sigma$. Suppose $A \in \mathcal{F}_P^\circ$. It suffices to show that

(a) $\sin d_{\mathbb{S}}(A, \Sigma) = 1 \Rightarrow \Delta(A) = 1$,
(b) $\sin d_{\mathbb{S}}(A, \Sigma) < d \Leftrightarrow \Delta(A) < d$ for all $0 < d < 1$.

The first case is easily established with the second part of Lemma 6.22. Thus, let $0 < d < 1$ be such that $\sin d_{\mathbb{S}}(A, \Sigma) < d$. Lemma 6.22 tells us that $d_{\mathbb{S}}(A, \Sigma) \le \frac{\pi}{2}$, hence $d_{\mathbb{S}}(A, \Sigma) < \arcsin d$. By (6.5) there exists $B = (b_1, \ldots, b_n) \notin \mathcal{F}_P^\circ$ such that $d_{\mathbb{S}}(A, B) < \arcsin d$. Additionally, we may assume that $\|b_i\| = 1$. Let $\theta_i = d_{\mathbb{S}}(a_i, b_i)$ (cf. Fig. 6.2).

Fig. 6.2 The definition of b_i

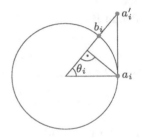

By definition, $d_{\mathbb{S}}(A, B) = \max_i \theta_i$; hence $\theta_i < \arcsin d$ for all i, and therefore

$$\|(\cos\theta_i)b_i - a_i\| = \sin\theta_i < d.$$

It follows from the definition of $\Delta(A)$ that $\Delta(A) < d$ (consider the matrix A' with the columns $(\cos\theta_i)b_i$).

Conversely, assume $\Delta(A) < d$ for $d < 1$. Then there exists $A' \notin \mathcal{F}_P^\circ$ such that $\max_i \|a_i' - a_i\| < d$. In particular, $a_i' \neq 0$. For $b_i := \frac{a_i'}{\|a_i'\|}$ we have $\theta_i := d_{\mathbb{S}}(a_i, b_i) < \frac{\pi}{2}$ and for all i,

$$\sin\theta_i = \min_{\lambda>0} \|\lambda b_i - a_i\| \leq \|a_i' - a\| < d$$

(cf. Fig. 6.2). Hence $d_{\mathbb{S}}(A, B) < \arcsin d$, and therefore we have $d_{\mathbb{S}}(A, \Sigma) = d_{\mathbb{S}}(A, (\mathbb{S}^{m-1})^n \setminus \mathcal{F}_P^\circ) < \arcsin d$.

The case $A \in \mathcal{F}_D^\circ$ is proved analogously. □

6.5 The GCC Condition Number and Spherical Caps

We provide here a characterization of the GCC condition number in terms of an optimization problem in spherical geometry.

For $p \in \mathbb{S}^{m-1}$ and $\alpha \in [0, 2\pi]$ recall that

$$\mathsf{cap}(p, \alpha) := \left\{ y \in \mathbb{S}^{m-1} \mid \langle p, y \rangle \geq \cos\alpha \right\}$$

denotes the spherical cap in $\mathbb{S}^{m-1}$ with center p and angular radius α.

Definition 6.24 A *smallest including cap* (SIC) for $A = (a_1, \ldots, a_n) \in (\mathbb{S}^{m-1})^n$ is a spherical cap $\mathsf{cap}(p, \alpha)$ of minimal radius containing the points $a_1, \ldots, a_n$. Its *blocking set* is defined as $\{i \in [n] \mid \langle a_i, p \rangle = \cos\alpha\}$ (which can be seen as the set of "active rows").

We remark that by a compactness argument, an SIC always exists. However, there may be several SICs (consider, for instance, three equidistant points on the circle). While an SIC for A might not be uniquely determined, its radius certainly is and will be denoted by $\rho(A)$.

Lemma 6.25 *We have $\rho(A) < \frac{\pi}{2}$ iff $A \in \mathcal{F}_D^\circ$. Moreover, $\rho(A) = \frac{\pi}{2}$ iff $A \in \Sigma$.*

Proof We have $\rho(A) < \frac{\pi}{2}$ iff $a_1, \ldots, a_n$ are contained in a spherical cap of radius less than $\frac{\pi}{2}$. This means that there exists $p \in \mathbb{S}^{m-1}$ such that $\langle a_1, -p \rangle < 0, \ldots, \langle a_n, -p \rangle < 0$. This is equivalent to $A \in \mathcal{F}_D^\circ$. By the same reasoning, $\rho(A) \leq \frac{\pi}{2}$ is equivalent to $A \in \mathcal{F}_D$. This proves the lemma. $\qquad\square$

Lemma 6.26 *Let $\mathrm{cap}(p, \rho)$ be an SIC for $A = (a_1, \ldots, a_n)$ with blocking set $[k]$. Write $t := \cos\rho$, so that*

$$\langle a_1, p \rangle = \cdots = \langle a_k, p \rangle = t, \qquad \langle a_{k+1}, p \rangle > t, \ldots, \langle a_n, p \rangle > t.$$

Then $tp \in \mathrm{conv}\{a_1, \ldots, a_k\}$.

Proof Suppose first that A is dual feasible, i.e., that $t \geq 0$. It suffices to show that $p \in \mathrm{cone}\{a_1, \ldots, a_k\}$. Indeed, if $p = \sum_{i=1}^k \lambda_i a_i$, $\lambda_i \geq 0$, then $tp = \sum_{i=1}^k t\lambda_i a_i$. Furthermore,

$$\sum_{i=1}^k t\lambda_i = \sum_{i=1}^k \lambda_i \langle a_i, p \rangle = \left\langle \sum_{i=1}^k \lambda_i a_i, p \right\rangle = \langle p, p \rangle = 1.$$

We argue by contradiction. If $p \notin \mathrm{cone}\{a_1, \ldots, a_k\}$, then by the separating hyperplane theorem (Theorem 6.1) there would exist a vector $v \in \mathbb{S}^{m-1}$ such that $\langle p, v \rangle < 0$ and $\langle a_i, v \rangle > 0$ for all i. For $\delta > 0$ we set

$$p_\delta := \frac{p + \delta v}{\|p + \delta v\|} = \frac{p + \delta v}{\sqrt{1 + 2\delta \langle p, v \rangle + \delta^2}}.$$

Then for $1 \leq i \leq k$ and sufficiently small δ we have

$$\langle a_i, p_\delta \rangle = \frac{t + \delta \langle a_i, v \rangle}{\sqrt{1 + 2\delta \langle p, v \rangle + \delta^2}} > t.$$

Moreover, by continuity we have $\langle a_i, p_\delta \rangle > t$ for all $i > k$ and δ sufficiently small. We conclude that for sufficiently small $\delta > 0$ there exists $t_\delta > 0$ such that $\langle a_i, p_\delta \rangle > t_\delta$ for all $i \in [n]$. Hence $\mathrm{cap}(p_\delta, \alpha_\delta)$ is a spherical cap containing all the a_i that have angular radius $\alpha_\delta = \arccos t_\delta < \alpha$, contradicting the minimality assumption.

In the case that A is dual infeasible ($t < 0$) one can argue analogously. $\qquad\square$

Theorem 6.27 *We have*

$$d_{\mathbb{S}}(A, \Sigma) = \begin{cases} \frac{\pi}{2} - \rho(A) & \text{if } A \in \mathcal{F}_D, \\ \rho(A) - \frac{\pi}{2} & \text{if } A \in (\mathbb{S}^{m-1})^n \setminus \mathcal{F}_D. \end{cases}$$

In particular, $d_{\mathbb{S}}(A, \Sigma) \leq \frac{\pi}{2}$ and

$$\mathscr{C}(A)^{-1} = \sin d_{\mathbb{S}}(A, \Sigma) = |\cos \rho(A)|.$$

Fig. 6.3 $A = (a_1, a_2, a_3) \in$
$\mathcal{F}_D$, $A' = (a_1', a_2', a_3') \in \Sigma$,
and $t = t(A)$

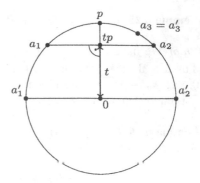

Proof We first assume that $A \in \mathcal{F}_D$. Let $\mathsf{cap}(p, \rho)$ be an SIC for A and put $t :=$ $\cos \rho$. Thus $\rho \leq \frac{\pi}{2}$ and hence $t \geq 0$. Let $A' \in (\mathbb{S}^{m-1})^n$ be such that $d_{\mathbb{S}}(A', A) \leq$ $\frac{\pi}{2} - \rho$. Since $d_{\mathbb{S}}(p, a_i) \leq \rho$ for all i, we get

$$d_{\mathbb{S}}(p, a_i') \leq d_{\mathbb{S}}(p, a_i) + d_{\mathbb{S}}(a_i, a_i') \leq \rho + \frac{\pi}{2} - \rho = \frac{\pi}{2}.$$

Hence $\langle p, a_i' \rangle \geq 0$ for all i, which implies $A' \in \mathcal{F}_D$. We have thus proved the implication

$$\forall A', \quad d_{\mathbb{S}}(A', A) \leq \frac{\pi}{2} - \rho \Rightarrow A' \in \mathcal{F}_D.$$

This implies

$$d_{\mathbb{S}}(A, \Sigma) = d_{\mathbb{S}}(A, (\mathbb{S}^{m-1})^n \setminus \mathcal{F}_D) \geq \frac{\pi}{2} - \rho.$$

For the other direction, without loss of generality, let $[k]$ be the blocking set of $\mathsf{cap}(p, \rho)$. We have $\langle a_i, p \rangle = t$ for $i \leq k$, $\langle a_i, p \rangle > t$ for $i > k$, and $tp \in$ $\mathsf{conv}\{a_1, \ldots, a_k\}$ by Lemma 6.26 (see Fig. 6.3). We assume that $a_i \neq tp$ for $i \in [k]$, since otherwise, $a_i = tp = p$ for all $i \in [n]$, and for this case the claim is easily established. Put

$$a_i' := \begin{cases} \frac{a_i - tp}{\|a_i - tp\|} & \text{for } i \leq k, \\ a_i & \text{for} i > k. \end{cases}$$

Then $\langle a_i', p \rangle \geq 0$ for all $i \in [n]$, $\langle a_i', p \rangle = 0$ for $i \leq k$, and $0 \in \mathsf{conv}\{a_1', \ldots, a_k'\}$. The characterization (6.3) (p. 131) implies that $A' = (a_1', \ldots, a_n') \in \Sigma$. Hence

$$d_{\mathbb{S}}(A, \Sigma) \leq d_{\mathbb{S}}(A, A') \leq \frac{\pi}{2} - \rho.$$

Altogether, we have shown that $d_{\mathbb{S}}(A, \Sigma) = \frac{\pi}{2} - \rho$, which proves the assertion in the case $A \in \mathcal{F}_D$.

Fig. 6.4 $A'q \leq 0$, $A \notin \mathcal{F}_D$, and $d_{\mathbb{S}}(a_{i_0}, a'_{i_0}) \geq \alpha - \frac{\pi}{2}$

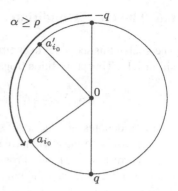

We assume now $A \notin \mathcal{F}_D$. Let $\mathsf{cap}(p, \rho)$ be an SIC for A. Note that for all $i \in [n]$ with $\langle a_i, p \rangle < 0$ we have $a_i \neq \langle a_i, p \rangle \cdot p$, since equality would yield a contradiction to the minimality of ρ, which is easily seen. We set

$$a'_i := \begin{cases} \frac{a_i - \langle a_i, p \rangle \cdot p}{\|a_i - \langle a_i, p \rangle \cdot p\|} & \text{if } a_i - \langle a_i, p \rangle \cdot p < 0, \\ a_i & \text{otherwise.} \end{cases}$$

As in the proof of the case $A \in \mathcal{F}_D$, we see that $A' = (a'_1, \ldots, a'_n) \in \Sigma$ and $d_{\mathbb{S}}(A', A) \leq \rho - \frac{\pi}{2}$. Hence

$$d_{\mathbb{S}}(A, \Sigma) \leq \rho - \frac{\pi}{2}.$$

For the other direction we need to prove that

$$\forall A' \quad \left(A' \in \mathcal{F}_D \Rightarrow d_{\mathbb{S}}(A', A) \geq \rho - \frac{\pi}{2} \right).$$

So let $A' \in \mathcal{F}_D$ and $q \in \mathbb{S}^{m-1}$ be such that $A'q \leq 0$. Consider the cap of smallest angular radius α with center $-q$ that contains all the points a_i. Then $\alpha \geq \rho$. Choose i_0 such that (see Fig. 6.4)

$$d_{\mathbb{S}}(a_{i_0}, -q) = \max_{1 \leq i \leq n} d_{\mathbb{S}}(a_i, -q) = \alpha.$$

It follows that

$$d_{\mathbb{S}}(A, A') \geq d_{\mathbb{S}}(a_{i_0}, a'_{i_0}) \geq d_{\mathbb{S}}(a_{i_0}, -q) - d_{\mathbb{S}}(-q, a'_{i_0}) \geq \alpha - \frac{\pi}{2} \geq \rho - \frac{\pi}{2}.$$

Therefore $d_{\mathbb{S}}(A, \Sigma) \geq \rho - \frac{\pi}{2}$, which completes the proof. $\square$

6.6 The GCC Condition Number and Images of Balls

The goal of this section is to exhibit a characterization of $\mathscr{C}(A)$ in the spirit of Proposition 1.11. This proposition (together with Theorem 1.7) tells us that for $A \in \mathbb{R}^{n \times n}$,

$$d_{12}(A, \Sigma) = \left\| A^{-1} \right\|_{21}^{-1} = \inf\{ \|y\| \mid y \in \{ Ax \mid \|x\|_1 = 1 \} \},$$

where Σ denotes here the set of singular matrices.

The positive orthant will have to play a role alongside the balls, and the statement of the corresponding result, Proposition 6.28 below, is far from apparent. To further motivate it we note the following fact, which follows easily from (6.2).

We assign to $A = [a_1, \ldots, a_n] \in \mathbb{R}^{m \times n}$ the convex hull

$$\mathcal{K} := \mathrm{conv}\{a_1, \ldots, a_n\} = \left\{ Ax \mid x \geq 0, \|x\|_1 = 1 \right\}.$$

Then $A \in \mathcal{F}_D^\circ$ implies $0 \notin \mathcal{K}$. Moreover, $A \in \mathcal{F}_P^\circ$, and $\mathrm{rank}\,A = m$ implies $0 \in \mathrm{int}(\mathcal{K})$. Proposition 6.28 is a quantitative version of this observation.

As usual, we will assume the matrix A to have columns a_i with unit norm, that is, $A \in (\mathbb{S}^{m-1})^n$. Recall from Corollary 1.3 that $\|A\|_{12} = \max_{i \leq n} \|a_i\|$. Therefore, for $S \in \{P, D\}$ and $A \in \mathcal{F}_S^\circ$, Definition 6.20 yields

$$\Delta(A) := \sup\{\delta > 0 \mid \forall E \in \mathbb{R}^{m \times n} (\|E\|_{12} < \delta \Rightarrow A + E \in \mathcal{F}_S^\circ)\}. \qquad (6.6)$$

Proposition 6.28 *Let* $A \in (\mathbb{S}^{m-1})^n$ *and* $\mathcal{K} := \mathrm{conv}\{a_1, \ldots, a_n\}$.

(a) *If* $A \in \mathcal{F}_D$, *then*

$$\Delta(A) = \inf\{ \|y\| \mid y \in \mathcal{K} \}.$$

(b) *If* $A \in \mathcal{F}_P$, *then*

$$\Delta(A) = \sup\{\delta \mid \|y\| \leq \delta \Rightarrow y \in \mathcal{K}\}.$$

Proof (a) Assume that the perturbation E is such that $A + E \in \mathcal{F}_P$. Then there exists $x \geq 0$, $x \neq 0$, such that $(A + E)x = 0$. Without loss of generality assume $\|x\|_1 = 1$. Then $y := -Ex = Ax \in \mathcal{K}$. Moreover, $\|y\| \leq \|E\|_{12} \|x\|_1 = \|E\|_{12}$. Therefore

$$\|E\|_{12} \geq \inf\{ \|y\| \mid y \in \mathcal{K} \}.$$

Since this holds for all E such that $A + E \in \mathcal{F}_P$, it follows from (6.6) that $\Delta(A) \geq \inf\{\|y\| \mid y \in \mathcal{K}\}$.

To see the reverse inequality, assume that $y = Ax$ with $x \geq 0$, $\|x\|_1 = 1$, is given. Consider the rank-one perturbation

$$E := -yu^{\mathrm{T}},$$

where $u \in \mathbb{R}^n$ satisfies $\|u\|_\infty = 1$ and $u^{\mathrm{T}}x = 1$ (use (1.3)). This perturbation satisfies $\|E\|_{12} = \|y\|$ and $(A+E)x = Ax + Ex = y - y = 0$ with $0 \neq x \geq 0$. In other words,

$A + E \in \mathcal{F}_P$. Therefore

$$\Delta(A) \leq \|E\|_{12} = \|y\|.$$

Since this holds for arbitrary $y \in \mathcal{K}$, we conclude, using (6.6) again, that $\Delta(A) \leq \inf\{\|y\| \mid y \in \mathcal{K}\}$ as well.

(b) We set $\Omega = \{\delta \mid \|y\| \leq \delta \Rightarrow y \in \mathcal{K}\}$ and first show that

$$\forall y \in \mathbb{R}^m \quad (\|y\| < \Delta(A) \Rightarrow y \in \mathcal{K}), \tag{6.7}$$

which implies $\Delta(A) \leq \sup \Omega$. By contradiction, suppose that there exists $y \notin \mathcal{K}$ with $\|y\| < \Delta(A)$. The separating hyperplane theorem (Theorem 6.1) applied to the closed convex set $\mathcal{K}$ shows that there exists $u \in \mathbb{R}^m$ with $\|u\| = 1$ and $\lambda \in \mathbb{R}$ such that

$$\forall i \in [n], \quad \langle u, y \rangle < \lambda < \langle u, a_i \rangle.$$

By the Cauchy–Schwarz inequality,

$$-\lambda < -u^{\mathsf{T}} y \leq \|y\| < \Delta(A),$$

whence $\lambda > -\Delta(A)$. Theorem 6.27 implies that $\Delta(A) = \sin d_{\mathbb{S}}(A, \Sigma) = \sin(\rho(A) - \frac{\pi}{2}) = -\cos\rho(A)$, since we assume that $A \in \mathcal{F}_P$. We have shown that

$$\forall i \in [n] \quad \cos\rho(A) = -\Delta(A) < \lambda < u^{\mathsf{T}} a_i.$$

It follows that there is a spherical cap centered at u containing all the a_i that has a radius strictly smaller than $\rho(A)$. This is a contradiction and proves (6.7).

To show that $\sup \Omega \leq \Delta(A)$, let $E \in \mathbb{R}^{m \times n}$ be such that $A + E \notin \mathcal{F}_P$. Then, $A + E \in \mathcal{F}_D^\circ$, and hence there exists $y \in \mathbb{R}^m$ such that $(A + E)^{\mathsf{T}} y \geq 0$ and $\|y\| = 1$. This implies that $E^{\mathsf{T}} y \geq -A^{\mathsf{T}} y$ and hence that

$$\text{for all } x \in \mathbb{R}^n, \ x \geq 0, \quad x^{\mathsf{T}} E^{\mathsf{T}} y \geq -x^{\mathsf{T}} A^{\mathsf{T}} y. \tag{6.8}$$

Consider now any $\delta \in \Omega$. By (1.3) there exists $\bar{y} \in \mathbb{R}^m$, $\|\bar{y}\| = \delta$, such that $\bar{y}^{\mathsf{T}} y = -\delta$. Since $\delta \in \Omega$ there exists $x \in \mathbb{R}^n$, $x \geq 0$, $\|x\|_1 = 1$, such that $Ax = \bar{y}$. Hence, using (6.8),

$$y^{\mathsf{T}} Ex = x^{\mathsf{T}} E^{\mathsf{T}} y \geq -x^{\mathsf{T}} A^{\mathsf{T}} y = -y^{\mathsf{T}} Ax = -y^{\mathsf{T}} \bar{y} = \delta,$$

which implies

$$\|E\|_{12} \geq \|Ex\| = \|Ex\|\|y\| \geq |y^{\mathsf{T}} Ex| \geq \delta.$$

This shows, using (6.6) a last time, that $\Delta(A) \geq \sup \Omega$. $\qquad\square$

It is possible to give other characterizations of $\mathscr{C}(A)$ in the spirit of Proposition 1.11. As an example, we state without proof the following result.

Proposition 6.29 *Let* $A = [a_1, \ldots, a_n] \in (\mathbb{S}^{m-1})^n$. *If* $A \in \mathcal{F}_D$, *then*

$$\Delta(A) = \sup\{\delta \mid \|\bar{x}\|_\infty \leq \delta \Rightarrow \bar{x} \in \{A^{\mathsf{T}} v + \mathbb{R}_+^n \mid \|v\| \leq 1\}\}. \qquad\square$$

Fig. 6.5 Understanding
$\Xi(A)$ for $A \in \mathcal{F}_D$

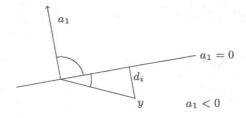

6.7 The GCC Condition Number and Well-Conditioned Solutions

The definition of $\mathscr{C}(A)$ given in Sect. 6.4 is in terms of a relativized distance to ill-posedness. Its characterization in Sect. 6.5 translates the space where the geometric property defining $\mathscr{C}(A)$ occurs from the space of data $(\mathbb{S}^{m-1})^n$—where $d_{\mathbb{S}}$ is defined—to the sphere $\mathbb{S}^{m-1}$—where smallest including caps are. With a little extra effort we can now look at $\mathbb{S}^{m-1}$ as the space of solutions for the problem $A^T y \leq 0$ and characterize $\mathscr{C}(A)$ in terms of the "best conditioned solution" (at least when $A \in \mathcal{F}_D$). This is the idea.

For $A \in \mathbb{R}^{m \times n}$ with nozero columns a_i we define

$$\Xi(A) := \min_{y \in \mathbb{S}^{m-1}} \max_{i \leq n} \frac{a_i^T y}{\|a_i\|}.$$

To understand $\Xi(A)$ assume $A \in \mathcal{F}_D$ and let $\mathsf{Sol}_D(A) = \{y \in \mathbb{S}^{m-1} \mid A^T y \leq 0\}$. This set is a polyhedral cone whose boundary is made of subsets of the hyperplanes $h_i := \{a_i^T y = 0\}$. Now consider $y \in \mathsf{Sol}_D(A)$. For each $i \in [n]$ we have $a_i^T y \leq 0$ and hence $\max_{i \leq n} \frac{a_i^T y}{\|a_i\|} \leq 0$. We claim that

$$-\max_{i \leq n} \frac{a_i^T y}{\|a_i\|} = \min d_i, \tag{6.9}$$

where d_i is the distance from y to the hyperplane h_i. Indeed, for each $i \in [n]$, we have (cf. Fig. 6.5)

$$d_i = \sin \angle(y, h_i) = -\cos \angle(y, a_i) = -\frac{a_i^T y}{\|a_i\|}$$

and consequently (6.9). Note that $\Xi_i \leq 0$ if and only if $A \in \mathcal{F}_D$.

Proposition 6.30 *For all $A \in \mathbb{R}^{m \times n}$ with nonzero columns, we have $|\Xi(A)| = \Delta(A)$.*

Proof By Theorem 6.27 it is enough to show that $\Xi(A) = -\cos \rho(A)$. To do so, we may assume in addition $\|a_i\| = 1$ for $i \in [n]$.

Let $\rho = \rho(A)$ and $p \in \mathbb{S}^{m-1}$ be such that $\mathsf{cap}(p, \rho)$ is an SIC for A. Take $\bar{y} = -p$. Then,

$$\mathcal{E}(A) \le \max_{i \le n} a_i^{\mathrm{T}} \bar{y} = - \min_{i \le n} a_i^{\mathrm{T}} p = - \cos \rho,$$

the last inequality resulting from $a_i \in \mathsf{cap}(p, \rho)$.

To prove the reverse inequality let $y_* \in \mathbb{S}^{m-1}$ be such that $\mathcal{E}(A) = \max_{i \le n} a_i^{\mathrm{T}} y_*$ and let $p = -y_*$ and $\alpha = \arccos(-\mathcal{E}(A))$. Then,

$$\min_{i \le n} a_i^{\mathrm{T}} p = - \max_{i \le n} a_i^{\mathrm{T}} y_* = -\mathcal{E}(A) = \cos \alpha.$$

It follows that $a_i \in \mathsf{cap}(p, \alpha)$ for all $i \in [n]$ and therefore that $\rho \le \alpha$. This implies $\mathcal{E}(A) = - \cos \alpha \ge - \cos \rho$. $\qquad\square$

6.8 Condition of Solutions and Condition Numbers

Proposition 6.30 introduces a new view for condition. In our first approach in the Overture we considered problems as functions $\varphi : \mathcal{D} \subseteq \mathbb{R}^m \to \mathbb{R}^q$. A number of natural problems, however, do not fit this pattern, since the desired output for a datum $a \in \mathcal{D}$ may not be uniquely specified, for instance, the problem of computing a complex root when given a univariate polynomial (which does not require any specific root to be returned), or the problem of, given a matrix $A \in \mathbb{R}^{m \times n}$, deciding whether $A \in \mathcal{F}_D$ and if so, returning a point $y \in \mathbb{R}^m \setminus \{0\}$ such that $A^{\mathrm{T}} y \le 0$.

For problems of this kind, we may approach conditioning from a different viewpoint. For an input a, let $\mathsf{Sol}(a)$ be its associated set of solutions (i.e., all the possible outputs for a). If for each $y \in \mathsf{Sol}(a)$ we have a number $\xi(a, y)$ quantifying the quality of the solution y, we may define the condition $\xi(a)$ of a by taking some function on the set $\{\xi(a, y) \mid y \in \mathsf{Sol}(a)\}$. Typical choices are

$$\xi(a) := \inf_{y \in \mathsf{Sol}(a)} \xi(a, y), \qquad \xi(a) := \mathop{\mathbb{E}}_{y \in \mathsf{Sol}(a)} \xi(a, y), \quad \text{and}$$

$$\xi(a) := \sup_{y \in \mathsf{Sol}(a)} \xi(a, y),$$

where the expectation in the middle expression is for some distribution on $\mathsf{Sol}(A)$. In the case of a matrix $A \in \mathcal{F}_D$ we have $\mathsf{Sol}_D(A) = \{y \in \mathbb{R}^m \setminus \{0\} \mid A^{\mathrm{T}} y \le 0\}$. If for $y \in \mathsf{Sol}_D(A)$, we define $\xi(A, y)$ by

$$\xi(A, y)^{-1} := \min_{i \le n} d_i = - \max_{i \le n} \frac{a_i^{\mathrm{T}} y}{\|a_i\| \|y\|}$$

(cf. (6.9)), then we have

$$\max_{y \in \mathsf{Sol}_D(A)} \xi(A, y)^{-1} = \max_{y \in \mathsf{Sol}_D(A)} - \max_{i \le n} \frac{a_i^{\mathrm{T}} y}{\|a_i\| \|y\|} = - \min_{y \in \mathsf{Sol}_D(A)} \max_{i \le n} \frac{a_i^{\mathrm{T}} y}{\|a_i\| \|y\|}$$

$$= -\mathcal{E}(A) = |\mathcal{E}(A)| = \Delta(A).$$

Therefore, $\mathscr{C}(A) = \min_{y \in \mathsf{Sol}_D(A)} \xi(A, y)$.

The quantity $\xi(A, y)^{-1}$ is the sine of the angular distance from y to the boundary of the cone $\mathsf{Sol}_D(A)$. The larger this distance, the better conditioned is the solution y. The equality $\mathscr{C}(A) = \min_{y \in \mathsf{Sol}_D(A)} \xi(A, y)$ thus expresses $\mathscr{C}(A)$ as the condition of the "best conditioned" point in $\mathsf{Sol}_D(A)$.

We finish this section by mentioning that we will encounter in Chaps. 17 and 18, in Part III, examples for the other two choices for $\xi(a)$, namely $\xi(a) := \mathbb{E}_{y \in \mathsf{Sol}(a)} \xi(a, y)$—the "average conditioned" solution—as well as $\xi(a) := \sup_{y \in \mathsf{Sol}(a)} \xi(a, y)$—the "worst conditioned" solution.

6.9 The Perceptron Algorithm for Feasible Cones

We close this chapter providing a first, simple, example of the use of $\mathscr{C}(A)$ in complexity analysis.

Assume we are given a matrix $A \in \mathbb{R}^{m \times n}$ such that $A \in \mathcal{F}_D^\circ$. Then, the set $\mathsf{Sol}_D^\circ(A) = \{y \in \mathbb{R}^m \mid A^{\mathrm{T}} y < 0\}$ is not empty, and we may be interested in finding a point in this set. Let us denote this problem by SLI (*system of linear inequalities*).

In what follows we describe an algorithm solving SLI, known as the *perceptron*, whose complexity is naturally analyzed in terms of $\mathscr{C}(A)$. One can devise an extension of this algorithm that actually decides whether $A \in \mathcal{F}_P^\circ$ or $A \in \mathcal{F}_D^\circ$, but we will proceed differently, postponing the issue to Chap. 9, where a different method to solve PCFP is described.

Let us denote by $a_1, \ldots, a_n$, the columns of A which, without loss of generality, we will assume to have norm one. That is, $a_j \in \mathbb{S}^{m-1}$ for $i = 1, \ldots, n$. The following is the perceptron algorithm.

Algorithm 6.1 Perceptron

Input: $a_1, \ldots, a_n \in \mathbb{S}^{m-1}$
Preconditions: $\{y \in \mathbb{R}^m \mid A^{\mathrm{T}} y < 0\} \neq \emptyset$

```
y := 0
repeat
        if A^T y < 0 then return y and halt
        else let j be the first index s.t. a_j^T y ≥ 0
            y := y - a_j
```

Output: $y \in \mathbb{R}^m$
Postconditions: $A^{\mathrm{T}} y < 0$

The role of $\mathscr{C}(A)$ in the analysis of Algorithm 6.1 is given in the following result

Theorem 6.31 *The number of iterations of Algorithm 6.1 is bounded by* $\mathscr{C}(A)^2$.

Proof Let $p \in \mathbb{S}^{m-1}$ and $\rho \in [0, \frac{\pi}{2})$ be such that an SIC for A is $\mathsf{cap}(p, \rho)$ (see Sect. 6.5). By Theorem 6.27, $\mathscr{C}(A) = (\cos \rho)^{-1}$. In addition, $w := -\frac{p}{\min_i a_i^\mathsf{T} p}$ is in $\mathsf{Sol}_D^\circ(A)$, and for all $j \leq n$ and $y \in \mathbb{R}^m$ such that $a_j^\mathsf{T} y \geq 0$, we have

$$\|y - a_j - w\|^2 = \|y - w\|^2 - 2a_j^\mathsf{T}(y - w) + 1 \leq \|y - w\|^2 - 2a_j^\mathsf{T} y + 2a_j^\mathsf{T} w + 1$$

$$\leq \|y - w\|^2 - 2\frac{a_j^\mathsf{T} p}{\min_i a_i^\mathsf{T} p} + 1 \leq \|y - w\|^2 - 2 + 1$$

$$= \|y - w\|^2 - 1.$$

A trivial induction shows that if y_k is the point produced at the kth iteration of the algorithm, then $\|y_k - w\|^2 \leq \|w\|^2 - k$. Hence, the algorithm stops after at most $\|w\|^2$ iterations. But

$$\|w\| = \frac{\|p\|}{\min_i a_i^\mathsf{T} p} = \frac{1}{\min_i a_i^\mathsf{T} p} = \frac{1}{\cos \rho},$$

with the last equality by Lemma 6.26. Since $\mathscr{C}(A) = \frac{1}{\cos \rho}$, we can conclude. □

Chapter 7
The Ellipsoid Method

In this chapter we describe an algorithm, known as the *ellipsoid method*, solving the problem SLI we described in Sect. 6.9. Its complexity analysis can also be done in terms of $\mathscr{C}(A)$, but in exchange for a loss of simplicity, we obtain bounds linear in $\ln \mathscr{C}(A)$ (instead of the quadratic dependence in $\mathscr{C}(A)$ of the perceptron algorithm).

We also introduce in this chapter, in its last section, a new theme: the use of condition numbers in the analysis of algorithms taking integer (as opposed to real) data. We will show that if the entries of $A \in \mathcal{F}_D^\circ$ are integer numbers, then one can return $y \in \mathrm{Sol}_D^\circ(A)$ with a cost—and since all our data are discrete, we mean bit cost (see Sect. 5.1)—polynomial in n, m and the bit-size of the largest entry in A.

7.1 A Few Facts About Ellipsoids

Definition 7.1 An *ellipsoid* in $\mathbb{R}^m$ with center $p \in \mathbb{R}^m$ is a set of the form

$$E = E(p, A) := \left\{ x \in \mathbb{R}^m \mid (x - p)^\mathsf{T} A^{-1} (x - p) \leq 1 \right\},$$

where $A \in \mathbb{R}^{m \times m}$ is a positive definite symmetric matrix.

In the special case that $p = 0$ and $A = \mathrm{diag}(\alpha_1^2, \ldots, \alpha_m^2)$ is a diagonal matrix, the ellipsoid $E(0, A)$ takes the special form

$$E(0, A) = \left\{ y \in \mathbb{R}^m \,\middle|\, \frac{y_1^2}{\alpha_1^2} + \cdots + \frac{y_m^2}{\alpha_m^2} \leq 1 \right\}. \tag{7.1}$$

The $\alpha_1, \ldots, \alpha_m$ can be interpreted as the lengths of the principal axes of $E(0, A)$.

It is straightforward to check that an invertible affine map $\phi \colon \mathbb{R}^m \to \mathbb{R}^m$, $x \mapsto Sx + v$, where $S \in \mathrm{GL}_m(\mathbb{R})$ and $v \in \mathbb{R}^m$, transforms ellipsoids into ellipsoids. More specifically, we have

$$\phi\big(E(p, A)\big) = E(q, B), \quad \text{where } q = Sp + v, \ B = SAS^\mathsf{T}. \tag{7.2}$$

P. Bürgisser, F. Cucker, *Condition*,
Grundlehren der mathematischen Wissenschaften 349,
DOI 10.1007/978-3-642-38896-5_7, © Springer-Verlag Berlin Heidelberg 2013

Fig. 7.1 The Löwner–John
ellipsoid E' of
$E(p, A) \cap \{x \in \mathbb{R}^2 \mid a^{\mathrm{T}}(x - p) \geq 0\}$

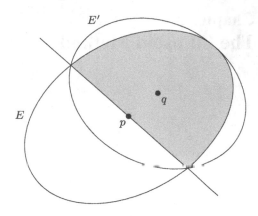

The volume of an ellipsoid can be expressed in terms of the determinant of the defining matrix A as follows.

Proposition 7.2 *We have* $\operatorname{vol} E(p, A) = \sqrt{\det A} \cdot \frac{\mathcal{O}_{m-1}}{m}$, *where* $\mathcal{O}_{m-1}$ *denotes the volume of the unit sphere* $\mathbb{S}^{m-1}$.

Proof There exists $S \in \mathrm{GL}_m(\mathbb{R})$ such that $B = SAS^{\mathrm{T}} = \mathrm{I}$ is the unit matrix. In particular, $1 = (\det S)^2 \det A$. By (7.2), the affine map ϕ corresponding to S and $v = -Sx_0$ maps $E(x_0, A)$ to the unit ball $E(0, \mathrm{I})$. Therefore,

$$\operatorname{vol} E(x_0, A) = (\det S)^{-1} \operatorname{vol} E(0, \mathrm{I}) = \sqrt{\det A} \operatorname{vol} E(0, \mathrm{I}).$$

In Corollary 2.20 it was shown that the volume of the unit ball in $\mathbb{R}^m$ equals $\operatorname{vol} E(0, \mathrm{I}) = \mathcal{O}_{m-1}/m$, which completes the proof. $\square$

Suppose we are given an ellipsoid $E = E(p, A)$ and a nonzero vector a in $\mathbb{R}^m$. We want to intersect E with the half-space $E \cap \{x \in \mathbb{R}^m \mid a^{\mathrm{T}}(x - p) \geq 0\}$ and enclose the resulting convex set in an ellipsoid E' of small volume (cf. Fig. 7.1).

The following result tells us how to do so.

Theorem 7.3 *Let* $E = E(p, A)$ *be an ellipsoid in* $\mathbb{R}^m$ *and* $a \in \mathbb{R}^m \setminus \{0\}$. *We define the symmetric matrix* A' *and the vector* p' *by*

$$p' := p + \frac{1}{m+1} \cdot \frac{1}{\sqrt{a^{\mathrm{T}}Aa}} \cdot Aa,$$

$$A' := \frac{m^2}{m^2 - 1}\left(A - \frac{2}{m+1} \cdot \frac{1}{a^{\mathrm{T}}Aa} \cdot Aaa^{\mathrm{T}}A^{\mathrm{T}}\right). \tag{7.3}$$

Then A' *is positive definite, and the ellipsoid* $E' = E(p', A')$ *satisfies*

$$E \cap \{x \in \mathbb{R}^m \mid a^{\mathsf{T}} x \geq a^{\mathsf{T}} p\} \subseteq E', \tag{7.4}$$

$$\mathrm{vol}\, E' < e^{-\frac{1}{2m}} \mathrm{vol}\, E. \tag{7.5}$$

Proof We first assume $p = 0$, $A = \mathrm{I}$, and $a = e_1 = (1, 0, \ldots, 0)$. Then $E = E(0, \mathrm{I})$ is the unit ball, and the half-space is described by $x_1 \geq 0$ (see Fig. 7.1). Equations (7.3) specialize to $p' := \frac{1}{m+1} e_1$ and

$$A' := \frac{m^2}{m^2 - 1}\left(\mathrm{I} - \frac{2}{m+1} \cdot e_1 e_1^{\mathsf{T}}\right) = \frac{m^2}{m^2 - 1} \mathrm{diag}\left(\frac{m-1}{m+1}, 1, \ldots, 1\right).$$

After some calculations we see that the ellipsoid $E' = E(p', A')$ is described by

$$\left(\frac{m+1}{m}\right)^2 \left(x_1 - \frac{1}{m+1}\right)^2 + \frac{m^2 - 1}{m^2} \sum_{i=2}^{m} x_i^2 \leq 1.$$

This can be easily rewritten as the inequality

$$\frac{m^2 - 1}{m^2} \sum_{i=1}^{m} x_i^2 + \frac{2(m+1)}{m^2} x_1(x_1 - 1) + \frac{1}{m^2} \leq 1. \tag{7.6}$$

We note that equality holds here for the standard basis vectors $e_1, \ldots, e_m$, which means that the boundary of E' contains the e_i, as suggested by Fig. 7.1. Using the characterization (7.6) of E', it is now easily checked that

$$\left\{x \in \mathbb{R}^m \;\middle|\; \sum_{i=1}^{m} x_i^2 \leq 1,\; x_1 \geq 0\right\} \subseteq E',$$

which proves (7.4). For proving (7.5), we note that by Proposition 7.2,

$$\frac{\mathrm{vol}\, E'}{\mathrm{vol}\, E} = \sqrt{\det A'} = \sqrt{\frac{m-1}{m+1}\left(\frac{m^2}{m^2 - 1}\right)^{\frac{m}{2}}} =: f(m). \tag{7.7}$$

For the function $f(m)$ on the right-hand side we have

$$f(m)^{-2m} = \left(1 + \frac{2}{m-1}\right)^{m-1}\left(1 + \frac{2}{m-1}\right)\left(1 - \frac{1}{m^2}\right)^{m^2},$$

which converges to $e^2 \cdot 1 \cdot e^{-1}$ for $m \to \infty$. A more detailed analysis shows that in fact, $f(m)^{-2m} \geq e$ for $m \geq 2$. This implies $(\frac{\mathrm{vol}\, E'}{\mathrm{vol}\, E})^{2m} \leq f(m)^{2m} \leq e^{-1}$ and hence (7.5).

The general case can be reduced to the special case we have just dealt with by the following considerations. We know that assertions (7.4)–(7.5) hold for $p = 0$, $A = \mathrm{I}$, and $a = e_1$. Let $\phi \colon \mathbb{R}^m \to \mathbb{R}^m$ be an affine transformation given by $S \in$

$GL_m(\mathbb{R})$ and $v \in \mathbb{R}^m$. By (7.2) we have $\phi(E(p, A)) = E(q, B)$, where $q = Sp + v$ and $B = SAS^T$. Defining $b = (S^{-1})^T a$, it is easily checked that

$$\phi(\{x \in \mathbb{R}^m \mid a^T(x - p) \geq 0\}) = \{y \in \mathbb{R}^m \mid b^T(y - q) \geq 0\}.$$

Now we define q' and B' as in (7.3) by

$$q' = q + \frac{1}{m+1} \cdot \frac{1}{\sqrt{b^T Bb}} \cdot Bb,$$

$$B' = \frac{m^2}{m^2 - 1}\left(B - \frac{2}{m+1} \cdot \frac{1}{b^T Bb} \cdot Bbb^T B^T\right).$$

We claim that the ellipsoid $E(q', B')$ satisfies

$$\phi(E(p', A')) = E(q', B'), \tag{7.8}$$

where p' and A' are as in the special case. Once this is proved, we can conclude from (7.4) for the triple (p, A, a), by applying ϕ, that

$$E(q, B) \cap \{y \in \mathbb{R}^m \mid b^T(y - q) \geq 0\} \subseteq E(q', B').$$

Moreover, from (7.5), also for the triple (p, A, a), it follows that $\text{vol}\, E(q', B') < e^{-\frac{1}{2m}}\,\text{vol}\, E(q, B)$, using that

$$\text{vol}\,\phi(E(p', A')) = \det S \cdot \text{vol}\, E(p, A), \qquad \text{vol}\,\phi(E(q', B')) = \det S \cdot \text{vol}\, E(q, B).$$

It therefore remains to verify the claim (7.8). A calculation yields

$$b^T Bb = a^T S^{-1} SAS^T (S^{-1})^T a = a^T Aa.$$

Moreover, we have $Bb = SAS^T(S^{-1})^T a = SAa$, and $Bbb^T B = SAaa^T A^T S^T$. From these observations one readily sees that $q' = Sq + v$ and $B' = SBS^T$, which proves the claim (7.8) and thus completes the proof of the theorem. $\square$

Remark 7.4

(a) The computation of p' and A' from p and A can be done with $\mathcal{O}(m^2)$ arithmetic operations (and one square root).

(b) It is known that for every convex compact set $K \subseteq \mathbb{R}^m$ there is a unique ellipsoid E of minimal volume containing K. One calls E the *Löwner–John ellipsoid*. It can be shown that the E' defined in Theorem 7.3 is the Löwner–John ellipsoid of $E \cap \{x \in \mathbb{R}^m \mid a^T(x - p) \geq 0\}$. See [114, §3.1] for more information.

7.2　The Ellipsoid Method

The goal of this section is to describe an algorithm finding a point in a nonempty convex closed subset $K \subseteq \mathbb{R}^m$. Before explaining the way the set K is specified, we recall from Theorem 6.1 that for a point $p \in \mathbb{R}^m$ not lying in K, there exists a *half-space H separating K from p*, that is, there exists $a \in \mathbb{R}^m$ such that $H = \{y \in \mathbb{R}^m \mid a^{\mathrm{T}}(y - p) \geq 0\}$ contains K (and the boundary of H contains p).

The algorithm to be described assumes the existence of (and makes calls to) a procedure that when given $p \in \mathbb{R}^m$, returns either the statement "$p \in K$" or a nonzero vector $a \in \mathbb{R}^m$ defining a half-space separating K from p. We call such a procedure a *separation oracle for K*. It also assumes that K is contained in the ball $B(p, R)$ and that both $p \in \mathbb{R}^m$ and $R > 0$ are given as input. Here is the general description of the ellipsoid method.

Algorithm 7.1 Ellip_Method

Input: $K \subseteq \mathbb{R}^m$ given by a separation oracle, $p \in \mathbb{R}^m$, $R > 0$

Preconditions: $K \neq \emptyset$ is convex and compact; $K \subseteq E_0 := B(p, R)$

```
y₀ := p
t := 0
repeat
      if  yₜ ∈ K  then return  yₜ  and halt
      if  yₜ ∉ K  use the separation oracle to find a
            separating half-space  H ⊇ K
      compute an ellipsoid  Eₜ₊₁  with center  yₜ₊₁
            containing  Eₜ ∩ H  by applying Theorem 7.3
      t := t + 1
```

Output: $y \in \mathbb{R}^m$

Postconditions: $y \in K$

In the case that K is a polyhedron, a separation oracle can be easily implemented. Indeed, suppose that $K \subseteq \mathbb{R}^m$ is given by a system of n linear inequalities ($a_i \in \mathbb{R}^m$, $b_i \in \mathbb{R}$)

$$a_1^{\mathrm{T}} y \leq b_1, \ldots, a_n^{\mathrm{T}} y \leq b_n.$$

Then a separation oracle can be implemented with $\mathcal{O}(mn)$ arithmetic operations. For a given $p \in \mathbb{R}^m$ one just checks the conditions $a_i^{\mathrm{T}} p \leq b_i$ for all i. If $p \notin K$, then one finds an index i such that $a_i^{\mathrm{T}} p > b_i$, and hence

$$H := \left\{ y \in \mathbb{R}^m \mid -a_i^{\mathrm{T}}(y - p) \geq 0 \right\}$$

is a half-space separating K from p.

Theorem 7.5 *The ellipsoid method works correctly. That is, it returns a point in K. Furthermore, for $m \geq 4$, the number of iterations it performs on input (K, R, p) is bounded by $\lceil 3(m + 1) \ln \frac{V}{v} \rceil$, where $V := \mathrm{vol}\, B(p, R)$ and $v = \mathrm{vol}\, K$.*

Proof Note that if the algorithm reaches iteration t, then $K \subseteq E_t$. Moreover, Theorem 7.3 implies that

$$\text{vol } E_t \leq e^{-\frac{t}{3(m+1)}} \text{ vol } E_0.$$

Hence, for $t > \lceil 3(m+1) \ln \frac{V}{\nu} \rceil$ we have $e^{-\frac{t}{3(m+1)}} \text{ vol } E_0 < \nu$ and therefore $\text{vol } K \leq \text{vol } E_t < \nu$, a contradiction. The claimed bound follows.   □

We can now proceed to analyze the complexity of the ellipsoid method for the problem SLI of finding a point in $\text{Sol}_D(A) = \{y \in \mathbb{R}^m \mid y \neq 0, A^T y \leq 0\}$ for given $A \in \mathcal{F}_D^o$ in terms of the GCC condition number $\mathscr{C}(A)$. Since we are dealing with cones, $\text{Sol}_D(A)$ is nonempty iff the convex compact set $K_A := \text{Sol}_D(A) \cap B(0,1)$ is nonempty.

We will apply Algorithm 7.1 to the set K_A given by the separation oracle explained before. (We note that even though the inequality $\sum_i y_i^2 \leq 1$ has to be added to the description of K_A, this inequality is never violated during the execution of the algorithm.) So we may take $p = 0$ and $R = 1$ in Algorithm 7.1. The next result shows that, in addition, we can replace the quotient of volumes V/ν by a simple function of the GCC condition number of the data A.

Lemma 7.6 *Let* $\text{cap}(p, \rho)$ *be an SIC for* $A \in \mathcal{F}_D^o$. *Then* $B(-p, \mathscr{C}(A)^{-1})$ *is contained in* K_A, *and consequently,* $\text{vol } B(0, \mathscr{C}(A)^{-1}) \leq \text{vol } K_A$.

Proof We first note that it is sufficient to show that $\text{cap}(-p, \frac{\pi}{2} - \rho) \subseteq K_A$, since $\mathscr{C}(A)^{-1} = \Delta(A) = \cos \rho$.

So assume that $y \in S^{m-1}$ satisfies $d_{\mathbb{S}}(y, -p) \leq \pi/2 - \rho$. Then we have $d_{\mathbb{S}}(y, p) \geq \pi/2 + \rho$. Moreover, since $d_{\mathbb{S}}(a_i, p) \leq \rho$, we conclude that for all i,

$$d_{\mathbb{S}}(y, a_i) \geq d_{\mathbb{S}}(y, p) - d_{\mathbb{S}}(p, a_i) \geq \pi/2.$$

This means that $a_i^T y \leq 0$ for all i and hence $y \in K_A$.   □

So in the analysis of Algorithm 7.1 we have

$$\frac{V}{\nu} \leq \frac{\text{vol } B(0,1)}{\text{vol } B(0, \mathscr{C}(A)^{-1})} = \mathscr{C}(A)^m.$$

Combining this observation with Theorem 7.5, we immediately get the following result.

Corollary 7.7 *Let* $A \in \mathbb{R}^{m \times n}$ *be such that* $A \in \mathcal{F}_D^o$. *The ellipsoid method, applied to* K_A, *finds a feasible point* $y \in K_A$ *with a number of iterations bounded by*

$$\lceil 3m(m+1) \ln(\mathscr{C}(A)) \rceil.$$

Hereby, each iteration step costs $\mathcal{O}(mn)$ *arithmetic operations for the implementation of the separation oracle for* K_A *and* $\mathcal{O}(m^2)$ *arithmetic operations (plus one* *square root) for the computation of the next ellipsoid. The total number of arithmetic operations can therefore be bounded by $\mathcal{O}(m^3 n \ln \mathscr{C}(A))$ operations.*   □

7.3   Polyhedral Conic Systems with Integer Coefficients

One of the facts giving historical relevance to the ellipsoid method is its protagonism in showing, in 1979, that linear programming problems with integer data can be solved in polynomial time. In this section we will show that this is the case for the solution of SLI with integer data matrices. The key result allowing us to do so is the following proposition relating condition and bit-size of data, which will be used in other contexts as well (cf. Remarks 9.18 and 10.5).

Proposition 7.8 (Hadamard's inequality)   *For $A = [a_1, \ldots, a_n] \in \mathbb{R}^{n \times n}$ we have*

$$|\det A| \leq \|a_1\| \|a_2\| \cdots \|a_n\|.$$

Proof Without loss of generality we assume that $\det A \neq 0$. Then the span S of $a_1, \ldots, a_{n-1}$ has the dimension $n - 1$, and by applying an orthogonal transformation we can achieve that $S = \mathbb{R}^{n-1} \times 0$. Since orthogonal transformations change neither the value of the determinant nor the lengths of the columns of A, it is sufficient to prove the assertion in the special case $S = \mathbb{R}^{n-1} \times 0$. We then have $a_i = (a_i', 0)$ with $a_i' \in \mathbb{R}^{n-1}$ for $1 \leq i \leq n - 1$. Therefore $\det A = a_{nn} \det A'$, where $A' = [a_1', \ldots, a_{n-1}']$. The assertion follows now by induction on n, using that $|a_{nn}| \leq \|a_n\|$.   □

Proposition 7.9 *Let $\widetilde{A} \in \mathbb{Z}^{m \times n}$ be an integer matrix with entries bounded in absolute value by U. We assume that the columns $\widetilde{a}_i$ of $\widetilde{A}$ are nonzero and form $A := [a_1, \ldots, a_n]$, where $a_i := \widetilde{a}_i / \|\widetilde{a}_i\|$. Then we have $\mathscr{C}(A) \leq (mU)^{\mathcal{O}(m)}$, provided $\mathscr{C}(A)$ is finite.*

Proof Let $p \in \mathbb{S}^{m-1}$ and $\rho \in [0, \pi]$ be such that $\mathsf{cap}(p, \rho)$ is an SIC for A with blocking set $[k]$ and put $t := \cos \rho$. We may assume $A \notin \Sigma$, so that $t \neq 0$. Lemma 6.26 implies $tp \in \mathsf{conv}\{a_1, \ldots, a_k\}$ and

$$a_1^{\mathsf{T}} p = \cdots = a_k^{\mathsf{T}} p = t. \tag{7.9}$$

Without loss of generality, let $a_1, \ldots, a_\ell$ be a basis of $\mathsf{span}\{a_1, \ldots, a_k\}$. Then the Gram matrix

$$G := [G_{ij}]_{1 \leq i, j \leq \ell} \quad \text{with } G_{ij} = a_i^{\mathsf{T}} a_j$$

is invertible.

Since $tp \in \mathsf{conv}\{a_1, \ldots, a_k\} \subseteq \mathsf{span}\{a_1, \ldots, a_\ell\}$, there exist ξ_j such that $p = \sum_{j=1}^{\ell} \xi_j a_j$. From $p^{\mathsf{T}} p = 1$ we deduce that $\xi^{\mathsf{T}} G \xi = 1$. On the other hand, (7.9) implies that $G\xi = t\mathbf{e}_\ell$, or $\xi = tG^{-1}\mathbf{e}_\ell$. Plugging this into the equality $\xi^{\mathrm{T}}G\xi = 1$, we obtain that

$$\mathscr{C}(A) = |t|^{-1} = \sqrt{\mathbf{e}_\ell^{\mathrm{T}}G^{-1}\mathbf{e}_\ell}.$$

It is therefore sufficient to show that the entries of G^{-1} are bounded as $(mU)^{\mathcal{O}(m)}$.

For this, we introduce the matrix $\tilde{G}$ with the entries $\tilde{a}_i^{\mathrm{T}}\tilde{a}_j$ and note that $\tilde{G} = \Delta G\Delta$ with the diagonal matrix $\Delta = \mathsf{diag}(\|a_1\|, \ldots, \|a_\ell\|)$. It follows that $G^{-1} = \Delta\tilde{G}^{-1}\Delta$, and hence it suffices to bound the entries of $\tilde{G}^{-1}$ by $(mU)^{\mathcal{O}(m)}$.

By Cramer's rule, we have

$$\left(\tilde{G}^{-1}\right)_{ij} = (-1)^{i+j} \det M_{ji} / \det \tilde{G},$$

where the minor M_{ji} is obtained from $\tilde{G}$ by omitting the jth row and the ith column. The assertion follows now from Hadamard's inequality (Proposition 7.8) and $\det \tilde{G} \geq 1$, which holds since $\tilde{G}$ is an invertible integer matrix. $\square$

Remark 7.10 Proposition 7.9 combined with Corollary 7.7 implies that for a matrix $\tilde{A} \in \mathbb{Z}^{m \times n}$ such that $A \in \mathcal{F}_D^\circ$, the ellipsoid method finds a feasible point $y \in \mathsf{Sol}_D(\tilde{A})$ with $\mathcal{O}(m^3 \log(mU))$ iterations. Furthermore, it can be shown that it is enough to implement the arithmetic operations and square roots to a precision of $(m \log U)^{\mathcal{O}(a)}$ digits. The overall number of bit operations is then polynomial in the bit-size of the input matrix $\tilde{A}$.

Chapter 8
Linear Programs and Their Solution Sets

The polyhedral cone feasibility problem PCFP that occupied us in the last two chapters, though fundamental, is better understood when regarded within the more general context of linear programming. Succinctly described, the latter is a family of problems that consist in optimizing (i.e., maximizing or minimizing) a linear function over a set defined by linear *constraints* (equalities and/or inequalities).

A first step towards the solution of such a problem requires one to decide whether the family of constraints is satisfiable, that is, whether it defines a nonempty set. The polyhedral cone feasibility problem is a particular case of such a requirement.

Interestingly, optimization and feasibility problems appear to reduce to one another. Thus, in Sect. 9.4, we will solve PCFP by recasting it as an optimization problem. Conversely, in Sect. 11.3.2, we will reduce the solution of optimization problems to a sequence of instances of PCFP.

Because of these considerations, before proceeding with the exposition of new algorithms, we make a pause and devote it to the understanding of linear programs and their sets of solutions. As usual, such an understanding will prove of the essence at the moment of defining condition.

8.1 Linear Programs and Duality

We start with a brief review of the basic concepts of linear programming. Because of the possible forms of the constraints of a linear program, as well as the choice maximization/minimization, linear programs occur in a variety of different shapes. They are all, however, equivalent in the sense that they can all be reduced to a single simple form. The most common such form, called *standard*, owes its widespread use to the fact that the first efficient algorithm developed to solve linear programs, the simplex method, applies to linear programs in this form. For use in subsequent chapters, we will consider in this section a slightly more general form, namely

P. Bürgisser, F. Cucker, *Condition*,
Grundlehren der mathematischen Wissenschaften 349,
DOI 10.1007/978-3-642-38896-5_8, © Springer-Verlag Berlin Heidelberg 2013

Table 8.1 Construction of the dual of a linear program

Maximization problem	← Dual →	Minimization problem
ith inequality ($\leq$) constraint		ith nonnegative variable
ith equality ($=$) constraint		ith unrestricted variable
jth nonnegative variable		jth inequality ($\geq$) constraint
jth unrestricted variable		jth equality ($=$) constraint
objective function coefficients		constant terms of constraints
constant terms of constraints		objective function coefficients
matrix of constraints A		matrix of constraints A^{T}

$$
\begin{aligned}
\min \quad & c^{\mathrm{T}}x + d^{\mathrm{T}}w \\
\text{s.t.} \quad & Ax + Gw = b, \\
& x \geq 0,
\end{aligned}
\tag{OP}
$$

where $A \in \mathbb{R}^{m \times n}$, $G \in \mathbb{R}^{m \times p}$, $b \in \mathbb{R}^m$, $c \in \mathbb{R}^n$, $d \in \mathbb{R}^p$, are the given data and we look for an optimal vector $(x, w) \in \mathbb{R}^{n+p}$. We say that (OP) is *feasible* if there exists $(x, w) \in \mathbb{R}^{n+p}$ such that $Ax + Gw = b$ and $x \geq 0$. The set of all such pairs is the *feasible set* of (OP).

The function $(x, w) \mapsto c^{\mathrm{T}}x + d^{\mathrm{T}}w$ is the *objective function*. A feasible linear program (OP) is called *bounded* if the minimum of the objective function is finite. Otherwise, it is called *unbounded*. In the first case this minimum is called the *optimal value*, and any feasible point (x, w) attaining it is an *optimal solution* (or an *optimizer*) of (OP).

Linear programming possesses a beautiful theory of duality. To any linear program one can associate another one, called its *dual*, which is obtained in a precise manner succinctly described in Table 8.1.

For the linear program (OP), given by the data A, G, b, c, d, we obtain as dual the following:

$$
\begin{aligned}
\max \quad & b^{\mathrm{T}}y \\
\text{s.t.} \quad & A^{\mathrm{T}}y \leq c, \\
& G^{\mathrm{T}}y = d
\end{aligned}
\tag{OD}
$$

where $y \in \mathbb{R}^m$. The notions of feasibility, boundedness, and optimality also apply here. Furthermore, the essence of duality theory is the interplay of these notions in both the primal and dual problems. We next elaborate on this interplay.

Feasible sets are intersections of hyperplanes and closed half-spaces. Therefore, they are closed convex subsets of Euclidean space. The following elementary lemma, whose proof we leave to the reader, recalls some facts of linear functions on convex sets.

Lemma 8.1

(a) *A linear function ℓ defined on a convex set $C \subseteq \mathbb{R}^q$ has no extrema in the relative interior of C unless it is constant on C.*

(b) *Under the hypothesis of (a), if C is closed and $\sup_{x \in C} \ell(x) < \infty$, then there exists $x^* \in C$ such that $\sup_{x \in C} \ell(x) = \ell(x^*)$. A similar statement holds for $\inf \ell$.*
(c) *If a linear function is constant on a set S, then it is constant on its convex hull* conv(S). $\square$

An important consequence of Lemma 8.1(b) is that if a linear program is feasible and bounded, then it has optimal solutions.

Suppose now that (OP) and (OD) are both feasible, say

$$Ax + Gw = b, \quad x \geq 0,$$

$$A^{\mathsf{T}} y \leq c, \qquad G^{\mathsf{T}} y = d, \quad \text{for some } x \in \mathbb{R}^n, \ w \in \mathbb{R}^p, \ y \in \mathbb{R}^m.$$

Introducing the vector $s := c - A^{\mathsf{T}} y$ of *slack variables*, we have $A^{\mathsf{T}} y + s = c$ and $s \geq 0$. Then

$$c^{\mathsf{T}} x + d^{\mathsf{T}} w - b^{\mathsf{T}} y = (s^{\mathsf{T}} + y^{\mathsf{T}} A)x + y^{\mathsf{T}} Gw - b^{\mathsf{T}} y$$

$$= s^{\mathsf{T}} x + y^{\mathsf{T}} (Ax + Gw - b) = s^{\mathsf{T}} x \geq 0. \tag{8.1}$$

In particular, for any feasible points (x, w) and y, we have

$$c^{\mathsf{T}} x + d^{\mathsf{T}} w \geq b^{\mathsf{T}} y. \tag{8.2}$$

It follows that if (OP) and (OD) are both feasible, then they are both bounded and $\max b^{\mathsf{T}} y \leq \min(c^{\mathsf{T}} x + d^{\mathsf{T}} w)$. The fundamental duality theorem of linear programming states that actually equality holds.

Theorem 8.2 (Duality theorem of linear programming)

(a) *The problem (OP) is bounded iff (OD) is bounded. In this case both problems have optimal solutions and their objective values are equal.*
(b) *If (OP) is unbounded, then (OD) is infeasible. If (OD) is unbounded, then (OP) is infeasible.*

Proof We have proved part (b) above. To prove part (a) we will show that if (OP) has an optimal solution then so has (OD), and in this case the optimal values of these problems coincide. The proof of the converse is similar.

Assume that (x^*, w^*) is an optimal solution of (OP) and let $v^* := c^{\mathsf{T}} x^* + d^{\mathsf{T}} w^*$ be the corresponding optimal value. For $\varepsilon \geq 0$ let $v_\varepsilon := v^* - \varepsilon$. Define

$$\mathcal{A} := \begin{bmatrix} A & G & -G \\ -c^{\mathsf{T}} & -d^{\mathsf{T}} & d^{\mathsf{T}} \end{bmatrix} \quad \text{and} \quad \mathbf{b}_\varepsilon := \begin{bmatrix} b \\ v_\varepsilon \end{bmatrix}.$$

Then, using that any real number can be written as the difference of two nonnegative reals, the system

$$\mathcal{A}(x, w', w'') = \mathbf{b}_\varepsilon, \quad x, w', w'' \geq 0,$$

is feasible when $\varepsilon = 0$ and infeasible when $\varepsilon > 0$. Farkas's lemma (Lemma 6.4) then implies that

$$\mathcal{A}^T(y, t) \leq 0, \qquad \mathbf{b}_\varepsilon^T(y, t) > 0$$

is infeasible when $\varepsilon = 0$ and feasible when $\varepsilon > 0$. This is equivalent to saying (now use that $z = 0$ if and only if $z \leq 0$ and $z \geq 0$) that the system

$$A^T y \leq ct, \qquad G^T y = dt, \qquad b^T y - v_\varepsilon t > 0 \tag{8.3}$$

is infeasible when $\varepsilon = 0$ and feasible when $\varepsilon > 0$.

For $\varepsilon > 0$ let $(y_\varepsilon, t_\varepsilon)$ be a solution of (8.3). Note that if $t_\varepsilon \leq 0$, then

$$0 < b^T y - v_\varepsilon t_\varepsilon = b^T y - v^* t_\varepsilon + \varepsilon t_\varepsilon \leq b^T y - v^* t_\varepsilon,$$

and hence a solution $(y_\varepsilon, t_\varepsilon)$ with $t_\varepsilon \leq 0$ would be a solution of the system for $\varepsilon = 0$ as well, which is a contradiction. We conclude that $t_\varepsilon > 0$. Dividing by t_ε, it follows that the system

$$A^T y \leq c, \qquad G^T y = d, \qquad b^T y - v_\varepsilon > 0$$

is infeasible when $\varepsilon = 0$ and feasible when $\varepsilon > 0$. That is, the linear function $\ell\colon y \mapsto b^T y$ is bounded above by v^* on the feasible set $\mathcal{S}_D$ of (OD) and its image on this set contains points arbitrarily close to v^*. Hence, $\sup_{y \in \mathcal{S}_D} b^T y = v^*$. Lemma 8.1(b) allows one to conclude that the maximum of ℓ is attained on the feasible set of (OD) and has the value v^*.   □

It is rewarding to consider (OP) and (OD) simultaneously. We define the *polyhedral set $\mathcal{S}$ of (primal–dual) feasible solutions* to be the set of points $z = (x, w, y, s) \in \mathbb{R}^{n+p+m+n}$ satisfying

$$Ax + Gw = b, \qquad A^T y + s = c, \qquad G^T y = d, \quad x \geq 0, \ s \geq 0. \tag{8.4}$$

We note that $\mathcal{S}$ is convex. We further note the following fundamental result.

Theorem 8.3 (Complementary slackness) *Let $(x, w, y, s) \in \mathcal{S}$. Then (x, w) is an optimal solution of (OP) and y is an optimal solution of (OD) if and only if*

$$x_1 s_1 = 0, \ldots, x_n s_n = 0. \tag{8.5}$$

Proof It follows from (8.1) and Theorem 8.2.   □

The equality (8.5) is known as the *complementary slackness condition*. We call relations (8.4) together with (8.5) *optimality conditions*. For a point $(x, w, y, s) \in \mathcal{S}$, the value $c^T x + d^T w - b^T y = s^T x$ is called the *duality gap*. Interior-point methods, which will be the theme of the next chapter, work by starting with a point in $\mathcal{S}$ and iteratively constructing a sequence of points in $\mathcal{S}$ with a fast decrease in their duality gap.

We close this section by giving the *standard form* mentioned at the beginning of this section. This is the linear programming form that will occupy us for the rest of this chapter (and in some chapters to come). Both the primal and the dual are obtained by removing all terms in (OP) and (OD) in which any of G, w, and d occurs. Thus, in the primal case we obtain

$$\text{min } c^{\mathsf{T}}x \quad \text{subject to} \quad Ax = b, \quad x \geq 0, \tag{SP}$$

and in the dual,

$$\text{max } b^{\mathsf{T}}y \quad \text{subject to} \quad A^{\mathsf{T}}y \leq c. \tag{SD}$$

In what follows, we will consider linear programs in standard form, and we will systematically assume that $n \geq m$. The first result we prove for this form is the following strengthening of the complementary slackness condition (8.5).

Proposition 8.4 (Strict complementary theorem) *If* (SP) *and* (SD) *are both feasible, then there exist optimizers* (x^*, y^*, s^*) *such that*

$$x_i^* = 0 \iff s_i^* > 0 \quad \text{for } i = 1, \ldots, n.$$

Proof We will first show that for each $i = 1, \ldots, n$ there exists an optimal solution $(x^{(a)}, y^{(a)}, s^{(a)})$ such that either $x_i^{(a)} \neq 0$ or $s_i^{(a)} \neq 0$.

Let v^* be the optimal value of the pair (SP–SD) and consider an optimizer $\bar{x}$ of (SP). If there exists an optimizer (y, s) of (SD) with $s_i \neq 0$, we take $(y^{(a)}, s^{(a)}) := (y, s)$ and we are done.

If instead, $s_i = 0$ for every optimizer (y, s) of (SD), then the linear program

$$
\begin{aligned}
\text{max} \quad & e_i^{\mathsf{T}}s \\
\text{s.t.} \quad & A^{\mathsf{T}}y + s = c, \\
& -b^{\mathsf{T}}y = -v^*, \\
& s \geq 0,
\end{aligned}
$$

where $e_i = (0, \ldots, 1, \ldots, 0)$ is the ith coordinate vector, has optimal value 0. By Theorem 8.2, its dual

$$
\begin{aligned}
\text{min} \quad & c^{\mathsf{T}}x - v^*t \\
\text{s.t.} \quad & Ax - bt = 0, \\
& x \geq e_i, \\
& x, t \geq 0,
\end{aligned}
$$

has then a feasible solution $(x, t) \in \mathbb{R}^{n+1}$ with objective value $c^{\mathsf{T}}x - v^*t = 0$.

Assume that for this solution we have $t = 0$. Then $c^{\mathsf{T}}x = 0$, $Ax = 0$, and $x \geq e_i$. This implies that $x^{(a)} := \bar{x} + x$ is an optimizer of (SP) and $x_i^{(a)} \neq 0$.

Assume now that instead, $t > 0$. Then the point $x^{(a)} := \frac{x}{t}$ is an optimizer of (SP) and $x_i^{(a)} \neq 0$.

We have therefore proved our initial claim. It is now immediate to see that the points

$$x^* := \frac{1}{n} \sum_{i=1}^{n} x^{(a)} \quad \text{and} \quad (y^*, s^*) := \frac{1}{n} \sum_{i=1}^{n} (y^{(a)}, s^{(a)})$$

satisfy that for all $i \leq n$, either $x_i^* > 0$ or $s_i^* > 0$. In addition, they are optimizers of (SP) and (SD) respectively, since they are convex combinations of optimizers and the optimal sets of linear programs are convex (Proposition 8.7(b)). The fact that not both x_i^* and s_i^* are greater than zero is, finally, a consequence of complementary slackness (8.5).   ⊔

Remark 8.5 A word of caution is called for regarding names. We have used the expression "standard primal" (and the corresponding tag (SP)) and likewise for the "standard dual." This choice of words follows a long established tradition that has its roots in the fact that the simplex method runs (only) on linear programs with the form (SP). It must be noted, however, that there are no naturally primal (or naturally dual) problems. To any given problem we may associate its dual using the method implicit in Table 8.1. And the dual of the dual is the original linear program.

8.2 The Geometry of Solution Sets

We denote by $\mathcal{S}_P$ and $\mathcal{S}_D$ the primal and dual feasible sets for $d = (A, b, c)$ respectively, that is,

$$\mathcal{S}_P := \{x \in \mathbb{R}^n \mid Ax = b, x \geq 0\}, \qquad \mathcal{S}_D := \{y \in \mathbb{R}^m \mid A^T y \leq c\}.$$

We also denote by $\mathcal{Q}_P$ and $\mathcal{Q}_D$ the corresponding sets of optimal solutions.

Proposition 8.6

(a) *Both $\mathcal{S}_P$ and $\mathcal{S}_D$ are polyhedra.*
(b) *If $\mathcal{S}_P$ is nonempty, then it contain vertices. The same holds true for $\mathcal{S}_D$ if in addition, rank $A = m$.*

Proof Part (a) is trivial. For part (b) recall that by Corollary 6.13, if a polyhedron has no vertices, then it contains a line. The fact that $\mathcal{S}_P$ contains vertices is then clear, since the set $\mathcal{S}_P$ is included in the positive orthant $\{x \in \mathbb{R}^n \mid x \geq 0\}$, and this set does not contain lines.

For the dual, we use that if there is a line ℓ contained in $\mathcal{S}_D$, then by Lemma 6.11, rank $A < m$, a contradiction.   □

Proposition 8.7

(a) *The sets $\mathcal{S}_P$ and $\mathcal{S}_D$ are both nonempty if and only if $\mathcal{Q}_P$ and $\mathcal{Q}_D$ are both nonempty.*

(b) *If this is the case, then $\mathcal{Q}_P$ and $\mathcal{Q}_D$ are faces of $\mathcal{S}_P$ and $\mathcal{S}_D$, respectively. In particular, they are polyhedra as well.*
(c) *In addition, $\mathcal{Q}_P$ possesses vertices, and so does $\mathcal{Q}_D$ if* rank $A = m$.

Proof Part (a) is an immediate consequence of Theorem 8.2(a). We then proceed to part (b), which we will prove for the primal case (the dual being similar).

If the objective function is constant on $\mathcal{S}_P$, then $\mathcal{Q}_P = \mathcal{S}_P$ and we are done. Assume then that this is not the case. We claim that $\mathcal{Q}_P$ is a union of proper faces of $\mathcal{S}_P$. Indeed, because of Lemma 8.1(a) with $C = \mathcal{S}_P$, we must have $\mathcal{Q}_P \subseteq \partial\mathcal{S}_P$, that is, $\mathcal{Q}_P$ is included in the union of the proper faces of $\mathcal{S}_P$. The same lemma applied to each of these faces shows that either the whole face is in $\mathcal{Q}_P$ or its intersection with $\mathcal{Q}_P$ is in a lower-dimensional subface. Repeating this argument proves the claim, i.e., $\mathcal{Q}_P$ is a union of proper faces of $\mathcal{S}_P$. If this union consists of a single face, we are done. Assume the contrary, and let F be a face of $\mathcal{S}_P$ of maximal dimension among those included in $\mathcal{Q}_P$. By assumption, there exists a face F', also included in $\mathcal{Q}_P$, such that neither $F' \subseteq F$ nor $F \subseteq F'$. Lemma 8.1(c) implies that conv$(F \cup F') \subseteq \mathcal{Q}_P$. But by Lemma 6.10, dim conv$(F \cup F') > \dim F = \dim \mathcal{Q}_P$, in contradiction to this inequality.

For part (c), assume $\mathcal{S}_P \neq \emptyset$. Then, Proposition 8.6(b) ensures that $\mathcal{S}_P$ has vertices, which implies, by Corollary 6.13, that $\mathcal{S}_P$ does not contain lines. Therefore, neither does $\mathcal{Q}_P$, and the same corollary (together with part (b)) implies that $\mathcal{Q}_P$ possesses vertices. A similar argument applies to $\mathcal{Q}_D$. $\square$

It is a common convention to assign dimension -1 to the empty set. With this convention, (SP) is feasible and bounded if and only if dim $\mathcal{Q}_P \geq 0$, and likewise for (SD). We can further distinguish among linear programming data as follows.

We say that a triple $d = (A, b, c)$ is *heavy for* (SP) (or *primal-heavy* when dim $\mathcal{Q}_P \geq 1$), and that it is *light*, i.e., $\mathcal{Q}_P$ is a vertex of $\mathcal{S}_P$, otherwise. Similarly for (SD). We say that d is *heavy* when it is either primal-heavy or dual-heavy.

Figure 8.1 shows examples of light and heavy instances for (SD) (the arrow showing the optimization direction, the lines and points in bold, the sets of optimal solutions). At the left we see an instance corresponding to a light triple. Both at the center and at the right are instances corresponding to heavy data, but the optimal set $\mathcal{Q}_D$ in the former is compact and in the latter is not. Because of this, for the data at the right, arbitrarily small perturbations may make the problem unbounded and consequently its dual (SP) infeasible. This is not possible for the central situation.

Proposition 8.8 *Let d be primal-heavy such that $\mathcal{Q}_P$ is noncompact but* (SP) *is bounded. Then, there exist arbitrarily small perturbations $\tilde{d}$ of d for which* (SP) *is unbounded (and hence* (SD) *infeasible). A similar statement holds for $\mathcal{Q}_D$.*

Proof Since $\mathcal{Q}_P$ is noncompact, there exist $x, w \in \mathbb{R}^n$, $\|w\| = 1$, such that $x_\lambda := x + \lambda w \in \mathcal{Q}_P$, for all $\lambda \geq 0$. Because (SP) is bounded we must have $c^\mathsf{T} x_\lambda = c^\mathsf{T} x + \lambda c^\mathsf{T} w = v^*$, for all $\lambda \geq 0$ (here v^* is the optimal value of d). This implies $c^\mathsf{T} w = 0$.

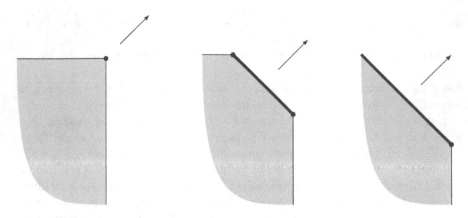

Fig. 8.1   Three situations for linear programs

Table 8.2   Possible optimal sets of a linear program

	dim $\mathcal{Q}$		
	-1	0	> 0
$\mathcal{Q}$ compact	infeasible	light	heavy
$\mathcal{Q}$ noncompact			heavy with dual nearly infeasible

Consider now, for any $\varepsilon > 0$, the point $\tilde{c} := c - \varepsilon w$ and the triple $\tilde{d} := (A, b, \tilde{c})$. Note that the set of feasible points of (SP) for $\tilde{d}$ coincides with that for d. In particular, it contains $\mathcal{Q}_P$. Also, for all $\lambda \geq 0$,

$$\tilde{c}^{\mathsf{T}} x_\lambda = c^{\mathsf{T}} x - \varepsilon w^{\mathsf{T}} x - \lambda \varepsilon.$$

Therefore, $\tilde{c}^{\mathsf{T}} x_\lambda \to -\infty$ when $\lambda \to \infty$, which shows that (SP) is unbounded for $\tilde{d}$. Since ε is arbitrarily small, the conclusion follows. $\square$

We can summarize the distinctions above in Table 8.2 (where empty boxes denote impossible situations, "dual" refers to the dual of the given problem, which may be either (SP) or (SD), and we used the expression *nearly infeasible* to denote that arbitrarily small perturbations may yield infeasibility).

We say that x^* is an *extremal optimal solution* of (SP) (or of the primal) when x^* is a vertex of $\mathcal{Q}_P$, and similarly for the dual problem.

8.3   The Combinatorics of Solution Sets

Proposition 8.7 ensures that if the primal–dual pair (SP–SD) is feasible, and rank $A = m$, then one may confine the search for optimizers to the vertices of the sets $\mathcal{Q}_P$ and $\mathcal{Q}_D$. But vertices are solutions of square systems of linear equations, an observation that suggests finding optimizers by solving this kind of system. This section pursues these ideas.

For any subset B of $\{1, 2, \ldots, n\}$, we denote by A_B the submatrix of A obtained by removing from A all the columns with index not in B. If $x \in \mathbb{R}^n$, x_B is defined analogously. Also, let $N := \{1, 2, \ldots, n\} \setminus B$. Then A_N and x_N are defined similarly.

Definition 8.9 By a *basis* B for a data triple $d = (A, b, c)$ we understand a subset $B \subseteq \{1, 2, \ldots, n\}$ with $|B| = m$ and such that A_B is invertible.

Let B be a basis for d. Then we may uniquely solve $A_B x = b$. Consider the point $x^* \in \mathbb{R}^n$ defined by $x_N^* = 0$ and $x_B^* - x$. Clearly, $Ax^* = b$. We say that x^* is the *primal basic solution* of (SP) associated with B. If, in addition, $x^* \geq 0$, which is equivalent to $x_B^* \geq 0$, then we say that x^* is a primal basic *feasible* solution.

Similarly, for any basis B for d we may uniquely solve $A_B^\mathsf{T} y = c_B$. The point y^* thus obtained is said to be the *dual basic solution* of (SD) associated with B. If, in addition, $A^\mathsf{T} y^* \leq c$, then y^* is said to be a dual basic *feasible* solution.

Definition 8.10 A basis B for a data triple d is called an *optimal basis* if both the primal and dual basic solutions associated with B are feasible, in which case the latter are called the *basic optimal solutions* of d associated to B.

There is a rationale for the word "optimal" in Definition 8.10.

Proposition 8.11 *Let B be an optimal basis for a triple d. If (x^*, y^*) is the basic optimal solution associated to B, then x^* and y^* are optimizers for the problems* (SP) *and* (SD), *respectively.*

Proof By construction, (x^*, y^*, s^*) is in the set $\mathcal{S}$ of feasible solutions (see (8.4)). In addition, one immediately checks that x^* and s^* satisfy the complementary slackness conditions (8.5). The optimality thus follows from Theorem 8.3. $\qquad\square$

It also follows from Definition 8.10 and Lemma 6.9 that if (x^*, y^*, s^*) is a basic optimal solution, then x^* and y^* are extremal optimal solutions of the primal and dual, respectively. The next example shows that the converse of this property does not necessarily hold. It is possible for a basis B that the associated basic solution for a linear program is optimal but the corresponding basic solution for the dual problem is not optimal (or even feasible). In other words, not all bases defining a vertex of $\mathcal{Q}$ are optimal bases.

Example 8.12 Consider the linear program

$$\begin{array}{rll}
\max & y_1 \\
\text{s.t.} & y_1 & \leq 0, \\
& -y_1 & \leq 0, \\
& y_2 \leq 0, \\
& -y_2 \leq 1,
\end{array}$$

with dual

$$\begin{array}{rll}
\min & x_4 \\
\text{s.t.} & x_1 - x_2 & = 1, \\
& x_3 - x_4 = 0, \\
& x_1, \ x_2, \ x_3, \ x_4 \geq 0.
\end{array}$$

The feasible set of the maximization problem is the interval with endpoints $(0, -1)$ and $(0, 0)$. Any point in this interval is an optimizer. The set of optimal points of its dual is a ray described by

$$\left\{ (\lambda + 1, \lambda, 0, 0) \mid \lambda \geq 0 \right\}.$$

Both problems have heavy sets of optimizers.

The set $\{2, 3\} \subseteq [4]$ is a basis. The associated solution of the maximization problem is $y^* = (0, 0)$, which is an optimizer. But the associated solution of the minimization problem is $(0, -1, 0, 0)$, which is infeasible.

Our next result, the highlight of this section, shows the existence of optimal bases.

Theorem 8.13 *Let $d = (A, b, c)$ be a data triple satisfying* rank $A = m$. *Then*:

(a) *There is an optimal basis for d if and only if both* (SP) *and* (SD) *are feasible.*
(b) *Let B be a basis for d. Then B is optimal if and only if both the primal and the dual basic solutions associated to B are optimizers for* (SP) *and* (SD) *respectively.*
(c) *If there is more than one optimal basis, say $B_1, \ldots, B_s$, $s \geq 2$, then the set of optimizers for* (SP) *contains the convex hull of $x_1^*, \ldots, x_s^*$, where $x_i^* \in \mathbb{R}^n$ is the primal basic solution associated to B_i, $i = 1, \ldots, s$. Likewise for the set of optimizers for* (SD).

Proof Clearly, if an optimal basis exists, then both primal and dual problems are feasible. To see the converse, assume that these problems are feasible. Then, by the Theorem 8.2(a), there exist optimal solutions x^* and y^* of (SP) and (SD), respectively. By Proposition 8.7(c), and since rank $A = m$, we may assume that y^* is a vertex. Therefore, by Lemma 6.9, there exists $B \subseteq [n]$ such that $|B| = m$, $A_B^{\mathsf{T}} y^* = c_B$, and rank $A_B = m$ (i.e., A_B is invertible). In other words, y^* is the dual basic solution associated to B.

Let $N := [n] \setminus B$ and assume that for all $i \in N$ we have $a_i^\mathsf{T} y^* < c_i$. Then, by complementary slackness (8.5), we must have that $x_i^* = 0$ for all $i \in N$. This implies that $A_B x_B^* = b$ and, consequently, that x^* is the primal basic solution associated to B. Since both x^* and y^* are feasible, we conclude that B is an optimal basis.

Assume now that instead, there exists $i \in N$ such that $a_i^\mathsf{T} y^* = c_i$, and let $D \subseteq N$ be the set of all such indices. For all $i \in D$, and since rank $A_B = m$, we can express a_i as a linear combination of the a_j for $j \in B$, say $a_i = \sum_j \lambda_j a_j$. Then

$$c_i = a_i^\mathsf{T} y^* = \sum_{j \in B} \lambda_j a_j^\mathsf{T} y^* = \sum_{j \in B} \lambda_j c_j.$$

It follows that $(a_i, c_i) \in \mathbb{R}^{m+1}$ is a linear combination of $\{(a_j, c_j) \mid j \in B\}$, for all $i \in D$. Consider the triple $d' = (\mathbf{A}, b, \mathbf{c})$, where $\mathbf{A}$ is obtained from A by removing its ith column for all $i \in D$ and likewise for $\mathbf{c}$. The set of feasible solutions of $\mathbf{A}^\mathsf{T} y \le \mathbf{c}$ is the same as that of $A^\mathsf{T} y \le c$. Therefore, the same holds for their sets of optimal solutions with respect to $y \mapsto b^\mathsf{T} y$, which we know is nonempty. By the duality theorem, the linear program $\min \mathbf{c}^\mathsf{T} x'$ subject to $\mathbf{A} x' = b$, $x' \ge 0$ (with now $x' \in \mathbb{R}^{n-|D|}$) also has a nonempty set of optimal solutions. We can therefore repeat the argument used above to show that B is an optimal basis for d', and padding with zeros the optimal basic solution x' of its primal, we obtain a primal basic feasible solution for the basis B of d. This finishes part (a).

The "only if" direction in part (b) is a consequence of Proposition 8.11. The other direction is trivial, since optimizers are, in particular, feasible points.

Part (c) is clear. $\square$

The following example shows a linear program with a unique optimal basis but a heavy set of optimizers for (SP). It also provides an instance for which the strict complementary guaranteed by Proposition 8.4 cannot be achieved at a basic optimal solution.

Example 8.14 Consider the linear program

$$\begin{aligned}
\min \quad & x_1 - x_2 \\
\text{s.t.} \quad & x_1 \qquad\quad - x_3 = 1, \\
& \qquad x_2 - x_3 = 1, \\
& x_1, \; x_2, \; x_3 \ge 0,
\end{aligned}$$

with dual

$$\begin{aligned}
\max \quad & y_1 - y_2 \\
\text{s.t.} \quad & y_1 \qquad\quad \le 1, \\
& \qquad y_2 \le -1, \\
& -y_1 - y_2 \le 0.
\end{aligned}$$

The feasible set of the primal is a ray with origin at $(1, 1, 0)$ and direction vector $(1, 1, 1)$. All points in this set are optimal solutions; hence, the datum is heavy for (SP). The feasible set of the dual reduces to the point $(1, -1)$.

The dual is nearly infeasible (it becomes infeasible if one replaces the third constraint by $-y_1 - y_2 \leq -\varepsilon$ for any $\varepsilon > 0$) and the primal is consequently nearly unbounded (it becomes so for the objective functions $x_1 - x_2 - \varepsilon x_3$).

We have $\mathcal{Q}_D = \{(1, -1)\}$, and the slackness at this point is $(0, 0, 0)$. Strict complementarity is achieved at primal solutions of the form $(r + 1, r + 1, r)$ for any $r > 0$. But these points are not basic solutions of (SP) (the only such solution corresponding to $r = 0$).

8.4 Ill-posedness and Degeneracy

We introduce in this section the notion of degeneracy, which links the algebra of a linear program with the geometry of its dual. Endowed with this notion, we discuss ill-posedness in the last part of this section.

8.4.1 Degeneracy

Proposition 8.4 (strict complementarity) imposes a constraint on the solutions of light data. If (x^*, y^*, s^*) is the only solution of a primal–dual pair of linear programs, then it must be a basic optimal solution by Theorem 8.13. If B is the associated optimal basis and $N := [n] \setminus B$, we must then have $x_N = 0$ and $s_B = 0$. Proposition 8.4 further implies that $s_N > 0$ and $x_B > 0$. This property motivates the following definition.

Definition 8.15 A feasible point x of (SP) is called *degenerate* when we have $|\{j \leq n \mid x_j = 0\}| > n - m$. Likewise, we say that a feasible point (y, s) of (SD) is *degenerate* when $|\{j \leq n \mid s_j = 0\}| > m$. This defines, by extension, the notions of *degenerate optimal solution*, *degenerate basic feasible point*, and *degenerate basic optimal solution*.

We say that a triple $d = (A, b, c)$ is *primal degenerate* if (SP) has degenerate optimal solutions and likewise for *dual degenerate*. We say that d is *degenerate* when it is either primal or dual degenerate.

Proposition 8.16 *The problem (SP) has a degenerate optimal solution if and only if it has a degenerate basic optimal solution. The same holds for (SD) if* rank $A = m$.

Proof We prove the result for (SD). The statement for (SP) admits a similar proof.

The "if" direction is trivial. For the converse, we note that in the course of the proof of Theorem 8.13(b) we started with an optimal solution (y^*, s^*) of (SD) and constructed a basic optimal solution for this problem. A new look at this proof reveals that in doing so, the number of nonzero components of s did not increase. Therefore, if (y^*, s^*) is a degenerate optimal solution of (SD), then so is the constructed basic optimal solution. This proves the second statement. □

The relationship between heaviness, duality, and degeneracy is captured in the following statement.

Proposition 8.17 *If one of (SP) or (SD) is heavy and has a nondegenerate extremal optimal solution, then all the optimal solutions of its dual are degenerate.*

Proof Assume that (SP) is heavy and let x^* be a nondegenerate extremal optimal solution with basis B, i.e., $B = \{ j \leq n \mid x_j^* > 0 \}$. Since (SP) is heavy, there exists another optimal solution $x' \neq x^*$ for (SP). Then there exists $i \notin B$ such that $x_i' > 0$. Otherwise, $A_B x_B' = A x' = b$, and it would follow that $x_B' = x_B^*$ and hence that $x' = x^*$.

Let $\bar{x} := \frac{1}{2}(x' + x^*)$. Then $\bar{x}$ is an optimizer, since it is a convex combination of two optimizers. Furthermore, since x^* is nondegenerate, we have $x_j^* > 0$ for all $j \in B$. This implies that $\bar{x}_j > 0$, for all $j \in B \cup \{i\}$. Now take any optimal solution (y^*, s^*) of (SD). Then, by complementary slackness (8.5), $s_j^* = 0$ for all $j \in B \cup \{i\}$. That is, (y^*, s^*) is degenerate.

The proof of the other case, i.e., (SD) heavy with a nondegenerate basic optimal solution, is similar. $\square$

Example 8.18 A linear program may be degenerate even if all its optimal solutions satisfy the strict complementarity condition. An example is the following primal–dual pair:

$$
\begin{array}{ll}
\min & x_1 + 2x_2 + 3x_3 \\
\text{s.t.} & x_1 + x_2 + x_3 = 1, \\
& x_1 + x_2 + x_3 = 1, \\
& x_1, x_2, x_3 \geq 0,
\end{array}
\qquad
\begin{array}{ll}
\max & y_1 \\
\text{s.t.} & y_1 + y_2 \leq 1, \\
& y_1 + y_2 \leq 2, \\
& y_1 + y_2 \leq 3.
\end{array}
$$

Example 8.19 A triple d may be infeasible (in the sense that either (SP) or (SD) is so) but have arbitrarily close feasible triples. An example is the infeasible primal–dual pair

$$
\begin{array}{ll}
\min & 2x_2 \\
\text{s.t.} & -x_1 + x_2 = 0, \\
& x_3 = -1, \\
& x_1, x_2, x_3 \geq 0,
\end{array}
\qquad
\begin{array}{ll}
\max & y_2 \\
\text{s.t.} & -y_1 \leq 0, \\
& y_1 \leq 2, \\
& -y_2 \leq 0,
\end{array}
$$

which is approximated (for $\varepsilon > 0$ small) by the following pairs:

$$
\begin{array}{ll}
\min & 2x_2 \\
\text{s.t.} & -x_1 + x_2 = 0, \\
& \varepsilon x_1 + \varepsilon x_2 - x_3 = 1, \\
& x_1, x_2, x_3 \geq 0,
\end{array}
\qquad
\begin{array}{ll}
\max & y_1 \\
\text{s.t.} & -y_1 + \varepsilon y_2 \leq 0, \\
& y_1 + \varepsilon y_2 \leq 2, \\
& -y_2 \leq 0,
\end{array}
$$

with optimal solutions $x_\varepsilon^* = (\frac{1}{2\varepsilon}, \frac{1}{2\varepsilon}, 0)$ and $y_\varepsilon^* = (1, \frac{1}{\varepsilon})$.

8.4.2 A Brief Discussion on Ill-posedness

The picture for the sets of optimal solutions emerging from the results in this chapter provides a framework to discuss ill-posedness for several problems in linear programming.

(a) Optimal Solution Problem   This is the problem of computing optimizers x^* and y^* for the linear programs (SP) and (SD), respectively. We want to identify the set of ill-posed data for this problem.

To do so, we first observe that infeasible triples should be considered as ill-posed if and only if they are like the one in Example 8.19, that is, if and only if arbitrarily small perturbations can make these triples feasible (and, consequently, create optimizers for them). We define

$$\Sigma_{\mathcal{I}} := \left\{ d \mid d \text{ is infeasible and } \forall \varepsilon > 0 \ \exists d' \text{ feasible with } \|d - d'\| \leq \varepsilon \right\}$$

and call this the set of *infeasible ill-posed triples*.

We next consider feasible triples, along with their sets of optimal solutions. Assume first that $\mathcal{Q}_P$ is heavy for some datum d. If it is compact, then arbitrarily small perturbations may turn $\mathcal{Q}_P$ into a singleton made by any of its vertices (cf. Fig. 8.1). If instead, $\mathcal{Q}_P$ is noncompact, then arbitrarily small perturbations of d may make (SD) infeasible (by Proposition 8.8). Similar statements hold for $\mathcal{Q}_D$. Therefore, we should consider data that are heavy for either (SP) or (SD) as ill-posed.

Assume now that instead, both $\mathcal{Q}_P$ and $\mathcal{Q}_D$ are light. Then Theorem 8.13 ensures that the only optimizers x^* and y^* for (SP) and (SD), respectively, are basic optimal solutions associated to a basis B. Furthermore, Proposition 8.4 implies that $s_N^* > 0$ (here $N := [n] \setminus B$). Therefore sufficiently small perturbations of d will still yield solutions for $A_B x_B = b$, $x_B > 0$, $A_B^\mathsf{T} y = c_B$, and $A_N^\mathsf{T} y < c_N$. In other words, we should consider data that are light for both (SP) and (SD) as well-posed.

We have thus identified the set of well-posed instances for the optimal solution problem as those having unique optimizers for both (SP) and (SD). Consequently, we define the set of ill-posed triples for this problem to be

$$\Sigma_{\text{opt}} := \left\{ d \mid d \text{ has at least two optimizers } (x, y) \right\} \cup \Sigma_{\mathcal{I}}.$$

The following result shows that rank-deficient triples are ill-posed.

Lemma 8.20 *Let $d = (A, b, c)$. If d is feasible and $\operatorname{rank} A < m$, then $d \in \Sigma_{\text{opt}}$.*

Proof Let x^*, y^* be optimizers for (SP) and (SD), respectively. Because of Proposition 8.4 we may assume that strict complementarity holds for this pair. We will show that other optimizers exist.

To do so, let $B := \{ j \leq n \mid x_j^* > 0 \}$. If $B = \emptyset$, then $A^\mathsf{T} y^* < c$, and consequently, sufficiently small perturbations of y^* will also be feasible points of (SD). Complementary slackness (8.5) ensures that they are actually optimizers. If instead, $B \neq \emptyset$, then

$$S := \left\{ x \in \mathbb{R}^{|B|} \mid A_B x = b \right\} = x_B^* + \ker A_B.$$

Since $\operatorname{rank} A < m$, we have $\operatorname{rank} A_B < m$ and hence $\dim \ker A_B > 0$. Since $x_B^* > 0$, the points x' given by $x_B' = x_B^* + x$ with $x \in \ker A_B$ and $x_N' = 0$ (here $N = [n] \setminus B$) will be, for sufficiently small $x \in \ker A_B$, feasible points of (SP). Complementary slackness ensures, again, that they are actually optimizers. $\square$

(b) Optimal Basis Problem This is the problem of computing an optimal basis. As for the optimal solution problem, we want to identify the set of ill-posed data. In contrast with the preceding problem, this one is discrete-valued: the output for any given datum d is an m-element subset in $[n]$. Therefore, the discussion in Sect. 6.1 applies, and the set Σ_{OB} of ill-posed triples for this problem should be taken as the boundary between these possible outputs. In other words, if $\mathrm{OB}(d)$ denotes the set of optimal bases for a triple d, we define

$$\Sigma_{\mathrm{OB}} := \left\{ d \mid \forall \varepsilon > 0 \ \exists d' \text{ s.t. } \mathrm{OB}(d) \neq \mathrm{OB}(d') \text{ and } \left\| d - d' \right\| \leq \varepsilon \right\}.$$

Again, feasible rank-deficient data are ill-posed for the optimal value problem.

Lemma 8.21 *Let $d = (A, b, c)$. If d is feasible and $\operatorname{rank} A < m$, then $d \in \Sigma_{\mathrm{OB}}$.*

Proof We begin as in the previous lemma with a pair x^*, y^* of optimizers for (SP) and (SD), respectively. Note that the hypothesis $\operatorname{rank} A < m$ implies that no $m \times m$ submatrix of A is invertible. Therefore, $\mathrm{OB}(d) = \emptyset$. We will show that there exist arbitrarily small perturbations $\overline{d}$ of d with $\mathrm{OB}(\overline{d}) \neq \emptyset$.

To do so, we need to fix a norm in the space of triples. Clearly, the norm is not relevant, so we may take $\|(A, b, c)\| := \max\{\|A\|, \|b\|, \|c\|\}$, where the first norm is the spectral and the other two, the Euclidean. Now let $R = \max\{1, \|x^*\|, \|y^*\|\}$. For any $\varepsilon > 0$ there exists a full-rank matrix $\overline{A}$ such that $\|A - \overline{A}\| \leq \frac{\varepsilon}{R}$. Let $\overline{b} := \overline{A} x^*$ and $\overline{c} \in \mathbb{R}^n$ be given by $\overline{c}_i := \max\{c_i, \overline{a}_i^{\mathsf{T}} y^*\}$. Then

$$\left\| b - \overline{b} \right\| = \left\| A x^* - \overline{A} x^* \right\| \leq \left\| A - \overline{A} \right\| \left\| x^* \right\| \leq \frac{\varepsilon}{R} \left\| x^* \right\| \leq \varepsilon.$$

Similarly, dividing by cases, $\|c - \overline{c}\| \leq \varepsilon$. It follows that if we take $\overline{d} = (\overline{A}, \overline{b}, \overline{c})$, we have $\|d - \overline{d}\| \leq \varepsilon$. But by construction, $\overline{A} x^* = \overline{b}$ and $\overline{A}^{\mathsf{T}} y^* \leq \overline{c}$. That is, x^* and y^* are feasible points for the primal and dual of $\overline{d}$, respectively. Theorem 8.13(a) now ensures that $\mathrm{OB}(\overline{d}) \neq \emptyset$. $\square$

Proposition 8.22 *Let $d = (A, b, c)$ be feasible. Then we have $d \in \Sigma_{\mathrm{opt}} \iff d \in \Sigma_{\mathrm{OB}}$.*

Proof Because of Lemmas 8.20 and 8.21 we know that the statement is true if $\operatorname{rank} A < m$. We therefore assume $\operatorname{rank} A = m$.

Suppose $d \notin \Sigma_{\mathrm{opt}}$. Then d has a unique optimal solution pair (x^*, y^*). Because of Theorem 8.13 this solution is basic. Furthermore, because of Proposition 8.4, we have A_B invertible, $x_B^* > 0$, and $A_N^{\mathsf{T}} y^* < c_N$ (here B is the optimal basis and $N := B \setminus [n]$). It is clear that sufficiently small perturbations d' of d will preserve these properties, so that $\mathrm{OB}(d') = \mathrm{OB}(d) = B$. Hence, $d \notin \Sigma_{\mathrm{OB}}$.

Suppose now $d \in \Sigma_{\mathrm{opt}}$. Then d is heavy either for (SP) or for (SD), say, without loss of generality, that for the former. Because of Theorem 8.13 there exists a basic optimal solution (x^*, y^*) of d with associated optimal basis B. If $\mathcal{Q}_P$ is non-compact, then because of Proposition 8.8, there exist arbitrarily small perturbations d' of d for which (SD) is infeasible. In particular, $\mathrm{OB}(d') = \emptyset$, and hence, since $B \in \mathrm{OB}(d)$, we deduce $d \in \Sigma_{\mathrm{OB}}$. If instead, $\mathcal{Q}_P$ is compact, then it contains at least two vertices. In particular, it contains a vertex $\tilde{x}$ different from x^*. Now arbitrarily small perturbations d' (just perturb c) may turn $\tilde{x}$ into the only optimizer of (SP) for d'. But then $B \notin \mathrm{OB}(d')$, and this shows that $d \in \Sigma_{\mathrm{OB}}$.   □

Corollary 8.23 *We have $\Sigma_{\mathrm{opt}} = \Sigma_{\mathrm{OB}}$. Furthermore, restricted to feasible data, this set coincides with $\{d \mid d \text{ is degenerate}\}$.*

Proof For an infeasible triple d we have $\mathrm{OB}(d) = \emptyset$. Hence, for such a triple, $d \in \Sigma_{\mathrm{OB}}$ if and only if there exist triples d' arbitrarily close to d for which $\mathrm{OB}(d') \neq \emptyset$, that is, if and only if $d \in \Sigma_{\mathcal{I}}$. The first statement therefore follows from Proposition 8.22.

To prove the second, let $d = (A, b, c)$ be feasible. If $d \notin \Sigma_{\mathrm{opt}}$, then d has a unique pair of optimizers (x^*, y^*). Also, because of Lemma 8.21, $\mathrm{rank}\, A = m$. Therefore, by Theorem 8.13, (x^*, y^*) is a basic optimal solution. By Proposition 8.4 this pair satisfies the strict complementarity condition. Therefore, it is nondegenerate.

We next prove the converse. For this, we assume that $d \in \Sigma_{\mathrm{opt}}$, that is, d is either primal heavy or dual heavy. We will then show that d is degenerate. We do so dividing by cases.

Assume first that $\mathrm{rank}\, A = m$. Then, by Theorem 8.13, there exists a basic optimal solution (x^*, y^*). If d is primal heavy, then either x^* is degenerate, in which case we are done, or it is nondegenerate, in which case y^* is degenerate by Proposition 8.17 and we are done as well. The same reasoning applies if d is dual heavy.

Assume now that $\mathrm{rank}\, A < m$ and consider an optimizer (x^*, y^*) satisfying the strict complementarity condition. Let $B := \{j \leq n \mid x_j^* > 0\}$. If $|B| < m$, then d is primal degenerate. If $|B| > m$, then d is dual degenerate. We are left with the case $|B| = m$. Since $A_B x^* = b$, we see that $b \in \mathbb{R}^m$ can be expressed as a linear combination of $\{a_i \mid i \in B\}$ with nonnegative coefficients, i.e., $b \in \mathrm{cone}\{a_i \mid i \in B\}$. Recall that $\dim \mathrm{span}\{a_i \mid i \in B\} < m$ by assumption. Hence, by Carathéodory's Theorem 6.5, there exist a set $I \subseteq B$, with $|I| \leq m - 1$, and nonnegative real numbers x_i', for $i \in I$, such that $A_I x' = b$ (here x' is the vector in $\mathbb{R}^{|I|}$ with entries x_i'). This shows that the point $\bar{x}$ given by $\bar{x}_I = x'$ and $\bar{x}_j = 0$ for $j \notin I$ is a feasible point for (SP). But the pair $(\bar{x}, y^*)$ satisfies the complementary slackness conditions. Therefore, x^* is an optimizer for the primal problem and it is degenerate.   □

The sharing of the set of ill-posed instances suggests that we should numerically solve the optimal solution problem by doing so for the optimal basis problem. We will do so in Chap. 11, where in passing, we will also define a condition number $\mathscr{H}(d)$ (the relativized inverse of the distance to Σ_{OB}) for the optimal basis problem.

(c) Feasibility Problem Both the optimal basis and the optimal solution problems require a previous (or simultaneous at least) solution of the *feasibility problem* for linear programming primal–dual pairs. This consists in deciding whether both $Ax = b$, $x \geq 0$, and $A^{T}y \leq c$ have feasible points. That is, the feasibility problem is a decision problem, and therefore, always following the discussion in Sect. 6.1, condition for this problem's data can be defined as the relativized distance to ill-posedness, with the latter defined as the boundary between the sets of feasible and infeasible triples. That is, letting

$$FP(d) = \begin{cases} 1 & \text{if } d \text{ is feasible,} \\ 0 & \text{if } d \text{ is infeasible,} \end{cases}$$

we define

$$\Sigma_{FP} := \left\{ d \mid \forall \varepsilon > 0 \; \exists d' \text{ s.t.} FP(d) \neq FP(d') \text{ and } \|d - d'\| \leq \varepsilon \right\}.$$

Clearly, $\Sigma_{FP} \subseteq \Sigma_{OB}$.

In Chap. 10, we will describe an algorithm solving the feasibility problem and analyze it in terms of a condition number $C(d)$ (the relativized inverse of the distance from d to Σ_{FP}). We will also show that $C(d)$ is closely related to $\mathscr{C}$.

(d) Optimal Value Problem A last problem that is worth mentioning here is the *optimal value problem*. This consists in computing the optimal value v^* (or occasionally, of computing an ε-approximation of it). A look at the three situations in Fig. 8.1 reveals a peculiar behavior. For the situation at the left ($\dim \mathcal{Q} = 0$), sufficiently small perturbations will produce only small perturbations of both the optimizer and the optimal value. That is, light triples are well-posed for the optimal value problem. For the situation at the right, arbitrarily small perturbations may drive the optimal value to ∞ (or $-\infty$ if it is a minimization problem). Hence we consider this situation to be ill-posed. But the middle situation (compact heavy data) appears to be well-posed for these problems: sufficiently small perturbations will neither affect feasibility nor drive the optimal value to $\pm\infty$. The optimal value problem appears to share the set of ill-posed inputs with the feasibility problem, and one could consequently expect to have algorithmic solutions analyzed in terms of $C(d)$. We will return to this problem in Sect. 11.5.

Chapter 9
Interior-Point Methods

The ellipsoid method presented in Chap. 7 has an undeniable historical relevance (due to its role in establishing polynomial time for linear programming with integer data). In addition, its underlying idea is simple and elegant. Unfortunately, it is not efficient in practice compared with both the simplex method and the more recent interior-point methods. In this chapter, we describe the latter in the context of linear programming.

Unlike the ellipsoid method, which seems tailored for feasibility problems, interior-point methods appear to be designed to solve optimization problems. In linear programming, however, it is possible to recast problems of one kind as problems of the other, and we will take advantage of this feature to present an algorithmic solution for the feasibility problem PCFP. We will see that again, the condition number $\mathscr{C}(A)$ of the data plays a role in the complexity of this solution.

9.1 Primal–Dual Interior-Point Methods: Basic Ideas

The most common method to solve linear programs is Dantzig's simplex method. This method relies on the geometry of the polyhedron of solutions and constructs a sequence of vertices on the boundary of this polyhedron leading to a basic optimal solution. By contrast, interior-point methods follow a path in the interior of the polyhedron, whence the name. The path is a nonlinear curve that is approximately followed by a variant of Newton's method.

In what follows we will consider primal–dual pairs of the form (OP)–(OD) we saw in Sect. 8.1. *Primal–dual interior-point methods* search for solutions of the optimality conditions for this pair, that is, for solutions of the system

$$Ax + Gw = b, \qquad A^T y + s = c, \qquad G^T y = d, \quad x \geq 0,\ s \geq 0,$$
$$x_1 s_1 = 0, \quad \ldots, \quad x_n s_n = 0, \tag{9.1}$$

P. Bürgisser, F. Cucker, *Condition*,
Grundlehren der mathematischen Wissenschaften 349,
DOI 10.1007/978-3-642-38896-5_9, © Springer-Verlag Berlin Heidelberg 2013

by following a certain curve in the *strictly (primal–dual) feasible set* $\mathcal{S}^\circ \subseteq \mathbb{R}^{n+p+m+n}$ defined by

$$\mathcal{S}^\circ := \big\{ (x, w, y, s) \mid Ax + Gw = b, A^\mathsf{T}y + s = c, G^\mathsf{T}y = d, x, s > 0 \big\}. \quad (9.2)$$

(Compare with the definition of the primal–dual feasible set $\mathcal{S}$ in (8.4).)

Note that (9.1) is only mildly nonlinear (quadratic equations $x_i s_i = 0$). It is the nonnegativity constraints that appear as the main source of difficulty. For a parameter $\mu > 0$ we add now to (9.2) the additional constraints

$$x_1 s_1 = \mu, \quad \ldots, \quad x_n s_n = \mu. \quad (9.3)$$

One calls μ the *duality measure*. Under mild genericity assumptions, there is exactly one strictly feasible solution $\zeta_\mu \in \mathcal{S}^\circ$ satisfying (9.3), and the limit $\zeta = (x, w, y, s) = \lim_{\mu \to 0} \zeta_\mu$ exists. Then it is clear that $\zeta \in \mathcal{S}$ and $x_i s_i = 0$ for all i. Hence ζ is a desired solution of the primal–dual optimization problem.

We postpone the proof of the next theorem to Sect. 9.2.

Theorem 9.1 *Suppose that $p \le m \le n$, rank $A = m$, rank $G = p$, and that there is a strictly feasible point, i.e., $\mathcal{S}^\circ \ne \emptyset$. Then for all $\mu > 0$ there exists a uniquely determined point $\zeta_\mu = (x^\mu, w^\mu, y^\mu, s^\mu) \in \mathcal{S}^\circ$ such that $x_i^\mu s_i^\mu = \mu$ for $i \in [n]$.*

Definition 9.2 The *central path* $\mathcal{C}$ of the primal–dual optimization problem given by A, G, b, c, d is the set

$$\mathcal{C} = \{ \zeta_\mu : \mu > 0 \}.$$

Suppose we know ζ_{μ_0} for some $\mu_0 > 0$. The basic idea of a *path-following method* is to choose a sequence of parameters $\mu_0 > \mu_1 > \mu_2 > \cdots$ converging to zero and to successively compute approximations z_k of $\zeta_k := \zeta_{\mu_k}$ for $k = 0, 1, 2, \ldots$ until a certain accuracy is reached (see Fig. 9.1). In most cases one chooses $\mu_k = \sigma^k \mu_0$ with a *centering parameter* $\sigma \in (0, 1)$.

It is useful to extend the duality measure to any $z = (x, w, y, s) \in \mathcal{S}^\circ$. We do so by taking

$$\mu(z) := \frac{1}{n} \sum_{i=1}^n x_i s_i = \frac{1}{n} s^\mathsf{T} x.$$

How can we compute the approximations z_k? This is based on *Newton's method*, one of the most fundamental methods in computational mathematics (which will occupy us in Part III).

Consider the map $F : \mathbb{R}^{n+p+m+n} \to \mathbb{R}^{n+p+m+n}$,

$$z = (x, w, y, s) \mapsto F(z) = \big(A^\mathsf{T}y + s - c, G^\mathsf{T}y - d, Ax + Gw - b, x_1 s_1, \ldots, x_n s_n \big).$$

Fig. 9.1 Central path $\mathcal{C}$

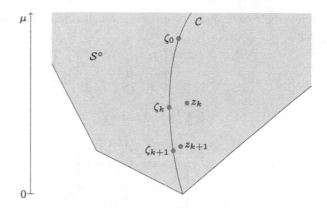

We note that by Theorem 9.1, $\{\zeta_\mu\} = F^{-1}(0, 0, 0, \mu \mathbf{e}_n)$, where $\mathbf{e}_n := (1, \ldots, 1)$ $\in \mathbb{R}^n$. The Jacobian matrix of F at z equals

$$DF(z) = \begin{bmatrix} 0 & 0 & A^{\mathrm{T}} & I \\ 0 & 0 & G^{\mathrm{T}} & 0 \\ A & G & 0 & 0 \\ S & 0 & 0 & X \end{bmatrix},$$

where here and in the following we set

$$S = \mathrm{diag}(s_1, \ldots, s_n), \qquad X = \mathrm{diag}(x_1, \ldots, x_n).$$

Depending on the context, z, $\mathbf{e}_n$, etc. should be interpreted as column vectors.

Lemma 9.3 *If $p \leq m \leq n$, rank $A = m$, and rank $G = p$, then $DF(z)$ is invertible, provided $s_i x_i \neq 0$ for all i.*

Proof By elementary column operations we can bring the matrix $DF(z)$ to the form

$$\begin{bmatrix} D & 0 & A^{\mathrm{T}} & I \\ 0 & 0 & G^{\mathrm{T}} & 0 \\ A & G & 0 & 0 \\ 0 & 0 & 0 & X \end{bmatrix},$$

where $D = \mathrm{diag}(-s_1 x_1^{-1}, \ldots, -s_n x_n^{-1})$. It is therefore sufficient to show that the matrix

$$\begin{bmatrix} D & 0 & A^{\mathrm{T}} \\ 0 & 0 & G^{\mathrm{T}} \\ A & G & 0 \end{bmatrix}$$

is invertible. Such matrices are of the so-called *Karush–Kuhn–Tucker* type. Suppose that

$$\begin{bmatrix} D & 0 & A^{\mathsf{T}} \\ 0 & 0 & G^{\mathsf{T}} \\ A & G & 0 \end{bmatrix} \begin{bmatrix} x \\ w \\ y \end{bmatrix} = 0,$$

that is, $Dx + A^{\mathsf{T}}y = 0$, $G^{\mathsf{T}}y = 0$ and $Ax + Gw = 0$. It follows that

$$0 - \begin{bmatrix} x^{\mathsf{T}} & w^{\mathsf{T}} & y^{\mathsf{T}} \end{bmatrix} \begin{bmatrix} D & 0 & A^{\mathsf{T}} \\ 0 & 0 & G^{\mathsf{T}} \\ A & G & 0 \end{bmatrix} \begin{bmatrix} x \\ w \\ y \end{bmatrix} = \begin{bmatrix} x^{\mathsf{T}} & w^{\mathsf{T}} & y^{\mathsf{T}} \end{bmatrix} \begin{bmatrix} Dx + A^{\mathsf{T}}y \\ G^{\mathsf{T}}y \\ 0 \end{bmatrix}$$

$$= x^{\mathsf{T}}Dx + (Ax + Gw)^{\mathsf{T}}y = x^{\mathsf{T}}Dx.$$

Since D is negative definite, it follows that $x = 0$. Hence $A^{\mathsf{T}}y = 0$. Therefore, $y = 0$, since $\operatorname{rank} A = m$. Also, since $Ax = 0$, we have $Gw = 0$ and therefore, since $\operatorname{rank} G = p$, $w = 0$.   □

We continue with the description of the basic algorithmic idea. Choose $\mu_k = \sigma^k \mu_0$ and set $\zeta_k = \zeta_{\mu_k}$. Then $F(\zeta_k) = (0, 0, 0, \mu_k \mathbf{e}_n)$ for all $k \in \mathbb{N}$. A first-order approximation gives

$$F(\zeta_{k+1}) \approx F(\zeta_k) + DF(\zeta_k)(\zeta_{k+1} - \zeta_k). \tag{9.4}$$

Suppose now that $z_k = (x, w, y, s) \in \mathcal{S}^\circ$ is an approximation of ζ_k. Then $F(z_k) = (0, 0, 0, x_1 s_1, \ldots, x_n s_n) = (0, 0, 0, XS\mathbf{e}_n)$. We obtain from (9.4), replacing the unknowns ζ_k by z_k,

$$(0, 0, 0, \mu_{k+1}\mathbf{e}_n) = F(\zeta_{k+1}) \approx F(z_k) + DF(z_k)(\zeta_{k+1} - z_k).$$

This leads to the definition

$$z_{k+1} := z_k + DF(z_k)^{-1}(0, 0, 0, \mu_{k+1}\mathbf{e}_n - XS\mathbf{e}_n) \tag{9.5}$$

of the approximation of ζ_{k+1}. This vector is well defined due to Lemma 9.3. Put $z_{k+1} = z_k + (\Delta x, \Delta w, \Delta y, \Delta s)$. Then

$$\left(A^{\mathsf{T}}\Delta y + \Delta s, G^{\mathsf{T}}\Delta y, A\Delta x + G\Delta w, S\Delta x + X\Delta s\right) = DF(z_k)(\Delta x, \Delta w, \Delta y, \Delta s)$$

$$= (0, 0, 0, \mu_{k+1}\mathbf{e}_n - XS\mathbf{e}_n),$$

and hence $A^{\mathsf{T}}\Delta y + \Delta s = 0$, $G^{\mathsf{T}}\Delta y = 0$, $A\Delta x + G\Delta w = 0$, which implies $A^{\mathsf{T}}(y + \Delta y) + (s + \Delta s) = c$, $G^{\mathsf{T}}(y + \Delta y) = d$, and $A(x + \Delta x) + G(w + \Delta w) = b$. We have shown that z_{k+1} satisfies the equalities in (9.2). By a suitable choice of the parameter σ we will see that one can achieve that z_{k+1} also satisfies the strict inequalities in (9.2), that is, $z_{k+1} \in \mathcal{S}^\circ$.

Summarizing, the framework for a primal–dual interior point method is the following algorithm.

Algorithm 9.1 Primal–Dual IPM

Input: $A \in \mathbb{R}^{m \times n}, G \in \mathbb{R}^{m \times p}, b \in \mathbb{R}^m, c \in \mathbb{R}^n, d \in \mathbb{R}^p,$
$z_0 = (x^0, w^0, y^0, s^0) \in \mathbb{R}^{2n+p+m}, \sigma \in (0, 1)$
Preconditions: rank $A = m \leq n$, rank $G = p \leq m$, $z_0 \in \mathcal{S}^\circ$

```
set  μ := μ(z₀)
repeat
        set  μ := σμ,   X := diag(x),   S := diag(s)
        solve
```

$$
\begin{bmatrix} 0 & 0 & A^{\mathrm{T}} & I \\ 0 & 0 & G^{\mathrm{T}} & 0 \\ A & G & 0 & 0 \\ S & 0 & 0 & X \end{bmatrix} \cdot \begin{bmatrix} \Delta x \\ \Delta w \\ \Delta y \\ \Delta s \end{bmatrix} = \begin{bmatrix} 0 \\ 0 \\ 0 \\ \mu\, e_n - X S e_n \end{bmatrix}
$$

```
        set
```

$$
(x, w, y, s) := (x, w, y, s) + (\Delta x, \Delta w, \Delta y, \Delta s)
$$

```
until a stopping criterion is satisfied by (x, w, y, s)
return (x, w, y, s) and halt
```

Output: $z = (x, w, y, s) \in \mathbb{R}^{2n+p+m}$
Postconditions: $z \in \mathcal{S}^0$ and z satisfies the stopping criterion

---

9.2 Existence and Uniqueness of the Central Path

We provide here the proof of the fundamental Theorem 9.1.

Lemma 9.4 *Suppose that* $\mathcal{S}^\circ \neq \emptyset$. *Then for all* $K \in \mathbb{R}$ *the set*

$$
\left\{ (x, s) \in \mathbb{R}^n \times \mathbb{R}^n \mid \exists w \in \mathbb{R}^p \exists y \in \mathbb{R}^m (x, w, y, s) \in \mathcal{S}, s^{\mathrm{T}} x \leq K \right\}
$$

is bounded.

Proof Let $(\bar{x}, \bar{w}, \bar{y}, \bar{s}) \in \mathcal{S}^\circ$. For any $(x, w, y, s) \in \mathcal{S}$ we have $A\bar{x} + G\bar{w} = b$ and $Ax + Gw = b$, hence $A(\bar{x} - x) + G(\bar{w} - w) = 0$. Similarly, $A^{\mathrm{T}}(\bar{y} - y) + (\bar{s} - s) = 0$ and $G^{\mathrm{T}}(\bar{y} - y) = 0$. This implies

$$
(\bar{s} - s)^{\mathrm{T}}(\bar{x} - x) = -(\bar{y} - y)^{\mathrm{T}} A(\bar{x} - x) = (\bar{y} - y)^{\mathrm{T}} G(\bar{w} - w) = 0.
$$

It follows, assuming $s^Tx \leq K$, that

$$s^T\bar{x} + \bar{s}^Tx = s^Tx + \bar{s}^T\bar{x} \leq K + \bar{s}^T\bar{x}.$$

The quantity $\xi := \min_i \min\{\bar{x}_i, \bar{s}_i\}$ is positive by assumption. We therefore get

$$\xi e_n^T(x+s) \leq K + \bar{s}^T\bar{x};$$

hence $\xi^{-1}(K + \bar{s}^T\bar{x})$ is an upper bound on x_i and s_i for all i.   □

Fix $\mu > 0$ and consider the *barrier function*

$$f : \mathcal{H}^\circ \to \mathbb{R}, \quad f(x,s) = \frac{1}{\mu}s^Tx - \sum_{j=1}^{n}\ln(x_js_j) \tag{9.6}$$

defined on the projection $\mathcal{H}^\circ$ of $\mathcal{S}^\circ$:

$$\mathcal{H}^\circ := \left\{(x,s) \in \mathbb{R}^n \times \mathbb{R}^n \mid \exists w \in \mathbb{R}^p \exists y \in \mathbb{R}^m (x,w,y,s) \in \mathcal{S}^\circ\right\}.$$

Note that $\mathcal{H}^\circ$ is convex because $\mathcal{S}^\circ$ is convex. Moreover, $f(x,s)$ approaches ∞ whenever any of the products x_js_j approaches zero.

Lemma 9.5

(a) *f is strictly convex.*
(b) *f is bounded from below.*
(c) *For all $\kappa \in \mathbb{R}$ there exist $0 < \alpha < \beta$ such that*

$$\left\{(x,s) \in \mathcal{H}^\circ \mid f(x,s) \leq \kappa\right\} \subseteq [\alpha, \beta]^n \times [\alpha, \beta]^n.$$

Proof (a) Consider the function $g : \mathbb{R}_+^n \times \mathbb{R}_+^n \to \mathbb{R}$, $g(x,s) = -\sum_{j=1}^{n}\ln(x_js_j)$. We have $\frac{\partial^2 g}{\partial x_j^2} = x_j^{-2}$, $\frac{\partial^2 g}{\partial s_j^2} = s_j^{-2}$, and all other second-order derivatives of g vanish. The Hessian of g is therefore positive definite and hence g is strictly convex. In particular, the restriction of g to $\mathcal{H}^\circ$ is strictly convex as well.

We claim that the restriction of s^Tx to $\mathcal{H}^\circ$ is linear. To show this, consider a fixed point $(\bar{x}, \bar{w}\bar{y}, \bar{s}) \in \mathcal{S}^\circ$. Then $A\bar{x} + G\bar{w} = b$, $A^T\bar{y} + \bar{s} = c$, and $G^T\bar{y} = d$. Now consider any $(x,s) \in \mathcal{H}^\circ$. There exist $w \in \mathbb{R}^p$ and $y \in \mathbb{R}^m$ such that $(x,w,y,s) \in \mathcal{S}^\circ$. Furthermore, by (8.1),

$$s^Tx = c^Tx + d^Tw - b^Ty = c^Tx + \bar{y}^TGw - \bar{x}^TA^Ty - \bar{w}^TG^Ty$$
$$= c^Tx + \bar{y}^T(b - Ax) - \bar{x}^T(c - s) - \bar{w}^Td$$
$$= c^Tx + \bar{y}^Tb - \bar{y}^TAx - \bar{x}^Tc + \bar{x}^Ts - \bar{w}^Td,$$

which is linear in (x,s). This proves the first assumption.

(b) We write

$$f(x,s) = \sum_{j=1}^{n} h\left(\frac{x_j s_j}{\mu}\right) + n - n\ln\mu,$$

where

$$h(t) := t - \ln t - 1.$$

It is clear that h is strictly convex on $(0, \infty)$ as well as that $\lim_{t\to 0} h(t) = \infty$, and $\lim_{t\to\infty} h(t) = \infty$. Moreover, $h(t) \geq 0$ for $t \in (0, \infty)$ with equality iff $t = 1$. Using this, we get

$$f(x,s) \geq n - n\ln\mu,$$

which proves the second assertion.

(c) Suppose $(x, s) \in \mathcal{H}^{\circ}$ with $f(x, s) \leq \kappa$ for some κ. Then, for all j,

$$h\left(\frac{x_j s_j}{\mu}\right) \leq \kappa - n + n\log\mu =: \bar{\kappa}.$$

From the properties of h it follows that there exist $0 < \alpha_1 < \beta_1$ such that

$$h^{-1}(-\infty, \bar{\kappa}] \subseteq [\alpha_1, \beta_1],$$

whence $\mu\alpha_1 \leq x_j s_j \leq \mu\beta_1$. Applying Lemma 9.4 with $K = n\mu\beta_1$ shows that there is some β such that $x_j \leq \beta$, $s_j \leq \beta$. Hence $x_j \geq \mu\alpha_1\beta^{-1}$, $s_j \geq \mu\alpha_1\beta^{-1}$, which proves the third assertion with $\alpha = \mu\alpha_1\beta^{-1}$. $\square$

Suppose that $\mathcal{S}^{\circ} \neq \emptyset$. Lemma 9.5(c) implies that f achieves its minimum in $\mathcal{H}^{\circ}$. Moreover, the minimizer is unique, since f is strictly convex. We shall denote this minimizer by (x^{μ}, s^{μ}). We note that if rank $A = m \leq n$, then y^{μ} is uniquely determined by the condition $A^{\mathsf{T}}y^{\mu} + s^{\mu} = c$, and similarly, if rank $G = p \leq m$, then w^{μ} is uniquely determined by the condition $Ax^{\mu} + Gw^{\mu} = b$.

To complete the argument, we will show that $x_i s_i = \mu$, $i = 1, 2, \ldots, n$, are exactly the first-order conditions characterizing local minima of the function f. (Note that a local minimum of f is a global minimum by the strict convexity of f.)

We recall a well-known fact about Lagrange multipliers from analysis. Let $g, h_1, \ldots, h_m \colon U \to \mathbb{R}$ be differentiable functions defined on the open subset $U \subseteq \mathbb{R}^n$. Suppose that $u \in U$ is a local minimum of g under the constraints $h_1 = 0, \ldots, h_m = 0$. Then, if the gradients $\nabla h_1, \ldots, \nabla h_m$ are linearly independent at u, there exist *Lagrange multipliers* $\lambda_1, \ldots, \lambda_m \in \mathbb{R}$ such that

$$\nabla g(u) + \lambda_1 \nabla h_1(u) + \cdots + \lambda_m \nabla h_m(u) = 0. \tag{9.7}$$

We apply this fact to the problem

$$\min \quad f(x, s) \quad \text{s.t.} \quad Ax + Gw = b, \quad A^{\mathsf{T}}y + s = c,$$
$$G^{\mathsf{T}}y = d, \quad x > 0, \ s > 0.$$

Suppose that (x, w, y, s) is a local minimum of f. The linear independence condition holds due to Lemma 9.3 and our assumptions $\operatorname{rank} A = m \leq n$ and $\operatorname{rank} G = p \leq m$. By (9.7) there are Lagrange multipliers $v \in \mathbb{R}^m$, $u \in \mathbb{R}^n$, $t \in \mathbb{R}^p$, such that

$$\mu^{-1}s - X^{-1}\mathbf{e}_n + A^{\mathsf{T}}v = 0, \qquad G^{\mathsf{T}}v = 0,$$

$$Au + Gt = 0, \qquad \mu^{-1}x - S^{-1}\mathbf{e}_n + u = 0. \tag{9.8}$$

(Here we have used that $\frac{\partial f}{\partial x} = \mu^{-1}s - X^{-1}\mathbf{e}_n$, $\frac{\partial f}{\partial w} = \frac{\partial f}{\partial y} = 0$, $\frac{\partial f}{\partial s} = \mu^{-1}x - S^{-1}\mathbf{e}_n$.) The last two equalities in (9.8) imply that

$$A\big(\mu^{-1}x - S^{-1}\mathbf{e}_n\big) - Gt = 0$$

and therefore that

$$\big(\mu^{-1}x - S^{-1}\mathbf{e}_n\big)^{\mathsf{T}}A^{\mathsf{T}}v - t^{\mathsf{T}}G^{\mathsf{T}}v = 0.$$

We now use the second equality in (9.8) to deduce $(\mu^{-1}Xe - S^{-1}\mathbf{e}_n)^{\mathsf{T}}A^{\mathsf{T}}v = 0$. Using the first equality we get

$$\big(\mu^{-1}X\mathbf{e}_n - S^{-1}\mathbf{e}_n\big)^{\mathsf{T}}\big(\mu^{-1}S\mathbf{e}_n - X^{-1}\mathbf{e}_n\big) = 0.$$

Therefore

$$0 = \big(\mu^{-1}X\mathbf{e}_n - S^{-1}\mathbf{e}_n\big)^{\mathsf{T}}\big(X^{-1/2}S^{1/2}\big)\big(X^{1/2}S^{-1/2}\big)\big(\mu^{-1}S\mathbf{e}_n - X^{-1}\mathbf{e}_n\big)$$

$$= \big\|\mu^{-1}(XS)^{1/2}\mathbf{e}_n - (XS)^{-1/2}\mathbf{e}_n\big\|^2.$$

This implies $XS\mathbf{e}_n = \mu\mathbf{e}_n$; hence (x, w, y, s) lies on the central path $\mathcal{C}$.

Conversely, suppose that $(x, w, y, s) \in \mathcal{S}^\circ$ satisfies $XS\mathbf{e}_n = \mu\mathbf{e}_n$. Put $v = 0$, $u = 0$, and $t = 0$. Then the first-order conditions (9.8) are satisfied. Since f is strictly convex, (x, s) is a global minimum of f. By the previously shown uniqueness, we have $(x, s) = (x^\mu, s^\mu)$. This completes the proof of Theorem 9.1. $\square$

9.3 Analysis of IPM for Linear Programming

Recall the following useful conventions: For a vector $u \in \mathbb{R}^d$ we denote by U the matrix $U = \operatorname{diag}(u_1, \ldots, u_d)$. Moreover, $\mathbf{e}_d$ stands for the vector $(1, \ldots, 1)$ of the corresponding dimension. Note that $U\mathbf{e}_d = u$.

Lemma 9.6 Let $u, v \in \mathbb{R}^d$ be such that $u^{\mathsf{T}}v \geq 0$. Then

$$\|UV\mathbf{e}_d\| \leq \frac{1}{2}\|u + v\|^2.$$

Proof We have

$$\|UV\mathbf{e}_d\| = \|Uv\| \le \|U\| \, \|v\| \le \|U\|_F \, \|v\| = \|u\| \, \|v\|.$$

Moreover, since $u^T v \ge 0$,

$$\|u\| \, \|v\| \le \frac{1}{2}\left(\|u\|^2 + \|v\|^2\right) \le \frac{1}{2}\left(\|u\|^2 + 2u^T v + \|v\|^2\right) = \frac{1}{2}\|u + v\|^2. \qquad \square$$

Let $A \in \mathbb{R}^{m \times n}$, $G \in \mathbb{R}^{m \times p}$, $b \in \mathbb{R}^m$, $c \in \mathbb{R}^n$, $d \in \mathbb{R}^p$, such that rank $A = m \le n$ and rank $G = p \le m$. Moreover, let $z = (x, w, y, s) \in \mathcal{S}^\circ$, that is, we have

$$Ax + Gw = b, \qquad A^T y + s = c, \qquad G^T y = d, \quad x > 0, \ s > 0.$$

We consider one step of Algorithm 9.1, the primal–dual IPM, with centering parameter $\sigma \in (0, 1)$. That is, we set $\mu := \mu(z) = \frac{1}{n}s^T x$, define $\Delta z = (\Delta x, \Delta w, \Delta y, \Delta s)$ by

$$\begin{bmatrix} 0 & 0 & A^T & I \\ 0 & 0 & G^T & 0 \\ A & G & 0 & 0 \\ S & 0 & 0 & X \end{bmatrix} \begin{bmatrix} \Delta x \\ \Delta w \\ \Delta y \\ \Delta s \end{bmatrix} = \begin{bmatrix} 0 \\ 0 \\ 0 \\ \sigma\mu\mathbf{e}_n - XS\mathbf{e}_n \end{bmatrix}, \tag{9.9}$$

and put

$$\tilde{z} = (\tilde{x}, \tilde{w}, \tilde{y}, \tilde{s}) = (x, w, y, s) + (\Delta x, \Delta w, \Delta y, \Delta s).$$

Lemma 9.7

(a) $\Delta s^T \Delta x = 0$.

(b) $\mu(\tilde{z}) = \sigma\mu(z)$.

(c) $\tilde{z} \in \mathcal{S}^\circ$ if $\tilde{x} > 0$, $\tilde{s} > 0$.

Proof (a) By definition of $\Delta z = (\Delta x, \Delta w, \Delta y, \Delta s)$ we have

$$\begin{aligned} A^T \Delta y + \Delta s &= 0, \\ G^T \Delta y &= 0, \\ A \Delta x + G \Delta w &= 0, \\ S \Delta x + X \Delta s &= \sigma\mu\mathbf{e}_n - XS\mathbf{e}_n. \end{aligned} \tag{9.10}$$

Therefore,

$$\Delta s^T \Delta x = -\Delta y^T A \Delta x = \Delta y^T G \Delta w = \Delta w^T G^T \Delta y = 0.$$

(b) The fourth equation in (9.10) implies $s^T \Delta x + x^T \Delta s = n\sigma\mu - x^T s$. Therefore,

$$\tilde{s}^T \tilde{x} = \left(s^T + \Delta s^T\right)(x + \Delta x) = s^T x + \Delta s^T x + s^T \Delta x + \Delta s^T \Delta x = n\sigma\mu.$$

This means that $\mu(\tilde{z}) = \frac{1}{n}\tilde{s}^T \tilde{x} = \sigma\mu$.

Fig. 9.2 Central path C and central neighborhood $\mathcal{N}(\beta)$

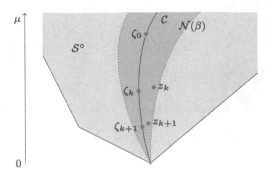

(c) We already verified at the end of Sect. 9.1 (by a straightforward calculation) that $\tilde{z}$ satisfies the equality constraints in (9.2). $\square$

A remaining issue is how to achieve $\tilde{x} > 0$, $\tilde{s} > 0$ by a suitable choice of the centering parameter σ.

Definition 9.8 Let $\beta > 0$. The *central neighborhood* $\mathcal{N}(\beta)$ is defined as the set of strictly feasible points $z = (x, w, y, s) \in \mathcal{S}^\circ$ such that

$$\left\| X S e_n - \mu(z) e_n \right\| \le \beta \mu(z).$$

The central neighborhood is a neighborhood of the central path C in $\mathcal{S}^\circ$ that becomes narrower as $\mu(z)$ approaches zero (see Fig. 9.2).

In the following we set $\beta = \frac{1}{4}$ and write $\mathcal{N} := \mathcal{N}(\frac{1}{4})$.

Lemma 9.9 *Let* $z = (x, w, y, s) \in \mathcal{N}$ *and* $\Delta z = (\Delta x, \Delta w, \Delta y, \Delta s)$ *be defined by* (9.9) *with respect to* $\sigma = 1 - \frac{\xi}{\sqrt{n}}$ *with* $0 < \xi \le \frac{1}{4}$. *Then* $\tilde{z} = z + \Delta z$ *satisfies* $\tilde{z} \in \mathcal{N}$.

Proof By (9.9) we have

$$X S e_n + X \Delta s + S \Delta x = \sigma \mu e_n, \tag{9.11}$$

which implies

$$\tilde{X} \tilde{S} e_n = X S e_n + X \Delta s + S \Delta x + \Delta X \Delta S e_n = \Delta X \Delta S e_n + \sigma \mu e_n.$$

Moreover, by Lemma 9.7(b), $\mu(\tilde{z}) = \sigma \mu$. We therefore need to show that

$$\| \tilde{X} \tilde{S} e_n - \sigma \mu e_n \| = \| \Delta X \Delta S e_n \| \le \beta \mu(\tilde{z}) = \beta \sigma \mu. \tag{9.12}$$

To do so, note first that $z \in \mathcal{N}$ implies $|x_i s_i - \mu| \le \beta \mu$ for all i, and hence

$$(1 - \beta)\mu \le x_i s_i \le (1 + \beta)\mu. \tag{9.13}$$

By (9.11) we have $X\Delta s + S\Delta x = \sigma\mu\mathbf{e}_n - XS\mathbf{e}_n$. Setting $D := X^{1/2}S^{-1/2}$, we get

$$D\Delta s + D^{-1}\Delta x = (XS)^{-1/2}(\sigma\mu\mathbf{e}_n - XS\mathbf{e}_n). \tag{9.14}$$

Because $(D^{-1}\Delta x)^{\mathsf{T}}(D\Delta s) = \Delta s^{\mathsf{T}}\Delta x = 0$ (cf. Lemma 9.7(a)), we can apply Lemma 9.6 with $u = D^{-1}\Delta x$ and $v = D\Delta s$ to obtain

$$\begin{aligned}
\|\Delta X\Delta S\mathbf{e}_n\| &= \left\|(D^{-1}\Delta X)(D\Delta S)\mathbf{e}_n\right\| \\
&\leq 2^{-1}\left\|D^{-1}\Delta x + D\Delta s\right\|^2 \\
&\leq 2^{-1}\left\|(XS)^{-1/2}(\sigma\mu\mathbf{e}_n - XS\mathbf{e}_n)\right\|^2 &&\text{by (9.14)} \\
&\leq \left(2\mu(1-\beta)\right)^{-1}\|\sigma\mu\mathbf{e}_n - XS\mathbf{e}_n\|^2 &&\text{by (9.13)} \\
&\leq \left(2\mu(1-\beta)\right)^{-1}\left(\|\mu\mathbf{e}_n - XS\mathbf{e}_n\| + \|\mu(\sigma-1)\mathbf{e}_n\|\right)^2 \\
&\leq \left(2\mu(1-\beta)\right)^{-1}\left(\beta\mu + (1-\sigma)\mu\sqrt{n}\right)^2 &&\text{by Def. 9.8} \\
&\leq \left(2(1-\beta)\right)^{-1}(\beta+\xi)^2\mu &&\text{by def. of } \sigma.
\end{aligned}$$

A small calculation shows that

$$\frac{1}{2(1-\beta)}(\beta+\xi)^2 \leq \beta(1-\xi) \leq \beta\left(1 - \frac{\xi}{\sqrt{n}}\right)$$

for $\beta = \frac{1}{4}$ and $0 \leq \xi \leq \frac{1}{4}$. This proves (9.12).

We still need to show that $\tilde{z} \in \mathcal{S}^\circ$. For this, by Lemma 9.7(c), it is sufficient to prove that $\tilde{x}, \tilde{s} > 0$. Inequality (9.12) implies $\tilde{x}_i\tilde{s}_i \geq (1-\beta)\sigma\mu > 0$. Suppose we had $\tilde{x}_i \leq 0$ or $\tilde{s}_i \leq 0$ for some i. Then $\tilde{x}_i < 0$ and $\tilde{s}_i < 0$, which implies $|\Delta x_i| > x_i$ and $|\Delta s_i| > s_i$. But then,

$$\beta\mu > \beta\sigma\mu \overset{(9.12)}{\geq} \|\Delta X\Delta S\mathbf{e}_n\| \geq |\Delta x_i\Delta s_i| > x_is_i \overset{(9.13)}{\geq} (1-\beta)\mu;$$

hence $\beta \geq \frac{1}{2}$, a contradiction. $\square$

Theorem 9.10 *On an input* $(A, G, b, c, d) \in \mathbb{R}^{m\times n} \times \mathbb{R}^{m\times p} \times \mathbb{R}^m \times \mathbb{R}^n \times \mathbb{R}^p$ *with* $\operatorname{rank} A = m \leq n$, $\operatorname{rank} G = p \leq m$, *and for the choice of the centering parameter* $\sigma = 1 - \frac{\xi}{\sqrt{n}}$ *with* $\xi \in (0, \frac{1}{4}]$, *Algorithm 9.1 produces, on a strictly feasible starting point* z_0 *in the central neighborhood* $\mathcal{N} = \mathcal{N}(\frac{1}{4})$, *a sequence of iterates* $z_k \in \mathcal{N}$ *such that* $\mu(z_k) = \sigma^k\mu(z_0)$, *for* $k \in \mathbb{N}$. *We therefore have, for all* $\varepsilon > 0$,

$$\mu(z_k) \leq \varepsilon \quad \text{for } k \geq \frac{\sqrt{n}}{\xi}\ln\frac{\mu(z_0)}{\varepsilon}.$$

Each iteration can be performed with $\mathcal{O}(n^3)$ *arithmetic operations.*

Proof It suffices to prove the displayed inequality. This follows from the implication

$$k \geq a^{-1} \ln B \implies (1-a)^k \leq \frac{1}{B} \tag{9.15}$$

for $0 < a < 1$, $B > 0$. (Use $\ln(1-a) \leq -a$ to show this.) $\square$

9.4 Condition-Based Analysis of IPM for PCFP

In the previous sections we described and analyzed an interior-point method (Algorithm 9.1) that approximates an optimal solution of the primal–dual pair (OP)–(OD). The goal of this section is to use this algorithm for solving the polyhedral cone feasibility problem (PCFP) and to analyze the resulting algorithm in terms of the GCC condition number.

Let $A \in (\mathbb{S}^{m-1})^n$ be given, $n > m$. Recall that the problem PCFP consists in deciding whether $A \in \mathcal{F}_P^\circ$ or $A \in \mathcal{F}_D^\circ$ (if $A \in \Sigma$, then A is an ill-posed instance of PCFP).

9.4.1 Reformulation

The first step is to reformulate the problem as a pair of primal–dual optimization problems by relaxation.

Without loss of generality assume $A\mathbf{e}_n \neq 0$, since otherwise the feasibility problem is trivial. Let $u := \frac{1}{\|A\mathbf{e}_n\|} A\mathbf{e}_n$. Notice that $\|u\| = 1$ by construction.

Instead of the primal feasibility problem $\exists x \in \mathbb{R}^n \setminus \{0\}$, $Ax = 0$, $x \geq 0$, we consider the relaxation (introducing one additional variable $x' \in \mathbb{R}$)

$$\min x' \quad \text{subject to} \quad Ax - ux' = 0,$$
$$\mathbf{e}_n^{\mathsf{T}} x = 1, \tag{9.16}$$
$$x \geq 0.$$

Also, the linear program dual of (9.16) (cf. Sect. 8.1) yields an analogous relaxation for the dual feasibility problem $\exists x \in \mathbb{R}^n \setminus \{0\}$, $A^{\mathsf{T}} y \leq 0$ (again, introducing one additional variable $y' \in \mathbb{R}$)

$$\max y' \quad \text{subject to} \quad A^{\mathsf{T}} y + \mathbf{e}_n y' \leq 0,$$
$$-u^{\mathsf{T}} y = 1. \tag{9.17}$$

We first note that the pair (9.16)–(9.17) has the form of the primal–dual pair (OP)–(OD) described in Sect. 8.1. We also note that these are both feasible problems. Indeed, the pair (x, x') with $x = \frac{\mathbf{e}_n}{n}$ and $x' = \frac{\|A\mathbf{e}_n\|}{n}$ satisfies the constraints of (9.16), and the pair (y, y') with $y = -u$ and $y' = -\|A^{\mathrm{T}} u\|_\infty$ satisfies those of (9.17). It follows from Theorem 8.2 that both have optimal solutions and that their optimal values are the same, which in the following we denote by v_*.

This optimal value is closely related to PCFP.

Lemma 9.11 *For $A \in \mathbb{R}^{m \times n}$ such that $A\mathbf{e}_n \neq 0$ we have*

$$A \in \mathcal{F}_P^\circ \iff v_* < 0,$$
$$A \in \mathcal{F}_D^\circ \iff v_* > 0,$$
$$A \in \Sigma, \quad \operatorname{rank} A = m \iff v_* = 0.$$

Proof Assume $A \in \mathcal{F}_P^\circ$. Then, there exists $x > 0$ such that $Ax = 0$. Let j be such that $x_j = \min_{i \leq n} x_i$ and put $\lambda := \mathbf{e}_n^{\mathrm{T}}(x - x_j \mathbf{e}_n)$. We have $\lambda > 0$, since $A\mathbf{e}_n \neq 0$. We define

$$\bar{x} := \lambda^{-1}(x - x_j \mathbf{e}_n), \qquad x' := -\lambda^{-1} x_j \|A\mathbf{e}_n\|.$$

Then, it is immediate to verify that $\bar{x} \geq 0$, $\mathbf{e}_n^{\mathrm{T}} \bar{x} = 1$, $x' < 0$, and $A\bar{x} - ux' = 0$. This shows that $v_* \leq x' < 0$.

Assume now that $A \in \mathcal{F}_D^\circ$. Then there exists $y \in \mathbb{R}^m$ such that $A^{\mathrm{T}} y < 0$. This implies $\mathbf{e}_n^{\mathrm{T}} A^{\mathrm{T}} y < 0$, or in other words, $u^{\mathrm{T}} y < 0$. Then $\bar{y} := -\frac{y}{u^{\mathrm{T}} y}$ satisfies $A^{\mathrm{T}} \bar{y} < 0$ and $-u^{\mathrm{T}} \bar{y} = 1$. In addition, for $y' = -\max_{i \leq n} a_i^{\mathrm{T}} \bar{y}$ we have $A^{\mathrm{T}} \bar{y} + \mathbf{e}_n y' \leq 0$ and $y' > 0$ and therefore $v_* \geq y' > 0$.

We next prove the converse of the two statements we just proved. To do so, consider optimal solutions (x_*, x_*') and (y_*, y_*') of (9.16) and (9.17), respectively. Because of Theorem 8.2 we have $x_*' = y_*' = v_*$.

If $v_* < 0$, then $x_*' < 0$, and we obtain a solution of $Ax = 0$, $x > 0$ by taking

$$x := x_* - \frac{x_*' \mathbf{e}_n}{\|A\mathbf{e}_n\|}.$$

This shows that $A \in \mathcal{F}_P^\circ$.

If instead, $v_* > 0$, then $y_*' > 0$, and we have $A^{\mathrm{T}} y_* \leq -y_*' \mathbf{e}_n < 0$, showing that $A \in \mathcal{F}_D^\circ$.

The reasoning above proves the first two equivalences in the statement. The third is now immediate from Theorem 6.14.   □

Remark 9.12 The rank condition in Lemma 9.11 is needed, since $\mathcal{F}_P^\circ \cap \Sigma = \{A \in \Sigma \mid \operatorname{rank} A < m\} \neq \emptyset$ (compare Theorem 6.14). While this condition may be confusing, it is quite irrelevant for our computational purposes.

By adding slack variables, we can recast (9.17) as

$$\max y' \quad \text{subject to} \quad A^{\mathrm{T}} y + \mathbf{e}_n y' + s = 0,$$
$$-u^{\mathrm{T}} y = 1, \tag{9.18}$$
$$s \geq 0.$$

Here $y \in \mathbb{R}^m$, $y' \in \mathbb{R}$, and $s \in \mathbb{R}^n$. In matrix form, (9.18) can be stated as

$$\max b^T \vec{y} \quad \text{subject to} \quad \mathcal{A}^T \vec{y} + s = c, \qquad G^T \vec{y} = d, \quad s \geq 0, \qquad (9.19)$$

where

$$\mathcal{A} = \begin{bmatrix} A \\ e_n^T \end{bmatrix} \in \mathbb{R}^{(m+1) \times n}, \qquad G = \begin{bmatrix} -u \\ 0 \end{bmatrix} \in \mathbb{R}^{m+1},$$

$$\vec{y} = \begin{pmatrix} y \\ y' \end{pmatrix} \in \mathbb{R}^{m+1}, \qquad b = \begin{pmatrix} 0 \\ 1 \end{pmatrix} \in \mathbb{R}^{m+1},$$

$$c = 0 \in \mathbb{R}^n, \qquad d = 1.$$

Likewise, (9.16) can be stated as

$$\min c^T x + dx' \quad \text{subject to} \quad Ax + Gx' = b, \quad x \geq 0. \qquad (9.20)$$

It is essential that (9.19) and (9.20) form a primal–dual pair. We note that $y' = b^T \vec{y} \leq c^T x + dx' = x'$ for any primal–dual feasible point $z = (x, x', \vec{y}, s)$, and therefore the duality gap $\mu(z)$ satisfies $n\mu(z) = x' - y'$.

9.4.2 Algorithmic Solution

We now run Algorithm 9.1 on input $(\mathcal{A}, G, b, c, d)$—together with an initial point z_0 given in Lemma 9.13 below—with the choice of the centering parameter $\sigma = 1 - \xi/\sqrt{n}$, $\xi = \frac{1}{4}$. This will solve PCFP for input A.

To see why, assume that we have at hand the starting point z_0 lying in the central neighborhood $\mathcal{N} = \mathcal{N}(\frac{1}{4})$. Because of Theorem 9.10, running Algorithm 9.1 on input $(\mathcal{A}, G, b, c, d)$ with the starting point z_0 produces a sequence of iterates $z_k = (x_k, x'_k, \vec{y}_k, s_k)$ in $\mathcal{N}$ such that $\mu(z_k) = \sigma^k \mu(z_0)$. The sequence z_k approaches an optimal solution of the primal–dual pair (9.19)–(9.20).

Suppose first that (9.19) has an optimal solution with $A^T y < 0$. Since $(\vec{y}_k, s_k)$ approaches an optimal solution of (9.19) for $k \to \infty$, we expect that for k sufficiently large, the first component y_k of $\vec{y}_k = (y_k, y'_k)$ will satisfy $A^T y_k < 0$ and certify the strict feasibility of the dual feasibility problem ($A \in \mathcal{F}_D^\circ$). This will be formally proved in Sect. 9.4.3 with an explicit bound on the number k of iterations required in terms of $\mathscr{C}(A)$.

Suppose instead that (9.20) has an optimal solution with $x > 0$. Since the iterates (x_k, x'_k) approach an optimal solution of (9.20), $Ax_k = ux'_k = \frac{x'_k}{\|Ae_n\|} Ae_n$ will be a negative multiple of Ae_n for k sufficiently large. Consequently, $x_k - \frac{x'_k}{\|Ae_n\|} e_n$ will solve $Ax = 0$, $x > 0$, thereby certifying the strict feasibility of the primal problem ($A \in \mathcal{F}_P^\circ$). This will be formally proved in Sect. 9.4.3. Again, a bound on how large k needs to be will be shown in terms of $\mathscr{C}(A)$.

It is not hard to find an explicit starting point z_0 in the central neighborhood $\mathcal{N}(\frac{1}{4})$.

Lemma 9.13 *The point $z_0 = (x, x', \vec{y}, s)$ defined by*

$$x = \frac{1}{n}\mathbf{e}_n, \qquad x' = \frac{\|A\mathbf{e}_n\|}{n},$$

$$y = -u, \qquad y' = -4\sqrt{n}, \qquad s = -A^{\mathrm{T}}y - \mathbf{e}_n y' \tag{9.21}$$

lies in the central neighborhood $\mathcal{N}(\frac{1}{4})$ and $\mu(z_0) \leq \frac{4\sqrt{n}+1}{n}$.

Proof It is straightforward to verify that

$$Ax - ux' = 0, \qquad A^{\mathrm{T}}y + \mathbf{e}_n y' + s = 0, \qquad -u^{\mathrm{T}}y = 1, \quad x, s > 0,$$

so that z_0 satisfies the constraints in (9.16) and (9.17). In addition, using $\mathbf{e}_n^{\mathrm{T}} A^{\mathrm{T}} u = \mathbf{e}_n^{\mathrm{T}} A^{\mathrm{T}} \frac{A\mathbf{e}_n}{\|A\mathbf{e}_n\|} = \|A\mathbf{e}_n\|$,

$$x^{\mathrm{T}}s = \frac{1}{n}\mathbf{e}_n^{\mathrm{T}}(-A^{\mathrm{T}}y - \mathbf{e}_n y') = \frac{1}{n}\|A\mathbf{e}_n\| - y'.$$

Thus, using $\|A\mathbf{e}_n\| = \|a_1 + \cdots + a_n\| \leq n$,

$$\mu(z_0) = \frac{x^{\mathrm{T}}s}{n} = -\frac{y'}{n} + \frac{\|A\mathbf{e}_n\|}{n^2} \in \left[-\frac{y'}{n}, \frac{1}{n}\right] \subseteq \left[\frac{4}{\sqrt{n}}, \frac{1+4\sqrt{n}}{n}\right],$$

or equivalently, $-n\mu(z_0) \in [-1-4\sqrt{n}, -4\sqrt{n}]$. On the other hand, using $|a_i^{\mathrm{T}}y| \leq 1$, we have for $i \in [n]$,

$$x_i s_i = \frac{s_i}{n} = \frac{-a_i^{\mathrm{T}}y - y'}{n} \in \left[\frac{-1 - y'}{n}, \frac{1 - y'}{n}\right],$$

or equivalently, $nx_i s_i \in [-1+4\sqrt{n}, 1+4\sqrt{n}]$. Therefore $n(x_i s_i - \mu(z_0)) \in [-2, 1]$, for each $i \in [n]$, and consequently $\|XSe_n - \mu\mathbf{e}_n\|^2 \leq \frac{4}{n} \leq \frac{1}{4}\mu(z_0)^2$. Thus $z_0 = (x, x', \vec{y}, s)$ is in the central neighborhood $\mathcal{N}(\frac{1}{4})$ of (9.19)–(9.20) with $\mu(z_0) \leq \frac{4\sqrt{n}+1}{n}$. $\square$

We are now ready to describe our algorithm for PCFP (see Algorithm 9.2 below).

The main result of this section is the following condition-based analysis of Algorithm 9.2.

Theorem 9.14 *Algorithm 9.2 returns, on input $A \notin \Sigma$, either a strictly feasible primal solution, i.e., a point $x \in \mathbb{R}^n$ such that $Ax = 0$, $x > 0$, or a strictly feasible dual solution, i.e., a point $y \in \mathbb{R}^m$ such that $A^{\mathrm{T}}y < 0$. In both cases, the algorithm halts after at most*

$$\mathcal{O}\big(\sqrt{n}(\log n + \log \mathscr{C}(A))\big)$$

iterations. The total number of arithmetic operations is bounded by

$$\mathcal{O}\big(n^{3.5}(\log n + \log \mathscr{C}(A))\big).$$

The proof of this theorem will be provided in the next subsection, and it relies on the characterization of $\mathscr{C}(A)$ given in Proposition 6.28.

Algorithm 9.2 Primal-Dual_IPM_for_PCFP

Input: $A \in \mathbb{R}^{m \times n}$
Preconditions: rank $A = m < n$

```
if  Aeₙ = 0 then return "Primal Feasible with
      solution eₙ" and halt
if  rank[Aᵀ eₙ] = m then
      compute y ∈ ℝᵐ, λ ∈ ℝ such that eₙ = λAᵀy
      if λ > 0 then return "Dual Feasible with
         solution −y" and halt
      if λ < 0 then return "Dual Feasible with
         solution y" and halt
σ := 1 − 1/(4√n)
set z := (x, x′, ȳ, s) with x, x′, ȳ, s as defined in (9.21)
μ := μ(z)
repeat
      if y′ > 0 then return "Dual Feasible with
         solution y" and halt
      if x′ < 0 then return "Primal Feasible with
         solution x̄ := x − x′/‖Aeₙ‖ eₙ" and halt
      set μ := σμ,  X = diag(x),  S = diag(s)
      compute Δz = (Δx, Δx′, Δȳ, Δs̄) by solving
```

$$
\begin{bmatrix} 0 & 0 & A^{\mathrm{T}} & I_n \\ 0 & 0 & G^{\mathrm{T}} & 0 \\ A & G & 0 & 0 \\ S & 0 & 0 & X \end{bmatrix} \begin{bmatrix} \Delta x \\ \Delta x' \\ \Delta \vec{y} \\ \Delta s \end{bmatrix} = \begin{bmatrix} 0 \\ 0 \\ 0 \\ \mu e_n - X S e_n \end{bmatrix},
$$

```
      set z := z + Δz
```

Output: either $x \in \mathbb{R}^n$ or $y \in \mathbb{R}^m$
Postconditions: The algorithm halts if $A \notin \Sigma$. In this case, $Ax = 0$ and $x > 0$ (if x is returned) or $A^{\mathrm{T}}y < 0$ (if y is returned)

9.4.3 Analysis

A key result in our analysis is the following quantitative complement of Lemma 9.11.

Lemma 9.15 *We have* $\Delta(A) \leq |v_*|$ *for* $A \in \mathbb{R}^{m \times n}$.

Proof If $A \in \Sigma$, then $\Delta(A) = 0$, and the result is trivial. We may then assume $A \notin \Sigma$.

Consider first the case $A \in \mathcal{F}_D^\circ$. Let (x_*, x_*') be an optimal solution of (9.16). Then $\|x_*\|_1 = 1$ and $Ax_* = ux_*'$. By Lemma 9.11 we have $x_*' = v_* > 0$. Proposition 6.28(a) implies that

$$\Delta(A) \leq \|Ax_*\| = \|ux_*'\| = x_*'.$$

Consider now the case $A \in \mathcal{F}_P^\circ$. Let $0 < \eta < \Delta(A)$ and $y = -\eta u$. Then $\|y\| = \eta < \Delta(A)$, and by Proposition 6.28(b) there exists $x \geq 0$ such that $\|x\|_1 = 1$ and $Ax = y$. Let $x' = -\eta$. Then $Ax - ux' = y + u\eta = 0$. Hence, (x, x') is a feasible point of (9.16), and it follows that the optimal value v_* of the pair (9.16)–(9.17) is at most $x' = -\eta$. Since this is true for all $\eta < \Delta(A)$, it follows that $v_* \leq -\Delta(A)$. $\square$

The goal is to prove the following result, from which Theorem 9.14 will easily follow.

Proposition 9.16 *Let $A \in \mathbb{R}^{m \times n}$ and $z = (x, x', \vec{y}, s) \in \mathcal{N}(\frac{1}{4})$.*

(a) *For $A \in \mathcal{F}_D^\circ$,*

$$\mu(z) < \frac{1}{n\mathscr{C}(A)} \quad \Rightarrow \quad y' > 0.$$

In particular, $A^T y < 0$.

(b) *For $A \in \mathcal{F}_P^\circ$,*

$$\mu(z) < \frac{1}{n\mathscr{C}(A)} \quad \Rightarrow \quad x' < 0.$$

In particular, the point $\bar{x} := x - \frac{x'}{\|Ae_n\|}e_n$ satisfies $A\bar{x} = 0, \bar{x} > 0$.

Proof Assume $A \in \mathcal{F}_D^\circ$. It follows from Lemma 9.15 that for any feasible point $z = (x, x', \vec{y}, \vec{s})$,

$$n\mu(z) = c^T x + dx' - b^T \vec{y} = x' - y' \geq x_*' - y' \geq \Delta(A) - y'.$$

Since $n\mu(z) < \mathscr{C}(A)^{-1} = \Delta(A)$, it follows that

$$y' \geq \Delta(A) - n\mu(z) > 0.$$

Now assume $A \in \mathcal{F}_P^\circ$. Using again Lemma 9.15, we deduce that for any feasible point $z = (x, x', \vec{y}, \vec{s})$,

$$n\mu(z) = x' - y' \geq x' - y_*' \geq x' + \Delta(A).$$

Since $n\mu(z) < \Delta(A)$, it follows that

$$x' \leq -\Delta(A) + n\mu(z) < 0. \qquad \square$$

Proof of Theorem 9.14 Assume $A \in \mathcal{F}_D^\circ$. By Lemma 9.13 and Theorem 9.10, the sequence of iterates z_k produced by Algorithm 9.2 stays in the central neighborhood $\mathcal{N}(\frac{1}{4})$ and $\mu(z_k) = \sigma^k \mu(z_0) \leq \sigma^k (4\sqrt{n} + 1)/n$ for all $k \in \mathbb{N}$. If

$$\frac{4\sqrt{n} + 1}{n} \left(1 - \frac{1}{4\sqrt{n}}\right)^k < \frac{1}{n \mathscr{C}(A)}, \tag{9.22}$$

then Proposition 9.16 implies $A^\mathsf{T} y < 0$, and hence Algorithm 9.2 halts and returns y. Inequality (9.22) holds if (compare (9.15))

$$k > 4\sqrt{n} \ln\left((4\sqrt{n} + 1)\mathscr{C}(A)\right).$$

The case $A \in \mathcal{F}_P^\circ$ is dealt with similarly. $\square$

Remark 9.17 In the preceding chapter we described the ellipsoid method, an elegant theoretical framework for showing the existence of efficient algorithms in convex optimization. We used such a method to find a point y such that $A^\mathsf{T} y < 0$ for a matrix $A \in \mathcal{F}_D^\circ$. Even though the task done by Algorithm 9.2 is more general than that performed by Algorithm 7.1, it makes sense to compare their theoretical complexity bounds. The total number of arithmetic operations is bounded by $\mathcal{O}(n^{3.5}(\log n + \log \mathscr{C}(A)))$ in the interior-point Algorithm 9.2 and by $\mathcal{O}(m^3 n \ln \mathscr{C}(A))$ for the ellipsoid Algorithm 7.1. This last bound can be smaller that the former when $n \gg m$, but for comparable m and n (say m a fraction of n) it is larger.

Remark 9.18 Since the complexity bound in Theorem 9.14 is in terms of $\mathscr{C}(A)$, we may use again Proposition 7.9 to deduce, as in Remark 7.10, that if the data matrix A has integer entries, then (a suitable modification of) Algorithm 9.2 works in time polynomial in n, m and the bit-size of the largest entry of A. Note that the only way for Algorithm 9.2 not to have halted when reaching this time bound is if $A \in \Sigma$. Hence, by "clocking" the algorithm, one can decide whether $A \in \mathcal{F}_P^\circ$, $A \in \mathcal{F}_D^\circ$, or $A \in \Sigma$ in polynomial time in the bit-size of the given integer matrix A.

9.5 Finite Precision for Decision and Counting Problems

We may now return to an issue pointed out in Sect. 6.1. We mentioned there that for a decisional problem such as PCFP, a different form of finite-precision analysis is needed. Recall that the obstruction is that the notion of loss of precision hardly makes sense for a problem whose possible outputs are the tags "primal strictly feasible" and "dual strictly feasible." It is apparent, however, that the value of the machine precision ϵ_{mach} has an influence on the computed solution for the problem. One feels that the smaller ϵ_{mach} is, the more likely the returned tag will be the correct one for the input matrix $A \in (\mathbb{S}^{m-1})^n$. This is clear if errors occur only in reading A. In this case—assuming that the error satisfies $d_{\sin}(A, \tilde{A}) \leq \epsilon_{\text{mach}}$—one has that the computed tag is guaranteed to be the correct one whenever $\epsilon_{\text{mach}} < \mathscr{C}(A)$.

If instead, errors occur all along the computation, then similar results can be sought with a more elaborate right-hand side.

Indeed, one can actually think of two template results for a (discrete-valued) problem $\varphi : \mathcal{D} \subseteq \mathbb{R}^m \to V$ for which a condition number cond^φ has been defined. In the first one, the machine precision ϵ_{mach} is fixed.

Template Result 1 *The cost of (fixed-precision) algorithm $\mathscr{A}$—computing the function $\mathscr{A}^\varphi$—satisfies, for all $a \in \mathcal{D}$, a bound of the form*

$$f\big(\text{dims}(a), \text{cond}^\varphi(a)\big).$$

Moreover, $\mathscr{A}^\varphi(a) = \varphi(a)$ if

$$\epsilon_{\text{mach}} \leq \frac{1}{g(\text{dims}(a), \text{cond}^\varphi(a))}.$$

Here f and g are functions of the dimensions of a and its condition. □

In the second template result, the finite-precision algorithm has the capacity to adaptively modify its machine precision. This leads to outputs that are guaranteed to be correct but do not have a priori bounds on the precision required for a given computation.

Template Result 2 *The cost of variable-precision algorithm $\mathscr{A}$—computing the function $\mathscr{A}^\varphi$—satisfies, for all $a \in \mathcal{D}$, a bound of the form*

$$f\big(\text{dims}(a), \text{cond}^\varphi(a)\big).$$

Moreover, for all $a \in \mathcal{D}$, $\mathscr{A}^\varphi(a) = \varphi(a)$, and the finest precision required satisfies the bound

$$\epsilon_{\text{mach}} \geq \frac{1}{g(\text{dims}(a), \text{cond}^\varphi(a))}.$$

Here f and g are functions of the dimensions of a and its condition. □

In practice, one may want to limit both the running time and the precision of $\mathscr{A}$. If this is the case, one may stop the execution of $\mathscr{A}$ on input a after a certain number of steps if the computation has not succeeded by then, and return a message of the form

```
The condition of the data is larger than K.
```

The value of K can be obtained by solving f for $\text{cond}^\varphi(a)$.

To give an example, we return to PCFP. For this problem, note that the assumption of finite precision sets some limitations on the solutions (feasible points) we may obtain. If system A belongs to $\mathcal{F}_D^\circ$, then we will obtain, after sufficiently refining the precision, a point $y \in \mathbb{R}^m$ such that $A^\mathsf{T} y < 0$. On the other hand, if $A \in \mathcal{F}_P^\circ$, then there is no hope of computing a point $x \in \mathbb{R}^n$ such that $Ax = 0, x > 0$, since the set of such points is thin in $\mathbb{R}^n$ (i.e., has empty interior). In such a case there is no way to ensure that the errors produced by the use of finite precision will not move any candidate solution out of this set. We may instead compute good approximations.

Let $\gamma \in (0, 1)$. A point $\hat{x} \in \mathbb{R}^n$ is a γ-forward solution of the system $Ax = 0$, $x \geq 0$, if $\hat{x} \geq 0$, $\hat{x} \neq 0$, there exists $\bar{x} \in \mathbb{R}^n$ such that

$$A\bar{x} = 0, \quad \bar{x} \geq 0,$$

and for $i = 1, \ldots, n$,

$$|\hat{x}_i - \bar{x}_i| \leq \gamma \hat{x}_i.$$

The point $\bar{x}$ is said to be an *associated solution* for $\hat{x}$. A point is a *forward-approximate solution* of $Ax = 0$, $x \geq 0$, if it is a γ-forward solution of the system for some $\gamma \in (0, 1)$. *Strict forward-approximate solutions* are defined, as usual, by replacing the inequalities by strict inequalities.

The main result in [69] is the following extension (in the form of Template Result 2) of Theorem 9.14.

Theorem 9.19 *There exists a round-off machine that with input a matrix $A \in \mathbb{R}^{m \times n}$ and a number $\gamma \in (0, 1)$ finds either a strict γ-forward solution $x \in \mathbb{R}^n$ of $Ax = 0$, $x \geq 0$, or a strict solution $y \in \mathbb{R}^m$ of the system $A^T y \leq 0$. The machine precision ϵ_{mach} varies during the execution of the algorithm. The finest required precision is*

$$\epsilon_{\mathsf{mach}} = \frac{1}{\mathbf{c} n^{12} \mathscr{C}(A)^2},$$

where $\mathbf{c}$ is a universal constant. The number of main (interior-point) iterations of the algorithm is bounded by

$$\mathcal{O}\big(n^{1/2}\big(\log n + \log(\mathscr{C}(A)) + |\log \gamma|\big)\big)$$

if $A \in \mathcal{F}_P^\circ$, and by the same expression without the $|\log \gamma|$ term if $A \in \mathcal{F}_D^\circ$. $\qquad\square$

Chapter 10
The Linear Programming Feasibility Problem

In Chap. 8 we introduced linear programming optimization problems. Then, in Chap. 9, we rephrased PCFP as one such problem. By doing so, we could apply an interior-point method to the latter and obtain an algorithm solving PCFP with a complexity bounded by a low-degree polynomial in n, m and $\log \mathscr{C}(A)$. A question conspicuously left open is the solution of the optimization problem itself. Theorem 9.10 provides a key step of this solution but leaves at least two aspects untouched: the initial feasible point is assumed to be given and there is no hint as how to deduce, at some moment of the process, the optimizers and optimal value.

In this chapter we deal with the first of the two aspects above. To fix a context, let us consider the primal–dual pair in standard form:

$$\min \ c^{\mathsf{T}}x \quad \text{subject to} \quad Ax = b, \quad x \geq 0, \tag{SP}$$

and

$$\max \ b^{\mathsf{T}}y \quad \text{subject to} \quad A^{\mathsf{T}}y \leq c. \tag{SD}$$

The problem at hand is, given the triple $d = (A, b, c)$ to decide whether the primal feasible set $\mathcal{S}_P = \{x \in \mathbb{R}^n \mid Ax = b, x \geq 0\}$ and dual feasible set $\mathcal{S}_D = \{y \in \mathbb{R}^m \mid A^{\mathsf{T}}y \leq c\}$ are both nonempty, and if this is so, to compute a pair (x, y) of points in these sets. If both primal and dual are feasible, we say that d is *feasible*. This is the feasibility problem we discussed in Sect. 8.4.2.

10.1 A Condition Number for Polyhedral Feasibility

We will call any system of linear equalities and inequalities a *polyhedral system*.

For any polyhedral system with data S we write $\mathbf{F}(S) = 0$ if the system is infeasible and $\mathbf{F}(S) = 1$ otherwise. Then, we define (assuming some norm in the space of data) the *distance to ill-posedness*

$$\rho(S) := \inf\{\|\Delta S\| : \mathbf{F}(S) \neq \mathbf{F}(S + \Delta S)\}$$

P. Bürgisser, F. Cucker, *Condition*,
Grundlehren der mathematischen Wissenschaften 349,
DOI 10.1007/978-3-642-38896-5_10, © Springer-Verlag Berlin Heidelberg 2013

as well as the *condition number*

$$C(S) := \frac{\|S\|}{\rho(S)}.$$

These definitions follow the lines of Sects. 1.3 and 6.4.

Consider now the data $d = (A, b, c)$ of the pair (SP–SD). For the primal polyhedral system $Ax = b, x \geq 0$, this yields a distance to ill-posedness $\rho_P(A, b)$ and a condition number $C_P(A, b)$; we similarly obtain $\rho_D(A, c)$ and $C_D(A, c)$ for the dual. Finally we define, for the triple $d = (A, b, c)$, the condition number

$$C(d) := \max\{C_P(A, b), C_D(A, c)\}.$$

As usual, if $C(d) = \infty$, we say that d is *ill-posed* and we say that it is *well-posed* otherwise.

We note that, following the discussion in Sect. 8.4.2(c), we could also define a condition number for the feasibility problem by taking

$$C_{\mathsf{FP}}(d) := \frac{\|d\|}{\mathsf{dist}(d, \Sigma_{\mathsf{FP}})}$$

with the set of ill-posed triples Σ_{FP} as defined there and where dist is the distance induced by $\| \ \|$. The condition number $C(d)$ is similar in spirit to C_{FP} but differs in the way the normalization of the inverse distance is made. In particular, we note that $C(d) = \infty$ if and only if $d \in \Sigma_{\mathsf{FP}}$.

To make these definitions precise, we need to fix a norm. It will be convenient to choose $\| \ \|_{12}$. That is, the norm of a pair (A, b) is the 12-norm of the matrix $[A, b]$, which we denote by $\|A, b\|_{12}$. Similarly for the dual, where for a pair (A, c), we consider the 12-norm of the matrix (A, c^{T}) (or equivalently, due to Lemma 1.2(c), the 2∞-norm of the matrix $[A^{\mathsf{T}}, c]$).

Remark 10.1 Note that $\rho_P(A, b) \leq \|A, b\|_{12}$ and $\rho_D(A, c) \leq \left\|\left(\begin{smallmatrix} A \\ c^{\mathsf{T}} \end{smallmatrix}\right)\right\|_{12}$.

We can now use $C(d)$ to state the main result of this chapter.

Theorem 10.2 *There exists an algorithm that given a triple $d = (A, b, c)$ with $A \in \mathbb{R}^{m \times n}$, $b \in \mathbb{R}^m$, and $c \in \mathbb{R}^n$, decides whether both S_P and S_D are nonempty with a cost bounded by*

$$\mathcal{O}\big(n^{3.5}\big(\log n + \log C(d)\big)\big).$$

Furthermore, if both sets are nonempty, the algorithm returns a pair (x, y) of strictly feasible points.

Note that the algorithm in Theorem 10.2 decides feasibility only for well-posed triples d.

10.2   Deciding Feasibility of Primal–Dual Pairs

The idea for proving Theorem 10.2 is simple: by homogenizing, we reduce the two feasibility problems to instances of PCFP. Indeed, given $d = (A, b, c)$, we have

$$\exists x > 0 \quad \text{s.t.} \quad Ax = b \quad \Longleftrightarrow \quad \exists x > 0, \ \exists t > 0 \quad \text{s.t.} \quad Ax - bt = 0 \quad (10.1)$$

and, for any $\alpha > 0$,

$$\exists y \quad \text{s.t.} \quad A^\mathsf{T} y < c \quad \Longleftrightarrow \quad \exists y \ \exists t \quad \text{s.t.} \quad \begin{bmatrix} A^\mathsf{T} & -c \\ 0 & -\alpha \end{bmatrix} \begin{bmatrix} y \\ t \end{bmatrix} < 0. \quad (10.2)$$

Therefore, we can decide the feasibility of d with two calls to Algorithm 9.2 (Primal-Dual_IPM_for_PCFP). In addition, we have freedom to choose $\alpha > 0$.

We say that d is *dual-normalized* when

$$\left\| \begin{pmatrix} A \\ c^\mathsf{T} \end{pmatrix} \right\|_{12} = 1.$$

From a computational viewpoint, this normalization can be straightforwardly achieved in a way that the dual feasible set S_D remains unchanged as does the condition number $C_D(d)$. These considerations lead to Algorithm 10.1 below.

The complexity bounds in Theorem 9.14 do not, however, directly apply to analyzing Algorithm 10.1, since they are expressed in terms of $\mathscr{C}$, a condition number for a homogeneous problem, and we want a bound in terms of $C(d)$ (i.e., in terms of C_P and C_D). Proposition 10.3 below shows that this is not a major obstacle, since these condition numbers are roughly the same. Indeed, the characterization of $\mathscr{C}$ in Proposition 6.21 shows that for $M \in \mathbb{R}^{m \times n}$,

$$\mathscr{C}(M) = \frac{\|M\|_{12}}{\rho^h(M)},$$

where we have written $\rho^h(M) := \mathrm{dist}_{12}(M, \Sigma)$ to emphasize the resemblance with the corresponding expressions for $C_P(A, b)$ and $C_D(A, c)$. Note actually that the only difference between $\rho^h\left(\begin{bmatrix} A & 0 \\ -c^\mathsf{T} & -1 \end{bmatrix} \right)$ and $\rho_D(A, c)$ is that in the former we are allowed to consider perturbations of the 0 and the -1 in the last column of the matrix, whereas in the latter we can perturb only the entries of A and c.

Proposition 10.3 *For any dual-normalized triple $d = (A, b, c)$,*

$$\mathscr{C}\left(\begin{bmatrix} A & 0 \\ -c^\mathsf{T} & -1 \end{bmatrix} \right) \geq C_D(A, c) \geq \frac{1}{5} \mathscr{C}\left(\begin{bmatrix} A & 0 \\ -c^\mathsf{T} & -1 \end{bmatrix} \right),$$

and for any triple $d = (A, b, c)$,

$$C_P(A, b) = \mathscr{C}([A, -b]).$$

Algorithm 10.1 FEAS_LP

Input: $A \in \mathbb{R}^{m \times n}, b \in \mathbb{R}^m, c \in \mathbb{R}^n$

Preconditions: rank $A = m < n$

```
run Primal-Dual_IPM_for_PCFP with input
      M := [A, -b]
if M(x, t) = 0, (x, t) > 0 is feasible
      return "primal feasible" with solution x
      else return "primal infeasible"
dual normalize (A, c)
run Primal-Dual_IPM_for_PCFP with input
```
$$M := \begin{bmatrix} A & 0 \\ -c^{\mathrm{T}} & -1 \end{bmatrix}$$
```
if M^T(y, t) < 0 is feasible
      return "dual feasible" with solution y
      else return "dual infeasible"
```

Output: $x \in \mathbb{R}^n$ or $y \in \mathbb{R}^m$ or both

Postconditions: The algorithm outputs x if $\mathcal{S}_P \neq \emptyset$ and $C_P(A, b) < \infty$ and it returns y if $\mathcal{S}_D \neq \emptyset$ and $C_P(A, c) < \infty$

Towards the proof of Proposition 10.3 we introduce some notation. Let

$$\mathcal{F}_D^{\#} = \left\{ (A, c) \in \mathbb{R}^{m \times n} \times \mathbb{R}^n \mid \exists y \in \mathbb{R}^m \text{ s.t. } A^{\mathrm{T}} y < c \right\}$$

and

$$\mathcal{F}_D = \left\{ B \in \mathbb{R}^{(m+1) \times (n+1)} \mid \exists z \in \mathbb{R}^{m+1} \text{ s.t. } B^{\mathrm{T}} z < 0 \right\}.$$

Lemma 10.4 *Let* $A \in \mathbb{R}^{m \times n}$ *and* $c \in \mathbb{R}^n$, *such that* $\left\| \left(\begin{smallmatrix} A \\ c^{\mathrm{T}} \end{smallmatrix} \right) \right\|_{12} = 1$. *Then*

$$\begin{bmatrix} A & 0 \\ -c^{\mathrm{T}} & -1 \end{bmatrix} \in \mathcal{F}_D \quad \Longleftrightarrow \quad (A, c) \in \mathcal{F}_D^{\#}$$

and

$$\rho^h \left(\begin{bmatrix} A & 0 \\ -c^{\mathrm{T}} & -1 \end{bmatrix} \right) \leq \rho_D(A, c) \leq 5\rho^h \left(\begin{bmatrix} A & 0 \\ -c^{\mathrm{T}} & -1 \end{bmatrix} \right).$$

Proof It is clear that

$$\begin{bmatrix} A & 0 \\ -c^{\mathrm{T}} & -1 \end{bmatrix} \in \mathcal{F}_D \quad \Longleftrightarrow \quad (A, c) \in \mathcal{F}_D^{\#}.$$

From here, the inequality

$$
\rho^h \left(\begin{bmatrix} A & 0 \\ -c^{\mathrm{T}} & -1 \end{bmatrix} \right) \leq \rho_D(A, c)
\tag{10.3}
$$

readily follows.

We now address the other inequality. By the definition of ρ^h, there exist $\Delta A, \Delta c, \Delta u, \Delta v$, such that

$$
\omega := \left\| \begin{pmatrix} \Delta A & \Delta u \\ \Delta c^{\mathrm{T}} & \Delta v \end{pmatrix} \right\|_{12} = \rho^h \left(\begin{bmatrix} A & 0 \\ -c^{\mathrm{T}} & -1 \end{bmatrix} \right)
$$

and

$$
\begin{bmatrix} A + \Delta A & \Delta u \\ -c^{\mathrm{T}} + \Delta c^{\mathrm{T}} & -1 + \Delta v \end{bmatrix} \in \mathcal{F}_D \quad \Longleftrightarrow \quad (A, c) \notin \mathcal{F}_D^\#.
\tag{10.4}
$$

Note that due to Remark 10.1 and (10.3), we have $\omega \leq 1$ and, in particular, $|\Delta v| \leq 1$. Assume $|\Delta v| < 1$ and consider

$$
M = \begin{bmatrix} (1 - \Delta v)I_m & \Delta u \\ 0 & 1 \end{bmatrix} \begin{bmatrix} A + \Delta A & \Delta u \\ -c^{\mathrm{T}} + \Delta c^{\mathrm{T}} & -1 + \Delta v \end{bmatrix}.
$$

Note that $M \in \mathbb{R}^{(m+1) \times (n+1)}$ and the matrix at the left in the product above is invertible. Therefore,

$$
M \in \mathcal{F}_D \quad \Longleftrightarrow \quad \begin{bmatrix} A + \Delta A & \Delta u \\ -c^{\mathrm{T}} + \Delta c^{\mathrm{T}} & -1 + \Delta v \end{bmatrix} \in \mathcal{F}_D.
$$

It now follows from (10.4) that $M \in \mathcal{F}_D$ if and only if $(A, c) \notin \mathcal{F}_D^\#$. In addition,

$$
M = \begin{bmatrix} (1 - \Delta v)(A + \Delta A) + \Delta u(-c^{\mathrm{T}} + \Delta c^{\mathrm{T}}) & 0 \\ -c^{\mathrm{T}} + \Delta c^{\mathrm{T}} & -1 + \Delta v \end{bmatrix}.
$$

Due to the form of M, and since $-1 + \Delta v < 0$, it follows, using (10.2), that $M \in \mathcal{F}_D$ if and only if

$$
\begin{bmatrix} (1 - \Delta v)(A + \Delta A) + \Delta u(-c^{\mathrm{T}} + \Delta c^{\mathrm{T}}) \\ -c^{\mathrm{T}} + \Delta c^{\mathrm{T}} \end{bmatrix} \in \mathcal{F}_D^\#.
$$

Therefore

$$
\rho_D(A, c) \leq \left\| \begin{pmatrix} A \\ -c^{\mathrm{T}} \end{pmatrix} - \begin{pmatrix} (1 - \Delta v)(A + \Delta A) + \Delta u(-c^{\mathrm{T}} + \Delta c^{\mathrm{T}}) \\ -c^{\mathrm{T}} + \Delta c^{\mathrm{T}} \end{pmatrix} \right\|_{12}
$$

$$
= \left\| \begin{pmatrix} -\Delta A + \Delta v A + \Delta v \Delta A + \Delta u c^{\mathrm{T}} - \Delta u \Delta c^{\mathrm{T}} \\ \Delta c^{\mathrm{T}} \end{pmatrix} \right\|_{12}
$$

$$\leq \left\| \begin{pmatrix} \Delta A \\ \Delta c^{\mathrm{T}} \end{pmatrix} \right\|_{12} + |\Delta v| \|A\|_{12} + |\Delta v| \|\Delta A\|_{12}$$

$$+ \|\Delta u c^{\mathrm{T}}\|_{12} + \|\Delta u \Delta c^{\mathrm{T}}\|_{12}$$

$$\leq \left\| \begin{pmatrix} \Delta A \\ \Delta c^{\mathrm{T}} \end{pmatrix} \right\|_{12} + |\Delta v| \|A\|_{12} + \|\Delta A\|_{12}$$

$$+ \|\Delta u\|_2 \|c\|_\infty + \|\Delta u\|_2 \|\Delta c\|_\infty,$$

where the last inequality is due to Lemma 1.2(a). Now use that

$$\left\| \begin{pmatrix} \Delta A \\ \Delta c^{\mathrm{T}} \end{pmatrix} \right\|_{12}, \|\Delta A\|_{12}, |\Delta v|, \|\Delta u\|_2, \|\Delta c\|_\infty \leq \left\| \begin{pmatrix} \Delta A & \Delta u \\ \Delta c^{\mathrm{T}} & \Delta v \end{pmatrix} \right\|_{12} = \omega$$

and that

$$\|A\|_{12}, \|c\|_\infty \leq \left\| \begin{pmatrix} A \\ c^{\mathrm{T}} \end{pmatrix} \right\|_{12} = 1$$

(together with $\omega^2 \leq \omega$) to obtain $\rho_D(A, c) \leq 5\omega$.

If instead, $|\Delta v| = 1$, then

$$\rho^h \left(\begin{bmatrix} A & 0 \\ -c^{\mathrm{T}} & -1 \end{bmatrix} \right) = \left\| \begin{pmatrix} \Delta A & \Delta u \\ \Delta c^{\mathrm{T}} & \Delta v \end{pmatrix} \right\|_{12} \geq 1. \tag{10.5}$$

In addition, since $\left\| \begin{pmatrix} A \\ c^{\mathrm{T}} \end{pmatrix} \right\|_{12} = 1$, $\rho_D(A, c) \leq 1$. Hence, using (10.3),

$$\rho^h \left(\begin{bmatrix} A & 0 \\ -c^{\mathrm{T}} & -1 \end{bmatrix} \right) \leq \rho_D(A, c) \leq 1. \tag{10.6}$$

Inequalities (10.5) and (10.6) yield $\rho^h \left(\begin{bmatrix} A & 0 \\ -c^{\mathrm{T}} & -1 \end{bmatrix} \right) = \rho_D(A, c) = 1$, and thus the statement. $\qquad\square$

Proof of Proposition 10.3 The equality in the primal case easily follows from (10.1). The inequalities for the dual case have been shown in Lemma 10.4. $\qquad\square$

Proof of Theorem 10.2 For both the primal and the dual systems, Theorem 9.14 (together with (10.1) and (10.2)) ensures that Algorithm 9.2 decides feasibility returning a strictly feasible point (if the system is well posed). Furthermore, this algorithm performs at most

$$\mathcal{O}\left(\sqrt{n} \left(\log n + \log \mathscr{C}(M) \right) \right)$$

iterations, where $M = [A, -b]$ in the primal case and $M = \begin{bmatrix} A & 0 \\ -c^{\mathrm{T}} & -1 \end{bmatrix}$ in the dual. Proposition 10.3 allows one to replace $\mathscr{C}(M)$ by $C(d)$ in both cases. $\qquad\square$

Remark 10.5 We can use Proposition 7.9 one more time to deduce, as in Remarks 7.10 and 9.18, that when restricted to data with integer entries, Algorithm 10.1 works in polynomial time.

Remark 10.6 Theorem 9.19 states that the problem PCFP can be accurately solved with finite precision. Since Algorithm 10.1 is, essentially, two calls to a PCFP-solver, a finite-precision version of this algorithm will work accurately as well.

Theorem 10.7 *There exists a finite-precision algorithm that, given a triple $d = (A, b, c)$ with $A \in \mathbb{R}^{m \times n}$, $b \in \mathbb{R}^m$, and $c \in \mathbb{R}^n$, decides whether both $\mathcal{S}_P$ and $\mathcal{S}_D$ are nonempty. The machine precision ϵ_{mach} varies during the execution of the algorithm. The finest required precision satisfies*

$$\epsilon_{\text{mach}} = \frac{1}{\mathcal{O}(n^{12} C(d)^2)}.$$

The total number of arithmetic operations is bounded by

$$\mathcal{O}\left(n^{3.5} \left(\log n + \log C(d)\right)\right).$$

Furthermore, if both sets are nonempty, and an additional input $\gamma \in (0, 1)$ is specified, the algorithm returns a pair (x, y), where y is a strictly feasible solution of $A^{\mathsf{T}} y \leq c$ and x is a strict γ-forward solution of $Ax = b$, $x \geq 0$. In this case the total number of arithmetic operations becomes

$$\mathcal{O}\left(n^{3.5} \left(\log n + \log C(d) + |\log \gamma|\right)\right). \qquad \square$$

Chapter 11
Condition and Linear Programming Optimization

In the previous chapter we analyzed an algorithm deciding feasibility for a triple $d = (A, b, c)$ specifying a pair of primal and dual linear programming problems in standard form,

$$\min c^{\mathrm{T}} x \quad \text{subject to} \quad Ax = b, \quad x \geq 0, \qquad \text{(SP)}$$

and

$$\max b^{\mathrm{T}} y \quad \text{subject to} \quad A^{\mathrm{T}} y \leq c. \qquad \text{(SD)}$$

If such an algorithm decides that a triple d is feasible, we may want to compute the optimizers x^* and y^*, as well as the optimal value v^*, of the pair (SP)–(SD). To do so is the goal of this chapter.

An approach to this problem is to apply the interior-point Algorithm 9.1 along with its basic analysis as provided in Theorem 9.10. A possible obstacle is the fact that the feasible point $z = (x, y, s)$ returned in Theorem 10.2 does not necessarily belong to the central neighborhood $\mathcal{N}(\frac{1}{4})$.

Another obstacle, now at the heart of this book's theme, is how to deduce, at some iteration of Algorithm 9.1, the optimizers x^* and y^*. Without doing so, Algorithm 9.1 will increasingly approach these optimizers without ever reaching them. It is not surprising that a notion of condition should be involved in this process. This notion follows lines already familiar to us. For almost all feasible triples d a small perturbation of d will produce a small change in x^* and y^*. For a thin subset of data, instead, arbitrarily small perturbations may substantially change these optimizers (recall the discussion in Sect. 8.4.2). The central character of this chapter, the condition number $\mathscr{K}(d)$, measures the relative size of the smallest perturbation that produces such a discontinuous change in the optimizers. We will formally define $\mathscr{K}(d)$ in Sect. 11.1. We will also show there a characterization of $\mathscr{K}(d)$ that, in line with the theme occupying Sect. O.5.2, makes its computation possible.

In Sect. 11.3 below we describe and analyze algorithms solving the optimal basis problem, which, we recall, consists in, given a feasible triple d, finding an optimal basis for it. The first main result concerning this goal is the following.

P. Bürgisser, F. Cucker, *Condition*,
Grundlehren der mathematischen Wissenschaften 349,
DOI 10.1007/978-3-642-38896-5_11, © Springer-Verlag Berlin Heidelberg 2013

Theorem 11.1 *There exists an algorithm that with input a full-rank matrix* $A \in \mathbb{R}^{m \times n}$, *vectors* $b \in \mathbb{R}^m$, $c \in \mathbb{R}^n$, *and a feasible point* $z_0 = (x_0, y_0, s_0)$ *in the central neighborhood* $\mathcal{N}(\frac{1}{4})$ *such that* $\mu(z_0) \leq (n \|d\|)^{\mathcal{O}(1)}$ *finds an optimal basis* B *for* d. *The number of iterations performed by the algorithm is bounded by*

$$\mathcal{O}\big(\sqrt{n}(\log n + \log \mathscr{K}(d))\big).$$

The total number of arithmetic operations is bounded by

$$\mathcal{O}\big(n^{3.5}(\log n + \log \mathscr{K}(d))\big).$$

Remark 11.2 There is no loss of generality in assuming $\mu(z_0) \leq (n\|d\|)^{\mathcal{O}(1)}$. In fact, a bit of observation shows that the feasible points x and (y, s) returned by Algorithm 10.1 satisfy $\max\{\|x\|, \|s\|\} = (n\|d\|)^{\mathcal{O}(1)}$. Therefore

$$\mu(z) = \frac{1}{n}x^{\mathsf{T}}s \leq \frac{1}{n}\|x\|\|s\| = \big(n\|d\|\big)^{\mathcal{O}(1)}.$$

Unfortunately, the assumption $z_0 \in \mathcal{N}(\frac{1}{4})$ appears to be more difficult to get rid of, and a discussion on the possible ways to deal with it would take us too far away from our main themes.

There is an alternative way to compute an optimal basis for d that does not require an initial point in $\mathcal{N}(\frac{1}{4})$. Instead, it proceeds by reducing the optimal basis problem to a sequence of polyhedral cone feasibility problems. The cost of this new approach is slightly larger than the cost in Theorem 11.1, but this is compensated by the simplicity in its overcoming the need for an initial point in the central neighborhood.

Theorem 11.3 *There exists an algorithm that with input a full-rank matrix* $A \in \mathbb{R}^{m \times n}$ *and vectors* $b \in \mathbb{R}^m$ *and* $c \in \mathbb{R}^n$, *finds an optimal basis* B *for* d. *The total number of arithmetic operations is bounded by*

$$\mathcal{O}\big(n^{3.5}(\log n + \log \mathscr{K}(d)) \log\log(\mathscr{K}(d) + 4)\big).$$

11.1 The Condition Number $\mathscr{K}(d)$

Definition 11.4 We say that d is *feasible well-posed* when there exists a unique optimal solution (x^*, y^*, s^*) for d. In this case, we write $d \in \mathcal{W}$. If more than one optimal solution exists, we say that d is *feasible ill-posed*.

Let $\mathcal{B} := \{B \subseteq [n] \mid |B| = m\}$. Theorem 8.13(c) implies that if $d \in \mathcal{W}$ is feasible well-posed with optimal solution (x^*, y^*, s^*), then there is a unique $B \in \mathcal{B}$ such that (x^*, y^*, s^*) is the basic optimal solution associated to B (see Definition 8.10). We called such B the *optimal basis for* d.

Fig. 11.1 The situation in the
space of data

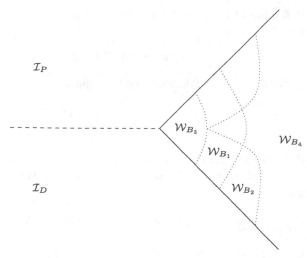

For any $B \in \mathcal{B}$, we write

$$\mathcal{W}_B := \{d \in \mathcal{W} \mid B \text{ is the optimal basis for } d\}.$$

The set $\mathcal{W}$ is thus partitioned by the family $\{\mathcal{W}_B \mid B \in \mathcal{B}\}$. Figure 11.1 schematically summarizes the situation in the space of data.

At the right of the diagram we have the set $\mathcal{W}$ of well-posed feasible triples, which is partitioned into the subsets $\mathcal{W}_B$. The dotted lines separating these subsets correspond to data with more than one optimal basis. These are degenerate triples. The boundary of this right part (made out of the two continuous segments) corresponds, using the notation of Sect. 8.4.2, to the set Σ_{FP}. The union of these lines—the dotted lines of degenerate triples and the continuous segments at the boundary of $\mathcal{W}$—forms the set $\Sigma_{\mathsf{opt}} = \Sigma_{\mathsf{OB}}$ of ill-posed triples for both the optimal solution and the optimal basis problems.

At the left of the diagram we have the sets $\mathcal{I}_P$ and $\mathcal{I}_D$ of triples that are primal, respectively dual, infeasible. The dashed line at their boundary correspond to triples that are both primal and dual infeasible. Away from this line, elements in $\mathcal{I}_P$ are primal infeasible (and hence dual feasible but unbounded), and similarly with $\mathcal{I}_D$.

To define condition we need a norm in the space of data triples. We fix $1 \leq r$, $s \leq \infty$ and define

$$\|d\|_{rs} = \left\| \begin{pmatrix} A & b \\ c^{\mathsf{T}} & 0 \end{pmatrix} \right\|_{rs}.$$

Since all the results in this section hold with respect to any of the norms $\| \ \|_{rs}$, we shall omit the indices r, s in order to simplify notation.

Definition 11.5 Let d be feasible well-posed. We define the *distance to ill-posedness* to be

$$\varrho(d) = \inf\{\|\Delta d\| : d + \Delta d \in \Sigma_{\text{opt}}\}.$$

Moreover, we define the *RCC condition number* $\mathscr{K}(d)$ of d by

$$\mathscr{K}(d) = \frac{\|d\|}{\varrho(d)}.$$

If d is feasible but not well-posed, we let $\varrho(d) = 0$ and $\mathscr{K}(d) = \infty$.

We shall write $\varrho_{rs}(d)$ and $\mathscr{K}_{rs}(d)$ in case we want to emphasize the choice of norm $\| \ \|_{rs}$.

Remark 11.6

(a) $\mathscr{K}(d)$ is undefined for infeasible d.

(b) We have $\varrho(d) \le \|d\|$ and hence $\mathscr{K}(d) \ge 1$.

(c) We saw in Corollary 8.23 that $\Sigma_{\text{opt}} = \Sigma_{\text{OB}}$. Also, it easily follows from its definition that Σ_{OB} is closed. Hence, the infimum in Definition 11.5 is actually attained. Also, $\varrho(d) > 0$ if and only if $d \in \mathcal{W}$.

(d) When d is feasible, we have $C_{rs}(d) \le \mathscr{K}_{rs}(d)$, where $C_{rs}(d)$ is the condition number for feasibility defined in Sect. 10.1. This follows immediately from the inclusion $\Sigma_{\text{FP}} \subseteq \Sigma_{\text{OB}}$.

Our next goal is to characterize $\mathscr{K}(d)$ in terms of distance to singularity for a certain family of matrices. To do so, for any square matrix S, we denote its *distance to singularity* by

$$\rho_{\text{sing}}(S) = \inf\{\|\Delta S\| \mid S + \Delta S \text{ is singular}\}.$$

Here the norm refers to $\| \ \|_{rs}$. The distance to singularity has been the core subject of Sect. 1.3. In the notation of that section we have $\rho_{\text{sing}}(S) = d_{rs}(S, \Sigma)$. For our characterization we need to consider this distance for a set of square matrices, which we next specify.

Let B be a basis, S_1 the set of all $m \times m$ submatrices of $[A_B, b]$, S_2 the set of all $(m+1) \times (m+1)$ submatrices of $\left[\begin{smallmatrix} A \\ c^T \end{smallmatrix}\right]$ containing A_B, and $S_B = S_1 \cup S_2$. Note that $|S_1| = m+1$ and $|S_2| = n - m$, so S_B has $n+1$ elements. Note that $A_B \in S_1$.

Theorem 11.7 *If d is feasible well-posed with optimal basis B, then*

$$\varrho(d) = \min_{S \in S_B} \rho_{\text{sing}}(S).$$

Remark 11.8 A consequence of Theorem 11.7 worth noting is an easy way to compute $\mathscr{K}(d)$ from an optimal basis B of d. Indeed, given such a basis, the $n+1$

matrices S in $\mathcal{S}_B$ are immediately written down, and for each such matrix S one can compute its inverse S^{-1}. Theorems 1.7 and 11.7 then give

$$\mathscr{K}_{rs}(d) = \frac{\|d\|_{rs}}{\min_{S \in \mathcal{S}_B} \rho_{\text{sing}}(S)} = \|d\|_{rs} \max_{S \in \mathcal{S}_B} \|S^{-1}\|_{sr}.$$

We now proceed with a few partial results towards the proof of Theorem 11.7. They rely on the well-known Cramer's rule for systems of linear equations.

Lemma 11.9 *Let $A \in \mathbb{R}^{m \times m}$ be invertible, $c \in \mathbb{R}^m$, and $y = A^{-T}c$. In addition, let $v \in \mathbb{R}^m$ and $c' \in \mathbb{R}$. Then*

$$v^T y = c' \quad \Leftrightarrow \quad \begin{pmatrix} A^T & c \\ v^T & c' \end{pmatrix} \text{ is singular.}$$

Proof Suppose that $v^T y = c'$. Then

$$\begin{pmatrix} A^T & c \\ v^T & c' \end{pmatrix} \begin{pmatrix} y \\ -1 \end{pmatrix} = 0,$$

and hence $\begin{pmatrix} A^T & c \\ v^T & c' \end{pmatrix}$ must be singular. The converse is shown similarly.   □

For the next result we introduce some notation. We defined degeneracy of solutions in Definition 8.15, and this notion was central to the content of Sect. 8.4. We now consider the slightly more restrictive notion of degeneracy for a basis.

Definition 11.10 We say that a basis B is *primal degenerate* for a triple d when its associated solution is degenerate (and hence, feasible). We similarly define *dual degenerate*. We say that B is *degenerate* for d when it is either primal or dual degenerate for this triple.

Let d be feasible well-posed with optimal basis B. We define

$$\rho_{\text{deg}}^P(d) := \min\{\|\Delta d\| : B \text{ is primal degenerate for } d + \Delta d\}$$

and

$$\rho_{\text{deg}}^D(d) := \min\{\|\Delta d\| : B \text{ is dual degenerate for } d + \Delta d\}.$$

Finally, let

$$\rho_{\text{deg}}(d) := \min\{\rho_{\text{deg}}^P(d), \rho_{\text{deg}}^D(d)\}.$$

The following characterization of ϱ, though less applicable than that in Theorem 11.7, will be useful to prove the latter result.

Proposition 11.11 *Let d be a feasible well-posed triple and B its optimal basis. Then $\varrho(d) = \min\{\rho_{\text{deg}}(d), \rho_{\text{sing}}(A_B)\}$.*

Proof Let $\Delta d = (\Delta A, 0, 0)$ be such that $(A + \Delta A)_B$ is singular. Then B is not a basis of $d + \Delta d$, and in particular, $\mathrm{OB}(d) \neq \mathrm{OB}(d + \Delta d)$. This shows that $\varrho(d) \leq \|\Delta d\| = \|\Delta A\|$ and hence $\varrho(d) \leq \rho_{\mathrm{sing}}(A_B)$. This inequality, together with the fact that triples d with degenerate solutions are ill-posed (Corollary 8.23), shows that $\varrho(d) \leq \min\{\rho_{\mathrm{deg}}(d), \rho_{\mathrm{sing}}(A_B)\}$. We next prove the reverse inequality.

Assume $\varrho(d) < \min\{\rho_{\mathrm{deg}}(d), \rho_{\mathrm{sing}}(A_B)\}$. Then, there is $\Delta d = (\Delta A, \Delta b, \Delta c)$ such that $\|\Delta d\| < \min\{\rho_{\mathrm{deg}}(d), \rho_{\mathrm{sing}}(A_B)\}$ and B is not an optimal basis for $d + \Delta d$.

For $t \in [0, 1]$ let $t\Delta d = (t\Delta A, t\Delta b, t\Delta c)$ and $A^{(t)} = A + t\Delta A$, $b^{(t)} = b + t\Delta b$, and $c^{(t)} = c + t\Delta c$. Then

$$\left\|(t\Delta A)_B\right\| \leq \|t\Delta d\| \leq \|\Delta d\| < \min\{\rho_{\mathrm{deg}}(d), \rho_{\mathrm{sing}}(A_B)\},$$

and therefore, $(A + t\Delta A)_B$ is invertible for all $t \in [0, 1]$.

Let $x^{(t)}$ and $y^{(t)}$ be the primal and dual basic solutions of $d^{(t)} := d + t\Delta d$, for $t \in [0, 1]$, i.e.,

$$x^{(t)} = \left(A_B^{(t)}\right)^{-1} b^{(t)}$$

and

$$y^{(t)} = \left(A_B^{(t)}\right)^{-\mathrm{T}} c_B^{(t)}.$$

Note that $x^{(0)}$ and $y^{(0)}$ are basic feasible solutions for d (since $d^{(0)} = d$) but either $x^{(a)}$ or $y^{(a)}$ is not a feasible basic solution for $d + \Delta d$, since B is not an optimal basis for $d + \Delta d$. Therefore,

$$\min\left\{\min_{j \in B} x_j^{(0)}, \min_{j \notin B} c_j - a_j^{\mathrm{T}} y^{(0)}\right\} > 0$$

and

$$\min\left\{\min_{j \in B} x_j^{(a)}, \min_{j \notin B} c_j - a_j^{\mathrm{T}} y^{(a)}\right\} < 0.$$

By continuity, there exists $t^* \in (0, 1)$ such that

$$\min\left\{\min_{j \in B} x_j^{(t^*)}, \min_{j \notin B} c_j - a_j^{\mathrm{T}} y^{(t^*)}\right\} = 0.$$

That is, B is a degenerate basis for $d + t^*\Delta d$ (it is even a degenerate optimal basis for this triple). But $\|t^*\Delta d\| < \rho_{\mathrm{deg}}(d)$, in contradiction to the definition of $\rho_{\mathrm{deg}}(d)$. Hence $\min\{\rho_{\mathrm{deg}}(d), \rho_{\mathrm{sing}}(A_B)\} \leq \varrho(d)$.   □

Lemma 11.12

$$\min_{S \in \mathcal{S}_1} \rho_{\mathrm{sing}}(S) \leq \rho_{\mathrm{deg}}^P(d) \quad \textit{and} \quad \min_{S \in \mathcal{S}_2} \rho_{\mathrm{sing}}(S) \leq \rho_{\mathrm{deg}}^D(d).$$

Proof We first prove the primal statement. Let $\Delta d = (\Delta A, \Delta b, \Delta c)$ be a perturbation such that B is primal degenerate for $d + \Delta d$. Then there exists $j \in B$ such that $x_j = 0$, where x denotes the primal basic solution of $d + \Delta d$ associated to B.

Cramer's rule implies that the matrix $(A + \Delta A)_B[j : b + \Delta b]$ is singular; see Notation 1.9. Therefore

$$\rho_{\text{sing}}\left(A_B[j : b]\right) \leq \left\| \Delta A_B[j : \Delta b] \right\| \leq \| \Delta d \|,$$

which implies $\min_{S \in \mathcal{S}_1} \rho_{\text{sing}}(S) \leq \| \Delta d \|$, since $A_B[j : b] \in \mathcal{S}_1$. The primal statement follows.

For the dual statement, let $\Delta d = (\Delta A, \Delta b, \Delta c)$ be a perturbation such that B is dual degenerate for $d + \Delta d$. Then there exists $j \notin B$ such that $(a_j + \Delta a_j)^{\mathsf{T}} y = (c + \Delta c)_j$, where y denotes the dual basic solution of $d + \Delta d$ associated to B. Note that $(A + \Delta A)_B$ is invertible, since by our assumption (the dual degeneracy of B for $d + \Delta d$), B is a basis of $d + \Delta d$.

By Lemma 11.9, the matrix

$$M := \begin{bmatrix} (A + \Delta A)_B^{\mathsf{T}} & (c + \Delta c)_B \\ (a_j + \Delta a_j)^{\mathsf{T}} & (c + \Delta c)_j \end{bmatrix}$$

is singular. Hence

$$\min_{S \in \mathcal{S}_2} \rho_{\text{sing}}(S) \leq \rho_{\text{sing}}\left(\begin{bmatrix} A_B^{\mathsf{T}} & c_B \\ a_j^{\mathsf{T}} & c_j \end{bmatrix} \right) \leq \left\| \begin{bmatrix} \Delta A_B^{\mathsf{T}} & \Delta c_B \\ \Delta a_j^{\mathsf{T}} & \Delta c_j \end{bmatrix} \right\| \leq \| d \|.$$

The dual assertion follows. $\square$

Proof of Theorem 11.7 We have $\min_{S \in \mathcal{S}_B} \rho_{\text{sing}}(S) \leq \rho_{\text{sing}}(A_B)$, since $A_B \in \mathcal{S}_1$. Also, it follows from Lemma 11.12 that $\min_{S \in \mathcal{S}_B} \rho_{\text{sing}}(S) \leq \rho_{\text{deg}}(d)$. Altogether and using Proposition 11.11, we obtain $\min_{S \in \mathcal{S}_B} \rho_{\text{sing}}(S) \leq \varrho(d)$.

To prove the reverse inequality, take any $S \in \mathcal{S}_B$. We need to show that

$$\varrho(d) \leq \rho_{\text{sing}}(S). \tag{11.1}$$

CASE I: $S = A_B$. Let ΔA be the smallest perturbation making A_B singular. Let $\tilde{d} = (A + \Delta A, b, c)$. Then $B \notin \text{OB}(\tilde{d})$. Hence, denoting by dist the distance induced by $\| \ \|$ and using Corollary 8.23,

$$\varrho(d) = \text{dist}(d, \Sigma_{\text{opt}}) = \text{dist}(d, \Sigma_{\text{OB}}) \leq \text{dist}(d, \tilde{d}) = \| \Delta A \| = \rho_{\text{sing}}(A_B).$$

CASE II: $S \in \mathcal{S}_1, S \neq A_B$. We assume that $B = \{1, 2, \dots, m\}$ and $S = [a_1, a_2, \dots, a_{m-1}, b]$ without loss of generality. There is a perturbation $\Delta S = [\Delta a_1, \Delta a_2, \dots, \Delta a_{m-1}, \Delta b]$ of S such that $S + \Delta S$ is singular and $\| \Delta S \| = \rho_{\text{sing}}(S)$. For $j \geq m$ we set $\Delta a_j := 0$ and thus have defined a matrix ΔA. Further, we set $\Delta d; = (\Delta A, \Delta b, 0)$. By construction, $\| \Delta d \| = \| \Delta S \|$. For proving (11.1), because of Proposition 11.11, it is sufficient to show that

$$\min\{\rho_{\text{deg}}^P(d), \rho_{\text{deg}}^D(d), \rho_{\text{rank}}(A)\} \leq \| \Delta S \|. \tag{11.2}$$

In the case that $(A + \Delta A)_B$ is singular, we have

$$\rho_{\mathrm{rank}}(A) \le \rho_{\mathrm{sing}}(A_B) \le \|\Delta A_B\| \le \|\Delta A\| \le \|\Delta S\|,$$

and (11.2) follows. So suppose that $(A + \Delta A)_B$ is invertible. It is now sufficient to show that B is primal degenerate for $d + \Delta d$. Let x^* be the primal basic solution of $d + \Delta d$ associated with B. Cramer's rule tells us that

$$x_m^* = \frac{\det(A + \Delta A)_B[m : b + \Delta b]}{\det(A + \Delta A)}.$$

But $(A + \Delta A)_B[m : b + \Delta b] = S + \Delta S$ is singular by assumption, hence $x_m^* = 0$. It follows that B is primal degenerate for $d + \Delta d$ and hence $\rho_{\mathrm{deg}}^P(d) \le \|\Delta d\| = \|\Delta S\|$, proving (11.2).

CASE III: $S \in \mathcal{S}_2$. Without loss of generality we assume $B = \{1, 2, \ldots, m\}$ and that the submatrix S is obtained by picking the first $m + 1$ columns of A. There is a perturbation ΔS of S such that $S + \Delta S$ is singular and $\|\Delta S\| = \rho_{\mathrm{sing}}(S)$. Now let $\Delta a_j = \Delta s_j$ for $j \le m + 1$ and $\Delta a_j = 0$ otherwise. Define $\Delta d = (\Delta A, 0, 0)$. As before, it is sufficient to prove the bound (11.2). In the case that $(A + \Delta A)_B$ is singular, we again have

$$\rho_{\mathrm{rank}}(A) \le \rho_{\mathrm{sing}}(A_B) \le \|\Delta A\| = \|\Delta S\|,$$

and (11.2) follows. So we may suppose that $(A + \Delta A)_B$ is invertible. It is now sufficient to show that B is dual degenerate for $d + \Delta d$. Let y^* be the dual basic solution of $d + \Delta d$ associated with B. That is, $(a_i + \Delta a_i^*)^{\mathsf{T}} y^* = c_i$ for $i = 1, 2, \ldots, m$. Since $(A + \Delta A)_B$ is invertible, the first m columns of $S + \Delta S$ are linearly independent. Hence, since $S + \Delta S$ is singular, the last column of $S + \Delta S$ must be a linear combination of the first m ones. We conclude that $(a_{m+1} + \Delta a_{m+1})^{\mathsf{T}} y^* = c_{m+1}$. Hence B is dual degenerate for $d + \Delta d$, and we obtain $\rho_{\mathrm{deg}}^D(d) \le \|\Delta S\|$, proving (11.2) and completing the proof.   □

11.2 $\mathscr{K}(d)$ and Optimal Solutions

In this section it will be convenient to fix particular norms in $\mathbb{R}^{n+1}$ and $\mathbb{R}^{m+1}$. We will actually endow these spaces with the norms $\|\ \|_1$ and $\|\ \|_2$, respectively. The distance to singularity we considered in the previous section now takes the form

$$\rho_{\mathrm{sing}}(S) = \min\{\|\Delta S\|_{12} \mid S + \Delta S \text{ is singular}\}.$$

It follows from the definition of $\|\ \|_{12}$ that $\|d\|$ is at least as large as the 2-norm of any of the columns or the ∞-norm of any of the rows of $\left[\begin{smallmatrix} A & b \\ c^{\mathsf{T}} & 0 \end{smallmatrix}\right]$; cf. Corollary 1.3. In particular, we will repeatedly use that with a_i denoting the ith column of A,

$$\max\{\|a_i\|, \|b\|, \|c\|_\infty\} \le \|d\|. \tag{11.3}$$

For a dual feasible solution $y \in \mathbb{R}^m$ we write, as usual, s for its slack. That is, $s = c - A^T y$.

Proposition 11.13 *Assume d is feasible well-posed. Let B be the optimal basis of d and (x^*, y^*, s^*) the associated basic optimal solution. Then*

$$\frac{1}{\mathscr{K}(d)} \leq \min \left\{ \min \{ x_i^* \mid i \in B \}, \frac{\min \{ s_j^* \mid j \notin B \}}{\|d\|} \right\}$$

and

$$\max \{ \|x^*\|_1, \|y^*\| \} \leq \mathscr{K}(d).$$

Proof For $i \in B$ consider $\Delta d = (0, \Delta b, 0)$, where $\Delta b = -x_i^* a_i$. Recall that $x^*(i : 0)$ denotes the vector obtained by substituting the ith entry of x^* by 0. Clearly, the point $(x^*(i : 0), y^*, s^*)$ is a feasible solution for $d + \Delta d$, and B is an optimal basis for $d + \Delta d$. By construction, B is primal degenerate for $d + \Delta d$. It follows that $\varrho(d) \leq \rho_{\text{deg}}^P(d) \leq \|\Delta d\| = \|\Delta b\| = x_i^* \|a_i\| \leq x_i^* \|d\|$. This proves that

$$\frac{\varrho(d)}{\|d\|} \leq \min \{ x_i^* \mid i \in B \}.$$

The bound $\varrho(d) \leq \min \{ s_j^* \mid j \notin B \}$ follows from a similar argument. For each $j \notin B$ we consider the triple $\Delta d = (0, 0, \Delta c)$, where $\Delta c := -s_j^* e_j$. Clearly, $(x^*, y^*, s^*(j : 0))$ is a feasible solution for $d + \Delta d$ and B is an optimal basis for this triple. Therefore, B is dual-degenerate for $d + \Delta d$. We deduce again that $\varrho(d) \leq \rho_{\text{deg}}^D(d) \leq \|\Delta d\| = \|\Delta c\|_1 = s_j^*$ and conclude that

$$\varrho(d) \leq \min \{ s_j^* \mid j \notin B \}.$$

The upper bounds on $\|x\|_1$ and $\|y\|$ follow from Theorem 11.7, since

$$\varrho(d) \leq \rho_{\text{sing}}(A_B) = \frac{1}{\|A_B^{-1}\|_{21}} \leq \frac{\|b\|}{\|A_B^{-1}b\|_1} \leq \frac{\|d\|}{\|x^*\|_1}$$

(we used Theorem 1.7 for the equality) and

$$\varrho(d) \leq \frac{1}{\|A_B^{-1}\|_{21}} = \frac{1}{\|A_B^{-T}\|_{\infty 2}} \leq \frac{\|c_B\|_\infty}{\|A_B^{-T}c_B\|} \leq \frac{\|d\|}{\|y^*\|},$$

where the equality follows from Lemma 1.2(c). $\square$

The next result gives a lower bound on changes in the objective function with respect to changes in either the primal or dual solution.

Theorem 11.14 *Let $d = (A, b, c)$ be a feasible well-posed triple.*

(a) *Let (y^*, s^*) be the optimal solution of the dual. Then, for any $y \in S_D$ with slack s,*

$$\frac{\|s - s^*\|_\infty}{\|d\|} \leq \|y - y^*\|$$

(b) *Let x^* be the optimal solution of the primal. Then, for any $x \in S_P$,*

$$\|x - x^*\|_1 \leq \frac{c^{\mathrm{T}}x - c^{\mathrm{T}}x^*}{\varrho(d)}.$$

Proof (a) Assume $y \neq y^*$, since otherwise, there is nothing to prove. Let $v \in \mathbb{R}^m$ be such that $\|v\| = 1$ and $v^{\mathrm{T}}(y - y^*) = \|y - y^*\|$ (we have used (1.3)). Now put

$$\Delta b := \frac{b^{\mathrm{T}}y^* - b^{\mathrm{T}}y}{\|y - y^*\|} v.$$

By construction, $(b + \Delta b)^{\mathrm{T}}(y - y^*) = 0$, i.e., both y^* and y have the same objective value for the triple $(A, b + \Delta b, c)$. We claim that the unique optimal basis B for d can no longer be a unique optimal basis for the perturbed data $d + \Delta d := (A, b + \Delta b, c)$. It follows from this claim that

$$\varrho(d) \leq \left\| \begin{pmatrix} 0 & \Delta b \\ 0 & 0 \end{pmatrix} \right\|_{12} = \|\Delta b\| = \frac{b^{\mathrm{T}}y^* - b^{\mathrm{T}}y}{\|y - y^*\|},$$

the last since $\|v\| = 1$, and

$$\|s - s^*\|_\infty = \|A^{\mathrm{T}}(y - y^*)\|_\infty \leq \|A^{\mathrm{T}}\|_{2\infty} \|y - y^*\| \leq \|d\| \|y - y^*\|.$$

Assertion (a) is a consequence of these inequalities.

We now prove the claim. To do so, note that y^* is a dual basic feasible solution for B and $d + \Delta d$ (the perturbation of b does not affect dual feasibility). If B is an optimal basis of $d + \Delta d$, then by Theorem 8.13(b), y^* is the dual optimal solution of $d + \Delta d$. Also, since y is dual feasible for d, it is dual feasible for $d + \Delta d$ as well. Finally, the equality $(b + \Delta b)^{\mathrm{T}}y = (b + \Delta b)^{\mathrm{T}}y^*$ implies that the objective value of y is the optimal value of $d + \Delta d$. We conclude that y is also a dual optimal solution of this triple. The claim now follows from Theorem 8.13(c), which implies that B is not the only optimal basis for $d + \Delta d$.

(b) The argument is similar to that in (a). Assume $x \neq x^*$, since otherwise, there is nothing to prove. Let $u \in \mathbb{R}^n$ be such that $\|u\|_\infty = 1$ and $u^{\mathrm{T}}(x - x^*) = \|x - x^*\|_1$. Now put

$$\Delta c := \frac{c^{\mathrm{T}}x^* - c^{\mathrm{T}}x}{\|x - x^*\|_1} u.$$

By construction, $(c + \Delta c)^{\mathrm{T}}(x - x^*) = 0$, i.e., both x^* and x have the same objective value for the triple $d + \Delta d := (A, b, c + \Delta c)$. Reasoning as in part (a), one shows that the unique optimal basis for d is no longer a unique optimal basis for $d + \Delta d$. Hence

$$\varrho(d) \leq \left\| \begin{pmatrix} 0 & 0 \\ (\Delta c)^T & 0 \end{pmatrix} \right\|_{12} = \|\Delta c\|_\infty = \frac{c^T x - c^T x^*}{\|x - x^*\|_1}. \qquad \square$$

11.3 Computing the Optimal Basis

In this section we will prove Theorems 11.1 and 11.3. To do so, we will exhibit and analyze two algorithms for computing optimal bases. In this context, it will be convenient to control the size of our data.

We say that d is *normalized* when

$$\|d\|_{12} = \left\| \begin{pmatrix} A & b \\ c^T & 0 \end{pmatrix} \right\|_{12} = 1.$$

From a computational viewpoint, this normalization can be straightforwardly achieved by multiplying the entries of d by $\|d\|^{-1}$. In addition, feasible sets and optimizers remain unchanged, as does the condition number $\mathscr{K}(d)$.

In the rest of this chapter we assume that d is feasible and has been normalized.

The general idea underlying our two algorithms relies on the following three ingredients:

(I)   a method to construct candidates $B \subseteq \{1, \ldots, n\}$ for the optimal basis,
(II) a criterion to check that a given candidate B is optimal,
(III) some bounds ensuring that the candidate in (I) eventually satisfies the criterion in (II) (and from which a complexity estimate can be deduced).

Let us begin with (I). If d is a well-posed feasible triple and x^* is the primal optimal solution, then

$$B^* = \left\{ j \leq n \mid x_j^* \neq 0 \right\}$$

is the optimal basis for d (otherwise, x^* would be degenerate and we would have $d \in \Sigma_{OB}$ by Corollary 8.23). In particular, B^* consists of the indices of the m largest components of x^*. By continuity, for a point $x \in \mathbb{R}^n$ sufficiently close to x^* we expect that the same choice will also produce the optimal basis. Therefore, for any point $x \in \mathbb{R}^n$ we define $B_1(x)$ to be the set of indices corresponding to the m largest components of x (ties are broken by taking the smallest index). Hence, $B_1(x)$ satisfies

$$B_1(x) \subseteq \{1, \ldots, n\}, \quad |B_1(x)| = m, \quad \text{and} \quad \max_{j \notin B_1(x)} x_j \leq \min_{j \in B_1(x)} x_j.$$

Similarly, if y^* is the optimal solution of the dual and $s^* = c - A^T y^*$, then $B^* = \{j \leq n \mid s_j^* = 0\}$ consists of the indices of the m smallest components of s^*. Again by continuity, for a point $y \in \mathbb{R}^m$ sufficiently close to y^* we expect that the same choice will also produce the optimal basis. Therefore, for any point $y \in \mathbb{R}^m$ we let $s = c - A^\mathrm{T} y$ and define $B_2(s)$ to be the set of indices corresponding to the m smallest components of s (ties are broken as above). Hence, $B_2(s)$ satisfies

$$B_2(s) \subseteq \{1, \ldots, n\}, \quad |B_2(s)| = m, \quad \text{and} \quad \max_{j \in B_2(s)} s_j \leq \min_{j \notin B_2(s)} s_j.$$

Given a point (x, y, s), we may take any of $B_1(x)$ and $B_2(s)$ as candidate for optimal basis.

We next look at (II): for this, we use the basic criterion that guarantees optimality when both the primal and dual solutions are feasible (that is, Definition 8.10).

As for (III), the desired conditions, for each of the algorithms, will be provided in Proposition 11.15.

11.3.1 An Interior-Point Algorithm

In this section we assume the availability of a point $z_0 = (x_0, y_0, y_0) \in \mathcal{N}(\frac{1}{4})$ such that $\mu(z_0) = n^{\mathcal{O}(1)}$, as described in the introduction to this chapter.

We can now describe the algorithm computing an optimal basis. It is just Algorithm 9.1, Primal-Dual IPM, enhanced with the ideas in ingredients (I) and (II) above.

Algorithm 11.1 OB

Input: $d = (A, b, c) \in \mathbb{R}^{m \times n} \times \mathbb{R}^m \times \mathbb{R}^n, z_0 = (x_0, y_0, s_0) \in \mathbb{R}^{n+m+n}$

Preconditions: d is feasible, normalized, rank $A = m < n$, $z_0 \in \mathcal{N}(\frac{1}{4})$ and $\mu(z_0) = n^{\mathcal{O}(1)}$

```
run Algorithm Primal-Dual_IPM with input (A, b, c) and z₀
at each iteration:
        compute B₁(x) and set B := B₁(x)
        if A_B is invertible then
            use B to compute x*, y*, s* given by x*_B = A_B⁻¹b,
               x*_N = 0,  y* = A_B⁻ᵀc_B,  and  s* = c − Aᵀy*
            if x*, s* ≥ 0 then Return B and Halt
```

Output: $B \subseteq \{1, \ldots, n\}$

Postconditions: B is an optimal basis for $d = (A, b, c)$

The correctness of Algorithm 11.1 is clear by the definition of optimal basis. To analyze its complexity we use the following result (recall that we are assuming $\|d\|_{12} = 1$).

Proposition 11.15 *Let $(x, y, s) \in \mathbb{R}^n \times \mathbb{R}^m \times \mathbb{R}^n$ be such that*

$$\|Ax - b\| \leq \frac{\varrho(d)}{2}, \qquad s = c - A^\mathrm{T} y, \quad x \geq 0, \ s \geq 0, \quad \text{and} \quad x^\mathrm{T} s < \frac{\varrho(d)^2}{12}.$$

Then $B_1(x) = B_2(s)$, and this is the optimal basis for d.

Proof Let $\Delta b := Ax - b$ and $\Delta d := (0, \Delta b, 0)$. Then

$$Ax = b + \Delta b, \qquad A^{\mathrm{T}}y + s = c, \quad x \geq 0, \text{ and } s \geq 0.$$

That is, x and (y, s) are feasible points for the primal and the dual for the triple $d + \Delta d$. Let x^* and (y^*, s^*) be the primal and dual optimal solutions of this triple. By (8.1),

$$c^{\mathrm{T}}x - c^{\mathrm{T}}x^* \leq c^{\mathrm{T}}x - (b + \Delta b)^{\mathrm{T}}y = x^{\mathrm{T}}s < \frac{\varrho(d)^2}{12}$$

and

$$(b + \Delta b)^{\mathrm{T}}y^* - (b + \Delta b)^{\mathrm{T}}y \leq c^{\mathrm{T}}x - (b + \Delta b)^{\mathrm{T}}y = x^{\mathrm{T}}s < \frac{\varrho(d)^2}{12}.$$

In addition, since $\|\Delta b\| \leq \varrho(d)/2$ and by Corollary 1.3,

$$\|\Delta d\|_{12} = \left\| \begin{pmatrix} 0 & \Delta b \\ 0 & 0 \end{pmatrix} \right\|_{12} = \|\Delta b\| \leq \varrho(d)/2. \tag{11.4}$$

Therefore, if B denotes the optimal basis for d, then by the definition of ϱ, B is also an optimal basis for the triple $d + \Delta d$ and $\varrho(d + \Delta d) \geq \varrho(d)/2$.

We now use Theorem 11.14 for $d + \Delta d$ to obtain

$$\|x^* - x\|_\infty \leq \|x^* - x\|_1 \leq \frac{c^{\mathrm{T}}x - c^{\mathrm{T}}x^*}{\varrho(d + \Delta d)} < \frac{\varrho(d)^2/12}{\varrho(d)/2} = \frac{\varrho(d)}{6}$$

and

$$\frac{\|s - s^*\|_\infty}{\|d + \Delta d\|} \leq \|y - y^*\| \leq \frac{(b + \Delta b)^{\mathrm{T}}y^* - (b + \Delta b)^{\mathrm{T}}y}{\varrho(d + \Delta d)} < \frac{\varrho(d)^2/12}{\varrho(d)/2} = \frac{\varrho(d)}{6}.$$

Note that inequality (11.4) and the normalization $\|d\| = 1$ imply $\|\Delta d\| \leq 1/2$ and therefore that $\|d + \Delta d\| \leq \frac{3}{2}$.

Now assume $B_1(x) \neq B$ and let $j_1 \in B_1(x) \setminus B$. Since x^* is an optimal solution for the triple $d + \Delta d$ and $j_1 \notin B$, we have $x_{j_1}^* = 0$. Let also $j_2 \in B \setminus B_1(x)$. By Proposition 11.13 applied to $d + \Delta d$,

$$x_{j_2}^* \geq \frac{\varrho(d + \Delta d)}{\|d + \Delta d\|} \geq \frac{2}{3}\varrho(d + \Delta d) \geq \frac{2}{3}\frac{\varrho(d)}{2} = \frac{\varrho(d)}{3}.$$

Since $\|x^* - x\|_\infty < \varrho(d)/6$, we have $x_{j_2} > \varrho(d)/6$ and $x_{j_1} < \varrho(d)/6$. This contradicts

$$\max_{j \notin B_1(x)} x_j \leq \min_{j \in B_1(x)} x_j.$$

The proof for B_2 is similar. Assume $B_2(s) \neq B$ and let $j_1 \in B \setminus B_2(s)$. Since y^* is an optimal solution for the triple $d + \Delta d$ and $j_1 \notin B$, we have $s_{j_1}^* = 0$. Now let $j_2 \in B_2(s) \setminus B$. By Proposition 11.13, $s_{j_2}^* \geq \varrho(d + \Delta d) \geq \varrho(d)/2$. Also, the bounds $\frac{\|s^* - s\|_\infty}{\|d + \Delta d\|} < \varrho(d)/6$ and $\|d + \Delta d\| \leq 3/2$ imply $\|s^* - s\|_\infty < \varrho(d)/4$. Therefore,

$s_{j_2} > \varrho(d)/4$ and $s_{j_1} < \varrho(d)/4$. This contradicts

$$\max_{j \in B_2(s)} s_j \leq \min_{j \notin B_2(s)} s_j. \qquad \Box$$

Proof of Theorem 11.1 Theorem 9.10 (take $\xi = \frac{1}{4}$) ensures that Algorithm 11.1 produces a sequence of iterates $(z_i)_{i \in \mathbb{N}}$ with $z_i = (x_i, y_i, s_i)$ feasible. Hence, the first four hypotheses in Proposition 11.15, namely $\|Ax - b\| \leq \frac{\varrho(d)}{2}$, $s = c - A^T y$, $x \geq 0$, and $s \geq 0$, are trivially satisfied by these z_i.

Theorem 9.10, along with the assumption $\mu(z_0) = n^{\mathcal{O}(1)}$, also ensures that

$$\mu(z_k) < \frac{\varrho(d)^2}{12n},$$

and consequently that $x_k^T s_k < \frac{\varrho(d)^2}{12}$, as soon as

$$k > 4\sqrt{n}\left(\ln \mu(z_0) + \ln \frac{12n}{\varrho(d)^2}\right) = \mathcal{O}\big(\sqrt{n}(\log n + \ln \mathcal{K}(d))\big).$$

Proposition 11.15 now finishes the proof. $\qquad \Box$

11.3.2 A Reduction to Polyhedral Feasibility Problems

Our second approach to finding an optimal basis also follows the roadmap based on ingredients (I–III) above but with a major deviation: the sequence of points $z = (x, y, s)$ used to construct B_1 or B_2 is obtained differently (and they are no longer feasible solutions of the pair (SP–SD)).

To see how, note that the optimal solution (x^*, y^*) is the only solution of the system

$$\begin{aligned} Ax &= b, \\ A^T y &\leq c, \\ c^T x - b^T y &\leq 0, \\ x &\geq 0. \end{aligned}$$

Therefore, points (x, y) close to (x^*, y^*) can be obtained as solutions of the relaxation (recall that $\mathbf{e}_m = (1, 1, \ldots, 1)$)

$$\begin{aligned} Ax &\leq b + \sigma_1 \mathbf{e}_m, \\ Ax &\geq b - \sigma_1 \mathbf{e}_m, \\ A^T y &\leq c, \\ c^T x - b^T y &\leq \sigma_2, \\ x &\geq 0, \end{aligned} \qquad (11.5)$$

where $\sigma = (\sigma_1, \sigma_2) \in \mathbb{R}_+^2$ has small components. To get solutions of such a system we can use any algorithm solving the polyhedral cone feasibility problem PCFP homogenizing the system above (with a new variable t) so that it becomes a polyhedral conic system (as we did in the preceding chapter). In our algorithm we will take σ_1, σ_2 to be functions of a single parameter $\varepsilon > 0$ as follows:

$$\sigma_1 := \frac{\varepsilon^3}{48m}, \qquad \sigma_2 := \frac{\varepsilon^2}{25}.$$

Furthermore, we will want to ensure that $t > 0$ and to control the magnitude of y/t so that

$$\|y/t\|_\infty \le \frac{2}{\varepsilon}.$$

The resulting extension of (11.5) is given by the polyhedral conic system

$$M_\varepsilon \begin{pmatrix} x \\ y \\ t \end{pmatrix} < 0, \tag{11.6}$$

where

$$M_\varepsilon := \begin{pmatrix} A & & -(b + \sigma_1 \mathbf{e}_m) \\ -A & & (b - \sigma_1 \mathbf{e}_m) \\ & A^{\mathrm{T}} & -c \\ c^{\mathrm{T}} & -b^{\mathrm{T}} & -\sigma_2 \\ -I_n & & 0 \\ & \varepsilon I_m & -2\mathbf{e}_m \\ & -\varepsilon I_m & -2\mathbf{e}_m \\ & & -1 \end{pmatrix}.$$

Note that $M_\varepsilon \in \mathbb{R}^{(4m+2n+2) \times (n+m+1)}$ and rank $M_\varepsilon = n + m + 1$ if rank $A = m$.

During the execution of the algorithm ε will decrease to zero. Therefore, successive pairs $(x/t, y/t)$ induced by the solutions of the system will be increasingly closer to (x^*, y^*).

Because of the lines we added in M_ε, it may happen that the linear conic system (11.6) has no solutions even though the system (11.5) has. The next result shows that for small enough ε this is not the case. Recall that data d are assumed to be normalized.

Proposition 11.16 *If d is feasible and $0 < \varepsilon \le \varrho(d)$, then the polyhedral conic system* (11.6) *is strictly feasible.*

Proof Let (x^*, y^*) be the optimal solution of d. Then, by Proposition 11.13, $\|y^*\| \le \mathcal{K}(d) \le \frac{1}{\varepsilon}$. Since $\|y^*\|_\infty \le \|y^*\|$, this shows that $(x^*, y^*, 1)$ satisfies the sixth and seventh lines of (11.6). The result follows since the other constraints are clear. $\square$

Hence, for ε sufficiently small the conic system (11.6) has solutions. Furthermore, continuity suggests that any point (x, y) such that $(x, y, 1)$ is such a solution will be close to (x^*, y^*). Therefore, we can construct our candidate for the optimal basis by taking either $B_1(x)$ or $B_2(s)$. Again, we need to prove that when ε becomes small enough, this candidate is indeed the optimal basis.

Proposition 11.17 *If* $0 < \varepsilon \leq \varrho(d)$ *and* $(x, y, t) \in \mathbb{R}^{n+m+1}$ *are such that*

$$M_\varepsilon(x, y, t) < 0,$$

then $B_1(x) = B_2(s)$, *and this is the optimal basis for* d.

A last basic ingredient is needed. In order to solve the feasibility problems $M_\varepsilon(x, y, t) < 0$, we will use Algorithm 9.2 (but we observe that we could perfectly well use, with appropriate modifications, an ellipsoid method or the perceptron algorithm). The number of iterations of Algorithm 9.2 for deciding the feasibility of (11.6) is a function of the GCC condition $\mathscr{C}(M_\varepsilon^{\mathrm{T}})$. We therefore need bounds on $\mathscr{C}(M_\varepsilon^{\mathrm{T}})$. The next result provides bounds for $\mathscr{C}(M_\varepsilon^{\mathrm{T}})$ for small enough ε.

Proposition 11.18 *If* $0 < \varepsilon \leq \varrho(d)$, *then*

$$\mathscr{C}(M_\varepsilon^{\mathrm{T}}) \leq B(n, m, \varepsilon) := 96m\sqrt{3(m + n + 1)}\varepsilon^{-4}.$$

We next describe our second algorithm for computing an optimal basis. Here $\mathbf{C}$ is any constant such that (recall Theorem 9.14) Algorithm 9.2 with input $M_\varepsilon^{\mathrm{T}}$ halts after at most

$$\mathbf{C}\sqrt{n}\left(\log_2 n + \log_2 \mathscr{C}(M_\varepsilon^{\mathrm{T}})\right)$$

iterations.

Algorithm 11.2 OB2

Input: $d = (A, b, c) \in \mathbb{R}^{m \times n} \times \mathbb{R}^m \times \mathbb{R}^n$

Preconditions: d is feasible well-posed, normalized, and rank $A = m < n$

```
set ε := 1/2
repeat
       write down M = Mε
       run at most C√n(log₂ n + log₂ B(n, m, ε))
              iterations of Algorithm 9.2 with input M
       if a solution of the system M(x, y, t) < 0 is found
              within the allowed number of iterations then
              compute B₁(x) and set B = B₁(x)
              use B to compute x*, y*, s* defined by
                  x*_B = A_B⁻¹b,  x*_N = 0,  y* = A_B⁻ᵀc_B,  s* = c − Aᵀy*
              if x*, s* ≥ 0 then return B and halt
       set ε := ε²
```

Output: $B \subseteq \{1, \ldots, n\}$

Postconditions: B is an optimal basis for $d = (A, b, c)$

We can now prove the second main result of this chapter.

Proof of Theorem 11.3 The correctness of Algorithm 11.2, as was that of Algorithm 11.1, is clear. For its cost, note that at the kth iteration the value of ε is 2^{-2^k}. Therefore, after $\log_2 \log_2(\mathscr{K}(d) + 4)$ iterations we have $\varrho(d)^2 < \varepsilon \leq \varrho(d)$. At this stage, Proposition 11.16 ensures that the system (11.6) is strictly feasible. Furthermore, Proposition 11.18 along with Theorem 9.14 guarantees that the clock we set for the execution of Algorithm 9.2 is generous enough to allow this procedure to find a strictly feasible solution of (11.6). Finally, Proposition 11.17 shows that the candidate basis constructed from this solution is the optimal basis for d.

Since the cost of each iteration of OB2 is at most

$$Cn^{3.5}\left(\log_2 n + \log_2 B(n, m, \varepsilon)\right),$$

a total bound of

$$\mathcal{O}\left(n^{3.5}\left(\log_2 n + \log_2 \mathscr{K}(d)\right)\log_2 \log_2(\mathscr{K}(d) + 4)\right)$$

follows for the total cost of Algorithm 11.2. □

To finish this section, the only task remaining is to prove Propositions 11.17 and 11.18.

Proposition 11.19 *Let $0 < \varepsilon \leq \varrho(d)$, and let $(x, y, s) \in \mathbb{R}^{n+2m}$ be such that*

$$\|Ax - b\| \leq \frac{\varepsilon^3}{48\sqrt{m}}, \qquad s = c - A^\mathsf{T} y, \qquad c^\mathsf{T} x - b^\mathsf{T} y \leq \frac{\varepsilon^2}{25},$$

$$\|y\| \leq \frac{2\sqrt{m}}{\varepsilon}, \quad x, s \geq 0.$$

Then $B_1(x) = B_2(s)$, and this is the optimal basis for d.

Proof Let $\Delta b = Ax - b$. Then $\|\Delta b\| \leq \frac{\varepsilon^3}{48\sqrt{m}}$, and we have

$$x^\mathsf{T} s = x^\mathsf{T} c - x^\mathsf{T} A^\mathsf{T} y = x^\mathsf{T} c - b^\mathsf{T} y - \Delta b^\mathsf{T} y$$

$$\leq \left(x^\mathsf{T} c - b^\mathsf{T} y\right) + \|y\| \|\Delta b\|$$

$$\leq \frac{\varepsilon^2}{25} + \frac{2\sqrt{m}}{\varepsilon}\frac{\varepsilon^3}{48\sqrt{m}} < \frac{\varepsilon^2}{12} \leq \frac{\varrho(d)^2}{12}.$$

In addition, since $\varepsilon \leq \varrho(d) \leq 1$,

$$\|Ax - b\| \leq \frac{\varepsilon^3}{48\sqrt{m}} \leq \frac{\varrho(d)}{2},$$

and the result follows from Proposition 11.15. □

Proof of Proposition 11.17 If (x, y, t) satisfy (11.6), then $M_\varepsilon(x/t, y/t, 1) \leq 0$. The first two lines in (11.6) yield $\|Ax - b\|_\infty \leq \frac{\varepsilon^3}{48m}$, which implies $\|Ax - b\| \leq \frac{\varepsilon^3}{48\sqrt{m}}$. In addition, the fourth line implies

$$c^T x - b^T y \leq \frac{\varepsilon^2}{25}.$$

Finally, letting $s := c - A^T y$ and now using the sixth, seventh, fifth, and third lines of (11.6), one has,

$$\|y\| \leq \sqrt{m}\|y\|_\infty \leq \frac{2\sqrt{m}}{\varepsilon} \quad \text{and} \quad x, s \geq 0.$$

The hypotheses of Proposition 11.19 are thus satisfied by (x, y, s), and therefore $B_1(x/t) = B_2(y/t)$ is the optimal basis of d. But $B_1(x) = B_1(x/t)$ and $B_2(y) = B_2(y/t)$. Thus $B_1(x)$ is the optimal basis for d.   □

Proof of Proposition 11.18 The proof will be based on the characterization of $\mathscr{C}(M_\varepsilon^T)$ given in Proposition 6.21. Let x^* and y^* be the optimal solutions of the primal and dual of d, respectively. Then

$$\begin{pmatrix} A & -(b + \sigma_1 e_m) \\ -A & (b - \sigma_1 e_m) \\ & A^T & -c \\ c^T & -b^T & -\sigma_2 \\ -I_n & & \end{pmatrix} \begin{pmatrix} x^* \\ y^* \\ 1 \end{pmatrix} \leq \begin{pmatrix} -\sigma_1 e_m \\ -\sigma_1 e_m \\ 0 \\ -\sigma_2 \\ 0 \end{pmatrix}.$$

In addition, by Proposition 11.13, $\|x^*\|_1, \|y^*\| \leq \mathscr{K}(d) \leq \frac{1}{\varepsilon}$, and the bound on $\|y^*\|$ implies (note that $\varepsilon \leq 1$)

$$\begin{pmatrix} \varepsilon I_m & -2e_m \\ -\varepsilon I_m & -2e_m \\ & -1 \end{pmatrix} \begin{pmatrix} x^* \\ y^* \\ 1 \end{pmatrix} \leq \begin{pmatrix} -e_m \\ -e_m \\ -\frac{1}{2} \end{pmatrix}.$$

Since $\min\{\sigma_1, \sigma_2\} = \sigma_1 = \frac{\varepsilon^3}{48m}$, it follows that $M_\varepsilon(x^*, y^*, 1) \leq -\frac{\varepsilon^3}{48m} e_{4m+2n+2}$.

Let E be any $(4m + 2n + 2) \times (m + n + 1)$ matrix such that $\|E\|_{2\infty} \leq \frac{\varepsilon^4}{48\sqrt{3m}}$ and let E_j be the jth row of E for $j = 1, 2, \ldots, 4m + 2n + 2$. Then, $\|E_j\| \leq \frac{\varepsilon^4}{48\sqrt{3m}}$. Similarly, let $M_{\varepsilon,j}$ be the jth row of M_ε. Then, for $j = 1, 2, \ldots, 4m + 2n + 2$,

$$(M_{\varepsilon,j} + E_j)(x^*, y^*, 1) = M_{\varepsilon,j}(x^*, y^*, 1) + E_j(x^*, y^*, 1)$$

$$< -\frac{\varepsilon^3}{48m} + \|E_j\| \|(x^*, y^*, 1)\|$$

$$\leq -\frac{\varepsilon^3}{48m} + \frac{\varepsilon^4}{48\sqrt{3m}} \frac{\sqrt{3}}{\varepsilon} = 0,$$

the last inequality following from $\|(x^*, y^*, 1)\| \leq \sqrt{3}/\varepsilon$. Therefore, $(x^*, y^*, 1)$ is also a solution of $(M_\varepsilon + E)(x, y, t) \leq 0$. We conclude that for Σ as defined in Sect. 6.3,

$$d_{12}(M_\varepsilon^T, \Sigma) = d_{2\infty}(M_\varepsilon, \Sigma) \geq \frac{\varepsilon^4}{48\sqrt{3}m}.$$

In addition, since $\|d\| = 1$, all the entries of M_ε are bounded in absolute value by 2, and we have $\|M_\varepsilon^T\|_{12} \leq 2\sqrt{m+n+1}$. Therefore, recalling Proposition 6.21,

$$\mathscr{C}(M_\varepsilon^T) = \frac{\|M_\varepsilon^T\|_{12}}{d_{12}(M_\varepsilon^T, \Sigma)} \leq 96m\sqrt{3(m+n+1)}\varepsilon^{-4}. \qquad \square$$

Remark 11.20 As we did in Remark 10.6, we observe now that since Algorithm 11.2 is essentially a sequence of calls to a PCFP-solver, it should come as no surprise that a finite-precision version of this algorithm will work accurately as well. Indeed, the main result in [58] (in the spirit of Template 2 in Sect. 9.5) is the following.

Theorem 11.21 *There exists a finite-precision algorithm that with input a full-rank matrix $A \in \mathbb{R}^{m \times n}$ and vectors $b \in \mathbb{R}^m$ and $c \in \mathbb{R}^n$ finds an optimal basis B for d. The machine precision ϵ_{mach} varies during the execution of the algorithm. The finest required precision satisfies*

$$\epsilon_{\mathsf{mach}} = \frac{1}{\mathcal{O}(n^{26} \mathscr{K}(d)^{16})}.$$

The total number of arithmetic operations is bounded by

$$\mathcal{O}\big(n^{3.5}(\log n + \log \mathscr{K}(d)) \log\log(\mathscr{K}(d) + 2)\big). \qquad \square$$

11.4 Optimizers and Optimal Bases: The Condition Viewpoint

Algorithms OB and OB2 in the preceding section compute optimal bases for (well-posed) triples $d = (A, b, c)$. Given such a basis B, we can obtain optimizers for primal and dual by taking

$$x_B^* := A_B^{-1}b, \qquad x_N^* := 0, \quad \text{and} \quad y^* := A_B^{-T}c_B.$$

Conversely, we note that any algorithm computing x^* and y^* for a well-posed triple d would produce (in an even simpler way) an optimal basis B by taking $B := \{j \leq n \mid x_j^* > 0\}$. Restricted to well-posed data and under infinite precision, these two problems are equivalent.

We can nonetheless abandon the infinite-precision hypothesis and consider at least the case of perturbed data. How do these two problems compare in this case? An answer to this question should involve a comparison of their condition numbers.

Fig. 11.2 A schematic
picture of $\mathscr{K}(d)$ and
$\text{cond}^{\text{opt}}(d)$

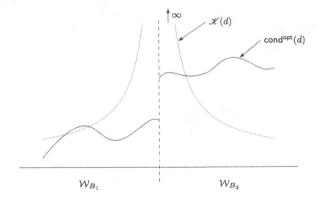

And we note at this point that we have never defined the condition number of the
optimal solution problem. Because of the continuously valued nature of this problem, one might be inclined to define condition following the lines laid down in the
Overture, that is, to take

$$\text{cond}^P(d) := \lim_{\delta \to 0} \sup_{\text{RelError}(d) \le \delta} \frac{\text{RelError}(x^*)}{\text{RelError}(d)},$$

and likewise for $\text{cond}^D(d)$, and finally define

$$\text{cond}^{\text{opt}}(d) := \max\{\text{cond}^P(d), \text{cond}^D(d)\}.$$

A moment's thought reveals, however, that the combinatorial structure of linear programming imposes on $\text{cond}^{\text{opt}}(d)$ the same shortcomings we discussed for discrete-
valued problems in Sect. 6.1. Indeed, if d is well-posed with optimal basis B, then
sufficiently small perturbations will also be well-posed with optimal basis B. By
Theorem 1.4, the (say dual) optimizer $\tilde{y}$ of such a perturbation $\tilde{d}$ will therefore satisfy

$$\frac{\|\tilde{y} - y^*\|}{\|y^*\|} \le 2\kappa\left(A_B^{\mathsf{T}}\right)\text{RelError}\left(A_B^{\mathsf{T}}, c\right) + o(1). \tag{11.7}$$

The key factor here is $\kappa(A_B^{\mathsf{T}})$; and this quantity may be moderate for data arbitrarily
close (or even belonging) to Σ_{opt}. The simplest example is an ill-posed triple $\bar{d}$ with
two optimal bases B_1 and B_2. Both matrices $A_{B_1}^{\mathsf{T}}$ and $A_{B_2}^{\mathsf{T}}$ may be well-conditioned
and yet $\bar{d}$ is ill-posed. What makes $\bar{d}$ ill-posed is the possibility to jump from one
extremal optimal solution to the other. But each of these extremal solutions is itself
well-posed as a function of the pair (A_B^{T}, c_B).

Figure 11.2 shows a schematic picture of this situation. The quantity $\text{cond}^{\text{opt}}(d)$
remains bounded around the boundary between $\mathcal{W}_{B_1}$ and $\mathcal{W}_{B_2}$ but it jumps to ∞ at
this boundary (in accordance with the discussion in Sect. 8.4.2). Figure 11.2 depicts
as well the condition number $\mathscr{K}(d)$, whose behavior appears to be more adapted to
its use in the analysis of algorithms.

We note that in addition, $\mathcal{K}(d)$ plays a central role in bounding the loss of precision for the computation of x^* and y^*. The following result gives a first idea.

Proposition 11.22 *Let d be feasible well-posed and B the optimal basis for d. Then, for all $r, s \geq 1$,*

$$\kappa_{rs}(A_B) \leq \mathcal{K}_{rs}(d) \quad \text{and} \quad \kappa_{s^*r^*}(A_B^{\mathsf{T}}) \leq \mathcal{K}_{rs}(d).$$

Proof Let $\Sigma \subseteq \mathbb{R}^{m \times m}$ denote the set of singular matrices. By Theorem 11.7 we have $\varrho(d) \leq d_{rs}(A_B, \Sigma)$. Hence, using Corollary 1.8, we get

$$\kappa_{rs}(A_B) = \frac{\|A_B\|_{rs}}{d_{rs}(A_B, \Sigma)} \leq \frac{\|d\|}{\varrho(d)} = \mathcal{K}_{rs}(d).$$

The second inequality follows from Lemma 1.2(c), according to which we have $\kappa_{s^*r^*}(A_B^{\mathsf{T}}) = \kappa_{rs}(A_B)$.   □

For $r, s \geq 1$ consider now the quantity

$$R(d) := \max_{B \in \mathcal{B}} \max \left\{ \frac{\|d\|_{rs}}{\|(A_B, b)\|_{rs}}, \frac{\|d\|_{rs}}{\|(A_B^{\mathsf{T}}, c_B)\|_{s^*r^*}} \right\},$$

which measures how balanced the norms of the square subsystems of d are for the different choices of basis. Then, by the primal version of (11.7) and Proposition 11.22,

$$\mathsf{RelError}(x^*) \leq 2\kappa_{rs}(A_B)\,\mathsf{RelError}(A_B, b) + o(1)$$

$$= 2\kappa_{rs}(A_B) \frac{\|(\tilde{A}_B, \tilde{b}) - (A_B, b)\|_{rs}}{\|(A_B, b)\|_{rs}} \frac{\|d\|_{rs}}{\|d\|_{rs}} + o(1)$$

$$\leq 2\kappa_{rs}(A_B) \frac{\|\tilde{d} - d\|_{rs}}{\|d\|_{rs}} \frac{\|d\|_{rs}}{\|(A_B, b)\|_{rs}} + o(1)$$

$$\leq 2\mathcal{K}_{rs}(d)\,\mathsf{RelError}(d)\,R(d) + o(1),$$

and a similar bound applies for $\mathsf{RelError}(y^*)$. We conclude that the loss of precision in the computation of the optimizers x^* and y^* is controlled by the product $\mathcal{K}(d)R(d)$.

11.5 Approximating the Optimal Value

We close this chapter with a few words on the problem of computing the optimal value v^*. We noted in Sect. 8.4.2 that the set of ill-posed triples for this problem is the same as that for the feasibility problem, and hence, one could expect to have algorithmic solutions analyzed in terms of $C(d)$ (as defined in Sect. 10.1). We cannot substantiate this last claim here, but we can nonetheless give some indications of the issues involved.

The first one is that as far as we know, there is no algorithm that will compute the optimal value without computing an optimal basis or an optimizer. This fact would appear to conflict with the picture above. But there is no such conflict. What happens is that to understand the role of condition in the computation of the optimal value, we need to consider the problem of approximating v^*, not of computing it exactly. We won't enter into the details, but we observe that given $\varepsilon > 0$ and a feasible point $z_0 = (x_0, y_0, s_0)$ in the central neighborhood $\mathcal{N}(\frac{1}{4})$ such that $\mu(z_0) \leq (n\|d\|)^{\mathcal{O}(1)}$—in what follows, we assume, without loss of generality that $\|d\| = 1$ we can compute a real number $\tilde{v}$ satisfying $|\tilde{v} - v^*| \leq \varepsilon$ using

$$\mathcal{O}\big(\sqrt{n}\,(\ln n + |\ln \varepsilon|)\big)$$

iterations (and n^3 times this bound for the number of arithmetic operations performed). This follows from Theorem 9.10 and the fact that if $z_k = (x_k, y_k, s_k)$ denotes the value of z at the kth iterate of Algorithm 9.1 and we take $v_k := c^\mathsf{T} x_k$ then

$$\big|v_k - v^*\big| \leq v_k - b^\mathsf{T} y_k = n\mu(z_k).$$

We therefore conclude that if we can compute a point z_0 in the central neighborhood $\mathcal{N}(\frac{1}{4})$ such that $\mu(z_0) \leq n^{\mathcal{O}(1)}$ with cost $\mathcal{O}(n^{3.5}(\ln n + \ln C(d)))$, then we can obtain the desired approximation of v^* with cost bounded by $\mathcal{O}(n^{3.5}(\ln n + \ln C(d) + |\ln \varepsilon|))$. As we pointed out in Remark 11.2, however, a discussion on the ways of doing the first computation would take us too far away from our main themes.

Chapter 12
Average Analysis of the RCC Condition Number

In Chap. 11 we considered the primal–dual pair of linear programming optimization problems

$$\min \ c^{\mathrm{T}}x \quad \text{subject to} \quad Ax = b, \quad x \geq 0, \qquad \text{(SP)}$$

and

$$\max \ b^{\mathrm{T}}y \quad \text{subject to} \quad A^{\mathrm{T}}y \leq c, \qquad \text{(SD)}$$

and analyzed two algorithms that in case both problems are feasible, return optimizers x^* and y^* for them, respectively. Recall that here $A \in \mathbb{R}^{m \times n}, b \in \mathbb{R}^m, c \in \mathbb{R}^n$, and $n \geq m \geq 1$.

To analyze these algorithms we introduced the condition number $\mathscr{K}_{rs}(d)$—here $d = (A, b, c)$ and the indices r, s refer to the underlying operator norm—and the main results in the previous chapter, Theorems 11.1 and 11.3, bound the cost of these algorithms by

$$\mathcal{O}\big(n^{3.5}\big(\log n + \log \mathscr{K}_{12}(d)\big)\big)$$

and

$$\mathcal{O}\big(n^{3.5}\big(\log n + \log \mathscr{K}_{12}(d)\big) \log \log \big(\mathscr{K}_{12}(d) + 2\big)\big),$$

respectively. Furthermore, Theorem 11.21 states that this task can be done with finite precision and the result is correct as long as the machine epsilon satisfies

$$\epsilon_{\mathsf{mach}} = \frac{1}{\mathcal{O}(n^{26}\mathscr{K}_{12}(d)^{16})}.$$

This means that the number of digits or bits necessary to perform the computation is bounded by $\mathcal{O}(\log n + \log \mathscr{K}_{12}(d))$.

The use of $\| \ \|_{12}$ in these bounds is irrelevant: the consideration of other norms will only change the constant in the $\mathcal{O}$ notation. The goal of this chapter, following a line of thought well established in our development, is to eliminate $\mathscr{K}(d)$ from these bounds via a probabilistic analysis.

P. Bürgisser, F. Cucker, *Condition*,
Grundlehren der mathematischen Wissenschaften 349,
DOI 10.1007/978-3-642-38896-5_12, © Springer-Verlag Berlin Heidelberg 2013

To be precise, we consider Gaussian triples d. That is, we assume that the entries of A, b, and c are i.i.d. random variables with standard normal distribution. Recall that we denoted by $\mathcal{W}$ the set of triples that are feasible well-posed. Our main result in this chapter is the following.

Theorem 12.1 *We have*

$$\mathop{\mathbb{E}}_{d \sim N(0,\mathrm{I})} \left(\log \mathscr{K}_2(d) \mid d \in \mathcal{W} \right) \leq \frac{5}{2} \log(n+1) + \log(m+1) + \log 24e$$

as well as

$$\mathop{\mathbb{E}}_{d \sim N(0,\mathrm{I})} \left(\log \log \mathscr{K}_2(d) \mid d \in \mathcal{W} \right) = \log \log n + \mathcal{O}(1)$$

and

$$\mathop{\mathbb{E}}_{d \sim N(0,\mathrm{I})} \left(\log \mathscr{K}_2(d) \log \log \mathscr{K}_2(d) \mid d \in \mathcal{W} \right) = \log n \log \log n + \mathcal{O}(1).$$

Theorems 11.1, 11.3, and 12.1 combine to yield the following average complexity results.

Corollary 12.2 *There exists an algorithm that with input a matrix $A \in \mathbb{R}^{m \times n}$, vectors $b \in \mathbb{R}^m$, $c \in \mathbb{R}^n$, and a feasible point $z_0 = (x_0, y_0, s_0)$ in the central neighborhood $\mathcal{N}(\frac{1}{4})$ finds an optimal basis B for d. The average number of iterations performed by the algorithm, on Gaussian data d, conditioned to d being feasible well-posed, is bounded by*

$$\mathcal{O}\left(\sqrt{n} \, \log n \right).$$

The average number of arithmetic operations is bounded by

$$\mathcal{O}\left(n^{3.5} \log n \right). \qquad \square$$

Corollary 12.3 *There exists an algorithm that with input a matrix $A \in \mathbb{R}^{m \times n}$ and vectors $b \in \mathbb{R}^m$ and $c \in \mathbb{R}^n$ finds an optimal basis B for d. The average cost of the algorithm, on Gaussian data d, conditioned to d being feasible well-posed, is bounded by*

$$\mathcal{O}\left(n^{3.5} \log n \log \log n \right). \qquad \square$$

In addition, a bound on the average maximum number of digits $\log \epsilon_{\mathrm{mach}}(d)$ required by the algorithm in Theorem 11.21 follows as well, namely,

$$\mathop{\mathbb{E}}_{d \sim N(0,\mathrm{I})} \log \epsilon_{\mathrm{mach}}(d) = \mathcal{O}(\log n).$$

The main difficulty in proving Theorem 12.1 is the conditioning over the event $d \in \mathcal{W}$. The idea of the proof involves rewriting the conditional expectation in the statement as an expectation over a Gaussian of some function (easier to deal with than $\mathcal{H}$). In order to do so, we will rely on an idea that will be central in Part III: the use of symmetry properties of functions and distributions expressed as invariance under the action of certain groups.

12.1 Proof of Theorem 12.1

Write $\mathcal{D} = \mathbb{R}^{mn+m+n}$ for the space of data inputs, and

$$\mathcal{B} = \{B \subseteq \{1, 2, \dots, n\} \mid |B| = m\}$$

for the family of possible bases.

12.1.1 The Group $\mathfrak{G}_n$ and Its Action

We consider the group $\mathfrak{G}_n = \{-1, 1\}^n$ with respect to componentwise multiplication. This group acts on $\mathcal{D}$ as follows. For $\mathsf{u} \in \mathfrak{G}_n$ let D_u be the diagonal matrix having u_j as its jth diagonal entry, and define

$$\mathsf{u}A := AD_\mathsf{u} = (\mathsf{u}_1 a_1, \mathsf{u}_2 a_2, \dots, \mathsf{u}_n a_n),$$

$$\mathsf{u}c := D_\mathsf{u}c = (\mathsf{u}_1 c_1, \mathsf{u}_2 c_2, \dots, \mathsf{u}_n c_n),$$

where a_i denotes the ith column of A. We define $\mathsf{u}d := (\mathsf{u}A, b, \mathsf{u}c)$. The group $\mathfrak{G}_n$ also acts on $\mathbb{R}^n$ by $\mathsf{u}x := (\mathsf{u}_1 x_1, \dots, \mathsf{u}_n x_n)$. It is immediate to verify that for all $A \in \mathbb{R}^{m \times n}$, all $x \in \mathbb{R}^n$, and all $\mathsf{u} \in \mathfrak{G}_n$ we have $\mathsf{u}A\,\mathsf{u}x = Ax$.

Now recall (from Sect. 11.1) the definition of S_B, for $B \in \mathcal{B}$, and consider the function

$$h_B : \quad \mathcal{D} \to [0, +\infty),$$

$$d \mapsto \min_{S \in S_B(d)} \rho_{\mathrm{sing}}(S).$$

These functions are important to us because for any $d \in \mathcal{W}$, Theorem 11.7 characterizes $\varrho(d)$ as $h_B(d)$, where B is the optimal basis of d. The functions ϱ and h_B are symmetric in a very precise sense.

Lemma 12.4 *The functions h_B are $\mathfrak{G}_n$-invariant. That is, for any $d \in \mathcal{D}$, $B \in \mathcal{B}$, and $\mathsf{u} \in \mathfrak{G}_n$,*

$$h_B(d) = h_B(\mathsf{u}d).$$

Proof Let S^* be any matrix in $\mathcal{S}_B(d)$ such that

$$\rho_{\text{sing}}(S^*) = \min_{S \in \mathcal{S}_B(d)} \rho_{\text{sing}}(S). \tag{12.1}$$

Let k be the number of rows (or columns) of S^* and let E be any matrix in $\mathbb{R}^{k \times k}$ such that $S^* + E$ is singular and

$$\|E\| = \rho_{\text{sing}}(S^*). \tag{12.2}$$

Then, there exists a nonzero $z \in \mathbb{R}^k$ such that

$$(S^* + E)z = 0. \tag{12.3}$$

Suppose S^* consists of the $j_1, j_2, \ldots, j_k$ columns of d (recall the definition of $\mathcal{S}_B$) and let $\bar{u} \in \mathfrak{G}_k$ be given by $\bar{u} = (u_{j_1}, u_{j_2}, \ldots, u_{j_k})$. Then, by the definition of $\mathcal{S}_B(d)$ and $\mathcal{S}_B(ud)$, we have $\bar{u}S^* \in \mathcal{S}_B(ud)$. Furthermore,

$$(\bar{u}S^* + \bar{u}E)\bar{u}z = \bar{u}(S^* + E)\bar{u}z = (S^* + E)(z) = 0,$$

the last equality by Eq. (12.3). That is, $(\bar{u}S^* + \bar{u}E)$ is also singular. By the definition of ρ_{sing},

$$\rho_{\text{sing}}(\bar{u}S^*) \le \|\bar{u}E\|. \tag{12.4}$$

Since operator norms are invariant under multiplication of arbitrary matrix columns by -1 we have $\|E\| = \|\bar{u}E\|$. Combining this equality with Eqs. (12.1), (12.2), and (12.4), we obtain

$$\rho_{\text{sing}}(\bar{u}S^*) \le \min_{S \in \mathcal{S}_B(d)} \rho_{\text{sing}}(S).$$

Since $\bar{u}S^* \in \mathcal{S}_B(ud)$, we obtain

$$\min_{S \in \mathcal{S}_B(ud)} \rho_{\text{sing}}(S) \le \min_{S \in \mathcal{S}_B(d)} \rho_{\text{sing}}(S).$$

The reverse inequality follows by exchanging the roles of uS and S. $\square$

Recall from Sect. 11.1 the partition $\{\mathcal{W}_B \mid B \in \mathcal{B}\}$ of the set $\mathcal{W}$ of well-posed feasible triples.

Lemma 12.5 *Let $d \in \mathcal{D}$ and $B \in \mathcal{B}$. If $h_B(d) > 0$, then there exists a unique $u \in \mathfrak{G}_n$ such that $ud \in \mathcal{W}_B$.*

Proof First observe that since $\min_{S \in \mathcal{S}_B(d)} \rho_{\text{sing}}(S) > 0$, the matrix A_B is invertible and therefore B is a basis for d. Let y^* and x^* be the dual and primal basic solutions of d for the basis B, i.e.,

$$y^* = A_B^{-\mathsf{T}} c_B, \qquad x_B^* = A_B^{-1} b, \quad \text{and} \quad x_j^* = 0, \quad \text{for all } j \notin B. \tag{12.5}$$

Similarly, let y^u and x^u be the dual and primal basic solutions of ud for the same basis. Then, using that $uA = AD_u$ and $uc = D_u c$,

$$y^u = (uA)_B^{-T}(uc)_B = A_B^{-T}(D_u)_B^{-T}(D_u)_B c_B = A_B^{-T} c_B = y^*, \tag{12.6}$$

the third equality by the definition of $(D_u)_B$. Similarly,

$$x_B^u = (uA)_B^{-1} b = (D_u)_B^{-1} A_B^{-1} b = (D_u)_B A_B^{-1} b = (D_u)_B x_B^* \tag{12.7}$$

and $x_j^u = 0$ for all $j \notin B$. Therefore,

$$B \text{ is optimal for } ud \quad \Leftrightarrow \quad x^u \text{ and } y^u \text{ are both feasible}$$

$$\Leftrightarrow \quad \begin{cases} x_B^u \geq 0 \\ (uA)_j^T y^u \leq (uc)_j, & \text{for } j \notin B \end{cases}$$

$$\Leftrightarrow \quad \begin{cases} (D_u)_B x_B^* \geq 0 \\ (u_j a_j)^T y^* \leq u_j c_j, & \text{for } j \notin B \end{cases}$$

$$\Leftrightarrow \quad \begin{cases} u_j x_j^* \geq 0, & \text{for } j \in B \\ u_j(c_j - a_j^T y) \geq 0, & \text{for } j \notin B, \end{cases} \tag{12.8}$$

the third equivalence by (12.6) and (12.7).

Since by hypothesis $h_B(d) > 0$, we have $\min_{S \in \mathcal{S}_B(d)} \rho_{\text{sing}}(S) > 0$ and hence

$$x_j^* \neq 0, \quad \text{for all } j \in B, \quad \text{and} \quad a_j^T y \neq c_j, \quad \text{for all } j \notin B. \tag{12.9}$$

Combining Eqs. (12.8) and (12.9), the statement follows for $u \in \mathfrak{G}_n$ given by $u_j = \frac{x_j^*}{|x_j^*|}$ if $j \in B$ and $u_j = \frac{c_j - a_j^T y}{|c_j - a_j^T y|}$ otherwise. Clearly, this u is unique. $\qquad\square$

For $B \in \mathcal{B}$ let

$$\Sigma_B := \{d \in \mathcal{D} \mid h_B(d) = 0\}$$

and $\mathcal{D}_B := \mathcal{D} \setminus \Sigma_B$. Lemma 12.4 implies that Σ_B and $\mathcal{D}_B$ are $\mathfrak{G}_n$-invariant, for all $B \in \mathcal{B}$. Lemma 12.5 immediately implies the following corollary.

Corollary 12.6 *For all $B \in \mathcal{B}$ the sets*

$$\mathcal{D}_u := \{d \in \mathcal{D}_B \mid ud \in \mathcal{W}_B\}, \quad \text{for } u \in \mathfrak{G}_n,$$

from a partition of $\mathcal{D}_B$. $\qquad\square$

Remark 12.7 The set of ill-posed feasible triples is included in the union of the sets Σ_B over $B \in \mathcal{B}$ (this follows from the proof of Theorem 11.7). The reverse inclusion, however, even restricted to feasible triples, does not hold. In other words,

Fig. 12.1 The situation in the
space of data, revisited

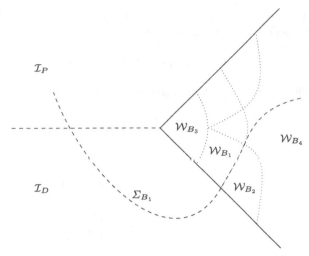

there are triples $d \in \mathcal{W}$ belonging to Σ_B for some $B \in \mathcal{B}$ (and actually, because of
an argument of symmetry, for all $B \in \mathcal{B}$). Needless to say, this B is not an optimal
basis for d.

As an example of the above, consider the matrix $A = [1, 0]$, and the vectors
$b = (1)$ and $c = (1, 1)$. Then the primal–dual pair of optimization problems is

$$
\begin{array}{ll}
\min \quad x_1 + x_2 & \max \quad y \\[4pt]
\text{s.t.} \quad [1, 0]\begin{pmatrix} x_1 \\ x_2 \end{pmatrix} = 1, & \text{s.t.} \quad \begin{bmatrix} 1 \\ 0 \end{bmatrix} y \le \begin{pmatrix} 1 \\ 1 \end{pmatrix}, \\[10pt]
\quad\quad x_1, x_2 \ge 0, &
\end{array}
$$

and it is immediate to check that $B_1 = \{1\}$ is an optimal basis and the corresponding
optimizers are $x_* = (1, 0)$ and $y_* = 1$. Furthermore,

$$
\mathcal{S}_{B_1} = \left\{ [1], [1], \begin{bmatrix} 1 & 0 \\ 1 & 1 \end{bmatrix} \right\},
$$

and hence $\varrho(d) = 1$. The other element in $\mathcal{B}$ is $B_2 = \{2\}$, for which we have

$$
\mathcal{S}_{B_2} = \left\{ [0], [1], \begin{bmatrix} 1 & 0 \\ 1 & 1 \end{bmatrix} \right\}
$$

and hence $h_{B_2}(d) = 0$. In summary, $d \in \mathcal{W}_{B_1}$ and $d \in \Sigma_{B_2}$.

Figure 12.1 revisits the situation discussed at the end of the introduction to
Chap. 11 and summarized in Fig. 11.1. We have added to the latter a dashed curve
corresponding to one of the sets Σ_B (say, B_1). We see that this set intersects both $\mathcal{I}_P$
and $\mathcal{I}_D$. Furthermore, it contains the boundary of $\mathcal{W}_{B_1}$, but it also contains triples
in the interior of other $\mathcal{W}_B$'s (in the figure, we have shown this for $\mathcal{W}_{B_4}$).

12.1.2 Probabilities

Although Theorem 12.1 is stated for the Gaussian distribution, a substantial part of its proof—a reduction from a conditional expectation to a nonconditional one—can be done more generally. Therefore, for a time to come, we fix a distribution $\mathscr{D}$ on the set of triples $d = (A, b, c)$, with density function f with respect to the Lebesgue measure $d\mathcal{D}$ on $\mathcal{D}$, satisfying the following conditions:

(Di) f is $\mathfrak{G}_n$-*invariant*; that is, if $d \sim \mathscr{D}$, then $ud \sim \mathscr{D}$ for all $u{\in}\mathfrak{G}_n$.

(Dii) for all $B \in \mathcal{B}$, $\mathrm{Prob}_{d\sim\mathscr{D}}\{h_B(d) = 0\} = 0$.

(Diii) for all $B_1, B_2 \in \mathcal{B}$ and all measurable functions $g : \mathbb{R} \to \mathbb{R}$,

$$\int_{d\in\mathcal{W}_{B_1}} g\big(h_{B_1}(d)\big)f(d)\,d\mathcal{D} = \int_{d\in\mathcal{W}_{B_2}} g\big(h_{B_2}(d)\big)f(d)\,d\mathcal{D}.$$

Note that condition (Dii) implies that the probability of having two optimal bases is zero. Also, condition (Diii) is satisfied whenever $\mathscr{D}$ comes from an i.i.d. distribution on $\mathcal{D}$. In particular, the standard Gaussian distribution $N(0, \mathrm{I})$ satisfies (Di)–(Diii) above.

Lemma 12.8 *For any* $u{\in}\mathfrak{G}_n$ *and* $B \in \mathcal{B}$,

$$\mathop{\mathrm{Prob}}_{d\sim\mathscr{D}}\{ud \in \mathcal{W}_B\} = \mathop{\mathrm{Prob}}_{d\sim\mathscr{D}}\{d \in \mathcal{W}_B\} = \frac{1}{2^n}.$$

Proof The equality between probabilities follows from (Di). Therefore, by Corollary 12.6 and condition (Dii), the probability of each of them is 2^{-n}. □

The following lemma tells us that for all $B \in \mathcal{B}$, the random variable $h_B(d)$ is independent of the event "$d \in \mathcal{W}_B$."

Lemma 12.9 *For all measurable* $g : \mathbb{R} \to \mathbb{R}$ *and* $B \in \mathcal{B}$,

$$\mathop{\mathbb{E}}_{d\sim\mathscr{D}}\big(g\big(h_B(d)\big) \mid d \in \mathcal{W}_B\big) = \mathop{\mathbb{E}}_{d\sim\mathscr{D}}\big(g\big(h_B(d)\big)\big).$$

Proof From the definition of conditional expectation and Lemma 12.8 we have

$$\mathop{\mathbb{E}}_{d\sim\mathscr{D}}\big(g\big(h_B(d)\big) \mid d \in \mathcal{W}_B\big) = \frac{\int_{d\in\mathcal{W}_B} g(h_B(d))f(d)\,d\mathcal{D}}{\mathrm{Prob}_{d\sim\mathscr{D}}\{d \in \mathcal{W}_B\}}$$

$$= 2^n \int_{d\in\mathcal{D}} \mathbb{1}_B(d)g\big(h_B(d)\big)f(d)\,d\mathcal{D},$$

where $\mathbb{1}_B$ denotes the indicator function of $\mathcal{W}_B$. Now, for any $u{\in}\mathfrak{G}_n$, the map $d \mapsto ud$ is a linear isometry on $\mathcal{D}$. Therefore

$$\int_{d\in\mathcal{D}} \mathbb{1}_B(d)g\big(h_B(d)\big)f(d)\,d\mathcal{D} = \int_{d\in\mathcal{D}} \mathbb{1}_B(ud)g\big(h_B(ud)\big)f(ud)\,d\mathcal{D}.$$

Using that $h_B(d) = h_B(ud)$ (by Lemma 12.4) and $f(d) = f(ud)$ (by the $\mathfrak{G}_n$-invariance of $\mathcal{D}$), it follows that

$$
\begin{aligned}
\underset{d\sim\mathcal{D}}{\mathbb{E}}\big(g(h_B(d)) \mid d \in \mathcal{W}_B\big) &= 2^n \int_{d\in\mathcal{D}} \mathbb{1}_B(d)g\big(h_B(d)\big)f(d)\,d\mathcal{D} \\
&= \sum_{u\in\mathfrak{G}_n} \int_{d\in\mathcal{D}} \mathbb{1}_B(ud)g\big(h_B(ud)\big)f(ud)\,d\mathcal{D} \\
&= \sum_{u\in\mathfrak{G}_n} \int_{d\in\mathcal{D}} \mathbb{1}_R(ud)g\big(h_R(d)\big)f(d)\,d\mathcal{D} \\
&= \int_{d\in\mathcal{D}} g\big(h_B(d)\big)f(d)\,d\mathcal{D} = \underset{d\sim\mathcal{D}}{\mathbb{E}}\big(g(h_B(d))\big),
\end{aligned}
$$

the last line by Corollary 12.6.   □

The following lemma is proved as Lemma 12.9.

Lemma 12.10 *For all $r, s \geq 1$ we have*

$$
\underset{d\sim\mathcal{D}}{\mathbb{E}}\,(\|d\|_{rs} \mid d \in \mathcal{W}) = \underset{d\sim\mathcal{D}}{\mathbb{E}}\,(\|d\|_{rs}). \qquad\qquad □
$$

Lemmas 12.9 and 12.10 eliminate the conditioning to "$d \in \mathcal{W}_B$" in the expectations we want to compute. A difficulty remains in the fact that $\varrho(d) = h_B(d)$ only when B is the optimal basis of d. Therefore, to compute $\mathbb{E}g(\varrho(d))$ we will have to compute $\mathbb{E}g(h_B(d))$, with B being a function of d. The next lemma solves this problem. Let $B^* = \{1, 2, \ldots, m\}$.

Lemma 12.11 *For all measurable $g : \mathbb{R} \to \mathbb{R}$,*

$$
\underset{d\sim\mathcal{D}}{\mathbb{E}}\big(g(\varrho(d)) \mid d \in \mathcal{W}\big) = \underset{d\sim\mathcal{D}}{\mathbb{E}}\big(g(h_{B^*}(d))\big).
$$

Proof By the definition of conditional expectation,

$$
\underset{d\sim\mathcal{D}}{\mathbb{E}}\big(g(\varrho(d)) \mid d \in \mathcal{W}\big) = \frac{\int_{d\in\mathcal{W}} g(\varrho(d))f(d)\,d\mathcal{D}}{\mathrm{Prob}_{d\sim\mathcal{D}}\{d \in \mathcal{W}\}}. \qquad (12.10)
$$

Because of (Dii), the probability that d has two optimal bases is 0. Using this and Lemma 12.8, we see that

$$
\underset{d\sim\mathcal{D}}{\mathrm{Prob}}\{d \in \mathcal{W}\} = \sum_{B\in\mathcal{B}} \underset{d\sim\mathcal{D}}{\mathrm{Prob}}\{d \in \mathcal{W}_B\} = \sum_{B\in\mathcal{B}} \frac{1}{2^n} = \binom{n}{m}\frac{1}{2^n}. \qquad (12.11)
$$

Combining Eqs. (12.10) and (12.11), we have

$$\binom{n}{m} \frac{1}{2^n} \mathop{\mathbb{E}}_{d \sim \mathcal{D}} \left(g\left(\varrho(d)\right) \mid d \in \mathcal{W} \right) = \int_{d \in \mathcal{W}} g\left(\varrho(d)\right) f(d) \, d\mathcal{D}$$

$$= \sum_{B \in \mathcal{B}} \int_{d \in \mathcal{W}_B} g\left(\varrho(d)\right) f(d) \, d\mathcal{D},$$

with the last equality from the fact that the probability that d has two optimal bases is 0. Using now this equality, condition (Diii), and the fact that $\varrho(d) = h_B(d)$ for $d \in \mathcal{W}_B$ (Theorem 11.7), we obtain

$$\frac{1}{2^n} \mathop{\mathbb{E}}_{d \sim \mathcal{D}} \left(g\left(\varrho(d)\right) \mid d \in \mathcal{W} \right) = \int_{d \in \mathcal{W}_{B*}} g\left(\varrho(d)\right) f(d) \, d\mathcal{D}$$

$$= \int_{d \in \mathcal{W}_{B*}} g\left(h_{B*}(d)\right) f(d) \, d\mathcal{D}.$$

Therefore, by Lemma 12.8 with $B = B^*$,

$$\mathop{\mathsf{Prob}}_{d \sim \mathcal{D}} \{ d \in \mathcal{W}_{B*} \} \mathop{\mathbb{E}}_{d \sim \mathcal{D}} \left(g\left(\varrho(d)\right) \mid d \in \mathcal{W} \right) = \int_{d \in \mathcal{W}_{B*}} g\left(h_{B*}(d)\right) f(d) \, d\mathcal{D}.$$

We conclude, since by the definition of conditional expectation and Lemma 12.9,

$$\mathop{\mathbb{E}}_{d \sim \mathcal{D}} \left(g\left(\varrho(d)\right) \mid d \in \mathcal{W} \right) = \mathop{\mathbb{E}}_{d \sim \mathcal{D}} \left(g\left(h_{B*}(d)\right) \mid d \in \mathcal{W}_{B*} \right) = \mathop{\mathbb{E}}_{d \sim \mathcal{D}} \left(g\left(h_{B*}(d)\right) \right). \qquad \square$$

Corollary 12.12 *For all $r, s \geq 1$ we have*

$$\mathop{\mathbb{E}}_{d \sim \mathcal{D}} \left(\log \mathscr{K}_{rs}(d) \mid d \in \mathcal{W} \right) = \mathop{\mathbb{E}}_{d \sim \mathcal{D}} \left(\log \|d\|_{rs} \right) - \mathop{\mathbb{E}}_{d \sim \mathcal{D}} \left(\log h_{B*}(d) \right).$$

Proof It is a trivial consequence of the definition of $\mathscr{K}(d)$ and Lemmas 12.10 and 12.11. $\qquad \square$

Corollary 12.12 reduces the computation of the conditional expectation of $\log \mathscr{K}_{rs}$ to those for the expectations of $\log \|d\|_{rs}$ and $\log h_{B*}(d)$. The reduction holds for any distribution $\mathcal{D}$ satisfying properties (Di)–(Diii). To proceed further and to give estimates of the latter two expectations, we need to choose a particular distribution $\mathcal{D}$ (and values for r and s). We next take $\mathcal{D} = N(0, \mathrm{I})$ and $r = s = 2$.

Lemma 12.13 *Let $B \in \mathcal{B}$ be fixed. Then,*

$$\mathop{\mathbb{E}}_{d \sim N(0, \mathrm{I})} \left(\sqrt{\frac{1}{h_B(d)}} \right) \leq 2\sqrt{e}(m + 1)^{1/2}(n + 1).$$

Proof Consider a random matrix $S \in \mathbb{R}^{p \times p}$. Using the tail bound in Corollary 4.20 (with $\overline{A} = 0$ and $\sigma = 1$) together with Proposition 2.27 (with $k = \frac{1}{2}$, $K = pe$,

$B = \sqrt{pe}$, and $\alpha = 1$) we obtain

$$\operatorname*{\mathbb{E}}_{S \sim N(0, I_{p^2})} \sqrt{\|S^{-1}\|} \leq 2\sqrt{pe}. \tag{12.12}$$

For any fixed $d \in \mathcal{D}$,

$$\sqrt{\frac{1}{h_B(d)}} = \max_{S \in \mathcal{S}_B} \sqrt{\frac{1}{\rho_{\mathrm{sing}}(S)}} \leq \sum_{S \in \mathcal{S}_B} \sqrt{\frac{1}{\rho_{\mathrm{sing}}(S)}} = \sum_{S \in \mathcal{S}_B} \sqrt{\|S^{-1}\|}.$$

Taking averages on both sides yields

$$\operatorname*{\mathbb{E}}_{d \sim \mathcal{D}} \left(\sqrt{\frac{1}{h_B(d)}} \right) \leq \operatorname*{\mathbb{E}}_{d \sim \mathcal{D}} \left(\sum_{S \in \mathcal{S}_B} \sqrt{\|S^{-1}\|} \right) = \sum_{S \in \mathcal{S}_B} \operatorname*{\mathbb{E}}_{d \sim \mathcal{D}} \left(\sqrt{\|S^{-1}\|} \right)$$

$$\leq \sum_{S \in \mathcal{S}_B} 2\sqrt{e}(m+1)^{1/2} \quad \text{by (12.12) with } p = m \text{ or } m+1$$

$$\leq 2\sqrt{e}(m+1)^{1/2}(n+1). \qquad \square$$

Proof of Theorem 12.1 Recall from Sect. 11.1 the definition of $\|d\|$. Since for a random Gaussian $M \in \mathbb{R}^{(m+1) \times (n+1)}$ we have $\mathbb{E}(\|M\|) \leq 6\sqrt{n+1}$ (Lemma 4.14), it follows from Jensen's inequality that

$$\operatorname*{\mathbb{E}}_{d \sim N(0, I)} \left(\log \|d\| \right) \leq \log \operatorname*{\mathbb{E}}_{M \sim N(0, I)} \left(\|M\| \right) \leq \frac{1}{2} \log(n+1) + \log 6.$$

In addition, using Lemma 12.13 and Jensen's inequality, we have

$$\operatorname*{\mathbb{E}}_{d \sim N(0, I)} \log\bigl(h_{B^*}(d)\bigr) = -2 \operatorname*{\mathbb{E}}_{d \sim N(0, I)} \log \sqrt{\frac{1}{h_{B^*}(d)}} \geq -2 \log \operatorname*{\mathbb{E}}_{d \sim N(0, I)} \sqrt{\frac{1}{h_{B^*}(d)}}$$

$$\geq -\log\bigl(4e(m+1)(n+1)^2\bigr).$$

Now use Corollary 12.12 with $\mathcal{D} = N(0, I)$ and $r = s = 2$ to obtain the first inequality in the statement. The remaining two inequalities follow from applying Jensen's inequality (Proposition 2.28) to the random variable $\log \mathscr{K}_2(d)$ and the concave functions $z \mapsto \log z$ and $z \mapsto z \log z$. $\qquad \square$

Chapter 13
Probabilistic Analyses of the GCC Condition Number

In Chap. 6 we identified the GCC condition number as the crucial parameter in the perturbation theory of the polyhedral conic feasibility problem PCFP. Later on, we saw that this quantity occurs in cost estimates for an ellipsoid method finding feasible points in a nonempty cone (Corollary 7.7) and for interior-point methods deciding feasibility of polyhedral conic systems (Theorem 9.14). Furthermore, the development in Chap. 10 showed that this condition number also plays a central role in cost estimates for deciding feasibility of primal–dual pairs in linear programming.

Continuing with one of the central themes in our exposition, we perform in this chapter probabilistic analyses of the GCC condition number, as was done in Chap. 2 for the condition number of linear equation solving. Our average-analysis result is the following.

Theorem 13.1 *For $A \in (\mathbb{S}^{m-1})^n$ chosen uniformly at random, $n > m$, we have*

$$\mathsf{Prob}\{\mathscr{C}(A) \geq t\} \leq cm^5 \frac{1}{t} \ln t \quad for \ t \geq e,$$

where c is a universal constant. Moreover, $\mathbb{E}(\ln \mathscr{C}(A)) = \mathcal{O}(\ln m)$.

Some average complexity results easily follow from Theorem 13.1. The following, which uses Theorem 9.14, is an example.

Corollary 13.2 *Let $\mathsf{cost}^{\mathsf{FEAS}}(A)$ denote the cost of Algorithm 9.2 on input $A \in \mathbb{R}^{m \times n}$. Then*

$$\mathop{\mathbb{E}}_{A \sim N(0,\mathrm{I})} \mathsf{cost}^{\mathsf{FEAS}}(A) = \mathcal{O}\big(n^{3.5}(\log n + \log m)\big).$$

$\square$

A glimpse at the right-hand side of the inequality in Corollary 13.2 shows that the contribution to average cost coming from conditioning is the humblest in the bound. For $n \gg m$ it is negligible.

P. Bürgisser, F. Cucker, *Condition*,
Grundlehren der mathematischen Wissenschaften 349,
DOI 10.1007/978-3-642-38896-5_13, © Springer-Verlag Berlin Heidelberg 2013

We also perform a uniform smoothed analysis of the GCC condition number in the sense of Sect. 2.4.3. That is, for $0 < \sigma \le 1$ we consider a random $A = (a_1, \ldots, a_n)$, with the points $a_i \in \mathbb{S}^{m-1}$ independently chosen at random from the uniform distribution in the $B(\overline{a}_i, \sigma)$ with center $\overline{a}_i$ and radius σ, with respect to $d_{\sin}$ (cf. Sect. 2.2.6). To simplify notation we write $\overline{A} = (\overline{a}_1, \ldots, \overline{a}_n) \in (\mathbb{S}^{m-1})^n$ and consider the σ-neighborhood of $\overline{A}$ in $(\mathbb{S}^{m-1})^n$, defined as

$$B(\overline{A}, \sigma) := B(\overline{a}_1, \sigma) \times \cdots \times B(\overline{a}_n, \sigma).$$

So we are assuming that A is chosen uniformly at random from $B(\overline{A}, \sigma)$. In this context our result is the following.

Theorem 13.3 *Let $0 < \sigma \le 1$ and $\overline{A} \in (\mathbb{S}^{m-1})^n$, $n > m$. Assume that $A \in B(\overline{A}, \sigma)$ is chosen uniformly at random. Then we have*

$$\mathsf{Prob}\{A \in \mathcal{F}_D, \ \mathscr{C}(A) \ge t\} \le \frac{13nm^2}{2\sigma} \frac{1}{t} \quad for \ t \ge \frac{2m^2}{\sigma}. \tag{13.1}$$

Moreover, we have for $t \ge 1$,

$$\mathsf{Prob}\{A \in \mathcal{F}_P, \ \mathscr{C}(A) \ge t\} \ \le \ \frac{845\,n^2 m^3}{8\sigma^2} \frac{1}{t} \ln t + \frac{65\,nm^3}{\sigma^2} \frac{1}{t}. \tag{13.2}$$

Combining the tail bounds of Theorem 13.3 with Proposition 2.26, e.g., using the rough estimate $t^{-1} \log t \le t^{-1/2}$, we obtain the following estimates for the expectation.

Corollary 13.4 *For $0 < \sigma \le 1$ and $n > m$ we have*

$$\sup_{\overline{A} \in (\mathbb{S}^{m-1})^n} \mathop{\mathbb{E}}_{A \in B(\overline{A}, \sigma)} \left(\log \mathscr{C}(A)\right) = \mathcal{O}\left(\log \frac{n}{\sigma}\right),$$

where the supremum is over all $\overline{A} \in (\mathbb{S}^{m-1})^n$. $\square$

We can derive from this result smoothed-complexity estimates. Again, as an example, we do so for polyhedral conic feasibility.

Corollary 13.5 *For $0 < \sigma \le 1$ and $n > m$ we have*

$$\sup_{\overline{A} \in (\mathbb{S}^{m-1})^n} \mathop{\mathbb{E}}_{A \in B(\overline{A}, \sigma)} \mathrm{cost}^{\mathsf{FEAS}}(A) = \mathcal{O}\left(n^{3.5} \log \frac{n}{\sigma}\right).$$

$\square$

This chapter, which completes Part II of this book, is technically somewhat more demanding than our previous developments.

13.1 The Probability of Primal and Dual Feasibility

A first step in the proof of our probabilistic estimates consists in computing the probability that a random $A \in \mathbb{R}^{m \times n}$ is dual (resp. primal) feasible. We begin by considering dual feasibility. That is we want to compute the probability $p(n, m)$ that

$$\exists y \in \mathbb{R}^m \quad \langle a_1, y \rangle < 0, \ldots, \langle a_n, y \rangle < 0$$

for independent standard Gaussian vectors $a_1, \ldots, a_n \in \mathbb{R}^m$.

Let us illustrate this problem by a simple example. In the case $m = 1$, we have $a_1, \ldots, a_n \in \mathcal{F}_D^\circ$ iff $a_1, \ldots, a_n$ have the same sign, either positive or negative. Since each sign occurs with the probability $1/2$, we obtain $p(n, 1) = 2/2^n$. The case $m = 2$ is already more challenging, and the reader may try to directly prove that $p(n, 2) = n/2^{n-1}$. The answer in the general case involves the binomial distribution. We shall take $\binom{n}{i} = 0$ if $i > n$.

Theorem 13.6 *For a standard Gaussian matrix $A \in \mathbb{R}^{m \times n}$ we have*

$$p(n, m) := \mathop{\mathsf{Prob}}_A \{A \in \mathcal{F}_D\} = \frac{1}{2^{n-1}} \sum_{i=0}^{m-1} \binom{n-1}{i}$$

and $\mathsf{Prob}_A\{A \in \mathcal{F}_P\} = 1 - \mathsf{Prob}_A\{A \in \mathcal{F}_D\}$.

Let us introduce some notation towards the proof. Fix nonzero vectors $a_1, \ldots, a_n \in \mathbb{R}^m$. To any $y \in \mathbb{R}^m$ we assign its *sign pattern* $\operatorname{sgn}(y) \in \{-1, 0, 1\}^n$ defined by $\operatorname{sgn}(y)_i := \operatorname{sgn}\langle a_i, y \rangle$. Moreover, for $\sigma \in \{-1, 0, 1\}$ we consider the *realization set* $R_A(\sigma) := \{y \in \mathbb{R}^m \mid \operatorname{sgn}(y) = \sigma\}$ of the sign pattern σ. We have a partition of $\mathbb{R}^m$ into the sets $R_A(\sigma)$. This partition is determined by the linear hyperplanes H_i given by $\langle a_i, y \rangle = 0$. The full-dimensional $R_A(\sigma)$'s shall be called *cells*. They correspond to the sign patterns $\sigma \in \{-1, 1\}$ with $R_A(\sigma) \neq \emptyset$. We say that the hyperplanes $H_1, \ldots, H_n$ of $\mathbb{R}^m$ are *in general position* if $\bigcap_{i \in I} H_i$ is of dimension $m - |I|$ for all $I \subseteq [n]$ with $|I| \leq m + 1$ (setting $\dim \emptyset = -1$). It is clear that this condition is satisfied by almost all $a_1, \ldots, a_n$.

Lemma 13.7 *A linear hyperplane arrangement $H_1, \ldots, H_n$ of $\mathbb{R}^m$ in general position has exactly $c(n, m) := 2 \sum_{i=0}^{m-1} \binom{n-1}{i}$ cells.*

Proof We have $c(n, 1) = c(1, m) = 2$, and hence the assertion is true for $m = 1$ or $n = 1$. Suppose now $m \geq 2$. We proceed by induction on n. Let $H_1, \ldots, H_{n+1}$ be hyperplanes of $\mathbb{R}^m$ in general position. By the induction hypothesis, the arrangement $\mathcal{H}$ given by $H_1, \ldots, H_n$ has exactly $c(n, m)$ cells. We now intersect this arrangement with H_{n+1}. If H_{n+1} intersects the interior of a cell C of $\mathcal{H}$, then this cell splits into two cells. This happens when $C \cap H_{n+1}$ is a cell of the arrangement of hyperplanes $H_1 \cap H_{n+1}, \ldots, H_n \cap H_{n+1}$ of $H_{n+1} \simeq \mathbb{R}^{m-1}$. By the induction hypothesis, this arrangement has exactly $c(n, m-1)$ cells. From this we may conclude that $H_1, \ldots, H_{n+1}$ has exactly $c(n, m) + c(n, m-1)$ cells, that is,

$$c(n, m) + c(n, m-1) = 2 + 2 \sum_{i=1}^{m-1} \binom{n-1}{i} + 2 \sum_{i=1}^{m-1} \binom{n-1}{i-1}$$

$$= 2 + 2 \sum_{i=1}^{m-1} \binom{n}{i} = c(n+1, m).$$

$\square$

Proof of Theorem 13.6 For $\sigma \in \Theta := \{-1, 1\}^n$ consider the event $\mathcal{E}_\sigma := \{A \in \mathbb{R}^{m \times n} \mid R_A(\sigma) \neq \emptyset\}$. Then, $\mathcal{F}_D$ coincides with $\mathcal{E}_\sigma$ for $\sigma = (-1, \ldots, -1)$. Moreover, all events $\mathcal{E}_\sigma$ have the same probability, since the standard Gaussian distribution is invariant under $a_i \mapsto -a_i$. We also note that $\sum_{\sigma \in \Theta} \mathbb{1}_{\mathcal{E}_\sigma}(A) = |\{\sigma \mid R_A(\sigma) \neq \emptyset\}|$ equals the number of cells of the hyperplane arrangement given by A. Now we conclude that

$$2^n \, \mathsf{Prob} \, \mathcal{F}_D = \sum_{\sigma \in \Theta} \mathsf{Prob} \, \mathcal{E}_\sigma = \sum_{\sigma \in \Theta} \mathbb{E}(\mathbb{1}_{\mathcal{E}_\sigma}) = \mathbb{E}\left(\sum_{\sigma \in \Theta} \mathbb{1}_{\mathcal{E}_\sigma} \right) = 2 \sum_{i=0}^{m-1} \binom{n-1}{i},$$

where the last equality is due to Lemma 13.7.

$\square$

By definition, $p(n, m)$ is the probability that n randomly chosen open hemispheres have a nonempty intersection. This is also the probability that the union of n randomly chosen closed hemispheres do not cover the whole sphere $\mathbb{S}^{m-1}$. More generally, let $p(n, m, \alpha)$ denote the probability that n randomly chosen spherical caps with centers $a_1, \ldots, a_n$ and angular radius α do not cover the sphere $\mathbb{S}^{m-1}$ (random meaning here that the centers a_i are independently chosen with respect to the uniform distribution of $\mathbb{S}^{m-1}$). Then it is clear that $p(n, m, \pi/2) = p(n, m)$.

The problem of determining the probabilities $p(n, m, \alpha)$ is arguably the central problem in the area of covering processes on spheres. Interestingly, there is a close connection between these probabilities and the probabilistic behavior of the GCC condition number. To explain this, recall from Sect. 6.4 that $\rho(A)$ denotes the angular radius of a smallest including cap of $a_1, \ldots, a_n \in \mathbb{S}^{m-1}$.

Proposition 13.8

(a) *We have for $0 \leq \alpha \leq \pi$,*

$$p(n, m, \alpha) = \mathsf{Prob}\{\rho(A) \leq \pi - \alpha\}.$$

(b) *Moreover, for $\pi/2 \leq \alpha \leq \pi$,*

$$p(n, m, \alpha) = \mathsf{Prob}\{A \in \mathcal{F}_D \text{ and } \mathscr{C}(A) \leq (-\cos \alpha)^{-1}\},$$

and for $0 \leq \alpha \leq \pi/2$,

$$p(n, m, \alpha) = p(n, m) + \mathsf{Prob}\{A \in \mathcal{F}_P \text{ and } \mathscr{C}(A) \geq (\cos \alpha)^{-1}\}.$$

Proof (a) The caps of radius α with center $a_1, \ldots, a_n$ do not cover $\mathbb{S}^{m-1}$ iff there exists $y \in S^{m-1}$ having distance greater than α from all a_i. The latter means that the cap of radius $\pi - \alpha$ centered at $-y$ contains all the a_i, which implies $\rho(A) \leq \pi - \alpha$ and vice versa. This proves the first claim.

(b) The following arguments are based on Theorem 6.27. Suppose first $\pi/2 \leq \alpha \leq \pi$. If $\rho(A) \leq \pi - \alpha$, then $\rho(A) \leq \pi/2$; hence $A \in \mathcal{F}_D$. Furthermore,

$$\cos \rho(A) \geq \cos(\pi - \alpha) = -\cos \alpha \geq 0,$$

whence $\mathscr{C}(A) = (\cos \rho(A))^{-1} \leq (-\cos \alpha)^{-1}$. On the other hand, if $A \in \mathcal{F}_D$ and $\mathscr{C}(A) \leq (-\cos \alpha)^{-1}$, we know that $\rho(A) \leq \pi/2$, and we can reverse the argument to infer $\rho(A) \leq \pi - \alpha$. Thus the asserted characterization of $p(n, m, \alpha)$ follows with part one.

Suppose now $0 \leq \alpha \leq \pi/2$. If $\rho(A) \leq \pi - \alpha$ then either $\rho(A) \leq \pi/2$ (meaning $A \in \mathcal{F}_D$), or $\pi/2 < \rho(A)$. In the latter case, $0 \leq -\cos \rho(A) \leq -\cos(\pi - \alpha) = \cos \alpha$, and hence $\mathscr{C}(A) = (-\cos \rho(A))^{-1} \geq (\cos \alpha)^{-1}$. Conversely, if $A \in \mathcal{F}_D$ and $\mathscr{C}(A) \geq (\cos \alpha)^{-1}$, then either $\rho(A) \leq \pi/2$ or $\pi/2 < \rho(A)$, in which case the above argument can be reversed to deduce that $\rho(A) \leq \pi - \alpha$. □

We conclude with a technical lemma about the asymptotic growth of $p(n, m)$, to be used later.

Lemma 13.9 *We have* $\sum_{n=4m}^{\infty} n \, p(n, m) = o(1)$ *for* $m \to \infty$.

Proof Let $n \geq 4m$. We have by Theorem 13.6, since $m - 1 \leq (n - 1)/2$,

$$np(n, m) \leq n \frac{m}{2^{n-1}} \binom{n-1}{m-1} \leq \frac{2nm}{2^n} \frac{(n-1)^{m-1}}{(m-1)!} \leq \frac{2m^2}{m!} \frac{n^m}{2^n}.$$

We also have $n^m 2^{-n} \leq 2^{-n/2}$ for $n \geq Cm \log m$, and sufficiently large m, where $C > 0$ is a suitable universal constant. Therefore, we get

$$\sum_{n \geq Cm \log m} n \, p(n, m) \leq \frac{2m^2}{m!} \sum_{n=0}^{\infty} \frac{1}{2^{n/2}} = o(1) \quad (m \to \infty).$$

We now deal with the case $n \in \{4m, \ldots, Cm \log m\}$. The function $x \mapsto x^m 2^{-x}$ is monotonically decreasing for $x \geq m/\ln 2$. Hence, using $n \geq 4m$ and $m! \geq (m/e)^m$, we get

$$\frac{1}{m!} \frac{n^m}{2^n} \leq \frac{1}{m!} \frac{(4m)^m}{2^{4m}} \leq \left(\frac{e}{4}\right)^m.$$

Since $e/4 < 1$, we conclude that

$$\sum_{n=4m}^{Cm \log m} np(n, m) \leq 2m^2 \left(\frac{e}{4}\right)^m Cm \log m = o(1) \quad (m \to \infty),$$

which completes the proof. □

13.2 Spherical Convexity

Section 6.2 gave a first introduction to the notion of convexity, through results such as Carathéodory's theorem, Helly's theorem, or the separating hyperplane theorem. In this section we further develop the theme of convexity by looking at convex sets in spheres. This amounts to studying convex cones C along with their duals $\check{C}$, which were already introduced previously, in Sect. 6.2.

A convex cone $C \subset \mathbb{R}^n$ is called *pointed* if $C \cap (-C) = \{0\}$. Suppose that $a_1, \ldots, a_k \in \mathbb{R}^n$ are nonzero. Then it is easy to check that $\mathrm{cone}\{a_1, \ldots, a_k\}$ is pointed iff 0 is not contained in the convex hull $\mathrm{conv}\{a_1, \ldots, a_k\}$. The following two lemmas give additional characterizations.

Lemma 13.10 *Let $C \subseteq \mathbb{R}^n$ be a convex cone. Then C is pointed iff $\check{C}$ has a nonempty interior.*

Proof $\check{C}$ has empty interior iff $\check{C}$ is contained in a hyperplane $H = (\mathbb{R}q)^{\perp}$ of $\mathbb{R}^n$. This implies by Proposition 6.3 that $\mathbb{R}q = \check{H} \subseteq \check{\check{C}} = C$; hence $0 \neq q \in C \cap (-C)$ and C is not pointed. The argument is reversible.   □

Lemma 13.11 *A convex cone C is pointed iff $C \setminus \{0\}$ is contained in an open half-space whose bounding hyperplane goes through the origin.*

Proof Suppose C is pointed. Then, by Lemma 13.10, there exists $q \in \mathrm{int}(\check{C})$. Let $x \in C \setminus \{0\}$. Then $\langle q, x \rangle \leq 0$. If we had $\langle q, x \rangle = 0$, then $\langle q', x \rangle > 0$ for some $q' \in \check{C}$ sufficiently close to q, which is a contradiction. Hence $\langle q, x \rangle < 0$ for all $x \in C \setminus \{0\}$. The converse direction is trivial.   □

We now define a notion of convexity for subsets of the sphere $\mathbb{S}^{m-1}$. Let $x, y \in \mathbb{S}^{m-1}$ be such that $x \neq \pm y$. We call $[x, y] := \mathrm{cone}\{x, y\} \cap \mathbb{S}^{m-1}$ the *great circle segment* connecting x and y.

Definition 13.12 A subset K of $\mathbb{S}^{m-1}$ is called *(spherically) convex* if we have $[x, y] \subseteq K$ for all $x, y \in K$ with $x \neq \pm y$. We call K *properly convex* if it is nonempty, convex, and does not contain a pair of antipodal points.

This notion of spherical convexity is closely related to convex cones. In fact, it is easy to see that a subset K of $\mathbb{S}^{m-1}$ is convex iff it is of the form $K = C \cap \mathbb{S}^{m-1}$ for some convex cone $C \subseteq \mathbb{R}^m$. In this case, we must have $C = \mathrm{cone}(K)$. Moreover, K is properly convex iff C is a pointed convex cone, i.e., $C \cap (-C) = \{0\}$. By the separating hyperplane theorem (Theorem 6.1) applied to $C = \mathrm{cone}(K)$, a convex subset K of $\mathbb{S}^{m-1}$ is contained in a closed half-space, unless $K = \mathbb{S}^{m-1}$. Moreover, by Lemma 13.11, a properly convex set K is always contained in an open half-space.

Example 13.13 A spherical cap $\mathrm{cap}(a, \alpha)$ of radius α is convex iff $\alpha \leq \pi/2$ or $\alpha = \pi$ (in which case the cap equals the whole sphere). The cap $\mathrm{cap}(a, \alpha)$ is properly convex iff $\alpha < \pi/2$.

We denote by $\mathrm{sconv}(M) := \mathrm{cone}(M) \cap \mathbb{S}^{m-1}$ the *(spherical) convex hull* of a subset M of $\mathbb{S}^{m-1}$, which is the smallest spherical convex set containing M. Clearly, M is convex iff $M = \mathrm{sconv}(M)$. Moreover, the closure of a convex set is convex as well.

Definition 13.14 The *dual set* of a convex set $K \subseteq \mathbb{S}^{m-1}$ is defined as

$$\check{K} := \left\{ a \in \mathbb{S}^{m-1} \mid \forall x \in K \ \langle a, x \rangle \leq 0 \right\}.$$

Clearly, if C is the convex cone generated by K and $\check{C}$ its dual cone, then $\check{K} = \check{C} \cap \mathbb{S}^{m-1}$. In particular, $\check{K}$ is a closed convex set disjoint from K. For example, the dual set of $\mathrm{cap}(a, \alpha)$ equals $\mathrm{cap}(-a, \pi/2 - \alpha)$, where $\alpha \leq \pi/2$.

By Proposition 6.3, the dual of $\check{K}$ equals K. Furthermore, by Lemma 13.10, a convex set $K \subseteq \mathbb{S}^{m-1}$ is properly convex iff $\check{K}$ has nonempty interior. Thus "nonempty interior" and "properly convex" are dual properties. We also note that $K_1 \subseteq K_2$ implies $\check{K}_1 \supseteq \check{K}_2$.

By a *convex body* K in $\mathbb{S}^{m-1}$ we will understand a closed convex set K such that both K and $\check{K}$ have nonempty interior. Therefore, the map $K \mapsto \check{K}$ is an involution of the set of convex bodies in $\mathbb{S}^{m-1}$.

We define the distance of $a \in \mathbb{S}^{m-1}$ to a nonempty set $K \subseteq \mathbb{S}^{m-1}$ by $d_{\mathbb{S}}(a, K) := \inf\{d_{\mathbb{S}}(a, x) \mid x \in K\}$ (recall that we defined $d_{\mathbb{S}}$ in Sect. 6.4). Then it is immediate that the dual set of a convex body K can be characterized in terms of distances as follows:

$$a \in \check{K} \iff d_{\mathbb{S}}(a, K) \geq \pi/2. \tag{13.3}$$

There is a simple relation between the distances of a to K and to $\check{K}$, respectively, if a lies outside of both K and $\check{K}$ (cf. Fig. 13.1).

Lemma 13.15 *Let K be a convex body in $\mathbb{S}^{m-1}$ and $a \in \mathbb{S}^{m-1} \setminus (K \cup \check{K})$. Then $d_{\mathbb{S}}(a, K) + d_{\mathbb{S}}(a, \check{K}) = \pi/2$.*

Proof Let $b \in K$ be such that $\theta := d_{\mathbb{S}}(a, b) = d_{\mathbb{S}}(a, K)$. Since $a \notin \check{K}$, we have $\theta < \pi/2$. The point $b^* := \langle a, b \rangle b = (\cos \theta) b$ is therefore nonzero and contained in $C := \mathrm{cone}(K)$. Put $p^* := a - b^*$. Then $\langle p^*, b \rangle = 0$, $\langle p^*, a \rangle = \sin^2 \theta$, and $\langle p^*, p^* \rangle = \sin^2 \theta$. In particular, $p^* \neq 0$.

By construction, b^* is the point of C closest to a. It follows that $\{x \in \mathbb{R}^{m+1} \mid \langle p^*, x \rangle = 0\}$ is a supporting hyperplane (cf. Theorem 6.1(b)) of C. Hence $\langle p^*, x \rangle \leq 0$ for all $x \in C$, and the point $p := p^*/\|p^*\|$ therefore belongs to $\check{K}$. Moreover, $\langle p, a \rangle = \sin \theta$, which implies $d_{\mathbb{S}}(a, p) = \pi/2 - \theta$. Hence

$$d_{\mathbb{S}}(a, K) + d_{\mathbb{S}}(a, \check{K}) \leq d_{\mathbb{S}}(a, b) + d_{\mathbb{S}}(a, p) = \pi/2.$$

To complete the proof it suffices to show that $d_{\mathbb{S}}(a, \check{K}) = d_{\mathbb{S}}(a, p)$. Suppose there exists $p' \in \check{K}$ such that $d_{\mathbb{S}}(a, p') < d_{\mathbb{S}}(a, p)$. Then $d_{\mathbb{S}}(b, p') \leq d_{\mathbb{S}}(b, a) + d_{\mathbb{S}}(a, p') < d_{\mathbb{S}}(b, a) + d_{\mathbb{S}}(a, p) = \pi/2$, which contradicts the fact that $b \in \check{K}$. $\quad\square$

Fig. 13.1 A cone K, its dual $\check{K}$, and a point $a \in \mathbb{S}^{m-1} \setminus (K \cup \check{K})$

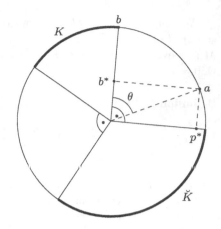

13.3 A Bound on the Volume of Tubes

In Sect. 2.2.6 we studied neighborhoods of special subsets of spheres and determined their volume. We will now look at this in more generality.

Definition 13.16

1. For $0 \le \varepsilon \le 1$, the ε-*neighborhood* of a nonempty subset U of $\mathbb{S}^{m-1}$ is defined as $T(U, \varepsilon) := \{x \in \mathbb{S}^{m-1} \mid d_{\mathbb{S}}(x, U) \le \arcsin \varepsilon\}$, where as usual $d_{\mathbb{S}}(x, U) := \inf_{y \in U} d_{\mathbb{S}}(x, y)$.
2. If U equals the boundary ∂K of a properly convex set K in $\mathbb{S}^{m-1}$, we call

$$T_o(\partial K, \varepsilon) := T(\partial K, \varepsilon) \setminus K \quad \text{and} \quad T_i(\partial K, \varepsilon) := T(\partial K, \varepsilon) \cap K$$

the *outer ε-neighborhood and inner ε-neighborhood of ∂K*, respectively.

Remark 13.17

(a) If U is symmetric, that is, $-U = U$, then

$$T(U, \varepsilon) = \left\{x \in \mathbb{S}^p \mid d_{\sin}(x, U) \le \varepsilon\right\}.$$

(b) We have $T(\partial K, \varepsilon) = T_o(\partial K, \varepsilon) \cup T_i(\partial K, \varepsilon)$.

For technical reasons, we define

$$B_{\sin}(\overline{a}, \sigma) := \left\{x \in \mathbb{S}^p \mid d_{\sin}(x, \overline{a}) \le \sigma\right\}, \tag{13.4}$$

which is the closed ball of radius σ, with respect to $d_{\sin}$, around $\overline{a}$ in $\mathbb{S}^p$. We note that $B_{\sin}(\overline{a}, \sigma) = B(\overline{a}, \sigma) \cup B(-\overline{a}, \sigma)$, where $B(\overline{a}, \sigma)$ denotes the spherical cap around $\overline{a}$ with angular radius $\arcsin \sigma$; compare Sect. 2.2.6.

For the probabilistic analyses in this chapter it will be crucial to effectively bound the volume of the neighborhoods $T(K, \varepsilon)$ of convex subsets K of a sphere. More specifically, such bounds suffice for an average analysis. For a uniform smoothed analysis as in Sect. 2.4.3 we need bounds on the volume of the intersection of $T(K, \varepsilon)$ with a ball $B_{\sin}(\overline{a}, \sigma)$, relative to the volume of $B_{\sin}(\overline{a}, \sigma)$.

We state such a result now, but postpone its proof to Sect. 21.6.

Theorem 13.18 *Let K be a properly convex subset of $\mathbb{S}^{m-1}$, let $\overline{a} \in \mathbb{S}^{m-1}$, and let $0 < \sigma, \varepsilon \le 1$. Then we have the following upper bound for the volume of the outer ε-neighborhood of ∂K:*

$$\frac{\mathrm{vol}(T_o(\partial K, \varepsilon) \cap B_{\sin}(\overline{a}, \sigma))}{\mathrm{vol}\, B_{\sin}(\overline{a}, \sigma)} \le e \frac{m\varepsilon}{\sigma} \quad \text{if } \varepsilon \le \frac{\sigma}{2m}.$$

The same upper bound holds for the relative volume of the inner ε-neighborhood of ∂K. For $\sigma = 1$ we obtain in particular,

$$\frac{\mathrm{vol}\, T_o(\partial K, \varepsilon)}{\mathrm{vol}\, \mathbb{S}^{m-1}} \le em\varepsilon \quad \text{if } \varepsilon \le \frac{1}{2m},$$

and the same upper bound holds for the relative volume of the inner ε-neighborhood of ∂K.

13.4 Two Essential Reductions

Recall from Sects. 6.3 and 6.4 the decomposition $(\mathbb{S}^{m-1})^n = \mathcal{F}_P \cup \mathcal{F}_D$ into the sets of primal and dual feasible instances of the polyhedral cone feasibility problem. The set of ill-posed instances equals $\Sigma = \mathcal{F}_P \cap \mathcal{F}_D$, and the GCC condition number $\mathscr{C}(A)$ of an instance $A = (a_1, \ldots, a_n) \in (\mathbb{S}^{m-1})^n$ is characterized as the inverse distance of A to Σ,

$$\mathscr{C}(A) = \frac{1}{d_{\sin}(A, \Sigma)};$$

see Proposition 6.23. We shall use the same symbols $\mathcal{F}_P$, $\mathcal{F}_D$, and Σ to denote the corresponding sets of primal feasible, dual feasible, and ill-posed instances in $(\mathbb{S}^{m-1})^n$ for different values of n.

We start with a simple observation.

Lemma 13.19 *Let $A = (a_1, \ldots, a_n) \in (\mathbb{S}^{m-1})^n$ and let K be the cone defined by $K := -\mathrm{sconv}\{a_1, \ldots, a_n\}$. Then, for any $b \in \mathbb{S}^{m-1}$,*

(a) *if $b \notin K$, then $(A, b) \in \mathcal{F}_D^\circ$,*
(b) *if $b \in \partial K$, then $(A, b) \in \Sigma$,*
(c) *if $b \in \mathrm{int}(K)$, then $(A, b) \in \mathcal{F}_P^\circ$.*

Proof (a) If $b \notin K$, then $-b \notin \mathrm{cone}\{a_1, \ldots, a_n\}$, and the separating hyperplane theorem (Theorem 6.1) yields the existence of $q \in \mathbb{S}^{m-1}$ such that $\langle a_i, q \rangle > 0$ for all i and $\langle -b, q \rangle < 0$. Hence $(A, b) \in \mathcal{F}_D^\circ$.

(b) If $b \in \partial K$, then $-b$ lies on the boundary of $\mathrm{cone}\{a_1, \ldots, a_n\}$. Hence there exists a supporting hyperplane with normal vector $q \in \mathbb{S}^{m-1}$ such that $\langle a_i, q \rangle \geq 0$ for all i and $\langle -b, q \rangle \leq 0$. Therefore, $(A, b) \in \mathcal{F}_D$. Moreover, since $-b \in \mathrm{cone}\{a_1, \ldots, a_n\}$, we have $0 = b + \sum_i \lambda_i a_i$ with some $\lambda \geq 0$, and hence $0 \in \mathrm{conv}\{a_1, \ldots, a_n\}$. Therefore, $(A, b) \in \Sigma$ by (6.3).

(c) If $b \in \mathrm{int}(K)$, then $-b$ lies in the interior of $\mathrm{cone}\{a_1, \ldots, a_n\}$. In particular, the latter is of dimension m. Hence $-b \in \mathrm{relint}(\mathrm{cone}\{a_1, \ldots, a_n\})$ and there exist $\lambda_i > 0$ such that $-b = \sum_i \lambda_i a_i$. It follows that $0 \in \mathrm{relint}(\mathrm{cone}\{a_1, \ldots, a_n, b\})$ and hence $(A, b) \in \mathcal{F}_P^\circ$ by (6.2). $\qquad\square$

A key idea for the probabilistic analyses of the matrix condition number in Sect. 2.4 was the following. Suppose that the square matrix $A \in \mathbb{R}^{n \times n}$ with columns $a_1, \ldots, a_n$ is ε-close to the set of singular matrices. This means that there exist linearly dependent vectors $b_1, \ldots, b_n$ such that $\|a_k - b_k\| \leq \varepsilon$ for all k. Then it is possible to pick a "pivot index" i and just perturb the ith component a_i to $\tilde{a}_i$ by at most $n\varepsilon$, i.e., $\|\tilde{a}_i - a_i\| \leq n\varepsilon$, such that $a_1, \ldots, a_{i-1}, \tilde{a}_i, a_{i+1}, \ldots, a_n$ are linearly dependent (cf. Proposition 2.44).

We try to employ a similar idea, but now we have to deal with convexity issues and inequalities. In a first step, assume that $A \in \mathbb{R}^{m \times n}$ is such that $(a_1, \ldots, a_n) \in \Sigma$. Hence a smallest including cap of these points has radius $\pi/2$; cf. Lemma 6.25. If q denotes the center of this cap, then we have, after a possible reordering,

$$\langle a_1, q \rangle = \cdots = \langle a_k, q \rangle = 0, \qquad \langle a_{k+1}, q \rangle > 0, \ldots, \langle a_n, q \rangle > 0,$$

and by Lemma 6.26 we must have $0 \in \mathrm{conv}\{a_1, \ldots, a_k\}$ (we called $[k]$ the blocking set of the cap). It follows that without loss of generality, $-a_1 \in \mathrm{conv}\{a_2, \ldots, a_k\}$. In particular, $-a_1$ lies in the convex set $K := \mathrm{sconv}\{a_2, \ldots, a_n\}$. Since K is contained in the half-space $\{x \mid \langle q, x \rangle \geq 0\}$ and $\langle a_1, q \rangle = 0$, we have $-a_1 \in \partial K$.

It is now plausible that this argument can be extended in the following way: suppose that $(a_1, \ldots, a_n) \in \mathcal{F}_D^\circ$ is ε-close to Σ. Then there exists a pivot index i, say $i = 1$, such that $-a_1$ is close to ∂K. The next result shows that this is indeed the case.

Proposition 13.20 *Let* $A = (a_1, \ldots, a_n) \in \mathcal{F}_D^\circ$ *and* $0 < \varepsilon \leq 1$. *If we have* $\mathscr{C}(A) \geq m\varepsilon^{-1}$, *then there exists* $i \in [n]$ *such that*

$$a_i \in T_o(\partial K_i, \varepsilon),$$

where $K_i := -\mathrm{sconv}\{a_1, \ldots, a_{i-1}, a_{i+1}, \ldots, a_n\}$.

Proof Note first that by Lemma 13.19 we have $a_i \notin K_i$ for all i. Hence $d_\mathbb{S}(a_i, \partial K_i) = d_\mathbb{S}(a_i, K_i)$. Put $\theta := \arcsin \varepsilon$.

We will prove the contrapositive of the assertion: assuming $d_\mathbb{S}(a_i, K_i) > \theta$ for all $i \in [n]$, we need to prove that $\sin d_\mathbb{S}(A, \Sigma_{n,m}) > m^{-1}\varepsilon$. Then we are done, since $\mathscr{C}(A)^{-1} = \sin d_\mathbb{S}(A, \Sigma)$ by Proposition 6.23.

In a first step we show that for each i there exists p_i close to a_i such that all a_j are contained in the open hemisphere with center p_i. More specifically, we claim that for every $i \in [n]$ there exists $p_i \in \mathbb{S}^{m-1}$ such that

$$\langle a_i, p_i \rangle > \varepsilon \quad \text{and} \quad \forall j \neq i \quad \langle a_j, p_i \rangle > 0. \tag{13.5}$$

To prove this, we distinguish two cases. If $a_i \in \check{K}_i$, we just choose $p_i \in \operatorname{int}(\check{K}_i)$ close enough to a_i such that $\langle a_i, p_i \rangle > \varepsilon$. If $a_i \notin \check{K}_i$, then Lemma 13.15 tells us that $d_{\mathbb{S}}(a_i, K_i) + d_{\mathbb{S}}(a_i, \check{K}_i) = \pi/2$. Hence $d_{\mathbb{S}}(a_i, \check{K}_i) < \pi/2 - \theta$. Choose $p_i \in \operatorname{int}(\check{K}_i)$ such that $d_{\mathbb{S}}(a_i, p_i) < \pi/2 - \theta$. This implies $\langle a_i, p_i \rangle > \cos(\pi/2 - \theta) = \varepsilon$. This completes the proof of the claim (13.5).

Let q be the center of a smallest included cap (SIC) of A. Then $q \in \operatorname{cone}(A)$ and hence $\langle q, p_i \rangle > 0$ for all i,

Consider now for $i \in [n]$ the following convex sets in $\mathbb{S}^{m-1}$:

$$C_i := \left\{ x \in \mathbb{S}^{m-1} \mid \langle a_i, x \rangle > \varepsilon/m \text{ and } \langle x, q \rangle > 0 \right\}.$$

It suffices to show that these sets have a nonempty intersection. Indeed, if $z \in \bigcap_{i=1}^n C_i$, then $d_{\mathbb{S}}(a_i, z) < \alpha$ for all i, where $\alpha := \arccos(\varepsilon/m)$. This implies that the spherical cap $\operatorname{cap}(z, \alpha)$ strictly contains all a_i. The radius $\rho(A)$ of the SIC of A is therefore strictly smaller than α. Hence, by Theorem 6.27, $\sin d_{\mathbb{S}}(A, \Sigma) = \cos \rho(A) > \cos \alpha = \varepsilon/m$, as claimed.

In fact, by Helly's theorem (Theorem 6.8), it suffices to show that any m of the sets C_i have a nonempty intersection. To see this, just use the bijective perspective map

$$\pi : \left\{ x \in \mathbb{S}^{m-1} \mid \langle x, q \rangle > 0 \right\} \to E, \quad x \mapsto \langle q, x \rangle^{-1} x,$$

to the affine hyperplane $E := \{ x \in \mathbb{R}^m \mid \langle x, q \rangle = 1 \} \simeq \mathbb{R}^{m-1}$ and note that $\pi(C_i)$ is well defined and convex.

Let now $I \subseteq [n]$ be of cardinality m and consider $p^* := \frac{1}{m} \sum_{j \in I} p_j$. Note that $\|p^*\| \leq 1$. We obtain for any $i \in I$, using (13.5),

$$\langle a_i, p^* \rangle = \frac{1}{m} \sum_{j \in I} \langle a_i, p_j \rangle \geq \frac{1}{m} \langle a_i, p_i \rangle > \frac{\varepsilon}{m}.$$

Moreover, $\langle p^*, q \rangle > 0$, and hence $p^* \neq 0$. It follows that $p := p^*/\|p^*\|$ is contained in C_i, for all $i \in I$. This completes the proof. $\square$

The next proposition describes the transition from the dual feasible to the primal feasible case. Suppose that $A = (a_1, \ldots, a_n)$ is strictly dual feasible, but after adding a further column vector b, (A, b) is not dual feasible anymore. By Lemma 13.19 this means that b lies in the convex set $K = -\operatorname{sconv}(A)$ and in fact, (A, b) is ill-posed iff $b \in \partial K$. It is now plausible that a large condition number of (A, b) may be caused by two reasons: first, b may be close to the boundary of K; second, A may have a large condition number itself. The following result turns this heuristic reasoning into a quantitative statement.

Proposition 13.21 *Let* $A = (a_1, \ldots, a_n) \in \mathcal{F}_D^\circ$, $K := -\text{sconv}(A)$, *and* $b \in \mathbb{S}^{m-1}$. *If* $(A, b) := (a_1, \ldots, a_n, b)$ *is not dual feasible, then* $b \in K$, *and we have*

$$\mathscr{C}(A, b) \leq \frac{5\mathscr{C}(A)}{\sin d_\mathbb{S}(b, \partial K)}.$$

Proof Lemma 13.19 implies that $b \in K$ and gives us the following characterization of $d_\mathbb{S}(b, \partial K)$ in terms of distances to Σ:

$$d_\mathbb{S}(b, \partial K) = \min\{d_\mathbb{S}(b, b') \mid b' \in \mathbb{S}^{m-1} \text{ such that } (a_1, \ldots, a_n, b') \in \Sigma\}.$$

By the characterization of the GCC condition number in Theorem 6.27, the assertion can be restated as

$$\sin d_\mathbb{S}((A, b), \Sigma) \geq \frac{1}{10} d_\mathbb{S}(b, \partial K) \sin d_\mathbb{S}(A, \Sigma). \tag{13.6}$$

Consider the convex set

$$C := -\check{K} = \{x \in \mathbb{S}^{m-1} \mid \langle a_1, x \rangle \leq 0, \ldots, \langle a_n, x \rangle \leq 0\}.$$

We claim that

$$s := \sin d_\mathbb{S}(b, \partial K) \leq \min_{x \in C} \langle b, x \rangle. \tag{13.7}$$

In order to establish this, suppose $x \in C$. Since $b \in K$, we have $\cos \omega := \langle b, x \rangle \geq 0$. We may assume that $\|b - x \cos \omega\|^2 = 1 - \cos^2 \omega$ is positive, since otherwise, $b = x$ and clearly $s \leq 1 = \langle b, x \rangle$. Therefore, $b' := (b - x \cos \omega)/\|b - x \cos \omega\|$ is a well-defined point in $b' \in \mathbb{S}^{m-1}$ and $\langle b', x \rangle = 0$. Note that $d_\mathbb{S}(b, b') = \pi/2 - \omega$. Therefore $(A, b') = (a_1, \ldots, a_n, b')$ is dual feasible. It is either strictly dual feasible, in which case $b' \notin K$, or (A, b') is ill-posed, in which case $b' \in \partial K$ (cf. Lemma 13.19). Since $b \in K$, we conclude that $d_\mathbb{S}(b, \partial K) \leq d_\mathbb{S}(b, b') = \pi/2 - \omega$. This implies $\sin d_\mathbb{S}(b, \partial K) \leq \cos \omega = \langle b, x \rangle$ and proves inequality (13.7).

Suppose now that $\text{cap}(p, \rho)$ is an SIC for A. Since we assume A to be strictly feasible, $t := \cos \rho$ is positive and we have $t = \sin d_\mathbb{S}(A, \Sigma)$; cf. Theorem 6.27.

We need to prove the assertion (13.6), which is equivalent to

$$\sin d_\mathbb{S}((A, b), \mathcal{F}_D) \geq \frac{1}{5} st, \tag{13.8}$$

since $(A, b) \notin \mathcal{F}_D$ by assumption. So let $(A', b') \in \mathcal{F}_D$ and put $\varphi := d_\mathbb{S}((A, b), (A', b'))$. We need to show that $\sin \varphi \geq st/10$. By Theorem 6.27, we may assume that $\varphi \leq \pi/2$. Since $(A', b') \in \mathcal{F}_D$, there exists $x' \in \mathbb{S}^{m-1}$ such that

$$\langle a_1', x' \rangle \leq 0, \quad \ldots, \quad \langle a_n', x' \rangle \leq 0, \quad \langle b', x' \rangle \leq 0.$$

Taking into account that $d_\mathbb{S}(a_i', a_i) \leq \varphi$, we see that $d_\mathbb{S}(a_i, x') \geq \pi/2 - \varphi$ and hence $\langle a_i, x' \rangle \leq \sin \varphi$.

We now put $\tilde{x} := x' - \lambda p$ with $\lambda := t^{-1} \sin \varphi$. Since $\langle a_i, p \rangle \geq t$, we have for all i,

$$\langle a_i, \tilde{x} \rangle = \langle a_i, x' \rangle - \lambda \langle a_i, p \rangle \leq \sin \varphi - \lambda t = 0.$$

Without loss of generality we may assume that $\tilde{x} \neq 0$. Otherwise, $t = \sin \theta$, and we are done, since $t \geq st/10$. So $\tilde{x}/\|\tilde{x}\|$ is well defined and lies in C.

Inequality (13.7) implies that (use $\|\tilde{x}\| \geq 1 - \lambda$)

$$s - \lambda \leq s(1 - \lambda) \leq \langle b, \tilde{x} \rangle.$$

Put $\Delta b := b' - b$. Then $\|\Delta b\| \leq 2\sin(\varphi/2) \leq 2$ by our assumption $d_S(b', b) \leq \varphi$. We obtain

$$\langle b, \tilde{x} \rangle = \langle b' - \Delta b, x' - \lambda p \rangle = \langle b', x' \rangle - \langle \Delta b, x' \rangle - \langle b', \lambda p \rangle + \langle \Delta b, \lambda p \rangle$$

$$\leq 0 + \|\Delta b\| + \lambda + \|\Delta b\| \lambda \leq 2\sin(\varphi/2) + 3\lambda.$$

Combining the above two estimates yields

$$s \leq 2\sin(\varphi/2) + 4\lambda.$$

Recalling $\lambda = t^{-1} \sin \varphi$ and using $t \leq 1$, we get

$$st \leq 2t \sin(\varphi/2) + 4\lambda t \leq 2\sin(\varphi/2) + 4\sin \varphi \leq 5\sin \varphi.$$

This proves (13.8) and thus completes the proof. $\square$

13.5   A Crash Course on Probability: III

Before continuing, we need to develop a few further facts from the theory of probability.

Suppose that X and Y are random variables on the data space M taking nonnegative values. We assume that the pushforward measure of μ_M with respect to $(X, Y): M \to \mathbb{R}^2$ has a density ρ. Associated with ρ are the marginal density $\rho_X(x) := \int_0^\infty \rho(x, y)\, dy$ and, for $x \in \mathbb{R}$ with $\rho_X(x) > 0$, the conditional density $\rho_Y(y \mid X = x) := \rho(x, y)/\rho_X(x)$; compare (2.8) and (2.9).

Proposition 13.22 *Suppose that $X, Y: M \to \mathbb{R}_+$ are random variables on the data space M taking nonnegative values such that (X, Y) has density ρ on $\mathbb{R}^2$. Then $Z := XY$ has the following density, for $z > 0$:*

$$\rho_Z(z) = \int_0^\infty x^{-1} \rho_X(x)\, \rho_Y(z/x \mid X = x)\, dx.$$

Moreover, the distribution function of Z is given by

$$\mathsf{Prob}\{Z \geq z\} = \int_0^\infty \rho_X(x)\, \mathsf{Prob}\{Y \geq z/x \mid X = x\}\, dx.$$

Proof Consider the diffeomorphism $\psi: (0, \infty)^2 \to (0, \infty)^2, (x, y) \mapsto (x, xy) = (x, z)$ having Jacobian $J\psi(x, y) = x$. By Proposition 2.11, the pushforward density δ of ρ under ψ is given by

$$\delta(x, z) = x^{-1} \rho(x, z/x), \quad \text{for } x, z > 0.$$

The distribution of the random variable Z is obtained as the marginal distribution of the distribution of δ, whence

$$\rho_Z(z) = \int_0^\infty \delta(x, z)\, dx = \int_0^\infty x^{-1} \rho(x, z/x)\, dx$$

$$= \int_0^\infty x^{-1} \rho_X(x)\, \rho_Y(z/x \mid X = x)\, dx.$$

This proves the first statement. To prove the second, note that we have

$$\mathsf{Prob}\{Z \geq z\} = \int_z^\infty \rho_Z(\zeta)\, d\zeta = \int_0^\infty \rho_X(x) \int_z^\infty x^{-1} \rho_Y(\zeta/x \mid X = x)\, d\zeta\, dx.$$

For fixed $x > 0$, the substitution $\zeta \mapsto y = \zeta/x$ yields

$$\int_z^\infty \rho_Y(\zeta/x \mid X = x)x^{-1}\, d\zeta = \int_{z/x}^\infty \rho_Y(y \mid X = x)\, dy$$

$$= \mathsf{Prob}\{Y \geq z/x \mid X = x\}.$$

Altogether, we get

$$\mathsf{Prob}\{Z \geq z\} = \int_0^\infty \rho_X(x)\, \mathsf{Prob}\{Y \geq z/x \mid X = x\}\, dx$$

as claimed.   □

The next result provides an upper bound for the tail of XY that may be easier to apply than Proposition 13.22.

Lemma 13.23 *Let X and Y be random variables on M taking nonnegative values such that (X, Y) has a density. Further, let $f, g: (0, \infty) \to (0, \infty)$ be piecewise differentiable functions such that for $x, y > 0$,*

$$\mathsf{Prob}\{X \geq x\} \leq f(x), \qquad \mathsf{Prob}\{Y \geq y \mid X = x\} \leq g(y).$$

We further assume that g is bounded and $\lim_{y \to \infty} g(y) = 0$. Then, for $z > 0$,

$$\mathsf{Prob}\{XY \geq z\} \leq \int_0^\infty f(z/y)\big(-g'(y)\big)\, dy.$$

Proof We apply Proposition 13.22 with $Z := XY$ to obtain for $z > 0$,

$$\text{Prob}\{Z \geq z\} = \int_0^\infty \rho_X(x)\, \text{Prob}\{Y \geq z/x \mid X = x\}\, dx$$

$$\leq \int_0^\infty \rho_X(x)\, g(z/x)\, dx = -\int_0^\infty \frac{d}{dx} \text{Prob}\{X \geq x\}\, g(z/x)\, dx$$

$$= \int_0^\infty \text{Prob}\{X \geq x\} \frac{d}{dx} g(z/x)\, dx.$$

For the last equality we used integration by parts together with

$$\lim_{x \to \infty} \text{Prob}\{X \geq x\} g(z/x)\, dx = \lim_{x \to 0} \text{Prob}\{X \geq x\} g(z/x) = 0,$$

which follows from $\lim_{x \to \infty} \text{Prob}\{X \geq x\} = 0$, $\lim_{y \to \infty} g(y) = 0$, and the assumption that g is bounded. Continuing, we get

$$\int_0^\infty \text{Prob}\{X \geq x\} \frac{d}{dx} g(z/x)\, dx \leq \int_0^\infty f(x)\, g'(z/x)\left(-\frac{z}{x^2}\right) dx$$

$$= \int_0^\infty f(z/y)\big(-g'(y)\big)\, dy. \qquad \square$$

Lemma 13.24 *Let X and Y be random variables on M taking nonnegative values such that (X, Y) has a density. Assume there are $x_0, y_0, c > 0$ such that for all $x, y > 0$,*

$$\text{Prob}\{X \geq x\} \leq \alpha x^{-c} \quad \text{if } x \geq x_0,$$

$$\text{Prob}\{Y \geq y \mid X = x\} \leq \beta y^{-c} \quad \text{if } y \geq y_0.$$

Then, for all $z > 0$,

$$\text{Prob}\{XY \geq z\} \leq c\alpha\beta z^{-c} \ln \max\left\{\frac{z}{x_0 y_0}, 1\right\} + \beta x_0^c z^{-c}.$$

Proof Lemma 13.23 with the functions f, g defined as

$$f(x) = \begin{cases} 1 & \text{if } x < x_0, \\ \alpha x^{-c} & \text{if } x \geq x_0, \end{cases} \qquad g(y) = \begin{cases} 1 & \text{if } y < y_0, \\ \beta y^{-c} & \text{if } y \geq y_0, \end{cases}$$

yields

$$\text{Prob}\{XY \geq z\} \leq \int_0^\infty f(z/y)\big(-g'(y)\big)\, dy. \tag{13.9}$$

If $z \geq x_0 y_0$, we estimate this by

$$\mathsf{Prob}\{XY \geq z\} \leq \int_{y_0}^{z/x_0} \alpha z^{-c} y^c c\beta y^{-c-1} \, dy + \int_{z/x_0}^{\infty} c\beta y^{-c-1} \, dy$$

$$= c\alpha\beta z^{-c} \ln\left(\frac{z}{x_0 x_1}\right) + \beta x_0^c z^{-c}.$$

If $z < x_0 y_0$, we have

$$\mathsf{Prob}\{XY \geq z\} \leq \int_{y_0}^{\infty} \left(-g'(y)\right) dy = g(y_0) = \beta y_0^{-c} \leq \beta x_0^c z^{-c}.$$

This completes the proof. $\square$

13.6 Average Analysis

In this section we prove Theorem 13.1. With this goal in mind, we first derive a tail bound for $\mathscr{C}(A)$, conditional on A being dual feasible. Recall from Sect. 13.1 that $p(n, m)$ stands for the probability of A being dual feasible.

Lemma 13.25 *For A chosen uniformly at random in $(\mathbb{S}^{m-1})^n$ we have*

$$\mathsf{Prob}\{A \in \mathcal{F}_D, \ \mathscr{C}(A) \geq t\} \leq \frac{13}{4} m^2 n p(n-1, m) \frac{1}{t} \quad \text{for } t \geq 2m^2.$$

Proof Let $t \geq 2m^2$ and put $\varepsilon := m/t$. Proposition 13.20 tells us that

$$\mathsf{Prob}\{A \in \mathcal{F}_D^\circ, \ \mathscr{C}(A) \geq t\} \leq \sum_{i=1}^{n} \mathsf{Prob}\{A \in \mathcal{F}_D^\circ, \ a_i \in T_o(\partial K_i, \varepsilon)\},$$

where $K_i := -\mathsf{sconv}\{a_1, \ldots, a_{i-1}, a_{i+1}, \ldots, a_n\}$. To bound the probabilities on the right-hand side we assume without loss of generality that $i = n$. We express the probability as an integral over $A' := (a_1, \ldots, a_{n-1})$ of probabilities conditioned on A'. Furthermore, we write $K_{A'} := K_n = -\mathsf{sconv}\{a_1, \ldots, a_{n-1}\}$ to emphasize the dependence on A'. By Fubini's theorem we obtain

$$\mathsf{Prob}\{A \in \mathcal{F}_D^\circ, \ a_n \in T_o(\partial K_{A'}, \varepsilon)\}$$

$$\leq \mathsf{Prob}\{A' \in \mathcal{F}_D^\circ, \ a_n \in T_o(\partial K_{A'}, \varepsilon)\}$$

$$= \frac{1}{\mathsf{vol}(\mathbb{S}^{m-1})^{n-1}} \int_{A' \in \mathcal{F}_D^\circ} \mathsf{Prob}\{a_n \in T_o(\partial K_{A'}, \varepsilon) \mid A'\} \, dA'.$$

For fixed $A' \in \mathcal{F}_D^{\circ}$ the set $K_{A'}$ in $\mathbb{S}^{m-1}$ is properly convex. Theorem 13.18 yields, since $\varepsilon \leq 1/(2m)$ by assumption,

$$\mathrm{Prob}\big\{a_n \in T_o(\partial K_{A'}, \varepsilon) \mid A'\big\} = \frac{\mathrm{vol}\, T_o(\partial K_{A'}, \varepsilon)}{\mathrm{vol}\, \mathbb{S}^{m-1}} \leq \frac{13m}{4}\, \varepsilon.$$

We therefore obtain

$$\mathrm{Prob}\big\{A \in \mathcal{F}_D^{\circ},\ a_n \in T_o(\partial K_{A'}, \varepsilon)\big\} \leq \frac{13m}{4}\, \varepsilon\, \mathrm{Prob}\big\{A' \in \mathcal{F}_D\big\}.$$

Note that $\mathrm{Prob}\{A' \in \mathcal{F}_D\} = p(n-1, m)$ and $\varepsilon = m/t$. The same bound holds for all $i \in [n]$; hence multiplying by n, we obtain the claimed upper bound on $\mathrm{Prob}\{A \in \mathcal{F}_D$ and $\mathscr{C}(A) \geq t\}$, since $\Sigma = \mathcal{F}_D \setminus \mathcal{F}_D^{\circ}$ has measure zero. $\qquad\square$

We study now the situation for A primal feasible. For $A = (a_1, \ldots, a_n) \in (\mathbb{S}^{m-1})^n$ and $1 \leq k \leq n$ we write $A_k := (a_1, \ldots, a_k)$. Again we shall use the same symbol $\mathcal{F}_D$ to denote the corresponding sets of dual feasible instances in $(\mathbb{S}^{m-1})^n$ and $(\mathbb{S}^{m-1})^k$, respectively.

Lemma 13.26 *Let $A \in (\mathbb{S}^{m-1})^n$ and $k \leq n$ be such that A_k is not dual feasible. Then $\mathscr{C}(A_k) \geq \mathscr{C}(A)$.*

Proof Let $A' = (a_1', \ldots, a_n') \in \Sigma$ be such that $d_{\mathbb{S}}(A, A') = d_{\mathbb{S}}(A, \Sigma)$. Since $A' \in \mathcal{F}_D$, we have $A_k' = (a_1', \ldots, a_k') \in \mathcal{F}_D$. But $A_k \notin \mathcal{F}_D$ by assumption. Hence, using $d_{\mathbb{S}}(A, \Sigma) \leq \pi/2$ and (6.5), we get

$$\sin d_{\mathbb{S}}(A_k, \Sigma) = \sin d_{\mathbb{S}}(A_k, \mathcal{F}_D) \leq \sin d_{\mathbb{S}}\big(A_k, A_k'\big) \leq \sin d_{\mathbb{S}}\big(A, A'\big)$$
$$= \sin d_{\mathbb{S}}(A, \Sigma),$$

and the assertion follows with Proposition 6.23. $\qquad\square$

Suppose that $A = A_n$ is not dual feasible. Since A_m is dual feasible, there exists a smallest index $k \geq m$ such that A_k is dual feasible and A_{k+1} is not dual feasible. Consider, for $t > 0$, the event

$$\mathcal{E}_k(t) := \big\{A \mid A_k \text{ dual feasible and } A_{k+1} \text{ not dual feasible and } \mathscr{C}(A_{k+1}) \geq t\big\}.$$

Using Lemma 13.26 we obtain

$$\mathrm{Prob}\big\{A \notin \mathcal{F}_D,\ \mathscr{C}(A) \geq t\big\} \leq \sum_{k=m}^{n-1} \mathrm{Prob}\,\mathcal{E}_k(t) \qquad (13.10)$$

for uniformly random $A \in (\mathbb{S}^{m-1})^n$.

Lemma 13.27 *We have, for a universal constant c,*

$$\text{Prob}\,\mathcal{E}_k(t) \leq cm^3kp(k-1,m)\frac{1}{t}\ln t \quad for\ t \geq e.$$

Proof We fix k and write from now on

$$A := (a_1,\ldots,a_k), \quad K_A := -\text{sconv}\{a_1,\ldots,a_k\}, \qquad b := a_{k+1}.$$

With this notation, we have (using that $\mathcal{F}_D \setminus \mathcal{F}_D^\circ = \Sigma$ has measure zero)

$$\Prob_{A,b}\,\mathcal{E}_k(t) = \Prob_{A,b}\big\{A \in \mathcal{F}_D^\circ,\ (A,b) \notin \mathcal{F}_D,\ \mathscr{C}(A,b) \geq t\big\}.$$

Proposition 13.21 implies that for $t > 0$,

$$\Prob_{A,b}\,\mathcal{E}_k(t) \leq \Prob_{A,b}\left\{A \in \mathcal{F}_D^\circ,\ b \in K_A,\ \frac{\mathscr{C}(A)}{\sin d_{\mathbb{S}}(b, K_A)} \geq \frac{t}{5}\right\}. \tag{13.11}$$

Introducing the random variables

$$U(A) := \mathbb{1}_{\mathcal{F}_D^\circ}(A)\,\mathscr{C}(A), \qquad V(A,b) := \mathbb{1}_{K_A}(b)\frac{1}{\sin d_{\mathbb{S}}(b, \partial K_A)},$$

where $\mathbb{1}_M$ denotes the indicator function of the set M, we may rewrite (13.11) as

$$\Prob_{A,b}\,\mathcal{E}_k(t) \leq \Prob_{A,b}\big\{U(A)\cdot V(A,b) \geq t/5\big\}. \tag{13.12}$$

Lemma 13.25 tells us that for $x \geq 2m^2$,

$$\Prob_A\big\{U(A) \geq x\big\} = \Prob_A\big\{A \in \mathcal{F}_D^\circ,\ \mathscr{C}(A) \geq x\big\} \leq \alpha x^{-1}, \tag{13.13}$$

where $\alpha := \frac{13}{4}m^2kp(k-1,m)$.

Moreover, for $A \in \mathcal{F}_D^\circ$, the set K_A is properly convex, and the bound in Theorem 13.18 on the inner neighborhood of ∂K_A implies

$$\Prob_b\big\{V(A,b) \geq y \mid A\big\} = \Prob_b\left\{b \in K_A,\ \frac{1}{\sin d_{\mathbb{S}}(b, \partial K_A)} \geq y \;\Big|\; A\right\}$$

$$= \Prob_b\big\{b \in T_i\big(\partial K_A, y^{-1}\big) \mid A\big\} \leq \beta\,y^{-1} \tag{13.14}$$

if $y \geq 2m$, where $\beta := \frac{13m}{4}$.

In Sect. 17.3 we will learn how to define the *conditional density* ρ_u of the map $U\colon \mathbb{R}^{m \times n} \setminus \Sigma \to \mathbb{R}$ on its fiber $U^{-1}(u)$ over $u \neq 0$ that is induced by the standard Gaussian density φ on $\mathbb{R}^{m \times n}$. (In fact, we will see that $\rho_u(A) :=$

$\int_{A \in U^{-1}(u)} \frac{\varphi}{\text{NJU}}(A) \, dA$, where NJU is the normal Jacobian of U.) For $u > 0$ we define the *conditional probability*

$$\Prob_{A,b}\{V(A,b) \geq y \mid U(A) = u\} := \int_{A \in \mathcal{F}_D^\circ} \Prob_b\{V(A,b) \geq y \mid A\} \rho_u(A) \, dA.$$

Using this, we deduce from (13.14) that for all $u > 0$ and $y \geq 2m$,

$$\Prob_{A,b}\{V(A,b) \geq y \mid U(A) = u\} \leq \beta y^{-1}. \tag{13.15}$$

We can combine the probability estimates (13.13) and (13.15) by applying Lemma 13.24 to the random variables U and V. This implies for $z > 0$,

$$\Prob_{A,b}\{U(A) \cdot V(A,b) \geq z\} \leq \frac{\alpha\beta}{z} \ln\max\{z, 1\} + 2m^2\beta\frac{1}{z}.$$

Setting $z = t/5$, we conclude that for $t \geq e$,

$$\Prob_{A,b} \mathcal{E}_k(t) \leq \Prob_{A,b}\{U(A) \cdot V(A,b) \geq t/5\} \leq cm^3 kp(k-1, m)\frac{1}{t} \ln t$$

with some universal constant $c > 0$. $\qquad\square$

Proof of Theorem 13.1 Combining equation (13.10) with Lemma 13.27, we obtain

$$\Prob\{A \notin \mathcal{F}_D, \mathscr{C}(A) \geq t\} \leq cm^3 \frac{1}{t} \ln t \sum_{k=m}^{n-1} kp(k-1, m).$$

In order to bound the sum, we use Lemma 13.9, which gives

$$\sum_{k=4m+1}^{n-1} k \, p(k-1, m) = \sum_{\ell=4m}^{n-2} \ell \, p(\ell, m) + \sum_{\ell=4m}^{n-2} p(\ell, m) = o(1) \quad (m \to \infty).$$

Therefore,

$$\sum_{k=m}^{n-1} k \, p(k-1, m) = \sum_{k=m}^{4m} k \, p(k-1, m) + \sum_{k=4m+1}^{n-1} k \, p(k-1, m) = \mathcal{O}(m^2).$$

We conclude that

$$\Prob\{A \notin \mathcal{F}_D, \mathscr{C}(A) \geq t\} \leq c' m^5 \frac{1}{t} \ln t$$

for some universal constant c'. Moreover, $\Prob\{A \in \mathcal{F}_D$ and $\mathscr{C}(A) \geq t\}$ can also be bounded in this way by Lemma 13.25. This proves the asserted tail estimate for $\mathscr{C}(A)$.

Finally, the claimed bound on the expectation of $\ln \mathscr{C}(A)$ follows from Proposition 2.26, e.g., using the rough estimate $t^{-1} \log t \leq t^{-1/2}$. $\qquad\square$

13.7 Smoothed Analysis

In this section we prove Theorem 13.3. We will see that this can be done by virtually the same method employed for the average analysis in the previous section, along with the use of Theorem 13.18.

Proof of Theorem 13.3 We first prove (13.1), proceeding exactly as in the proof of Lemma 13.25. Fix $\overline{A} \in (\mathbb{S}^{m-1})^n$ and $0 < \sigma \le 1$. Further, suppose that $t \ge 2m^2\sigma^{-1}$ and put $\varepsilon := mt^{-1}$. If we suppose that A is chosen uniformly at random in $B(\overline{A}, \sigma)$, then Proposition 13.20 tells us that

$$\Prob_{A \in B(\overline{A},\sigma)} \left\{ A \in \mathcal{F}_D^\circ, \; \mathscr{C}(A) \ge t \right\} \le \sum_{i=1}^{n} \Prob_{A \in B(\overline{A},\sigma)} \left\{ A \in \mathcal{F}_D^\circ, \; a_i \in T_o(\partial K_i, \varepsilon) \right\},$$

where $K_i := -\mathrm{sconv}\{a_1, \ldots, a_{i-1}, a_{i+1}, \ldots, a_n\}$. To bound the probabilities on the right-hand side, we assume without loss of generality that $i = n$. We will express the probability as an integral over $A' := (a_1, \ldots, a_{n-1})$ of probabilities conditioned on A' and write $K_{A'} := K_n = -\mathrm{sconv}\{a_1, \ldots, a_{n-1}\}$. Note that $B(\overline{A}, \sigma) = B(\overline{A}', \sigma) \times B(\overline{a}_n, \sigma)$, where $\overline{A}' := (\overline{a}_1, \ldots, \overline{a}_{n-1})$. Then the distribution of a_n conditional on A' is just the uniform distribution of $B(\overline{a}_n, \sigma)$. By Fubini's theorem we obtain

$$\Prob_{A \in B(\overline{A},\sigma)} \left\{ A \in \mathcal{F}_D^\circ, \; a_n \in T_o(\partial K_{A'}, \varepsilon) \right\}$$

$$\le \Prob_{\substack{A' \in B(\overline{A}',\sigma) \\ a_n \in B(\overline{a}_n,\sigma)}} \left\{ A' \in \mathcal{F}_D^\circ, \; a_n \in T_o(\partial K_{A'}, \varepsilon) \right\}$$

$$= \frac{1}{\mathrm{vol}\, B(\overline{A}', \sigma)} \int_{A' \in \mathcal{F}_D^\circ \cap B(\overline{A}',\sigma)} \Prob_{a_n \in B(\overline{a}_n,\sigma)} \left\{ a_n \in T_o(\partial K_{A'}, \varepsilon) \mid A' \right\} dA'.$$

For fixed $A' \in \mathcal{F}_D^\circ$ the set $K_{A'}$ in $\mathbb{S}^{m-1}$ is properly convex. Theorem 13.18 implies

$$\Prob_{a_n \in B(\overline{a}_n,\sigma)} \left\{ a_n \in T_o(\partial K_{A'}, \varepsilon) \mid A' \right\} = \frac{\mathrm{vol}(T_o(\partial K_{A'}, \varepsilon) \cap B(\overline{a}_n, \sigma))}{\mathrm{vol}\, B(\overline{a}_n, \sigma)}$$

$$\le \frac{13m}{2} \frac{\varepsilon}{\sigma}.$$

Note that we get an extra factor of two by considering $B(\overline{a}_n, \sigma)$ instead of $B_{\sin}(\overline{a}_n, \sigma)$. Note also that $\varepsilon \le \sigma/(2m)$ by our assumption $t \ge 2m^2\sigma^{-1}$. Hence, using $\varepsilon = mt^{-1}$, we conclude that

$$\Prob_{\substack{A' \in B(\overline{A}',\sigma) \\ a_n \in B(\overline{a}_n,\sigma)}} \left\{ A \in \mathcal{F}_D^\circ, \; a_n \in T_o(\partial K_{A'}, \varepsilon) \right\} \le \frac{13m^2}{2\sigma t} \Prob_{A' \in B(\overline{A}',\sigma)} \left\{ A' \in \mathcal{F}_D \right\}$$

$$\le \frac{13m^2}{2\sigma t}.$$

Note that in contrast to the average analysis, we do not have a good bound for $\text{Prob}\{A' \in \mathcal{F}_D\}$, so we had to bound this quantity by 1. Since the same bound holds for all $i \in [n]$, we obtain the claim (13.1) by multiplying by n.

We continue now with the proof of (13.2) and proceed as in Theorem 13.1 and Lemma 13.27. We fix k and write

$$A := (a_1, \ldots, a_k), \quad K_A := -\text{sconv}\{a_1, \ldots, a_k\}, \quad b := a_{k+1}.$$

As above, we suppose that (A, b) is chosen uniformly at random in $B(\overline{A}, \sigma) \times B(\overline{b}, \sigma)$ and consider, for $t > 0$, the event

$$\mathcal{E}_k(t) = \{(A, b) \mid A \text{ is dual feasible}, \ (A, b) \text{ is not dual feasible}, \ \mathcal{C}(A, b) \geq t\}.$$

Then, as in (13.12) and using the notation from there, we have

$$\text{Prob}_{A,b} \mathcal{E}_k(t) \leq \text{Prob}_{A,b}\{U(A) \cdot V(A, b) \geq t/5\}. \tag{13.16}$$

From (13.1) we know that for $x \geq 2m^2/\sigma$,

$$\text{Prob}_A\{U(A) \geq x\} = \text{Prob}_A\{A \in \mathcal{F}_D^\circ, \ \mathcal{C}(A) \geq x\} \leq \alpha x^{-1}, \tag{13.17}$$

where we have set here $\alpha := \frac{13\,km^2}{2\sigma}$.

Moreover, for $A \in \mathcal{F}_D^\circ$, the set K_A is properly convex, and the bound in Theorem 13.18 on the inner neighborhood of ∂K_A implies, for $y \geq 2m/\sigma$,

$$\text{Prob}_b\{V(A, b) \geq y \mid A\} = \text{Prob}_b\{b \in T_i(\partial K_A, y^{-1}) \mid A\}$$

$$= \frac{\text{vol}(T_i(\partial K_A, y^{-1}) \cap B(\overline{b}, \sigma))}{\text{vol}\, B(\overline{b}, \sigma)} \leq \beta y^{-1},$$

where we have set $\beta := \frac{13m}{2\sigma}$. Since $U(A) > 0$ implies $A \in \mathcal{F}_D^\circ$, we get, for all $u > 0$ and $y \geq 2m/\sigma$,

$$\text{Prob}_{A,b}\{V(A, b) \geq y \mid U(A) = u\} \leq \beta y^{-1}. \tag{13.18}$$

We can now combine the estimates (13.17) and (13.18) by applying Lemma 13.24 to the random variables U and V. This yields, for $z > 0$,

$$\text{Prob}_{A,b}\{U(A) \cdot V(A, b) \geq z\} \leq \frac{\alpha\beta}{z} \ln \max\{z, 1\} + \frac{2m^2}{\sigma}\beta\frac{1}{z}.$$

Setting $z = t/5$ and using (13.16), we conclude that for $t \geq 1$,

$$\text{Prob}_{A,b} \mathcal{E}_k(t) \leq \frac{845\,km^3}{4\sigma^2}\frac{1}{t}\ln t + \frac{65\,m^3}{\sigma^2}\frac{1}{t}.$$

As in (13.10) we have

$$\text{Prob}\{A \notin \mathcal{F}_D, \ \mathscr{C}(A) \geq t\} \leq \sum_{k=m}^{n-1} \text{Prob}\, \mathcal{E}_k(t).$$

Summing the bounds for $\text{Prob}_{A,b}\, \mathcal{E}_k(t)$ over all k and using $\sum_{k=1}^{n} k \leq n^2/2$, the assertion (13.2) follows. $\square$

Intermezzo II: The Condition of the Condition

How costly is it to compute a condition number? This question presents two aspects: computational cost and accuracy. We begin by briefly discussing the first of these aspects. To do so, we recall a few of the condition numbers we have met thus far.

Take matrix–vector multiplication. We analyzed this problem in Sect. O.4, where we proved (Proposition O.8) that the normwise condition number $\mathsf{cond}(A, x)$ for this problems satisfies

$$\mathsf{cond}(A, x) = \frac{\|A\|_\infty \|x\|_\infty}{\|Ax\|_\infty}.$$

The denominator on the right-hand side indicates that to compute $\mathsf{cond}(A, x)$, we need at least to compute Ax, that is, to solve the problem for which (A, x) is the input data.

Consider now matrix inversion. Its normwise condition number (for the norms $\|\ \|_{rs}$ and $\|\ \|_{sr}$ in data and solution space, respectively) is, as we proved in Theorem 1.5,

$$\kappa_{rs}(A) = \|A\|_{rs} \|A^{-1}\|_{sr}.$$

Again, it is apparent that for computing $\kappa_{rs}(A)$ one needs to solve the problem for which A is the data, i.e., inverting A.

Finally, consider the condition number $\mathscr{C}(A)$ for PCFP. All its characterizations, in Sect. 6.5 via smallest including caps, in Sect. 6.6 via images of balls, and in Sect. 6.7 via well-conditioned solutions, turn into computations of $\mathscr{C}(A)$ that require, among other things, the solution of PCFP for input A.

It would seem that invariably, to compute $\mathsf{cond}^\varphi(a)$ we need to compute $\varphi(a)$. This is not true. The function $\varphi(a) = a^k$ satisfies $\mathsf{cond}^\varphi(a) = k$ for all $a \neq 0$; it is thus trivially computed. Yet the cost of computing $\varphi(a)$ can be bounded by the cost of computing $\mathsf{cond}^\varphi(a)$ plus a constant. The emerging picture can be thus summarized as follows:

The cost of computing $\mathsf{cond}^\varphi(a)$ is, modulo an additive constant, at least the cost of computing $\varphi(a)$. That is, $\mathsf{cost}(\mathsf{cond}^\varphi) \geq \mathsf{cost}(\varphi) + \mathcal{O}(1)$.

P. Bürgisser, F. Cucker, *Condition*,
Grundlehren der mathematischen Wissenschaften 349,
DOI 10.1007/978-3-642-38896-5, © Springer-Verlag Berlin Heidelberg 2013

The nature of this statement makes it difficult to formally prove it. We will therefore refrain from continuing and leave the statement as an empirical conclusion.

We can now proceed with the second aspect mentioned above. The accuracy in the computation of $\mathsf{cond}^\varphi(a)$ depends on the algorithm used to compute $\mathsf{cond}^\varphi(a)$ as well as on the condition of a for the function $\mathsf{cond}^\varphi : \mathcal{D} \subseteq \mathbb{R}^m \to [0, \infty)$. Disregarding the former, the question is posed, what is the condition number of condition number computation? This "condition of the condition" is called *level-2 condition number*.

In this intermezzo we give an answer for a large class of condition numbers. We say that a condition number cond^φ is *à la Renegar* when there exists a $\Sigma \subset \mathbb{R}^m$, $\Sigma \neq \emptyset$, such that for all $a \in \mathcal{D} \subseteq \mathbb{R}^m$,

$$\mathsf{cond}^\varphi(a) = \frac{\|a\|}{\mathsf{dist}(a, \Sigma)}. \tag{II.1}$$

Here $\| \ \|$ is an arbitrary norm in $\mathbb{R}^m$ and dist is the distance induced by that norm. As we have seen, several condition numbers have this form (or are well approximated by expressions of this form). Furthermore (cf. Sect. 6.1), expression (II.1) is the definition of choice for condition numbers of discrete-valued problems (e.g., $\mathscr{C}(A)$) when the set of ill-posed inputs is clear.

Denote by $\mathsf{cond}^\varphi_{[2]}(a)$ the normwise (for the norm $\| \ \|$) condition number of the function cond^φ. Our main result is the following.

Theorem II.1 *Let φ be any problem and let cond^φ be given by (II.1). Then*

$$\mathsf{cond}^\varphi(a) - 1 \leq \mathsf{cond}^\varphi_{[2]}(a) \leq \mathsf{cond}^\varphi(a) + 1.$$

Proof To simplify notation, let $\varrho(a) = \mathsf{dist}(a, \Sigma)$. For all input data a,

$$
\begin{aligned}
\mathsf{cond}^\varphi_{[2]}(a) &= \lim_{\delta \to 0} \sup_{\|\Delta a\| \leq \delta \|a\|} \frac{|\mathsf{cond}^\varphi(a + \Delta a) - \mathsf{cond}^\varphi(a)| \|a\|}{\mathsf{cond}^\varphi(a) \|\Delta a\|} \\
&= \lim_{\delta \to 0} \sup_{\|\Delta a\| \leq \delta \|a\|} \frac{\left| \frac{\|a + \Delta a\|}{\varrho(a + \Delta a)} - \frac{\|a\|}{\varrho(a)} \right| \|a\|}{\frac{\|a\|}{\varrho(a)} \|\Delta a\|} \\
&= \lim_{\delta \to 0} \sup_{\|\Delta a\| \leq \delta \|a\|} \left| \frac{\|a + \Delta a\| \varrho(a) - \|a\| \varrho(a + \Delta a)}{\varrho(a + \Delta a) \|\Delta a\|} \right|.
\end{aligned}
\tag{II.2}
$$

To prove the upper bound, note that for every perturbation Δa,

$$\left| \|a + \Delta a\| - \|a\| \right| \leq \|\Delta a\|$$

and

$$\left| \varrho(a + \Delta a) - \varrho(a) \right| \leq \|\Delta a\|.$$

Therefore,

$$\left| \|a + \Delta a\| \varrho(a) - \|a\| \varrho(a) \right| \leq \|\Delta a\| \varrho(a)$$

and

$$\left| \|a\|\varrho(a + \Delta a) - \|a\|\varrho(a) \right| \leq \|a\| \|\Delta a\|.$$

It follows that

$$\left| \|a + \Delta a\|\varrho(a) - \|a\|\varrho(a + \Delta a) \right| \leq \|\Delta a\|\varrho(a) + \|a\| \|\Delta a\|$$

and consequently that for sufficiently small Δa,

$$\left| \frac{\|a + \Delta a\|\varrho(a) - \|a\|\varrho(a + \Delta a)}{\varrho(a + \Delta a)\|\Delta a\|} \right| \leq \frac{\|\Delta a\|\varrho(a) + \|a\| \|\Delta a\|}{(\varrho(a) - \|\Delta a\|)\|\Delta a\|} = \frac{\varrho(a) + \|a\|}{\varrho(a) - \|\Delta a\|}.$$

Now use this inequality together with (II.2) to obtain

$$\mathrm{cond}^{\varphi}_{[2]}(a) = \lim_{\delta \to 0} \sup_{\|\Delta a\| \leq \delta \|a\|} \left| \frac{\|a + \Delta a\|\varrho(a) - \|a\|\varrho(a + \Delta a)}{\varrho(a + \Delta a)\|\Delta a\|} \right|$$

$$\leq \lim_{\delta \to 0} \sup_{\|\Delta a\| \leq \delta \|a\|} \frac{\varrho(a) + \|a\|}{\varrho(a) - \|\Delta a\|}$$

$$= \frac{\varrho(a) + \|a\|}{\varrho(a)} = 1 + \frac{\|a\|}{\varrho(a)} = 1 + \mathrm{cond}^{\varphi}(a).$$

This proves the upper bound. We now proceed with the lower bound.

Let Δa^* be such that $\varrho(a) = \|\Delta a^*\|$ and $a + \Delta a^* \in \Sigma$. For any $\varepsilon \in \mathbb{R}$ satisfying $0 < \varepsilon < \|\Delta a^*\|$ let

$$\Delta a^*_{\varepsilon} = \frac{\varepsilon}{\varrho(a)} \Delta a^*.$$

Then, $\|\Delta a^*_{\varepsilon}\| = \varepsilon$ and $\varrho(a + \Delta a^*_{\varepsilon}) = \varrho(a) - \|\Delta a^*_{\varepsilon}\| = \varrho(a) - \varepsilon$ and therefore

$$\left| \frac{\|a + \Delta a^*_{\varepsilon}\|\varrho(a) - \|a\|\varrho(a + \Delta a^*_{\varepsilon})}{\varrho(a + \Delta a^*_{\varepsilon})\|\Delta a^*_{\varepsilon}\|} \right| = \left| \frac{\|a + \Delta a^*_{\varepsilon}\|\varrho(a) - \|a\|(\varrho(a) - \varepsilon)}{(\varrho(a) - \varepsilon)\varepsilon} \right|$$

$$\geq \frac{(\|a\| - \|\Delta a^*_{\varepsilon}\|)\varrho(a) - \|a\|(\varrho(a) - \varepsilon)}{(\varrho(a) - \varepsilon)\varepsilon}$$

$$= \frac{(\|a\| - \varepsilon)\varrho(a) - \|a\|(\varrho(a) - \varepsilon)}{(\varrho(a) - \varepsilon)\varepsilon}$$

$$= \frac{-\varepsilon\varrho(a) + \|a\|\varepsilon}{(\varrho(a) - \varepsilon)\varepsilon} = \frac{\|a\| - \varrho(a)}{\varrho(a) - \varepsilon}.$$

Again, use this inequality together with (II.2) to obtain

$$\mathrm{cond}^{\varphi}_{[2]}(a) = \lim_{\delta \to 0} \sup_{\|\Delta a\| \leq \delta \|a\|} \left| \frac{\|a + \Delta a\|\varrho(a) - \|a\|\varrho(a + \Delta a)}{\varrho(a + \Delta a)\|\Delta a\|} \right|$$

$$\geq \lim_{\delta \to 0} \frac{\|a\| - \varrho(a)}{\varrho(a) - \delta\|a\|} = \frac{\|a\| - \varrho(a)}{\varrho(a)} = \mathrm{cond}^{\varphi}(a) - 1.$$

This proves the lower bound. □

Remark II.2 The bounds in Theorem II.1 are sharp, as shown by the following toy example. Consider φ to be the problem of deciding whether a point $x \in \mathbb{R}$ is greater than a fixed value $\xi > 0$. Then $\Sigma = \{\xi\}$, and for $x \in \mathbb{R}$, $x > 0$, Eq. (II.1) yields

$$\mathrm{cond}^\varphi(x) = \begin{cases} \frac{x}{x-\xi} & \text{if } x > \xi, \\ \frac{x}{\xi-x} & \text{if } x < \xi, \\ \infty & \text{if } x = \xi. \end{cases}$$

Since cond^φ is differentiable at x for $x \neq \xi$, we have (compare Proposition 14.1)

$$\mathrm{cond}_{[2]}(x) = \left| \frac{d}{dx} \mathrm{cond}^\varphi(x) \right| \frac{x}{|\mathrm{cond}^\varphi(x)|} = \begin{cases} \frac{\xi}{x-\xi} & \text{if } x > \xi, \\ \frac{\xi}{\xi-x} & \text{if } x < \xi. \end{cases}$$

Now note that $\frac{x}{x-\xi} = \frac{\xi}{x-\xi} + 1$ and $\frac{x}{\xi-x} = \frac{\xi}{\xi-x} - 1$.

Another simple example shows that a result like Theorem II.1 (actually, even a version with multiplicative constants) may fail to hold for condition numbers not having a characterization of the form (II.1). Consider the problem $\varphi : \mathbb{R} \to \mathbb{R}$ given by $\varphi(x) = x^2 + x + c$, for some $c \in \mathbb{R}$. For $x \in \mathbb{R}$, let $\mathrm{cond}^\varphi(x)$ be its condition number, as defined in (O.1). Since φ is differentiable on $\mathbb{R}$, we have

$$\mathrm{cond}^\varphi(x) = \frac{|x\varphi'(x)|}{|\varphi(x)|}$$

and, assuming $x\varphi'(x), \varphi(x) > 0$,

$$\mathrm{cond}^\varphi_{[2]}(x) = \left| \left(\frac{x\varphi'(x)}{\varphi(x)} \right)' \right| \frac{|\varphi(x)|}{|\varphi'(x)|} = \frac{|x\varphi''(x)\varphi(x) + \varphi'(x)\varphi(x) - x(\varphi'(x))^2|}{|\varphi(x)\varphi'(x)|}.$$

Now take $x = 1$ and $c > -2$ (so that $x, \varphi(x), \varphi'(x) > 0$). Then

$$\mathrm{cond}^\varphi(a) = \frac{3}{2+c}$$

and

$$\mathrm{cond}^\varphi_{[2]}(a) = \frac{|5c+1|}{3(2+c)}.$$

When $c \to \infty$ we have $\mathrm{cond}^\varphi(a) \to 0$ and $\mathrm{cond}^\varphi_{[2]}(a) \to \frac{5}{3}$, while for $c = -\frac{1}{5}$ we have $\mathrm{cond}^\varphi(a) = \frac{5}{3}$ and $\mathrm{cond}^\varphi_{[2]}(a) = 0$.

Part III
Condition in Polynomial Equation Solving (*Allegro con brio*)

Chapter 14
A Geometric Framework for Condition Numbers

Solving equations—linear, algebraic, differential, difference, analytic, Diophantine ...—is arguably the most central problem in mathematics. A case of this problem that can be efficiently tackled is that of linear systems of equations. What could be considered the level of difficulty immediately above that for linear systems, the case of quadratic, or more generally, polynomial equations, is substantially more complicated. Even for polynomials in one variable, classical results of Abel and Galois deprive us of any hope to actually compute their zeros. The best we can do is to approximate them (and a number of algorithms compute these approximations quite efficiently).

For systems of multivariate polynomials we need to add complexity obstructions. The first that meets the eye is the possibly large number of solutions. A system of n quadratic equations in n variables has (generically) 2^n solutions in complex space $\mathbb{C}^n$. But each polynomial in the system has $\frac{1}{2}(n^2 + 3n + 2)$ coefficients, and therefore the whole system is specified with $\Theta(n^3)$ coefficients. If we were to compute approximations for all its zeros, the size of the output would be exponential in the input size!

A focal theme in this third part of the book is that of systems of polynomial equations and algorithms that approximate solutions of these systems. These algorithms have a "numeric" character, and it goes without saying that their analyses strongly rely on appropriate condition numbers. But the nature of these systems and their solutions suggests a view of their condition numbers within a more general framework than the one underlying Sect. O.2. The present chapter introduces this framework and provides some motivating (but also interesting per se) examples.

14.1 Condition Numbers Revisited

Let us reexamine the general definition of condition number given at the very beginning of this book. Our goal in this section is to bring this concept closer to calculus, so that it will become apparent how to extend it to the more general framework of manifolds.

P. Bürgisser, F. Cucker, *Condition*,
Grundlehren der mathematischen Wissenschaften 349,
DOI 10.1007/978-3-642-38896-5_14, © Springer-Verlag Berlin Heidelberg 2013

We begin by assuming that X and Y are finite-dimensional normed real vector spaces and consider a function

$$\varphi : X \supseteq \mathcal{D} \to Y$$

defined on an open subset $\mathcal{D}$ of X. (Everything we say immediately extends to finite-dimensional complex normed vector spaces.) In Sect. O.2 we defined the *relative normwise condition number* $\text{cond}^\varphi(x)$ of φ at a nonzero input $x \in \mathcal{D}$ satisfying $\varphi(x) \neq 0$ by

$$\text{cond}^\varphi(x) = \lim_{\delta \to 0} \sup_{\text{RelError}(x) \leq \delta} \frac{\text{RelError}(\varphi(x))}{\text{RelError}(x)}.$$

More specifically, the supremum is over all $\tilde{x} \in X$ such that

$$\text{RelError}(x) := \frac{\|\tilde{x} - x\|}{\|x\|} \leq \delta,$$

where we used the abbreviation $\text{RelError}(\varphi(x)) := \frac{\|\varphi(\tilde{x}) - \varphi(x)\|}{\|\varphi(x)\|}$. We can as well define an *absolute normwise condition number* by

$$\text{acond}^\varphi(x) := \lim_{\delta \to 0} \sup_{\|\tilde{x} - x\| \leq \delta} \frac{\|\varphi(\tilde{x}) - \varphi(x)\|}{\|\tilde{x} - x\|}.$$

It is clear that $\text{cond}^\varphi(x) = \text{acond}^\varphi(x) \frac{\|x\|}{\|\varphi(x)\|}$.

In the case that φ is differentiable, condition numbers turn out to be a familiar concept from calculus. Indeed, the absolute condition number of φ at x is nothing but the operator norm of the derivative $D\varphi(x) : X \to Y$ of φ at x,

$$\|D\varphi(x)\| := \max_{\|\dot{x}\| = 1} \|D\varphi(x)(\dot{x})\|.$$

Let us explicitly state this important insight.

Proposition 14.1 *If φ is differentiable at x, then*

$$\text{acond}^\varphi(x) = \|D\varphi(x)\|, \quad \text{cond}^\varphi(x) = \text{acond}^\varphi(x) \frac{\|x\|}{\|\varphi(x)\|}.$$

Proof It suffices to prove the assertion about the absolute condition number. The proof is basically a rewriting of the definition of differentiability. We fix x and write $\varphi(x + y) = \varphi(x) + D\varphi(x)y + \|y\|r(y)$ with a function r defined in a neighborhood of 0 such that $\lim_{y \to 0} \|r(y)\| = 0$. For $\varepsilon > 0$ there exists $\delta_\varepsilon > 0$ such that $\sup_{\|y\| \leq \delta_\varepsilon} \|r(y)\| \leq \varepsilon$. For any y satisfying $\|y\| \leq \delta_\varepsilon$ we get

$$\frac{\|D\varphi(x)y\|}{\|y\|} - \varepsilon \leq \frac{\|\varphi(x + y) - \varphi(x)\|}{\|y\|} \leq \frac{\|D\varphi(x)y\|}{\|y\|} + \varepsilon,$$

and hence we obtain for any $0 < \delta \le \delta_\varepsilon$,

$$\sup_{\|y\| \le \delta} \frac{\|D\varphi(x)y\|}{\|y\|} - \varepsilon \le \sup_{\|y\| \le \delta} \frac{\|\varphi(x+y) - \varphi(x)\|}{\|y\|} \le \sup_{\|y\| \le \delta} \frac{\|D\varphi(x)y\|}{\|y\|} + \varepsilon.$$

But $\sup_{\|y\| \le \delta} \frac{\|D\varphi(x)y\|}{\|y\|} = \|D\varphi(x)\|$. Now take the limit for $\delta \to 0$. The claim follows since ε was arbitrary. $\qquad\square$

To illustrate Proposition 14.1, let us briefly review the proof of Theorem 1.5 on the condition number of matrix inversion.

Example 14.2 Consider the map $\varphi \colon \mathrm{GL}_n(\mathbb{R}) \to \mathbb{R}^{n \times n}$ given by $\varphi(A) = A^{-1}$, where, we recall, $\mathrm{GL}_n(\mathbb{R}) = \{A \in \mathbb{R}^{n \times n} \mid \det A \ne 0\}$. The argument at the beginning of the proof of Theorem 1.4 shows that $D\varphi(A)(\dot{A}) = -A^{-1}\dot{A}A^{-1}$. We choose the norm $\| \ \|_{rs}$ on the input space $X = \mathbb{R}^{n \times n}$ and the norm $\| \ \|_{sr}$ on the output space $Y = \mathbb{R}^{n \times n}$, for $r, s \ge 1$. Then we have by (1.5),

$$\left\| A^{-1}\dot{A}A^{-1} \right\|_{sr} \le \left\| A^{-1} \right\|_{sr} \|\dot{A}\|_{rs} \left\| A^{-1} \right\|_{sr} = \left\| A^{-1} \right\|_{sr}^2$$

for $\dot{A}$ with $\|\dot{A}\|_{rs} = 1$. The argument at the end of the proof of Theorem 1.5, which we shall not repeat here, shows that equality holds for some $\dot{A}$. Therefore,

$$\mathsf{acond}^\varphi(A) = \left\| D\varphi(A) \right\|_{rs,sr} = \left\| A^{-1} \right\|_{sr}^2.$$

Finally,

$$\mathsf{cond}^\varphi(a) = \mathsf{acond}^\varphi(A)\|A\|_{rs} \left\| A^{-1} \right\|_{sr}^{-1} = \|A\|_{rs} \left\| A^{-1} \right\|_{sr} = \kappa_{rs}(A).$$

14.1.1 Complex Zeros of Univariate Polynomials

In many situations of interest, the map φ is only implicitly given. For example, consider the problem of finding a complex zero of a univariate polynomial $f = \sum_{j=0}^{d} a_j Z^j$, $a_j \in \mathbb{C}$. The zeros ζ are given implicitly by the nonlinear equation $f(\zeta) = 0$, and in general, there are d zeros by the fundamental theorem of algebra.

Consider the input space $\mathcal{P}_d := \{\sum_{j=0}^{d} a_j Z^j \mid a_j \in \mathbb{C}\} \simeq \mathbb{C}^{d+1}$, let $f_0 \in \mathcal{P}_d$, and suppose that $\zeta_0 \in \mathbb{C}$ is a simple zero of f_0, that is, $f_0(\zeta_0) = 0$ and $f_0'(\zeta_0) \ne 0$.

Consider the map $F \colon \mathcal{P}_d \times \mathbb{C} \to \mathbb{C}$, $F(f, \zeta) := f(\zeta)$ and note that $\frac{\partial F}{\partial \zeta}(f_0, \zeta_0) = f_0'(\zeta_0) \ne 0$. The implicit function theorem (Theorem A.1) applied to F implies that for all f sufficiently close to f_0, there is a unique zero ζ of f close to ζ_0, and moreover, ζ is a differentiable function of f. More specifically, there are open neighborhoods $U \subseteq \mathcal{P}_d$ of f_0 and $V \subseteq \mathbb{C}$ of ζ_0, and there is a differentiable function $\varphi \colon U \to V$ such that for all $f \in U$, $\varphi(f)$ is the only zero of f in V.

The derivative $D\varphi(\zeta_0)\colon \mathcal{P}_d \to \mathbb{C}$ at ζ_0 is a linear map that can be calculated by the following general method. Consider a smooth curve $\mathbb{R} \to U$, $t \mapsto f(t) = \sum_{j=0}^{d} a_j(t)Z^j$, such that $f(0) = f_0$ and write

$$\dot{f} = \frac{df}{dt}(0) = \sum_{j=0}^{d} \frac{da_j}{dt}(0)Z^j =: \sum_{j=0}^{d} \dot{a}_j Z^j.$$

Let $\mathbb{R} \to \mathbb{C}$, $t \mapsto \zeta(t) := \varphi(f(t))$ be the corresponding curve of solutions and write $\dot{\zeta} = \frac{d\zeta}{dt}(0)$. Then we have $D\varphi(f_0)(\dot{f}) = \dot{\zeta}$ by the chain rule. Differentiating the equality

$$0 = f(t)\big(\zeta(t)\big) = \sum_j a_j(t)\zeta(t)^j$$

with respect to t at zero yields

$$0 = \sum_j \dot{a}_j \zeta_0^j + \sum_j a_j j \zeta_0^{j-1} \dot{\zeta} = \dot{f}(\zeta_0) + f_0'(\zeta_0)\dot{\zeta}. \qquad (14.1)$$

Since $f_0'(\zeta_0) \neq 0$, we get $D\varphi(f_0)(\dot{f}) = \dot{\zeta} = -f_0'(\zeta_0)^{-1} \dot{f}(\zeta_0)$.

To simplify notation we write from now on $f = f_0$ and $\zeta = \zeta_0$. Once we fix a norm $\|\ \|$ on $\mathcal{P}_d$ (and take the absolute value as the norm on $\mathbb{C}$), the condition number of φ at f is defined and hence given by

$$\mathrm{cond}^{\varphi}(f) = \frac{\|f\|}{|\zeta|} \|D\varphi(f)\| = \frac{\|f\|}{|\zeta|\,|f'(\zeta)|} \max_{\|\dot{f}\|=1} \big|\dot{f}(\zeta)\big|.$$

The standard choice of a norm on $\mathcal{P}_d$ is

$$\|\dot{f}\|_{\mathrm{st}} := \left(\sum_{j=0}^{d} |\dot{a}_j|^2 \right)^{1/2},$$

which comes from the isomorphism $\mathcal{P}_d \simeq \mathbb{C}^{d+1}$ and the standard Hermitian inner product $\langle\ ,\ \rangle$ on $\mathbb{C}^{d+1}$. We shall denote the corresponding condition number by $\mathrm{cond}_{\mathrm{st}}^{\varphi}(f)$. Since $\dot{f}(\zeta) = \langle \dot{a}, (\zeta^j) \rangle$, the Cauchy–Schwarz inequality yields

$$\mathrm{cond}_{\mathrm{st}}^{\varphi}(f) = \frac{\|f\|}{|\zeta|}\frac{1}{|f'(\zeta)|} \left(\sum_{j=0}^{d} |\zeta|^{2j} \right)^{1/2}.$$

Another choice of norm on $\mathcal{P}_d$ is given by

$$\|\dot{f}\|_W := \left(\sum_{j=0}^{d} \binom{d}{j}^{-1} |\dot{a}_j|^2 \right)^{1/2},$$

and we shall denote the corresponding condition number by $\mathrm{cond}^\varphi_W(f)$. In Sect. 16.1 we will learn that $\|\dot{f}\|_W$ is a natural choice when our aim is a unitarily invariant theory. Writing $\dot{a} = (\sqrt{\binom{d}{j}}\, \dot{b}_j)_j$ with $\sum_j |\dot{b}_j|^2 = 1$, we obtain with the Cauchy-Schwarz inequality

$$|\dot{f}(\zeta)| = \left|\sum_{j=0}^{d} \dot{b}_j \sqrt{\binom{d}{j}}\, \zeta^j\right| \le \left(\sum_{j=0}^{d} \binom{d}{j} |\zeta|^{2j}\right)^{1/2} = (1 + |\zeta|^2)^{d/2}.$$

Clearly, the right-hand side is attained at some $\dot{a}$, so that

$$\mathrm{cond}^\varphi_W(f) = \frac{\|f\|_W}{|\zeta|}\, \frac{1}{|f'(\zeta)|}\, (1 + |\zeta|^2)^{d/2}.$$

We can specialize the content of the previous example to particular polynomials, for instance, to cyclotomic polynomials.

Example 14.3 Let $f = Z^d - 1$ and let ζ be a dth root of unity, i.e., $\zeta^d = 1$. Then

$$\mathrm{cond}^\varphi_{st}(f) = \frac{\sqrt{2(d+1)}}{d}, \qquad \mathrm{cond}^\varphi_W(f) = \frac{2^{\frac{d+1}{2}}}{d}.$$

Note the exponential difference in these results: while $\mathrm{cond}^\varphi_{st}(f)$ goes to zero as $d \to \infty$, $\mathrm{cond}^\varphi_W(f)$ grows exponentially with d. So the choice of the norm on $\mathcal{P}_d$ may make a huge difference in the corresponding condition.

14.1.2 A Geometric Framework

The previous discussion is just a special case of a general geometric framework. Let X and Y be finite-dimensional real vector spaces. Suppose $F: X \times Y \to \mathbb{R}^n$ is a smooth (C^∞) map (which can be defined on an open subset only) and consider its zero set

$$V := \big\{(x, y) \in X \times Y \mid F(x, y) = 0\big\}.$$

We shall interpret X as the space of inputs, Y as the space of outputs, and $(x, y) \in V$ as meaning that y is a "solution" to input x. We shall suppose $n = \dim Y$ and that the derivative $\frac{\partial F}{\partial y}(x, y)$ has full rank n for all $(x, y) \in V$. Then the implicit function theorem implies that V is a submanifold of $X \times Y$ of dimension $\dim X$. We shall call V the *solution manifold*. Consider the subset

$$\Sigma' := \left\{(x, y) \in V \mid \mathrm{rank}\, \frac{\partial F}{\partial y}(x, y) < n\right\}. \tag{14.2}$$

For reasons to be clear soon, we call the elements of Σ' *ill-posed* and the elements of $V \setminus \Sigma'$ *well-posed*. Let (x_0, y_0) be well-posed. Then the implicit function theorem tells us that there exist open neighborhoods $U' \subseteq X \times Y$ of (x_0, y_0) and $U \subseteq X$ of x_0 such that the projection $U' \cap V \to U$, $(x, y) \mapsto x$ is bijective and has a smooth inverse $U \to U' \cap V$, $x \mapsto (x, G(x))$ given by some function $G : U \to Y$. Thus locally around (x_0, y_0), V is the graph of G. Note that $y_0 = G(x_0)$. We call G the *solution map*, since $y = G(x)$ is the unique solution for input $x \in U$ such that $(x, y) \in U'$. Moreover, we call the derivative $DG(x) : X \to Y$ the *condition map* of the problem at input x. (Note that G, and hence DG, depends on the initial choice of (x_0, y_0).) After choosing bases of X and Y, the condition map determines the *condition matrix* $DG(x_0)$.

We may, in addition, fix norms on the space X of inputs and the space Y of outputs. Then, according to Sect. 14.1, the solution map $G : U \to Y$ has well-defined absolute and relative normwise condition numbers. By Proposition 14.1 they take the following form:

$$\mathrm{acond}^G(x) = \| DG(x) \|, \qquad \mathrm{cond}^G(x) = \frac{\|x\|}{\|G(x)\|} \| DG(x) \|.$$

For an ill-posed $(x, y) \in \Sigma'$ we define the (absolute) condition to be infinity.

Even though G is only implicitly given, it is easy to find an explicit formula for $DG(x)$. Indeed, differentiating the equation $F(x, G(x)) = 0$ yields

$$\frac{\partial F}{\partial x}(x, y) + \frac{\partial F}{\partial y}(x, y) \, DG(x) = 0. \tag{14.3}$$

Hence

$$DG(x) = -\left(\frac{\partial F}{\partial y}(x, y) \right)^{-1} \frac{\partial F}{\partial x}(x, y). \tag{14.4}$$

Lemma 14.4 *The tangent space $T_{(x,y)}V$ of V at (x, y) is given by*

$$T_{(x,y)}V = \left\{ (\dot{x}, \dot{y}) \in X \times Y \; \middle| \; \frac{\partial F}{\partial x}(x, y) \dot{x} + \frac{\partial F}{\partial y}(x, y) \dot{y} = 0 \right\}.$$

Moreover, denoting by $\pi : V \to X$, $(x, y) \mapsto x$ the projection to the first component,

$$\Sigma' = \left\{ (x, y) \in V \mid \mathrm{rank} \, D\pi(x, y) < \dim X \right\}. \tag{14.5}$$

Proof The linear space $T_{(x,y)}V$ equals the kernel of $DF(x, y)$, which is determined by

$$DF(x, y)(\dot{x}, \dot{y}) = \frac{\partial F}{\partial x}(x, y) \dot{x} + \frac{\partial F}{\partial y}(x, y) \dot{y} = 0.$$

Moreover, $D\pi(x, y)$ equals the projection $T_{(x,y)}V \to X$, $(\dot{x}, \dot{y}) \mapsto \dot{x}$. This projection has a nontrivial kernel iff the matrix $\partial F/\partial y(x, y)$ is singular, which by definition means that $(x, y) \in \Sigma'$. □

14.1.3 Linear Equation Solving

We take up the example of linear equation solving, whose condition was already discussed in Sect. 1.2. Consider $X = \mathbb{R}^{n \times n} \times \mathbb{R}^n$, $Y = \mathbb{R}^n$, and the map $F : X \times Y \to \mathbb{R}^n$, $(A, b, y) \mapsto Ay - b$. We make X and Y normed spaces by considering, for fixed $r, s \geq 1$, the norm

$$\|(A, b)\| := \max\{\|A\|_{rs}, \|b\|_s\} \tag{14.6}$$

on X and the norm $\|y\|_r$ on Y. If the matrix A is invertible, then the input (A, b) has the unique solution $G(A, b) = A^{-1}b = y$, with G denoting the solution map.

We want to compute $\mathsf{acond}^G(A, b) = \|DG(A, b)\|$, and from it, $\mathsf{cond}^G(A, b)$. In principle, we could expect the same bounds we obtained in Theorem 1.5 (which would make $\kappa_{rs}(A) \leq \mathsf{cond}^G(A, b) \leq 2\kappa_{rs}(A)$). A more careful look at the hypotheses in Theorem 1.5 shows that the relative error in (A, b) considered there is the maximum of the normwise relative errors A and b. This introduces a minor, but not negligible, "componentwise viewpoint" that does not fit into our present geometric framework. The latter is entirely normwise.

The derivative of F at (A, b, y) is given by

$$DF(A, b, y)(\dot{A}, \dot{b}, \dot{y}) = A\dot{y} + \dot{A}y - \dot{b},$$

which clearly has full rank for all (A, b, y). So

$$V = F^{-1}(0) = \{(A, b, y) \in X \times Y \mid Ay = b\}$$

is a smooth submanifold of $X \times Y$ of dimension $\dim X = n^2 + n$. Moreover,

$$\frac{\partial F}{\partial(A, b)}(A, b, y)(\dot{A}, \dot{b}) = DF(A, b, y)(\dot{A}, \dot{b}, 0) = \dot{A}y - \dot{b},$$

$$\frac{\partial F}{\partial y}(A, b, y)(\dot{y}) = DF(A, b, y)(0, 0, \dot{y}) = A\dot{y}.$$

By Eq. (14.4), the condition map equals

$$DG(A, b)(\dot{A}, \dot{b}) = -A^{-1}(\dot{A}y - \dot{b}).$$

Let $r, s \geq 1$ and consider the norm $\|(A, b)\|$ on X defined in (14.6) and the norm $\|y\|_r$ on Y. We have

$$\left\|A^{-1}(\dot{A}y - \dot{b})\right\|_r \leq \left\|A^{-1}\right\|_{sr}(\|\dot{A}\|_{rs}\|y\|_r + \|\dot{b}\|_s).$$

This implies for the corresponding operator norm

$$\|DG(A, b)\| = \max_{\substack{\|\dot{A}\|_{rs} \leq 1 \\ \|\dot{b}\|_s \leq 1}} \left\|A^{-1}(\dot{A}y - \dot{b})\right\|_r \leq \left\|A^{-1}\right\|_{sr}(\|y\|_r + 1),$$

and it is straightforward to check that equality holds. Therefore,

$$\mathsf{acond}^G(A, b) = \|A^{-1}\|_{sr}(\|y\|_r + 1).$$

From this we obtain, recalling $\kappa_{rs}(A) = \|A\|_{rs}\|A^{-1}\|_{sr}$,

$$\mathsf{cond}^G(A, b) = \mathsf{acond}^G(A, b)\,\frac{\|(A, b)\|}{\|y\|_r}$$

$$= \|A^{-1}\|_{sr}(\|y\|_r + 1)\,\frac{1}{\|y\|_r}\,\max\{\|A\|_{rs}, \|b\|_s\}$$

$$= \max\{\kappa_{rs}(A), \|A^{-1}\|_{sr}\|b\|_s\}\cdot(1 + \|y\|_r^{-1}).$$

Hence, $\mathsf{cond}^G(A, b) \geq \kappa_{rs}(A)$. In addition, using $\|b\|_s \leq \|A\|_{rs}\|y\|_r$, it follows that

$$\mathsf{cond}^G(A, b) \leq \kappa_{rs}(A)\cdot\max\{1, \|y\|_r\}\cdot(1 + \|y\|_r^{-1}).$$

Putting these bounds together, we obtain

$$\kappa_{rs}(A) \leq \mathsf{cond}^G(A, b) \leq \kappa_{rs}(A)\big(1 + \max\{\|y\|_r\|y\|_r^{-1}\}\big).$$

As already discussed, this result is different from the one in Theorem 1.5. As an exercise, the reader may check that if we take the norm

$$\|(A, b)\| := \big(\|A\|_{rs}^2 + \|b\|_s^2\big)^{1/2},$$

then we obtain yet another result, namely

$$\mathsf{acond}^G(A, b) = \|A^{-1}\|_{sr}\sqrt{1 + \|y\|_r^2}.$$

For the analysis of certain problems, a further generalization of the geometric framework described in this section is necessary. In the following it is convenient to use the notation $W_* := W \setminus \{0\}$ for any vector space W. For instance, let us consider the problem of computing the eigenvalues and eigenvectors of a given matrix $A \in \mathbb{C}^{n \times n}$. A first attempt to formalize this problem would be to consider the set of solutions

$$V := \big\{(A, v, \lambda) \in \mathbb{C}^{n \times n} \times (\mathbb{C}^n)_* \times \mathbb{C} \mid Av = \lambda v\big\}.$$

However, even if A has only simple eigenvalues, its eigenvectors v are determined only up to scaling. It is therefore natural to replace $(\mathbb{C}^n)_*$ by the *complex projective space* $\mathbb{P}^{n-1}$, which is defined as the set of one-dimensional linear subspaces of $\mathbb{C}^n$. This geometric object will also be of paramount importance for our analysis of polynomial equation solving.

The space $\mathbb{P}^{n-1}$ is a Riemannian manifold, and we will see in Sect. 14.3 that the geometric framework discussed so far naturally extends to this more general setting. We shall also continue there the discussion of the eigenvalue and eigenvector problems.

Before doing so, we give a short introduction to the geometry of $\mathbb{P}^{n-1}$.

14.2 Complex Projective Space

Let V be a finite-dimensional complex vector space and recall $V_* := V \setminus \{0\}$. For $v \in V_*$ we write $[v] := \mathbb{C}v$ for the one-dimensional linear subspace spanned by v.

Definition 14.5 The *complex projective space* $\mathbb{P}(V)$ is defined as

$$\mathbb{P}(V) := \{[v] \mid v \in V_*\}.$$

One writes $\mathbb{P}^{n-1} := \mathbb{P}(\mathbb{C}^n)$.

The space $\mathbb{P}(V)$ comes with a topology. Consider the canonical map $\pi \colon V_* \to \mathbb{P}(V)$, $v \mapsto [v]$. We say that $U \subseteq \mathbb{P}(V)$ is open if $\pi^{-1}(U)$ is open in V with respect to the standard topology (induced by the Euclidean topology via a linear isomorphism $V \simeq \mathbb{C}^n$).

We argue now that $\mathbb{P}(V)$ is compact: Let $\langle \ , \ \rangle$ be a Hermitian inner product on V. Then $v \in V$ has the norm $\|v\| := \sqrt{\langle v, v \rangle}$, and we can define the sphere

$$\mathbb{S}(V) := \{v \in V \mid \|v\| = 1\},$$

which is compact. Consider the restriction $\pi_\mathbb{S} \colon \mathbb{S}(V) \to \mathbb{P}(V)$, $w \mapsto [w]$ of π. This map is surjective, and its fibers are given by $\pi_\mathbb{S}^{-1}([v]) = \{\lambda v \mid |\lambda| = 1\}$. Since $\pi_\mathbb{S}$ is continuous, it follows that $\mathbb{P}(V)$ is compact.

In the next subsections we shall explain that $\mathbb{P}(V)$ carries the structure of a Riemannian manifold. For a brief introduction to these concepts, see Appendix A.2.

14.2.1 Projective Space as a Complex Manifold

We show here that $\mathbb{P}(V)$ is a complex manifold by exhibiting an atlas for it.

Fix a Hermitian inner product $\langle \ , \ \rangle$ on $V \simeq \mathbb{C}^n$. For $v \in V_*$, we consider the orthogonal complement of $\mathbb{C}v$,

$$T_v := \{z \in V \mid \langle z, v \rangle = 0\}.$$

Clearly, T_v is a linear subspace of V of complex codimension one, and we have $V = \mathbb{C}v \oplus T_v$. Consider the open subsets

$$\mathbb{A}_v := \{L \in \mathbb{P}(V) \mid L \not\subseteq T_v\} = \{[v + w] \mid w \in T_v\}$$

of $\mathbb{P}(V)$. It is easy to check that $\mathbb{P}(V) = \mathbb{A}_{v_1} \cup \cdots \cup \mathbb{A}_{v_n}$ when $v_1, \ldots, v_n$ is a basis of V. We can parameterize $\mathbb{A}_v$ by the bijective map

$$\Psi_v \colon T_v \to \mathbb{A}_v, \qquad w \mapsto [v + w]. \tag{14.7}$$

Note that $\Psi_v(0) = [v]$.

The next lemma shows that Ψ_v^{-1} is a chart and the collection $\{\Psi_v^{-1} \mid v \in V_*\}$ is a holomorphic atlas for $\mathbb{P}(V)$.

Lemma 14.6

(a) *We have* $\Psi_v^{-1}([x]) = \varphi_v(x)$, *where*

$$\varphi_v : V \setminus T_v \to T_v, \quad x \mapsto \frac{\|v\|^2}{\langle x, v \rangle} x - v. \tag{14.8}$$

(b) *The derivative of* φ_v *at* $x \in V \setminus T_v$ *is given by* $\dot{y} = D\varphi_v(x)(\dot{x})$, *where*

$$\dot{y} = \frac{\|v\|^2}{\langle x, v \rangle^2} \left(\langle x, v \rangle \dot{x} - \langle \dot{x}, v \rangle x \right). \tag{14.9}$$

(c) Ψ_v *is a homeomorphism.*
(d) *The change of coordinates map*

$$T_u \supseteq \Psi_u^{-1}(\mathbb{A}_u \cap \mathbb{A}_v) \to \Psi_v^{-1}(\mathbb{A}_u \cap \mathbb{A}_v) \subseteq T_v, \quad w \mapsto \Psi_v^{-1}\Psi_u(w)$$

   is a complex differentiable map.

Proof (a, b) These are verified by a direct computation.
   (c) The map Ψ_v is the composition of $T_v \to V \setminus T_v$, $w \mapsto v + w$ with the canonical map $V \setminus T_v \to \mathbb{A}_v$, $x \mapsto [x]$, and hence Ψ_v is continuous. By part (a), the inverse Ψ_v^{-1} factors over the continuous map $\varphi_v : V \setminus T_v \to T_v$ and hence Ψ_v^{-1} is continuous. We have thus shown that Ψ_v is a homeomorphism.
   (d) By part (a), the change of coordinates map is given by $\Psi_v^{-1}\Psi_u(w) = \varphi_v(u + w)$. It is straightforward to check that this is a complex differentiable map. $\square$

   In the following we view $\mathbb{P}(V)$ as a complex manifold with respect to the above atlas. We therefore have a well-defined abstract notion of the *tangent space* $T_{[v]}\mathbb{P}(V)$ at $[v]$; see Appendix A.2 (and Sect. A.3.2). We make this now more concrete.
   By Lemma 14.6(c), the map Ψ_v is a complex diffeomorphism (i.e., biholomorphism). Since the tangent space of the vector space T_v at 0 can be identified with T_v, the derivative $D\Psi_v(0)$ of Ψ_v at 0 provides a $\mathbb{C}$-linear isomorphism

$$D\Psi_v(0) : T_v \to T_{[v]}\mathbb{P}(V). \tag{14.10}$$

In the following, we shall identify $T_{[v]}\mathbb{P}(V)$ with T_v via this map. A little care has to be taken here, because there is a choice of the representative v of $[v]$. Suppose that a vector in $T_{[v]}\mathbb{P}(V)$ is represented by $w \in T_v$ via $D\Psi_v(0)$. Then this same vector is represented by $\lambda w \in T_{\lambda v} = T_v$ when the representative v is replaced by λv. This fact is a consequence of the following commutative diagram

$$
\begin{array}{ccc}
T_{\lambda v} & \xrightarrow{\Psi_{\lambda v}} & \mathbb{A}_{\lambda v} \\
\lambda \cdot \uparrow & & \| \\
T_v & \xrightarrow{\Psi_v} & \mathbb{A}_v
\end{array}
$$

where the vertical arrow $\lambda \cdot$ stands for the multiplication by λ.

Remark 14.7 A more invariant, yet concrete, description of $T_{[v]}\mathbb{P}(V)$ is obtained by replacing T_v by the isomorphic vector space $\mathscr{L}(\mathbb{C}v; T_v)$ of linear maps $\mathbb{C}v \to T_v$. The isomorphism is $\mathscr{L}(\mathbb{C}v; T_v) \to T_v,\ \alpha \mapsto \alpha(e)$.

In the following, we will mostly forget the complex structure and view $\mathbb{P}(V)$ as a smooth (C^∞) manifold. Here is a useful result for concrete computations.

Lemma 14.8 *Let $\gamma\colon \mathbb{R} \to V_*$ be a smooth map and let $\gamma_\mathbb{P}\colon [0,1] \to \mathbb{P}(V)$ be defined by $\gamma_\mathbb{P}(t) = [\gamma(t)]$. Then, writing $\dot\gamma(t) := d\gamma(t)/dt$, we have*

$$\frac{d\gamma_\mathbb{P}(t)}{dt} = \frac{p_t(\dot\gamma(t))}{\|\gamma(t)\|},$$

where $p_t\colon V \to T_{\gamma(t)}$ denotes the orthogonal projection.

Proof Fix $t_0 \in [0,1]$ and put $v := \gamma(t_0)$. Since the statement is local, we may assume that $\gamma_\mathbb{P}\colon \mathbb{R} \to \mathbb{A}_v$. By scale invariance we may assume that $\|v\| = 1$. Further, by choosing a suitable orthonormal basis of V, we may assume that $V = \mathbb{C}^n$ and $v = e_n$. We express now the curve γ in the coordinates provided by the chart $\Psi_{e_n}\colon T_{e_n} \to \mathbb{A}_{e_n}$; cf. (14.7). The composition $g := \Psi_{e_n}^{-1} \circ \gamma\colon \mathbb{R} \to T_{e_n}$ is given by (cf. (14.8))

$$g(t) = \frac{1}{\gamma_n(t)}\big(\gamma_1(t), \ldots, \gamma_{n-1}(t), 0\big).$$

The derivative $\dot g(t_0) \in T_{e_n}$ represents $d\gamma_\mathbb{P}(t_0)/dt$ in the chosen coordinates. Taking the derivative at t_0 and using $\gamma(t_0) = e_n$ implies

$$\dot g(t_0) = \big(\dot\gamma_1(t_0), \ldots, \dot\gamma_{n-1}(t_0), 0\big).$$

This completes the proof. $\qquad\square$

14.2.2 Distances in Projective Space

We again fix a Hermitian inner product $\langle\,,\,\rangle$ on V. The real and imaginary parts of a complex number $z \in \mathbb{C}$ shall be denoted by $\Re z$ and $\Im z$, respectively. Setting $\langle v, w\rangle_\mathbb{R} := \Re\langle v, w\rangle$ defines an associated inner product $\langle\,,\,\rangle_\mathbb{R}$ on V. This inner product defines the same norm as $\langle\,,\,\rangle$, since $\langle v, v\rangle_\mathbb{R} = \langle v, v\rangle$ for $v \in V$. Moreover, $\langle iv, v\rangle_\mathbb{R} = 0$ for all $v \in V$.

The sphere $\mathbb{S}(V) = \{v \in V \mid \|v\| = 1\}$ is a submanifold of V, and its tangent space at $v \in \mathbb{S}(V)$ is given by the real subspace

$$T_v\mathbb{S}(V) = \big\{a \in V \mid \langle a, v\rangle_\mathbb{R} = 0\big\}. \tag{14.11}$$

Recall the projection $\pi_\mathbb{S}\colon \mathbb{S}(V) \to \mathbb{P}(S),\ w \mapsto [w]$.

Lemma 14.9 *For all $v \in \mathbb{S}(V)$ we have the orthogonal decomposition $T_v\mathbb{S}(V) = T_v \oplus \mathbb{R}iv$, which is orthogonal with respect to $\langle \, , \, \rangle_{\mathbb{R}}$. Moreover, the derivative $D\pi_{\mathbb{S}}(e)\colon T_v\mathbb{S}(V) \to T_v$ is the orthogonal projection onto T_v.*

Proof It is clear that $T_v \subseteq T_v\mathbb{S}(V)$ and moreover $iv \in T_v\mathbb{S}(V)$, since $\langle iv, v \rangle_{\mathbb{R}} = \Re i \langle v, v \rangle = 0$. The first statement follows by comparing the dimensions. For the second statement take a smooth curve $\gamma\colon \mathbb{R} \to \mathbb{S}(V)$ and consider $\gamma_{\mathbb{P}} := \pi_{\mathbb{S}} \circ \gamma$. Then $\dot\gamma_{\mathbb{P}}(t) = D\pi_{\mathbb{S}}(\gamma(t))(\dot\gamma(t))$. Now use Lemma 14.8. $\qquad\square$

A result similar to Lemma 14.8 holds for spheres.

Lemma 14.10 *Let $\gamma\colon \mathbb{R} \to V_*$ be a smooth map and let $\gamma_{\mathbb{S}}\colon [0, 1] \to \mathbb{S}(V)$ be defined by $\gamma_{\mathbb{S}}(t) = \frac{\gamma(t)}{\|\gamma(t)\|}$. Then we have*

$$\frac{d\gamma_{\mathbb{S}}(t)}{dt} = \frac{P_t(\dot\gamma(t))}{\|\gamma(t)\|},$$

where $P_t\colon V \to T_{\gamma(t)}\mathbb{S}(V)$ denotes the orthogonal projection.

Proof A straightforward calculation shows that

$$\dot\gamma_{\mathbb{S}} = \frac{\dot\gamma}{\|\gamma\|} - \frac{\langle \gamma, \dot\gamma \rangle}{\|\gamma\|^3}\gamma = \frac{1}{\|\gamma\|}P(\dot\gamma),$$

where

$$P(\dot\gamma) = \dot\gamma - \frac{\langle \gamma, \dot\gamma \rangle}{\|\gamma\|^2}\gamma$$

equals the orthogonal projection of $\dot\gamma$ onto $T_{\gamma(t)}\mathbb{S}(V)$. $\qquad\square$

The inner product $\langle \, , \, \rangle_{\mathbb{R}}$ on V induces an inner product on the subspace $T_v\mathbb{S}(V)$ of V, which turns the sphere $\mathbb{S}(V)$ into a Riemannian manifold.

As in any Riemannian manifold, we have a well-defined notion of Riemannian distance $d_{\mathbb{S}}(v, w)$ between points $v, w \in \mathbb{S}(V)$; cf. Appendix A.2. It is a well-known fact that $d_{\mathbb{S}}(v, w)$ equals the *angle* between v and w, that is,

$$d_{\mathbb{S}}(v, w) = \arccos\langle v, w \rangle_{\mathbb{R}}. \qquad (14.12)$$

Similarly, we define an inner product on the tangent space $T_{[v]}\mathbb{P}(V)$ of the projective space $\mathbb{P}(V)$ by setting, for $a, b \in T_v$,

$$\langle a, b \rangle_v := \frac{\langle a, b \rangle_{\mathbb{R}}}{\|v\|^2}. \qquad (14.13)$$

The reader should note that this is a well-defined notion, independent of the choice of the representative v of $[v]$. Clearly, if $v \in \mathbb{S}(V)$, this coincides with the inner product defined on $T_v\mathbb{S}(V)$.

The next lemma is a precise formulation of the fact that the inner product $\langle\,,\,\rangle_v$ "varies smoothly" with the base point v. It implies that $\mathbb{P}(V)$ is also a *Riemannian manifold*. The corresponding metric is called *Fubini–Study metric*.

Lemma 14.11 *Fix $v \in V_*$ and recall $\Psi_v \colon T_v \to \mathbb{A}_v$, $p, w \mapsto [v + w]$. Consider its derivative $D\Psi_v(w) \colon T_v \to T_{[v+w]}\mathbb{P}(V)$ at $w \in T_v$. Then, for fixed $a, b \in T_v$,*

$$T_v \to \mathbb{R}, \quad w \mapsto \big\langle D\Psi_v(w)(a), D\Psi_v(w)(b)\big\rangle_{[v+w]},$$

is a smooth map.

Proof Consider the derivative of Ψ_v at $w \in T_v$,

$$D\Psi_v(w) \colon T_v \to T_{[x]}\mathbb{A}_v \simeq T_x, \quad \dot{y} \mapsto \dot{x},$$

where we write $x = v + w$ and recall that $[x] = \Psi_v(w)$. Lemma 14.6 implies that $\dot{x}$ and $\dot{y}$ are related according to Eq. (14.9).

Assume now without loss of generality that $\|v\| = 1$ (scaling). Let $e_1, \ldots, e_n = v$ be an orthonormal basis of V. Without loss of generality, we may assume that $V = \mathbb{C}^n$ and that e_i is the standard basis. Then (14.9) becomes

$$\dot{y}_k = \frac{1}{x_n^2}(x_n \dot{x}_k - \dot{x}_n x_k).$$

Fix $i < n$ and let $\dot{y} = e_i \in T_{e_n}$ be the ith standard basis vector. Solving the above equation for $\dot{x}$ under the constraint $\dot{x} \in T_x$, that is, $\langle \dot{x}, x \rangle = 0$, yields

$$\dot{x} = e_i - \frac{\bar{x}_i}{\|x\|^2}x,$$

as is easily verified. Now taking $a = e_i$ and $b = e_j$ in T_{e_n}, for fixed $i, j < n$, we obtain

$$\big\langle D\Psi_{e_n}(x)(a), D\Psi_{e_n}(x)(b)\big\rangle_{\Psi_{e_n}(x)} = \frac{1}{\|x\|^2}\Big\langle e_i - \frac{\bar{x}_i}{\|x\|^2}x, e_j - \frac{\bar{x}_j}{\|x\|^2}x\Big\rangle_{\mathbb{R}}$$

$$= \frac{1}{\|x\|^2}\Big(\delta_{ij} - \frac{1}{\|x\|^2}\Re(x_i \bar{x}_j)\Big).$$

Clearly, this depends smoothly on $x_1, \ldots, x_{n-1}$, which completes the proof. $\square$

We denote by $d_{\mathbb{P}}$ the *Riemannian distance* of $\mathbb{P}(V)$, cf. Appendix A.2. It turns out that $d_{\mathbb{P}}([u], [v])$ equals the angle between the complex lines $\mathbb{C}u$ and $\mathbb{C}v$. More specifically, we have the following result.

Proposition 14.12 *We have for $v, w \in \mathbb{S}(V)$;*

$$d_{\mathbb{P}}([v], [w]) = \min_{\lambda \in \mathbb{S}(\mathbb{C})} d_{\mathbb{S}}(v, \lambda w) = \arccos\big|\langle v, w \rangle\big|.$$

Proof The right-hand equality follows from the definition (14.12) of $d_{\mathbb{S}}$ and the fact $\max_{\lambda \in \mathbb{S}(\mathbb{C})} \Re(\lambda z) = |z|$, for $z \in \mathbb{C}$.

For the left-hand equality take a smooth curve $\gamma \colon [0, 1] \to \mathbb{S}(V)$ connecting v with λw. Then $\gamma_{\mathbb{P}} \colon [0, 1] \to \mathbb{P}(V)$ defined as $\gamma_{\mathbb{P}}(t) = [\gamma(t)]$ connects $[v]$ with $[w]$. Lemma 14.8 implies that $\|\dot\gamma_{\mathbb{P}}(t)\| \leq \|\dot\gamma(t)\|$. By the definition (A.3) of the length of curves we obtain $L(\gamma_{\mathbb{P}}) \leq L(\gamma)$. This shows that

$$d_{\mathbb{P}}([v], [w]) \leq \min_{\lambda \in \mathbb{S}(\mathbb{C})} d_{\mathbb{S}}(v, \lambda w).$$

In order to prove the reverse inequality, take a smooth curve $\gamma_{\mathbb{P}} \colon [0, 1] \to \mathbb{P}(V)$. Using charts as in the proof of Lemma 14.8, it is easy to see that $\gamma_{\mathbb{P}}$ can be lifted to $\mathbb{S}(V)$, that is, there exists a smooth curve $\gamma \colon [0, 1] \to \mathbb{S}(V)$ such that $\gamma_{\mathbb{P}} = \pi_{\mathbb{S}} \circ \gamma$. Hence $\dot\gamma_{\mathbb{P}}(t) = D\pi_{\mathbb{S}}(\gamma(t))(\dot\gamma(t))$. If we have

$$\langle \dot\gamma(t), \gamma(t) \rangle = 0, \tag{14.14}$$

then $\dot\gamma_{\mathbb{P}}(t) = \dot\gamma(t)$, since $D\pi_{\mathbb{S}}(\gamma(t))$ is the orthogonal projection onto $T_{\gamma(t)}$; see Lemma 14.9. It follows that $L(\gamma_{\mathbb{P}}) = L(\gamma)$, and we are done.

In order to achieve (14.14) we multiply γ by a smooth function $\lambda \colon [0, 1] \to \mathbb{C}^\times$. A short calculation shows that $\langle \frac{d(\lambda\gamma)}{dt}, \lambda\gamma \rangle = 0$ iff $\dot\lambda = -\langle \dot\gamma, \gamma \rangle \lambda$. This linear differential equation has the solution $\lambda(t) = \exp(\alpha(t))$, where $\alpha(t)$ is a primitive function of $-\langle \dot\gamma(t), \gamma(t) \rangle$. Note that since $\langle \dot\gamma, \gamma \rangle_{\mathbb{R}} = 0$, we have $\alpha(t) \in i\mathbb{R}$ and hence $|\lambda(t)| = 1$.   □

We define the *sine distance* on $\mathbb{P}(V)$ by setting $d_{\sin}([v], [w]) := \sin d_{\mathbb{P}}([v], [w])$. Recall that we have already introduced a similar notion $d_{\sin}$ for spheres in Definition 2.32. As for spheres, one can show that this defines a metric on $\mathbb{P}(V)$ (cf. Remark 2.33).

For later use we present the following result.

Lemma 14.13

(a) *Let $v, w \in \mathbb{S}(V)$. Then*

$$d_{\sin}(v, w) = \min_{\mu \in \mathbb{R}} \|v - \mu w\|.$$

If the minimum is attained at μ_0, then $\langle v - \mu_0 w, w \rangle = 0$ and $|\mu_0| \leq 1$. Moreover, if $\langle v, w \rangle \neq 0$, then $\mu_0 \neq 0$.

(b) *Let $v, w \in V_*$. Then*

$$d_{\sin}([v], [w]) = \min_{\lambda \in \mathbb{C}} \frac{\|v - \lambda w\|}{\|v\|}.$$

If the minimum is attained at λ_0, then $\langle v - \lambda_0 w, w \rangle = 0$ and $\|\lambda_0 w\| \leq \|v\|$. Moreover, if $\langle v, w \rangle \neq 0$, then $\lambda_0 \neq 0$.

Proof (a) The corresponding statement for two points on the circle $\mathbb{S}^1$ can be proved by elementary geometry. This already implies the first assertion.

(b) For the second assertion we may assume $v, w \in \mathbb{S}(V)$ without loss of generality. Then we have by Proposition 14.12 and part (a);

$$\sin d_{\mathbb{P}}([v], [w]) = \min_{|\lambda|=1} \sin d_{\mathbb{S}}(v, \lambda w) = \min_{|\lambda|=1} \min_{\mu \in \mathbb{R}} \|v - \mu \lambda w\| = \min_{\lambda \in \mathbb{C}} \|v - \lambda w\|.$$

The claim about the minimum attained at λ_0 is an immediate consequence of the corresponding statement for $\mathbb{S}(V)$ in part (a). $\square$

14.3   Condition Measures on Manifolds

We return now to the main theme of this chapter, the definition of condition in a general geometric framework. Let X be a manifold of inputs, Y a manifold of outputs, and let $V \subseteq X \times Y$ be a submanifold of "solutions" to some computational problem. We assume that X and V have the same dimension n to guarantee the local uniqueness of solutions. Consider the projection $\pi_1 \colon V \to X$, $(x, y) \mapsto x$, and its derivative $D\pi_1(x, y) \colon T_{(x,y)}V \to T_x X$, which is the restriction of the projection $T_x X \times T_y Y \to T_x X$, $(\dot{x}, \dot{y}) \mapsto \dot{x}$, to the subspace $T_{(x,y)}V$. Following (14.5), we define the *set of ill-posed solutions* as

$$\Sigma' := \left\{ (x, y) \in V \mid \operatorname{rank} D\pi_1(x, y) < \dim X \right\}. \tag{14.15}$$

If $(x_0, y_0) \in V \setminus \Sigma'$, then $D\pi_1(x_0, y_0) \colon T_{(x_0, y_0)}V \to T_{x_0}X$ is a linear isomorphism. The implicit function theorem tells us that $\pi_1 \colon V \to X$ can be locally inverted around (x_0, y_0). Its inverse $x \mapsto (x, G(x))$ is given by the *solution map* $G \colon X \supseteq U \to Y$ defined on an open neighborhood U of x_0. So we have $(x, G(x)) \in V$ for all $x \in U$. The derivative

$$DG(x_0) \colon T_{x_0}X \to T_{y_0}Y$$

will again be called the *condition map*. Clearly, the inverse of $D\pi_1(x_0, y_0)$ is given by

$$T_{x_0}X \to T_{(x_0, y_0)}V, \qquad \dot{x} \mapsto \left(\dot{x}, DG(x_0)(\dot{x}) \right). \tag{14.16}$$

If V is given as the zero set of a smooth map $F \colon X \times Y \to \mathbb{R}^n$, then, as in Lemma 14.4 and (14.2), we have the following characterization of Σ':

$$\Sigma' = \left\{ (x, y) \in V \mid \operatorname{rank} \partial F / \partial y(x, y) < n \right\}. \tag{14.17}$$

Here the partial derivative $\partial F / \partial y(x, y) \colon T_y Y \to \mathbb{R}^n$ is defined as the restriction of $DF(x, y)$ to $T_y Y$.

Now suppose that X and Y are Riemannian manifolds. That is, we have an inner product on each tangent space $T_x X$ and $T_y Y$ that varies smoothly with x and y, respectively. In particular, $T_x X$ and $T_y Y$ are normed vector spaces. In this case, we may define the (absolute) normwise condition number

$$\text{acond}^G(x_0) := \| DG(x_0) \| = \max_{\|\dot{x}\|=1} \| DG(x_0)(\dot{x}) \|$$

as the operator norm of $DG(x_0)$. We note that in this general framework, it does not directly make sense to define relative condition numbers. However, implicitly, we can model relative notions of condition numbers by choosing the manifolds appropriately. For instance, working with projective spaces means to study ratios, which accounts for a relative notion.

Remark 14.14 The discussion above ties in with the theme of Sect. 6.8. For an element x_0 in the manifold of inputs X we have a finite number of points $(x_0, y_1), \ldots, (x_0, y_s)$ in the fiber $\pi_1^{-1}(x_0)$. Each of them has a corresponding solution map G_j, $j = 1, \ldots, s$, and a condition number $\text{acond}^{G_j}(x_0)$. The condition of x_0 will depend on the computational problem we are considering associated to the geometric situation $V \subseteq X \times Y$. As described in Sect. 6.8, the three typical choices are

$$\text{acond}(x_0) := \inf_{j \leq s} \text{acond}^{G_j}(x_0), \qquad \text{acond}(x_0) := \mathbb{E}_{j \leq s} \, \text{acond}^{G_j}(x_0),$$

and

$$\text{acond}(x_0) := \sup_{j \leq s} \text{acond}^{G_j}(x_0).$$

Note that the last two choices force one to define as *set of ill-posed inputs* the set $\Sigma := \pi_1(\Sigma')$, whereas for the first, one should take instead

$$\Sigma := \left\{ x \in X \mid (x, y) \in \Sigma' \text{ for all } (x, y) \in \pi^{-1}(x) \right\}.$$

14.3.1 Eigenvalues and Eigenvectors

The computation of eigenvalues and eigenvectors can be modeled as follows. Consider the manifold $X = \mathbb{C}^{n \times n}$ of inputs, the manifold $Y = \mathbb{P}(\mathbb{C}^n) \times \mathbb{C}$ of outputs, and the solution manifold

$$V := \left\{ (A, [v], \lambda) \in X \times Y \mid Av = \lambda v \right\}.$$

(We will see shortly that V is indeed a smooth submanifold of $X \times Y$; compare Lemma 14.17.) If λ is a simple eigenvalue of A with eigenvector v, then we can locally invert the projection $\pi_1 \colon V \to X$ around $(A, [v], \lambda)$ and thus have a well-defined solution map $G \colon X \supseteq U \to Y$ defined on an open neighborhood U of A. We may decompose the map G via $G(A) = (G_1(A), G_2(A))$, where $G_1 \colon U \to \mathbb{P}^{n-1}$ is the solution map for the computation of eigenvectors and $G_2 \colon U \to \mathbb{C}$ is the solution map for the computation of eigenvalues. We may thus interpret the operator norms of the derivatives

$$DG_1(A): \quad T_A X = \mathbb{C}^{n \times n} \to T_{[v]}\mathbb{P}^{n-1},$$

$$DG_2(A): \quad T_A X = \mathbb{C}^{n \times n} \to \mathbb{C}$$

as the (absolute) condition numbers of the corresponding computational problems. Clearly, a choice of norms on $\mathbb{C}^{n \times n}$ and $T_{[v]}\mathbb{P}^{n-1}$ has to be made. From a geometric point of view, it is natural to take the norms coming from the inner products on $T_A X$ and $T_{[v]}\mathbb{P}^{n-1}$ induced by the structure of the Riemannian manifolds in X and $\mathbb{P}^{n-1}$, respectively. Note that on $T_A X$, this would amount to considering the Frobenius norm. However, we may as well choose other norms. As in Sect. 1.1, we may fix $r, s \geq 1$ and consider the corresponding operator norm $\| \|_{rs}$ on $T_A X = \mathbb{C}^{n \times n}$. On the space $T_{[v]}\mathbb{P}^{n-1} = T_v$ we shall consider the norm $\frac{1}{\|v\|_r} \|\dot{v}\|_r$ for $\dot{v} \in T_v$.

Within this context, we can compute the condition numbers of the eigenvector $[v]$ and the eigenvalue λ,

$$\mathsf{acond}^{G_1}\big(A, [v]\big) = \big\|DG_1(A)\big\| \quad \text{and} \quad \mathsf{acond}^{G_2}(A, \lambda) = \big\|DG_2(A)\big\|.$$

Before stating the result we need to introduce the notions of left and right eigenvectors. Suppose that λ is a simple eigenvalue of $A \in \mathbb{C}^{n \times n}$, i.e., $\ker(\lambda I - A)$ is one-dimensional. Let $v \in \mathbb{C}^n_*$ be a corresponding (right) eigenvector, so $Av = \lambda v$. Consider the characteristic polynomial $\chi_A(z) = \det(zI - A) = (z - \lambda)g(z)$ with $g(\lambda) \neq 0$. Taking complex conjugates, we get

$$\det\big(\bar{z}I - A^*\big) = \det(\bar{z}I - \overline{A}) = (\bar{z} - \bar{\lambda})\bar{g}(\bar{z}).$$

Hence $\chi_{A^*}(z) = \det(zI - \overline{A}) = (z - \bar{\lambda})\bar{g}(z)$, and we see that $\bar{\lambda}$ is a simple eigenvalue of A^*. Let $u \in \mathbb{C}^n_*$ be a corresponding eigenvector of A^*, that is, $A^*u = \bar{\lambda}u$, or equivalently $u^*A = \lambda u^*$. One calls u a *left eigenvector* of A. We note that for $\dot{v} \in \mathbb{C}^n$ we have

$$\big\langle u, (\lambda I - A)\dot{v}\big\rangle = \langle u, \lambda \dot{v}\rangle - \langle u, A\dot{v}\rangle = \bar{\lambda}\langle u, \dot{v}\rangle - \big\langle A^*u, \dot{v}\big\rangle = \bar{\lambda}\langle u, \dot{v}\rangle - \langle \bar{\lambda}u, \dot{v}\rangle = 0.$$

So the image of $\lambda I - A$ is contained in T_u and hence equals T_u for dimensional reasons.

Let $P \colon \mathbb{C}^n \to T_v$ denote the orthogonal projection, which is given by $P(z) = z - \|v\|^{-2}\langle z, v\rangle v$. We will see shortly that $\langle u, v\rangle \neq 0$, so that the restriction of P induces an isomorphism $T_u \xrightarrow{\sim} T_v$. Thus $P(\lambda I - A) \colon \mathbb{C}^n \to T_v$ is surjective with kernel $\mathbb{C}v$. We can thus take its Moore–Penrose inverse $(P(\lambda I - A))^\dagger$, which provides an isomorphism from T_v onto itself. The next proposition tells us that the norm of this map equals $\|DG_1(A)\|$.

Proposition 14.15 *Choosing the norm $\| \|_{rs}$ on $T_A X = \mathbb{C}^{n \times n}$ and $\frac{1}{\|v\|_r} \| \|_r$ on T_v, the condition maps DG_1 for the eigenvector problem and DG_2 for the eigenvalue* *problem have the following operator norms*:

$$\mathsf{acond}^{G_1}\left(A,[v]\right) = \left\|DG_1(A)\right\| = \left\|\left(P(\lambda I - A)\right)^\dagger\right\|_{sr},$$

$$\mathsf{acond}^{G_2}(A,\lambda) = \left\|DG_2(A)\right\| = \frac{\|u\|_{s^*}\|v\|_r}{|\langle u, v\rangle|},$$

where $\| \ \|_{s^*}$ *denotes the dual norm of* $\| \ \|_s$; *cf.* (1.3).

Let us first illustrate this result in a special case of interest.

Example 14.16 Suppose that $A \in \mathbb{C}^{n \times n}$ is Hermitian, i.e., $A^* = A$, with distinct eigenvalues $\lambda_1, \ldots, \lambda_n$ and corresponding left eigenvectors $v_1, \ldots, v_n$. Then λ_i is real and $u_i = v_i$ is a right eigenvector of λ_i. Suppose that $r = s = 2$. Proposition 14.15 easily implies that

$$\mathsf{acond}^{G_2}(A, \lambda_i) = 1, \qquad \mathsf{acond}^{G_1}\left(A, [v_i]\right) = \frac{1}{\min_{j \neq i} |\lambda_i - \lambda_j|}.$$

So, in accordance with our intuition, a large condition $\mathsf{acond}^{G_1}(A, [v_i])$ means that λ_i is close to other eigenvalues. By contrast, $\mathsf{acond}^{G_2}(A, \lambda_i)$ always equals 1.

For the proof of Proposition 14.15 we first need to compute the derivative $DG(A)\colon \mathbb{C}^{n \times n} \to T_v \times \mathbb{C}$.

Lemma 14.17

(a) *V is a smooth submanifold of $X \times Y$ and* $\dim V = \dim X$.

(b) *We have* $\langle v, u\rangle \neq 0$.

(c) *If λ is a simple eigenvalue of $A \in \mathbb{C}^{n \times n}$ with right eigenvector v and left eigenvector u, then the derivative of the solution map is given by* $DG(A)(\dot{A}) = (\dot{v}, \dot{\lambda})$, *where*

$$\dot{\lambda} = \frac{\langle \dot{A}v, u\rangle}{\langle v, u\rangle}, \qquad \dot{v} = \left(P(\lambda I - A)\right)^\dagger P \dot{A} v.$$

Proof Consider the map

$$F\colon \mathbb{C}^{n \times n} \times \left(\mathbb{C}_*^n\right) \times \mathbb{C} \to \mathbb{C}^n, \qquad (A, v, \lambda) \mapsto Av - \lambda v.$$

For all (A, v, λ), the derivative of F at (A, v, λ) is given by

$$DF(A, v, \lambda)(\dot{A}, \dot{v}, \dot{\lambda}) = \dot{A}v + A\dot{v} - \dot{\lambda}v - \lambda\dot{v}.$$

Since $DF(A, v, \lambda)$ has rank n, the zero set

$$\hat{V} := \left\{(A, v, \lambda) \in \mathbb{C}^{n \times n} \times \mathbb{C}_*^n \times \mathbb{C} \mid F(A, v, \lambda) = 0\right\}$$

is a submanifold of $\mathbb{C}^{n \times n} \times \mathbb{C}^n_* \times \mathbb{C} \to \mathbb{C}^n$ of real dimension $2(n^2 + 1)$. Since V is obtained as the image of $\hat{V}$ under the canonical map $(A, v, \lambda) \mapsto (A, [v], \lambda)$, it follows that V is a manifold of dimension $2n^2$. This proves part (a).

Let $\hat{G}: X \supseteq U \to \mathbb{C}^n_* \times \mathbb{C}$ be a local lifting of the solution map $G: U \to \mathbb{P}^{n-1} \times \mathbb{C}$. If $D\hat{G}(A)(\dot{A}) = (\dot{v}, \dot{\lambda})$, then $DG(A)(\dot{A}) = (P\dot{v}, \dot{\lambda})$, where $P: \mathbb{C}^n \to T_v$ is the orthogonal projection and we have identified $T_{[v]}\mathbb{P}^{n-1}$ with T_v. It is thus sufficient to compute the derivative of $\hat{G}$.

We have $F(A, \hat{G}(A)) = 0$ for $A \in U$, and taking the derivative at A yields (compare (14.3))

$$\frac{\partial F}{\partial A}(A, v, \lambda)A + \frac{\partial F}{\partial(v, \lambda)}(A, v, \lambda)D\hat{G}(A) = 0. \tag{14.18}$$

Furthermore,

$$\frac{\partial F}{\partial A}(A, v, \lambda)\dot{A} = DF(A, v, \lambda)(\dot{A}, 0, 0) = \dot{A}v,$$

$$\frac{\partial F}{\partial(v, \lambda)}(A, v, \lambda)(\dot{v}, \dot{\lambda}) = DF(A, v, \lambda)(0, \dot{v}, \dot{\lambda}) = (A - \lambda I)\dot{v} - \dot{\lambda}v.$$

Thus setting $(\dot{v}, \dot{\lambda}) = D\hat{G}(A)(\dot{A})$, Eq. (14.18) yields

$$\dot{A}v + (A - \lambda I)\dot{v} - \dot{\lambda}v = 0. \tag{14.19}$$

Recall that $\operatorname{im}(\lambda I - A) = T_u$. Taking the inner product of (14.19) with u, we thus obtain

$$\langle \dot{A}v, u \rangle - \dot{\lambda}\langle v, u \rangle = 0.$$

This implies that $\langle v, u \rangle \neq 0$, since $u, v \neq 0$ and $\dot{A}$ was arbitrary. Part (b) and the stated formula for $\dot{\lambda}$ follow.

For part (c) we apply the orthogonal projection $P: \mathbb{C}^n \to T_v$ to (14.19), to get for $\dot{v} \in T_v$,

$$P\dot{A}v = P(\lambda I - A)\dot{v} = P(\lambda I - A)P\dot{v},$$

noting that $P\dot{v} = \dot{v}$. This implies, since the kernel of $P(\lambda I - A)$ equals $\mathbb{C}v$,

$$P\dot{v} = \left(P(\lambda I - A)\right)^{\dagger} P\dot{A}v.$$

The proof is now complete, since $D\hat{G}(A)(\dot{A})$ is obtained from $(\dot{v}, \dot{\lambda}) = D\hat{G}(A)(\dot{A})$ by projecting $\dot{v}$ orthogonally onto T_v. But we already assumed $\dot{v} \in T_v$. $\qquad \square$

Proof of Proposition 14.15 For all $\dot{A} \in \mathbb{C}^{n \times n}$ we have by Hölder's inequality (1.3),

$$\left| \langle \dot{A}v, u \rangle \right| \leq \|\dot{A}v\|_s \|u\|_{s^*} \leq \|\dot{A}\|_{rs} \|v\|_r \|u\|_{s^*}.$$

Moreover, by Lemma 1.2, there exists $\dot{A}$ such that $\|\dot{A}\|_{rs} = 1$ and $\dot{A}v/\|v\|_r$ $= u/\|u\|_s$. For this choice of $\dot{A}$ we have equality above. This implies with Lemma 14.17 that

$$\|DG_2(A)\| = \max_{\|\dot{A}\|_{rs}=1} \frac{|\langle \dot{A}v, u \rangle|}{|\langle v, u \rangle|} = \frac{\|v\|_r \|u\|_{s*}}{|\langle v, u \rangle|},$$

as claimed.

For the assertion on $\|DG_1(A)\|$ we note that for all $\dot{A}$,

$$\|P\dot{A}v\|_s \leq \|P\|_{ss} \|\dot{A}\|_{rs} \|v\|_r \leq \|\dot{A}\|_{rs} \|v\|_r.$$

Let $w \in T_v$ be such that $\|w\|_s = \|v\|_r$. By Lemma 1.2 there exists $\dot{A}$ such that $\|\dot{A}\|_{rs} = 1$ and $\dot{A}v/\|v\|_r = w/\|w\|_s$; hence $P\dot{A}v = w$. This observation implies

$$\max_{\|\dot{A}\|_{rs}=1} \left\| \left(P(\lambda I - A) \right)^{\dagger} P\dot{A}v \right\|_r = \max_{\substack{w \in T_v \\ \|w\|_s=\|v\|_r}} \left\| \left(P(\lambda I - A) \right)^{\dagger} w \right\|_r$$

$$= \|v\|_r \left\| \left(P(\lambda I - A) \right)^{\dagger} \right\|_{sr}.$$

The assertion follows with Lemma 14.17 (recall the norm $\|v\|_r^{-1} \| \|_r$ on T_v).   □

14.3.2 Computation of the Kernel

The goal of this short section is to show that in our geometric framework, the condition number $\kappa(A) = \|A\| \|A^{\dagger}\|$ of a rectangular matrix A (cf. Sect. 1.6) has a natural interpretation as the condition to compute the kernel of A. For this we require a certain understanding of the notion of a Grassmann manifold. This section is not required for the understanding of the remainder of the book and may be skipped.

Fix $1 \leq r \leq m \leq n$ and consider the input space $X := \{A \in \mathbb{R}^{m \times n} \mid \text{rank } A = r\}$, which is a smooth submanifold by Proposition A.5. As the output space Y we take the *Grassmann manifold* consisting of the k-dimensional linear subspaces of $\mathbb{R}^n$, where $k := n - r$. The solution map is $G: X \to Y$, $A \mapsto \ker A$.

Computations in the Grassmann manifold are best performed in the *Stiefel manifold* $\text{St}_{n,k}$, which is defined as the set of $n \times k$ matrices M satisfying $M^{\mathsf{T}}M = I_k$. According to Proposition A.4, $\text{St}_{n,k}$ is indeed a submanifold of $\mathbb{R}^{n \times k}$. Note the $\text{St}_{n,k}$ is invariant under the right action of the orthogonal group $\mathcal{O}(k)$ on $\mathbb{R}^{n \times k}$. We have a natural surjective map $\pi: \text{St}_{n,k} \to Y$, which maps M to its image $\text{Im } M$. Clearly, this map is constant on $\mathcal{O}(k)$-orbits $M\mathcal{O}(k) := \{Mg \mid g \in \mathcal{O}(k)\}$.

Lemma 14.18 *The orbit $M\mathcal{O}(k)$ is a submanifold of $\mathbb{R}^{n \times k}$. The derivative $D\pi(M)$ is surjective and $\ker D\pi(M) = T_M M\mathcal{O}(k)$. Moreover, its orthogonal complement $(\ker D\pi(M))^{\perp}$ in $T_M \text{St}_{n,k}$ consists of the matrices $\dot{M} \in \mathbb{R}^{n \times k}$ such that $\text{Im } \dot{M} = (\text{Im } M)^{\perp}$. In particular, the orthogonal projection of $T_M \text{St}_{n,k}$ onto $T_M M\mathcal{O}(k)$ is given by $V \mapsto PV$, where P is the orthogonal projection onto $(\text{Im } M)^{\perp}$.*

Proof We leave it to the reader to verify that π has local sections. More specifically, for each $L \in Y$ and each $M \in \mathrm{St}_{n,k}$ such that $\pi(M) = L$ there exist an open neighborhood U of L and a smooth map $\iota \colon U \to \mathrm{St}_{n,k}$ such that $\pi \circ \iota = \mathrm{I}_U$. This implies that $D\pi(M)D\iota(L) = \mathrm{I}$. Hence $D\pi(M)$ is surjective.

It follows that all $L \in Y$ are regular values of π. Theorem A.9 implies therefore that $M\mathcal{O}(k) = \pi^{-1}(\pi(M))$ is a submanifold and $\ker D\pi(M) = T_M M\mathcal{O}(k)$.

For the remaining statements, we may assume that $M = (\mathrm{I}_k, 0)^\mathrm{T}$. This is possible by the singular value decomposition (see Sect. 1.5) and the orthogonal invariance of the statement. Proposition A.4 states that $T_M \mathrm{St}_{n,k}$ consists of the matrices $(\dot{B}, \dot{C})^\mathrm{T}$, where $\dot{B} \in \mathbb{R}^{k \times k}$ is skew-symmetric, i.e., $\dot{B} + \dot{B}^\mathrm{T} = 0$, and $\dot{C} \in \mathbb{R}^{k \times r}$. Similarly, it follows that the tangent space of $M\mathcal{O}(k)$ at M consists of the matrices $(\dot{B}, 0)^\mathrm{T}$, where $\dot{B} + \dot{B}^\mathrm{T} = 0$. Hence $(\ker D\pi(M))^\perp$ equals the set of matrices $(0, \dot{C})^\mathrm{T}$, as stated. $\qquad\square$

By this lemma, $D\pi(M)$ provides an isomorphism of the orthogonal complement of $T_M M\mathcal{O}(k)$ with the tangent space $T_{\pi(M)} Y$ of Y at $\mathrm{Im}\, M$. In the following, we shall identify $T_{\pi(M)} Y$ with this subspace of $T_M \mathrm{St}_{n,k}$ to have a concrete model to work with. This also defines a Riemannian metric on Y.

The following result shows that $\kappa(A) = \|A\|\, \mathrm{acond}^G(A)$, and hence $\kappa(A)$ can be interpreted as the relative condition number for computing the kernel of A from a given A.

Proposition 14.19 *We have* $\mathrm{acond}^G(A) = \|A^\dagger\|$ *for* $A \in X$.

Proof From the existence of local sections for π it follows that G can be locally lifted to a smooth map $\hat{G} \colon X \to \mathrm{St}_{n,k}$ such that $G = \pi \circ \hat{G}$. Let $A(t)$ be a smooth curve in X and put $M(t) := \hat{G}(A(t))$. Since $\ker A(t) = \mathrm{Im}\, M(t)$, we have $A(t)M(t) = 0$. Taking derivatives, we get $\dot{A}M + A\dot{M} = 0$, hence $A^\dagger A\dot{M} = -A^\dagger \dot{A}M$.

Recall that $A^\dagger A$ is the orthogonal projection onto $(\ker A)^\perp = (\mathrm{Im}\, M)^\perp$; cf. Lemma 1.24. Lemma 14.18 and our interpretation of the tangent spaces of Y imply now

$$\frac{d}{dt}\pi\big(M(t)\big) = A^\dagger A\dot{M} = -A^\dagger \dot{A}M.$$

Taking norms, we obtain

$$\big\|A^\dagger A\dot{M}\big\|_F \le \big\|A^\dagger\big\| \cdot \big\|\dot{A}M\big\|_F \le \big\|A^\dagger\big\| \cdot \big\|\dot{A}\big\|_F.$$

Here we have used that $\|M\| = 1$ for $M \in \mathrm{St}_{n,k}$, as well as the easily verified general facts $\|PQ\|_F \le \|P\| \cdot \|Q\|_F$ and $\|PQ\|_F \le \|P\|_F \cdot \|Q\|$ for matrices P, Q of compatible formats. It follows that

$$\|DG(A)\| = \sup_{\|\dot{A}\|_F = 1} \left\|\frac{d}{dt}\pi\big(M(t)\big)\right\|_F \le \|A^\dagger\|.$$

In order to see that equality holds, using the singular value decomposition, one may assume without loss of generality that $A = \sum_{i=1}^{r} \sigma_i E_{ii}$, where E_{ij} stands for the matrix with entry 1 at position (i, j) and 0 elsewhere. We assume that σ_i is the smallest positive singular value, so that $\|A^\dagger\| = \sigma_1^{-1}$. Now we choose the curve $A(t) = A + t E_{1,r+1}$ in X and take $M := \sum_{i=1}^{k} E_{r+i,i}$. Then it is easily verified that $\|A^\dagger \dot{A} M\|_F = \sigma_1^{-1} = \|A^\dagger\|$. $\qquad\square$

Chapter 15
Homotopy Continuation and Newton's Method

A general approach to solving a problem consists in reducing it to another problem for which a solution can be found. The first section in this chapter is an example of this approach for the zero-finding problem. Yet, in most occurrences of this strategy, this auxiliary problem is different from the original one, as in the reduction of a nonlinear problem to one or more linear ones. In contrast with this, the treatment we will consider reduces the situation at hand to the consideration of a number of instances of *the same* problem with different data. The key remark is that for these instances, either we know the corresponding solution or we can compute it with little effort.

We mentioned in the introduction of the previous section that even for functions as simple as univariate polynomials, there is no hope of computing their zeros, and the best we can do is to compute accurate approximations. A goal of the second section in this chapter is to provide a notion of approximation (of a zero) that does not depend on preestablished accuracies. It has an intrinsic character. In doing so, we will rely on a pearl of numerical analysis, Newton's method, and on the study of it pioneered by Kantorovich and Smale.

15.1 Homotopy Methods

Homotopy (or continuation) methods are a family of algorithms to compute zeros of a given function, say f, belonging to a class $\mathcal{F}$ of functions defined on a domain Y. They require a pair (g, ζ) at hand with $g \in \mathcal{F}$ and $\zeta \in Y$ a zero of g.

The general idea of the method is to consider a path

$$\gamma : [0, 1] \to \mathcal{F}, \quad t \mapsto q_t,$$

such that $q_0 = g$ and $q_1 = f$. This path needs to be computable (in the sense that q_t may be computed from f, g, and t).

Under certain conditions the path γ can be lifted to a path Γ in the solution manifold $V \subset \mathcal{F} \times Y$ such that $\Gamma(0) = (q_0, \zeta)$. If this is the case, by projecting onto Y we obtain a path $\{\zeta_t\}_{t \in [0,1]}$ on Y such that $\zeta_0 = \zeta$ and $f(\zeta_1) = 0$. The goal of homo-

P. Bürgisser, F. Cucker, *Condition*,
Grundlehren der mathematischen Wissenschaften 349,
DOI 10.1007/978-3-642-38896-5_15, © Springer-Verlag Berlin Heidelberg 2013

topy methods is to "follow" (or "continue") the path on V to eventually obtain an approximation of ζ_1. A nonalgorithmic instance of this continuation is at the heart of our proof of Bézout's theorem in the next chapter (see Sect. 16.5). The algorithmic scheme, broadly described (and this includes just an informal description of the postcondition satisfied by the output) is the following:

Algorithm 15.1 Homotopy_Continuation

Input: $f, g \in \mathcal{F}, \zeta \in Y, k \in \mathbb{N}$

Preconditions: $g(\zeta) = 0$

```
set a partition t₀ = 0 < t₁ < ··· < t_{k-1} < t_k = 1
set z₀ := ζ
for i = 0,..., k − 1 do
      compute an approximation z_{i+1} of ζ_{t_{i+1}}
      from z_i and q_{t_{i+1}}
end for
Return z_k
```

Output: $z \in Y$

Postconditions: z is an approximate zero of f

Figure 15.1 below depicts the process.

To turn this broad description into a working algorithm, a number of issues need to be clarified. For instance: how are the points t_i computed and how many of them

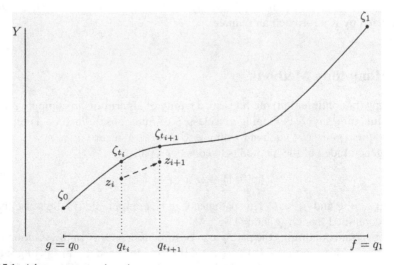

Fig. 15.1 A homotopy continuation

are considered (i.e., which k should be given as input)? What is "an approximation" of a zero ζ_t of q_t? How does one compute one such approximation?

These questions are too general for an all-encompassing answer to be possible. Furthermore, they are not independent, and the answer given to one of them may affect the way we approach the others. The next section provides a first answer to the last two questions above. Prior to proceeding with such an answer, however, we stress a couple of issues leading to an improved version of Algorithm 15.1.

The number k of iterations in Algorithm 15.1 is given as input and presumably needs to be estimated in some way such that it is appropriate for the triple (f, g, ζ). Such an estimation does not appear to be simple. A way to avoid the issue altogether is to compute, at the ith iteration, the point t_{i+1} as a function of t_i, q_{t_i}, and z_i. The underlying idea is the following. Assume that z_i is a "strong approximation" of ζ_{t_i} in the sense that for all t reasonably close to t_i, z_i is an approximation of ζ_t. Assume, in addition, that the computation of z_{i+1} in Algorithm 15.1 is such that if z_i is an approximation of $\zeta_{t_{i+1}}$, then z_{i+1} is a strong approximation of $\zeta_{t_{i+1}}$. Then the good functioning of the homotopic continuation will be guaranteed—by a trivial induction argument—as long as (a) z_0 is a strong approximation of ζ, and (b) the point t_{i+1} is chosen reasonably close (in the sense above) of t_i.

Requirement (a) allows for a relaxation. We no longer need a zero of g at hand. It is enough to have a strong approximation of one such zero. Requirement (b) introduces condition into the scenario. Intuitively, for z_i to be an approximation of $\zeta_{t_{i+1}}$ it has to be close to this zero. For this to occur, we expect t_{i+1} to be close to t_i, but how close exactly—as shown by a look at Fig. 15.1—will depend on how fast ζ_t moves apart from ζ_{t_i} when t increases from t_i. That is, on how large $\mu(g_{t_i}, \zeta_{t_i})$ is.

The discussion above introduces one more notion to be made precise (what exactly we mean by strong approximation) but allows us to (again, broadly) describe an adaptive version of homotopy.

We do so in Algorithm 15.2.

Algorithm 15.2 Adaptive_Homotopy

Input: $f, g \in \mathcal{F}, z \in Y$

Preconditions: $\exists \zeta \in Y$ such that $g(\zeta) = 0$ and z strongly approximates ζ

```
set i := 0,  z_i := z,  and  t_i := 0
while t_i < 1 do
      compute t_{i+1} from t_i, z_i, and q_{t_i}
      compute an approximation z_{i+1} of ζ_{t_{i+1}}
      from z_i and q_{t_{i+1}}
end while
Return z_k
```

Output: $z \in Y$

Postconditions: z is an approximate zero of f

We may now turn to the notions of approximation.

15.2 Newton's Method

Newton's method is doubtless one the most widely used algorithms in numerical analysis. To understand the idea behind it, consider an analytic function $f : \mathbb{C} \to \mathbb{C}$. Given an initial point $z_0 \in \mathbb{C}$, the method constructs a sequence of iterates $z_0, z_1, z_2, \ldots$, where

$$z_{i+1} = N_f(z_i) := z_i - \frac{f(z_i)}{f'(z_i)}.$$

Here f' is the derivative of f (which we assume well defined for all $i \geq 0$). An immediate property of N_f is the fact that $N_f(z) = z$ if and only if $f(z) = 0$ and $f'(z) \neq 0$. Also, for a point $z \in \mathbb{C}$,

$$N'_f(z) = 1 - \frac{f'(z)^2 - f(z)f''(z)}{f'(z)^2} = \frac{f(z)f''(z)}{f'(z)^2}.$$

In particular, for a simple zero ζ of f we have $N'_f(\zeta) = 0$, and the Taylor expansion of N_f at ζ is given by

$$N_f(z) - \zeta = \frac{1}{2} N''_f(\zeta)(z - \zeta)^2 + \mathcal{O}\big((z - \zeta)^3\big). \tag{15.1}$$

This implies that if the initial point z_0 is close enough to ζ, the sequence of points generated by Newton's method approaches ζ quadratically fast. Newton's method does not necessarily find a zero of f, but starting with a modest approximation of such a zero, it refines its accuracy in a very fast manner.

The above reasoning can be extended to many variables. Let $f : \mathbb{C}^n \to \mathbb{C}^n$ be analytic. Newton's method is an iteration based on the map

$$N_f(z) = z - Df(z)^{-1} f(z),$$

where $Df(z)$ is the derivative of f at z. This formula is defined if $Df(z)$ is invertible. As for the one-dimensional case (15.1), one shows that for $z \to \zeta$,

$$\|N_f(z) - \zeta\| = \mathcal{O}\big(\|z - \zeta\|^2\big).$$

Hence, for all initial points z_0 sufficiently close to ζ, the distance from the iterates $z_{i+1} := N_f(z_i)$ to ζ decreases quadratically.

This property inspired Steve Smale to introduce a notion of approximate zero that does not depend on prescribed accuracies: a point is an approximate zero of a function when Newton's method starting at this point converges to a zero of the function immediately, quadratically fast. In what follows we make this idea precise (which, we note, gives a possible answer to the second question at the end of the previous section).

Definition 15.1 We say that $z \in \mathbb{C}^n$ is an *approximate zero* of f if the sequence given by $z_0 = z$ and $z_{i+1} = N_f(z_i)$ is defined for all natural numbers i, and there is a ζ with $f(\zeta) = 0$ such that for all $i \in \mathbb{N}$,

$$\|z_i - \zeta\| \le \left(\frac{1}{2}\right)^{2^i - 1} \|z - \zeta\|.$$

We say that ζ is the *associated zero* of z.

Remark 15.2 An approximate zero in the sense above yields approximations to any desired accuracy of a zero of f. Indeed, for any $\varepsilon > 0$ and approximate zero z of f with associated zero ζ, we may compute the kth iterate $z_k = N_f^k(z)$. If $\|z - \zeta\| \le M$, then $\|z_k - \zeta\| < \varepsilon$ if $k \ge \log\log\frac{M}{\varepsilon} + 1$.

Let's go back to the discussion on homotopy methods. Definition 15.1 provides a notion of approximate zero. We can take as "strong approximate zero" the image under Newton's operator of an approximate zero. That is, if z is an approximate zero of f, then $N_f(z)$ is a strong approximate zero. An immediate advantage of this is that we can make the second computation in the while loop in Algorithm 15.2 precise. It is

$$\texttt{compute } z_{i+1} := N_{q_{t_{i+1}}}(z_i).$$

Having made precise what we understand by approximation and strong approximation sheds light as well on the meaning of the condition (b) of the previous section, namely, to choose t_{i+1} reasonably close to t_i. Indeed, this means choosing t_{i+1} such that z_i is an approximate zero of $q_{t_{i+1}}$.

To provide an algorithmic procedure for this choice of t_{i+1} will take us some time (and we will do that only in the specific context of homogeneous polynomial systems). Yet, it is apparent that a first step in this endeavor is to have an estimate, for a zero ζ of an analytic function f, of how large its *basin of quadratic attraction* is, that is, how large the set

$$\{z \in \mathbb{C}^n \mid z \text{ is an approximate zero of } f \text{ with associated zero } \zeta\}$$

is. An answer to this question is naturally given in terms of a quantity $\gamma(f, \zeta)$ that we shall define in a moment.

Before doing so, we recall some basic facts from multivariate calculus. For finite-dimensional vector spaces $V_1, \ldots, V_k, W$ we denote by $\mathscr{L}_k(V_1, \ldots, V_k; W)$ the space of k-linear maps from $V_1 \times \cdots \times V_k$ to W. In case $k = 1$, we omit the index. Also, if $V_1 = \cdots = V_k = V$, we simply write $\mathscr{L}_k(V; W)$. If $V_1, \ldots, V_k, W$ are normed vector spaces, then one defines the induced norm

$$\|\varphi\| := \max_{\|v_1\| = \cdots = \|v_k\| = 1} \|\varphi(v_1, \ldots, v_k)\|$$

for $\varphi \in \mathscr{L}_k(V_1, \ldots, V_k; W)$. In this way, $\mathscr{L}_k(V_1, \ldots, V_k; W)$ becomes a normed vector space as well. We do have a canonical isomorphism

$$\mathscr{L}_{k-1}(V_1, \ldots, V_{k-1}; \mathscr{L}(V_k; W)) \simeq \mathscr{L}_k(V_1, \ldots, V_k; W), \qquad (15.2)$$

which is an isometry if $V_1, \ldots, V_k, W$ are normed vector spaces.

For $z \in \mathbb{C}^n$, we denote by $Df(z)$ the derivative of f at z. This is a linear map $Df(z) : \mathbb{C}^n \to \mathbb{C}^n$ so that $Df : \mathbb{C}^n \to \mathscr{L}(\mathbb{C}^n; \mathbb{C}^n)$. Taking the derivative of Df at $z \in \mathbb{C}^n$, we obtain that this second derivative is a linear map $D^2 f(z) : \mathbb{C}^n \to \mathscr{L}(\mathbb{C}^n; \mathscr{L}(\mathbb{C}^n; \mathbb{C}^n))$. That is, using (15.2), $D^2 f(z) \in \mathscr{L}_2(\mathbb{C}^n, \mathbb{C}^n)$. Repeating this argument we find, for all $k \geq 1$ and $z \in \mathbb{C}^n$, that the kth derivative of f at z satisfies $D^k f(z) \in \mathscr{L}_k(\mathbb{C}^n; \mathbb{C}^n)$. It is known that $D^k f(z)$ is a symmetric k-linear map. By abuse of notation, we shall abbreviate $D^k f(z)(y - z, \ldots, y - z)$ by $D^k f(z)(y - z)^k$ for $y \in \mathbb{C}^n$. The definition of the norm of $D^k f(z)$ implies that

$$\left\| D^k f(z)(y - z)^k \right\| \leq \left\| D^k f(z) \right\| \, \|y - z\|^k. \qquad (15.3)$$

Recall that an analytic function $f : \mathbb{C}^n \to \mathbb{C}^n$ can be expanded in a *Taylor series* around a point $\zeta \in \mathbb{C}^n$,

$$f(z) = \sum_{k=0}^{\infty} \frac{1}{k!} D^k f(\zeta)(z - \zeta)^k,$$

and we have absolute convergence for $z \in \mathbb{C}^n$ such that $\|z - \zeta\| < \rho$, where the radius of convergence $\rho > 0$ is given by *Hadamard's formula*

$$\rho^{-1} = \limsup_{k \to \infty} \left\| \frac{D^k f(z)}{k!} \right\|^{\frac{1}{k}}. \qquad (15.4)$$

We can now define the quantity $\gamma(f, \zeta)$.

Definition 15.3 For an analytic function $f : \mathbb{C}^n \to \mathbb{C}^n$ and $z \in \mathbb{C}^n$ such that $Df(z)$ is invertible, we define

$$\gamma(f, z) = \sup_{k \geq 2} \left\| \frac{Df(z)^{-1} D^k f(z)}{k!} \right\|^{\frac{1}{k-1}}.$$

We set $\gamma(f, z) = \infty$ if $Df(z)$ is not invertible.

Remark 15.4 The supremum exists so that $\gamma := \gamma(f, z)$ is well defined. Indeed, by (15.4), the supremum

$$C := \sup_{k} \left\| \frac{D^k f(z)}{k!} \right\|^{\frac{1}{k}}$$

is finite. Therefore,

$$\left\| \frac{Df(z)^{-1}D^k f(z)}{k!} \right\|^{\frac{1}{k-1}} \leq \left(\|Df(z)^{-1}\| C^k \right)^{\frac{1}{k-1}},$$

and the right-hand side converges to C for $k \to \infty$ and hence is bounded.

On the other hand, we have

$$\left\| \frac{D^k f(z)}{k!} \right\|^{\frac{1}{k}} \leq \|Df(z)\|^{\frac{1}{k}} \left\| \frac{Df(z)^{-1}D^k f(z)}{k!} \right\|^{\frac{1}{k}} \leq \|Df(z)\|^{\frac{1}{k}} \gamma^{1-\frac{1}{k}}.$$

Therefore, by (15.4), $\rho^{-1} \leq \gamma$. So γ^{-1} is a lower bound for the radius of convergence ρ.

Theorem 15.5 (Smale's γ-theorem) *Suppose that $f : \mathbb{C}^n \to \mathbb{C}^n$ is analytic, $f(\zeta) = 0$, and $Df(\zeta)$ is invertible. If, for $z \in \mathbb{C}^n$,*

$$\|z - \zeta\| \leq \frac{3 - \sqrt{7}}{2\gamma(f,\zeta)},$$

then z is an approximate zero of f with associated zero ζ.

For the proof of this theorem we use the following stepping stones.

Lemma 15.6 *For $-1 < u < 1$ we have*

$$\sum_{k=0}^{\infty} u^k = \frac{1}{1 - u}, \qquad \sum_{k=1}^{\infty} k u^{k-1} = \frac{1}{(1 - u)^2}.$$

Proof The first equation is the well-known geometric series. The second equation is obtained from the first by (termwise) differentiation. □

Lemma 15.7 *Let $A \in \mathcal{L}(\mathbb{C}^n; \mathbb{C}^n)$ be invertible and let $\Delta \in \mathcal{L}(\mathbb{C}^n; \mathbb{C}^n)$ be such that $\|\Delta\| \cdot \|A^{-1}\| < 1$. Then $A + \Delta$ is invertible and*

$$\left\| (A + \Delta)^{-1} \right\| \leq \frac{\|A^{-1}\|}{1 - \|\Delta\| \|A^{-1}\|}.$$

Proof We have $\|B\| < 1$ for $B := \Delta A^{-1}$. The geometric series converges and yields $(I + B)^{-1} = \sum_{k=0}^{\infty} (-B)^k$. In particular, $I + B$ is invertible. Bounding the norms, we get $\|(I + B)^{-1}\| \leq \sum_{k=0}^{\infty} \|B\|^k = (1 - \|B\|)^{-1}$. Finally, we note that $A + \Delta = (I + B)A$; hence $(A + \Delta)^{-1} = A^{-1}(I + B)^{-1}$ and the assertion follows. □

The following simple quadratic polynomial plays an important role in our estimates:

$$\psi(u) = 1 - 4u + 2u^2. \tag{15.5}$$

The proof of the following properties of ψ is left to the reader.

Lemma 15.8 *The function* $\psi(u) = 1 - 4u + 2u^2$ *is monotonically decreasing and nonnegative in* $[0, 1 - \frac{\sqrt{2}}{2}]$ *and satisfies*

$$\frac{u}{\psi(u)} < 1 \quad for\ 0 \le u < \frac{5 - \sqrt{17}}{4},$$

$$\frac{u}{\psi(u)} \le \frac{1}{2} \quad for\ 0 \le u \le \frac{3 - \sqrt{7}}{2}.$$

$\square$

The following crucial lemma gives an estimate on how much $Df(z)$ changes when we perturb z a little.

Lemma 15.9 *Let* $f: \mathbb{C}^n \to \mathbb{C}^n$ *be analytic and* $y, z \in \mathbb{C}^n$ *such that* $Df(z)$ *is invertible. We put* $u := \|y - z\|\gamma(f, z)$. *If* $u < 1 - \frac{\sqrt{2}}{2}$, *then* $Df(y)$ *is invertible and we have:*

(a) $Df(z)^{-1}Df(y) = I + \Delta$, *where* $\Delta \in \mathcal{L}(\mathbb{C}^n; \mathbb{C}^n)$, $\|\Delta\| \le \frac{1}{(1-u)^2} - 1 < 1$.

(b) $\|Df(y)^{-1}Df(z)\| \le \frac{(1-u)^2}{\psi(u)}$.

Proof Put $g := Df$ and consider the Taylor expansion of $g: \mathbb{C}^n \to \mathcal{L}(\mathbb{C}^n; \mathbb{C}^n)$ around $z \in \mathbb{C}^n$:

$$g(y) = g(z) + \sum_{\ell=1}^{\infty} \frac{1}{\ell!} D^\ell g(z)(y - z)^\ell.$$

Composition with $g(z)^{-1}$ yields $g(z)^{-1}g(y) = I + \Delta$, where $\Delta \in \mathcal{L}(\mathbb{C}^n; \mathbb{C}^n)$ is given by

$$\Delta := \sum_{\ell=1}^{\infty} \frac{1}{\ell!} g(z)^{-1} D^\ell g(z)(y - z)^\ell.$$

Setting $k = \ell + 1$, using (15.3) as well as the isometric identification (15.2) along with $D^{k-1}g(z) = D^k f(z)$, we can bound as follows:

$$\frac{1}{k!}\|g(z)^{-1}D^{k-1}g(z)(y - z)^{k-1}\| \le \frac{1}{k!}\|g(z)^{-1}D^{k-1}g(z)\|\ \|y - z\|^{k-1}$$

$$= \frac{1}{k!}\|g(z)^{-1}D^k f(z)\|\ \|y - z\|^{k-1}$$

$$\le \gamma(f, z)^{k-1}\|y - z\|^{k-1},$$

where we have used the definition of $\gamma(f, z)$ for the last inequality. Therefore, by Lemma 15.6,

$$\|\Delta\| \le \sum_{k=2}^{\infty} k\big(\gamma(f,z)\|y - z\|\big)^{k-1} = \frac{1}{(1 - u)^2} - 1 < 1,$$

where the strict inequality on the right is due to our assumption $u < 1 - \sqrt{2}/2$. Lemma 15.7 (with $A = \mathrm{I}$) implies that $\mathrm{I} + \Delta$ is invertible. Hence $Df(y) = g(y) = g(z)(\mathrm{I} + \Delta)$ is invertible as well. We have thus proved part (a).

Part (b) follows from the relations

$$\big\|g(y)^{-1}g(z)\big\| = \big\|\big(g(z)^{-1}g(y)\big)^{-1}\big\| = \big\|(\mathrm{I} + \Delta)^{-1}\big\|$$

$$\le \frac{1}{1 - \|\Delta\|} \le \frac{1}{1 - (\frac{1}{(1-u)^2} - 1)} = \frac{(1 - u)^2}{\psi(u)},$$

where we have again used Lemma 15.7 for the first inequality. $\qquad\square$

We now apply the previous lemma to the analysis of the Newton iteration. We shall denote by N_f^k the k-fold iterate of the Newton operator N_f.

Proposition 15.10 *Let $f : \mathbb{C}^n \to \mathbb{C}^n$ be analytic, $f(\zeta) = 0$, and let $Df(\zeta)$ be invertible. Let $z \in \mathbb{C}^n$ be such that*

$$u := \|z - \zeta\|\gamma(f,\zeta) < \frac{1}{4}\big(5 - \sqrt{17}\big).$$

Then $\frac{u}{\psi(u)} < 1$ and

(a) $\|N_f(z) - \zeta\| < \frac{u}{\psi(u)}\|z - \zeta\| = \frac{\gamma(f,\zeta)}{\psi(u)}\|z - \zeta\|^2$.

(b) $\|N_f^k(z) - \zeta\| \le \big(\frac{u}{\psi(u)}\big)^{2^k - 1}\|z - \zeta\|$ *for all $k \ge 0$.*

Proof The bound $\frac{u}{\psi(u)} < 1$ was established in Lemma 15.8.

Towards proving part (a), we expand f around ζ to obtain in $\mathbb{C}^n$,

$$f(z) = \sum_{k=1}^{\infty} \frac{1}{k!} D^k f(\zeta)(z - \zeta)^k.$$

Similarly, expanding $g := Df$ around ζ, we obtain in $\mathscr{L}(\mathbb{C}^n; \mathbb{C}^n)$,

$$g(z) = \sum_{k=1}^{\infty} \frac{1}{(k - 1)!} D^{k-1} g(\zeta)(z - \zeta)^{k-1}.$$

Evaluating both sides of the last equality at $z - \zeta$, we obtain

$$Df(z)(z - \zeta) = \sum_{k=1}^{\infty} \frac{1}{(k - 1)!} D^k f(\zeta)(z - \zeta)^k,$$

which is an equality in $\mathbb{C}^n$. We may now subtract from it the first equality above to obtain

$$Df(z)(z - \zeta) - f(z) = \sum_{k=1}^{\infty} \left(\frac{1}{(k-1)!} - \frac{1}{k!} \right) D^k f(\zeta)(z - \zeta)^k$$

$$= \sum_{k=1}^{\infty} (k-1) \frac{D^k f(\zeta)}{k!} (z - \zeta)^k. \tag{15.6}$$

From this it follows that

$$N_f(z) - \zeta = z - \zeta - Df(z)^{-1}\big(f(z)\big) = Df(z)^{-1}\big(Df(z)(z - \zeta) - f(z)\big)$$

$$= Df(z)^{-1} Df(\zeta) \sum_{k=1}^{\infty} (k-1) \frac{Df(\zeta)^{-1} D^k f(\zeta)}{k!} (z - \zeta)^k.$$

We can bound the norm of this as follows, recalling $u = \gamma(f, z)\|z - \zeta\|$:

$$\big\| N_f(z) - \zeta \big\| \le \big\| Df(z)^{-1} Df(\zeta) \big\| \sum_{k=1}^{\infty} (k-1) \left\| \frac{Df(\zeta)^{-1} D^k f(\zeta)}{k!} \right\| \|z - \zeta\|^k$$

$$\le \big\| Df(z)^{-1} Df(\zeta) \big\| \|z - \zeta\| \sum_{k=1}^{\infty} (k-1) u^{k-1}.$$

Lemma 15.6 implies

$$\sum_{k=1}^{\infty} (k-1) u^{k-1} = \sum_{k=1}^{\infty} k u^{k-1} - \sum_{k=1}^{\infty} u^{k-1} = \frac{1}{(1-u)^2} - \frac{1}{1-u} = \frac{u}{(1-u)^2}.$$

Combining this with Lemma 15.9(b), we conclude that

$$\big\| N_f(z) - \zeta \big\| \le \frac{(1-u)^2}{\psi(u)} \frac{u}{(1-u)^2} \|z - \zeta\| = \frac{u}{\psi(u)} \|z - \zeta\|,$$

which proves (a).

We next prove (b). For $k = 0$ this is trivial. For $k \ge 1$ we assume by induction that

$$\big\| N_f^{k-1}(z) - \zeta \big\| < \left(\frac{u}{\psi(u)} \right)^{2^{k-1}-1} \|z - \zeta\|.$$

Part (a) implies that $\|N_f(z) - \zeta\| < \|z - \zeta\|$, since $u/\psi(u) < 1$. Applying this $k - 1$ times, it follows that

$$\bar{u} := \big\| N_f^{k-1}(z) - \zeta \big\| \gamma(f, \zeta) < u.$$

Furthermore, since ψ is decreasing in $[0, \frac{5-\sqrt{17}}{4})$, we have $\psi(\bar{u}) > \psi(u)$. So, by part (a) and the induction hypothesis,

$$\left\| N_f^k(z) - \zeta \right\| = \left\| N_f\left(N_f^{k-1}(z)\right) - \zeta \right\| \leq \frac{\gamma(f,\zeta)}{\psi(\bar{u})} \left\| N_f^{k-1}(z) - \zeta \right\|^2$$

$$< \frac{\gamma(f,\zeta)}{\psi(u)} \left(\frac{u}{\psi(u)}\right)^{2^k-2} \|z - \zeta\|^2 = \left(\frac{u}{\psi(u)}\right)^{2^k-1} \|z - \zeta\|. \qquad \square$$

Proof of Theorem 15.5 By Lemma 15.8, $\frac{u}{\psi(u)} < \frac{1}{2}$ if $u < \frac{3-\sqrt{7}}{2}$. Now Proposition 15.10(b) finishes the proof. $\qquad \square$

A bound for the separation of zeros of an analytic function easily follows as a consequence of the previous results.

Corollary 15.11 *If ζ, ξ are distinct zeros of f, then*

$$\|\zeta - \xi\| \geq \frac{5-\sqrt{17}}{4} \frac{1}{\min\{\gamma(f,\zeta), \gamma(f,\xi)\}}.$$

Proof Assume, without loss of generality, that $\min\{\gamma(f,\zeta), \gamma(f,\xi)\} = \gamma(f,\zeta)$. If $\|\zeta - \xi\| < \frac{5-\sqrt{17}}{4} \frac{1}{\gamma(f,\zeta)}$, then by Proposition 15.10(b) with $z = \xi$ we deduce that $N_f^k(\xi) \to \zeta$ as $k \to \infty$. However, since $f(\xi) = 0$, we have $N_f^k(\xi) = \xi$ for all k. Hence $\xi = \zeta$. $\qquad \square$

Remark 15.12 (A fresh view of interior-point methods) The content of this chapter allows us to look at the interior-point method described in Sect. 9.1 with new eyes. The reader may recall (or have a look at the Eqs. (9.1)) that the idea was to find a solution (x, w, y, s) of the function F given by

$$(x, w, y, s) \mapsto \begin{bmatrix} A^\mathsf{T}y + s - c \\ G^\mathsf{T}y - d \\ Ax + Gw - b \\ x_1 s_1 \\ \vdots \\ x_n s_n \end{bmatrix}.$$

To this end we considered a curve C of functions F_μ parameterized by $\mu \in [0, \mu_0]$ with F_μ given by

$$(x, w, y, s) \mapsto \begin{bmatrix} A^\mathsf{T}y + s - c \\ G^\mathsf{T}y - d \\ Ax + Gw - b \\ x_1 s_1 - \mu \\ \vdots \\ x_n s_n - \mu \end{bmatrix}$$

and such that we had at hand a point ζ_{μ_0} with $F_{\mu_0}(\zeta_{\mu_0}) = 0$. Note that by construction, $F_0 = F$. The central path $\mathcal{C}$ we defined in Sect. 9.1 in fact is obtained from the lifting of the curve C containing $(F_{\mu_0}, \zeta_{\mu_0})$ by projecting on the space of solutions.

But there is more. A look at Algorithm 9.1 (or, probably more compelling, at Fig. 9.1) reveals that the central path is "followed" from the initial point ζ_{μ_0} by a sequence of points z_i, each of them approximating a point ζ_{μ_i} on $\mathcal{C}$. And a somehow closer look at the way the point z_{i+1} is computed reveals that a Newton's step is used to approximate $\zeta_{\mu_{i+1}}$, as a zero of $f_{\mu_{i+1}}$, starting at z_i; compare Eq. (9.5).

Thus interior-point methods, as described in Sect. 9.1, turn out to be an illustration of the main ideas of this chapter.

Chapter 16
Homogeneous Polynomial Systems

We finished the preceding chapter with a notion of approximate zero of a function and an algorithmic scheme to compute these approximate zeros, the adaptive homotopy.

Within this scheme, we identified as critical the issue of determining the step length at each iteration of the continuation process, and as a first step towards this goal, we estimated the size of the basin of quadratic attraction of a proper zero ζ for a given analytic function f: this basin contains a ball of radius $\frac{3-\sqrt{7}}{2\gamma(f,\zeta)}$ centered at ζ.

At this stage we perceive two weaknesses in this estimate. Firstly, the computation of $\gamma(f,\zeta)$ appears to require the computation of the norm of all the higher order derivatives of f at ζ. Even if we deal with polynomials (for which the number of such computations is finite), this can be very costly. Secondly, we can hardly compute these derivatives without having ζ at our disposal. And the whole idea of the adaptive homotopy relies on not having resource to the zeros ζ_t in the lifted path Γ.

In this chapter we provide solutions for these shortcomings. To do so, we narrow the context we are working on and focus on a specific class of functions, namely homogeneous multivariate polynomial functions $f : \mathbb{C}^{n+1} \to \mathbb{C}^n$. Homogenization is a common approach to the study of zeros for not necessarily homogeneous polynomial systems: given one such system, one homogenizes its component polynomials and considers the zeros of the resulting homogeneous system, which are now sets of lines through the origin, as points in projective space $\mathbb{P}^n$. In doing so, one avoids the distortions produced by having "large" zeros or, in the limit, zeros at infinity. We denote by $\mathcal{H}_{\mathbf{d}}$ the linear space of homogeneous polynomial systems with degree pattern $\mathbf{d} = (d_1, \ldots, d_n)$ (more details in Sect. 16.1 below).

Newton's method as described in the previous chapter can be modified to work in this setting (i.e., acting on $\mathbb{P}^n$ and with underlying function in $\mathcal{H}_{\mathbf{d}}$); we will do that in Sect. 16.6. With a few natural modifications we recover both the notion of approximate zero and a version $\gamma_{\text{proj}}(f,\zeta)$ of the γ invariant. Furthermore—and gratifyingly, also with only a few minor modifications—we show that the size of the basin of quadratic attraction of a zero is controlled by γ_{proj} in about the same manner as what we saw in Theorem 15.5 (see Theorem 16.38 below).

P. Bürgisser, F. Cucker, *Condition*,
Grundlehren der mathematischen Wissenschaften 349,
DOI 10.1007/978-3-642-38896-5_16, © Springer-Verlag Berlin Heidelberg 2013

The invariant $\gamma_{\mathrm{proj}}(f, \zeta)$ is also defined in terms of higher-order derivatives and therefore shares the first weakness mentioned above. Condition proves helpful to overcoming it. The solution in $\mathbb{P}^n$ of systems in $\mathcal{H}_\mathbf{d}$ fits within the framework described in Sect. 14.3 and therefore, to a pair $(f, \zeta) \in \mathcal{H}_\mathbf{d} \times \mathbb{P}^n$ with $f(\zeta) = 0$ we may associate a condition number $\mathsf{acond}^G(f)$ (here G denotes the solution map corresponding to (f, ζ)). It is common to denote $\mathsf{acond}^G(f)\|f\|^{-1}$ by $\mu(f, \zeta)$. Shub and Smale introduced a normalization of $\mu(f, \zeta)$—denoted by $\mu_{\mathrm{norm}}(f, \zeta)$—whose value is close to $\mu(f, \zeta)$ and is computed with the same cost. This normalized condition number allows for some elegant statements, such as a condition number theorem; see Theorem 16.19. To follow what is an already established tradition, we will base our exposition in $\mu_{\mathrm{norm}}(f, \zeta)$.

Since we shall only be able to compute approximations z of a true zero ζ of f, we will extend the definition of $\mu_{\mathrm{norm}}(f, z)$ (or $\mu(f, z)$ for that matter) to any pair $f \in \mathcal{H}_\mathbf{d}$ and $z \in \mathbb{P}^n$ as long as $Df(z)|_{T_z}$ is invertible. It is an important feature that the quantity $\mu_{\mathrm{norm}}(f, z)$ depends only on the equivalence class of f in $\mathbb{P}(\mathcal{H}_\mathbf{d})$. We may therefore view μ_{norm} as a function defined (almost everywhere) on $\mathbb{P}(\mathcal{H}_\mathbf{d}) \times \mathbb{P}^n$.

A main result in this exposition is the following. Let $D = \max_{i \leq n} \deg f_i$.

Theorem 16.1 *For all nonzero $f \in \mathcal{H}_\mathbf{d}$ and $z \in \mathbb{P}^n$ we have*

$$\gamma_{\mathrm{proj}}(f, z) \leq \frac{1}{2} D^{3/2} \mu_{\mathrm{norm}}(f, z).$$

Theorem 16.1 allows one to use $\mu_{\mathrm{norm}}(f, \zeta)$ instead of $\gamma_{\mathrm{proj}}(f, \zeta)$ to estimate sizes of basins of quadratic attraction. This solves the first of the two shortcomings above.

To solve the second shortcoming, a key step will be the observation that the condition μ_{norm} satisfies a *Lipschitz property* that allows one to estimate $\mu_{\mathrm{norm}}(g, y)$ in terms of $\mu_{\mathrm{norm}}(f, z)$ for pairs (g, y) close to (f, z). Hereby we measure both distances $d_\mathbb{P}(f, g)$ and $d_\mathbb{P}(y, z)$ in the Riemannian metric of the complex projective spaces $\mathbb{P}(\mathcal{H}_\mathbf{d})$ and $\mathbb{P}^n$, respectively. (Recall Sect. 14.2 for the definition and some properties of this metric.)

Proposition 16.2 *Fix $0 \leq \varepsilon \leq \frac{1}{7}$. Let $f, g \in \mathbb{P}(\mathcal{H}_\mathbf{d})$ and $y, z \in \mathbb{P}^n$ be such that*

$$\mu_{\mathrm{norm}}(f, z) \max\left\{ D^{1/2} d_\mathbb{P}(f, g), D^{3/2} d_\mathbb{P}(y, z) \right\} \leq \frac{\varepsilon}{4}.$$

Then

$$\frac{1}{1 + \varepsilon} \mu_{\mathrm{norm}}(f, z) \leq \mu_{\mathrm{norm}}(g, y) \leq (1 + \varepsilon) \mu_{\mathrm{norm}}(f, z).$$

The way this proposition solves the second shortcoming can be briefly stated if we leave technical details aside. At the ith iteration of the adaptive homotopy, we compute $\mu_{\mathrm{norm}}(q_{t_i}, z_i)$. Since z_i is a strong approximation of ζ_{t_i}, this quantity yields $\mu_{\mathrm{norm}}(q_{t_i}, \zeta_{t_i})$ up to a factor of $1 + \varepsilon$. Having this estimate in hand allows us to chose

t_{i+1}, so that $d_{\mathbb{P}}(q_{t_i}, q_{t_{i+1}}) \leq \frac{\varepsilon}{4D^{1/2}\mu_{\text{norm}}(q_{t_i}, \zeta_{t_i})}$. This ensures that $\mu_{\text{norm}}(q_{t_i}, \zeta_{t_i})$ yields $\mu_{\text{norm}}(q_{t_{i+1}}, \zeta_{t_{i+1}})$, again up to a factor $1 + \varepsilon$, and therefore that $\mu_{\text{norm}}(q_{t_i}, z_i)$ does so up to a factor of $(1 + \varepsilon)^2$. We will see that it also ensures that z_i is an approximate zero of $q_{t_{i+1}}$ and therefore $z_{i+1} := N_{q_{t_{i+1}}}(z_i)$ is a strong approximate zero of $\zeta_{t_{i+1}}$. This allows one to iterate the procedure. We will provide the details of the resulting (fully specified) version of the adaptive homotopy in the next chapter (Sect. 17.1). In the meanwhile, we devote this chapter to proving the two results stated above.

16.1   A Unitarily Invariant Inner Product

Let $\mathcal{H}_d = \mathbb{C}[X_0, \ldots, X_n]$ be the complex vector space of homogeneous polynomials of degree d in $n + 1$ variables. A basis of $\mathcal{H}_d$ is given by the system of monomials $X^\alpha := X_0^{\alpha_0} \cdots X_n^{\alpha_n}$ running over all multi-indices $\alpha = (\alpha_0, \ldots, \alpha_n) \in \mathbb{N}^{n+1}$ such that $|\alpha| = \sum_i \alpha_i$. The dimension of $\mathcal{H}_d$ equals $N_d = \binom{d+n}{d}$. It will be essential to rescale this basis of monomials by considering the basis $\binom{d}{\alpha}^{1/2} X^\alpha$ of $\mathcal{H}_d$ defined with the *multinomial coefficients*

$$\binom{d}{\alpha} := \frac{d!}{\alpha_0! \cdots \alpha_n!}.$$

We call the basis $\{\binom{d}{\alpha}^{1/2} X^\alpha\}_{|\alpha|=d}$ of $\mathcal{H}_d$ *Weyl's basis*. The dot product in this basis defines a Hermitian inner product $\langle\,,\,\rangle$ on $\mathcal{H}_d$. More specifically,

$$\langle f, g \rangle := \sum_\alpha f_\alpha \overline{g}_\alpha$$

for $f = \sum_\alpha \binom{d}{\alpha}^{\frac{1}{2}} f_\alpha X^\alpha$ and $g = \sum_\alpha \binom{d}{\alpha}^{\frac{1}{2}} g_\alpha X^\alpha$ in $\mathcal{H}_d$. We shall call $\langle\,,\,\rangle$ *Weyl's inner product*. It defines a norm on $\mathcal{H}_d$ that we shall denote by $\|\ \|$. The reason to consider this inner product is its invariance under the *unitary group* $\mathscr{U}(n+1)$, which, we recall, is defined as

$$\mathscr{U}(n+1) := \{u \in \mathrm{GL}_{n+1}(\mathbb{C}) \mid uu^* = \mathrm{I}_{n+1}\},$$

where u^* denotes the *adjoint* of u, i.e., $(u^*)_{ij} = \bar{u}_{ji}$.

Every unitary transformation $u \in \mathscr{U}(n+1)$ induces a transformation on $\mathcal{H}_d$ by setting $uf := f \circ u^{-1}$. The invariance just mentioned is stated in the following theorem.

Theorem 16.3 *For all $f, g \in \mathcal{H}_d$ and all $u \in \mathscr{U}(n+1)$ we have*

$$\langle uf, ug \rangle = \langle f, g \rangle.$$

Proof Consider the function

$$K: \mathbb{C}^{n+1} \times \mathbb{C}^{n+1} \to \mathbb{C}, \quad (x, y) \mapsto \langle x, y \rangle^d,$$

where $\langle x, y \rangle := \sum_k x_k \overline{y}_k$ denotes the standard Hermitian inner product on $\mathbb{C}^{n+1}$. It is clear that $K(ux, uy) = K(x, y)$ for $u \in \mathscr{U}(n+1)$ and $x, y \in \mathbb{C}^{n+1}$. Moreover,

$$K_y(x) := K(x, y) = (x_0 \overline{y}_0 + \cdots + x_n \overline{y}_n)^d = \sum_{|\alpha|=d} \binom{d}{\alpha} \overline{y}^\alpha x^\alpha,$$

and hence $K_y \in \mathcal{H}_d$. We conclude that for $f = \sum_\alpha \binom{d}{\alpha}^{1/2} f_\alpha X^\alpha$,

$$\langle f, K_y \rangle = \sum_{|\alpha|=d} f_\alpha \binom{d}{\alpha}^{1/2} y^\alpha = f(y). \tag{16.1}$$

We also note the following transformation behavior:

$$(u K_y)(X) = K_y(u^{-1} X) = \langle u^{-1} X, y \rangle^d = \langle X, uy \rangle^d = K_{uy}(X),$$

and therefore $u K_y = K_{uy}$.

Hence, if $f \in \mathcal{H}_d$ satisfies $\langle f, K_y \rangle = f(y) = 0$ for all y, we have $f = 0$. It follows that the set $\{K_y \mid y \in \mathbb{C}^{n+1}\}$ generates the vector space $\mathcal{H}_d$. So, it is enough to prove the assertion for the functions in $\{K_y \mid y \in \mathbb{C}^{n+1}\}$. We can now conclude, since for $x, y \in \mathbb{C}^{n+1}$,

$$\begin{aligned}
\langle u K_x, u K_y \rangle &= \langle K_{ux}, K_{uy} \rangle \\
&= K_{ux}(uy) \quad \big(\text{by } (16.1)\big) \\
&= \langle ux, uy \rangle^d = \langle x, y \rangle^d = K_y(x) \\
&= \langle K_x, K_y \rangle,
\end{aligned}$$

where the last equality is again obtained by (16.1). $\qquad\square$

Remark 16.4

(a) The function K in the proof of Theorem 16.3 is a kernel, and the space H constructed in this proof (in a more general, infinite-dimensional, case one would have to take a completion) is the reproducing kernel Hilbert space associated to K. The reproducing property justifying the name is the equality in (16.1).

(b) Up to scaling, Weyl's inner product is the only unitarily invariant Hermitian inner product on $\mathcal{H}_d$. This can be readily deduced from the fact that $\mathcal{H}_d$ is an irreducible $\mathscr{U}(n+1)$-module, that is, $\mathcal{H}_d$ does not contain a nontrivial $\mathscr{U}(n+1)$-invariant linear subspace. Remarkably, this uniqueness property fails for the space of real homogeneous polynomials of degree d when one considers orthogonal invariance.

(c) Weyl's inner product can also be described by the integral

$$\langle f, g \rangle = c \int_{\mathbb{S}(\mathcal{H}_d)} f \, \overline{g} \, d\mathbb{S}(\mathcal{H}_d)$$

with respect to the volume measure of the sphere $\mathbb{S}(\mathcal{H}_d)$, where $f, g \in \mathcal{H}_d$ and c denotes a constant. This follows immediately from the uniqueness stated in (b).

(d) We briefly encountered Weyl's norm in the univariate case and denoted it by $\| \ \|_W$ in Sect. 14.1.1.

We proceed with a few observations regarding Weyl's inner product. Throughout, we denote by $\|x\|$ the Euclidean norm of $x \in \mathbb{C}^{n+1}$. We first determine the norm of the evaluation map at x, defined as $\mathrm{eval}_x : \mathcal{H}_d \to \mathbb{C}, \ f \mapsto f(x)$.

Lemma 16.5 *For all $x \in \mathbb{C}^{n+1}$ we have*

$$\|\mathrm{eval}_x\| = \max_{\substack{f \in \mathcal{H}_d \\ \|f\|=1}} \left| f(x) \right| = \|x\|^d.$$

Proof Note that $|f(e_0)| \leq \|f\|$ for $f \in \mathcal{H}_d$, since $f(e_0)$ equals the coefficient of X_0^d in f. Let $u \in \mathcal{U}(n+1)$ be such that $u(x) = \|x\|e_0$. For any $f \in \mathcal{H}_d$,

$$\left| f(x) \right| = \left| uf\big(u(x)\big) \right| = \|x\|^d \left| uf(e_0) \right| \leq \|x\|^d \|uf\| = \|x\|^d \|f\|.$$

This shows that $\max_{\substack{f \in \mathcal{H}_d \\ \|f\|=1}} |f(x)| \leq \|x\|^d$. The reverse inequality is obtained by taking $f = u^{-1} X_0^d$. $\qquad\square$

We next extend the development above to polynomial systems. Given a *degree pattern*, $\mathbf{d} = (d_1, \ldots, d_n)$ we consider the space $\mathcal{H}_{\mathbf{d}} = \mathcal{H}_{d_1} \times \cdots \times \mathcal{H}_{d_n}$. We make $\mathcal{H}_{\mathbf{d}}$ an inner product space by defining, for $f, g \in \mathcal{H}_{\mathbf{d}}$,

$$\langle f, g \rangle = \langle f_1, g_1 \rangle + \cdots + \langle f_n, g_n \rangle, \tag{16.2}$$

and call $\langle \ , \ \rangle$ *Weyl's inner product* on $\mathcal{H}_{\mathbf{d}}$. It defines a norm on $\mathcal{H}_{\mathbf{d}}$ that we again denote by $\| \ \|$. The unitary group $\mathcal{U}(n+1)$ naturally acts on $\mathcal{H}_{\mathbf{d}}$ componentwise, that is, $uf := (uf_1, \ldots, uf_n)$, and as a consequence of Theorem 16.3, Weyl's inner product is invariant under this action.

Summarizing, we have a linear action of the group $\mathcal{U}(n+1)$ on the vector space $\mathcal{H}_{\mathbf{d}}$ that leaves Weyl's inner product invariant. This symmetry property has a profound impact on the design and analysis of algorithms developed here and in the chapters to follow.

Lemma 16.6 *Let $x \in \mathbb{C}^{n+1}$. The linear map*

$$\mathcal{H}_{\mathbf{d}} \to \mathbb{C}^n, \quad f \mapsto \big(f_1(x)/\|x\|^{d_1}, \ldots, f_n(x)/\|x\|^{d_n} \big),$$

maps the unit ball in $\mathcal{H}_{\mathbf{d}}$ onto the Euclidean unit ball in $\mathbb{C}^n$.

Proof Lemma 16.5 implies that $|f_i(x)|/\|x\|^{d_i} \leq \|f_i\|$, for $x \neq 0$ and $f \in \mathcal{H}_\mathbf{d}$. Therefore

$$\sum_i |f_i(x)|^2/\|x\|^{2d_i} \leq \sum_i \|f_i\|^2 = \|f\|^2.$$

The assertion follows immediately from this observation.   □

We also note the following basic observation.

Corollary 16.7 *For all* $u \in \mathcal{U}(n+1)$, $f \in \mathcal{H}_\mathbf{d}$, *and* $\zeta, w \in \mathbb{C}^{n+1}$,

$$D(uf)(u\zeta)(uw) = Df(\zeta)(w).$$

Proof By the chain rule, $D(f \circ u^{-1})(u\zeta) = Df(\zeta) \circ u^{-1}$. The assertion follows by applying this to uw.   □

There is a straightforward relation between a polynomial $f \in \mathcal{H}_d$ and its first-order partial derivatives, known as *Euler's formula*:

$$d \cdot f = \sum_{i=0}^n X_i \frac{\partial f}{\partial X_i}. \tag{16.3}$$

The following is an immediate consequence of this relation.

Corollary 16.8 *If* $f(\zeta) = 0$, *then* $Df(\zeta)(\zeta) = 0$, *for* $f \in \mathcal{H}_\mathbf{d}$ *and* $\zeta \in \mathbb{C}^{n+1}$. *Hence* $\mathbb{C}\zeta = \ker Df(\zeta)$, *and assuming* $\mathrm{rank}\, Df(\zeta) = n$, *we have for all* $w \in \mathbb{C}^n$,

$$Df(\zeta)|_{T_\zeta}^{-1} w = Df(\zeta)^\dagger w.$$ □

16.2 A Unitarily Invariant Condition Number

We are now in position to fit the context of homogeneous polynomial systems into the framework for condition we developed in Sect. 14.3.

We shall consider the vector space $\mathcal{H}_\mathbf{d}$ as the space of inputs and complex projective space $\mathbb{P}^n = \mathbb{P}(\mathbb{C}^{n+1})$ as the space of outputs. The solution manifold is defined as

$$V := \left\{ (f, \zeta) \in \mathcal{H}_\mathbf{d} \times \mathbb{P}^n \mid f(\zeta) = 0 \right\}$$

and comes with the two projections $\pi_1 \colon V \to \mathcal{H}_\mathbf{d}$ and $\pi_2 \colon V \to \mathbb{P}^n$ onto the first and second components, respectively.

In Theorem 16.3 we saw that the unitary group $\mathcal{U}(n+1)$ acts on $\mathcal{H}_\mathbf{d}$ by unitary transformations $f \mapsto f \circ u^{-1}$, for all $u \in \mathcal{U}(n+1)$. Combined with the natural action of $\mathcal{U}(n+1)$ on $\mathbb{P}^n$, we have the action

$$u(f, \zeta) = \left(f \circ u^{-1}, u\zeta \right)$$

of $\mathcal{U}(n+1)$ on the product $\mathcal{H}_{\mathbf{d}} \times \mathbb{P}^n$. It is immediate that V is invariant under this action. Moreover, both π_1 and π_2 are $\mathcal{U}(n+1)$-*equivariant*; that is, $\pi_1(u(f, \zeta)) = u\pi_1(f, \zeta)$ for $u \in \mathcal{U}$ and $(f, \zeta) \in V$, and similarly for π_2 (the proof is obvious).

The group $\mathbb{S}^1 = \{\lambda \in \mathbb{C} \mid |\lambda| = 1\}$ acts on the sphere $\mathbb{S}(\mathbb{C}^{n+1})$ by scalar multiplication. We may interpret $\mathbb{P}^n$ as the set of orbits of the sphere $\mathbb{S}(\mathbb{C}^{n+1})$ under this action. When looking at representatives in the sphere $\mathbb{S}(\mathbb{C}^{n+1})$ of points in projective space $\mathbb{P}^n$, one is led to study the related solution set

$$\hat{V} := \{(f, \zeta) \in \mathcal{H}_{\mathbf{d}} \times \mathbb{S}(\mathbb{C}^{n+1}) \mid f(\zeta) = 0\}.$$

Clearly, V is the image of $\hat{V}$ under the canonical map $\mathcal{H}_{\mathbf{d}} \times \mathbb{S}(\mathbb{C}^{n+1}) \to \mathcal{H}_{\mathbf{d}} \times \mathbb{P}^n$. We may interpret V as the quotient of $\hat{V}$ under the $\mathbb{S}^1$-action.

The next lemma summarizes some important geometric properties of $\hat{V}$ and V.

Lemma 16.9

(a) $\hat{V}$ is a connected smooth submanifold of $\mathcal{H}_{\mathbf{d}} \times \mathbb{S}(\mathbb{C}^{n+1})$ of real dimension $\dim_{\mathbb{R}} \hat{V} = \dim_{\mathbb{R}} \mathcal{H}_{\mathbf{d}} + 1$.

(b) *The tangent space of* $\hat{V}$ *at* (f, ζ) *equals*

$$T_{(f, \zeta)}\hat{V} = \{(\dot{f}, \dot{\zeta}) \in \mathcal{H}_{\mathbf{d}} \times T_{\zeta}\mathbb{S}(\mathbb{C}^{n+1}) \mid \dot{f}(\zeta) + Df(\zeta)\dot{\zeta} = 0\}.$$

(c) V *is a connected complex submanifold of* $\mathcal{H}_{\mathbf{d}} \times \mathbb{P}^n$ *of complex dimension* $\dim_{\mathbb{C}} V = \dim_{\mathbb{C}} \mathcal{H}_{\mathbf{d}}$.

(d) *The tangent space of* V *at* (f, ζ) *equals*

$$T_{(f, \zeta)}V = \{(\dot{f}, \dot{\zeta}) \in \mathcal{H}_{\mathbf{d}} \times T_{\zeta} \mid \dot{f}(\zeta) + Df(\zeta)\dot{\zeta} = 0\}.$$

Here we fixed a representative $\zeta \in \mathbb{C}^{n+1}$ *(denoted by the same symbol) and identified* $T_{\zeta}\mathbb{P}^n$ *with* T_{ζ}.

Proof Write $(\mathbb{C}^{n+1})_* := \mathbb{C}^{n+1} \setminus \{0\}$ and consider the evaluation map

$$F : \mathcal{H}_{\mathbf{d}} \times (\mathbb{C}^{n+1})_* \to \mathbb{C}^n, \quad (f, \zeta) \mapsto f(\zeta).$$

Computing its derivative at (f, ζ) as in the univariate case, cf. (14.1), we obtain

$$DF(f, \zeta) : \mathcal{H}_{\mathbf{d}} \times \mathbb{C}^{n+1} \to \mathbb{C}^n, \quad DF(f, \zeta)(\dot{f}, \dot{\zeta}) = \dot{f}(\zeta) + Df(\zeta)\dot{\zeta}. \quad (16.4)$$

Note that $DF(f, \zeta)$ is surjective, even when restricted to $\mathcal{H}_{\mathbf{d}} \times \{0\}$. If $\hat{F}$ denotes the restriction of F to $\mathcal{H}_{\mathbf{d}} \times \mathbb{S}(\mathbb{C}^{n+1})$, we have $\hat{V} = \hat{F}^{-1}(0)$. Since $D\hat{F}(f, \zeta)$ equals the restriction of $DF(f, \zeta)$ to the tangent space of $\mathcal{H}_{\mathbf{d}} \times \mathbb{S}(\mathbb{C}^{n+1})$ at (f, ζ), the latter is surjective and Theorem A.9 implies that $\hat{V} = F^{-1}(0)$ is a smooth submanifold of $\mathcal{H}_{\mathbf{d}} \times \mathbb{S}(\mathbb{C}^{n+1})$ with

$$\dim_{\mathbb{R}} \hat{V} = \dim_{\mathbb{R}} \mathcal{H}_{\mathbf{d}} + \dim_{\mathbb{R}}(\mathcal{H}_{\mathbf{d}} \times \mathbb{S}(\mathbb{C}^{n+1})) - \dim_{\mathbb{R}} \mathbb{C}^n = \dim_{\mathbb{R}} \mathcal{H}_{\mathbf{d}} + 1.$$

Moreover, by Theorem A.9, the tangent space $T_{(f,\zeta)}\hat{V}$ equals the kernel of $D\hat{F}(f,\zeta)$. This proves the first two assertions except for the connectedness of $\hat{V}$.

To establish the latter, let $(f,\zeta),(\tilde{f},\tilde{\zeta}) \in \hat{V}$. Clearly, (f,ζ) can be connected with $(0,\zeta)$ by a continuous path in $\hat{V}$. Analogously, $(\tilde{f},\tilde{\zeta})$ can be connected with $(0,\tilde{\zeta})$. Since $\mathbb{S}(\mathbb{C}^{n+1})$ is connected, one can connect $(0,\zeta)$ with $(0,\tilde{\zeta})$ by a continuous path in $\hat{V}$.

We leave it to the reader to verify that V is a complex submanifold of $\mathcal{H}_{\mathbf{d}} \times \mathbb{P}^n$. The manifold V is connected, since it is the image of $\hat{V}$ under the canonical map $\mathrm{I} \times \pi_{\mathbb{S}} \colon \mathcal{H}_{\mathbf{d}} \times \mathbb{S}(\mathbb{C}^{n+1}) \to \mathcal{H}_{\mathbf{d}} \times \mathbb{P}^n$. Moreover, the tangent space $T_{(f,\zeta)}V$ equals the image of $T_{(f,\zeta)}\hat{V}$ under the derivative of $\mathrm{I} \times \pi_{\mathbb{S}}$ at (f,ζ), which equals $\mathrm{I} \times D\pi_{\mathbb{S}}(\zeta)$; cf. Lemma 14.9 . This proves the fourth assertion. In addition, $\dim_{\mathbb{R}} V = \dim_{\mathbb{R}} T_{(f,\zeta)}V = \dim_{\mathbb{R}} T_{(f,\zeta)}\hat{V} - 1 = \dim_{\mathbb{R}} \hat{V} - 1 = \dim_{\mathbb{R}} \mathcal{H}_{\mathbf{d}}$. $\square$

As in (14.15) we define the *set of ill-posed solutions* as

$$\Sigma' := \big\{(f,\zeta) \in V \mid \operatorname{rank} D\pi_1(f,\zeta) < \dim \mathcal{H}_{\mathbf{d}}\big\}.$$

If $(f,\zeta) \in V \setminus \Sigma'$, then we say that ζ is a *simple zero* of f. By the implicit function theorem, if ζ is a simple zero of f, there are neighborhoods U_1 and U_2 of f and ζ, respectively, such that for all $f' \in U_1$, there is exactly one zero of f' in U_2. One can show that this is not true if $(f,\zeta) \in \Sigma'$. Hence the name *multiple zero* of f for ζ in this case. We also define the set $\Sigma := \pi_1(\Sigma')$ consisting of those systems $f \in \mathcal{H}_{\mathbf{d}}$ that have some multiple zero ζ.

We may also characterize Σ' as follows:

$$\Sigma' = \big\{(f,\zeta) \in V \mid \operatorname{rank} Df(\zeta) < n\big\}. \tag{16.5}$$

This follows from Eq. (14.17) applied to the evaluation map $F(f,\zeta) = f(\zeta)$, noting that $\partial F/\partial \zeta(f,\zeta) = Df(\zeta)$ by (16.4).

Suppose now that $(f_0,\zeta_0) \in V \setminus \Sigma'$. According to the general principles explained in Sect. 14.3, we can locally invert the projection map $\pi_1 \colon V \to \mathcal{H}_{\mathbf{d}}$ around (f_0,ζ_0). Its inverse $f \mapsto (f,\zeta)$ is defined in terms of a solution map $G \colon U \to \mathbb{P}^n$, $G(f) = \zeta$, that is defined on an open neighborhood U of f in $\mathcal{H}_{\mathbf{d}}$. We consider the derivative $DG(f) \colon \mathcal{H}_{\mathbf{d}} \to T_\zeta \mathbb{P}^n$ of G at f and define the condition number as follows:

$$\mu(f,\zeta) := \|f\| \cdot \|DG(f)\|, \tag{16.6}$$

where the operator norm is defined with respect to Weyl's inner product (16.2) on $\mathcal{H}_{\mathbf{d}}$ and the inner product on $T_\zeta \mathbb{P}^n$ coming from the standard inner product on $\mathbb{C}^{n+1}$.

We shall see next that $\mu(f,\zeta)$ can be expressed in terms of the derivative $Df(\zeta) \colon \mathbb{C}^{n+1} \to \mathbb{C}^n$ of $f \colon \mathbb{C}^{n+1} \to \mathbb{C}^n$. Corollary 16.8 implies that ζ lies in the kernel of $Df(\zeta)$. So, if $Df(\zeta)$ is of full rank, its kernel is $\mathbb{C}\zeta$ and T_ζ equals its orthogonal complement. The inverse of the restriction $Df(\zeta)_{|T_\zeta}$ is described by the Moore–Penrose inverse $Df(\zeta)^{\dagger}$.

Proposition 16.10 *For* $(f, \zeta) \in V \setminus \Sigma'$ *we have*

$$\mu(f, \zeta) = \|f\| \cdot \left\| Df(\zeta)^\dagger \, \mathsf{diag}\big(\|\zeta\|^{d_i - 1}\big) \right\|.$$

Proof We fix a representative ζ and identify $T_\zeta \mathbb{P}^n$ with T_ζ. According to (14.16), the derivative $DG(f)$ can be described in terms of the inverse of the derivative $D\pi_1(f, \zeta)$ of the projection π_1 as follows: we have $\dot{\zeta} = DG(f)(\dot{f})$ iff $(\dot{f}, \dot{\zeta}) \in T_{(f, \zeta)}V$. By Lemma 16.9, this can be restated as $Df(\zeta)(\dot{\zeta}) = -\dot{f}(\zeta)$ with $\dot{\zeta} \in T_\zeta$. Equivalently, $\dot{\zeta} = -Df(\zeta)^\dagger(\dot{f}(\zeta))$. So we have for all $\dot{f} \in \mathcal{H}_{\mathbf{d}}$,

$$DG(f)(\dot{f}) = \dot{\zeta} = -Df(\zeta)^\dagger\big(\dot{f}(\zeta)\big).$$

The operator norm of $DG(\zeta)$ is defined as the maximum of

$$\frac{\|\dot{\zeta}\|}{\|\zeta\|} = \left\| Df(\zeta)^\dagger \frac{1}{\|\zeta\|} \dot{f}(\zeta) \right\| = \left\| Df(\zeta)^\dagger \, \mathsf{diag}\big(\|\zeta\|^{d_i - 1}\big) \, \mathsf{diag}\big(\|\zeta\|^{-d_i}\big) \, \dot{f}(\zeta) \right\|$$

over the $\dot{f}$ in the unit ball in $\mathcal{H}_{\mathbf{d}}$ (compare the definition of the norm on T_ζ in (14.13)). Lemma 16.6 states that $\dot{f} \mapsto \mathsf{diag}(\|\zeta\|^{-d_i}) \dot{f}(\zeta)$ maps the unit ball in $\mathcal{H}_{\mathbf{d}}$ onto the Euclidean unit ball in $\mathbb{C}^n$. We conclude that

$$\|DG(f)\| = \max_{\substack{w \in \mathbb{C}^n \\ \|w\| \leq 1}} \left\| Df(\zeta)^\dagger \, \mathsf{diag}\big(\|\zeta\|^{d_i - 1}\big) w \right\| = \left\| Df(\zeta)^\dagger \, \mathsf{diag}\big(\|\zeta\|^{d_i - 1}\big) \right\|. \qquad \square$$

Remark 16.11 We note that $\mu(f, \zeta)$ defined in (16.6) should be interpreted as a relative normwise condition number. With respect to the input f, the relative nature of $\mu(f, \zeta)$ is obvious (and this is why we multiplied $\mathsf{acond}^G(f) = \|DG(f)\|$ by $\|f\|$ in (16.6)). With respect to the output, the relative nature of $\mu(f, \zeta)$ is built into the choice of the output's space, which is a projective space.

We could also have considered the solution manifold $V_{\mathbb{P}}$ as a subset of $\mathbb{P}(\mathcal{H}_{\mathbf{d}}) \times \mathbb{P}^n$ with the corresponding solution maps $G_{\mathbb{P}} \colon \mathbb{P}(\mathcal{H}_{\mathbf{d}}) \supseteq U \to \mathbb{P}^n$. Had we done so, it would have turned out that

$$\left\| DG_{\mathbb{P}}\big([f]\big) \right\| = \|f\| \cdot \|DG(f)\| = \mu(f, \zeta). \tag{16.7}$$

We leave the straightforward proof of this fact to the reader. Also, it is a good exercise to directly check that $\|Df(\zeta)^\dagger \, \mathsf{diag}(\|\zeta\|^{d_i - 1})\|$ is invariant under the scaling of ζ.

Corollary 16.12 *The condition number* μ *is invariant under the action of* $\mathcal{U}(n+1)$. *That is,* $\mu(u(f, \zeta)) = \mu(f, \zeta)$, *for all* $f \in \mathcal{H}_{\mathbf{d}}$ *and* $u \in \mathcal{U}(n+1)$.

Proof Corollary 16.7 tells us that $D(uf)(u\zeta) = Df(\zeta) \circ u^{-1}$ for $u \in \mathcal{U}(n+1)$. The invariance of μ under $\mathcal{U}(n+1)$ is thus a consequence of Proposition 16.10 combined with the fact that $\mathcal{U}(n+1)$ acts unitarily on $\mathcal{H}_{\mathbf{d}}$ and $\mathbb{C}^{n+1}$. $\qquad \square$

We finish this section with a useful observation, which also illustrates the advantages of working with the projective space of $\mathcal{H}_{\mathbf{d}}$. Suppose that $(f, \zeta) \in V \setminus \Sigma'$ and let $G_{\mathbb{P}} \colon U \to \mathbb{P}^n$ be the solution map defined on an open subset U of the projective space $\mathbb{P}(\mathcal{H}_{\mathbf{d}})$ such that $G_{\mathbb{P}}([f]) = \zeta$. Suppose further that $t \mapsto f_t$ is a smooth curve in $\mathcal{H}_{\mathbf{d}}$ with $f_0 = f$ and put $\zeta_t := G_{\mathbb{P}}([f_t])$.

Lemma 16.13 *We have*

$$\|\dot{\zeta}_t\| \leq \mu(f_t, \zeta_t) \left\| \frac{d}{dt}[f_t] \right\| = \mu(f_t, \zeta_t) \frac{\|P(\dot{f}_t)\|}{\|f_t\|},$$

where P denotes the orthogonal projection of $\mathcal{H}_{\mathbf{d}}$ onto T_{f_t}.

Proof Differentiating $\zeta_t = G_{\mathbb{P}}([f_t])$, we get

$$\dot{\zeta}_t = DG_{\mathbb{P}}([f_t]) \frac{d}{dt}[f_t],$$

which implies by (16.7),

$$\|\dot{\zeta}_t\| \leq \left\| DG_{\mathbb{P}}([f_t]) \right\| \left\| \frac{d}{dt}[f_t] \right\| = \mu(f_t, \zeta_t) \left\| \frac{d}{dt}[f_t] \right\|. \tag{16.8}$$

Lemma 14.8 yields $\frac{d}{dt}[f_t] = \|f_t\|^{-1} P(\dot{f}_t)$ (as usual identifying $T_{[f_t]}\mathbb{P}(\mathcal{H}_{\mathbf{d}})$ with T_{f_t}). This implies the second equality. $\square$

Corollary 16.14 *Let $[0, 1] \to V, t \mapsto (f_t, \zeta_t) \in V$, be a smooth curve such that $f_t \in \mathbb{S}(\mathcal{H}_{\mathbf{d}})$ for all t. Then we have $\|\dot{\zeta}_t\| \leq \mu(f_t, \zeta_t)\|\dot{f}_t\|$.*

Proof We combine Lemma 16.13 with the observation $\|\frac{d}{dt}[f_t]\| \leq \|\frac{d}{dt}f_t\|$, which follows from Lemma 14.9. $\square$

16.3 Orthogonal Decompositions of $\mathcal{H}_{\mathbf{d}}$

We identify here a family of orthogonal decompositions[1] of $\mathcal{H}_{\mathbf{d}}$, parameterized by $\zeta \in \mathbb{P}^n$.

For $\zeta \in (\mathbb{C}^{n+1})_*$ we consider the subspace R_ζ of $\mathcal{H}_{\mathbf{d}}$ consisting of all systems h that vanish at ζ to higher order:

$$R_\zeta := \{h \in \mathcal{H}_{\mathbf{d}} \mid h(\zeta) = 0, Dh(\zeta) = 0\}.$$

---

[1] In fact, we have an orthogonal decomposition of the trivial vector bundle $\mathcal{H}_{\mathbf{d}} \times \mathbb{P}^n \to \mathbb{P}^n$, but we won't use this bundle structure.

We further decompose the orthogonal complement $R_\zeta^\perp$ of R_ζ in $\mathcal{H}_\mathbf{d}$ (defined with respect to Weyl's inner product). Let L_ζ denote the subspace of $R_\zeta^\perp$ consisting of the systems vanishing at ζ and let C_ζ denote its orthogonal complement in $R_\zeta^\perp$. Then we have an orthogonal decomposition

$$\mathcal{H}_\mathbf{d} = C_\zeta \oplus L_\zeta \oplus R_\zeta \tag{16.9}$$

parameterized by $\zeta \in (\mathbb{C}^{n+1})_*$. Note that the spaces C_ζ, L_ζ, and R_ζ depend only on $[\zeta] \in \mathbb{P}^n$. We next show that the above orthogonal decomposition is compatible with the unitary action. Before doing so, recall that the *stabilizer* of ζ is the subgroup $\mathscr{U}_\zeta := \{u \in \mathscr{U}(n+1) \mid u\zeta = \zeta\}$ of $\mathscr{U}(n+1)$.

Lemma 16.15 *Let $\zeta \in (\mathbb{C}^{n+1})_*$. We have $uC_\zeta = C_{u\zeta}$, $uL_\zeta = L_{u\zeta}$, $uR_\zeta = R_{u\zeta}$ for $u \in \mathscr{U}(n+1)$. In particular, the decomposition $\mathcal{H}_\mathbf{d} = C_\zeta \oplus L_\zeta \oplus R_\zeta$ is invariant under the action of the stabilizer $\mathscr{U}_\zeta$ of ζ.*

Proof We first prove the inclusion $uR_\zeta \subseteq R_{u\zeta}$. Suppose $h \in R_\zeta$. Then $(uh)(u\zeta) = h(\zeta) = 0$. Corollary 16.7 implies $D(uh)(u\zeta)(uw) = Dh(\zeta)(w) = 0$ for all $w \in \mathbb{C}^{n+1}$; hence $D(uh)(u\zeta) = 0$. Altogether, $uh \in R_{u\zeta}$. The inclusion shown above implies $u^{-1}R_{u\zeta} \subseteq R_\zeta$ and hence $uR_\zeta = R_{u\zeta}$.

Since u acts unitarily on $\mathcal{H}_\mathbf{d}$, we deduce that $uR_\zeta^\perp = R_{u\zeta}^\perp$. This immediately gives $uL_\zeta = L_{u\zeta}$ and hence $uC_\zeta = C_{u\zeta}$. $\qquad \square$

Let us now have a concrete look at this orthogonal decomposition in the special case $\zeta = e_0 = (1, 0, \ldots, 0)$. Expanding f_i according to the powers of X_0 with respect to decreasing degree, we can write

$$f_i = c_i X_0^{d_i} + X_0^{d_i - 1} \sqrt{d_i} \sum_{j=1}^n a_{ij} X_j + h_i. \tag{16.10}$$

A simple calculation shows that

$$f_i(e_0) = c_i, \quad \partial_{X_0} f_i(e_0) = d_i c_i, \quad \partial_{X_j} f_i(e_0) = \sqrt{d_i}\, a_{ij} \quad \text{for } j \geq 1.$$

Therefore, $f \in R_{e_0}$ iff $c_i = 0$ and $a_{ij} = 0$ for all i, j, which means that $f_i = h_i$.

Suppose now $f \in R_{e_0}^\perp$, which means that $h_i = 0$ for all i. In this case, we have $f \in L_\zeta$ iff $c_i = 0$ for all i. Similarly, $f \in C_\zeta$ iff $a_{ij} = 0$ for all i, j. Furthermore, for $f \in L_\zeta$, by the definition of Weyl's inner product we have $\|f_i\|^2 = \sum_j |a_{ij}|^2$. Indeed, note that $\binom{d_i}{\alpha} = d_i$, where α denotes the exponent vector corresponding to the monomial $X_0^{d_i - 1} X_j$. This observation is the reason to introduce the factors $\sqrt{d_i}$ in (16.10). We also note that $Df(e_0)(w) = (\sqrt{d_i} \sum_{j=1}^n a_{ij} w_j)$ for $w \in T_{e_0}$.

Combining these findings with the unitary invariance of the orthogonal decompositions expressed in Lemma 16.15, we arrive at the following result.

Proposition 16.16

(a) *The space C_ζ consists of the systems $(c_i \langle X, \zeta \rangle^{d_i})$ with $c_i \in \mathbb{C}$. We have $Dk(\zeta)|_{T_\zeta} = 0$ for $k \in C_\zeta$.*
The space L_ζ consists of the systems

$$g = \left(\sqrt{d_i} \, \langle X, \zeta \rangle^{d_i - 1} \ell_i \right),$$

where ℓ_i is a linear form vanishing at ζ. If $\ell_i = \sum_{j=0}^{n} m_{ij} X_j$ with $M = (m_{ij})$, then we may characterize the matrix $M \in \mathbb{C}^{n \times (n+1)}$ by

$$M = \Delta^{-1} Dg(\zeta), \quad \text{where } \Delta = \text{diag}\left(\sqrt{d_i} \|\zeta\|^{d_i - 1} \right).$$

Moreover, we have $M\zeta = 0$ and $\|g\| = \|M\|_F$.

(b) *Given $f \in \mathcal{H}_d$ and $\zeta \in (\mathbb{C}^{n+1})_*$ we have $f = k + g + h \in C_\zeta \oplus L_\zeta \oplus R_\zeta$ with, for $1 = 1, \ldots, n$,*

$$k_i = f_i(\zeta)\langle X, \zeta \rangle^{d_i}, \qquad g_i = \sqrt{d_i}\langle X, \zeta \rangle^{d_i - 1} \sum_{j=0}^{n} m_{ij} X_j,$$

where $m_{ij} = d_i^{-1/2}(\partial_{X_j} f_i(\zeta) - d_i f_i(\zeta)\bar{\zeta}_j)$.

Proof The reasoning above proves both assertions for the special point $\zeta = e_0$. For multiples of e_0, the result follows by scaling appropriately. For a general ζ, it follows by unitary invariance using Lemma 16.15. $\qquad \square$

We determine now the "best conditioned" pairs among all $(f, \zeta) \in V \setminus \Sigma'$, i.e., those for which $\mu(f, \zeta)$ is minimal. The result itself will not be essential in our development, but its proof will. Recall that $P_\zeta : \mathbb{C}^{n+1} \to T_\zeta$ denotes the orthogonal projection onto T_ζ.

Proposition 16.17 *We have*

$$\min_{(f,\zeta)\in V\setminus\Sigma'} \mu(f, \zeta) = \left(\sum_{i=1}^{n} \frac{1}{d_i} \right)^{1/2}.$$

Moreover, the pair (f, ζ) minimizes μ iff $f \in L_\zeta$ and $Df(\zeta) = \sigma P_\zeta$ for some $\sigma > 0$.

Proof Let $\zeta \in \mathbb{S}(\mathbb{C}^{n+1})$ and $f = g + h$ with $g \in L_\zeta$ and $h \in R_\zeta$. Note that $\|f\| \geq \|g\|$ and $Df(\zeta) = Dg(\zeta)$. By Proposition 16.10 we have $\mu(f, \zeta) \geq \mu(g, \zeta)$, and equality holds iff $f = g$. It remains to find the minimum of $\mu(g, \zeta) = \|g\| \|N^\dagger\|$ over $g \in L_\zeta$.

The description of L_ζ in Proposition 16.16(a) tells us that the group $\mathcal{U}(n) \times \mathcal{U}_\zeta$ acts on L_ζ via $(v, u)g := v \circ g \circ u^{-1}$ for $(v, u) \in \mathcal{U}(n) \times \mathcal{U}_\zeta$ and $g \in L_\zeta$. Moreover, $D(v \circ g \circ u^{-1})(\zeta) = vDg(\zeta)u^{-1}$.

By unitary invariance, we may assume that $\zeta = e_0$. Then the first column of $N :=$ $Dg(\zeta)$ equals zero, due to $N\zeta = 0$. Further, using the singular value decomposition (cf. Sect. 1.5), we may assume that N, after the first column has been removed, equals a diagonal matrix $\mathrm{diag}(\sigma_1, \ldots, \sigma_n)$, with the singular values $\sigma_1 \geq \sigma_2 \geq \cdots \geq \sigma_n > 0$. By Proposition 16.16 this implies that

$$g_i = X_0^{d_i - 1} \sigma_i X_i,$$

and we have

$$\|g\|^2 = \sum_{i=1}^{n} \|g_i\|^2 = \sum_{i=1}^{n} \frac{\sigma_i^2}{d_i}, \qquad \|N^\dagger\| = \frac{1}{\sigma_n}.$$

Hence

$$\mu(g, \zeta) = \left(\sum_{i=1}^{n} \frac{\sigma_i^2}{d_i \sigma_n^2} \right)^{1/2} \geq \left(\sum_{i=1}^{n} \frac{1}{d_i} \right)^{1/2}.$$

Equality holds iff $\sigma_j = \sigma_n$ for all j. This means that $N = Dg(\zeta)$ is a multiple of the orthogonal projection P_{e_0}. $\square$

Remark 16.18 Proposition 16.17 identifies the system $\overline{g} \in \mathcal{H}_{\mathbf{d}}$ given by

$$\overline{g}_i = X_0^{d_i - 1} X_i$$

as the only one, up to scaling and unitary invariance, having a zero that is best possibly conditioned, namely $e_0 = (1, 0, \ldots, 0)$. As if by divine justice, all other zeros of $\overline{g}$ are ill-posed.

16.4 A Condition Number Theorem

Proposition 16.16 suggests that we modify the definition of the condition number $\mu(f, \zeta)$ by introducing additional scaling factors $\sqrt{d_i}$. We will see that this leads to an elegant characterization of condition as an inverse distance to ill-posedness.

We define the *normalized condition number* $\mu_{\mathrm{norm}}(f, \zeta)$ for $(f, \zeta) \in V$ by

$$\mu_{\mathrm{norm}}(f, \zeta) := \|f\| \cdot \big\| Df(\zeta)^\dagger \, \mathrm{diag}\big(\sqrt{d_i} \|\zeta\|^{d_i - 1}\big) \big\|. \tag{16.11}$$

Note that the introduction of the $\sqrt{d_i}$ factors is the only change compared with $\mu(f, \zeta)$; cf. Proposition 16.10. As for $\mu(f, \zeta)$, we note that $\mu_{\mathrm{norm}}(f, \zeta)$ does not depend on the choice of a representative of ζ and it is thus well defined. Moreover, as in Corollary 16.12, one can show that $\mu_{\mathrm{norm}}(f, \zeta)$ is $\mathcal{U}(n + 1)$-invariant.

Setting $D := \max_i d_i$, we clearly have

$$\mu(f, \zeta) \leq \mu_{\mathrm{norm}}(f, \zeta) \leq \sqrt{D} \mu(f, \zeta). \tag{16.12}$$

We shall now prove a condition number theorem that expresses $\mu_{\text{norm}}(f, \zeta)$ as an inverse distance to ill-posedness.

Fix $\zeta \in (\mathbb{C}^{n+1})_*$ and consider the vector space $V_\zeta = \{f \in \mathcal{H}_{\mathbf{d}} : f(\zeta) = 0\}$, that is, $V_\zeta = L_\zeta \oplus R_\zeta$. We focus here on the corresponding projective space $\mathbb{P}(V_\zeta)$ and, as in Sect. 14.2.2, denote by d_{sin} the sine distance on $\mathbb{P}(V_\zeta)$.

Now consider the set $\Sigma_\zeta := \{[\tilde{f}] \mid (\tilde{f}, \zeta) \in \Sigma'\}$ of systems for which ζ is a multiple zero and let $d_{\text{sin}}([f], \Sigma_\zeta)$ denote the minimum sine distance of $[f]$ to Σ_ζ. According to Lemma 14.13, this quantity may be characterized by

$$d_{\text{sin}}\big([f], \Sigma_\zeta\big) = \min_{\tilde{f} \in \Sigma_\zeta \backslash 0} \frac{\|f - \tilde{f}\|}{\|f\|}. \tag{16.13}$$

Theorem 16.19 *For* $(f, \zeta) \in V$ *we have*

$$\mu_{\text{norm}}(f, \zeta) = \frac{1}{d_{\text{sin}}([f], \Sigma_\zeta)}.$$

Proof We decompose $f \in V_\zeta$ and $\tilde{f} \in \Sigma_\zeta$ as

$$f = g + h, \qquad \tilde{f} = \tilde{g} + \tilde{h} \quad \text{with } g, \tilde{g} \in L_\zeta \text{ and } h, \tilde{h} \in R_\zeta.$$

Consider the diagonal matrix $\Delta = \text{diag}(\sqrt{d_i} \|\zeta\|^{d_i - 1})$ and define the following matrices in $\mathbb{C}^{n \times (n+1)}$:

$$M := \Delta^{-1} Df(\zeta), \qquad \tilde{M} := \Delta^{-1} D\tilde{f}(\zeta).$$

We note that rank $\tilde{M} < n$, since ζ is a multiple zero of $\tilde{f}$. Since $g - \tilde{g} \in L_\zeta$ and

$$M - \tilde{M} = \Delta^{-1} D(f - \tilde{f})(\zeta) = \Delta^{-1} D(g - \tilde{g})(\zeta),$$

Proposition 16.16 implies that

$$\|g - \tilde{g}\| = \|M - \tilde{M}\|_F.$$

The characterization of the (Moore–Penrose) matrix condition number as relativized inverse distance to singularity in Corollary 1.27 implies

$$\|M - \tilde{M}\|_F \geq \|M^\dagger\|^{-1}. \tag{16.14}$$

By the orthogonality of the decomposition $V_\zeta = L_\zeta + R_\zeta$ and the Pythagorean theorem we have

$$\|f - \tilde{f}\|^2 = \|g - \tilde{g}\|^2 + \|h - \tilde{h}\|^2 \geq \|g - \tilde{g}\|^2 = \|M - \tilde{M}\|_F^2.$$

Altogether, we obtain

$$\frac{\|f - \tilde{f}\|}{\|f\|} \geq \frac{\|M - \tilde{M}\|_F}{\|f\|} \geq \frac{1}{\|f\| \|M^\dagger\|} = \frac{1}{\mu_{\text{norm}}(f, \zeta)}.$$

With Eq. (16.13) this implies

$$d_{\sin}\bigl([f], \Sigma_\zeta\bigr) \geq \frac{1}{\mu_{\mathrm{norm}}(f, \zeta)}.$$

In order to show that equality holds, it suffices to trace back the proof. According to Corollary 1.27 there exists a singular matrix $\tilde{M} \in \mathbb{C}^{n \times (n+1)}$ such that equality holds in (16.14). Let $\tilde{g}$ be the corresponding system in L_ζ such that $\tilde{M} = \Delta^{-1} D\tilde{g}(\zeta)$, and put $\tilde{f} := \tilde{g}$, so that $\tilde{h} = 0$. Then we have

$$\|f - \tilde{f}\|^2 = \|g - \tilde{g}\|^2 = \|M - \tilde{M}\|_F^2. \qquad \square$$

We remark that Theorem 16.19 again implies that the condition number μ_{norm} is unitarily invariant.

Example 16.20 Consider now the following particular system $\bar{U} \in \mathcal{H}_{\mathbf{d}}$ defined as

$$\bar{U}_1 = \frac{1}{\sqrt{2n}}\bigl(X_0^{d_1} - X_1^{d_1}\bigr), \quad \ldots, \quad \bar{U}_n = \frac{1}{\sqrt{2n}}\bigl(X_0^{d_n} - X_n^{d_n}\bigr), \qquad (16.15)$$

where the scaling factor guarantees that $\|\bar{U}\| = 1$. This system will serve as the starting system in a homotopy continuation algorithm studied in Chap. 18.

Denote by $z_{(a)}$ a d_ith primitive root of unity. The zeros of $\bar{U} = (\bar{U}_1, \ldots, \bar{U}_n)$ are then the points $\mathbf{z}_j = [(1, z_{(a)}^{j_1}, \ldots, z_{(n)}^{j_n})] \in \mathbb{P}^n$ for all the possible tuples $j = (j_1, \ldots, j_n)$ with $j_i \in \{0, \ldots, d_i - 1\}$. Clearly, each $\mathbf{z}_j$ can be obtained from $\mathbf{z}_1 := [(1, 1, \ldots, 1)]$ by a unitary transformation u_j that leaves $\bar{U}$ invariant, that is,

$$u_j \mathbf{z}_1 = \mathbf{z}_j \quad \text{and} \quad u_j \bar{U} = \bar{U}.$$

The following lemma results from the unitary invariance of our setting. The proof is immediate.

Lemma 16.21 *Let $g \in \mathcal{H}_{\mathbf{d}}$, $\zeta \in \mathbb{P}^n$ a zero of g, and $u \in \mathcal{U}(n + 1)$. Then $\mu_{\mathrm{norm}}(g, \zeta) = \mu_{\mathrm{norm}}(ug, u\zeta)$.* $\qquad \square$

Note that Lemma 16.21 implies $\mu_{\mathrm{norm}}(\bar{U}, \mathbf{z}_j) = \mu_{\mathrm{norm}}(\bar{U}, \mathbf{z}_1)$ for all j. The following result gives an upper bound for these condition numbers.

Lemma 16.22 *Let $D := \max_i d_i$. Then*

$$\mu_{\mathrm{norm}}^2(\bar{U}, \bar{\zeta}) \leq 2n \max_{i \leq n} \frac{1}{d_i}(n + 1)^{d_i - 1}$$

with equality if $d_i = D$ for all i. In particular, $\mu_{\mathrm{norm}}^2(\bar{U}, \bar{\zeta}) \leq 2(n + 1)^D$.

Proof Put $M := \text{diag}(d_i^{-\frac{1}{2}} \|\bar{\zeta}\|^{1-d_i}) D\bar{U}(\bar{\zeta}) \in \mathbb{C}^{n \times (n+1)}$. By definition (16.11) we have

$$\mu_{\text{norm}}(\bar{U}, \bar{\zeta}) = \|\bar{U}\| \, \|M^\dagger\| = \|M^\dagger\| = \frac{1}{\sigma_{\min}(M)},$$

where $\sigma_{\min}(M)$ denotes the smallest singular value of M. It can be characterized as a constrained minimization problem as follows:

$$\sigma_{\min}^2(M) = \min_{w \in \mathbb{C}^{n+1}} \|Mw\|^2 \quad \text{subject to} \quad w \in (\ker M)^\perp, \quad \|w\|^2 = 1.$$

In our situation, $\ker M = \mathbb{C}(1, \ldots, 1)$ and $D\bar{U}(\bar{\zeta})$ is given by the following matrix:

$$D\bar{U}(\bar{\zeta}) = \frac{1}{\sqrt{2n}} \begin{bmatrix} d_1 & -d_1 & 0 & \cdots & 0 \\ d_2 & 0 & -d_2 & \cdots & 0 \\ \vdots & \vdots & \vdots & \ddots & 0 \\ d_n & 0 & 0 & & -d_n \end{bmatrix}.$$

Hence for $w = (w_0, \ldots, w_n) \in \mathbb{C}^{n+1}$,

$$\|Mw\|^2 = \frac{1}{2n} \sum_{i=1}^{n} \frac{d_i}{(n+1)^{d_i-1}} |w_i - w_0|^2 \geq \frac{1}{2n} \min_i \frac{d_i}{(n+1)^{d_i-1}} \cdot \sum_{i=1}^{n} |w_i - w_0|^2,$$

with equality holding if $d_i = D$. A straightforward calculation shows that

$$\sum_{i=1}^{n} |w_i - w_0|^2 \geq 1 \quad \text{if} \quad \sum_{i=0}^{n} w_i = 0, \quad \sum_{i=0}^{n} |w_i|^2 = 1.$$

The assertion follows by combining these observations.   □

16.5 Bézout's Theorem

Let us further study the solution manifold V with its two projections π_1, π_2. The fiber $V_\zeta := \{ f \in \mathcal{H}_\mathbf{d} \mid (f, \zeta) \in V \}$ of $\pi_2 \colon V \to \mathbb{P}^n$ over any $\zeta \in \mathbb{P}^n$ is clearly a linear subspace[2] of $\mathcal{H}_\mathbf{d}$ with complex codimension n. It decomposes as $V_\zeta = L_\zeta \oplus R_\zeta$.

We now have a look at the fiber $\pi_1^{-1}(f)$, which can be identified with the set $Z_\mathbb{P}(f) := \{ \zeta \in \mathbb{P}^n \mid f_1(\zeta) = 0, \ldots, f_n(\zeta) = 0 \}$ of common zeros of $f_1, \ldots, f_n$. Recall that $\Sigma \subset \mathcal{H}_\mathbf{d}$ denotes the set of systems $f \in \mathcal{H}_\mathbf{d}$ having a multiple zero.

The following result is the celebrated *Bézout's theorem*. It states that the fibers $\pi_1^{-1}(f)$ are finite with $\mathcal{D} := d_1 \cdots d_n$ elements, provided $f \in \mathcal{H}_\mathbf{d} \setminus \Sigma$. One calls $\mathcal{D}$ the *Bézout number*. We shall prove this result using a non-algorithmic version of the

---

[2]One can even show that $\pi_2 \colon V \to \mathbb{P}^n$ is a vector bundle, but again, we will not need this here.

homotopy continuation we saw in Sect. 15.1. In the next two chapters we will see that with considerable more effort, the idea underlying this existence proof can be converted into an efficient numerical algorithm. A main goal of the third part of this book is the analysis of this algorithm and its variations.

Theorem 16.23 *The zero set* $Z_{\mathbb{P}}(f) = \{\zeta \in \mathbb{P}^n \mid f_1(\zeta) = 0, \ldots, f_n(\zeta) = 0\}$ *of a system of homogeneous polynomials* $(f_1, \ldots, f_n) \in \mathcal{H}_\mathbf{d} \setminus \Sigma$ *is finite and has exactly* $\mathcal{D} = d_1 \cdots d_n$ *elements. Recall that* $d_i = \deg f_i$.

The proof relies on a few concepts and results from algebraic geometry; cf. Sect. A.3.

Lemma 16.24 Σ' *is the zero set of finitely many polynomial functions of* $(f, \zeta) \in \mathcal{H}_\mathbf{d} \times \mathbb{C}^{n+1}$ *that are homogeneous in the arguments* f *and* ζ.

Proof We have rank $Df(\zeta) < n$ iff the determinant of all of the $n \times n$ submatrices of $Df(\zeta)$ vanish. Since the entries of $Df(\zeta)$ are linear in f and homogeneous in ζ, the assertion follows. $\square$

Proposition 16.25 *The image* Σ *of* Σ' *under the projection* $\pi_1 \colon V \to \mathcal{H}_\mathbf{d}$ *is an algebraic variety, closed under multiplication by complex scalars. It is called the discriminant variety.*

Proof Recall that Σ equals the image of Σ' under the projection $\pi_1 \colon V \to \mathcal{H}_\mathbf{d}$. The assertion is a consequence of Lemma 16.24 combined with the main theorem of elimination theory; cf. Theorem A.39. $\square$

Remark 16.26 One can show that Σ is the zero set of a single polynomial, called the *multivariate discriminant* in N variables with integer coefficients (cf. Sect. A.3.5). This implies that Σ is a complex hypersurface in $\mathcal{H}_\mathbf{d}$. There exists a well-defined notion of dimension for algebraic varieties (which are not necessarily submanifolds); cf. Sect. A.3.2. It is known that Σ is of complex codimension one and of real codimension two in $\mathcal{H}_\mathbf{d}$. This makes it intuitively plausible that $\mathcal{H}_\mathbf{d} \setminus \Sigma$ is connected (in the Euclidean topology). The next result provides a formal proof of this important fact.

Corollary 16.27 *The complement* $\mathcal{H}_\mathbf{d} \setminus \Sigma$ *of the discriminant variety is connected.*

Proof By Proposition 16.25, Σ is the zero set of a system $F_1, \ldots, F_s$ of homogeneous polynomials. Let $f, g \in \mathcal{H}_\mathbf{d} \setminus \Sigma$. We may assume that f and g are $\mathbb{C}$-linearly independent and denote by E the complex span of f, g. Then $E \cap \Sigma$ is the zero set of the restrictions $F_{1|E}, \ldots, F_{s|E}$. There exists i such that $F_{i|E} \neq 0$, since $f \in E \setminus \Sigma$; hence $E \cap \Sigma \neq E$. We need to show that f and g can be connected by a continuous path in $E \setminus (E \cap \Sigma)$.

In order to see this, we note that the image of $E \cap \Sigma$ under the canonical projection $E \setminus 0 \to \mathbb{P}(E), q \mapsto [q]$, is contained in the zero set of the (homogeneous bivariate) polynomial $F_{i|E_i} \neq 0$, which thus consists of finitely many points in $\mathbb{P}(E)$. Moreover, it is known that $\mathbb{P}(E)$ is homeomorphic to a (Riemann) sphere. Removing finitely many points from $\mathbb{P}(E)$ cannot destroy connectedness. Hence $[f]$ and $[g]$ can be connected by a continuous path in $\mathbb{P}(E)$ avoiding these points. This path can be lifted to $E \setminus (E \cap \Sigma)$ as in the proof of Proposition 14.12. $\square$

Proof of Theorem 16.23 The system $\bar{U}$ from Example 16.20 (omitting the scaling factors),

$$\bar{U}_1 = X_1^{d_1} - X_0^{d_1}, \quad \ldots, \quad \bar{U}_n = X_n^{d_n} - X_0^{d_n},$$

has exactly $\mathcal{D} = d_1 \cdots d_n$ zeros. They are of the form $(\zeta_1, \ldots, \zeta_n)$, where ζ_j runs through all d_jth roots of unity $\exp(\frac{2\pi i k}{d_j})$ for $0 \leq k \leq d_j - 1$. It is straightforward to check that all these zeros ζ are simple and hence $g \notin \Sigma$.

We consider the following restriction of the projection π_1:

$$\varphi \colon V \setminus \pi_1^{-1}(\Sigma) \to \mathcal{H}_{\mathbf{d}} \setminus \Sigma, \quad (f, \zeta) \mapsto f.$$

By Corollary 16.27 we know that $\mathcal{H}_{\mathbf{d}} \setminus \Sigma$ is connected. We shall prove that the function $\chi \colon V \setminus \Sigma' \to \mathbb{N}, f \mapsto |\varphi^{-1}(f)|$, is well defined and locally constant. Then the theorem follows, since $\chi(\bar{U}) = |\varphi^{-1}(\bar{U})| = \mathcal{D}$.

We first argue that the fibers of φ are finite. Note that $\varphi^{-1}(f)$ is a closed subset of $\{f\} \times \mathbb{P}^n$ and thus compact. The inverse function theorem implies that $\varphi^{-1}(f)$ is a discrete set. (This means that for each f, there exists a neighborhood W' such that $W' \cap \varphi^{-1}(f)$ consists of f only.) However, a compact discrete set must be finite. It follows that the fibers of φ are finite.

Pick now any $\tilde{f} \in V \setminus \pi^{-1}(\Sigma)$ and let $\varphi^{-1}(\tilde{f}) = \{\zeta_1, \ldots, \zeta_k\}$. By the implicit function theorem, there exists an open neighborhood $W \subseteq \mathcal{H}_{\mathbf{d}} \setminus \Sigma$ of $\tilde{f}$ and there exist pairwise disjoint open subsets $W'_1, \ldots, W'_k$ of $V \setminus \pi_1^{-1}(\Sigma)$ with $(\tilde{f}, \zeta_i) \in W'_i$ such that $\varphi^{-1}(W) = W'_1 \cup \cdots \cup W'_k$ and such that for each i, the projection $W'_i \to W, (f, \zeta) \mapsto f$, is bijective (actually, a diffeomorphism). It follows that $\varphi^{-1}(f)$ has exactly k elements for all $f \in W$, and hence χ is locally constant. $\square$

We finish this section with a further result that will be of great relevance in Chap. 17. Recall that $\mathrm{aff}(f, g)$ denotes the *real* line passing through two distinct points f and g in $\mathcal{H}_{\mathbf{d}}$.

Lemma 16.28 *For all $f \in \mathcal{H}_{\mathbf{d}} \setminus \Sigma$ the set $\{g \in \mathcal{H}_{\mathbf{d}} \setminus \{f\} \mid \mathrm{aff}(f, g) \cap \Sigma \neq \emptyset\}$ has measure zero.*

Proof By Proposition 16.25, Σ is an algebraic variety. Since $\Sigma \neq \mathcal{H}_{\mathbf{d}}$, we have $\dim_{\mathbb{C}} \Sigma \leq \dim_{\mathbb{C}} \mathcal{H}_{\mathbf{d}} - 1$. Hence $\dim_{\mathbb{R}} \Sigma = 2 \dim_{\mathbb{C}} \Sigma \leq \dim_{\mathbb{R}} \mathcal{H}_{\mathbf{d}} - 2$. The assertion is now an immediate consequence of Corollary A.36. $\square$

16.6 A Projective Newton's Method

In this section we extend Newton's method to projective space. More precisely, for $f \in \mathcal{H}_\mathbf{d}$ having at least one simple zero, we shall define a map $N_f : \mathbb{P}^n \to \mathbb{P}^n$ (defined almost everywhere) with properties similar to those of Newton's method in $\mathbb{C}^n$ we saw in Sect. 15.2. In particular, we prove a projective version of Smale's γ-theorem.

Let $(f, z) \in \mathcal{H}_\mathbf{d} \times (\mathbb{C}^{n+1})_*$. The derivative of f at z is a linear map $Df(z) : \mathbb{C}^{n+1} \to \mathbb{C}^n$. Suppose that its restriction $Df(z)|_{T_z}$ to the subspace T_z is invertible. Then we can define the value of the projective Newton operator associated to f, at z, by

$$N_f(z) := z - Df(z)|_{T_z}^{-1} f(z).$$

We next verify that N_f can be interpreted as mapping points from $\mathbb{P}^n$ to $\mathbb{P}^n$.

Lemma 16.29 *We have* $Df(\lambda z)|_{T_z}^{-1} f(\lambda z) = \lambda Df(z)|_{T_z}^{-1} f(z)$ *for* $\lambda \in \mathbb{C}_*$. *Hence* $N_f(\lambda z) = \lambda N_f(z)$.

Proof The assertion is a consequence of

$$f(\lambda z) = \mathrm{diag}(\lambda^{d_i}) f(z), \quad Df(\lambda z) = \mathrm{diag}(\lambda^{d_i - 1}) Df(z),$$

which follows from the homogeneity of f. $\qquad\square$

Example 16.30 Since T_z is defined in terms of the Hermitian inner product, the definition of N_f involves not only operations of the field $\mathbb{C}$, but also complex conjugation. So N_f is not a rational map over $\mathbb{C}$. We illustrate this in the case $n = 1$. It is easy to check that T_z is spanned by $(-\bar{z}_1, \bar{z}_0)$. Solving the equation $Df(z)\lambda(-\bar{z}_1, \bar{z}_0)^T = f(z)$ for λ yields

$$N_f(z) = \begin{bmatrix} z_0 \\ z_1 \end{bmatrix} - \frac{f(z)}{-\bar{z}_1 \partial_{z_0} f + \bar{z}_0 \partial_{z_1} f} \begin{bmatrix} -\bar{z}_1 \\ \bar{z}_0 \end{bmatrix}.$$

We next investigate the cost of one Newton step.

Lemma 16.31 *Let* $z \in \mathbb{C}^{n+1}$. *A homogeneous polynomial* $f \in \mathcal{H}_d$ *can be evaluated at* z *with* $3\binom{n+d}{n} + d - 3$ *arithmetic operations. For a system* $f \in \mathcal{H}_\mathbf{d}$, *assuming* $d_i \geq 2$ *for all* i, *we can compute both* $f(z)$ *and* $\|f(z)\|$ *with* $\mathcal{O}(N)$ *operations.*

Proof Let $T(n, d)$ be the number of additions and multiplications sufficient to compute any fixed $f \in \mathcal{H}_d$ from the powers $X_0, X_0^2, \ldots, X_0^d$, the variables $X_1, \ldots, X_n$, and complex numbers. Any linear form $f = \sum_{i=0}^n a_i X_i$ can be evaluated with $n + 1$ multiplications and n additions, whence $T(n, 1) \leq 2n + 1$.

It is easy to see that any $f \in \mathcal{H}_d$ can be written as $f = aX_0^d + \sum_{i=1}^n f_i X_i$, where $a \in \mathbb{C}$ and f_i is a homogeneous polynomial of degree $d - 1$ in the variables $X_0, \ldots, X_i$. This implies the following recurrence

$$T(n,d) \le \sum_{i=1}^n T(i, d-1) + 2n + 1.$$

Induction on d proves that $T(n,d) \le 3\binom{n+d}{n} - 2$. Since $X_0^2, \ldots, X_0^d$ can be computed with $d - 1$ further multiplications, the first assertion follows.

For $f \in \mathcal{H}_\mathbf{d}$ write $N_i = \binom{n+d_i}{n}$ and $N := \sum_{i=1}^n N_i$. We have just seen that we can compute $f(z)$ from the coefficients of f and z with $\sum_{i=1}^n (3N_i - 2) + d - 1 = 3N - 2n + d - 1 = \mathcal{O}(N)$ arithmetic operations. The computation of $\|f(z)\|$ from $f(z)$ has cost $\mathcal{O}(n)$. $\square$

Proposition 16.32 *One Newton step, i.e., the evaluation of $N_f(z)$ from the coefficients of $f \in \mathcal{H}_\mathbf{d}$ and $z \in \mathbb{C}_*^{n+1}$, can be performed with $\mathcal{O}(N + n^3)$ arithmetic operations. If $d_i \ge 2$ for all i, then this is $\mathcal{O}(N)$.*

Proof Based on Lemma 16.31, the Jacobian matrix $Df(z)$ can be computed with $\mathcal{O}(nN)$ arithmetic operations. By a more sophisticated reasoning, based on a general transformation of straight-line programs, one can show that in fact, $\mathcal{O}(N)$ operations are sufficient for this. Instead of proving this fact, we refer to the Notes for references.

Further, by linear algebra over $\mathbb{R}$, one can compute $Df(z)|_{T_z}^{-1} f(z)$ from $Df(z)$ and $f(z)$ with $\mathcal{O}(n^3)$ operations. Hence $\mathcal{O}(N + n^3)$ arithmetic operations are sufficient for evaluating $N_f(z)$. Moreover, if we assume that $d_i \ge 2$ for all i, then we have $n^2 = \mathcal{O}(N_i)$ and hence $n^3 = \mathcal{O}(N)$. $\square$

The *projective Newton operator associated with* f is the map

$$N_f \colon \mathbb{P}^n \setminus \Lambda_f \to \mathbb{P}^n, \quad N_f(z) = z - Df(z)|_{T_z}^{-1} f(z)$$

defined on the complement of the following subset of $\mathbb{P}^n$:

$$\Lambda_f := \left\{ z \in \mathbb{P}^n \mid Df(z)|_{T_z} \text{ not invertible} \right\}. \tag{16.16}$$

Note also that $N_{\lambda f} = N_f$ for $\lambda \in \mathbb{C}_*$, so that N_f depends on f only as an element of $\mathbb{P}(\mathcal{H}_\mathbf{d})$. Moreover, for $\zeta \notin \Lambda_f$, we have $N_f(\zeta) = \zeta$ iff $f(\zeta) = 0$.

The following result tells us that N_f is defined almost everywhere.

Lemma 16.33 *If $f \in \mathcal{H}_\mathbf{d}$ has a simple zero ζ, then Λ_f has measure zero in $\mathbb{P}^n$.*

Proof First, by (16.5), $(f, \zeta) \notin \Sigma'$ means that $\text{rank} \, Df(\zeta) = n$. Because of Corollary 16.8, we see that $\zeta \notin \Lambda_f$. Hence Λ_f is properly contained in $\mathbb{P}^n$.

Let $\hat{\Lambda}_f \subseteq (\mathbb{C}^{n+1})_*$ denote the cone corresponding to Λ_f. We shall view $\mathbb{C}^{n+1}$ as the real vector space $\mathbb{R}^{2n+2}$.

Claim $\hat{\Lambda}_f$ is the zero set of a system of homogeneous real polynomials.

In other words, $\hat{\Lambda}_f$ corresponds to a real projective variety. Corollary A.36 implies that $\hat{\Lambda}_f$ has measure zero in $\mathbb{C}^{n+1}$, which will complete the proof (cf. Sect. A.2.4).

In order to prove the claim, consider the orthogonal projection onto T_z:

$$P_z \colon \mathbb{C}^{n+1} \to T_z, \quad P_z(w) = w - \|z\|^{-2} \langle w, z \rangle z.$$

We have $z \in \hat{\Lambda}_f$ iff $\mathrm{rank}(Df(z)\|z\|^2 P_z) < n$. The latter means that the determinant of all of the $n \times n$ submatrices $A(z)$ of $Df(z)\|z\|^2 P_z$ vanish. Now note that (with e_j denoting the standard basis) $\|z\|^2 P_z(e_j) = \|z\|^2 e_j - \bar{z}_j z_i$. Hence the real and imaginary parts of $\|z\|^2 P_z$ are homogeneous quadratic polynomials in the real and imaginary parts of the z_j. It follows that the real and the imaginary parts of $\det A(z)$ are homogeneous polynomials in the real and imaginary parts of the z_j as well. This proves the claim. $\square$

It is now natural to extend the notion of approximate zero (Definition 15.1) from $\mathbb{C}^n$ to $\mathbb{P}^n$. We shall measure distances in $\mathbb{P}^n$ using the Riemannian distance $d_{\mathbb{P}}$, i.e., by the angle, as defined in Proposition 14.12.

Definition 16.34 We say that $z \in \mathbb{P}^n$ is an *approximate zero* of $f \in \mathcal{H}_{\mathbf{d}}$ if the sequence given by $z_0 = z$ and $z_{i+1} = N_f(z_i)$ is defined for all natural numbers i, and there exists $\zeta \in \mathbb{P}^n$ with $f(\zeta) = 0$ such that for all i,

$$d_{\mathbb{P}}(z_i, \zeta) \le \left(\frac{1}{2}\right)^{2^i - 1} d_{\mathbb{P}}(z, \zeta).$$

We say that ζ is the *associated zero* of z.

Note that if z is an approximate zero of f with associated zero ζ, then one Newton step reduces the distance to ζ by a factor of two: $d_{\mathbb{P}}(z_1, \zeta) \le \frac{1}{2} d_{\mathbb{P}}(z, \zeta)$.

We define now a projective version of the invariant γ introduced in Definition 15.3 for Euclidean space.

Definition 16.35 For $(f, z) \in \mathcal{H}_{\mathbf{d}} \times \mathbb{C}^{n+1}_*$ such that $Df(z)|_{T_z}$ is invertible we define

$$\gamma_{\mathrm{proj}}(f, z) := \|z\| \sup_{k \ge 2} \left\| Df(z)|_{T_z}^{-1} \frac{D^k f(z)}{k!} \right\|^{\frac{1}{k-1}}.$$

If $Df(z)|_{T_z}$ is not invertible, we set $\gamma_{\mathrm{proj}}(f, z) := \infty$.

Fig. 16.1 Graph of ψ_δ for, from top to bottom, $\delta = 0, \frac{\pi}{6}, \frac{\pi}{3}$, and $\frac{\pi}{2}$

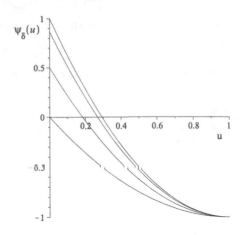

Note that the existence of the supremum follows as for γ; cf. Definition 15.3.

Lemma 16.36 *For all* $\lambda \in \mathbb{C}_*$ *we have* $\gamma_{\text{proj}}(f, \lambda z) = \gamma_{\text{proj}}(f, z)$ *and* $\gamma_{\text{proj}}(\lambda f, z) = \gamma_{\text{proj}}(f, z)$. *In particular,* γ_{proj} *induces a function* $V \setminus \Sigma' \to \mathbb{R}$.

Proof By homogeneity we have

$$D^k f(\lambda z) = \text{diag}\left(\lambda^{d_i - k}\right) D^k f(z).$$

In particular, $Df(\lambda z) = \text{diag}(\lambda^{d_i - 1}) Df(z)$. This implies

$$\left\| Df(\lambda z)\big|_{T_z}^{-1} D^k f(\lambda z) \right\| = \frac{1}{|\lambda|^{k-1}} \left\| Df(z)\big|_{T_z}^{-1} D^k f(z) \right\|,$$

and it follows that $\gamma_{\text{proj}}(f, \lambda \zeta) = \gamma_{\text{proj}}(f, \zeta)$. The second assertion is trivial. $\qquad\square$

For the statement of the projective γ-theorem below we need to define certain numerical quantities.

For $0 \le \delta < \pi/2$ let us consider the following family of quadratic functions:

$$\psi_\delta(u) := (1 + \cos \delta)(1 - u)^2 - 1 = (1 + \cos \delta)u^2 - 2(1 + \cos \delta)u + \cos \delta; \quad (16.17)$$

cf. Fig. 16.1.

For $\delta = 0$ we retrieve the function ψ introduced in Sect. 15.2. We note that $\psi_\delta(0) = \cos \delta$ and $\psi_\delta(a) = -1$. Moreover, $\psi_{\delta'}(u) \ge \psi_\delta(u)$ for $\delta' \le \delta$.

For $\frac{2}{\pi} \le r \le 1$ we define $\delta(r)$ as the smallest nonnegative real number δ such that $r\delta = \sin \delta$. Then we have

$$r\delta < \sin \delta, \quad \text{for } 0 \le \delta < \delta(r).$$

For example, taking $r = 2/\pi$, we get $\delta(r) = \pi/2$.

We also define $u(r)$ as the smallest nonnegative number u satisfying the equation

$$\frac{u}{r\psi_{\delta(r)}(u)} = \frac{1}{2}.$$

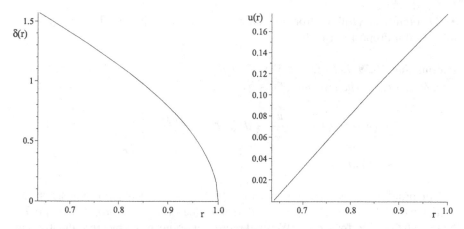

Fig. 16.2 The values of $\delta(r)$ (*left*) and $u(r)$ (*right*) as functions of $r \in [\frac{2}{\pi}, 1]$

Table 16.1 Some examples of r, $\delta(r)$, and $u(r)$

r	$\delta(r)$	$u(r)$
1	0	$\frac{3-\sqrt{7}}{2}$
$0.99991\ldots$	$\frac{1}{45}$	$0.17708\ldots$
$0.99500\ldots$	$0.17333\ldots$	$0.17486\ldots$
$0.88800\ldots$	$0.83415\ldots$	$0.12469\ldots$

Then we have, for $0 < \delta \le \delta(r)$ and $0 \le u < u(r)$,

$$\frac{u}{r\psi_\delta(u)} < \frac{u}{r\psi_{\delta(r)}(u)} = \frac{1}{2}. \tag{16.18}$$

Figure 16.2 displays the functions $\delta(r)$ and $u(r)$. An approximation of $\delta(r)$ and $u(r)$ for a few values of r is shown in Table 16.1.

The following trivial result will be repeatedly used.

Lemma 16.37 *For $\delta \le \delta(r)$ and $u \le u(r)$ we have $\psi_\delta(u) > 0$.*

Proof It follows from (16.18) that $\psi_\delta(u) > \frac{2u}{r} \ge 0$. $\square$

We can now state the main result of this section.

Theorem 16.38 (Projective γ-theorem) *Fix $\frac{2}{\pi} \le r < 1$. Let $(f, \zeta) \in V \setminus \Sigma'$ and $z \in \mathbb{P}^n$ be such that*

$$d_{\mathbb{P}}(z, \zeta) \le \delta(r), \quad d_{\mathbb{P}}(z, \zeta)\gamma_{\mathrm{proj}}(f, \zeta) \le u(r).$$

Then z is an approximate zero of f with associated zero ζ.

Theorem 16.38 follows from the following proposition just as Theorem 15.5 followed from Proposition 15.10.

Proposition 16.39 *Let* $(f, \zeta) \in V \setminus \Sigma'$ *and* $z \in \mathbb{P}^n \setminus \Lambda_f$. *Put* $\delta := d_{\mathbb{P}}(z, \zeta)$ *and* $u := \delta \gamma_{\mathrm{proj}}(f, \zeta)$. *Then we have, for* $\frac{2}{\pi} \leq r < 1$,

$$d_{\mathbb{P}}(N_f(z), \zeta) \leq \frac{u}{r \psi_{\delta(r)}(u)} \, d_{\mathbb{P}}(z, \zeta) = \frac{\gamma_{\mathrm{proj}}(f, \zeta)}{r \psi_{\delta(r)}(u)} \, d_{\mathbb{P}}(z, \zeta)^2,$$

provided $\delta \leq \delta(r)$ *and* $u \leq u(r)$.

The proof will be similar to Proposition 15.10 and proceeds in several steps.

For $z \in (\mathbb{C}^{n+1})_*$ let $P_z : \mathbb{C}^{n+1} \to T_z$ denote the orthogonal projection onto T_z. Note that $P_{\lambda z} = P_z$ for $\lambda \in \mathbb{C}_*$. We shall represent points in $\mathbb{P}^n$ by representatives in the sphere $\mathbb{S}(\mathbb{C}^{n+1}) := \{x \in \mathbb{C}^{n+1} \mid \|x\| = 1\}$. Their angular distance is denoted by $d_{\mathbb{S}}(y, z)$. It follows from Proposition 14.12 that $d_{\mathbb{P}}(y, z) \leq d_{\mathbb{S}}(y, z)$.

The easy proof of the following observation is left to the reader.

Lemma 16.40 *Let* $z, y \in (\mathbb{C}^{n+1})_*$ *and assume* $\delta := d_{\mathbb{S}}(z, y) < \pi/2$. *Then* $P_y|_{T_z} : T_z \to T_y$ *is invertible and* $\|P_y|_{T_z}^{-1}\| \leq (\cos \delta)^{-1}$. □

We prove now a variant of Lemma 15.9 for homogeneous maps $\mathbb{C}^{n+1} \to \mathbb{C}$.

Lemma 16.41 *Let* $f \in \mathcal{H}_{\mathbf{d}}$ *and* $y \in \mathbb{S}(\mathbb{C}^{n+1})$ *be such that* $Df(y)|_{T_y}$ *is invertible. Let* $z \in \mathbb{S}(\mathbb{C}^{n+1})$ *and put* $\delta := d_{\mathbb{S}}(z, y)$, $u := \delta \gamma_{\mathrm{proj}}(f, y)$. *If* $\psi_\delta(u) > 0$, *then* $Df(z)|_{T_z}$ *is invertible and we have*

(a) $Df(y)|_{T_y}^{-1} Df(z)|_{T_z} = P_y|_{T_z} + B$, *with* $B \in \mathcal{L}(T_z; T_y)$, $\|B\| \leq \frac{1}{(1-u)^2} - 1$.

(b) $\|Df(z)|_{T_z}^{-1} Df(y)|_{T_y}\| \leq \frac{(1-u)^2}{\psi_\delta(u)}$.

Proof (a) We proceed as in Lemma 15.9 and can therefore be brief. Note first that $Df(y)|_{T_y}^{-1} Df(y)|_{T_z} = P_y|_{T_z}$. Taking the Taylor expansion of $Df : \mathbb{C}^{n+1} \to \mathcal{L}(\mathbb{C}^{n+1}; \mathbb{C}^n)$ around $y \in \mathbb{C}^{n+1}$, restricting to T_z, and then composing from the left with $Df(y)|_{T_y}^{-1}$ yields

$$Df(y)|_{T_y}^{-1} Df(z)|_{T_z} = Df(y)|_{T_y}^{-1} \left(Df(y)|_{T_z} + \sum_{k=2}^{\infty} \frac{D^k f(y)(z-y)^{k-1}|_{T_z}}{(k-1)!} \right),$$

$$= P_y|_{T_z} + B,$$

where

$$B = \sum_{k=2}^{\infty} k \, \frac{Df(y)|_{T_y}^{-1} D^k f(y)(z-y)^{k-1}|_{T_z}}{k!}.$$

We can now bound $\|B\| \leq \frac{1}{(1-u)^2} - 1$ as in the proof of Lemma 15.9 using that $\|z - y\| \leq d_{\mathbb{S}}(z, y)$.

(b) Put $P := P_y|_{T_z}$. According to part (a) we need to show that when $\psi_\delta(u) > 0$, $P + B$ is invertible (and hence $Df(z)|_{T_z}$ is invertible as well) and $\|(P + B)^{-1}\| \leq \frac{(1-u)^2}{\psi_\delta(u)}$.

Towards this end, first note that by Lemma 16.40,

$$\left\|P^{-1}B\right\| \leq \left\|P^{-1}\right\|\|B\| \leq \frac{1}{\cos\delta}\left(\frac{1}{(1-u)^2} - 1\right) < 1$$

due to our assumption $\psi_\delta(u) > 0$. Lemma 15.7 implies that $I + P^{-1}B$ is invertible. Now $(P + B)^{-1} = (I + P^{-1}B)^{-1}P^{-1}$. Bounding the norms with Lemma 15.7 and using part (a) yields

$$\left\|(P + B)^{-1}\right\| \leq \left\|I + P^{-1}B\right\|^{-1}\|P\|^{-1}$$

$$\leq \frac{\|P^{-1}\|}{1 - \|P^{-1}B\|} \leq \frac{\|P^{-1}\|}{1 - \|P^{-1}\|\|B\|}$$

$$= \frac{1}{\cos\delta - \|B\|} \leq \frac{1}{\cos\delta - (\frac{1}{(1-u)^2} - 1)}$$

$$= \frac{(1-u)^2}{(1+\cos\delta)(1-u)^2 - 1} = \frac{(1-u)^2}{\psi_\delta(u)},$$

where we have again used Lemma 16.40 for the penultimate equality. $\square$

Proof of Proposition 16.39   We choose representatives $z, \zeta \in \mathbb{S}(\mathbb{C}^{n+1})$ such that $\delta := d_{\mathbb{P}}(z, \zeta) = d_{\mathbb{S}}(z, \zeta)$. Note that $\|z - \zeta\| \leq \delta$.

Note as well that $\psi_\delta(u) \geq 0$ by Lemma 16.37. Also, since $(f, \zeta) \notin \Sigma'$, we have that $Df(\zeta)|_{T_\zeta}$ is invertible. We are therefore in the hypothesis of Lemma 16.41 with $y = \zeta$.

We can therefore proceed similarly as for Proposition 15.10. As we did for (15.6), we get

$$Df(z)(z - \zeta) - f(z) = \sum_{k=1}^{\infty}(k - 1)\frac{D^k f(\zeta)}{k!}(z - \zeta)^k.$$

Applying $Df(z)|_{T_z}^{-1}$ to this equation and inserting $Df(\zeta)|_{T_\zeta}Df(\zeta)|_{T_\zeta}^{-1} = I_{\mathbb{C}^n}$, we obtain

$$N_f(z) - \zeta = (z - \zeta) - Df(z)|_{T_z}^{-1}f(z)$$

$$= \sum_{k=1}^{\infty}(k - 1)Df(z)|_{T_z}^{-1}Df(\zeta)Df(\zeta)|_{T_\zeta}^{-1}\frac{D^k f(\zeta)}{k!}(z - \zeta)^k.$$

By bounding the norm, using $\|z - \zeta\|\gamma_{\mathrm{proj}}(f, \zeta) \le \delta\gamma_{\mathrm{proj}}(f, \zeta) = u$, and recalling the definition of $\gamma_{\mathrm{proj}} := \gamma_{\mathrm{proj}}(f, \zeta)$, we get

$$\|N_f(z) - \zeta\| \le \|Df(z)|_{T_z}^{-1} Df(\zeta)\| \delta \sum_{k=1}^{\infty} (k - 1)(\gamma_{\mathrm{proj}}\delta)^{k-1}$$

$$\le \frac{(1 - u)^2}{\psi_\delta(u)} \delta \left(\frac{1}{(1 - u)^2} - \frac{1}{(1 - u)} \right)$$

$$- \frac{u}{\psi_\delta(u)} \delta,$$

where we used Lemma 15.9 for the second inequality. By Lemma 14.13 we have

$$\sin d_{\mathbb{P}}\big(N_f(z), \zeta\big) \le \|N_f(z) - \zeta\|.$$

Hence we obtain

$$\sin d_{\mathbb{P}}\big(N_f(z), \zeta\big) \le \frac{u}{\psi_\delta(u)} d_{\mathbb{P}}(z, \zeta). \qquad (16.19)$$

This implies, since

$$\frac{u}{r\psi_\delta(u)} \le \frac{u}{r\psi_{\delta(r)}(u)} = \frac{1}{2}$$

for $\delta \le \delta(r)$ and $u \le u(r)$,

$$\frac{2}{\pi} d_{\mathbb{P}}\big(N_f(z), \zeta\big) \le \sin d_{\mathbb{P}}\big(N_f(z), \zeta\big) \le \frac{r}{2} d_{\mathbb{P}}(z, \zeta) \le \frac{1}{2} d_{\mathbb{P}}(z, \zeta).$$

Here we have used that $\frac{2}{\pi} \varphi \le \sin \varphi$ for $0 \le \varphi \le \pi/2$ for the left-hand inequality. Hence $d_{\mathbb{P}}(N_f(z), \zeta) \le d_{\mathbb{P}}(z, \zeta) \le \delta(r)$. We can now conclude from (16.19) that

$$d_{\mathbb{P}}\big(N_f(z), \zeta\big) \le \frac{1}{r} \sin d_{\mathbb{P}}\big(N_f(z), \zeta\big) \le \frac{u}{r\psi_{\delta(r)}(u)} d_{\mathbb{P}}(z, \zeta),$$

where the first inequality follows from the definition of $\delta(r)$. This completes the proof. $\qquad\square$

One can deduce from Theorem 16.38 bounds for the separation of zeros of $f \in \mathcal{H}_{\mathbf{d}}$, just as we did for Corollary 15.11. We leave the straightforward proof to the reader.

Corollary 16.42 *Let $\zeta, \xi \in \mathbb{P}^n$ be two distinct zeros of $f \in \mathcal{H}_{\mathbf{d}}$. Then we have for any $\frac{2}{\pi} \le r < 1$,*

$$d_{\mathbb{P}}(\zeta, \xi) \ge \min\left\{ \delta(r), \frac{u(r)}{\gamma_{\mathrm{proj}}(f, \zeta)}, \frac{u(r)}{\gamma_{\mathrm{proj}}(f, \xi)} \right\}. \qquad\square$$

16.7 A Higher Derivative Estimate

Since our algorithms work with approximations of zeros only, it will be convenient to extend the notion of condition number $\mu_{\mathrm{norm}}(f, z)$ to the case that z is not a zero of f.

Definition 16.43 For $f \in \mathcal{H}_\mathbf{d}$ and $z \in (\mathbb{C}^{n+1})_*$ we define the *normalized condition number* $\mu_{\mathrm{norm}}(f, z)$ as

$$\mu_{\mathrm{norm}}(f, z) := \|f\| \cdot \left\| Df(z)|_{T_z}^{-1} \operatorname{diag}\left(\sqrt{d_i} \, \|z\|^{d_i - 1}\right) \right\|$$

if $Df(z)|_{T_z} : T_z \to \mathbb{C}^n$ is invertible. Otherwise, we set $\mu_{\mathrm{norm}}(f, z) := \infty$.

By Corollary 16.8, if $f(z) = 0$, this is consistent with the original definition (16.11):

$$\mu_{\mathrm{norm}}(f, z) = \|f\| \cdot \left\| Df(z)^\dagger \operatorname{diag}\left(\sqrt{d_i} \, \|z\|^{d_i - 1}\right) \right\|.$$

We note, however, that using this formula in the case $f(z) \neq 0$ would lead to a different notion of $\mu_{\mathrm{norm}}(f, z)$ (although the difference is small if z is close to ζ).

Again, μ_{norm} is invariant under scaling. That is, for $\lambda_1, \lambda_2 \in \mathbb{C}_*$, we have $\mu_{\mathrm{norm}}(\lambda_1 f, \lambda_2 z) = \mu_{\mathrm{norm}}(f, z)$. Moreover, $\mu_{\mathrm{norm}}(f, z)$ is as well invariant under the action of $\mathcal{U}(n + 1)$ in this more general setting.

Lemma 16.44 *We have* $\mu_{\mathrm{norm}}(f, z) \geq \sqrt{n}$ *for all* $(f, z) \in \mathcal{H}_\mathbf{d} \times (\mathbb{C}^{n+1})_*$.

Proof Let $f = k + g + h$ with $k \in C_z$, $g \in L_z$, $h \in R_z$. By the orthogonality of the decomposition (16.9) we have $\|f\|^2 = \|k\|^2 + \|g\|^2 + \|h\|^2 \geq \|g\|^2$. Moreover, $Df(z)|_{T_z} = Dg(z)|_{T_z}$, since $Dh(z) = 0$ and $Dk(z)|_{T_z} = 0$; cf. Proposition 16.16. Therefore, $\mu_{\mathrm{norm}}(f, z) \geq \mu_{\mathrm{norm}}(g, z)$. We now argue as for Proposition 16.17. Instead of $N := Dg(z)$ we consider the scaled matrix $M := \operatorname{diag}(\sqrt{d_i})^{-1} N$, and similarly, we obtain

$$\|g\| = \|M\|_F = \left(\sum_{j=1}^n \frac{\sigma_j^2}{d_j} \right)^{1/2}, \qquad \|M^\dagger\| = \max_i \frac{\sqrt{d_i}}{\sigma_i}.$$

Then we get

$$\mu_{\mathrm{norm}}(f, z) \geq \mu_{\mathrm{norm}}(g, z) = \|M\|_F \|M^\dagger\| \geq \left(\sum_{j=1}^n \frac{\sigma_j^2}{d_j} \frac{d_j}{\sigma_j^2} \right)^{1/2} = \sqrt{n},$$

thus finishing the proof. $\qquad \square$

As in the proof of Proposition 16.32, we obtain the following estimate on the cost of evaluating $\mu_{\mathrm{norm}}(f, x)$.

Proposition 16.45 *The computation of $\mu_{\text{norm}}(f, x)$ from the coefficients of $f \in \mathcal{H}_\mathbf{d}$ and $x \in (\mathbb{C}^{n+1})_*$ can be performed with $\mathcal{O}(N + n^3)$ arithmetic operations and square roots. If $d_i \geq 2$ for all i, then this is $\mathcal{O}(N)$.* $\qquad\square$

The goal of this section is to prove Theorem 16.1, which, we recall, states that for $f \in \mathcal{H}_\mathbf{d}$ and $z \in \mathbb{P}^n$,

$$\gamma_{\text{proj}}(f, z) \leq \frac{1}{2} D^{3/2} \mu_{\text{norm}}(f, z).$$

We begin with some inequalities relating norms of polynomials, points in $\mathbb{C}^{n+1}$, and function values.

For fixed $w \in \mathbb{C}^{n+1}$, consider the derivative evaluated at w as a map $D_w : \mathcal{H}_d \to \mathcal{H}_{d-1}$ given by $D_w f := Df(X)(w) = \sum_{j=0}^n w_j \partial_{X_j} f$. Similarly, for $k \geq 2$ and $w_1, \ldots, w_k \in \mathbb{C}^{n+1}$, we consider $D_{\bar{w}}^k : \mathcal{H}_d \to \mathcal{H}_{d-k}$ given by $D_{\bar{w}}^k f := D^k f(X)(w_1, \ldots, w_k)$. Here $\bar{w}$ denotes the k-tuple $(w_1, \ldots, w_k) \in (\mathbb{C}^{n+1})^k$.

Lemma 16.46 *For any $f \in \mathcal{H}_d$ and any $w \in \mathbb{C}^{n+1}$,*

$$\|D_w f\| \leq d \, \|f\| \, \|w\|.$$

Proof By homogeneity we may suppose $\|w\| = 1$. Moreover, by unitary invariance and Corollary 16.7 we may further suppose that $w = e_0$.

If $f = \sum a_\alpha X^\alpha$, then

$$Df(X)(e_0) = \partial_{X_0} f = \sum_{\alpha \mid \alpha_0 \neq 0} \alpha_0 a_\alpha X_0^{\alpha_0 - 1} X_1^{\alpha_1} \cdots X_n^{\alpha_n}.$$

Then, by the definition of the Weyl norm,

$$\|D_{e_0} f\|^2 = \sum_{\alpha \mid \alpha_0 \neq 0} \alpha_0^2 |a_\alpha|^2 \frac{(\alpha_0 - 1)! \alpha_1! \cdots \alpha_n!}{(d-1)!}$$

$$= d \sum_{\alpha \mid \alpha_0 \neq 0} \alpha_0 |a_\alpha|^2 \frac{\alpha_0! \cdots \alpha_n!}{d!}$$

$$\leq d^2 \sum_\alpha |a_\alpha|^2 \frac{\alpha_0! \cdots \alpha_n!}{d!} = d^2 \|f\|. \qquad\square$$

We extend now the previous result to higher order derivatives.

Lemma 16.47 *For $f \in \mathcal{H}_d$ and $w_1, \ldots, w_k \in \mathbb{C}^{n+1}$,*

$$\left\| D_{\bar{w}}^k f \right\| \leq \frac{d!}{(d-k)!} \|f\| \|w_1\| \cdots \|w_k\|,$$

where $\bar{w} = (w_1, \ldots, w_k)$.

Proof We proceed by induction. The case $k = 0$ is trivial. For $k \geq 1$ let $\tilde{w} = (w_1, \ldots, w_{k-1})$ and $g = D_{\tilde{w}}^{k-1} f \in \mathcal{H}_{d-k+1}$, so that $D_{\tilde{w}}^k f = D_{w_k} g$. Hence, by Lemma 16.46,

$$\left\| D_{\tilde{w}}^k f \right\| = \| D_{w_k} g \| \leq (d - k + 1) \, \| g \| \, \| w_k \|.$$

But

$$\| g \| = \left\| D_{\tilde{w}}^{k-1} f \right\| \leq \frac{d!}{(d - k + 1)!} \, \| f \| \, \| w_1 \| \cdots \| w_{k-1} \|$$

by the induction hypothesis, so that

$$\left\| D_{\tilde{w}}^k f \right\| \leq \frac{d!}{(d - k)!} \| f \| \| w_1 \| \cdots \| w_k \|$$

and we are finished. $\qquad\qquad\qquad\qquad\qquad\qquad\qquad\qquad\qquad\qquad\qquad\square$

Proposition 16.48 *Let $f \in \mathcal{H}_d$. For all $x, w_1, \ldots, w_k \in \mathbb{C}^{n+1}$, the kth derivative of f satisfies*

$$\left| D^k f(x)(w_1, \ldots, w_k) \right| \leq d(d - 1) \cdots (d - k + 1) \| f \| \| x \|^{d-k} \| w_1 \| \cdots \| w_k \|.$$

Proof This is an immediate consequence of Lemmas 16.5 and 16.47. $\qquad\qquad\qquad\square$

Lemma 16.49 *Let $d \geq k \geq 2$ be positive integers and put*

$$A_k := \left(\frac{d(d - 1) \cdots (d - k + 1)}{d^{1/2} k!} \right)^{\frac{1}{k-1}}.$$

Then $\max_{k>1} A_k$ *is attained at* $k = 2$.

Proof It is sufficient to show that $A_{k+1} < A_k$ for $k \geq 2$. This amounts to

$$\frac{d(d - 1) \cdots (d - k)}{d^{1/2} (k + 1)!} < \left(\frac{d(d - 1) \cdots (d - k + 1)}{d^{1/2} k!} \right)^{\frac{k}{k-1}},$$

which follows from

$$\frac{(d - 1) \cdots (d - k)}{(k + 1)!} < \left(\frac{(d - 1) \cdots (d - k + 1)}{k!} \right)^{1 + \frac{1}{k-1}},$$

which in turn is equivalent to

$$\frac{d - k}{k + 1} < \left(\frac{(d - 1) \cdots (d - k + 1)}{k!} \right)^{\frac{1}{k-1}}.$$

The last inequality is clear. $\qquad\qquad\qquad\qquad\qquad\qquad\qquad\qquad\qquad\qquad\qquad\square$

Lemma 16.50 *Let* $f \in \mathcal{H}_\mathbf{d}$, $x \in \mathbb{C}^{n+1}$, *and* $k \geq 2$. *Then*

$$\left\| \frac{1}{\|f\|} \, \mathrm{diag}(d_i^{1/2} \, \|x\|^{d_i-k})^{-1} \frac{D^k f(x)}{k!} \right\|^{\frac{1}{k-1}} \leq \frac{1}{2} D^{3/2}.$$

Proof By Proposition 16.48 we have

$$\left\| D^k f_i(x) \right\| \leq d_i(d_i - 1) \cdots (d_i - k + 1) \|f_i\| \|x\|^{d_i-k}.$$

Hence,

$$\frac{\|D^k f_i(x)\|}{d_i^{1/2} \|x\|^{d_i-k} k! \|f_i\|} \leq \left(\frac{d_i(d_i - 1) \cdots (d_i - k + 1)}{d_i^{1/2} k!} \right)^{\frac{1}{k-1}} \leq \frac{1}{2} d_i^{3/2}, \qquad (16.20)$$

the last inequality resulting from Lemma 16.49.

Note that for all $\varphi = (\varphi_1, \ldots, \varphi_n) \in \mathcal{L}_k(V; \mathbb{C}^n)$ with $\varphi_i \in \mathcal{L}_k(V; \mathbb{C})$, we have $\|\varphi\| \leq (\sum_i \|\varphi_i\|^2)^{1/2}$. Therefore,

$$\left(\frac{\|\mathrm{diag}(d_i^{1/2} \|x\|^{d_i-k})^{-1} D^k f(x)\|}{k! \|f\|} \right)^{\frac{1}{k-1}} \leq \left(\sum_{i=1}^{n} \left(\frac{\|D^k f_i(x)\|}{d_i^{1/2} \|x\|^{d_i-k} k! \|f\|} \right)^2 \right)^{\frac{1}{2(k-1)}}.$$

From (16.20) we obtain

$$\frac{\|D^k f_i(x)\|}{d_i^{1/2} k! \|f\| \|x\|^{d_i-k}} \leq \left(\frac{1}{2} D^{3/2} \right)^{k-1} \frac{\|f_i\|}{\|f\|},$$

from which the claim follows.   □

We can now prove Theorem 16.1.

Proof of Theorem 16.1 By Definition 16.35 we have

$$\gamma_{\mathrm{proj}}(f, z)^{k-1} = \max_{k \geq 2} \left\| \|z\|^{k-1} Df(z)|_{T_z}^{-1} \frac{D^k f(z)}{k!} \right\|.$$

Using Definition (16.11) of μ_{norm} and Lemma 16.50, we estimate as follows

$$\left\| \|z\|^{k-1} Df(z)|_{T_z}^{-1} \frac{D^k f(z)}{k!} \right\|$$

$$\leq \|f\| \left\| Df(z)|_{T_z}^{-1} \, \mathrm{diag}(d_i^{1/2} \|z\|^{d_i-1}) \right\| \cdot \left\| \frac{1}{\|f\|} \, \mathrm{diag}(d_i^{1/2} \|z\|^{d_i-k})^{-1} \frac{D^k f(z)}{k!} \right\|$$

$$\leq \mu_{\mathrm{norm}}(f, z) \cdot \left(\frac{1}{2} D^{3/2} \right)^{k-1} \leq \mu_{\mathrm{norm}}(f, z)^{k-1} \left(\frac{1}{2} D^{3/2} \right)^{k-1}.$$

For the last inequality note that $\mu_{\mathrm{norm}}(f, z) \geq 1$ by Lemma 16.44. The assertion is now immediate.   □

16.8 A Lipschitz Estimate for the Condition Number

The goal of this section is to prove the Lipschitz property stated in Proposition 16.2.

The invariance of μ_{norm} under scaling suggests that we think of (nonzero) inputs $f \in \mathcal{H}_{\mathbf{d}}$ as elements of the corresponding complex projective space $\mathbb{P}(\mathcal{H}_{\mathbf{d}})$. We denote by $d_{\mathbb{P}}(f, g)$ the Riemannian distance of the corresponding points in $\mathbb{P}(\mathcal{H}_{\mathbf{d}})$; compare Sect. 14.2.2.

We shall proceed in several steps. First we only perturb the system. Recall $D := \max_i d_i$.

Lemma 16.51 *Let $f, g \in \mathbb{P}(\mathcal{H}_{\mathbf{d}})$ and $z \in \mathbb{P}^n$. Then*

$$\mu_{\mathrm{norm}}(g, z) \leq \frac{\mu_{\mathrm{norm}}(f, z)}{1 - D^{1/2}\mu_{\mathrm{norm}}(f, z) \sin d_{\mathbb{P}}(f, g)},$$

provided $\mu_{\mathrm{norm}}(f, z) < \infty$ and the denominator is positive.

Proof We choose representatives and denote them by the same symbol $f, g \in \mathcal{H}_{\mathbf{d}}$. Note that the assumption $D^{1/2}\mu_{\mathrm{norm}}(f, z) \sin d_{\mathbb{P}}(f, g) < 1$ implies $d_{\mathbb{P}}(f, g) < \pi/2$ since $\mu_{\mathrm{norm}}(f, z) \geq 1$ by Lemma 16.44. Hence $\langle f, g \rangle \neq 0$. Also, to simplify notation, we may choose a representative z in the sphere $\mathbb{S}(\mathbb{C}^{n+1})$ by the invariance of μ_{norm} under scaling.

By Definition 16.43,

$$\mu_{\mathrm{norm}}(f, z) = \|f\| \, \|A^{-1}\|, \quad \text{where } A := \mathrm{diag}\big(d_i^{-1/2}\big) Df(z)|_{T_z}.$$

We shall apply Lemma 15.7 with $\Delta := \mathrm{diag}(d_i^{-1/2}) D(g - f)(z)|_{T_z} \in \mathscr{L}(T_z; \mathbb{C}^n)$. First we prove that

$$\|\Delta\| \leq D^{1/2} \|g - f\|. \tag{16.21}$$

Indeed, Lemma 16.46 implies that for all $w \in \mathbb{C}^{n+1}$,

$$\big\|D_w(g_i - f_i)\big\| \leq d_i \|g_i - f_i\| \|w\|,$$

where $D_w(g_i - f_i) = \sum_{j=0}^n w_j \partial_{X_j}(g_i - f_i)$. Evaluating the polynomial $D_w(g_i - f_i)$ at z and using Lemma 16.5, we get

$$\big|D(g_i - f_i)(z)(w)\big| = \big|D_w(g_i - f_i)(z)\big| \leq \big\|D_w(g_i - f_i)\big\|,$$

since $\|z\| = 1$. Therefore,

$$\big|d_i^{-1/2} D(g_i - f_i)(z)(w)\big| \leq d_i^{1/2} \|g_i - f_i\| \|w\| \leq D^{1/2} \|g_i - f_i\| \|w\|,$$

and the claim (16.21) follows.

From (16.21) we obtain

$$\|\Delta\|\,\|A^{-1}\| \le D^{1/2}\,\frac{\|g-f\|}{\|f\|}\,\mu_{\mathrm{norm}}(f,z).$$

Let $\lambda_0 \in \mathbb{C}$ be such that

$$\frac{\|\lambda_0 g - f\|}{\|f\|} = \min_{\lambda \in \mathbb{C}} \frac{\|\lambda g - f\|}{\|f\|} = \sin d_{\mathbb{P}}(g,f).$$

Lemma 14.13 ensures that $\lambda_0 \ne 0$ and $\|\lambda_0 g\| \le \|f\|$, since $\langle f,g\rangle \ne 0$. Replacing g by $\lambda_0 g$, we may assume that

$$\frac{\|g-f\|}{\|f\|} = \sin d_{\mathbb{P}}(g,f), \quad \|g\| \le \|f\|,$$

since the assertion of Proposition 16.51 is invariant under scaling of g. Therefore, we conclude from the above that

$$\|\Delta\|\,\|A^{-1}\| \le D^{1/2}\,\sin d_{\mathbb{P}}(g,f)\,\mu_{\mathrm{norm}}(f,z).$$

Lemma 15.7 implies now, using $\mathrm{diag}(d_i^{-1/2})\,Dg(z)|_{T_z} = A + \Delta$,

$$\mu_{\mathrm{norm}}(g,z) = \|g\|\,\|(A+\Delta)^{-1}\| \le \|f\|\,\|(A+\Delta)^{-1}\|$$

$$\le \frac{\|f\|\,\|A^{-1}\|}{1-\|\Delta\|\,\|A^{-1}\|} \le \frac{\mu_{\mathrm{norm}}(f,z)}{1-D^{1/2}\sin d_{\mathbb{P}}(g,f)\mu_{\mathrm{norm}}(f,z)}. \qquad \square$$

Corollary 16.52 *Let* $f,g \in \mathbb{P}(\mathcal{H}_{\mathbf{d}})$ *and* $z \in \mathbb{P}^n$ *be such that*

$$D^{1/2}\mu_{\mathrm{norm}}(f,z)\sin d_{\mathbb{P}}(f,g) \le \varepsilon < 1.$$

Then we have

$$(1-\varepsilon)\,\mu_{\mathrm{norm}}(f,z) \le \mu_{\mathrm{norm}}(g,z) \le \frac{1}{1-\varepsilon}\,\mu_{\mathrm{norm}}(f,z).$$

Proof The right-hand inequality follows from Lemma 16.51. The left-hand inequality is clear if $\mu_{\mathrm{norm}}(f,z) \le \mu_{\mathrm{norm}}(g,z)$. If, on the other hand, $\mu_{\mathrm{norm}}(g,z) < \mu_{\mathrm{norm}}(f,z)$, we obtain $D^{1/2}\mu_{\mathrm{norm}}(g,z)\sin d_{\mathbb{P}}(f,g) \le \varepsilon$, and the left-hand inequality follows from the right-hand inequality by exchanging f and g. $\qquad \square$

Now we investigate what happens when we perturb the point $z \in \mathbb{P}^n$. Recall the family of functions $\psi_\delta(u)$ introduced in (16.17).

Lemma 16.53 *Let* $f \in \mathcal{H}_{\mathbf{d}}$ *and* $z \in \mathbb{P}^n$ *be such that* $\gamma_{\mathrm{proj}}(f,z) < \infty$. *For* $y \in \mathbb{P}^n$ *put* $\delta := d_{\mathbb{P}}(y,z)$ *and* $u := \delta\gamma_{\mathrm{proj}}(f,z)$. *If* $\psi_\delta(u) > 0$, *then*

$$\mu_{\mathrm{norm}}(f,y) \le \frac{(1-u)^2}{\psi_\delta(u)}\,\mu_{\mathrm{norm}}(f,z).$$

Proof We choose representatives $y, z \in \mathbb{S}(\mathbb{C}^{n+1})$ such that $\delta = d_{\mathbb{P}}(y, z) = d_{\mathbb{S}}(y, z)$. Lemma 16.41 tells us that $Df(y)|_{T_y}$ is invertible. We have

$$Df(y)|_{T_y}^{-1} \, \mathrm{diag}(\sqrt{d_i}) = Df(y)|_{T_y}^{-1} Df(z)|_{T_z} \, Df(z)|_{T_z}^{-1} \, \mathrm{diag}(\sqrt{d_i}),$$

and hence

$$\left\| Df(y)|_{T_y}^{-1} \, \mathrm{diag}(\sqrt{d_i}) \right\| \leq \left\| Df(y)|_{T_y}^{-1} Df(z)|_{T_z} \right\| \cdot \left\| Df(z)|_{T_z}^{-1} \, \mathrm{diag}(\sqrt{d_i}) \right\|$$

$$\leq \frac{(1-u)^2}{\psi_\delta(u)} \left\| Df(z)|_{T_z}^{-1} \, \mathrm{diag}(\sqrt{d_i}) \right\|,$$

where the last inequality follows from Lemma 16.41(b). Multiplying by $\|f\|$ and using the definition

$$\mu_{\mathrm{norm}}(f, y) = \|f\| \left\| Df(y)|_{T_y}^{-1} \, \mathrm{diag}(\sqrt{d_i}) \right\|,$$

the assertion follows. $\square$

Corollary 16.54 *Let* $0 \leq \varepsilon \leq 1/4$. *For all* $f \in \mathbb{P}(\mathcal{H}_{\mathbf{d}})$ *and all* $y, z \in \mathbb{P}^n$ *the following is true: if* $D^{3/2} \mu_{\mathrm{norm}}(f, z) d_{\mathbb{P}}(y, z) \leq \varepsilon$, *then*

$$(1 - 2\varepsilon) \, \mu_{\mathrm{norm}}(f, z) \leq \mu_{\mathrm{norm}}(f, y) \leq \frac{1}{1 - 2\varepsilon} \mu_{\mathrm{norm}}(f, z).$$

Proof It suffices to prove the right-hand inequality, since the left-hand inequality then follows by exchanging the roles of f and g as in the proof of Corollary 16.52.

Our assumption combined with Theorem 16.1 implies, setting $\delta := d_{\mathbb{P}}(y, z)$,

$$u := \gamma_{\mathrm{proj}}(f, z) \delta \leq \frac{1}{2} D^{3/2} \mu_{\mathrm{norm}}(f, z) \delta \leq \frac{\varepsilon}{2}.$$

Moreover, by Lemma 16.44,

$$\delta \leq D^{3/2} \mu_{\mathrm{norm}}(f, z) \delta \leq \varepsilon.$$

According to Lemma 16.53 it suffices to show that

$$\frac{\psi_\delta(u)}{(1-u)^2} \geq 1 - 2\varepsilon \quad \text{for all } 0 \leq \delta \leq \varepsilon, \ 0 \leq u \leq \varepsilon/2. \tag{16.22}$$

By definition (16.17), $\psi_\delta(u) = (1 + \cos\delta)(1 - u)^2 - 1$, whence

$$\frac{\psi_\delta(u)}{(1-u)^2} = 1 + \cos\delta - \frac{1}{(1-u)^2} = \cos\delta - \frac{u(2-u)}{(1-u)^2}.$$

Using $\cos \delta \geq 1 - \delta^2/2$ and that $u \mapsto \frac{2u}{(1-u)^2}$ is monotonically increasing, we see that the inequality (16.22) is a consequence of

$$\frac{\delta^2}{2} + \frac{2u}{(1-u)^2} \leq 2\varepsilon \quad \text{for } \delta = 2u = \varepsilon. \tag{16.23}$$

We are now going to check this inequality. Using $(1-u)^{-1} \leq 1 + 2u$ for $0 \leq u \leq 1/2$, we get

$$u^2 + \frac{u}{(1-u)^2} \leq u^2 + u(1+2u)^2 = 4u^3 + 5u^2 + u$$

$$\leq 4 \cdot \frac{1}{2} u^2 + 5u^2 + u = 7u^2 + u.$$

But, since $u = \varepsilon/2$ and $\varepsilon \leq 1/4$, we have

$$7u^2 + u = 7\frac{\varepsilon^2}{4} + \frac{\varepsilon}{2} \leq 2\varepsilon^2 + \frac{\varepsilon}{2} \leq \frac{\varepsilon}{2} + \frac{\varepsilon}{2} = \varepsilon,$$

and hence (16.23) follows. □

Proposition 16.55 *Fix $0 \leq \varepsilon \leq \frac{7}{10}$. Let $f, g \in \mathbb{P}(\mathcal{H}_\mathbf{d})$ and $x, y \in \mathbb{P}^n$ be such that*

$$\mu_{\mathrm{norm}}(f, z) \max\{D^{1/2} d_\mathbb{P}(f, g), D^{3/2} d_\mathbb{P}(y, z)\} \leq \frac{2\varepsilon}{7}.$$

Then

$$(1 - \varepsilon) \mu_{\mathrm{norm}}(f, z) \leq \mu_{\mathrm{norm}}(g, y) \leq \frac{1}{1-\varepsilon} \mu_{\mathrm{norm}}(f, z).$$

Proof Let $0 \leq \varepsilon \leq \frac{7}{10}$ and put $\varepsilon' := \frac{2}{7}\varepsilon$. By hypothesis,

$$\mu_{\mathrm{norm}}(f, z) \max\{D^{1/2} d_\mathbb{P}(f, g), D^{3/2} d_\mathbb{P}(y, z)\} \leq \varepsilon'.$$

Corollary 16.52 implies

$$\left(1 - \varepsilon'\right) \mu_{\mathrm{norm}}(f, z) \leq \mu_{\mathrm{norm}}(g, z) \leq \frac{1}{1-\varepsilon'} \mu_{\mathrm{norm}}(f, z).$$

Therefore,

$$D^{3/2} \mu_{\mathrm{norm}}(g, z) d_\mathbb{P}(y, z) \leq \frac{1}{1-\varepsilon'} D^{3/2} \mu_{\mathrm{norm}}(f, z) d_\mathbb{P}(y, z) \leq \frac{\varepsilon'}{1-\varepsilon'} =: \varepsilon''.$$

We have $\varepsilon'' \leq \frac{1}{4}$, since $\varepsilon' \leq \frac{1}{5}$. Corollary 16.54 now implies

$$\mu_{\mathrm{norm}}(g, y) \leq \frac{1}{1-2\varepsilon''} \mu_{\mathrm{norm}}(g, z) \leq \frac{1}{(1-2\varepsilon'')(1-\varepsilon')} \mu_{\mathrm{norm}}(f, z),$$

and in the same way,

$$(1 - 2\varepsilon'')(1 - \varepsilon')\,\mu_{\text{norm}}(f, z) \le \mu_{\text{norm}}(g, y).$$

Note that $(1 - \varepsilon')^{-1} \le 1 + \frac{5}{4}\varepsilon'$ for $0 \le \varepsilon' \le \frac{1}{5}$. Therefore

$$\varepsilon' + 2\varepsilon'' = \varepsilon' + \frac{2\varepsilon'}{1 - \varepsilon'} \le \varepsilon' + 2\varepsilon'\left(1 + \frac{5}{4}\varepsilon'\right) = 3\varepsilon' + \frac{5}{2}\varepsilon'^2 \le 3\varepsilon' + \frac{5}{2}\varepsilon'\frac{1}{5} = \varepsilon.$$

Hence $(1 - 2\varepsilon'')(1 - \varepsilon') \ge 1 - \varepsilon' - 2\varepsilon'' \ge 1 - \varepsilon$, which completes the proof.   □

We can finally prove Proposition 16.2.

Proof of Proposition 16.2   Let $1 + \bar{\varepsilon} = \frac{1}{1-\varepsilon}$. Then $0 \le \varepsilon \le \frac{7}{10}$ corresponds to $0 \le \bar{\varepsilon} \le \frac{7}{3}$. Moreover, when assuming $\bar{\varepsilon} \le \frac{1}{7}$, we have $\frac{\bar{\varepsilon}}{4} \le \frac{2\varepsilon}{7} = \frac{2}{7}\frac{\bar{\varepsilon}}{1+\bar{\varepsilon}}$. Thus Proposition 16.2 follows from Proposition 16.55. For ease of notation we renamed $\bar{\varepsilon}$ by ε in the final statement.   □

Chapter 17
Smale's 17th Problem: I

In 1998, at the request of the International Mathematical Union, Steve Smale published a list of mathematical problems for the twenty-first century. The 17th problem in the list reads as follows:

Can a zero of n complex polynomial equations in n unknowns be found approximately, on the average, in polynomial time with a uniform algorithm?

Smale pointed out that "it is reasonable" to homogenize the polynomial equations by adding a new variable and to work in projective space. That is, he considered as input a system $f \in \mathcal{H}_\mathbf{d}$ to which he associated its zeros in $\mathbb{P}^n$. Smale also stressed that the word "approximately" refers to the computation of an approximate zero in the sense of Definition 16.34 and that "average" refers to expectation with respect to f after endowing $\mathcal{H}_\mathbf{d}$ with a standard Gaussian measure. This amounts to considering the coefficients of a system f—with respect to the Weyl basis—as independent and identically distributed complex standard Gaussian variables. We will denote this distribution by $N(0, \mathrm{I})$ (instead of the more cumbersome $N(0, \mathrm{I}_{2N})$). Finally, Smale used the expression "uniform algorithm" to refer to a numerical algorithm like those we have seen thus far and "time" to refer to the running time, or cost, of this algorithm as we defined in Sect. 5.1.

As of today, there is no conclusive answer to the question above. But a number of partial results towards such an answer have been obtained in recent years. We will devote this and the next chapter to the exposition of these results. The core of this is an algorithm, proposed by Carlos Beltrán and Luis Miguel Pardo, that finds an approximate zero in average polynomial time but makes random choices (flips coins, so to speak) during the computation. The result of the computation is not affected by these choices, but its cost, for any given input $f \in \mathcal{H}_\mathbf{d}$, is a random variable. For such an input one is forced to replace cost by *expected* (also called *randomized*) cost, and the average time that Smale wants to consider is the average over f of this expected cost. We will describe these notions in some detail in Sect. 17.2. We can nonetheless state here the main result in this chapter.

P. Bürgisser, F. Cucker, *Condition*,
Grundlehren der mathematischen Wissenschaften 349,
DOI 10.1007/978-3-642-38896-5_17, © Springer-Verlag Berlin Heidelberg 2013

Fig. 17.1 The family q_τ,
$\tau \in [0, 1]$

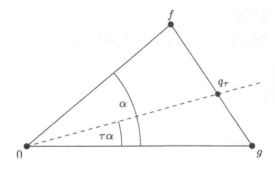

Theorem 17.1 *We exhibit a randomized algorithm that on input $f \in \mathcal{H}_\mathbf{d} \setminus \Sigma$ returns an approximate zero of f. The average of the expected (randomized) cost of this algorithm is bounded by $\mathcal{O}(D^{3/2} n N^2)$.*

Theorem 17.1 provides a probabilistic solution to Smale's 17th problem.

17.1 The Adaptive Linear Homotopy for $\mathcal{H}_\mathbf{d}$

Suppose that we are given an input system $f \in \mathcal{H}_\mathbf{d}$ and an initial pair (g, ζ) in the solution variety $V \subseteq \mathcal{H}_\mathbf{d} \times \mathbb{P}^n$ such that f and g are $\mathbb{R}$-linearly independent. Let $\alpha := d_\mathbb{S}(f, g) \in (0, \pi)$ denote the *angle* between the rays $\mathbb{R}_+ f$ and $\mathbb{R}_+ g$. Consider the line segment $E_{f,g}$ in $\mathcal{H}_\mathbf{d}$ with endpoints f and g. We parameterize this segment by writing

$$E_{f,g} = \{ q_\tau \in \mathcal{H}_\mathbf{d} \mid \tau \in [0, 1] \}$$

with q_τ being the only point in $E_{f,g}$ such that $d_\mathbb{S}(g, q_\tau) = \tau \alpha$ (see Fig. 17.1).

Recall the discriminant variety from Proposition 16.25. If the line segment $E_{f,g}$ does not intersect the discriminant variety Σ, then starting at the zero ζ of g, the map $[0, 1] \to \mathcal{H}_\mathbf{d}, \tau \mapsto q_\tau$, can be uniquely extended to a continuous map

$$[0, 1] \to V, \quad \tau \mapsto (q_\tau, \zeta_\tau),$$

such that $\zeta_0 = \zeta$, We call this map the *lifting* of $E_{f,g}$ with origin (g, ζ). In fact, the formal argument for the existence of such a lifting was the basis of our proof of Bézout's theorem (Theorem 16.23). We shall also call $\tau \mapsto (q_\tau, \zeta_\tau)$ the *solution path* in V corresponding to the input system f and initial pair (g, ζ).

In order to find an approximation of the zero ζ_1 of $f = q_1$ we may start with the zero $\zeta = \zeta_0$ of $g = q_0$ and numerically follow the path (q_τ, ζ_τ) by subdividing $[0, 1]$ into points $0 = \tau_0 < \tau_1 < \cdots < \tau_K = 1$ and by successively computing approximations z_i of ζ_{τ_i} by Newton's method. The main result of this section states that the number K of Newton steps sufficient to follow the path can be bounded by a constant times the integral $\int_0^1 \mu_{\mathrm{norm}}^2(q_\tau, \zeta_\tau) \, d\tau$ of the square of the condition number μ_{norm}.

This integral can be given a more invariant interpretation, which will be essential in the proofs to follow. We associate with the solution path in V the following curve in $\mathbb{S}(\mathcal{H}_d) \times \mathbb{P}^n$:

$$[0, 1] \to V, \quad \tau \mapsto (p_\tau, \zeta_\tau) := \left(\frac{q_\tau}{\|q_\tau\|}, \zeta_\tau \right),$$

where we recall $\mathbb{S}(\mathcal{H}_d) := \{q \in \mathcal{H}_d \mid \|q\| = 1\}$. (In fact, we could also associate with the solution path a corresponding curve in $\mathbb{P}(\mathcal{H}_d) \times \mathbb{P}^n$, but in view of the homotopy algorithm to be discussed in a moment, the spherical viewpoint is more natural.) Recall that $\alpha = d_{\mathbb{S}}(f, g)$. The meaning of the parameterization by τ is that $\alpha\tau$ is the parameterization of $\tau \mapsto p_\tau$ by arc length, which means that $\|\frac{dp_\tau}{d\tau}\| = \alpha$.

Let now $[0, 1] \to [0, 1]$, $\tau \mapsto t(\tau)$, be any smooth bijective map such that $dt/d\tau > 0$. Then we have

$$\left\| \frac{dp_{\tau(t)}}{dt} \right\| = \left\| \frac{dp_\tau}{d\tau} \right\| \left| \frac{d\tau}{dt} \right| = \alpha \frac{d\tau}{dt},$$

and hence, by variable transformation,

$$\alpha \int_0^1 \mu_{\mathrm{norm}}^2(p_\tau, \zeta_\tau) \, d\tau = \int_0^1 \mu_{\mathrm{norm}}^2(p_{\tau(t)}, \zeta_{\tau(t)}) \left\| \frac{dp_{\tau(t)}}{dt} \right\| \, dt. \qquad (17.1)$$

In fact, for the probabilistic analysis later on, it will be essential to consider a specific parameterization of $E_{f,g}$ different from τ.

Proposition 17.2 *For all $\tau \in [0, 1]$ we have $q_\tau = tf + (1 - t)g$, where $t = t(\tau)$ is given by*

$$t(\tau) = \frac{\|g\|}{\|f\| \sin \alpha \cot(\tau\alpha) - \|f\| \cos \alpha + \|g\|}.$$

Proof We use some elementary geometry. For this, we introduce Cartesian coordinates (x, y) in the plane spanned by f and g and assume that g has the coordinates $(s, 0)$ and f has the coordinates $(r \cos \alpha, r \sin \alpha)$ (see Fig. 17.1), so that $r = \|f\|$ and $s = \|g\|$.

Then, the lines determining q_τ have the equations

$$x = y \frac{\cos(\tau\alpha)}{\sin(\tau\alpha)} \quad \text{and} \quad x = y \frac{r \cos \alpha - s}{r \sin \alpha} + s,$$

from which it follows that the coordinate y of q_τ is

$$y = \frac{rs \sin \alpha \sin(\tau\alpha)}{r \sin \alpha \cos(\tau\alpha) - r \cos \alpha \sin(\tau\alpha) + s \sin(\tau\alpha)}. \qquad (17.2)$$

Since $t(\tau) = \frac{y}{r \sin \alpha}$, we conclude that

$$t(\tau) = \frac{s}{r \sin \alpha \cot(\tau\alpha) - r \cos \alpha + s}. \qquad \square$$

We now explicitly describe the path-following algorithm ALH (adaptive linear homotopy) by specifying the step size in an adaptive way. For the step size parameter we chose $\lambda = 0.008535284254$ (cf. Remark 17.4).

Algorithm 17.1 ALH

Input: $f, g \in \mathcal{H}_\mathbf{d}$ and $\zeta \in \mathbb{P}^n$
Preconditions: $g(\zeta) = 0$

$\alpha := d_\mathbb{S}(f, g), \quad r := \|f\|, \quad s := \|g\|$
$\tau := 0, \quad q := g, \quad z := \zeta$
repeat
$\qquad \Delta\tau := \dfrac{\lambda}{\alpha D^{3/2} \mu_{\mathrm{norm}}^2(q.z)}$
$\qquad \tau := \min\{1, \tau + \Delta\tau\}$
$\qquad t := \dfrac{s}{r \sin\alpha \cot(\tau\alpha) - r \cos\alpha + s}$
$\qquad q := tf + (1 - t)g$
$\qquad z := N_q(z)$
until $\tau = 1$
return z and halt

Output: $z \in (\mathbb{C}^{n+1})_*$
Postconditions: The algorithm halts if the lifting of $E_{f,g}$ at ζ does not cut Σ'. In this case, $[z] \in \mathbb{P}^n$ is an approximate zero of f.

The following result estimates the number of iterations performed by algorithm ALH.

Theorem 17.3 *Suppose that $E_{f,g}$ does not intersect the discriminant variety Σ. Then the algorithm* ALH *stops after at most K steps with*

$$K \leq 188 \, D^{3/2} d_\mathbb{S}(f, g) \int_0^1 \mu_{\mathrm{norm}}^2(q_\tau, \zeta_\tau) \, d\tau.$$

The returned point z is an approximate zero of f with associated zero ζ_1. Furthermore, the bound above is optimal up to a constant: we have

$$K \geq 74 \, D^{3/2} d_\mathbb{S}(f, g) \int_0^1 \mu_{\mathrm{norm}}^2(q_\tau, \zeta_\tau) \, d\tau.$$

Proof For $0 \leq \varepsilon \leq \frac{1}{7}$ put $C := \frac{\varepsilon}{4}$. Proposition 16.2 on the Lipschitz continuity of μ_{norm} implies that for all $f, g \in \mathbb{S}(\mathcal{H}_\mathbf{d})$ and all $y, z \in \mathbb{P}^n$ such that

$$\mu_{\mathrm{norm}}(f, z) \max\left\{ D^{1/2} d_\mathbb{S}(f, g), D^{3/2} d_\mathbb{P}(y, z) \right\} \leq C$$

we have

$$\frac{1}{1+\varepsilon}\,\mu_{\text{norm}}(f,z) \le \mu_{\text{norm}}(g,y) \le (1+\varepsilon)\,\mu_{\text{norm}}(f,z).$$

(Note that $d_{\mathbb{P}}(f,g) \le d_{\mathbb{S}}(f,g)$ by Proposition 14.12.)

The analysis of ALH is based on this Lipschitz property with the choice $\varepsilon := \frac{1}{8}$ and $C := \frac{\varepsilon}{4} = \frac{1}{32} = 0.03125$. Furthermore, we set $\lambda = \frac{\varepsilon(1-\varepsilon)}{8(1+\varepsilon)^4} = \frac{2^3 \cdot 7}{3^8} = 0.008535284254$. (See Remark 17.4(a) below for an explanation of these choices.)

As before, we consider the curve $p_\tau := q_\tau/\|q_\tau\|$ in the sphere $\mathbb{S}(\mathcal{H}_{\mathbf{d}})$. Let $0 = \tau_0 < \tau_1 < \cdots < \tau_K = 1$ and $\zeta_0 = z_0, z_1, \ldots, z_K$ be the sequences of τ-values and points in $\mathbb{P}^n$ generated by the algorithm ALH. To simplify notation we write p_i instead of p_{τ_i} and ζ_i instead of ζ_{τ_i}.

We claim that for $i = 0, \ldots, K-1$, the following statements are true:

(a) $d_{\mathbb{P}}(z_i, \zeta_i) \le \dfrac{C}{D^{3/2}\mu_{\text{norm}}(p_i,\zeta_i)}$.

(b) $\dfrac{\mu_{\text{norm}}(p_i,z_i)}{1+\varepsilon} \le \mu_{\text{norm}}(p_i,\zeta_i) \le (1+\varepsilon)\mu_{\text{norm}}(p_i,z_i)$.

(c) $d_{\mathbb{S}}(p_i, p_{i+1}) \le \dfrac{C}{D^{3/2}\mu_{\text{norm}}(p_i,\zeta_i)}$.

(d) $d_{\mathbb{P}}(\zeta_i, \zeta_{i+1}) \le \dfrac{C}{D^{3/2}\mu_{\text{norm}}(p_i,\zeta_i)}\dfrac{1-\varepsilon}{1+\varepsilon}$.

(e) $d_{\mathbb{P}}(z_i, \zeta_{i+1}) \le \dfrac{2C}{(1+\varepsilon)D^{3/2}\mu_{\text{norm}}(p_i,\zeta_i)}$.

(f) z_i is an approximate zero of p_{i+1} with associated zero ζ_{i+1}.

We proceed by induction, showing that

$$(\mathbf{a}, i) \Rightarrow (\mathbf{b}, i) \Rightarrow \big((\mathbf{c}, i) \text{ and } (\mathbf{d}, i)\big) \Rightarrow (\mathbf{e}, i) \Rightarrow \big((\mathbf{f}, i) \text{ and } (\mathbf{a}, i+1)\big).$$

Inequality (a) for $i = 0$ is trivial.

Assume now that (a) holds for some $i \le K-1$. Then, Proposition 16.2 (with $f = g = p_i$) implies

$$\frac{\mu_{\text{norm}}(p_i, z_i)}{1+\varepsilon} \le \mu_{\text{norm}}(p_i, \zeta_i) \le (1+\varepsilon)\mu_{\text{norm}}(p_i, z_i)$$

and thus (b). We now prove (c) and (d). To do so, let $\tau_* > \tau_i$ be such that

$$\int_{\tau_i}^{\tau_*} \big(\|\dot{p}_\tau\| + \|\dot{\zeta}_\tau\|\big)\,d\tau = \frac{C}{D^{3/2}\mu_{\text{norm}}(p_i, \zeta_i)}\,\frac{1-\varepsilon}{1+\varepsilon}$$

or $\tau_* = 1$, whichever is smaller. Then, for all $t \in [\tau_i, \tau_*]$,

$$d_{\mathbb{P}}(\zeta_i, \zeta_t) = \int_{\tau_i}^{t} \|\dot{\zeta}_\tau\|\,d\tau \le \int_{\tau_i}^{\tau_*} \big(\|\dot{p}_\tau\| + \|\dot{\zeta}_\tau\|\big)\,d\tau$$

$$\le \frac{C}{D^{3/2}\mu_{\text{norm}}(p_i, \zeta_i)}\,\frac{1-\varepsilon}{1+\varepsilon}. \tag{17.3}$$

Similarly,

$$d_{\mathbb{S}}(p_i, p_t) = \int_{\tau_i}^{t} \|\dot{p}_\tau\| \, d\tau \le \int_{\tau_i}^{\tau_*} \left(\|\dot{p}_\tau\| + \|\dot{\zeta}_\tau\| \right) d\tau$$

$$\le \frac{C}{D^{3/2} \mu_{\mathrm{norm}}(p_i, \zeta_i)} \frac{1 - \varepsilon}{1 + \varepsilon} \le \frac{C}{D^{3/2} \mu_{\mathrm{norm}}(p_i, \zeta_i)}. \tag{17.4}$$

It is therefore enough to show that $\tau_{i+1} \le \tau_*$. This is trivial if $\tau_* = 1$. We therefore assume $\tau_* < 1$. The two bounds above allow us to apply Proposition 16.2 and to deduce, for all $\tau \in [\tau_i, \tau_*]$,

$$\mu_{\mathrm{norm}}(p_\tau, \zeta_\tau) \le (1 + \varepsilon) \mu_{\mathrm{norm}}(p_i, \zeta_i).$$

Corollary 16.14 implies that (using $\mu \le \mu_{\mathrm{norm}}$)

$$\|\dot{\zeta}_\tau\| \le \mu_{\mathrm{norm}}(p_\tau, \zeta_\tau) \|\dot{p}_\tau\|.$$

It follows that using $\mu_{\mathrm{norm}} \ge 1$,

$$\|\dot{p}_\tau\| + \|\dot{\zeta}_\tau\| \le 2\mu_{\mathrm{norm}}(p_\tau, \zeta_\tau) \|\dot{p}_\tau\|.$$

We now deduce that

$$\frac{C}{D^{3/2} \mu_{\mathrm{norm}}(p_i, \zeta_i)} \frac{1 - \varepsilon}{1 + \varepsilon} = \int_{\tau_i}^{\tau_*} \left(\|\dot{p}_\tau\| + \|\dot{\zeta}_\tau\| \right) d\tau$$

$$\le \int_{\tau_i}^{\tau_*} 2\mu_{\mathrm{norm}}(p_\tau, \zeta_\tau) \|\dot{p}_\tau\| \, d\tau$$

$$\le 2(1 + \varepsilon) \mu_{\mathrm{norm}}(p_i, \zeta_i) \int_{\tau_i}^{\tau_*} \|\dot{p}_\tau\| \, d\tau$$

$$\le 2(1 + \varepsilon) \mu_{\mathrm{norm}}(p_i, \zeta_i) \, d_{\mathbb{S}}(p_i, p_{\tau_*}).$$

Consequently, using (b), we obtain

$$d_{\mathbb{S}}(p_i, p_{\tau_*}) \ge \frac{C(1 - \varepsilon)}{2(1 + \varepsilon)^2 D^{3/2} \mu_{\mathrm{norm}}^2(p_i, \zeta_i)} \ge \frac{C(1 - \varepsilon)}{2(1 + \varepsilon)^4 D^{3/2} \mu_{\mathrm{norm}}^2(p_i, z_i)}.$$

Recall that the parameter λ in ALH was chosen as $\lambda = \frac{C(1-\varepsilon)}{2(1+\varepsilon)^4}$. By the definition of $\tau_{i+1} - \tau_i$ in ALH we have $\alpha(\tau_{i+1} - \tau_i) = \frac{\lambda}{D^{3/2} \mu_{\mathrm{norm}}^2(p_i, z_i)}$. So we obtain

$$d_{\mathbb{S}}(p_i, p_{\tau_*}) \ge \alpha(\tau_{i+1} - \tau_i) = d_{\mathbb{S}}(p_i, p_{i+1}).$$

This implies $\tau_{i+1} \le \tau_*$ as claimed, and hence inequalities (c) and (d) follow from (17.4) and (17.3), respectively. With them, we may apply Proposition 16.2 to deduce, for all $\tau \in [\tau_i, \tau_{i+1}]$,

$$\frac{\mu_{\mathrm{norm}}(p_i, \zeta_i)}{1 + \varepsilon} \le \mu_{\mathrm{norm}}(p_\tau, \zeta_\tau) \le (1 + \varepsilon) \mu_{\mathrm{norm}}(p_i, \zeta_i). \tag{17.5}$$

Next we use the triangle inequality, (a), and (d) to obtain

$$d_\mathbb{P}(z_i, \zeta_{i+1}) \le d_\mathbb{P}(z_i, \zeta_i) + d_\mathbb{P}(\zeta_i, \zeta_{i+1})$$

$$\le \frac{C}{D^{3/2}\mu_{\mathrm{norm}}(p_i, \zeta_i)} + \frac{C}{D^{3/2}\mu_{\mathrm{norm}}(p_i, \zeta_i)}\frac{1-\varepsilon}{1+\varepsilon}$$

$$= \frac{2C}{(1+\varepsilon)D^{3/2}\mu_{\mathrm{norm}}(p_i, \zeta_i)}, \tag{17.6}$$

which proves (e). Now note that since $D \ge 2$ and $\mu_{\mathrm{norm}}(p_i, \zeta_i) \ge 1$, we have

$$d_\mathbb{P}(z_i, \zeta_{i+1}) \le \frac{2C}{(1+\varepsilon)2^{3/2}} \le \frac{1}{45}.$$

For $r = 0.99991\ldots$ we have that $\delta(r) = \frac{1}{45}$ (recall Table 16.1) and $u(r) = 0.17708\ldots$. Inequality (17.6) combined with (17.5) for $\tau = \tau_{i+1}$ yields

$$\frac{1}{2}D^{3/2}\mu_{\mathrm{norm}}(p_{i+1}, \zeta_{i+1})\, d_\mathbb{P}(z_i, \zeta_{i+1}) \le \frac{C}{1+\varepsilon}\frac{\mu_{\mathrm{norm}}(p_{i+1}, \zeta_{i+1})}{\mu_{\mathrm{norm}}(p_i, \zeta_i)} \le C.$$

Together with Theorem 16.1 and $C = \frac{1}{32} < u(r)$, this implies

$$\gamma_{\mathrm{proj}}(p_{i+1}, \zeta_{i+1})\, d_\mathbb{P}(z_i, \zeta_{i+1}) \le u(r).$$

We can therefore apply Theorem 16.38 for this value of r to deduce that z_i is an approximate zero of p_{i+1} associated with its zero ζ_{i+1}, and hence (f) holds.

It follows from (f) that $z_{i+1} = N_{p_{i+1}}(z_i)$ satisfies

$$d_\mathbb{P}(z_{i+1}, \zeta_{i+1}) \le \frac{1}{2}d_\mathbb{P}(z_i, \zeta_{i+1}).$$

Using (e) and the right-hand inequality in (17.5) with $\tau = \tau_{i+1}$, we obtain from (17.6)

$$d_\mathbb{P}(z_{i+1}, \zeta_{i+1}) \le \frac{C}{(1+\varepsilon)D^{3/2}\mu_{\mathrm{norm}}(p_i, \zeta_i)} \le \frac{C}{D^{3/2}\mu_{\mathrm{norm}}(p_{i+1}, \zeta_{i+1})},$$

which proves (a) for $i + 1$. The claim is thus proved.

Note that (f) for $K - 1$ shows that z_{K-1} is an approximate zero of $q_K = f$ with associated zero ζ_1 and consequently, so is the returned point $z_K = N_f(z_{K-1})$.

Consider now any $i \in \{0, \ldots, K - 1\}$. Using (17.5), (b), and by the choice of the step size $\Delta\tau$ in Algorithm 17.1, we obtain

$$\int_{\tau_i}^{\tau_{i+1}} \mu_{\mathrm{norm}}^2(p_\tau, \zeta_\tau)\, d\tau \ge \int_{\tau_i}^{\tau_{i+1}} \frac{\mu_{\mathrm{norm}}^2(p_i, \zeta_i)}{(1+\varepsilon)^2}\, d\tau = \frac{\mu_{\mathrm{norm}}^2(p_i, \zeta_i)}{(1+\varepsilon)^2}(\tau_{i+1} - \tau_i)$$

$$\ge \frac{\mu_{\mathrm{norm}}^2(p_i, z_i)}{(1+\varepsilon)^4}(\tau_{i+1} - \tau_i)$$

$$= \frac{\mu_{\text{norm}}^2(p_i, z_i)}{(1+\varepsilon)^4} \frac{\lambda}{\alpha D^{3/2} \mu_{\text{norm}}^2(p_i, z_i)}$$

$$= \frac{\lambda}{(1+\varepsilon)^4 \alpha D^{3/2}} = \frac{\varepsilon(1-\varepsilon)}{8(1+\varepsilon)^8} \frac{1}{\alpha D^{3/2}}$$

$$\geq \frac{1}{188} \frac{1}{\alpha D^{3/2}}.$$

This implies

$$\int_0^1 \mu_{\text{norm}}^2(p_\tau, \zeta_\tau) \, d\tau \geq \frac{K}{188} \frac{1}{\alpha D^{3/2}},$$

which proves the stated upper bound on K. The lower bound follows from

$$\int_{\tau_i}^{\tau_{i+1}} \mu_{\text{norm}}^2(p_\tau, \zeta_\tau) \, d\tau \leq \int_{\tau_i}^{\tau_{i+1}} \mu_{\text{norm}}^2(p_i, \zeta_i)(1+\varepsilon)^2 \, d\tau$$

$$= \mu_{\text{norm}}^2(p_i, \zeta_i)(1+\varepsilon)^2(\tau_{i+1} - \tau_i)$$

$$\leq \mu_{\text{norm}}^2(p_i, z_i)(1+\varepsilon)^4(\tau_{i+1} - \tau_i)$$

$$= \frac{\lambda(1+\varepsilon)^4}{\alpha D^{3/2}} = \frac{\varepsilon(1-\varepsilon)}{8} \frac{1}{\alpha D^{3/2}} \leq \frac{1}{74} \frac{1}{\alpha D^{3/2}}. \qquad \square$$

Remark 17.4

(a) The proof of Theorem 17.3 gives a rationale for the choice of the value ε. It is the one minimizing the expression $F(\varepsilon) := \frac{8(1+\varepsilon)^8}{\varepsilon(1-\varepsilon)}$ on the interval $[0, 1/7]$ that produces the constant 188. A computation shows that F is minimized at $\varepsilon_m = \frac{3}{4} - \frac{1}{12}\sqrt{57} = 0.120847\ldots$ and $F(\varepsilon_m) = 187.568\ldots$ We have approximated ε_m by $\varepsilon = 1/8 = 0.125$, which yields $F(\varepsilon) = 187.668\ldots < 188$.

(b) Algorithm 17.1 requires the computation of μ_{norm}, which, in turn, requires the computation of the operator norm of a matrix. This cannot be done exactly with rational operations and square roots only. We can do, however, with a sufficiently good approximation of $\mu_{\text{norm}}^2(q, z)$, and there exist several numerical methods efficiently computing such an approximation. We will therefore neglect this issue, pointing out, however, for the skeptical reader that another course of action is possible. Indeed, one may replace the operator by the Frobenius norm in the definition of μ_{norm} and use the bounds $\|M\| \leq \|M\|_F \leq \sqrt{\text{rank}(M)}\|M\|$ to show that this change preserves the correctness of Algorithm 17.1 and adds a multiplicative factor n to the right-hand side of Theorem 17.3. A similar comment applies to the computation of α and $\cot(\tau\alpha)$ in Algorithm 17.1, which cannot be done exactly with rational operations.

For applying Theorem 17.3, it will be central in our development to calculate the integral (17.1) of the squared condition number with respect to the parameterization t of $E_{f,g}$ introduced in Proposition 17.2. Abusing notation, we shall write

Fig. 17.2 An elementary
geometric argument

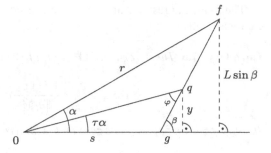

$q_t = (1 - t)g + tf$. For this parameterization we have the following bound on the norm of the speed of the spherical curve $t \mapsto p_t := \frac{q_t}{\|q_t\|}$.

Lemma 17.5 *We have*

$$\left\| \frac{dp_t}{dt} \right\| \leq \frac{\|f\| \|g\|}{\|q_t\|^2}.$$

Proof Note that $\frac{dq_t}{dt} = f - g$. Hence, if P denotes the orthogonal projection of $\mathcal{H}_\mathbf{d}$ onto the tangent space $T_{q_t}\mathbb{S}(\mathcal{H}_\mathbf{d})$, we have by Lemma 14.10,

$$\frac{dp_t}{dt} = \frac{1}{\|q_t\|} P(f - g).$$

We show now by some elementary geometry that $\|P(f - g)\| \leq \|f\| \|g\|$. For this, as for Proposition 17.2, we introduce Cartesian coordinates in the plane spanned by f and g and assume that g has the coordinates $(s, 0)$ and f has the coordinates $(r \cos \alpha, r \sin \alpha)$; see Fig. 17.2.

We write $q := q_t$ and $L := \|f - g\|$. Then $\|q - g\| = tL$, and trigonometry tells us that

$$\frac{\sin \varphi}{\sin(\tau \alpha)} = \frac{s}{tL}.$$

Hence

$$\|P(f - g)\| = L \sin \varphi = \frac{s}{t} \sin(\tau \alpha) = \frac{s}{t} \frac{y}{\|q\|}.$$

We have

$$\frac{y}{t} = L \sin \beta \leq r,$$

and therefore

$$\frac{\|P(f - g)\|}{\|q\|} = \frac{1}{\|q\|^2} \frac{y}{t} s \leq \frac{\|f\| \|g\|}{\|q_t\|^2}$$

as claimed.   □

The following result is an immediate consequence of Theorem 17.3, (17.1), and Lemma 17.5.

Corollary 17.6 *The algorithm* ALH *stops after at most K steps with*

$$K \leq 188 \, D^{3/2} \int_0^1 \frac{\|f\| \|g\|}{\|q_t\|^2} \mu_{\text{norm}}^2(q_t, \zeta_t) \, dt.$$

Its output z is an approximate zero of f with associated zero ζ_1. $\square$

Algorithm 17.1 together with Theorem 17.3, or Corollary 17.6, provides the details of how a linear path is followed in V and how many iterations are needed to do so. It now becomes imperative to deal with an issue we have neglected thus far: the choice of the initial pair.

17.2 Interlude: Randomization

17.2.1 Randomized Algorithms

We start this section, for a change, with a problem in algorithmic number theory—primality testing—which does not appear to bear any relation with conditioning. It consists in, given an integer $n \geq 3$, deciding whether n is prime. The most obvious algorithm to do so checks for all numbers d from 2 to $\lfloor \sqrt{n} \rfloor$ whether d divides n. If such a d is found, the algorithm halts and returns COMPOSITE. Otherwise, it returns PRIME. The simplicity of this algorithm is shadowed by its poor complexity. The size $s = \text{size}(n)$ of the input n is the number of bits needed to write n, which is approximately $\log n$. And the number of candidate divisors we may need to consider is, in the worst case, about $\lfloor \sqrt{n} \rfloor$, i.e., about $2^{\frac{s}{2}}$. By the 1970s, variants of this naive approach had been proposed that improved this behavior but not in any substantial manner: the cost was still exponential.

It is at this time that a new idea entered the stage, proposed by Robert Solovay and Volker Strassen. To understand this idea let us return to the naive algorithm. If a number d, $2 \leq d \leq \lfloor \sqrt{n} \rfloor$, divides n, then d is a "certificate" of n's compositeness. Given n and d, to decide whether d is such a certificate (i.e., whether d divides n) can be quickly done. The shortcoming of the naive algorithm is the possible large number of certifications to be checked. What Solovay and Strassen came up with was a different manner by which a number $a \in \{2, \ldots, n-1\}$ could certify that n is composite, namely, to check the identity

$$a^{\frac{n-1}{2}} \not\equiv \left(\frac{a}{n} \right) \pmod{n}, \tag{17.7}$$

where $\left(\frac{a}{n} \right)$ denotes the Jacobi symbol. We write certif_C(a, n) when (17.7) holds. Again, if n is prime, then there is no $a \in \{2, \ldots, n-1\}$ such that certif_C(a, n)

holds. In contrast with the naive certification, however, if n is composite, at least half of the candidates a in $\{2, \ldots, n - 1\}$ are certificates for that. Furthermore, Jacobi's symbol can be quickly computed (with cost $\mathcal{O}(\log^2 n) = \mathcal{O}(s^2)$). Solovay and Strassen therefore proposed the following algorithm.

Algorithm 17.2 Randomized_Primality_Testing

Input: $n, k \in \mathbb{N}$

Preconditions: $n \geq 3$ odd, $k \geq 1$

```
repeat k times
     draw a at random from {2,...,n-1}
     if certif_C(a,n) then return COMPOSITE and halt
return PRIME and halt
```

Output: a tag in $\{\mathtt{PRIME}, \mathtt{COMPOSITE}\}$

Postconditions: if the tag is COMPOSITE, then n is composite; if the tag is PRIME, then n is prime with probability at least $1 - 2^{-k}$

Algorithm 17.2 presents some features that are new in our exposition. The most noticeable is the presence of the instruction `draw...at random`. Up to now, all algorithms we have described rely on the basic arithmetic operations, on comparisons, and on occasionally taking a square root. Algorithm 17.2 uses a new tool: *randomization*. To be precise, it assumes at hand a function random_bit( ) returning an element in $\{0, 1\}$, each of them with probability $\frac{1}{2}$. Note that the number a in the algorithm can be obtained with s calls to this function (corresponding to the n first bits in the binary expansion of a). It is out of our scope to describe how this function is implemented. Suffice it for our purposes to note that many implementations exist (usually called *pseudorandom number generators*) and are widely accepted as appropriate for their task.

A second new feature in Algorithm 17.2 is the possibility of a wrong answer for some composite numbers. Indeed, if the algorithm returns COMPOSITE then its input n is so. But there is a possibility of returning PRIME on a composite input n. Yet, since the k draws of a are independent, this happens with a probability of at most $\frac{1}{2^k}$. And for moderate values of k, say around 100, this probability of a mistaken output is certainly negligible.

In 1992, Leonard Adleman and Ming-Deh Huang devised a new randomized algorithm (this is how algorithms making random draws are called) that differed from Algorithm 17.2 in an important aspect: the certificate now was for primality. Consequently, the possibility of a wrong output was now associated with returning COMPOSITE. Let us denote by certif_P(b, n) the fact that b is a certificate of primality for n and assume that for a random $b \in \{0, 1\}^{q(s)}$ the probability that b certifies n's primality, if n is prime, is at least $\frac{1}{2}$. Here q is some low-degree polynomial. Consider now the following algorithm.

Algorithm 17.3 Randomized_Primality_Testing_2

Input: $n \in \mathbb{N}$
Preconditions: $n \geq 3$ odd

```
repeat
      draw a at random from {2, ..., n − 1}
      if certif_C(a, n) then return COMPOSITE and halt
      draw b at random from {0, 1}^{q(s)}
      if certif_P(b, n) then return PRIME and halt
```

Output: a tag in $\{\text{PRIME}, \text{COMPOSITE}\}$
Postconditions: the tag is PRIME iff n is prime

This algorithm never gives a wrong output. But its running time is no longer bounded by a function of s. Each iteration is done in time polynomial in s, but the number of iterations itself is a random variable. The probability of performing more than k iterations is at most $\frac{1}{2^k}$. Consequently, the expectation of the number of iterations performed is (use Lemma 15.6)

$$\sum_{k=1}^{\infty} \frac{k}{2^k} = 2.$$

Algorithm 17.3 belongs to a class commonly referred to as *Las Vegas algorithms*, as opposed to *Monte Carlo algorithms*. In the latter, the running time is bounded by a function of the input size but incorrect outputs occur with a small probability. In the former, it is the opposite. Outputs are always correct, but the running time is a random variable.

Because of this, we consider for Las Vegas algorithms a notion of *randomized cost*, which consists of the expectation of the cost over all possible random draws.

17.2.2 A Las Vegas Homotopy Method

With this new set of ideas in mind, let us return to the problem of computing an approximate zero of a system $f \in \mathcal{H}_\mathbf{d}$.

All the efforts to couple linear homotopies (such as Algorithm 17.1) with some deterministically constructed initial pair (g, ζ) to produce zero-finding algorithms working on average polynomial time have, as of today, failed. A way out to deal with the manifest difficulty of the problem, recently proposed by Carlos Beltrán and Luis Miguel Pardo, is to randomly draw the initial pair (g, ζ). For this, they endowed V with a probability distribution ρ_{st} and described an efficient procedure for drawing a pair from ρ_{st}. With such a procedure at hand, the following Las Vegas algorithm is a natural way of proceeding.

Algorithm 17.4 LV

Input: $f \in \mathcal{H}_\mathbf{d}$

Preconditions: $f \neq 0$

---

draw $(g, \zeta) \in V$ from ρ_{st}
run ALH on input (f, g, ζ)

Output: $z \in (\mathbb{C}^{n+1})_*$

Postconditions: The algorithm halts if the lifting of $E_{f,g}$ at ζ does not cut Σ'. In this case, $[z] \in \mathbb{P}^n$ is an approximate zero of f.

Due to our analysis of ALH we know that for an input $f \in \mathcal{H}_\mathbf{d}$, algorithm LV either outputs an approximate zero z of f or loops forever (in case the lifting of the segment $E_{f,g}$ intersects Σ'). Furthermore, the number of iterations performed by ALH depends on the initial pair (g, ζ). The analysis of LV will therefore pass through the notion of randomized cost described above.

At this moment it becomes apparent that the probabilistic framework we have been using thus far, based on Euclidean spaces, spheres, and their products, is too narrow to accommodate the measure ρ_{st}, supported on V. A new installment of our crash course is called for.

17.3 A Crash Course on Probability: IV

In Sect. 2.1 we gave a brief introduction to integration on "data spaces," which were defined in an ad hoc manner as open subsets of a finite product of Euclidean spaces and spheres. The study of these particular spaces turned out to be sufficient for the purposes of the first two parts of this book. Now we need to extend the scope of this theory to the framework of Riemannian manifolds. For background information on this concept we refer to Sect. A.2. Note that a data space by definition is a submanifold of a Euclidean space and thus inherits the structure of a Riemannian manifold, i.e., an inner product in each of its tangent spaces from the inner product of the ambient space.

It is important that on a Riemannian manifold M there is a well-defined measure vol_M obtained by integrating the indicator functions $\mathbb{1}_A$ of Borel-measurable subsets $A \subseteq M$ against the volume form dM of M:

$$\mathop{\mathrm{vol}}_M(A) = \int_M \mathbb{1}_A \, dM$$

(for the definition of the volume form see Sect. A.2.5). This is clearly an extension of the natural measure vol_M for data spaces encountered in Sect. 2.1, and dividing $\mathbb{1}$ by $\mathrm{vol}_M(M)$ if $\mathrm{vol}_M(M) < \infty$, it leads to a natural notion of *uniform distribution* on M. More generally, we will call any measurable function $f : M \to [0, \infty]$ such that $\int_M f \, dM = 1$ a *probability density* on M.

The most fundamental tool encountered in Sect. 2.1 was the transformation formula in Theorem 2.1 for diffeomorphisms between data spaces. The extension of this result to not necessarily bijective smooth maps between Riemannian manifolds, called the coarea formula, is of paramount importance for us. In order to state this result, we first need to generalize the notion of Jacobians.

Suppose that M, N are Riemannian manifolds of dimensions m, n, respectively, such that $m > n$. Let $\psi : M \to N$ be a smooth map. By definition, the derivative $D\psi(x) : T_x M \to T_{\psi(x)} N$ at a regular point $x \in M$ is surjective. Hence the restriction of $D\psi(x)$ to the orthogonal complement of its kernel yields a linear isomorphism. The absolute value of its determinant is called the *normal Jacobian* of ψ at x and denoted by $NJ\psi(x)$. We set $NJ\psi(x) := 0$ if x is not a regular point.

Remark 17.7 In the special case that $m = n$, the kernel of $D\psi(x)$ reduces to zero and its orthogonal complement is therefore all of $T_x M$. Therefore, $NJ\psi(x) = J\psi(x)$.

If y is a regular value of ψ, then the fiber $F_y := \psi^{-1}(y)$ is a Riemannian submanifold of M of dimension $m - n$ (see Theorem A.9). Sard's lemma states that almost all $y \in N$ are regular values.

We can now state the *coarea formula*.

Theorem 17.8 (Coarea formula) *Suppose that M, N are Riemannian manifolds of dimensions m, n, respectively, and let $\psi : M \to N$ be a surjective smooth map. Put $F_y = \psi^{-1}(y)$. Then we have for any function $\chi : M \to \mathbb{R}$ that is integrable with respect to the volume measure of M that*

$$\int_M \chi \, dM = \int_{y \in N} \left(\int_{F_y} \frac{\chi}{NJ\psi} \, dF_y \right) dN.$$

It should be clear that this result contains the transformation formula (Theorem 2.1). as a special case. Moreover, if we apply the coarea formula to the projection $\pi_2 : M \times N \to N$, $(x, y) \mapsto y$, we retrieve Fubini's equality (2.2), since $NJ\pi_2 = 1$. For this reason, the coarea formula is sometimes also called *Fubini's theorem for Riemannian manifolds*. It tells us how probability distributions on Riemannian manifolds transform.

Example 17.9 The natural projection $\mathbb{R}^{2n+2} \setminus \{0\} \cong \mathbb{C}^{n+1} \setminus \{0\} \to \mathbb{P}^n$ factors through a projection $\pi_{\mathbb{S}} : \mathbb{S}^{2n+1} \to \mathbb{P}^n$ with fibers isometric to $\mathbb{S}^1$. Theorem 17.8 allows us to reduce the computation of integrals on $\mathbb{P}^n$ to the computation of integrals on $\mathbb{S}^{2n+1}$. In Lemma 14.9 we showed that the derivative $D\pi_{\mathbb{S}}(x) : T_x \mathbb{S}^{2n+1} \to T_{[v]} \mathbb{P}^n$ equals the orthogonal projection onto $T_v = T_{[v]} \mathbb{P}^n$. Hence the normal Jacobian of $\pi_{\mathbb{S}}$ equals 1. By Theorem 17.8, we have for any integrable function

$f : \mathbb{P}^n \to \mathbb{R}$ and measurable $U \subseteq \mathbb{P}^n$,

$$\int_U f \, d\mathbb{P}^n = \frac{1}{2\pi} \int_{\pi_{\mathbb{S}}^{-1}(U)} (f \circ \pi) \, d\mathbb{S}^{2n+1}. \tag{17.8}$$

Taking $f = 1$ and $U = \mathbb{P}^n$ yields the volume of complex projective space,

$$\text{vol} \, \mathbb{P}^n = \frac{1}{2\pi} \text{vol} \, \mathbb{S}^{2n+1} = \frac{\mathcal{O}_{2n+1}}{2\pi} = \frac{\pi^n}{n!}. \tag{17.9}$$

For later use we note the following immediate consequence of Theorem 17.8.

Corollary 17.10 *Let M, N be Riemannian manifolds of the same dimension and let $\psi : M \to N$ be a surjective smooth map. Suppose that $\int_M |\det D\psi| \, dM$ is finite. Then the fiber $\psi^{-1}(y)$ is finite for almost all $y \in N$, and we have*

$$\int_M |\det D\psi| \, dM = \int_{y \in N} \#\big(\psi^{-1}(y)\big) \, dN(y).$$

Here and in what follows, # denotes cardinality. $\qquad\square$

In Sect. 2.2 we studied probability densities on data spaces and looked at the concepts of marginal and conditional distributions for densities defined on a product space $M \times N$. We shall now see how these notions generalize to the setting of probability densities on Riemannian manifolds.

Suppose that we are in the situation described in the statement of Proposition 17.8 and we have a probability measure on M with density ρ_M. For a regular value $y \in N$ we set

$$\rho_N(y) := \int_{F_y} \frac{\rho_M}{\text{NJ}\psi} \, dF_y. \tag{17.10}$$

The coarea formula implies that for all measurable sets $B \subseteq N$ we have

$$\int_{\psi^{-1}(B)} \rho_M \, dM = \int_B \rho_N \, dN.$$

Hence ρ_N is a probability density on N. We call it the *pushforward* of ρ_M with respect to ψ. Note that this generalizes Proposition 2.11.

Further, for a regular value $y \in N$ and $x \in F_y$ we define

$$\rho_{F_y}(x) := \frac{\rho_M(x)}{\rho_N(y)\text{NJ}\psi(x)}. \tag{17.11}$$

Clearly, this defines a probability density on F_y. In the special case that $\psi : M \times N \to N$, $(x, y) \mapsto y$, is the projection, we have $\text{NJ}\psi = 1$, and we retrieve the formula (2.9) for the conditional density.

The coarea formula implies that for all measurable functions $\chi : M \to \mathbb{R}$,

$$\int_M \chi \, \rho_M \, dM = \int_{y \in N} \left(\int_{F_y} \chi \, \rho_{F_y} \, dF_y \right) \rho_N(y) \, dN,$$

provided the left-hand integral exists. Therefore, we can interpret ρ_{F_y} as the *density of the conditional distribution* of x on the fiber F_y and briefly express the formula above in probabilistic terms as

$$\mathop{\mathbb{E}}_{x \sim \rho_M} \chi(x) = \mathop{\mathbb{E}}_{y \sim \rho_N} \mathop{\mathbb{E}}_{x \sim \rho_{F_y}} \chi(x). \tag{17.12}$$

Remark 17.11 In the context of a map $\psi : M \to N$, we started with a probability density ρ_M on M and derived both its pushforward density ρ_N on N and, for every $y \in N$, the conditional density ρ_{F_y} on the fiber $F_y \subseteq M$.

Conversely, we can start with a density ρ_N on N and densities ρ_{F_y} on the fibers F_y of ψ. The operational process of first drawing y from ρ_N and then x from ρ_{F_y} determines a distribution ρ_M on M, which, following (17.11), has the form

$$\rho_M(x) := \rho_N(y) \rho_{F_y}(x) \mathrm{NJ}\psi(x).$$

These two processes are inverse to each other and hence, for instance, the push-forward of the derived $\rho_M(x)$ above is the original ρ_N. In order to emphasize its possible primary character, and by analogy with the case of product spaces, we will call it the *marginal density* on N.

In summary, any density on M "decomposes" as a marginal density on N and conditional densities on the fibers, and we can recover the distribution on M from this decomposition.

17.4 Normal Jacobians of Projections

We shall determine here the normal Jacobians of various projection maps. Let us start with a few general comments. The $\mathbb{R}$-linear map $\mathbb{C} \to \mathbb{C}, z \mapsto \lambda z$, with $\lambda \in \mathbb{C}$ has determinant $|\lambda|^2$.

Later on, we will need the following observation, whose easy proof is left to the reader.

Lemma 17.12 *For some fixed nonzero* $\lambda \in \mathbb{R}$ *let* $G \subseteq \mathbb{C} \times \mathbb{C}$ *denote the graph of the linear map* $\mathbb{C} \to \mathbb{C}, z \mapsto \lambda z$. *Then the* $\mathbb{R}$-*linear isomorphism* $\mathbb{C} \to G, z \mapsto (z, \lambda z)$, *has determinant* $1 + \lambda^2$. $\qquad\square$

We shall distinguish points in $\mathbb{P}^n$ from their representatives ζ in the sphere $\mathbb{S}(\mathbb{C}^{n+1}) := \{\zeta \in \mathbb{C}^{n+1} \mid \|\zeta\| = 1\}$. The lifting

$$\hat{V} := \left\{ (f, \zeta) \in \mathcal{H}_{\mathbf{d}} \times \mathbb{S}(\mathbb{C}^{n+1}) \mid f(\zeta) = 0 \right\} \tag{17.13}$$

of the solution variety V is a smooth submanifold of $\mathcal{H}_\mathbf{d} \times \mathbb{S}(\mathbb{C}^{n+1})$ by Lemma 16.9. Our goal is to determine the normal Jacobians of the projections

$$\pi_1 \colon \hat{V} \mapsto \mathcal{H}_\mathbf{d}, \quad (q, \zeta) \mapsto q, \quad \text{and} \quad \pi_2 \colon \hat{V} \mapsto \mathbb{S}(\mathbb{C}^{n+1}), \quad (q, \zeta) \mapsto \zeta.$$

(No confusion should arise from the fact that in Sect. 16.5 we denoted the projections $V \to \mathcal{H}_\mathbf{d}$ and $V \to \mathbb{P}^n$ by the same symbols π_1, π_2.)

Recall from Sect. 16.2 that the unitary group $\mathcal{U}(n+1)$ acts on $\mathcal{H}_\mathbf{d} \times \mathbb{S}(\mathbb{C}^{n+1})$, leaving invariant the solution variety $\hat{V}$. It is clear that the projections π_1 and π_2 are $\mathcal{U}(n+1)$-equivariant. This implies that the normal Jacobians of π_1 and π_2 are constant on $\mathcal{U}(n+1)$-orbits. Let us explicitly state this important insight.

Lemma 17.13 *For all $(q, \zeta) \in \hat{V}$ and all $u \in \mathcal{U}(n+1)$ we have*

$$\mathrm{NJ}\pi_1(q, \zeta) = \mathrm{NJ}\pi_1(uq, u\zeta), \qquad \mathrm{NJ}\pi_2(q, \zeta) = \mathrm{NJ}\pi_2(uq, u\zeta). \qquad \square$$

We shall first investigate the special case in which all the degrees d_i equal 1. Consider the vector space $\mathcal{M} := \mathbb{C}^{n \times (n+1)}$ of matrices and define

$$\hat{W} := \left\{ (M, \zeta) \in \mathcal{M} \times \mathbb{S}(\mathbb{C}^{n+1}) \mid M\zeta = 0 \right\}. \tag{17.14}$$

Note that in the special case $d_i = 1$, we can indeed identify $\mathcal{H}_\mathbf{d}$ with $\mathcal{M}$ and $\hat{V}$ specializes to $\hat{W}$. In particular, $\hat{W}$ is a smooth manifold. If $M \in \mathcal{M}$ has rank n, then the linear system $M\zeta = 0$ has a unique solution $\zeta \in \mathbb{S}(\mathbb{C}^{n+1})$ up to scaling by a complex number of modulus 1. That is, the fiber of the projection

$$p_1 \colon \hat{W} \to \mathcal{M}, \quad (M, \zeta) \mapsto M,$$

over M then equals the unit circle $\{(M, e^{i\theta}\zeta) \mid \theta \in \mathbb{R}\}$. We also note that the fibers of the projection $p_2 \colon \hat{W} \to \mathbb{P}^n$, $(M, \zeta) \mapsto \zeta$, are vector spaces of complex dimension n^2.

The group $\mathcal{U} = \mathcal{U}(n) \times \mathcal{U}(n+1)$ acts on $\mathcal{M} = \mathbb{C}^{n \times (n+1)}$ via $(v, u)M := vMu^{-1}$, and it acts on $\mathbb{S}(\mathbb{C}^{n+1})$ by $(v, u)\zeta := u\zeta$. Hence, by the complex singular value decomposition, every $(M, \zeta) \in \hat{W}$ can be transformed into the special form in which $\zeta = e_0 = (1, 0, \ldots, 0)$ and M consists of the zero column and the diagonal matrix $\mathrm{diag}(\sigma_1, \ldots, \sigma_n)$, where $\sigma_1, \ldots, \sigma_n$ are the singular values of M.

It is clear that the projections p_1 and p_2 are $\mathcal{U}$-equivariant. This implies that the normal Jacobians of p_1 and p_2 are constant on $\mathcal{U}$-orbits. Therefore, $\mathrm{NJ}p_1$ and $\mathrm{NJ}p_2$ must be functions of the singular values $\sigma_1, \ldots, \sigma_n$ of M only. We now determine these functions.

Lemma 17.14 *Let $\sigma_1, \ldots, \sigma_n$ be the singular values of $M \in \mathcal{M}$ of full rank. Then we have*

$$\mathrm{NJ}p_1(M, \zeta) = \prod_{i=1}^{n} \frac{\sigma_i^2}{1 + \sigma_i^2}, \qquad \mathrm{NJ}p_2(M, \zeta) = \prod_{i=1}^{n} \frac{1}{1 + \sigma_i^2},$$

and

$$\frac{\mathrm{NJ}p_1}{\mathrm{NJ}p_2}(M, \zeta) = \det(MM^*).$$

Proof The tangent space to the sphere $\mathbb{S}(\mathbb{C}^{n+1})$ at ζ is given by $T_\zeta \mathbb{S}(\mathbb{C}^{n+1}) = \{\dot{\zeta} \in \mathbb{C}^{n+1} \mid \mathrm{Re}\langle \zeta, \dot{\zeta}\rangle = 0\}$; compare (14.11). Lemma 16.9 implies that the tangent space $T_{(M,\zeta)}\hat{W}$ consists of the $(\dot{M}, \dot{\zeta}) \in \mathcal{M} \times T_\zeta \mathbb{S}(\mathbb{C}^{n+1})^n$ such that $\dot{M}\zeta + M\dot{\zeta} = 0$.

As already explained before, by unitary invariance, we may assume that $\zeta = (1, 0, \ldots, 0)$. Then the first column of M vanishes, and using the singular value decomposition, we may assume that the remaining part $A \in \mathbb{C}^{n \times n}$ of M equals $A = \mathrm{diag}(\sigma_1, \ldots, \sigma_n)$.

Let $\dot{u} \in \mathbb{C}^n$ denote the first column of $\dot{M}$ and $\dot{A} \in \mathbb{C}^{n \times n}$ its remaining part. We may thus identify $T_{(M,\zeta)}\hat{W}$ with the product $E \times \mathbb{C}^{n \times n}$ via $(\dot{M}, \dot{\zeta}) \mapsto ((\dot{u}, \dot{\zeta}), \dot{A})$, where E denotes the subspace

$$E := \{(\dot{u}, \dot{\zeta}) \in \mathbb{C}^n \times \mathbb{C}^{n+1} \mid \dot{u}_i + \sigma_i \dot{\zeta}_i = 0, 1 \le i \le n, \dot{\zeta}_0 \in i\mathbb{R}\}.$$

We also note that $E \simeq \mathrm{graph}(-A) \times i\mathbb{R}$. The derivative of p_1 is described by the following commutative diagram:

$$
\begin{array}{ccc}
T_{(M,\zeta)}\hat{W} & \xrightarrow{\simeq} & (\mathrm{graph}(-A) \times i\mathbb{R}) \times \mathbb{C}^{n \times n} \\
\Big\downarrow {\scriptstyle Dp_1(M,\zeta)} & & \Big\downarrow {\scriptstyle \mathrm{pr}_1 \times I} \\
\mathcal{M} & \xrightarrow{\simeq} & \mathbb{C}^n \times \mathbb{C}^{n \times n},
\end{array}
$$

where $\mathrm{pr}_1(\dot{u}, \dot{\zeta}) = \dot{u}$. Note that pr_1 has kernel $i\mathbb{R}$. Since $A = \mathrm{diag}(\sigma_1, \ldots, \sigma_n)$, the pseudoinverse of the projection pr_1 is given by the linear map

$$\varphi \colon \mathbb{C}^n \to \mathrm{graph}(-A), \quad (\dot{u}_1, \ldots, \dot{u}_n) \mapsto (\dot{u}_1, \ldots, \dot{u}_n, -\sigma_1^{-1}\dot{u}_1, \ldots, -\sigma_n^{-1}\dot{u}_n).$$

Lemma 17.12 implies that $\det \varphi = \prod_{i=1}^n (1 + \sigma_i^{-2})$, where the determinant refers to φ as an $\mathbb{R}$-linear map. Noting that $1/\mathrm{NJ}p_1(M, \zeta) = \det \varphi$, the first assertion follows.

For the second assertion we consider the following commutative diagram:

$$
\begin{array}{ccc}
T_{(M,\zeta)}\hat{W} & \xrightarrow{\simeq} & (\mathrm{graph}(-A) \times i\mathbb{R}) \times \mathbb{C}^{n \times n} \\
\Big\downarrow {\scriptstyle Dp_2(M,\zeta)} & & \Big\downarrow {\scriptstyle \mathrm{pr}_2} \\
T_\zeta \mathbb{S}(\mathbb{C}^{n+1}) & \xrightarrow{\simeq} & \mathbb{C}^n \times i\mathbb{R},
\end{array}
$$

where $\mathrm{pr}_2(\dot{u}, \dot{\zeta}, \dot{A}) = \dot{\zeta}$. The map pr_2 has the kernel $\mathbb{C}^{n \times n}$, and its pseudoinverse is given by

$$\psi \colon \mathbb{C}^n \times i\mathbb{R} \to \mathrm{graph}(-A) \times i\mathbb{R}, \quad (\dot{\zeta}_1, \ldots, \dot{\zeta}_n, \dot{\zeta}_0) \mapsto (-\sigma_1 \dot{\zeta}_1, \ldots, -\sigma_n \dot{\zeta}_n, \dot{\zeta}_0).$$

As before, we conclude that $1/\mathrm{NJ}p_2(M, \zeta) = \det \psi = \prod_{j=1}^n (1 + \sigma_j^2)$, proving the second assertion.

The third assertion follows immediately from the first and second. $\square$

We next show that the normal Jacobians of the projections π_j can be expressed in terms of the normal Jacobians of the projections p_j that we just determined.

Lemma 17.15 *For $(q, \zeta) \in \hat{V}$ and $N := Dq(\zeta)$ we have*

$$\mathrm{NJ}\pi_1(q, \zeta) = \mathrm{NJ}p_1(N, \zeta), \qquad \mathrm{NJ}\pi_2(q, \zeta) = \mathrm{NJ}p_2(N, \zeta).$$

Proof By unitary invariance we may assume without loss of generality that $\zeta = (1, 0, \ldots, 0)$. If we write $N = (n_{ij}) = Dq(\zeta) \in \mathcal{M}$, we must have $n_{i0} = 0$, since $N\zeta = 0$. Moreover, according to the orthogonal decomposition (16.9) and Proposition 16.16, we have for $1 \leq i \leq n$,

$$q_i = X_0^{d_i - 1} \sum_{j=1}^{n} n_{ij} X_j + h_i$$

for some $h = (h_1, \ldots, h_n) \in R_\zeta$. We express $\dot{q}_i \in T_q \mathcal{H}_\mathbf{d} = \mathcal{H}_\mathbf{d}$ as

$$\dot{q}_i = \dot{u}_i X_0^{d_i} + \sqrt{d_i} X_0^{d_i - 1} \sum_{j=1}^{n} \dot{a}_{ij} X_j + \dot{h}_i$$

in terms of the coordinates $\dot{u} = (\dot{u}_i) \in \mathbb{C}^n$, $\dot{A} = (\dot{a}_{ij}) \in \mathbb{C}^{n \times n}$, and $\dot{h} = (\dot{h}_i) \in R_\zeta$. The reason to put the factor $\sqrt{d_i}$ here is that

$$\|\dot{q}\|^2 = \sum_i |\dot{u}_i|^2 + \sum_{ij} |\dot{a}_{ij}|^2 + \sum_i \|\dot{h}_i\|^2 \tag{17.15}$$

by the definition of Weyl's inner product.

The tangent space $T_{(q, \zeta)} \hat{V}$ consists of the $(\dot{q}, \dot{\zeta}) \in \mathcal{H}_\mathbf{d} \times T_\zeta \mathbb{S}(\mathbb{C}^{n+1})^n$ such that $\dot{q}(\zeta) + N\dot{\zeta} = 0$; see Lemma 16.9. This condition can be expressed in coordinates as

$$\dot{u}_i + \sum_{j=1}^{n} n_{ij} \dot{\zeta}_j = 0, \quad i = 1, \ldots, n. \tag{17.16}$$

By (17.15) the inner product on $T_{(q, \zeta)} \hat{V}$ is given by the standard inner product in the chosen coordinates $\dot{u}_i, \dot{a}_{ij}, \dot{\zeta}_j$ if $h_i = 0$. Thinking of the description of $T_{(N, \zeta)} W$ given in the proof of Lemma 17.14, we may therefore isometrically identify $T_{(q, \zeta)} V$ with the product $T_{(N, \zeta)} W \times R_\zeta$ via $(\dot{q}, \dot{\zeta}) \mapsto ((\dot{u}, \dot{A}, \dot{\zeta}), \dot{h})$. The derivative of π_1 is then described by the commutative diagram

$$
\begin{array}{ccc}
T_{(q, \zeta)} \hat{V} & \xrightarrow{\cong} & T_{(N, \zeta)} \hat{W} \times R_\zeta \\
{\scriptstyle D\pi_1(q, \zeta)} \downarrow & & \downarrow {\scriptstyle Dp_1(N, \zeta) \times I} \\
\mathcal{H}_\mathbf{d} & \xrightarrow{\cong} & \mathcal{M} \times R_\zeta.
\end{array}
\tag{17.17}
$$

The claim $\text{NJ}\pi_1(q, \zeta) = \text{NJ}p_1(N, \zeta)$ is now immediate.

Similarly, we have the commutative diagram

$$
\begin{array}{ccc}
T_{(q,\zeta)}\hat{V} & \xrightarrow{\simeq} & T_{(N,\zeta)}\hat{W} \times R_\zeta \\
D\pi_2(q,\zeta) \downarrow & & \downarrow Dp_2(N,\zeta) \times \text{zero} \\
T_\zeta \mathbb{S}(\mathbb{C}^{n+1}) & \xrightarrow{\simeq} & T_\zeta \mathbb{S}(\mathbb{C}^{n+1}),
\end{array}
\tag{17.18}
$$

where zero: $R_\zeta \to 0$ is the zero map. Hence $\text{NJ}\pi_2(q, \zeta) = \text{NJ}p_2(N, \zeta)$. $\qquad\square$

The following corollary will be crucial for the proof of the main result of this chapter.

Corollary 17.16 *For $(q, \zeta) \in \hat{V}$ we have*

$$
\frac{\text{NJ}\pi_1}{\text{NJ}\pi_2}(q, \zeta) = \mathcal{D} \det(MM^*),
$$

where $M := \text{diag}(\sqrt{d_i})^{-1} Dq(\zeta)$ and $\mathcal{D} = d_1 \cdots d_n$.

Proof Lemma 17.15 implies that

$$
\frac{\text{NJ}\pi_1}{\text{NJ}\pi_2}(g, \zeta) = \frac{\text{NJ}p_1}{\text{NJ}p_2}(N, \zeta),
$$

where $N := Dq(\zeta)$. Moreover, Lemma 17.14 says that

$$
\frac{\text{NJ}p_1}{\text{NJ}p_2}(N, \zeta) = \det(NN^*).
$$

If we put $\Delta := \text{diag}(\sqrt{d_i})$, then $N = \Delta M$ and hence

$$
\det(NN^*) = \det(\Delta MM^* \Delta) = \det(\Delta^2) \det(MM^*) = \mathcal{D} \det(MM^*).
$$

Combining these equations, the assertion follows. $\qquad\square$

Remark 17.17 One obtains the same formulas for the normal Jacobians of the projections $V \to \mathcal{H}_\mathbf{d}$ and $V \to \mathbb{P}^n$.

17.5 The Standard Distribution on the Solution Variety

We may now return to the probability distribution ρ_{st}. The most immediate approach to ρ_{st} defines it via the following procedure (recall that N denotes the complex dimension of $\mathcal{H}_\mathbf{d}$, so that $2N$ equals its real dimension):

- draw $g \in \mathcal{H}_\mathbf{d}$ from $N(0, I_{2N})$;
- draw one of the $\mathcal{D}$ zeros of g from the uniform distribution on $\{1, \ldots, \mathcal{D}\}$.

The goal of this section is to provide an explicit density function for ρ_{st} and to prove some properties that will simplify the computation of expectations for this measure. To do so it will be more convenient to work with a standard distribution $\hat{\rho}_{st}$ on the lifted solution variety $\hat{V} \subseteq \mathcal{H}_{\mathbf{d}} \times \mathbb{S}(\mathbb{C}^{n+1})$ considered in (17.13). This distribution arises from drawing $(q, [\zeta]) \in V$ from the standard distribution as described above and further drawing a representative ζ uniformly at random from the circle $[\zeta] \cap \mathbb{S}(\mathbb{C}^{n+1})$.

Recall that $Z_{\mathbb{P}}(q)$ denotes the set of zeros in $\mathbb{P}^n$ of $q \in \mathcal{H}_{\mathbf{d}}$. Bézout's theorem tells us that $Z_{\mathbb{P}}(q)$ is finite of cardinality $\mathcal{D} = d_1 \cdots d_n$ if q does not lie in the discriminant variety Σ. This implies that the fiber over $q \notin \Sigma$,

$$\hat{V}(q) := \left\{ \zeta \in \mathbb{S}(\mathbb{C}^{n+1}) \mid (q, \zeta) \in \hat{V} \right\},$$

of the projection $\pi_1 \colon \hat{V} \to \mathcal{H}_{\mathbf{d}}$, $(q, \zeta) \mapsto q$, consists of $\mathcal{D}$ disjoint circles. Hence the volume of such fibers is $2\pi \mathcal{D}$.

With the help of the coarea formula, we can now give a formal definition of these standard distributions by specifying their densities. If $\varphi_{\mathcal{H}_{\mathbf{d}}}$ denotes the density of the standard Gaussian distribution on $\mathcal{H}_{\mathbf{d}}$, we have (cf. Sect. 2.2.2)

$$\varphi_{\mathcal{H}_{\mathbf{d}}}(q) := \frac{1}{(2\pi)^N} e^{-\frac{\|q\|^2}{2}}.$$

A naive attempt to define the density $\hat{\rho}_{st}$ would be to take the product $\frac{1}{2\pi \mathcal{D}} \varphi_{\mathcal{H}_{\mathbf{d}}}$. However, this function is not even a density function on $\hat{V}$, since its integral over $\hat{V}$ differs from 1. As it happens, we have to take into account the normal Jacobian of the projection π_1. We define the density $\hat{\rho}_{st}$ of the *standard distribution on $\hat{V}$* as follows:

$$\hat{\rho}_{st}(q, \zeta) := \frac{1}{2\pi \mathcal{D}} \varphi_{\mathcal{H}_{\mathbf{d}}}(q) \, \mathrm{NJ}\pi_1(q, \zeta). \qquad (17.19)$$

This definition is justified by the following lemma.

Lemma 17.18

(a) *The function $\hat{\rho}_{st}$ is a probability density on $\hat{V}$.*

(b) *The pushforward of $\hat{\rho}_{st}$ with respect to $\pi_1 \colon \hat{V} \to \mathcal{H}_{\mathbf{d}}$ equals $\varphi_{\mathcal{H}_{\mathbf{d}}}$.*

(c) *The pushforward of $\hat{\rho}_{st}$ with respect to the projection $\pi_2 \colon \hat{V} \to \mathbb{S}(\mathbb{C}^{n+1})$ equals the density of the uniform distribution on $\mathbb{S}(\mathbb{C}^{n+1})$.*

(d) *For $q \notin \Sigma$, the conditional distribution on the fiber $\hat{V}(q)$ is the uniform distribution on $\hat{V}(q)$.*

(e) *The expectation of a function $F \colon \hat{V} \to \mathbb{R}$ with respect to $\hat{\rho}_{st}$ can be expressed as*

$$\mathop{\mathbb{E}}_{(q,\zeta) \sim \hat{\rho}_{st}} F(q, \zeta) = \mathop{\mathbb{E}}_{q \sim \varphi_{\mathcal{H}_{\mathbf{d}}}} F_{\widehat{av}}(q),$$

where $F_{\widehat{av}}(q) := \frac{1}{2\pi \mathcal{D}} \int_{\zeta \in \hat{V}(q)} F(q, \zeta) \, d\zeta$.

Proof The coarea formula (Theorem 17.8) applied to $\pi_1 : \hat{V} \to \mathcal{H}_\mathbf{d}$ implies

$$\int_{\hat{V}} F \hat{\rho}_{st} \, d\hat{V} = \int_{q \in \mathcal{H}_\mathbf{d}} \left(\int_{\zeta \in \hat{V}(q)} F(q, \zeta) \frac{\hat{\rho}_{st}(q, \zeta)}{\mathrm{NJ}\pi_1(q, \zeta)} \, d\hat{V}(q) \right) d\mathcal{H}_\mathbf{d}$$

$$= \int_{q \in \mathcal{H}_\mathbf{d}} F_{\widehat{av}}(q) \, \varphi_{\mathcal{H}_\mathbf{d}}(q) \, d\mathcal{H}_\mathbf{d},$$

where $F : \hat{V} \to \mathbb{R}$ is a function that is integrable with respect to the volume measure on $\hat{V}$. Taking $F = 1$ reveals that $\hat{\rho}_{st}$ is a density, proving the first assertion. The above formula also proves the fifth assertion.

By eq. (17.10) the pushforward density ρ_1 of $\hat{\rho}_{st}$ with respect to π_1 satisfies

$$\rho_1(q) = \int_{\zeta \in \hat{V}(q)} \frac{\hat{\rho}_{st}(q, \zeta)}{\mathrm{NJ}\pi_1(q, \zeta)} \, d\hat{V}(q) = \varphi_{\mathcal{H}_\mathbf{d}}(q).$$

This establishes the second assertion.

For the third assertion we first note that by its definition and Lemma 17.13, $\hat{\rho}_{st}$ is unitarily invariant. Since π_2 is an equivariant map, it follows that the pushforward density of $\hat{\rho}_{st}$ is unitarily invariant on $\mathbb{S}(\mathbb{C}^{n+1})$. Hence it must be the uniform distribution.

Finally, by (17.11) the conditional density satisfies

$$\rho_{\hat{V}(q)}(\zeta) = \frac{\hat{\rho}_{st}(q, \zeta)}{\varphi_{\mathcal{H}_\mathbf{d}}(q) \, \mathrm{NJ}\pi_1(q, \zeta)} = \frac{1}{2\pi \mathcal{D}},$$

which proves the fourth assertion.   □

We may now recover the density ρ_{st} on the original solution variety V as the pushforward of $\hat{\rho}_{st}$ under the canonical map $\hat{V} \to V$. As in the proof of Lemma 17.18, one shows that

$$\rho_{st}(q, \zeta) = 2\pi \hat{\rho}_{st}(q, \zeta) = \frac{1}{\mathcal{D}} \varphi_{\mathcal{H}_\mathbf{d}}(q) \, \mathrm{NJ}\pi_1(q, \zeta).$$

Moreover, the expectation of an integrable function $F : V \to \mathbb{R}$ with respect to ρ_{st} can be expressed as

$$\mathop{\mathbb{E}}_{(g, \zeta) \sim \rho_{st}} F = \mathop{\mathbb{E}}_{g \sim N(0, \mathrm{I})} F_{av}, \qquad (17.20)$$

where $F_{av}(q) := \frac{1}{\mathcal{D}} \sum_{\zeta \mid g(\zeta) = 0} F(q, \zeta)$.

Recall from (17.14) the manifold $\hat{W}$, which naturally arises as a special case of $\hat{V}$ in the case $d_i = 1$. We put $\Delta := \mathrm{diag}(d_i^{1/2})$ and consider the *linearization map*

$$\Psi : \hat{V} \to \hat{W}, \quad (q, \zeta) \mapsto (M, \zeta), \quad \text{where } M := \Delta^{-1} Dq(\zeta). \qquad (17.21)$$

The proof of the following result is postponed to Sect. 18.1.3 in the next chapter, where it will be a consequence of more general results.

Lemma 17.19 *The pushforward density of the standard distribution $\hat{\rho}_{\text{st}}$ on $\hat{V}$ with respect to the map Ψ equals the standard distribution on $\hat{W}$.*

17.6 Beltrán–Pardo Randomization

Our next goal is to describe an efficient sampling procedure for the standard distribution ρ_{st} on the solution variety $V \subseteq \mathcal{H}_{\mathbf{d}} \times \mathbb{P}^n$.

An immediate difference with the context of Sect. 17.2.1 is the inadequacy of random_bit( ) as the fundamental building block for randomized algorithms dealing with continuous distributions. We will instead rely on the basic procedure rand_Gaussian( ), which returns, with no input and with unit cost, a real number z drawn from $N(0, 1)$. It is obvious how to draw $z \in \mathbb{C}$ from $N(0, \mathrm{I}_2)$ using this procedure. And it is equally easy to use these draws to draw systems in $\mathcal{H}_{\mathbf{d}}$ from $N(0, \mathrm{I}_{2N})$ (recall that N denotes the complex dimension of $\mathcal{H}_{\mathbf{d}}$, so that $2N$ equals its real dimension). See Algorithm 17.5 below.

Algorithm 17.5 random_system

Input: $d_1, \ldots, d_n \in \mathbb{N}$
Preconditions: $n \geq 1$ and $d_i \geq 1$ for $i = 1, \ldots, n$

```
for i = 1,...,n do
      for α ∈ ℕⁿ⁺¹ with |α| = dᵢ do
          draw fᵢ,α ∈ ℂ from N(0, I₂)
      fᵢ := ∑|α|=dᵢ fᵢ,α (dᵢ α)^(1/2) Xᵅ
return f := (f₁,...,fₙ) and halt
```

Output: $f \in \mathcal{H}_{\mathbf{d}}$
Postconditions: $f \sim N(0, \mathrm{I})$

Recall that the standard distribution arises as follows: we first draw $q \in \mathcal{H}_{\mathbf{d}}$ at random from the standard Gaussian distribution $N(0, \mathrm{I}_{2N})$ on $\mathcal{H}_{\mathbf{d}}$, and then uniformly draw one of the (almost surely) $\mathcal{D}$ zeros of q. Algorithm 17.5 allows one to do the first task. But to do the second once q has been obtained appears to be difficult, since we do not have the zeros of q at hand. Actually, computing one such zero is the problem we wanted to solve in the first place!

Beltrán and Pardo's idea to turn around this obstruction is very elegant. We have shown in Sect. 16.3 that for any $[\zeta] \in \mathbb{P}^n$, the space $\mathcal{H}_{\mathbf{d}}$ is written as a direct sum $C_\zeta \oplus L_\zeta \oplus R_\zeta$ and any system q correspondingly decomposes as $k_\zeta + g_\zeta + h_\zeta$. If $[\zeta]$ is going to be a zero of q, then k_ζ needs to equal 0. Furthermore, Proposition 16.16(a) shows that L_ζ is isometrically bijected with the space $\mathcal{M}_\zeta = \{M \in$

$\mathbb{C}^{n\times(n+1)} \mid M[\zeta] = 0\}$. More precisely, given a representative $\zeta \in (\mathbb{C}^{n+1})_*$ of $[\zeta]$ and $M \in \mathcal{M}_\zeta$, we compute $g_\zeta \in L_\zeta$ by taking

$$g_\zeta = g_{M,\zeta} := \left(\sqrt{d_i}\, \langle X, \zeta \rangle^{d_i-1} \sum_{j=0}^{n} m_{ij} X_j \right). \tag{17.22}$$

To draw $(q, \zeta) \in V$, we can therefore first draw $M \in \mathcal{M} = \mathbb{C}^{n\times(n+1)}$ from a standard Gaussian distribution, then compute $[\zeta] \in \mathbb{P}^n$ such that $M[\zeta] = 0$, then g_ζ using (17.22), and finally draw $h_\zeta \in R_\zeta$ also from a Gaussian distribution. The system $q = g_\zeta + h_\zeta$ satisfies $q(\zeta) = 0$—that is $(q, \zeta) \in V$—and is certainly random (we have randomized both M and h_ζ). The somehow surprising fact is that the resulting distribution on V is precisely ρ_{st}.

The following is a high-level description of the Beltrán–Pardo randomization scheme.

Algorithm 17.6 BP_Randomization_scheme

Input: $d_1, \ldots, d_n \in \mathbb{N}$

Preconditions: $n \geq 1$ and $d_i \geq 1$ for $i = 1, \ldots, n$

```
draw M ∈ ℳ from the standard Gaussian distribution
      # almost surely M has rank n #
compute the unique [ζ] ∈ ℙⁿ such that M[ζ] = 0
choose ζ uniformly at random in [ζ]∩𝕊(ℂⁿ⁺¹)
compute gₘ,ζ according to (17.22)
draw h ∈ Rζ from the standard Gaussian distribution
compute q = gₘ,ζ + h
return (q, ζ) and halt
```

Output: $(q, \zeta) \in \mathcal{H}_\mathbf{d} \times (\mathbb{C}^{n+1})_*$

Postconditions: $(q, [z]) \in V$, $(q, [\zeta]) \sim \rho_{\mathrm{st}}$

It is obvious how to draw $M \in \mathcal{M}$ in the first line of Algorithm 17.6 using $2(n^2 + n)$ calls to rand_Gaussian( ). A representative of the class $[\zeta] \in \mathbb{P}^n$ such that $M[\zeta] = 0$ can be computed by standard algorithms in linear algebra. The drawing of ζ from the uniform distribution in $[\zeta] \cap \mathbb{S}(\mathbb{C}^{n+1})$ is done by drawing $z \in \mathbb{C}$ from $N(0, I_2)$ and then multiplying the representative of $[\zeta]$ obtained above by $\frac{z}{|z|}$.

The drawing of $h \in R_\zeta$ requires more thought but is nonetheless simple. The idea is to draw $f \in \mathcal{H}_\mathbf{d}$ from $N(0, I)$ and then compute the image h of f under the orthogonal projection $\mathcal{H}_\mathbf{d} \to R_\zeta$. Since the orthogonal projection of a standard Gaussian is a standard Gaussian, this amounts to drawing h from a standard Gaussian in R_ζ. For computing the projection h we use the orthogonal decomposition $f = k_\zeta + g_\zeta + h$ with $k_\zeta \in C_\zeta$, and $g_\zeta \in L_\zeta$ given by Proposition 16.16(b). A precise description is given in Algorithm 17.7 (random_h) below.

Algorithm random_h returns a system $h \in R_\zeta$ randomly drawn from a Gaussian in this space performing $2N$ calls to rand_Gaussian( ). Furthermore, its overall cost is low.

Lemma 17.20 *Algorithm* random_h *can be implemented such that it uses only* $\mathcal{O}(DnN)$ *arithmetic operations.*

Proof First recall that by Lemma 16.31, a polynomial $f_i \in \mathcal{H}_{d_i}$ can be evaluated with $\mathcal{O}(N_i)$ arithmetic operations, where $N_i = \binom{n+d_i}{n}$ equals the number of its coefficients. This implies that one can evaluate f_i and all of its first-order partial derivatives with $\mathcal{O}(nN_i)$ arithmetic operations. This implies that the entries of the matrix M can be computed with $\mathcal{O}(nN)$ arithmetic operations (recall $N = \sum_i N_i$).

Let $f_i \in \mathcal{H}_{d_i}$ and let $\ell \in \mathcal{H}_1$ be a linear form. Then the coefficients of the product $\ell \cdot f_i$ can be obtained from the coefficients of ℓ and f_i with $\mathcal{O}(nN_i)$ arithmetic operations. It follows that the coefficients of $\langle X, \zeta \rangle^k$ for $k = 1, 2, \ldots, d_i$ can be computed with a total of $\mathcal{O}(d_i nN_i)$ arithmetic operations. This implies that we can compute the coefficients of the polynomials $k_1, \ldots, k_n$ as well as those of $g_{M,\zeta}$ with $\mathcal{O}(DnN)$ arithmetic operations, where we recall $D = \max_i d_i$. $\square$

Algorithm 17.7 random_h

Input: $d_1, \ldots, d_n \in \mathbb{N}$ and $\zeta \in (\mathbb{C}^{n+1})_*$

Preconditions: $n \geq 1$, $d_i \geq 1$ for $i = 1, \ldots, n$, and $\|\zeta\| = 1$

```
draw f ∈ H_d from N(0, I)
for i = 1, ..., n do
      k_i := f_i(ζ)⟨X, ζ⟩^{d_i}
      for j = 0, ..., n do
            m_{ij} := d_i^{-1/2}(∂_{X_j} f_i(ζ) − d_i f_i(ζ)ζ̄_j)
      (g_{M,ζ})_i := √d_i ⟨X, ζ⟩^{d_i−1} Σ_{j=0}^{n} m_{ij} X_j
h :=: f − k − g_{M,ζ}
return h and halt
```

Output: $h \in \mathcal{H}_d$

Postconditions: $h \in R_\zeta$, $h \sim N(0, I)$

More importantly, we have the following result for the overall behavior of Algorithm 17.6.

Proposition 17.21

(a) *Algorithm 17.6 returns a random pair* $(g, \zeta) \in \hat{V}$ *according to the density* $\hat{\rho}_{\text{st}}$.
(b) *The routine in Algorithm 17.7 performs* $2(N + n^2 + n + 1)$ *draws of random real numbers from the standard Gaussian distribution and can be implemented*

with $\mathcal{O}(DnN + n^3)$ *arithmetic operations (including square roots of positive numbers).*

Proof We delay the proof of part (a) to Sect. 18.1.3 in the next chapter, where it will follow from more general results.

For part (b), we note that the total number of calls to rand_Gaussian( ) is $2(N + n^2 + n + 1)$, of which $2(n^2 + n)$ are to draw M, 2 to draw z, and the remaining $2N$ to draw h. The claim on the operation count follows from Lemma 17.20, noting that $\mathcal{O}(n^3)$ operations suffice for the computation of a representative of $[\zeta]$ in solving $M[\zeta] = 0$. $\qquad\square$

17.7 Analysis of Algorithm LV

Recall from Sect. 5.1 that $\mathrm{cost}^{\mathsf{ALH}}(f, g, \zeta)$ is the number of elementary operations (i.e., arithmetic operations, elementary functions, and comparisons) performed by algorithm ALH with input (f, g, ζ). The *randomized cost* $\mathrm{r\_cost}^{\mathsf{LV}}(f)$ of LV on input $f \in \mathcal{H}_{\mathbf{d}}$ is given by

$$\mathrm{r\_cost}^{\mathsf{LV}}(f) := \mathcal{O}\big(DnN + n^3\big) + \mathop{\mathbb{E}}_{(g,\zeta)\sim\rho_{\mathrm{st}}} \mathrm{cost}^{\mathsf{ALH}}(f, g, \zeta),$$

where the first term is the cost of drawing a pair (g, ζ) from ρ_{st} (Proposition 17.21). We next focus on the second term.

For all f, g, ζ_0, the quantity $\mathrm{cost}^{\mathsf{ALH}}(f, g, \zeta)$ is given by the *number of iterations* $K(f, g, \zeta)$ of ALH with input this triple times the cost of an iteration. The latter is dominated by the computation of one Newton iterate (which is $\mathcal{O}(N)$ independently of the triple (f, g, ζ); see Proposition 16.32). It therefore follows that analyzing the expected cost of LV amounts to doing so for the expected value—over $(g, \zeta) \in V$ drawn from ρ_{st}—of $K(f, g, \zeta)$. We denote this expectation by

$$K(f) := \mathop{\mathbb{E}}_{(g,\zeta)\sim\rho_{\mathrm{st}}} \big(K(f, g, \zeta)\big).$$

To compute bounds for $K(f)$, the following quantity (suggested by the form of F_{av} in (17.20)) will be of the essence. For $q \in \mathcal{H}_{\mathbf{d}} \setminus \Sigma$ we define its *mean square condition number* by

$$\mu_{\mathrm{av}}^2(q) := \frac{1}{D} \sum_{\zeta \mid q(\zeta)=0} \mu_{\mathrm{norm}}^2(q, \zeta). \tag{17.23}$$

If $q \in \Sigma$, then we set $\mu_{\mathrm{av}}(q) := \infty$.

Remark 17.22 Note that μ_{av}^2 is F_{av} for $F = \mu_{\mathrm{norm}}^2$. In this sense, we should write $(\mu_{\mathrm{norm}}^2)_{\mathrm{av}}$. But we will use μ_{av}^2 for the sake of simplicity, and we may even abuse notation and write μ_{av} for $\sqrt{\mu_{\mathrm{av}}^2}$.

The definition of $\mu_{\text{av}}^2(q)$ as an average is an example for the discussion in Sect. 6.8 (see also Remark 14.14).

The use of μ_{av}, together with Corollary 17.6, yields an upper bound for $K(f)$.

Proposition 17.23 *The expected number of iterations of* ALH *on input* $f \in \mathcal{H}_{\mathbf{d}} \setminus \Sigma$ *is bounded as*

$$K(f) \leq 188 \, D^{3/2} \mathop{\mathbb{E}}_{g \sim N(0,\mathrm{I})} \int_0^1 \frac{\|f\| \|g\|}{\|q_t\|^2} \mu_{\text{av}}^2(q_t) \, dt.$$

Proof Fix $f \in \mathcal{H}_{\mathbf{d}} \setminus \Sigma$. Consider any $g \in \mathcal{H}_{\mathbf{d}}$ such that the segment $E_{f,g}$ does not intersect the discriminant variety Σ. By Lemma 16.28, this is the case for almost all $g \in \mathcal{H}_{\mathbf{d}}$. To each of the zeros $\zeta^{(a)}$ of g there corresponds a lifting $[0, 1] \to V, \tau \mapsto (q_t, \zeta_t^{(a)})$, of $E_{f,g}$ such that $\zeta_0^{(a)} = \zeta^{(a)}$. Corollary 17.6 states that

$$K\big(f, g, \zeta^{(a)}\big) \leq 188 \, D^{3/2} \int_0^1 \frac{\|f\| \|g\|}{\|q_t\|^2} \mu_{\text{norm}}^2\big(q_t, \zeta_t^{(a)}\big) \, dt.$$

Since $\zeta_t^{(a)}, \ldots, \zeta_t^{(\mathcal{D})}$ are the zeros of q_t, we have, by the definition (17.23),

$$\frac{1}{\mathcal{D}} \sum_{i=1}^{\mathcal{D}} K\big(f, g, \zeta^{(a)}\big) \leq 188 \, D^{3/2} \int_0^1 \frac{\|f\| \|g\|}{\|q_t\|^2} \mu_{\text{av}}^2(q_t) \, dt.$$

The assertion follows now from (17.20), since

$$K(f) = \mathop{\mathbb{E}}_{(g,\zeta) \sim \rho_{\text{st}}} \big(K(f, g, \zeta)\big) = \mathop{\mathbb{E}}_{g \sim N(0,\mathrm{I})} \left(\frac{1}{\mathcal{D}} \sum_{i=1}^{\mathcal{D}} K\big(f, g, \zeta^{(a)}\big) \right). \qquad \square$$

Remark 17.24 Let $\mathcal{H}_{\mathbf{d}}^{\mathbb{R}}$ denote the subspace of $\mathcal{H}_{\mathbf{d}}$ with real coefficients and let $f \in \mathcal{H}_{\mathbf{d}} \setminus \Sigma$. Were we to try to take the average of $K(f, g, \zeta)$ over all real standard Gaussian $g \in \mathcal{H}_{\mathbf{d}}^{\mathbb{R}}$ and its zeros $\zeta \in \mathbb{P}(\mathbb{R}^{n+1})$, then the argument of Proposition 17.23 would break down. The reason is that $\Sigma \cap \mathcal{H}_{\mathbf{d}}^{\mathbb{R}}$ has codimension one. Hence, for random $g \in \mathcal{H}_{\mathbf{d}}^{\mathbb{R}}$, the line segment $E_{f,g}$ intersects $\Sigma \cap \mathcal{H}_{\mathbf{d}}^{\mathbb{R}}$ with positive probability. (Compare Lemma 16.28.) Therefore, ALH will fail with positive probability.

We can further take the expectation of $K(f)$ for $f \sim N(0, \mathrm{I})$ to obtain the *average expected cost* (or *average randomized cost*) of LV. Because of Proposition 17.23, this quantity is bounded as

$$\mathop{\mathbb{E}}_{f \sim N(0,\mathrm{I})} K(f) \leq 188 \, D^{3/2} \mathop{\mathbb{E}}_{f \sim N(0,\mathrm{I})} \mathop{\mathbb{E}}_{g \sim N(0,\mathrm{I})} \int_0^1 \frac{\|f\| \|g\|}{\|q_t\|^2} \mu_{\text{av}}^2(q_t) \, dt. \qquad (17.24)$$

At this point it is perhaps befitting to stress a difference between the two expectations in the formula above. From a technical point of view, they have exactly the same nature: both f and g are drawn (independently) from $N(0, \mathrm{I})$. Yet, the two drawings play very different roles. In the case of g or, more precisely, of (g, ζ), the nature of the underlying probability distribution is irrelevant as long as one can efficiently draw elements from it. In contrast, in the case of f, the underlying distribution is supposed to model the (elusive) notion of "frequency in practice," and the appropriateness of the Gaussian for this purpose, recall the discussion in Sect. 2.2.7, is not without contention.

But let us return to the bound (17.24). It is tempting to swap the integral and the expectations in this expression because, for a fixed $t \in [0, 1]$, q_t is Gaussian and we know its mean and variance (by Proposition 2.17). We could then replace the two expectations for a single one in q_t. An obstruction to doing so is the presence of $\|f\| \|g\|$, but this obstruction can be easily overcome.

We consider, for $T, \sigma > 0$, the *truncated Gaussian* $N_T(0, \sigma^2 \mathrm{I})$ on $\mathcal{H}_\mathbf{d}$ given by the density (recall Sect. 2.2.2)

$$\rho_T^\sigma(f) = \begin{cases} \frac{\varphi_{2N}^\sigma(f)}{P_{T,\sigma}} & \text{if } \|f\| \leq T, \\ 0 & \text{otherwise,} \end{cases} \tag{17.25}$$

where $P_{T,\sigma} := \mathsf{Prob}_{f \sim N(0,\sigma^2 \mathrm{I})}\{\|f\| \leq T\}$, and as usual, φ_{2N}^σ is the density of $N(0, \sigma^2 \mathrm{I}_{2N})$. In the following we set the threshold $T := \sqrt{2N}$.

Lemma 17.25 *We have $P_{T,\sigma} \geq \frac{1}{2}$ for all $0 < \sigma \leq 1$.*

Proof Clearly it suffices to assume $\sigma = 1$. The statement follows from Proposition 2.22 and the fact that the random variable $\|f\|^2$ is chi-square distributed with $2N$ degrees of freedom. $\square$

Proposition 17.26 *The average randomized number of iterations of* LV *satisfies*

$$\mathop{\mathbb{E}}_{f \sim N(0,\mathrm{I})} K(f) \leq 752\pi \, D^{3/2} N \mathop{\mathbb{E}}_{q \sim N(0,\mathrm{I})} \frac{\mu_{\mathrm{av}}^2(q)}{\|q\|^2}.$$

Proof By (17.24) we have

$$\mathop{\mathbb{E}}_{f \sim N(0,\mathrm{I})} K(f) \leq 188 \, D^{3/2} \mathop{\mathbb{E}}_{f \sim N(0,\mathrm{I})} \mathop{\mathbb{E}}_{g \sim N(0,\mathrm{I})} \int_0^1 \frac{\|f\| \|g\|}{\|q_t\|^2} \mu_{\mathrm{av}}^2(q_t) \, dt$$

$$= 188 \, D^{3/2} \mathop{\mathbb{E}}_{f \sim N_T(0,\mathrm{I})} \mathop{\mathbb{E}}_{g \sim N_T(0,\mathrm{I})} \int_0^1 \frac{\|f\| \|g\|}{\|q_t\|^2} \mu_{\mathrm{av}}^2(q_t) \, dt.$$

The equality follows from the fact that since both $\frac{\|f\| \|g\|}{\|q_t\|^2}$ and $\mu_{\mathrm{av}}^2(q_t)$ are homogeneous of degree 0 in both f and g, we may replace the standard Gaussian by any rotationally invariant distribution on $\mathcal{H}_\mathbf{d}$, in particular by the centered truncated Gaussian $N_T(0, \mathrm{I})$. The last expression can be bounded (we use, as usual, φ to denote the density of $N(0, \mathrm{I})$) as follows:

$$188D^{3/2}\frac{T^2}{P_{T,1}^2}\int_{\|f\|\le T}\int_{\|g\|\le T}\int_0^1\frac{\mu_{\mathrm{av}}^2(q_t)}{\|q_t\|^2}\,dt\,\varphi(g)\,\varphi(f)\,dg\,df$$

$$\le 188D^{3/2}\frac{T^2}{P_{T,1}^2}\mathop{\mathbb{E}}_{f\sim N(0,\mathrm{I})}\mathop{\mathbb{E}}_{g\sim N(0,\mathrm{I})}\int_0^1\frac{\mu_{\mathrm{av}}^2(q_t)}{\|q_t\|^2}\,dt$$

$$= 188D^{3/2}\frac{T^2}{P_{T,1}^2}\int_0^1\left(\mathop{\mathbb{E}}_{q_t\sim N(0,(t^2+(1-t)^2)\mathrm{I})}\frac{\mu_{\mathrm{av}}^2(q_t)}{\|q_t\|^2}\right)dt,$$

where the last equality follows from the fact that for fixed t, the random polynomial system $q_t = tf + (1-t)g$ has a Gaussian distribution with law $N(0,\sigma_t^2\mathrm{I})$, where $\sigma_t^2 := t^2 + (1-t)^2$ (by Proposition 2.17). Note that we deal with nonnegative integrands, so the interchange of integrals is justified by Tonelli's theorem (cf. Sect. 2.1). We next note that by Lemma 17.25, we have $\frac{T^2}{P_{T,1}^2}\le 8N$, and we use the homogeneity (of degree -2) of $\frac{\mu_{\mathrm{av}}^2(q)}{\|q\|^2}$ to obtain

$$\mathop{\mathbb{E}}_{f\sim N(0,\mathrm{I})}K(f)\le 1504D^{3/2}N\int_0^1\left(\mathop{\mathbb{E}}_{q_t\sim N(0,(t^2+(1-t)^2)\mathrm{I})}\frac{\mu_{\mathrm{av}}^2(q_t)}{\|q_t\|^2}\right)dt$$

$$= 1504D^{3/2}N\mathop{\mathbb{E}}_{q\sim N(0,\mathrm{I})}\frac{\mu_{\mathrm{av}}^2(q)}{\|q\|^2}\int_0^1\frac{1}{t^2+(1-t)^2}\,dt$$

$$= 1504D^{3/2}N\mathop{\mathbb{E}}_{q\sim N(0,\mathrm{I})}\frac{\mu_{\mathrm{av}}^2(q)}{\|q\|^2}\frac{\pi}{2}.\qquad(17.26)$$

$\square$

We can now complete the average analysis of LV. The remaining step is achieved in the following result.

Proposition 17.27 *We have*

$$\mathop{\mathbb{E}}_{q\sim N(0,\mathrm{I})}\frac{\mu_{\mathrm{av}}^2(q)}{\|q\|^2}\le\frac{e(n+1)}{2}.$$

Proof By the definition (17.23) of $\mu_{\mathrm{av}}^2(q)$ we have

$$\mathop{\mathbb{E}}_{\mathcal{H}_{\mathbf{d}}}\frac{\mu_{\mathrm{av}}^2(q)}{\|q\|^2}=\int_{q\in\mathcal{H}_{\mathbf{d}}}\frac{\mu_{\mathrm{av}}^2(q)}{\|q\|^2}\varphi_{\mathcal{H}_{\mathbf{d}}}(q)\,dq$$

$$=\int_{q\in\mathcal{H}_{\mathbf{d}}}\frac{1}{D}\sum_{[\zeta]\in Z(q)}\frac{\mu_{\mathrm{norm}}^2(q,\zeta)}{\|q\|^2}\varphi_{\mathcal{H}_{\mathbf{d}}}(q)\,dq$$

$$=\int_{q\in\mathcal{H}_{\mathbf{d}}}\frac{1}{2\pi D}\left(\sum_{[\zeta]\in Z(q)}\int_{\mathbb{S}^1}\frac{\mu_{\mathrm{norm}}^2(q,\zeta)}{\|q\|^2}\,d\theta\right)\varphi_{\mathcal{H}_{\mathbf{d}}}(q)\,dq$$

$$= \int_{q \in \mathcal{H}_{\mathbf{d}}} \frac{1}{2\pi \mathcal{D}} \left(\int_{(q,\zeta) \in \pi_1^{-1}(q)} \frac{\mu_{\text{norm}}^2(q,\zeta)}{\|q\|^2} \, d\pi_1^{-1}(q) \right) \varphi_{\mathcal{H}_{\mathbf{d}}}(q) \, dq$$

$$= \int_{(q,\zeta) \in \hat{V}} \frac{\mu_{\text{norm}}^2(q,\zeta)}{\|q\|^2} \frac{\text{NJ}\pi_1(q,\zeta)}{2\pi \mathcal{D}} \, \varphi_{\mathcal{H}_{\mathbf{d}}}(q) \, d\hat{V},$$

the last equality by the coarea formula applied to $\pi_1: \hat{V} \to \mathcal{H}_{\mathbf{d}}$. We next apply the coarea formula to the projection $\pi_2: \hat{V} \to \mathbb{S}(\mathbb{C}^{n+1})$ and obtain that the last expression above equals

$$\int_{\zeta \in \mathbb{S}(\mathbb{C}^{n+1})} \frac{1}{2\pi \mathcal{D}} \int_{(q,\zeta) \in \pi_2^{-1}(\zeta)} \frac{\mu_{\text{norm}}^2(q,\zeta)}{\|q\|^2} \frac{\text{NJ}\pi_1(q,\zeta)}{\text{NJ}\pi_2(q,\zeta)} \, \varphi_{\mathcal{H}_{\mathbf{d}}}(q) \, d\pi_2^{-1}(\zeta) \, d\mathbb{S}(\mathbb{C}^{n+1}).$$

Recall the orthogonal decompositions $\mathcal{H}_{\mathbf{d}} = C_\zeta \oplus L_\zeta \oplus R_\zeta$ from (16.9). For fixed $\zeta \in \mathbb{S}(\mathbb{C}^{n+1})$, the fiber $\pi_2^{-1}(\zeta)$ can be identified with the linear subspace

$$\hat{V}_\zeta := \{q \in \mathcal{H}_{\mathbf{d}} \mid q(\zeta) = 0\} = L_\zeta \oplus R_\zeta.$$

For $q \in \hat{V}_\zeta$ let us write $q = 0 + g + h$ corresponding to the orthogonal sum above. Factoring the standard Gaussian density as in (2.12), we obtain

$$\varphi_{\mathcal{H}_{\mathbf{d}}}(q) = \varphi_{C_\zeta}(0) \, \varphi_{L_\zeta}(g) \, \varphi_{R_\zeta}(h) = \frac{1}{(2\pi)^n} \, \varphi_{L_\zeta}(g) \, \varphi_{R_\zeta}(h),$$

since $C_\zeta \simeq \mathbb{C}^n \simeq \mathbb{R}^{2n}$. Furthermore, we put

$$M := \text{diag}(\sqrt{d_i})^{-1} Dq(\zeta) = \text{diag}(\sqrt{d_i})^{-1} Dg(\zeta). \qquad (17.27)$$

Note that $M\zeta = 0$. By the definition (16.11) of the condition number μ_{norm}, we have

$$\frac{\mu_{\text{norm}}^2(q,\zeta)}{\|q\|^2} = \|M^\dagger\|^2. \qquad (17.28)$$

Moreover, Corollary 17.16 tells us that

$$\frac{\text{NJ}\pi_1}{\text{NJ}\pi_2}(q,\zeta) = \mathcal{D} \det(MM^*).$$

Fubini's theorem (Theorem 2.2) implies now, using $\int_{h \in R_\zeta} \varphi_{R_\zeta}(h) \, dh = 1$,

$$\int_{(q,\zeta) \in \pi_2^{-1}(\zeta)} \frac{\mu_{\text{norm}}^2(q,\zeta)}{\|q\|^2} \frac{\text{NJ}\pi_1}{\text{NJ}\pi_2}(q,\zeta) \, \varphi_{\mathcal{H}_{\mathbf{d}}}(q) \, d\pi_2^{-1}(\zeta)$$

$$= \frac{\mathcal{D}}{(2\pi)^n} \int_{g \in L_\zeta} \|M^\dagger\|^2 \det(MM^*) \varphi_{L_\zeta}(g) \, dg.$$

By unitary invariance, this expression is independent of $\zeta \in \mathbb{S}(\mathbb{C}^{n+1})$, so that we may assume that $\zeta = e_0$; cf. Lemma 16.15. Hence we obtain that

$$\mathop{\mathbb{E}}_{\mathcal{H}_{\mathbf{d}}} \frac{\mu_{\text{av}}^2(q)}{\|q\|^2} = \frac{\text{vol}\,\mathbb{S}(\mathbb{C}^{n+1})}{(2\pi)^{n+1}} \int_{g \in L_{e_0}} \|M^\dagger\|^2 \det(MM^*) \varphi_{L_{e_0}}(g)\, dg, \qquad (17.29)$$

where M is determined by (17.27). Proposition 16.16 reveals that

$$L_{e_0} \to \mathcal{M}_{e_0}, \; g \mapsto M = \text{diag}(\sqrt{d_i})^{-1} Dg(\zeta)$$

is a linear isometry, where $\mathcal{M}_{e_0} := \{M \in \mathcal{M} \mid Me_0 = 0\}$. This space can be identified with $\mathbb{C}^{n \times n}$, since it consists of the matrices in $\mathcal{M}$ whose first column is zero. Corollary 4.23 states that

$$\int_{A \in \mathbb{C}^{n \times n}} \|A^{-1}\|^2 |\det(A)|^2 \varphi_{2n^2}(A)\, dA \le 2^n n! \frac{e(n+1)}{2}. \qquad (17.30)$$

Moreover, cf. Proposition 2.19,

$$\text{vol}\,\mathbb{S}(\mathbb{C}^{n+1}) = \text{vol}\,\mathbb{S}^{2n+1} = \frac{2\pi^{n+1}}{n!}. \qquad (17.31)$$

Combining (17.29)–(17.31), we get

$$\mathop{\mathbb{E}}_{\mathcal{H}_{\mathbf{d}}} \frac{\mu_{\text{av}}^2(q)}{\|q\|^2} \le \frac{e(n+1)}{2},$$

as claimed.   □

We close this section by proving the main result in this chapter.

Proof of Theorem 17.1 We already know that the algorithm LV described in Sect. 17.2 returns an approximate zero of its input f, since ALH does so. In addition, it follows from Propositions 17.26 and 17.27, and the $\mathcal{O}(N)$ cost of each iteration of ALH established in Proposition 16.32 that the average cost of ALH over random $f \sim N(0, \mathrm{I})$ and random $(g, \zeta) \sim \rho_{\text{st}}$ is bounded as

$$\mathop{\mathbb{E}}_{f \sim N(0,\mathrm{I})} \mathop{\mathbb{E}}_{(g,\zeta) \sim \rho_{\text{st}}} \text{cost}^{\text{ALH}}(f, g, \zeta) \le \mathcal{O}(D^{3/2} n N^2).$$

Proposition 17.21 further ensures that the cost of the initial randomization in LV, i.e., $\mathcal{O}(DnN + n^3)$, is dominated by this bound.   □

17.8 Average Analysis of μ_{norm}, μ_{av}, and μ_{max}

It is clear that for successfully applying the algorithm ALH, one needs a starting pair (g, ζ) in the solution variety V having small condition $\mu_{\text{norm}}(g, z)$. It is therefore of interest to understand the distribution of μ_{norm} on V. For instance, what is the order of magnitude of the expectation $\mathbb{E}\mu_{\mathrm{norm}}$ with respect to the standard distribution?

In order to analyze this, recall from (17.14) the special case

$$\hat{W} := \left\{ (M, \zeta) \in \mathscr{M} \times \mathbb{S}(\mathbb{C}^{n+1}) \mid M\zeta = 0 \right\}$$

of the solution manifold, where $\mathscr{M} := \mathbb{C}^{n \times (n+1)}$. We put $\Delta := \mathrm{diag}(d_i^{1/2})$ and consider the linearization map

$$\Psi : \hat{V} \rightarrow \hat{W}, \quad (q, \zeta) \mapsto (M, \zeta), \quad \text{where } M := \Delta^{-1} Dq(\zeta),$$

already introduced in (17.21). We claim that if $(q, \zeta) \in \hat{V}$ is random following $\hat{\rho}_{\mathrm{st}}$, then $M \in \mathscr{M}$ is a standard Gaussian matrix. Indeed, Lemma 17.19 states that the pushforward of the standard distribution $\hat{\rho}_{\mathrm{st}}$ on $\hat{V}$ under the map Ψ equals the standard distribution on $\hat{W}$. Moreover, Lemma 17.18(b), applied to the projection $p_1 : \hat{W} \to \mathscr{M}$, implies that the pushforward of the standard distribution on $\hat{W}$ under p_1 equals the standard Gaussian on $\mathscr{M}$.

According to (16.11), the condition number $\mu_{\mathrm{norm}}(q, \zeta)$ can be described in terms of Ψ as follows:

$$\frac{\mu_{\mathrm{norm}}(q, \zeta)}{\|q\|} = \|M^\dagger\|, \quad \text{where } (M, \zeta) = \Psi(q, \zeta). \tag{17.32}$$

In Sect. 4.4 we already analyzed the distribution of $\|M^\dagger\|$ for a standard Gaussian. Putting things together, it is now easy to prove the following result.

Theorem 17.28 *For $t \geq \sqrt{nN}$ we have*

$$\Prob_{(q,\zeta)\sim\rho_{\mathrm{st}}} \left\{ \mu_{\mathrm{norm}}(q, \zeta) \geq t \right\} \leq 24n^2 N^2 \frac{\ln^2 t}{t^4}.$$

Moreover,

$$\mathop{\mathbb{E}}_{(q,\zeta)\sim\rho_{\mathrm{st}}} \mu_{\mathrm{norm}}(q, \zeta) \leq 5\sqrt{nN} \, \ln^2(nN)$$

and

$$\mathop{\mathbb{E}}_{(q,\zeta)\sim\rho_{\mathrm{st}}} \mu_{\mathrm{norm}}^2(q, \zeta) = \mathcal{O}\left(nN \ln^2(nN)\right).$$

Proof Proposition 4.27 implies that for any $\varepsilon > 0$,

$$\Prob_{M\sim N(0,\mathrm{I})} \left\{ \|M^\dagger\| \geq \frac{n^{\frac{1}{2}}}{(8e)^{\frac{1}{4}}} \varepsilon^{-\frac{1}{4}} \right\} \leq \varepsilon.$$

Moreover, Corollary 4.6 implies that for any $\varepsilon > 0$,

$$\Prob_{q\sim N(0,\mathrm{I})} \left\{ \|q\| \geq \sqrt{2N} + \sqrt{2\ln\frac{1}{\varepsilon}} \right\} \leq \varepsilon.$$

Using the observation (4.17), we combine the above two tail estimates to obtain

$$\mathrm{Prob}\{\|M^{\dagger}\|\,\|q\| \geq t(\varepsilon)\} \leq 2\varepsilon, \qquad (17.33)$$

where

$$t(\varepsilon) := \frac{n^{\frac{1}{2}}\varepsilon^{-\frac{1}{4}}}{(8e)^{\frac{1}{4}}}\left(\sqrt{2N} + \sqrt{2\ln\frac{1}{\varepsilon}}\right).$$

Clearly, the function $t(\varepsilon)$ is strictly monotonically decreasing on $(0,1)$ and $\lim_{\varepsilon \to 0} t(\varepsilon) = \infty$. Hence for all $t \geq t(a)$ there exists a unique $\varepsilon = \varepsilon(t)$ such that $t = t(\varepsilon)$. In order to bound $\varepsilon(t)$ from above in terms of ε, we note that

$$t(\varepsilon) \geq \frac{n^{\frac{1}{2}}\varepsilon^{-\frac{1}{4}}}{(8e)^{\frac{1}{4}}}\sqrt{2N}$$

and hence

$$\frac{1}{\varepsilon} \leq \frac{2e}{(nN)^2}\, t(\varepsilon)^4 \leq t(\varepsilon)^4.$$

Using this bound in (17.33), we get for $\varepsilon = \varepsilon(t)$,

$$t \leq \frac{n^{\frac{1}{2}}}{(8e)^{\frac{1}{4}}}\varepsilon^{-\frac{1}{4}}\left(\sqrt{2N} + \sqrt{2\ln t^4}\right) \leq \frac{n^{\frac{1}{2}}}{(8e)^{\frac{1}{4}}}\varepsilon^{-\frac{1}{4}}4\sqrt{N\ln t},$$

where we have used that $a + b \leq ab$ for $a, b \geq 2$. This implies

$$\varepsilon \leq \frac{4^4}{8e}n^2N^2\frac{\ln^2 t}{t^4}.$$

Since $t(a) = \frac{\sqrt{2nN}}{(8e)^{\frac{1}{4}}} \leq \sqrt{nN}$, this bound is valid for any $t \geq \sqrt{nN}$. Hence we obtain from (17.33) that

$$\mathrm{Prob}\{\|M^{\dagger}\|\,\|q\| \geq t\} \leq 2\varepsilon \leq 24\,n^2N^2\frac{\ln^2 t}{t^4},$$

proving the tail estimate.

The bound on the expectation follows from

$$\mathbb{E}\,\mu_{norm} = \int_1^\infty \mathrm{Prob}\{\mu_{norm} \geq t\}\,dt \leq \sqrt{nN} + 24n^2N^2\int_{\sqrt{nN}}^\infty \frac{\ln^2 t}{t^4}\,dt,$$

noting that

$$\int_{t_0}^\infty \frac{\ln^2 t}{t^4}\,dt = \frac{\ln^2 t_0}{3t_0^3} + \frac{2\ln t_0}{9t_0^3} + \frac{2}{27t_0^3} \leq \frac{17}{27}\frac{\ln^2 t_0}{t_0^3}.$$

Therefore,

$$\mathbb{E}\,\mu_{\text{norm}} \leq \sqrt{nN} + \frac{24 \cdot 17}{27 \cdot 4}(nN)^{\frac{1}{2}}\ln^2(nN) \leq 5(nN)^{\frac{1}{2}}\ln^2(nN).$$

We can argue similarly for the expectation of $\mu_{\text{norm}}^2(q)$, where $q \in \mathcal{H}_\mathbf{d}$ is standard Gaussian.   □

The previous result easily implies information on the distribution of μ_{av}.

Corollary 17.29 *For $t \geq 1$ we have*

$$\Prob_{q \sim N(0,\mathrm{I})} \{\mu_{\text{av}}(q) \geq t\} = \mathcal{O}\left(\frac{nN}{t^2}\ln^2(nN)\right)$$

and

$$\mathbb{E}_{q \sim N(0,\mathrm{I})} \mu_{\text{av}}^2(q) = \mathcal{O}(nN\ln^2(nN)).$$

Proof Lemma 17.18(e) implies $\mathbb{E}\,\mu_{\text{av}}^2 = \mathbb{E}\,\mu_{\text{norm}}^2$. Hence the bound on the expectation of μ_{av}^2 follows from Theorem 17.28. The tail bound is now a consequence of Markov's inequality (Corollary 2.9).   □

Remark 17.30 The t^{-2} tail decay in Corollary 17.29 results from a simple application of Markov's inequality and does not describe the true behavior. This can be seen be comparing the tail bound with such a bound in Theorem 17.28 in the case $d_i = 1$ for all i, in which $\mu_{\text{norm}} = \mu_{\text{av}}$.

We finish this chapter with a brief discussion of the distribution of the *maximum condition number* μ_{max} defined, for $q \in \mathcal{H}_\mathbf{d} \setminus \Sigma$, by

$$\mu_{\text{max}}(q) := \max_{i \leq \mathcal{D}} \mu_{\text{norm}}(q, \zeta_i(q)),$$

where $\zeta_1(q), \ldots, \zeta_{\mathcal{D}}(q)$ are the zeros of q. By definition, $\mu_{\text{max}}(q) \geq t$ iff there exists $j \leq \mathcal{D}$ such that $\mu_{\text{norm}}(q, \zeta_j(q)) \geq t$. Hence, for standard Gaussian q,

$$\Prob_{q \sim N(0,\mathrm{I})} \{\mu_{\text{max}}(q) \geq t\} \leq \sum_{j=1}^{\mathcal{D}} \Prob\{\mu_{\text{norm}}(q, \zeta_j(q)) \geq t\}.$$

If we could assume that $(q, \zeta_j(q))$ follows the standard distribution, for all j, then we could deduce from Theorem 17.28 that

$$\Prob_{q \sim N(0,\mathrm{I})} \{\mu_{\text{max}}(q) \geq t\} = \mathcal{O}\left(\mathcal{D}n^2N^2\frac{\ln^2 t}{t^4}\right).$$

While it is not clear that the latter assumption is in fact true, the following result can be rigorously proven. We omit the proof.

Proposition 17.31 *In the case $n > 1$ we have for $t \geq \sqrt{n}$,*

$$\Prob_{q \sim N(0,\mathrm{I})} \{ \mu_{\mathrm{max}}(q) \geq t \} = \mathcal{O}\left(\frac{Dn^3 N^2}{t^4} \right)$$

and

$$\mathop{\mathbb{E}}_{q \sim N(0,\mathrm{I})} \mu_{\mathrm{max}}^2(q) = \mathcal{O}\left(D^{\frac{1}{2}} n^{\frac{3}{2}} N \right).$$

In the case $n = 1$ we have for $t \geq 1$,

$$\Prob_{q \sim N(0,\mathrm{I})} \{ \mu_{\mathrm{max}}(q) \geq t \} \leq d \left(1 - \left(1 - \frac{1}{t^2} \right)^{d-1} \left(1 + \frac{d-1}{t^2} \right) \right). \qquad \square$$

Remark 17.32 In the special case $n = 1$, Proposition 17.31 implies that for a random q of degree d we have $\mu_{\mathrm{max}}(q) = \mathcal{O}(d)$ with probability at least $1/2$. Remarkably, as of today, no deterministic polynomial-time procedure is known to produce a family (q_d) of univariate polynomials such that $\deg q_d = d$ and $\mu_{\mathrm{max}}(q) = \mathcal{O}(d)$.

Chapter 18
Smale's 17th Problem: II

In the preceding chapter we described Smale's 17th problem and a probabilistic solution for it, namely, a randomized algorithm whose average cost is polynomially
bounded. The present chapter continues with this theme, adding further understanding of the behavior of Algorithm LV (introduced in Sect. 17.2.2). Also, getting closer
to the original formulation of Smale's problem, this chapter exhibits a deterministic
algorithm whose average cost is nearly polynomial.

Our first result here is an extension of Theorem 17.1 providing a smoothed
analysis for the randomized cost of Algorithm LV. For this result we use a trun-
cated Gaussian as defined in (17.25) but noncentered. For $\overline{f} \in \mathcal{H}_{\mathbf{d}}$ we define
$N(\overline{f}, \sigma^2 I) := \overline{f} + N(0, \sigma^2 I)$.

Theorem 18.1 *For any* $0 < \sigma \leq 1$*, Algorithm* LV *satisfies*

$$\sup_{\overline{f} \in \mathbb{S}(\mathcal{H}_{\mathbf{d}})} \underset{f \sim N_T(\overline{f}, \sigma^2 I)}{\mathbb{E}} \text{r\_cost}^{\text{LV}}(f) = \mathcal{O}\left(D^{3/2} n N^2 \frac{1}{\sigma}\right).$$

Our second result is a condition-based cost analysis of Algorithm LV. We are
here interested in estimating $K(f)$ for a fixed input system $f \in \mathcal{H}_{\mathbf{d}} \setminus \Sigma$. Such an
estimate will have to depend on, besides D, n, and N, the condition of f. We take
for the latter the maximum condition number (which we met in Sect. 17.8)

$$\mu_{\max}(f) := \max_{\zeta \mid f(\zeta)=0} \mu_{\text{norm}}(f, \zeta), \tag{18.1}$$

which, we note in passing, provides an example for the third (and last) form of
measuring condition in terms of solutions discussed in Sect. 6.8.

Our condition-based analysis of LV is summarized in the following statement.

Theorem 18.2 *The randomized cost of Algorithm* LV *with input* $f \in \mathcal{H}_{\mathbf{d}} \setminus \Sigma$ *is
bounded as*

$$\text{r\_cost}^{\text{LV}}(f) = \mathcal{O}\left(D^3 n N^2 \mu_{\max}^2(f)\right).$$

P. Bürgisser, F. Cucker, *Condition,*
Grundlehren der mathematischen Wissenschaften 349,
DOI 10.1007/978-3-642-38896-5_18, © Springer-Verlag Berlin Heidelberg 2013

We finally want to consider *deterministic* algorithms for finding zeros of polynomial systems. One such algorithm with polynomial average cost would provide a positive answer to Smale's 17th problem. As of today, this answer has proved to be elusive. The last main result in this chapter is just a step in this direction.

Theorem 18.3 *There is a deterministic real-number algorithm that on input* $f \in \mathcal{H}_\mathbf{d}$ *computes an approximate zero of* f *in average time* $N^{\mathcal{O}(\log \log N)}$. *Moreover, if we restrict data to polynomials satisfying*

$$D \leq n^{\frac{1}{1+\varepsilon}} \quad or \quad D \geq n^{1+\rho},$$

for some fixed $\varepsilon > 0$, *then the average time of the algorithm is polynomial in the input size* N.

A common characteristic of the contexts of the three results above is the fact that the systems q_t occurring in their corresponding homotopy paths, while still random because of the randomness of either f or g, follow distributions that are no longer centered at 0. Gaussianity remains, but centrality doesn't. Consequently, variance plays a role. This characteristic adds a degree of intricacy to the arguments of the preceding chapter. In particular, it requires the consideration of extensions of the measure ρ_{st}. We therefore begin this chapter with some probability results including, among others, the postponed proof of Proposition 17.21.

18.1 The Main Technical Result

The technical heart of the proof of the results in this chapter is the following smoothed analysis of the mean square condition number μ_{av}. Note that this result extends Proposition 17.27.

Theorem 18.4 *For* $\overline{q} \in \mathcal{H}_\mathbf{d}$ *and* $\sigma > 0$ *we have*

$$\mathop{\mathbb{E}}_{q \sim N(\overline{q}, \sigma^2 \mathrm{I})} \left(\frac{\mu_{\mathsf{av}}^2(q)}{\|q\|^2} \right) \leq \frac{e(n+1)}{2\sigma^2}.$$

We note that no bound on the norm of $\overline{q}$ is required here. Indeed, using $\mu_{\mathsf{av}}(\lambda q) = \mu_{\mathsf{av}}(q)$, it is easy to see that the assertion for a pair $(\overline{q}, \sigma)$ implies the assertion for $(\lambda \overline{q}, \lambda \sigma)$, for any $\lambda > 0$.

18.1.1 Outline of the Proof

We shall now give an outline of the proof of Theorem 18.4. Let $\rho_{\mathcal{H}_\mathbf{d}}$ denote the density of the Gaussian $N(\overline{q}, \sigma^2 \mathrm{I})$ on $\mathcal{H}_\mathbf{d}$, where $\overline{q} \in \mathcal{H}_\mathbf{d}$ and $\sigma > 0$. For fixed $\zeta \in$

$\mathbb{S}(\mathbb{C}^{n+1})$ we decompose the mean $\overline{q}$ as

$$\overline{q} = \overline{k}_\zeta + \overline{g}_\zeta + \overline{h}_\zeta \in C_\zeta \oplus L_\zeta \oplus R_\zeta$$

according to the orthogonal decomposition (16.9). If we denote by ρ_{C_ζ}, ρ_{L_ζ}, and ρ_{R_ζ} the densities of the Gaussian distributions in the spaces C_ζ, L_ζ, and R_ζ with covariance matrices $\sigma^2 I$ and means $\overline{k}_\zeta$, $\overline{M}_\zeta$, and $\overline{h}_\zeta$, respectively, then the density $\rho_{\mathcal{H}_\mathbf{d}}$ factors as

$$\rho_{\mathcal{H}_\mathbf{d}}(k + g + h) = \rho_{C_\zeta}(k) \cdot \rho_{L_\zeta}(g) \cdot \rho_{R_\zeta}(h); \tag{18.2}$$

compare (2.12).

Recall from (17.14) the manifold

$$\hat{W} := \{(M, \zeta) \in \mathcal{M} \times \mathbb{S}(\mathbb{C}^{n+1}) \mid M\zeta = 0\},$$

whose projection $p_2 \colon \hat{W} \to \mathbb{S}(\mathbb{C}^{n+1})$, $(M, \zeta) \mapsto \zeta$, has the fibers

$$W_\zeta := \{M \in \mathcal{M} \mid M\zeta = 0\}.$$

Proposition 16.16 tells us that we have the isometric linear maps

$$W_\zeta \to L_\zeta, \quad M \mapsto g_{M,\zeta} := \left(\sqrt{d_i} \, \langle X, \zeta \rangle^{d_i-1} \sum_j m_{ij} X_j \right). \tag{18.3}$$

Thus the Gaussian density ρ_{L_ζ} on L_ζ induces a Gaussian density ρ_{W_ζ} on the fiber W_ζ with covariance matrix $\sigma^2 I$ via this map, so that $\rho_{W_\zeta}(M) = \rho_{L_\zeta}(g_{M,\zeta})$.

We derive now from the given Gaussian distribution $\rho_{\mathcal{H}_\mathbf{d}}$ on $\mathcal{H}_\mathbf{d}$ a probability distribution on the solution variety V as follows (naturally extending ρ_{st} introduced in Sect. 17.5). Think of choosing (q, ζ) at random from $\hat{V}$ by first choosing $q \in \mathcal{H}_\mathbf{d}$ from $N(\overline{q}, \sigma^2 I)$, then choosing one of its $\mathcal{D}$ zeros $[\zeta] \in \mathbb{P}^n$ at random from the uniform distribution on $\{1, \ldots, \mathcal{D}\}$, and finally choosing a representative ζ in the unit circle $[\zeta] \cap \mathbb{S}^n$ uniformly at random. (An explicit expression of the corresponding probability density $\rho_{\hat{V}}$ on $\hat{V}$ is given in (18.16); compare the special case (17.19).)

The road map to proving Theorem 18.4 is as follows. By the definition of $\rho_{\mathcal{H}_\mathbf{d}}$ and as in Lemma 17.18, we have

$$\underset{q \sim N(\overline{q}, \sigma^2 I)}{\mathbb{E}} \left(\frac{\mu_{\mathrm{av}}^2(q)}{\|q\|^2} \right) = \underset{(q,\zeta) \sim \rho_{\hat{V}}}{\mathbb{E}} \left(\frac{\mu_{\mathrm{norm}}^2(q, \zeta)}{\|q\|^2} \right). \tag{18.4}$$

Recall from (17.13) the lifting $\hat{V} \subseteq \mathcal{H}_\mathbf{d} \times \mathbb{S}(\mathbb{C}^{n+1})$ of the solution variety $V \subseteq \mathcal{H}_\mathbf{d} \times \mathbb{P}^n$. Put $\Delta := \mathrm{diag}(d_i^{1/2})$. In (17.32) we already noted that the condition number $\mu_{\mathrm{norm}}(q, \zeta)$ can be described in terms of the linearization map

$$\Psi \colon \hat{V} \to \hat{W}, \quad (q, \zeta) \mapsto (M, \zeta), \quad \text{where } M := \Delta^{-1} Dq(\zeta),$$

as follows:

$$\frac{\mu_{\text{norm}}(q,\zeta)}{\|q\|} = \|M^\dagger\|, \quad \text{where } (M,\zeta) = \Psi(q,\zeta).$$

Hence

$$\mathop{\mathbb{E}}_{(q,\zeta)\sim\rho_{\hat{V}}} \left(\frac{\mu_{\text{norm}}^2(q,\zeta)}{\|q\|^2} \right) = \mathop{\mathbb{E}}_{M\sim\rho_{\mathcal{M}}} (\|M^\dagger\|^2), \tag{18.5}$$

where $\rho_{\mathcal{M}}$ denotes the pushforward density of $\rho_{\hat{V}}$ with respect to the map $p_1 \circ \Psi \colon \hat{V} \to \mathcal{M}$.

It will turn out that the density $\rho_{\mathcal{M}}$ has the following explicit description. For $M \in \mathcal{M}$ of rank n and $\zeta \in \mathbb{S}(\mathbb{C}^{n+1})$ with $M\zeta = 0$ we have

$$\rho_{\mathcal{M}}(M) = \rho_{C_\zeta}(0) \cdot \frac{1}{2\pi} \int_{\lambda\in S^1} \rho_{W_{\lambda\zeta}}(M)\, d\mathbb{S}^1. \tag{18.6}$$

By (17.12) we have

$$\mathop{\mathbb{E}}_{M\sim\rho_{\mathcal{M}}} (\|M^\dagger\|^2) = \mathop{\mathbb{E}}_{\zeta\sim\rho_{\mathbb{S}(\mathbb{C}^{n+1})}} \left(\mathop{\mathbb{E}}_{M\sim\tilde{\rho}_{W_\zeta}} (\|M^\dagger\|^2) \right), \tag{18.7}$$

where $\rho_{\mathbb{S}(\mathbb{C}^{n+1})}$ is the pushforward density of ρ_V with respect to $p_2 \circ \Psi \colon V \to \mathbb{S}(\mathbb{C}^{n+1})$ and $\tilde{\rho}_{W_\zeta}$ denotes the conditional density on the fiber W_ζ. This conditional density will turn out to be of the form

$$\tilde{\rho}_{W_\zeta}(M) = c_\zeta^{-1} \cdot \det(MM^*)\, \rho_{W_\zeta}(M), \tag{18.8}$$

with c_ζ denoting a normalization factor possibly depending on ζ. In the case $\zeta = (1, 0, \ldots, 0)$ we can identify W_ζ with $\mathbb{C}^{n\times n}$, and $\tilde{\rho}_{W_\zeta}$ takes the form (4.21) studied in Sect. 4.4. Proposition 4.22 and unitary invariance imply that for all $\zeta \in \mathbb{S}(\mathbb{C}^{n+1})$,

$$\mathop{\mathbb{E}}_{M\sim\tilde{\rho}_{W_\zeta}} (\|M^\dagger\|^2) \le \frac{e(n+1)}{2\sigma^2}. \tag{18.9}$$

This implies by (18.7) that

$$\mathop{\mathbb{E}}_{M\sim\rho_{\mathcal{M}}} (\|M^\dagger\|^2) \le \frac{e(n+1)}{2\sigma^2},$$

and this bound, replaced in (18.5) and back in (18.4), completes the outline of the proof of Theorem 18.4.

The formal proof of the stated facts (18.6) and (18.8) is quite involved and will be given in the remainder of this section.

18.1.2 Normal Jacobians of Linearizations

In (17.32) we saw that the condition number $\mu_{norm}(q, \zeta)$ can be described in terms of the *linearization map* $\Psi \colon \hat{V} \to \hat{W}$. As a stepping stone towards the analysis of the normal Jacobian of Ψ we introduce now the related linearization map

$$\Phi \colon \hat{V} \to \hat{W}, \quad (q, \zeta) \mapsto (N, \zeta) = \big(Dq(\zeta), \zeta\big). \tag{18.10}$$

Lemma 18.5 *The derivative of* $D\Phi(q, \zeta) \colon T_{(q,\zeta)}V \to T_{(N,\zeta)}W$ *is determined by*

$$D\Phi(q, \zeta)(\dot{q}, \dot{\zeta}) = (\dot{N}, \dot{\zeta}), \quad \text{where } \dot{N} = D\dot{q}(\zeta) + D^2 q(\zeta)(\dot{\zeta}, \cdot).$$

Proof Consider a smooth curve $t \mapsto (q(t), \zeta(t))$ in $\hat{V}$ with $(q(0), \zeta(0)) = (q, \zeta)$ and put $N(t) = Dq(t)(\zeta(t))$. In coordinates, $n_{ij}(t) = \partial_{X_j} q_i(t)(\zeta(t))$. Differentiating this with respect to t at zero, we obtain

$$\dot{n}_{ij} = \partial_{X_j}\dot{q}_i(\zeta) + \sum_{k=0}^{n} \partial_{X_k}\partial_{X_j}q_i(\zeta)\,\dot{\zeta}_k.$$

This is nothing but the claimed assertion, written in coordinates.   □

It is a crucial observation that the normal Jacobian of Φ is constant.

Proposition 18.6 *We have* $\mathrm{NJ}\Phi(q, \zeta) = \mathcal{D}^n$ *for all* $(q, \zeta) \in \hat{V}$.

Proof We adopt the notation from the proof of Lemma 17.15. Using the shorthand $\partial_k q$ for the partial derivative $\partial_{X_k} q$, etc., a short calculation yields, for $j > 0$,

$$\partial_0\dot{q}_i(\zeta) = d_i\dot{u}_i, \qquad \partial_j\dot{q}_i(\zeta) = \sqrt{d_i}\,\dot{a}_{ij}, \qquad \partial_{0j}^2 q_i(\zeta) = (d_i - 1)\,n_{ij}. \tag{18.11}$$

Similarly, we obtain $\partial_0 q_i(\zeta) = 0$ and $\partial_j q_i(\zeta) = n_{ij}$ for $j > 0$.

Introducing the coordinates $\dot{N} = (\dot{n}_{ij})$, the formula of Lemma 18.5 can be written as

$$\dot{n}_{ij} = \partial_j\dot{q}_i(\zeta) + \sum_{k=1}^{n} \partial_{jk}^2\dot{q}_i(\zeta)\,\dot{\zeta}_k. \tag{18.12}$$

For $j > 0$ this gives, using (18.11),

$$\dot{n}_{ij} = \sqrt{d_i}\,\dot{a}_{ij} + \sum_{k=1}^{n} \partial_{jk}^2\dot{q}_i(\zeta)\,\dot{\zeta}_k. \tag{18.13}$$

For $j = 0$ we obtain from (18.12), using (18.11) and (17.16),

$$\dot{n}_{i0} = \partial_0\dot{q}_i(\zeta) + \sum_{k=1}^{n} \partial_{0k}^2\dot{q}_i(\zeta)\,\dot{\zeta}_k = d_i\dot{u}_i + (d_i - 1)\sum_{k=1}^{n} n_{ik}\dot{\zeta}_k = \dot{u}_i. \tag{18.14}$$

Note the crucial cancellation taking place here!

From (18.13) and (18.14) we see that the kernel K of $D\Phi(q, \zeta)$ is determined by the conditions $\dot{\zeta} = 0$, $\dot{u} = 0$, $\dot{A} = 0$. Hence, recalling $T_{(q,\zeta)}V \simeq T_{(N,\zeta)}W \times R_\zeta$ from the proof of Lemma 17.15, we have $K \simeq 0 \times R_\zeta$ and $K^\perp \simeq T_{(N,\zeta)}W \times 0$. Moreover, as in the proof of Lemma 17.14 (but replacing M by N), we write

$$E := \left\{ (\dot{u}, \dot{\zeta}) \in \mathbb{C}^n \times \mathbb{C}^{n+1} \mid \dot{u}_i + \sum_{j=1}^{n} n_{ij}\dot{\zeta}_j = 0, 1 \le i \le n, \dot{\zeta}_0 \in i\mathbb{R} \right\}$$

and identify $T_{(N,\zeta)}W$ with $E \times \mathbb{C}^{n \times n}$. Using this identification of spaces, (18.13) and (18.14) imply that $D\Phi(q, \zeta)_{K^\perp}$ has the following structure:

$$D\Phi(q, \zeta)_{K^\perp}: \quad E \times \mathbb{C}^{n \times n} \to E \times \mathbb{C}^{n \times n},$$

$$((\dot{u}, \dot{\zeta}), \dot{A}) \mapsto ((\dot{u}, \dot{\zeta}), \lambda(\dot{A}) + \rho(\dot{\zeta})),$$

where the linear map $\lambda: \mathbb{C}^{n \times n} \to \mathbb{C}^{n \times n}$, $\dot{A} \mapsto (\sqrt{d_i}\,\dot{a}_{ij})$, multiplies the ith row of $\dot{A}$ by $\sqrt{d_i}$ and $\rho: \mathbb{C}^{n+1} \to \mathbb{C}^{n \times n}$ is given by $\rho(\dot{\zeta})_{ij} = \sum_{k=1}^{n} \partial_{jk}^2 \dot{q}_i(\zeta)\dot{\zeta}_k$.

By definition we have $\mathrm{NJ}\Phi(q, \zeta) = |\det D\Phi(q, \zeta)_{|K^\perp}|$. The triangular form of $D\Phi(q, \zeta)_{|K^\perp}$ shown above implies that $|\det D\Phi(q, \zeta)_{|K^\perp}| = \det \lambda$. Finally, using the diagonal form of λ, we obtain $\det \lambda = \prod_{i=1}^{n} \prod_{j=1}^{n} \sqrt{d_i}^2 = \mathcal{D}^n$, which completes the proof.   □

Remark 18.7 Denote by $\mathcal{H}_{\mathbf{d}}^{\mathbb{R}}$ the linear space of homogeneous polynomial systems with degree pattern $\mathbf{d} = (d_1, \ldots, d_n)$ and real coefficients. The real solution variety $V_{\mathbb{R}} \subseteq \mathcal{H}_{\mathbf{d}}^{\mathbb{R}} \times \mathbb{S}^n$ is defined in the obvious way and so is $W_{\mathbb{R}} \subseteq \mathcal{M}_{\mathbb{R}} \times \mathbb{S}^n$, where $\mathcal{M}_{\mathbb{R}} = \mathbb{R}^{n \times (n+1)}$. The same proof as for Proposition 18.6 shows that the normal Jacobian of the map $\Phi_{\mathbb{R}}: V_{\mathbb{R}} \to W_{\mathbb{R}}$, $(q, \zeta) \mapsto (Dq(\zeta), \zeta)$, has the constant value $\mathcal{D}^{n/2}$. As the only change in the proof we notice that the $\mathbb{R}$-linear map $\mathbb{C} \to \mathbb{C}$, $z \mapsto \sqrt{d_i}z$, has determinant d_i, while the map $\mathbb{R} \to \mathbb{R}$, $x \mapsto \sqrt{d_i}x$, has determinant $\sqrt{d_i}$.

The normal Jacobian of the map $\Psi: \hat{V} \to \hat{W}$ is not constant and takes a more complicated form in terms of the normal Jacobians of the projection $p_1: \hat{W} \to \mathcal{M}$. For obtaining an expression for $\mathrm{NJ}\Psi$ we need the following lemma.

Lemma 18.8 *The scaling map* $\gamma: \hat{W} \to \hat{W}$, $(N, \zeta) \mapsto (M, \zeta)$, *with* $M = \Delta^{-1}N$ *of rank n satisfies*

$$\det D\gamma(N, \zeta) = \frac{1}{\mathcal{D}^{n+1}} \cdot \frac{\mathrm{NJ}p_1(N, \zeta)}{\mathrm{NJ}p_1(M, \zeta)}.$$

Proof Note that $T_{(M,\zeta)}\hat{W} = T_{(M,\zeta)}W \oplus \mathbb{R}i\zeta$, where W denotes the solution variety in $\mathcal{M} \times \mathbb{P}^n$. Let $p_1': W \to \mathcal{M}$ denote the projection. The derivative $D\gamma_{\mathbb{P}}(N, \zeta)$ of the corresponding scaling map $\gamma_{\mathbb{P}}: W \to W$ is determined by the commutative diagram

$$
\begin{CD}
T_{(N,\zeta)}W @>{D\gamma_{\mathbb{P}}(N,\zeta)}>> T_{(M,\zeta)}W \\
@V{Dp_1'(N,\zeta)}VV @VV{Dp_1'(M,\zeta)}V \\
\mathcal{M} @>{\mathrm{sc}}>> \mathcal{M},
\end{CD}
$$

where the vertical arrows are linear isomorphisms. The assertion follows by observing that $\mathrm{NJ}p_1(N,\zeta) = \det Dp_1'(N,\zeta)$, $\mathrm{NJ}\gamma(N,\zeta) = \det D\gamma_{\mathbb{P}}(N,\zeta)$, and using that the $\mathbb{R}$-linear map $\mathrm{sc}\colon \mathcal{M} \to \mathcal{M}, N \mapsto M = \Delta^{-1}N$, has determinant $1/\mathcal{D}^{n+1}$. $\qquad\square$

Proposition 18.6 combined with Lemma 18.8 immediately gives

$$
\mathrm{NJ}\Psi(q,\zeta) = \frac{1}{\mathcal{D}} \cdot \frac{\mathrm{NJ}p_1(N,\zeta)}{\mathrm{NJ}p_1(M,\zeta)} \tag{18.15}
$$

for $N = Dq(\zeta)$, $M = \Delta^{-1}N$.

Remark 18.9 Here is a sketch of an alternative proof of Proposition 18.6. For given $(q,\zeta) \in \hat{V}$ we decompose $q = g + h$ with $g \in L_\zeta$ and $h \in R_\zeta$ according to the orthogonal decomposition (16.9). It turns out that $\mathrm{NJ}\Phi(q,\zeta)$ depends only on the component g, so that we may assume that $h = 0$ and $q = g$.

The map Φ is equivariant under the action of $\mathcal{U}(n+1)$. Hence $\mathrm{NJ}\Phi$ is constant on $\mathcal{U}$-orbits. We may therefore assume that $\zeta = e_0 = (1, 0, \ldots, 0)$. The elements $g = (g_i)$ of L_{e_0} are of the form $g_i = X_0^{d_i - 1} \sum_{j=1}^n n_{ij} X_j$ and in bijection with the matrices $N = [n_{ij}] \in \mathcal{M}$ having zero as their first column. The action of the stabilizer $\mathcal{U}_{e_0} \simeq \mathcal{U}(n)$ of e_0 corresponds to the multiplication of N by unitary matrices from the right. However, we also have an action of $\mathcal{U}(n)$ on $\mathcal{M}$ given by multiplication from the left. This defines a $\mathcal{U}(n)$-action on L_{e_0}. (Note that this action does not extend to $\mathcal{H}_{\mathbf{d}}$ unless all the degrees d_i are equal.) It can be shown that Φ is also equivariant with respect to this action. As a consequence, $\mathrm{NJ}\Phi$ is constant on $\mathcal{U}(n) \times \mathcal{U}_{e_0}$-orbits. By the singular value decomposition of the matrix N, we may therefore assume that N consists of the zero column and the diagonal matrix $\mathrm{diag}(\sigma_1, \ldots, \sigma_n)$, where $\sigma_1, \ldots, \sigma_n$ are the singular values of N. Summarizing, (g, ζ) is of the special form

$$
g_1 = \sigma_1 X_0^{d_i - 1} X_1, \ldots, g_n = \sigma_n X_0^{d_i - 1} X_n, \qquad \zeta = (1, 0, \ldots, 0),
$$

that we already encountered in Remark 16.18. A closer look then reveals that $\mathrm{NJ}\Phi(g, \zeta)$ does not depend on the singular values σ_i. Using this fact, it is possible to derive the actual value $\mathrm{NJ}\Phi$ by an indirect argument as follows. As in the proof of Theorem 19.2, one can derive the following complex analogue of (19.11):

$$
\int_{q \in \mathcal{H}_{\mathbf{d}}} \#_{\mathbb{C}}(q)\, \varphi_{\mathcal{H}_{\mathbf{d}}}(q)\, d\mathcal{H}_{\mathbf{d}} = \frac{\mathcal{D}^{n+1}}{\mathrm{NJ}\Phi},
$$

where $\#_{\mathbb{C}}(q)$ denotes the number of zeros of q in $\mathbb{P}^n(\mathbb{C})$ and $\varphi_{\mathcal{H}_{\mathbf{d}}}$ is the standard Gaussian distribution on $\mathcal{H}_{\mathbf{d}}$. Bézout's theorem implies that $\#_{\mathbb{C}}(q) = \mathcal{D}$ for almost all $q \in \mathcal{H}_{\mathbf{d}}$. Therefore, $\mathrm{NJ}\Phi = \mathcal{D}^n$.

18.1.3 Induced Probability Distributions

By Bézout's theorem, the fiber $\hat{V}(q)$ of the projection $\pi_1 \colon \hat{V} \to \mathcal{H}_{\mathbf{d}}$ at $q \in \mathcal{H}_{\mathbf{d}} \setminus \Sigma$ is a disjoint union of $\mathcal{D} = d_1 \cdots d_n$ unit circles and therefore has volume $2\pi \mathcal{D}$.

Recall that $\rho_{\mathcal{H}_{\mathbf{d}}}$ denotes the density of the Gaussian distribution $N(\overline{q}, \sigma^2 I)$ for fixed $\overline{q} \in \mathcal{H}_{\mathbf{d}}$ and $\sigma > 0$. We associate with $\rho_{\mathcal{H}_{\mathbf{d}}}$ the function $\rho_{\hat{V}} \colon \hat{V} \to \mathbb{R}$ defined by

$$\rho_{\hat{V}}(q, \zeta) := \frac{1}{2\pi \mathcal{D}} \, \rho_{\mathcal{H}_{\mathbf{d}}}(q) \, \mathrm{NJ}\pi_1(q, \zeta). \tag{18.16}$$

By the same proof as for the standard distribution dealt with in Lemma 17.18 we can prove the following.

Lemma 18.10

(a) *The function $\rho_{\hat{V}}$ is a probability density on $\hat{V}$.*

(b) *The expectation of a function $F \colon \hat{V} \to \mathbb{R}$ with respect to $\rho_{\hat{V}}$ can be expressed as*

$$\mathop{\mathbb{E}}_{(q,\zeta) \sim \rho_{\hat{V}}} F(q, \zeta) = \mathop{\mathbb{E}}_{q \sim \rho_{\mathcal{H}_{\mathbf{d}}}} F_{\mathsf{sav}}(q),$$

where $F_{\mathsf{sav}}(q) := \frac{1}{2\pi \mathcal{D}} \int_{\hat{V}(q)} F \, d\hat{V}(q)$.

(c) *The pushforward of $\rho_{\hat{V}}$ with respect to $\pi_1 \colon \hat{V} \to \mathcal{H}_{\mathbf{d}}$ equals $\rho_{\mathcal{H}_{\mathbf{d}}}$.*

(d) *For $q \notin \Sigma$, the conditional density on the fiber $\hat{V}(q)$ is the density of the uniform distribution on $\hat{V}(q)$.* $\square$

We can now determine the various probability distributions induced by $\rho_{\hat{V}}$.

Proposition 18.11 *Let $\zeta \in \mathbb{C}^{n+1}_*$. For $h \in R_\zeta$ we have*

$$\frac{\rho_{\hat{V}}}{\mathrm{NJ}\Psi}(g_{M,\zeta} + h, \zeta) = \rho_{\hat{W}}(M, \zeta) \cdot \rho_{R_\zeta}(h),$$

where the pushforward density $\rho_{\hat{W}}$ of $\rho_{\hat{V}}$ with respect to $\Psi \colon \hat{V} \to \hat{W}$ satisfies

$$\rho_{\hat{W}}(M, \zeta) = \frac{1}{2\pi} \, \rho_{C_\zeta}(0) \cdot \rho_{W_\zeta}(M) \cdot \mathrm{NJ}p_1(M, \zeta).$$

Proof Using the factorization of Gaussians (18.2) and Lemma 17.15, the density $\rho_{\hat{V}}$ can be written as

$$\rho_{\hat{V}}(g_{M,\varsigma} + h, \zeta) = \frac{1}{2\pi\mathcal{D}} \rho_{C_\zeta}(0)\,\rho_{W_\varsigma}(M)\,\rho_{R_\zeta}(h)\,\mathrm{NJ}p_1(N, \zeta),$$

where $N = \Delta M$. It follows from (18.15) that

$$\frac{\rho_{\hat{V}}}{\mathrm{NJ}\Psi}(g_{M,\varsigma} + h, \zeta) = \frac{1}{2\pi} \rho_{C_\zeta}(0)\,\rho_{W_\varsigma}(M)\,\rho_{R_\zeta}(h)\,\mathrm{NJ}p_1(M, \zeta). \qquad (18.17)$$

This implies, using (17.10) for $\Psi : V \to W$ and the isometry $\Psi^{-1}(M, \zeta) \simeq R_\zeta$ for the fiber at ζ, that

$$\rho_{\hat{W}}(M, \zeta) = \int_{h \in R_\zeta} \frac{\rho_{\hat{V}}}{\mathrm{NJ}\Psi}(g_{M,\varsigma} + h, \zeta)\,dR_\zeta$$

$$= \frac{1}{2\pi} \rho_{C_\zeta}(0) \cdot \rho_{W_\varsigma}(M) \cdot \mathrm{NJ}p_1(M, \zeta) \int_{h \in R_\zeta} \rho_{R_\zeta}(h)\,dR_\zeta$$

$$= \frac{1}{2\pi} \rho_{C_\zeta}(0) \cdot \rho_{W_\varsigma}(M) \cdot \mathrm{NJ}p_1(M, \zeta)$$

as claimed. Replacing in (18.17), we therefore obtain

$$\frac{\rho_{\hat{V}}}{\mathrm{NJ}\Psi}(g_{M,\varsigma} + h, \zeta) = \rho_{\hat{W}}(M, \zeta)\,\rho_{R_\zeta}(h). \qquad \square$$

Lemma 18.12 *Let c_ζ denote the expectation of $\det(MM^*)$ with respect to ρ_{W_ς}. We have*

$$\frac{\rho_{\hat{W}}}{\mathrm{NJ}p_2}(M, \zeta) = \rho_{\mathbb{S}(\mathbb{C}^{n+1})}(\zeta) \cdot \tilde{\rho}_{W_\varsigma}(M),$$

where $\rho_{\mathbb{S}(\mathbb{C}^{n+1})}(\zeta) = \frac{c_\zeta}{2\pi}\rho_{C_\zeta}(0)$ is the pushforward density of $\rho_{\hat{W}}$ with respect to $p_2 : W \to \mathbb{S}(\mathbb{C}^{n+1})$, and where the conditional density $\tilde{\rho}_{W_\varsigma}$ on the fiber W_ς of p_2 is given by

$$\tilde{\rho}_{W_\varsigma}(M) = c_\zeta^{-1} \cdot \det(MM^*)\rho_{W_\varsigma}(M).$$

Proof Lemma 17.16 states that

$$\frac{\mathrm{NJ}p_1}{\mathrm{NJ}p_2}(M, \zeta) = \det(MM^*).$$

Combining this with Proposition 18.11, we get

$$\frac{\rho_{\hat{W}}}{\mathrm{NJ}p_2}(M, \zeta) = \frac{1}{2\pi} \rho_{C_\zeta}(0) \cdot \rho_{W_\varsigma}(M) \cdot \det(MM^*).$$

Integrating over W_ζ we get $\rho_{\mathbb{S}(\mathbb{C}^{n+1})}(\zeta) = \frac{1}{2\pi} \rho_{C_\zeta}(0) \cdot c_\zeta$, and finally (cf. (17.11))

$$\tilde{\rho}_{W_\zeta}(M) = \frac{\rho_{\hat{W}}(M, \zeta)}{\rho_{\mathbb{S}(\mathbb{C}^{n+1})}(\zeta)\,\mathrm{NJ}p_2(M, \zeta)} = c_\zeta^{-1} \cdot \rho_{W_\zeta}(M) \cdot \det(MM^*),$$

as claimed.   □

We can finally complete the proof of the main technical result of this chapter.

Proof of Theorem 18.4 The claimed formula (18.6) for the pushforward density $\rho_{\mathscr{M}}$ of $\rho_{\hat{W}}$ with respect to $p_1 \colon \hat{W} \to \mathscr{M}$ immediately follows from Proposition 18.11 by integrating $\rho_{\hat{W}}/\mathrm{NJ}p_1$ over the fibers of p_1; compare (17.10).

Moreover, Lemma 18.12 shows that the conditional density $\tilde{\rho}_{W_\zeta}$ has the form stated in (18.8). We have thus filled the two gaps in the outline of the proof given in Sect. 18.1.1.   □

We close this section by providing a proof of Lemma 17.19 as well as of Proposition 17.21(a), both tasks that we postponed in the previous chapter.

We begin by noting that the fibers of Ψ allow for a transparent description. Indeed, for $(q, \zeta) \in \hat{V}$ we have the decomposition $q = g + h$ with $g \in L_\zeta$ and $h \in R_\zeta$ according to (16.9). If $\Psi(q, \zeta) = (M, \zeta)$, then g is uniquely determined by (M, ζ) (by (18.3)). It follows that the fiber of Ψ over (M, ζ) is in bijective correspondence with the space R_ζ as follows:

$$R_\zeta \to \Psi^{-1}(M, \zeta), \quad h \mapsto (g_{M,\zeta} + h, \zeta). \tag{18.18}$$

Proposition 18.13 *The conditional distribution on the fiber of Ψ over $(M, \zeta) \in \hat{W}$ is induced from the Gaussian in R_ζ via the bijection* (18.18).

Proof By the definition (17.11) of the conditional distributions on fibers we have that the density $\tilde{\rho}_{\hat{V}_\zeta}$ satisfies, for $(q, \zeta) \in \psi^{-1}(M, \zeta) = \hat{V}_\zeta$,

$$\tilde{\rho}_{\hat{V}_\zeta}(q, \zeta) = \frac{\rho_{\hat{V}}(q, \zeta)}{\rho_{\hat{W}}(M, \zeta)\,\mathrm{NJ}\Psi(q, \zeta)} = \rho_{R_\zeta}(h),$$

where q decomposes as $q = g_{M,\zeta} + h \in L_\zeta \oplus R_\zeta$, the last equality following by Proposition 18.11. This proves (a).   □

Proof of Lemma 17.19 Since we assume here $\rho_{\mathcal{H}_d}$ to be standard Gaussian, the induced distributions on C_ζ, L_ζ, and R_ζ are standard Gaussian as well. Hence ρ_{W_ζ} equals the standard Gaussian distribution on the fiber W_ζ. Moreover, $\rho_{C_\zeta}(0) =$

$(\sqrt{2\pi})^{-2n}$. Therefore, using the second statement in Proposition 18.11, we get

$$\rho_{\hat{W}}(M, \zeta) = \frac{1}{2\pi} \rho_{C_\zeta}(0) \cdot \rho_{W_\zeta}(M) \cdot \mathrm{NJ}p_1(M, \zeta)$$

$$= \frac{1}{2\pi} \frac{1}{(2\pi)^n} \frac{1}{(2\pi)^{n^2}} \exp\left(-\frac{1}{2}\|M\|_F^2\right) \cdot \mathrm{NJ}p_1(M, \zeta)$$

$$= \frac{1}{2\pi} \varphi_{\mathscr{M}}(M) \cdot \mathrm{NJ}p_1(M, \zeta),$$

where $\varphi_{\mathscr{M}}$ denotes the standard Gaussian distribution on $\mathscr{M}$. It follows from the definition (17.19) of standard distribution (taking $\mathcal{D} = 1$ there, since we are dealing with $\hat{W}$) that $\rho_{\hat{W}}$ is the density of this distribution on $\hat{W}$.   □

Proof of Proposition 17.21(a) Let us denote by ρ_{BP} the density for the distribution of the pairs $(q, \zeta) \in \hat{V}$ returned by Algorithm 17.6.

Pairs are drawn from ρ_{BP} by first drawing (M, ζ) from a marginal distribution $\rho_{\mathrm{BP}}^{\hat{W}}$ on $\hat{W}$ and then drawing (q, ζ) from the conditional distribution on the fiber $\Psi^{-1}(M, \zeta)$ (recall Remark 17.11). Draws from the marginal $\rho_{\mathrm{BP}}^{\hat{W}}$ are likewise obtained by first drawing M from the standard Gaussian $\varphi_{\mathscr{M}}$ on $\mathscr{M}$ and then drawing ζ from the uniform distribution on $M^{-1}(0) \cap \mathbb{S}(\mathbb{C}^{n+1})$ (which is almost certainly $\mathbb{S}^1$). From here it follows that $\rho_{\mathrm{BP}}^{\hat{W}}$ is the standard distribution on $\hat{W}$. Indeed, Lemma 17.18(b) applied to $\hat{W}$ states that the pushforward of the standard distribution on $\hat{W}$ with respect to $p_1 \colon \hat{W} \to \mathscr{M}$ equals the standard Gaussian distribution on $\mathscr{M}$, and part (d) of the same lemma ensures that the conditional distribution on the fiber $p_1^{-1}(M)$ equals the uniform distribution on $M^{-1}(0) \cap \mathbb{S}(\mathbb{C}^{n+1})$. Hence the standard distribution on $\hat{W}$ decomposes with respect to p_1 in the same manner as $\rho_{\mathrm{BP}}^{\hat{W}}$.

A similar argument shows that ρ_{st} and ρ_{BP} decompose in the same manner with respect to the linearization $\Psi \colon \hat{V} \to \hat{W}$. Indeed, the pushforward of ρ_{st} with respect to Ψ is, by Lemma 17.19, the standard distribution on $\hat{W}$, showing that the marginals coincide. The conditionals for pairs (M, ζ) coincide as well, since in the case of ρ_{BP} these are the standard Gaussian in R_ζ by construction and in the case of ρ_{st} they are the same distribution by Proposition 18.13.   □

18.2   Smoothed Analysis of LV

The smoothed analysis of LV, that is, Theorem 18.1, is shown similarly to its average-case analysis.

Proof of Theorem 18.1 Fix $\overline{f} \in \mathbb{S}(\mathcal{H}_{\mathbf{d}})$. Reasoning as in the proof of Proposition 17.26 and using $\|f\| \le \|\overline{f}\| + \|f - \overline{f}\| \le 1 + T$, we show that

$$\underset{f \sim N_T(\overline{f}, \sigma^2 I)}{\mathbb{E}} K(f) \leq 188 D^{3/2} \frac{(T+1)T}{P_{T,\sigma} P_{T,1}} \underset{f \sim N(\overline{f}, \sigma^2 I)}{\mathbb{E}} \underset{g \sim N(0,I)}{\mathbb{E}} \left(\int_0^1 \frac{\mu_2^2(q_t)}{\|q_t\|} dt \right)$$

$$= 188 D^{3/2} \frac{(T+1)T}{P_{T,\sigma} P_{T,1}} \int_0^1 \underset{q_t \sim N(\overline{q}_t, \sigma_t^2 I)}{\mathbb{E}} \left(\frac{\mu_2^2(q_t)}{\|q_t\|} \right) dt$$

with $\overline{q}_t = t\overline{f}$ and $\sigma_t^2 = (1-t)^2 + \sigma^2 t^2$. We now apply Theorem 18.4 to deduce

$$\int_0^1 \underset{q_t \sim N(\overline{q}_t, \sigma_t^2 I)}{\mathbb{E}} \left(\frac{\mu_2^2(q_t)}{\|q_t\|^2} \right) dt \leq \frac{e(n+1)}{2} \int_0^1 \frac{dt}{(1-t)^2 + \sigma^2 t^2} = \frac{e\pi(n+1)}{4\sigma}.$$

Consequently, using Lemma 17.25, we get

$$\mathbb{E}_{f \sim N_T(\overline{f}, \sigma^2 I)} K(f) \leq 188 \, D^{3/2} \cdot 4 \cdot \left(2N + \sqrt{2N} \right) \frac{e\pi(n+1)}{4\sigma},$$

which, combined with the $\mathcal{O}(N)$ cost of each iteration in LV, proves the assertion. $\square$

18.3   Condition-Based Analysis of LV

The last two results stated in the introduction of this chapter involve homotopies on which one endpoint of the homotopy segment is fixed, not randomized. The following result provides the major stepping stone in their proofs.

Theorem 18.14 *For $g \in \mathbb{S}(\mathcal{H}_\mathbf{d}) \setminus \Sigma$ we have*

$$\underset{f \sim N(0,I)}{\mathbb{E}} \left(d_{\mathbb{S}}(f, g) \int_0^1 \mu_{av}^2(q_\tau) d\tau \right) \leq 639 \, D^{3/2} N(n+1) \mu_{max}^2(g) + 0.02.$$

The idea for proving Theorem 18.14 is simple. For small values of τ the system q_τ is close to g, and therefore, the value of $\mu_{av}^2(q_\tau)$ can be bounded by a small multiple of $\mu_{max}^2(g)$. For the remaining values of τ, the corresponding $t = t(\tau)$ is bounded away from 0, and therefore, so is the variance σ_t^2 in the distribution $N(\overline{q}_t, \sigma_t^2 I)$ for q_t. This allows one to control the denominator on the right-hand side of Theorem 18.4 when using this result. Here are the precise details.

In the following fix $g \in \mathbb{S}(\mathcal{H}_\mathbf{d}) \setminus \Sigma$. First note that we may again replace the Gaussian distribution of f on $\mathcal{H}_\mathbf{d}$ by the truncated Gaussian $N_T(0, I)$. As in Sect. 17.7 we choose $T := \sqrt{2N}$. Recall also from this section the probability $P_{T,1}$, which is at least $1/2$; cf. Lemma 17.25. We therefore need to bound the quantity

$$Q_g := \mathop{\mathbb{E}}_{f \sim N_T(0,I)} \left(d(f,g) \int_0^1 \mu_{av}^2(q_\tau) \, d\tau \right).$$

To simplify notation, we set $\varepsilon := \frac{1}{8}$, $C := \frac{1}{32}$, $\lambda := 0.00853\ldots$ as in the proof of Theorem 17.3 and define

$$\delta_0 := \frac{\lambda}{D^{3/2} \mu_{\max}^2(g)}, \qquad t_T := \frac{1}{1 + T + 1.00001 \frac{T}{\delta_0}}.$$

Proposition 18.15 *We have*

$$Q_g \le (1+\varepsilon)^2 \delta_0 \, \mu_{\max}^2(g) + \frac{T}{P_{T,1}} \int_{t_T}^1 \mathop{\mathbb{E}}_{q_t \sim N(\overline{q}_t, t^2 I)} \left(\frac{\mu_{av}^2(q_t)}{\|q_t\|^2} \right) dt,$$

where $\overline{q}_t = (1-t)g$.

Proof Let $\zeta^{(a)}, \ldots, \zeta^{(D)}$ be the zeros of g and denote by $(q_\tau, \zeta_\tau^{(j)})_{\tau \in [0,1]}$ the lifting of $E_{f,g}$ in V corresponding to the initial pair $(g, \zeta^{(j)})$ and final system $f \in \mathcal{H}_\mathbf{d} \setminus \Sigma$.

Equation (17.5) for $i = 0$ in the proof of Theorem 17.3 shows the following: for all j and all $\tau \le \frac{\lambda}{d_\mathbb{S}(f,g) D^{3/2} \mu_{\text{norm}}^2(g,\zeta^{(j)})}$ we have

$$\mu_{\text{norm}}(q_\tau, \zeta_\tau^{(j)}) \le (1+\varepsilon) \mu_{\text{norm}}(g, \zeta^{(j)}) \le (1+\varepsilon) \mu_{\max}(g).$$

In particular, this inequality holds for all j and all $\tau \le \frac{\delta_0}{d_\mathbb{S}(f,g)}$, and hence for all such τ, we have

$$\mu_{av}(q_\tau) \le (1+\varepsilon) \mu_{\max}(g). \tag{18.19}$$

Splitting the integral in Q_g at $\tau_0(f) := \min\{1, \frac{\delta_0}{d_\mathbb{S}(f,g)}\}$, we obtain

$$Q_g = \mathop{\mathbb{E}}_{f \sim N_T(0,I)} \left(d_\mathbb{S}(f,g) \int_0^{\tau_0(f)} \mu_{av}^2(q_\tau) \, d\tau \right)$$

$$+ \mathop{\mathbb{E}}_{f \sim N_T(0,I)} \left(d_\mathbb{S}(f,g) \int_{\tau_0(f)}^1 \mu_{av}^2(q_\tau) \, d\tau \right).$$

Using (18.19) we bound the first term on the right-hand side as follows:

$$\mathop{\mathbb{E}}_{f \sim N_T(0,I)} \left(d_\mathbb{S}(f,g) \int_0^{\tau_0(f)} \mu_{av}^2(q_\tau) \, d\tau \right) \le (1+\varepsilon)^2 \delta_0 \mu_{\max}(g)^2.$$

For bounding the second term, we assume without loss of generality that $\tau_0(f) \le 1$. It then follows from (17.1) and Lemma 17.5 that for a fixed f,

$$d_\mathbb{S}(f,g) \int_{\tau_0(f)}^1 \mu_{av}^2(q_\tau) \, d\tau \le \int_{t_0(f)}^1 \|f\| \frac{\mu_{av}^2(q_t)}{\|q_t\|^2} \, dt,$$

where $t_0(f)$ is given by

$$t_0(f) = \frac{1}{1 + \|f\|(\sin\alpha \cot\delta_0 - \cos\alpha)}, \quad \alpha := d_{\mathbb{S}}(f, g).$$

Now note that $\|f\| \leq T$, since we draw f from $N_T(0, \mathrm{I})$. This will allow us to bound $t_0(f)$ from below by a quantity independent of f. For $\|f\| \leq T$ we have

$$0 \leq \sin\alpha \cot\delta_0 - \cos\alpha \leq \frac{1}{\sin\delta_0} - \cos\alpha \leq \frac{1}{\sin\delta_0} + 1,$$

and moreover, $\sin\delta_0 \geq 0.9999978\,\delta_0$, since $\delta_0 \leq 2^{-3/2}\lambda \leq 0.0037$. We can therefore bound $t_0(f)$ as

$$t_0(f) \geq \frac{1}{1 + T + \frac{T}{\sin\delta_0}} \geq \frac{1}{1 + T + 1.00001\frac{T}{\delta_0}} = t_T.$$

We can now bound the second term in Q_g as follows:

$$\mathop{\mathbb{E}}_{f\sim N_T(0,\mathrm{I})}\left(d_{\mathbb{S}}(f, g)\int_{\tau_0(f)}^1 \mu_{\mathrm{av}}^2(q_\tau)\,d\tau\right)$$

$$\leq \mathop{\mathbb{E}}_{f\sim N_T(0,\mathrm{I})}\left(T\int_{t_T}^1 \frac{\mu_{\mathrm{av}}^2(q_t)}{\|q_t\|^2}\,dt\right)$$

$$= T\int_{t_T}^1 \mathbb{E}_{f\sim N_T(0,\mathrm{I})}\left(\frac{\mu_{\mathrm{av}}^2(q_t)}{\|q_t\|^2}\right)dt \leq \frac{T}{P_{T,1}}\int_{t_T}^1 \mathop{\mathbb{E}}_{f\sim N(0,\mathrm{I})}\left(\frac{\mu_{\mathrm{av}}^2(q_t)}{\|q_t\|^2}\right)dt.$$

To conclude, note that for fixed t and when f is distributed following $N(0, \mathrm{I})$, the variable $q_t = (1-t)g + tf$ follows the Gaussian $N(\bar{q}_t, t^2\mathrm{I})$, where $\bar{q}_t = (1-t)g$. $\square$

Proof of Theorem 18.14 We only need to estimate Q_g, for which we use the right-hand side of Proposition 18.15. In order to bound the first term there, we note that

$$(1+\varepsilon)^2\delta_0\,\mu_{\max}^2(g) = (1+\varepsilon)^2\lambda D^{-3/2} \leq (1+\varepsilon)^2\lambda \leq 0.02.$$

For bounding the second term we apply Theorem 18.4 to deduce that

$$\int_{t_T}^1 \mathop{\mathbb{E}}_{q_t\sim N(\bar{q}_t, t^2\mathrm{I})}\left(\frac{\mu_{\mathrm{av}}^2(q_t)}{\|q_t\|^2}\right)dt \leq \int_{t_T}^1 \frac{e(n+1)}{2t^2}\,dt = \frac{e(n+1)}{2}\left(\frac{1}{t_T} - 1\right)$$

$$= \frac{e(n+1)T}{2}\left(1 + \frac{1.00001}{\delta_0}\right).$$

Replacing this bound in Proposition 18.15, we obtain

$$Q_g \leq \frac{eT^2(n+1)}{2P_{T,1}}\left(1 + \frac{1.00001}{\lambda}D^{3/2}\mu_{\max}^2(g)\right) + 0.02$$

$$\leq 2eN(n+1)D^{3/2}\mu_{\max}^2(g)\left(\frac{1}{D^{3/2}}+\frac{1.00001}{\lambda}\right)+0.02$$

$$\leq 639\,N(n+1)D^{3/2}\mu_{\max}^2(g)+0.02,$$

where we used $D \geq 2$ for the last inequality. $\square$

Proof of Theorem 18.2 The result follows immediately by combining Proposition 17.23 with Theorem 18.14, with the roles of f and g swapped. $\square$

18.4 A Near-Solution to Smale's 17th Problem

In this section we prove the last main result of this chapter, namely Theorem 18.3.

18.4.1 A Deterministic Homotopy Continuation

The analysis of the previous section allows one to eliminate the randomness from the system in one of the extremes of the homotopy segment. Unfortunately, though, it does not, in principle, allow one to do so for the choice of the zero (or, equivalently, of the lifting of this segment in V) of this system. Therefore, it cannot be directly used to analyze the average complexity of a homotopy with a given initial pair (g, ζ).

There is one particular case, however, in which this analysis can be used. Recall the system $\bar{U}$ introduced in Example 16.20,

$$\bar{U}_i = \frac{1}{\sqrt{2n}}(X_0^{d_i} - X_i^{d_i}), \quad i = 1, \ldots, n,$$

along with its zeros $\mathbf{z}_1, \ldots, \mathbf{z}_{\mathcal{D}} \in \mathbb{P}^n$, where the ith components of the $\mathbf{z}$'s run through all possible d_ith roots of unity. We set $\mathbf{z}_1 = [(1, \ldots, 1)]$.

The various invariances we saw for this system are now helpful. Denote by $K_{\bar{U}}(f)$ the number of iterations performed by ALH with input $(f, \bar{U}, \mathbf{z}_1)$. The following result is an immediate consequence of Lemma 16.21.

Lemma 18.16 *Let $g \in \mathcal{H}_{\mathbf{d}}, \zeta \in \mathbb{P}^n$ be a zero of g, and $u \in \mathcal{U}(n+1)$. Then, for all $f \in \mathcal{H}_{\mathbf{d}}$, we have $K(f, g, \zeta) = K(uf, ug, u\zeta)$.* $\square$

Proposition 18.17 $K_{\bar{U}}(f) = K(f, \bar{U}, \mathbf{z}_1)$ *satisfies*

$$\underset{f \sim N(0,\mathrm{I})}{\mathbb{E}}\,K_{\bar{U}}(f) = \underset{f \sim N(0,\mathrm{I})}{\mathbb{E}}\,\frac{1}{\mathcal{D}}\sum_{j=1}^{\mathcal{D}} K(f, \bar{U}, \mathbf{z}_j).$$

Proof Let $u_j \in \mathscr{U}(n+1)$ be such that $\mathbf{z}_j = u_j \mathbf{z}_1$. Then $u_j \bar{U} = \bar{U}$ and Lemma 18.16 implies that

$$K(f, \bar{U}, \mathbf{z}_1) = K(u_j f, u_j \bar{U}, u_j \mathbf{z}_1) = K(u_j f, \bar{U}, \mathbf{z}_j).$$

It follows that

$$K_{\bar{U}}(f) = K(f, \bar{U}, \mathbf{z}_1) = \frac{1}{\mathcal{D}} \sum_{j=1}^{\mathcal{D}} K(u_j f, \bar{U}, \mathbf{z}_j).$$

The assertion follows now, since for all integrable functions $F \colon \mathcal{H}_{\mathbf{d}} \to \mathbb{R}$ and all $u \in \mathscr{U}(n+1)$, we have

$$\mathop{\mathbb{E}}_{f \sim N(0, \mathrm{I})} F(f) = \mathop{\mathbb{E}}_{f \sim N(0, \mathrm{I})} F(uf),$$

due to the unitary invariance of $N(0, \mathrm{I})$.   □

We consider the following algorithm MD (moderate degree).

Algorithm 18.1 MD

Input: $f \in \mathcal{H}_{\mathbf{d}}$

Preconditions: $f \neq 0$

---

run ALH on input $(f, \bar{U}, \mathbf{z}_1)$

Output: $z \in (\mathbb{C}^{n+1})_*$

Postconditions: The algorithm halts if the lifting of $E_{f, \bar{U}}$ at $\mathbf{z}_1$ does not cut Σ'. In this case, $[z] \in \mathbb{P}^n$ is an approximate zero of f.

Proposition 18.17, together with the bound for $\mu_{\mathrm{norm}}(\bar{U}, \mathbf{z}_1)$ we derived in Sect. 16.4, yields bounds for the average cost of MD.

Theorem 18.18 *Let* $\mathrm{cost}^{\mathrm{MD}}(f)$ *denote the cost of Algorithm MD with input* $f \in \mathcal{H}_{\mathbf{d}}$. *Then*

$$\mathop{\mathbb{E}}_{f \sim N(0, \mathrm{I})} \mathrm{cost}^{\mathrm{MD}}(f) = \mathcal{O}\big(D^3 N^2 n^{D+1}\big).$$

Proof Theorem 17.3, together with the definition of μ_{av}^2, implies for $g = \bar{U}$ that

$$\frac{1}{\mathcal{D}} \sum_{i=1}^{\mathcal{D}} K(f, \bar{U}, \mathbf{z}_i) \leq 188 \, D^{3/2} \, d_{\mathbb{S}}(f, \bar{U}) \int_0^1 \mu_{\mathrm{av}}^2(q_\tau) \, d\tau.$$

Using Proposition 18.17 we get

$$\operatorname*{\mathbb{E}}_{f\sim N(0,\mathrm{I})} K_{\bar{U}}(f) \leq 188\, D^{3/2} \operatorname*{\mathbb{E}}_{f\sim N(0,\mathrm{I})} \left(d_{\mathbb{S}}(f,\bar{U}) \int_0^1 \mu_{\mathrm{av}}^2(q_\tau)\, d\tau \right).$$

Applying Theorem 18.14 with $g = \bar{U}$ we obtain

$$\operatorname*{\mathbb{E}}_{f\sim N(0,\mathrm{I})} K_{\bar{U}}(f) = \mathcal{O}\big(D^3 N n\, \mu_{\max}^2(\bar{U})\big).$$

We now plug in the bound $\mu_{\max}^2(\bar{U}) \leq 2(n+1)^D$ of Lemma 16.22 to obtain

$$\operatorname*{\mathbb{E}}_{f\sim N(0,\mathrm{I})} K_{\bar{U}}(f) = \mathcal{O}\big(D^3 N n^{D+1}\big).$$

Multiplying by $\mathcal{O}(N)$ to take into account the cost of each iteration completes the proof. □

Algorithm MD is efficient when D is small, say, when $D \leq n$. Otherwise, it has a cost exponential in D. This is an unusual feature. The common cost of zero-finding algorithms is polynomial in D but exponential in n. We will take advantage of this fact to use, for $D > n$, a different approach with this kind of complexity bound. The combination of both procedures yields the desired near-polynomial cost.

18.4.2 An Elimination Procedure for Zero-Finding

For our second procedure we will rely on an algorithm due to Jim Renegar.

Before giving the specification of Renegar's algorithm, we need to fix some notation. We shall identify $\mathbb{P}_0^n := \{[(z_0,\dots,z_n)] \in \mathbb{P}^n \mid z_0 \neq 0\}$ with $\mathbb{C}^n$ via the bijection $[(z_0,\dots,z_n)] \mapsto \underline{z} := (z_1/z_0,\dots,z_n/z_0)$. For $z \in \mathbb{P}_0^n$ we shall denote by $\|z\|_{\mathrm{aff}}$ the Euclidean norm of $\underline{z} \in \mathbb{C}^n$, that is,

$$\|z\|_{\mathrm{aff}} := \|\underline{z}\| = \left(\sum_{i=1}^n \left| \frac{z_i}{z_0} \right|^2 \right)^{\frac{1}{2}},$$

and we put $\|z\|_{\mathrm{aff}} = \infty$ if $z \in \mathbb{P}^n \setminus \mathbb{P}_0^n$. Furthermore, for $z, y \in \mathbb{P}_0^n$ we shall write $d_{\mathrm{aff}}(z,y) := \|\underline{z} - \underline{y}\|$, and we set $d_{\mathrm{aff}}(z,y) := \infty$ otherwise. An elementary argument shows that

$$d_{\mathbb{P}}(z,y) \leq d_{\mathrm{aff}}(z,y) \quad \text{for all } z, y \in \mathbb{P}_0^n.$$

By a δ-*approximation* of a zero $\zeta \in \mathbb{P}_0^n$ of $f \in \mathcal{H}_{\mathbf{d}}$ we understand a $z \in \mathbb{P}_0^n$ such that $d_{\mathrm{aff}}(z,\zeta) \leq \delta$.

Renegar's algorithm Ren takes as input $f \in \mathcal{H}_{\mathbf{d}}$ and $R, \delta \in \mathbb{R}$ with $R \geq \delta > 0$, decides whether its zero set $Z_{\mathbb{P}}(f) \subseteq \mathbb{P}^n$ is finite, and if so, computes δ-approximations z to at least all zeros ζ of f satisfying $\|\zeta\|_{\mathrm{aff}} \leq R$. We may formally specify this algorithm as follows.

Algorithm 18.2 Ren

Input: $f \in \mathcal{H}_{\mathbf{d}}, R, \delta \in \mathbb{R}$

Preconditions: $f \neq 0, R \geq \delta > 0$

Output: Either $\ell \in \mathbb{N}$ and $z_1, \dots, z_\ell \in \mathbb{P}_0^n$ or tag INFINITE

Postconditions: If tag INFINITE is returned, then $Z_{\mathbb{P}}(f)$ is infinite. Otherwise, we have $\{\zeta \in Z_{\mathbb{P}}(f) \mid \|\zeta\|_{\mathrm{aff}} \leq R\} = \{\zeta_1, \dots, \zeta_\ell\}, d_{\mathrm{aff}}(\zeta_i, z_i) \leq \delta$ for $i = 1, \dots, \ell$.

It is known that the cost of Ren on input (f, R, δ) is bounded by

$$\mathcal{O}\left(n\mathcal{D}^4 (\log \mathcal{D}) \left(\log \log \frac{R}{\delta} \right) + n^2 \mathcal{D}^4 \left(\frac{1 + \sum_i d_i}{n} \right)^4 \right). \tag{18.20}$$

Algorithm Ren finds δ-approximations, not necessarily approximate zeros in the sense of Definition 16.34. This is not a hindrance; the following result relates these two forms of approximation.

Proposition 18.19 *Let $z \in \mathbb{P}_0^n$ be a δ-approximation of a zero $\zeta \in \mathbb{P}_0^n$ of f. If $D^{3/2} \mu_{\mathrm{norm}}(f, z)\delta \leq \frac{1}{28}$, then z is an approximate zero of f.*

Proof From the hypothesis and Proposition 16.2 with $g = f$ we obtain that $\mu_{\mathrm{norm}}(f, \zeta) \leq (1 + \varepsilon)\mu_{\mathrm{norm}}(f, z)$ with $\varepsilon = \frac{1}{7}$. We are going to apply Theorem 16.38 with $r = 0.99500, \delta(r) = 0.17333\dots, u(r) = 0.17486\dots$ (see Table 16.1). Writing $C := \frac{\varepsilon}{4} = \frac{1}{28}$, we can bound $d := d_{\mathbb{P}}(z, \zeta) \leq d_{\mathrm{aff}}(z, \zeta) \leq \delta$ by

$$\delta \leq \frac{C}{D^{3/2} \mu_{\mathrm{norm}}(f, z)} \leq \frac{C}{2^{3/2}} < 0.0127 \leq \delta(r).$$

Moreover,

$$\frac{1}{2} D^{3/2} \mu_{\mathrm{norm}}(f, \zeta) d \leq \frac{1}{2}(1 + \varepsilon) D^{3/2} \mu_{\mathrm{norm}}(f, z) d \leq \frac{1}{2}(1 + \varepsilon)C < 0.021 \leq u(r).$$

Hence $\gamma_{\mathrm{proj}}(f, \zeta) d_{\mathbb{P}}(z, \zeta) \leq u(r)$ by Theorem 16.1. It follows from Theorem 16.38 that z is an approximate zero of f. $\qquad\square$

To find an approximate zero of f we may therefore use $\mathrm{Ren}(R, \delta)$ iteratively for $R = 4^k$ and $\delta = 2^{-k}$ for $k = 1, 2, \dots$ until we are successful. More precisely, we consider the following algorithm (here, and for the rest of this section, $\varepsilon = \frac{1}{7}$ and $C = \frac{1}{28}$):

Algorithm 18.3 ItRen

Input: $f \in \mathcal{H}_\mathbf{d}$
Preconditions: $f \neq 0$

```
for k = 1, 2, ... do
        run Ren(4^k, 2^{-k}) on input f
        for all δ-approximations z found
            if D^{3/2} μ_norm(f, z)δ ≤ C return z and halt
```

Output: $z \in (\mathbb{C}^{n+1})_*$
Postconditions: The algorithm halts if $f \notin \Sigma$ and $Z_{\mathbb{P}}(f) \cap \mathbb{P}_0^n \neq \emptyset$. In this case $[z] \in \mathbb{P}^n$ is an approximate zero of f.

Let $\Sigma_0 := \Sigma \cup \{f \in \mathcal{H}_\mathbf{d} \mid Z(f) \cap \mathbb{P}_0^n = \emptyset\}$. It is obvious that ItRen stops on inputs $f \notin \Sigma_0$. In particular, ItRen stops almost surely. We next show that it does so, on average, with cost polynomial in N and $\mathcal{D}$.

Proposition 18.20 *Let* $\mathrm{cost}^{\mathrm{ItRen}}(f)$ *denote the running time of algorithm* ItRen *on input* f. *Then,*

$$\mathop{\mathbb{E}}_{f \sim N(0, \mathrm{I})} \mathrm{cost}^{\mathrm{ItRen}}(f) = (\mathcal{D} N n)^{\mathcal{O}(1)}.$$

Towards the proof of Proposition 18.20 we first bound the probability Probfail that the main loop of ItRen, with parameters R and δ, fails to output an approximate zero for a standard Gaussian input $f \in \mathcal{H}_\mathbf{d}$. We do so in a sequence of lemmas.

Lemma 18.21 *Let* $\mathcal{E}$ *denote the set of* $f \in \mathcal{H}_\mathbf{d}$ *such that at least one* z *on the output list of* $\mathrm{Ren}(R, \delta)$ *on input* f *satisfies* $D^{3/2} \mu_{\mathrm{norm}}(f, z)\delta > C$. *Then*

$$\mathrm{Probfail} \leq \mathrm{Prob}\left\{ \min_{f \in \mathcal{H}_\mathbf{d}} \|\zeta\|_{\mathrm{aff}} \geq R \right\} + \mathrm{Prob}\,\mathcal{E}.$$

Proof We may assume that $Z(f)$ is finite. Let $z_1, \ldots, z_\ell$ be the output of $\mathrm{Ren}(R, \delta)$ on input f. If $\ell = 0$, then by the specification of ItRen, all the zeros ζ of f satisfy $\|\zeta\|_{\mathrm{aff}} > R$. Otherwise, $\ell \geq 1$. If ItRen fails, then all z_i fail the test, so that $D^{3/2} \mu_{\mathrm{norm}}(f, z_i)\delta > C$ for $i = 1, \ldots, \ell$. In particular, at least one z on the output list satisfies this, and hence $f \in \mathcal{E}$. $\square$

Lemma 18.22 *For* $R > 0$ *and standard Gaussian* $f \in \mathcal{H}_\mathbf{d}$ *we have*

$$\mathrm{Prob}\left\{ \min_{f \in \mathcal{H}_\mathbf{d}} \|\zeta\|_{\mathrm{aff}} \geq R \right\} \leq \frac{n}{R^2}.$$

Proof Choose $f \in \mathcal{H}_\mathbf{d}$ standard Gaussian and pick one of the $\mathcal{D}$ zeros $\zeta_f^{(a)}$ of f uniformly at random; call it ζ. Then the resulting distribution of (f, ζ) in V has density ρ_{st}. Lemma 17.18 (adapted to V) implies that ζ is uniformly distributed in $\mathbb{P}^n$. Therefore,

$$\Prob_{f \in \mathcal{H}_d}\left\{\min_i \|\zeta_f^{(a)}\|_{\text{aff}} \geq R\right\} \leq \Prob_{\zeta \in \mathbb{P}^n}\left\{\|\zeta\|_{\text{aff}} \geq R\right\}.$$

To estimate the right-hand-side probability we set $\mathbb{P}^{n-1} := \{z \in \mathbb{P}^n \mid z_0 = 0\}$, and we define θ by $R = \tan\theta$. It is straightforward to check that

$$\|\zeta\|_{\text{aff}} \geq R \iff d_{\mathbb{P}}(\zeta, \mathbb{P}^{n-1}) \leq \frac{\pi}{2} - \theta.$$

Therefore,

$$\Prob_{\zeta \in \mathbb{P}^n}\left\{\|\zeta\|_{\text{aff}} \geq R\right\} = \frac{\text{vol}\{z \in \mathbb{P}^n \mid d_{\mathbb{P}}(z, \mathbb{P}^{n-1}) \leq \frac{\pi}{2} - \theta\}}{\text{vol}(\mathbb{P}^n)}.$$

In Lemma 20.8 we shall provide bounds on the volume of the tubes in $\mathbb{P}^n$ around $\mathbb{P}^{n-1}$. Using this and $\text{vol}(\mathbb{P}^n) = \pi^n/n!$, cf. (17.9), we see that

$$\Prob_{\zeta \in \mathbb{P}^n}\left\{\|\zeta\|_{\text{aff}} \geq R\right\} \leq \frac{\text{vol}(\mathbb{P}^{n-1})\,\text{vol}(\mathbb{P}^1)}{\text{vol}(\mathbb{P}^n)}\sin^2\left(\frac{\pi}{2} - \theta\right)$$

$$= n\cos^2\theta = \frac{n}{1 + R^2} \leq \frac{n}{R^2}. \qquad \square$$

Lemma 18.23 *We have* $\Prob\,\mathcal{E} = \mathcal{O}(\mathcal{D}N^2n^3D^6\delta^4)$.

Proof Assume that $f \in \mathcal{E}$. Then, there exist $\zeta, z \in \mathbb{P}_0^n$ such that $f(\zeta) = 0$, $\|\zeta\|_{\text{aff}} \leq R$, $d_{\text{aff}}(\zeta, z) \leq \delta$, Ren returns z, and $D^{3/2}\mu_{\text{norm}}(f, z)\delta > C$.

We proceed by cases. Suppose first that $\delta \leq \frac{C}{D^{3/2}\mu_{\text{norm}}(f,\zeta)}$. Then, by Proposition 16.2 (with $\varepsilon = 1/7$, $C = 1/28$),

$$(1 + \varepsilon)^{-1}C < (1 + \varepsilon)^{-1}D^{3/2}\mu_{\text{norm}}(f, z)\delta \leq D^{3/2}\mu_{\text{norm}}(f, \zeta)\delta,$$

and hence

$$\mu_{\max}(f) \geq \mu_{\text{norm}}(f, \zeta) \geq (1 + \varepsilon)^{-1}CD^{-3/2}\delta^{-1}.$$

If, on the other hand, $\delta > \frac{C}{D^{3/2}\mu_{\text{norm}}(f,\zeta)}$, then we have

$$\mu_{\max}(f) \geq \mu_{\text{norm}}(f, \zeta) \geq CD^{-3/2}\delta^{-1}.$$

Therefore, for any $f \in \mathcal{E}$,

$$\mu_{\max}(f) \geq (1 + \varepsilon)^{-1}CD^{-3/2}\delta^{-1} = \frac{1}{32}D^{-3/2}\delta^{-1}.$$

Proposition 17.31 shows that $\mathsf{Prob}_f\{\mu_{max}(f) \geq \rho^{-1}\} = \mathcal{O}(\mathcal{D}N^2n^3\rho^4)$ for all $\rho > 0$. Therefore, we get

$$\mathsf{Prob}\,\mathcal{E} \leq \mathop{\mathsf{Prob}}_{f\in\mathcal{H}_\mathbf{d}}\left\{\mu_{max}(f) \geq \frac{1}{32}D^{-3/2}\delta^{-1}\right\} = \mathcal{O}(\mathcal{D}N^2n^3D^6\delta^4),$$

as claimed.   □

From Lemma 18.22 and Lemma 18.23 we immediately obtain the following.

Lemma 18.24 *We have* $\mathsf{Probfail} = \mathcal{O}(\mathcal{D}N^2n^3D^6\delta^4 + nR^{-2})$.   □

Proof of Proposition 18.20 The probability that ItRen stops in the $(k+1)$th loop is bounded above by the probability p_k that $\mathsf{Ren}(4^k, 2^{-k})$ fails to produce an approximate zero. Lemma 18.24 tells us that

$$p_k = \mathcal{O}(\mathcal{D}N^2n^3D^6\,16^{-k}).$$

If A_k denotes the running time of the $(k+1)$th loop, we conclude that

$$\mathop{\mathbb{E}}_{f\sim N(0,\mathrm{I})}\,\mathsf{cost}^{\mathsf{ItRen}}(f) \leq \sum_{k=0}^{\infty} A_k p_k.$$

According to (18.20), A_k is bounded by

$$\mathcal{O}\left(n\mathcal{D}^4(\log\mathcal{D})(\log k) + n^2\mathcal{D}^4\left(\frac{1+\sum_i d_i}{n}\right)^4 + (N+n^3)\mathcal{D}\right),$$

where the last term accounts for the cost of the tests. The assertion now follows by distributing the products $A_k p_k$ and using that the series $\sum_{k\geq1} 16^{-k}$, and $\sum_{k\geq1} 16^{-k}\log k$ have finite sums.   □

18.4.3   Some Inequalities of Combinatorial Numbers

Theorem 18.18 and Proposition 18.20 yield bounds (exponential in D and n, respectively) for the cost of computing an approximate zero. We next relate these bounds to bounds purely in terms of the input size N.

Lemma 18.25

(a) *For* $D \leq n, n \geq 4$, *we have*

$$n^D \leq \binom{n+D}{D}^{\ln n}.$$

(b) *For $D^2 \geq n \geq 1$ we have*

$$\ln n \leq 2 \ln \ln \binom{n+D}{n} + 4.$$

(c) *For $0 < c < 1$ there exists K such that for all n, D,*

$$D \leq n^{1-c} \implies n^D \leq \binom{n+D}{n}^K.$$

(d) *For $D \leq n$ we have*

$$n^D \leq N^{2 \ln \ln N + \mathcal{O}(1)}.$$

(e) *For $n \leq D$ we have*

$$D^n \leq N^{2 \ln \ln N + \mathcal{O}(1)}.$$

Proof Stirling's formula states that $n! = \sqrt{2\pi} n^{n+\frac{1}{2}} e^{-n} e^{\frac{\Theta_n}{12n}}$ with $\Theta_n \in (0, 1)$. Let $H(x) = x \ln \frac{1}{x} + (1-x) \ln \frac{1}{1-x}$ denote the binary entropy function, defined for $0 < x < 1$. By a straightforward calculation we get from Stirling's formula the following asymptotics for the binomial coefficient: for any $0 < m < n$ we have

$$\ln \binom{n}{m} = n H\left(\frac{m}{n}\right) + \frac{1}{2} \ln \frac{n}{m(n-m)} - 1 + \varepsilon_{n,m}, \tag{18.21}$$

where $-0.1 < \varepsilon_{n,m} < 0.2$. This formula holds as well for the extension of binomial coefficients on which m is not necessarily integer.

(a) The first claim is equivalent to $e^D \leq \binom{n+D}{D}$. The latter is easily checked for $D \in \{1, 2, 3\}$ and $n \geq 4$. So assume $n \geq D \geq 4$. By monotonicity it suffices to show that $e^D \leq \binom{2D}{D}$ for $D \geq 4$. Equation (18.21) implies

$$\ln \binom{2D}{D} > 2D \ln 2 + \frac{1}{2} \ln \frac{2}{D} - 1.1,$$

and the right-hand side is easily checked to be at least D, for $D \geq 4$.

(b) Put $m := \sqrt{n}$. If $D \geq m$, then $\binom{n+D}{n} \geq \binom{n+\lceil m \rceil}{n}$, so it is enough to show that $\ln n \leq 2 \ln \ln \binom{n+\lceil m \rceil}{n} + 4$. Equation (18.21) implies

$$\ln \binom{n + \lceil m \rceil}{n} \geq \ln \binom{n+m}{n} \geq (n+m) H\left(\frac{m}{n+m}\right) + \frac{1}{2} \ln \frac{1}{m} - 1.1.$$

The entropy function can be bounded as

$$H\left(\frac{m}{n+m}\right) \geq \frac{m}{n+m} \ln \left(1 + \frac{n}{m}\right) \geq \frac{m}{n+m} \ln m.$$

It follows that

$$\ln\binom{n+\lceil m\rceil}{n} \geq \frac{1}{2}\sqrt{n}\,\ln n - \frac{1}{4}\ln n - 1.1 \geq \frac{1}{4}\sqrt{n}\,\ln n,$$

the right-hand inequality holding for $n \geq 10$. Hence

$$\ln\ln\binom{n+\lceil m\rceil}{n} \geq \frac{1}{2}\ln n + \ln\ln n - \ln 4 \geq \frac{1}{2}\ln n - 2,$$

the right-hand inequality holding for $n \geq 2$. This proves the second claim for $n \geq 10$. The cases $n \leq 9$ are easily directly checked.

(c) Writing $D = n\delta$, we obtain from (18.21),

$$\ln\binom{n+D}{n} = (n+D)H\left(\frac{\delta}{1+\delta}\right) - \frac{1}{2}\ln D + \mathcal{O}(1).$$

Estimating the entropy function yields

$$H\left(\frac{\delta}{1+\delta}\right) \geq \frac{\delta}{1+\delta}\ln\left(1+\frac{1}{\delta}\right) \geq \frac{\delta}{2}\ln\frac{1}{\delta} = \frac{\delta\varepsilon}{2}\ln n,$$

where ε is defined by $\delta = n^{-\varepsilon}$. By assumption, $\varepsilon \geq c$. From the last two lines we get

$$\frac{1}{D\ln n}\ln\binom{n+D}{n} \geq \frac{c}{2} - \frac{1-c}{2D} + \mathcal{O}\left(\frac{1}{\ln n}\right).$$

In the case $c \leq \frac{3}{4}$ we have $D \geq n^{1/4}$, and we bound the above by

$$\frac{c}{2} - \frac{1}{2n^{1/4}} + \mathcal{O}\left(\frac{1}{\ln n}\right),$$

which is greater than $c/4$ for sufficiently large n. In the case $c \geq \frac{3}{4}$ we bound as follows:

$$\frac{1}{D\ln n}\ln\binom{n+D}{n} \geq \frac{c}{2} - \frac{1-c}{2} + \mathcal{O}\left(\frac{1}{\ln n}\right) = c - \frac{1}{2} + \mathcal{O}\left(\frac{1}{\ln n}\right) \geq \frac{1}{5}$$

for sufficiently large n.

We have shown that for $0 < c < 1$ there exists n_c such that for $n \geq n_c$, $D \leq n^{1-c}$, we have

$$n^D \leq \binom{n+D}{n}^{K_c},$$

where $K_c := \max\{4/c, 5\}$. By increasing K_c we can achieve that the above inequality holds for all n, D with $D \leq n^{1-c}$.

(d) Clearly, $N \geq \binom{n+D}{n}$. If $D \leq \sqrt{n}$, then by part (c), there exists K such that

$$n^D \leq \left(\frac{n+D}{n}\right)^K \leq N^K.$$

Otherwise, $D \in [\sqrt{n}, n]$, and the desired inequality is an immediate consequence of parts (a) and (b).

(e) Use $\binom{n+D}{n} = \binom{n+D}{D}$ and swap the roles of n and D in part (d) above. $\square$

We finally proceed to the proof of Theorem 18.3.

Proof of Theorem 18.3 We use Algorithm MD if $D \leq n$ and Algorithm ItRen if $D > n$.

Theorem 18.18 combined with Lemma 18.25(d) implies that

$$\underset{f \sim N(0,\mathrm{I})}{\mathbb{E}} \mathrm{cost}^{\mathrm{MD}}(f) = N^{2\ln\ln N + \mathcal{O}(1)} \quad \text{if } D \leq n. \tag{18.22}$$

Note that this bound is nearly polynomial in N. Moreover, if $D \leq n^{1-c}$ for some fixed $0 < c < 1$, then Lemma 18.25(c) implies

$$\underset{f \sim N(0,\mathrm{I})}{\mathbb{E}} (f) = N^{\mathcal{O}(1)}. \tag{18.23}$$

In this case, the average cost is polynomially bounded in the input size N.

For the case $D > n$ we use Proposition 18.20 together with the inequality $\mathcal{D}^{\mathcal{O}(1)} \leq D^{\mathcal{O}(n)} \leq N^{\mathcal{O}(\log\log N)}$, which follows from Lemma 18.25(e). Moreover, in the case $D \geq n^{1+\varepsilon}$, Lemma 18.25(c) implies $\mathcal{D} \leq D^n \leq N^{\mathcal{O}(1)}$. $\square$

Chapter 19
Real Polynomial Systems

The development of the preceding three chapters focused on complex systems of homogeneous polynomial equations. The main algorithmic results in these chapters were satisfying: we can compute an approximate zero of a system f in average (and even smoothed) randomized polynomial time. Central in these results were the consideration of complex numbers for both the coefficients of the input system and the components of the computed approximate zero.

For a variety of purposes, however, one is interested in real zeros of systems with real coefficients. An observation previous to any consideration about the computation of any such zero is that in this context there are systems having no zeros at all. For instance, the polynomial $X_0^2 + X_1^2 + X_2^2$ has no zeros in $\mathbb{P}(\mathbb{R}^3)$. Furthermore, this absence of zeros is not a phenomenon occurring almost nowhere. The simplest example is given by the quadratic polynomials

$$aX_1^2 + bX_0X_1 + cX_0^2$$

with $a, b, c \in \mathbb{R}$, not all three of them zero. Such a polynomial has two zeros in $\mathbb{P}(\mathbb{R}^2)$ if $b^2 > 4ac$, one zero if $b^2 = 4ac$, and no zeros at all if $b^2 < 4ac$. Therefore—and this is a situation we have already met when dealing with linear programming—the issue of feasibility precedes that of computing zeros.

For systems of n homogeneous polynomials in $n + 1$ variables one can consider a problem more demanding than feasibility, namely, to count how many zeros the system has. Let us denote by $\mathcal{H}_{\mathbf{d}}^{\mathbb{R}}$ the linear space of these systems for a fixed degree pattern $\mathbf{d} = (d_1, \ldots, d_n)$. The goal of this chapter is to exhibit and analyze an algorithm for zero-counting. Even though we will not pursue the issue here, the motivating idea for this algorithm was the possibility to implement it with finite precision (see Remark 19.28 at the end of the chapter). A measure of conditioning was therefore a must, and not unexpectedly, this measure appears in the complexity analysis of the algorithm as well.

This measure follows a pattern we have already studied. Recall the discussion in Sect. 6.1 on conditioning for problems with a discrete set of values. In accordance with it, we say that a system $f \in \mathcal{H}_{\mathbf{d}}^{\mathbb{R}}$ is *ill-posed* when arbitrary small perturbations

P. Bürgisser, F. Cucker, *Condition*,
Grundlehren der mathematischen Wissenschaften 349,
DOI 10.1007/978-3-642-38896-5_19, © Springer-Verlag Berlin Heidelberg 2013

of f can change its number of real zeros. We observe that this is the case if and only if f has multiple real zeros in $\mathbb{P}(\mathbb{R}^{n+1})$. Let $\Sigma_{\mathbb{R}} \subset \mathcal{H}_{\mathbf{d}}^{\mathbb{R}}$ be the set of ill-posed systems. We define

$$\kappa(f) := \frac{\|f\|}{d(f, \Sigma_{\mathbb{R}})}. \tag{19.1}$$

In Sect. 19.2 below we will relate $\kappa(f)$ to the quantities $\mu_{\text{norm}}(f, x)$ via a characterization of the former akin to a condition number theorem.

Otherwise, the main result of this chapter is the following.

Theorem 19.1 *There exists an iterative algorithm that given an input $f \in \mathcal{H}_{\mathbf{d}}^{\mathbb{R}} \setminus \Sigma_{\mathbb{R}}$:*

(a) *Returns the number of real zeros of f in $\mathbb{P}(\mathbb{R}^{n+1})$.*
(b) *Performs $\mathcal{O}(\log_2(nD\kappa(f)))$ iterations and has a total cost (number of arithmetic operations) of*

$$\mathcal{O}\left(\left(\mathbf{C}(n+1)D^2\kappa(f)^2\right)^{2(n+1)} N \log_2\left(nD\kappa(f)\right)\right)$$

for some universal constant $\mathbf{C}$.
(c) *It can be modified to return, in addition, at the same cost, and for each real zero $\zeta \in \mathbb{P}(\mathbb{R}^{n+1})$ of f, an approximate zero x of f with associated zero ζ.*

In addition to Theorem 19.1 we present in this chapter some additional results related to real polynomial systems. Firstly, we profit from the tools developed in Sects. 17.5 and 18.1 to give a short proof of a well-known result of Shub and Smale giving the expected value for the output of the counting problem.

Theorem 19.2 *The average number of zeros of a standard Gaussian random $f \in \mathcal{H}_{\mathbf{d}}^{\mathbb{R}}$ (with respect to Weyl's basis) in real projective space $\mathbb{P}(\mathbb{R}^{n+1})$ equals $\sqrt{\mathcal{D}}$.*

Secondly, we briefly describe and analyze an algorithm to decide feasibility of underdetermined systems of real polynomials.

19.1   Homogeneous Systems with Real Coefficients

We will use for real systems of polynomials the same notation we used for complex systems. Furthermore, we observe that a number of the notions and results we proved for the latter carry over, with only natural modifications, to the real setting. In particular, we may endow $\mathcal{H}_{d}^{\mathbb{R}}$ with the Weyl inner product defined in Sect. 16.1 and consider for $f \in \mathcal{H}_{d}^{\mathbb{R}}$ and $x \in \mathbb{R}^{n+1}$ the quantity $\mu_{\text{norm}}(f, x)$ defined in Sect. 16.7. The arguments used to show unitary invariance for both $\langle \, , \, \rangle$ and μ_{norm} carry over to show invariance, now under the action of the orthogonal group $\mathcal{O}(n + 1)$.

Fig. 19.1 Newton's operator
on $\mathbb{S}^n$

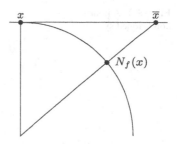

Newton's method can be defined to act on the sphere $\mathbb{S}^n$. Indeed, for $f \in \mathcal{H}_\mathbf{d}^\mathbb{R}$ and $x \in \mathbb{S}^n$ we let

$$\overline{x} := x - Df(x)|_{T_x \mathbb{S}^n}^{-1} f(x)$$

and put $N_f(x) := \frac{\overline{x}}{\|\overline{x}\|}$; see Fig. 19.1. In this way we get the operator $N_f : \mathbb{S}^n \setminus \Lambda_f^\mathbb{R} \to \mathbb{S}^n$, where $\Lambda_f^\mathbb{R} := \{x \in \mathbb{S}^n \mid Df(x)|_{T_x \mathbb{S}^n}$ not invertible$\}$.

The invariant $\gamma_{\text{proj}}(f, x)$ is defined as in Sect. 16.6, and Theorem 16.1 carries over unchanged. The notion of approximate zero is defined in the same manner (part (c) of Theorem 19.1 above refers to this notion), and the projective γ-theorem (Theorem 16.38) holds as well.

The covering map $\mathbb{S}^n \to \mathbb{P}(\mathbb{R}^{n+1})$ has degree 2. Hence, real projective zeros of polynomial systems $f \in \mathcal{H}_\mathbf{d}^\mathbb{R}$ correspond to pairs of zeros $(-\zeta, \zeta)$ of the restriction $f_{|\mathbb{S}^n}$ of f to $\mathbb{S}^n$. We will thus consider a system $f \in \mathcal{H}_\mathbf{d}^\mathbb{R}$ to be a (centrally symmetric) mapping of $\mathbb{S}^n$ into $\mathbb{R}^n$. In particular, the number $\#_\mathbb{R}(f)$ of real projective zeros of $f \in \mathcal{H}_\mathbf{d}^\mathbb{R}$ is equal to half the number of zeros of f in $\mathbb{S}^n$. That is,

$$\#_\mathbb{R}(f) := \left|Z_\mathbb{P}(f)\right| = \frac{1}{2}\left|Z_\mathbb{S}(f)\right|.$$

Our algorithm will thus compute $\#_\mathbb{R}(f)$ by counting the number of points in $Z_\mathbb{S}(f)$.

The same reason is behind the use of $d_\mathbb{S}(x, y)$ instead of $d_\mathbb{P}(x, y)$, a choice that has no consequences as long as the angle between x and y is at most $\frac{\pi}{2}$.

The tangent spaces of $\mathbb{P}^n$ and $\mathbb{S}^n$ at x can be identified, and it will be convenient to denote them by $T_x := x^\perp = T_x \mathbb{S}^n = T_x \mathbb{P}^n$.

For the rest of this chapter all systems $f \in \mathcal{H}_\mathbf{d}^\mathbb{R}$ considered are different from 0.

19.2   On the Condition for Real Zero-Counting

The goal of this section is to provide an explicit characterization of $\kappa(f)$ that will be useful in calculations. We have similarly done so for $\mathcal{C}(A)$ (Theorems 6.27 and Propositions 6.28 and 6.30) and for $\mathcal{K}(d)$ (Theorem 11.7).

The development in the previous chapters suggests that the condition numbers $\mu_{\text{norm}}(f, \zeta)$ for the zeros $\zeta \in \mathbb{S}^n$ of f should play a role. But it is apparent that these quantities cannot be the only ingredient. For in the first place, it may happen that

Fig. 19.2 A poorly
conditioned system f

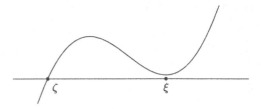

$Z_\mathbb{S}(f) = \emptyset$, in which case there would be no ingredient at all, but also because the
poor conditioning of a system f may be caused by the behavior of f away from
its zeros. Figure 19.2 gives an idea. Here f has only one zero ζ, and it is well
conditioned (has a small value of μ_{norm}). Yet, a small perturbation of f will make
ξ a zero as well. That is, $d(f, \Sigma_\mathbb{R})$ is small, or equivalently, $\kappa(f)$ is large.

The situation is clear: f is poorly conditioned when there are points x for which
both $\|f(x)\|$ is small and $\mu_{norm}(f, x)$ is large. This inspires the following measure
of condition.

We define $v(f, x) \in (0, \infty]$ by

$$v(f, x) := \frac{\|f\|}{(\|f\|^2 \mu_{norm}(f, x)^{-2} + \|f(x)\|_2^2)^{1/2}},\qquad(19.2)$$

the ∞ corresponding to division by 0, and

$$v(f) := \max_{x \in \mathbb{S}^n} v(f, x).$$

Note that $v(f, x) \geq \mu_{norm}(f, x)$ with $v(f, x) = \mu_{norm}(f, x)$ if and only if $f(x) = 0$.
Furthermore, $v(f) = \infty$ if and only if there exists a zero ζ of f with $\mu_{norm}(f, \zeta)$
$= \infty$. The fact that we allow $v(f, x)$ to take the value ∞ is what allows us to use a
maximum in the definition of $v(f)$. For ill-posed systems this maximum is attained
at an ill-posed zero of f.

Our characterization is the following.

Theorem 19.3 *We have* $\kappa(f) = v(f)$ *for all* $f \in \mathcal{H}_\mathbf{d}^\mathbb{R}$.

A first consequence of Theorem 19.3 is the following result.

Corollary 19.4 *For all* $f \in \mathcal{H}_\mathbf{d}^\mathbb{R}$, $v(f) \geq 1$.

Proof Since $\Sigma_\mathbb{R}$ has points arbitrarily close to 0, we have $d(f, \Sigma_\mathbb{R}) \leq \|f\|$ and
hence $\kappa(f) \geq 1$. Now use Theorem 19.3.   □

Remark 19.5 It is worth noting that although $v(f)$ is somehow related to the condition number $\mu_{max}(f)$ or $\mu_{av}(f)$ for complex polynomial systems, a result like
Theorem 19.3 does not hold for the latter. As we have seen (in Theorem 16.19),
a result of this kind holds for μ_{norm} on the fibers of the zeros. It can be shown,
however, that it does not hold globally.

Consider a point $x \in \mathbb{S}^n$ and a system $f \in \mathcal{H}_\mathbf{d}^\mathbb{R}$. One may define a notion of ill-posedness relative to the point x by taking

$$\Sigma_\mathbb{R}(x) = \{f \in \mathcal{H}_\mathbf{d}^\mathbb{R} \mid x \text{ is a multiple zero of } f\},$$

the set of systems that are ill-posed at x. Note that $\Sigma_\mathbb{R}(x) \neq \emptyset$ for all $x \in \mathbb{S}^n$ and that

$$\Sigma_\mathbb{R} = \{f \in \mathcal{H}_\mathbf{d}^\mathbb{R} \mid f \text{ has a multiple zero in } \mathbb{S}^n\} = \bigcup_{x \in \mathbb{S}^n} \Sigma_\mathbb{R}(x).$$

Note that for all $\lambda \neq 0$, $\nu(\lambda f) = \nu(f)$ and $d(\lambda f, \Sigma_\mathbb{R}) = |\lambda| d(f, \Sigma_\mathbb{R})$. The same is true relative to a point $x \in \mathbb{S}^n$. We will therefore assume, without loss of generality, that $\|f\| = 1$, and denote by $\mathbb{S}(\mathcal{H}_\mathbf{d}^\mathbb{R})$ the unit sphere in $\mathcal{H}_\mathbf{d}^\mathbb{R}$.

Proposition 19.6 *For all $x \in \mathbb{S}^n$ and $f \in \mathbb{S}(\mathcal{H}_\mathbf{d}^\mathbb{R})$,*

$$\nu(f, x) = \frac{1}{d(f, \Sigma_\mathbb{R}(x))}.$$

Proof For $0 \leq i \leq n$, let $e_i = (0, \ldots, 0, 1, 0, \ldots, 0)$ denote the ith standard basis vector. The group $\mathcal{O}(n+1)$ acts on $\mathcal{H}_\mathbf{d}^\mathbb{R} \times \mathbb{S}^n$ and leaves $\mu_{\text{norm}}, \kappa$ and the distance to $\Sigma_\mathbb{R}(\ )$ invariant. Therefore, we may assume without loss of generality that $x = e_0$. This implies that $T_{e_0}\mathbb{S}^n \simeq \text{span}\{e_1, \ldots, e_n\}$, and we may write the singular value decomposition

$$\text{diag}\left(\frac{1}{\sqrt{d_i}}\right)Df(e_0)|_{T_{e_0}\mathbb{S}^n} = \underbrace{[u_1 \quad \cdots \quad u_n]}_{U} \begin{bmatrix} \sigma_1 & & \\ & \ddots & \\ & & \sigma_n \end{bmatrix} V^\mathsf{T}$$

with U and V orthogonal and $\sigma_1 \geq \sigma_2 \geq \cdots \geq \sigma_n \geq 0$. Recall from Sect. 16.3 that the stabilizer $\mathcal{O}_{e_0}$ is the subgroup of $\mathcal{O}(n+1)$ leaving e_0 invariant. Since $\mathcal{O}_{e_0}$ is isomorphic to $\mathcal{O}(n)$ and acts on $T_{e_0}\mathbb{S}^n$, we may as well assume that $V = \mathrm{I}$. Note that $\mu_{\text{norm}}(f, e_0) = \sigma_n^{-1}$, and therefore $\kappa(f, e_0) = (\sigma_n^2 + \|f(e_0)\|_2^2)^{-1/2}$.

In what follows, for the sake of simplicity, we write $Dg_i(e_0)$ instead of $Dg_i(e_0)|_{T_{e_0}\mathbb{S}^n}$ and we denote by Δ the matrix $\text{diag}(\sqrt{d_i})$.

For $i = 1, \ldots, n$, let $g_i(x) := f_i(x) - f_i(e_0)x_0^{d_i} - \sqrt{d_i}\sigma_n u_{in} x_0^{d_i-1} x_n$, where $u_n = (u_{1n}, \ldots, u_{nn})^\mathsf{T}$.

Clearly, $g_i(e_0) = 0$ and $Dg_i(e_0)e_n = 0$, since $\partial g_i/\partial x_n(e_0) = \partial f_i/\partial x_n(e_0) - \sqrt{d_i}u_{in}\sigma_n = 0$. Thus, $g = (g_1, \ldots, g_n) \in \Sigma_\mathbb{R}(e_0)$. Moreover,

$$\|f_i - g_i\|^2 = \left(\frac{d_i}{d_i}\right)^{-1} f_i(e_0)^2 + \left(\frac{d_i}{d_i - 1, 1}\right)^{-1} (\sqrt{d_i}\sigma_n u_{in})^2 = f_i(e_0)^2 + \sigma_n^2 u_{in}^2,$$

and hence, using $\|u_n\| = 1$,

$$\|f - g\|^2 = \|f(e_0)\|_2^2 + \sigma_n^2 = \kappa(f, e_0)^{-2}.$$

It follows that

$$d\big(f, \Sigma_{\mathbb{R}}(e_0)\big) \le \|f - g\| = \kappa(f, e_0)^{-1}.$$

For the reverse inequality, let $g \in \Sigma_{\mathbb{R}}(e_0)$. Then, $g(e_0) = 0$ and $Dg(e_0)$ is singular. We want to show that $\|f - g\| \ge \kappa(f, e_0)^{-1}$. To this end, we write

$$f_i(x) = f_i(e_0)x_0^{d_i} + \frac{\partial f_i}{\partial x_1}(e_0)x_0^{d_i-1}x_1 + \cdots + \frac{\partial f_i}{\partial x_n}(e_0)x_0^{d_i-1}x_n + Q_i(x)$$

with $\deg_{x_0} Q_i \le d_i - 2$ and, similarly,

$$g_i(x) = \frac{\partial g_i}{\partial x_1}(e_0)x_0^{d_i-1}x_1 + \cdots + \frac{\partial g_i}{\partial x_n}(e_0)x_0^{d_i-1}x_n + R_i(x).$$

Then

$$\|f_i - g_i\|^2 \ge f_i(e_0)^2 + \frac{1}{d_i}\big\|Df_i(e_0) - Dg_i(e_0)\big\|_2^2$$

and hence

$$\|f - g\|^2 \ge \big\|f(e_0)\big\|_2^2 + \big\|\Delta^{-1}Df(e_0) - \Delta^{-1}Dg(e_0)\big\|_F^2.$$

By assumption, $\Delta^{-1}Dg(e_0)$ is singular. Hence, denoting by Sing_n the set of singular $n \times n$ matrices and by d_F the Frobenius distance on this set of matrices, we get

$$d_F\big(\Delta^{-1}Df(e_0), \Delta^{-1}Dg(e_0)\big) \ge d_F\big(\Delta^{-1}Df(e_0), \mathsf{Sing}_n\big) = \sigma_n,$$

the equality holding by Corollary 1.19. It follows that

$$\|f - g\|^2 \ge \big\|f(e_0)\big\|_2^2 + \sigma_n^2 = \kappa(f, e_0)^{-2}. \qquad \square$$

Proof of Theorem 19.3 Again we can assume $f \in \mathbb{S}(\mathcal{H}_{\mathbf{d}}^{\mathbb{R}})$. Note that

$$d(f, \Sigma_{\mathbb{R}}) = \min_{g \in \Sigma_{\mathbb{R}}} d(f, g) = \min_{x \in \mathbb{S}^n} d\big(f, \Sigma_{\mathbb{R}}(x)\big),$$

since $\Sigma_{\mathbb{R}} = \bigcup_{x \in \mathbb{S}^n} \Sigma_{\mathbb{R}}(x)$. Therefore, using Proposition 19.6,

$$\nu(f) = \max_{x \in \mathbb{S}^n} \nu(f, x) = \max_{x \in \mathbb{S}^n} \frac{1}{d(f, \Sigma_{\mathbb{R}}(x))} = \frac{1}{\min_{x \in \mathbb{S}^n} d(f, \Sigma_{\mathbb{R}}(x))} = \frac{1}{d(f, \Sigma_{\mathbb{R}})}.$$

$$\square$$

19.3 Smale's α-Theory

The projective γ-theorem shows that the quantity $\gamma_{\mathrm{proj}}(f, \zeta)$ can be used to estimate the size of the basin of quadratic attraction of a zero ζ of f. It cannot, however, be directly used to check that a point z is an approximate zero of f associated to ζ. For checking this, Steve Smale introduced another quantity $\alpha_{\text{proj}}(f, x)$. We define

$$\beta_{\text{proj}}(f, x) = \left\| Df(x)|_{T_x\mathbb{S}^n}^{-1} f(x) \right\|,$$

$$\alpha_{\text{proj}}(f, x) = \beta_{\text{proj}}(f, x)\, \gamma_{\text{proj}}(f, x).$$

Recall that $N_f(x) = \frac{\overline{x}}{\|\overline{x}\|}$, where $\overline{x} = x - Df(x)|_{T_x\mathbb{S}^n}^{-1} f(x)$. Then, by construction, $\beta_{\text{proj}}(f, x) = \|\overline{x} - x\|$ is the tangent of the Newton step $d_{\mathbb{S}}(x, N_f(x))$; cf. Fig. 19.1. In particular, $d_{\mathbb{S}}(x, N_f(x)) \le \beta_{\text{proj}}(f, x)$. We have, in addition, the following bound.

Proposition 19.7 *Let $f \in \mathcal{H}_{\mathbf{d}}^{\mathbb{R}}$ and $x, \zeta \in \mathbb{S}^n$. If x is an approximate zero of f with associated zero ζ, then $d_{\mathbb{S}}(x, \zeta) \le 2\beta_{\text{proj}}(f, x)$.*

Proof Let $x_1 := N_f(x)$. By the definition of approximate zero, $d_{\mathbb{S}}(x_1, \zeta) \le \frac{1}{2} d_{\mathbb{S}}(x, \zeta)$. This implies

$$d_{\mathbb{S}}(x, \zeta) \le d_{\mathbb{S}}(x, x_1) + d_{\mathbb{S}}(x_1, \zeta) \le d_{\mathbb{S}}(x, x_1) + \frac{1}{2} d_{\mathbb{S}}(x, \zeta),$$

from which it follows that $d_{\mathbb{S}}(x, \zeta) \le 2 d_{\mathbb{S}}(x, x_1) \le 2\beta_{\text{proj}}(f, x)$. $\square$

We should next turn to the meaning of α_{proj}, which, roughly speaking, guarantees the existence of zeros near points x where $\alpha_{\text{proj}}(f, x)$ is small enough. For ease of computation, though, we will use slight modifications of the quantities α_{proj} and γ_{proj}. We set

$$\overline{\beta}(f, x) := \mu_{\text{norm}}(f, x) \frac{\|f(x)\|}{\|f\|},$$

$$\overline{\gamma}(f, x) := \frac{1}{2} D^{3/2} \mu_{\text{norm}}(f, x),$$

$$\overline{\alpha}(f, x) := \overline{\beta}(f, x)\overline{\gamma}(f, x) = \frac{1}{2} D^{3/2} \mu_{\text{norm}}^2(f, x) \frac{\|f(x)\|}{\|f\|}.$$

We will also use, for technical reasons, the quantity

$$\widetilde{\alpha}(f, x) := \beta_{\text{proj}}(f, x)\overline{\gamma}(f, x).$$

Lemma 19.8 *For $f \in \mathcal{H}_{\mathbf{d}}^{\mathbb{R}}$ and $x \in \mathbb{S}^n$ we have*

(a) $\alpha_{\text{proj}}(f, x) \le \widetilde{\alpha}(f, x) \le \overline{\alpha}(f, x)$, $\beta_{\text{proj}}(f, x) \le \overline{\beta}(f, x)$, $\gamma_{\text{proj}}(f, x) \le \overline{\gamma}(f, x)$,
(b) $\overline{\beta}(f, x) \le \mu_{\text{norm}}(f, x)$ *and* $\overline{\gamma}(f, x) \ge \frac{1}{2}\sqrt{n} D^{3/2}$,
(c) $\left\| Df(x)|_{T_x}^{-1} Df(x)(x) \right\| \le \sqrt{D}\,\overline{\beta}(f, x)$.

Proof (a) The inequality $\gamma_{\mathrm{proj}}(f,x) \le \overline{\gamma}(f,x)$ is just a restatement of the higher derivative estimate (Theorem 16.1). Further,

$$\beta_{\mathrm{proj}}(f,x) = \left\| Df(x)|_{T_x\mathbb{S}^n}^{-1} f(x) \right\| \le \left\| Df(x)|_{T_x\mathbb{S}^n}^{-1} \right\| \left\| f(x) \right\|$$

$$\le \mu_{\mathrm{norm}}(f,x)\frac{\|f(x)\|}{\|f\|} = \overline{\beta}(f,x).$$

The inequalities $\alpha_{\mathrm{proj}}(f,x) \le \widetilde{\alpha}(f,x) \le \overline{\alpha}(f,x)$ are then immediate.

(b) This is a consequence of $\|f(x)\| \le \|f\|$ (see Lemma 16.6) and of Lemma 16.44.

(c) Euler's formula (16.3) implies $Df(x)(x) = \mathrm{diag}(d_i)f(x)$. Writing

$$Df(x)|_{T_x}^{-1} Df(x)(x) = Df(x)|_{T_x}^{-1}\mathrm{diag}\left(\sqrt{d_i}\right)\mathrm{diag}\left(\sqrt{d_i}\right)f(x)$$

and bounding norms, the third assertion follows.   □

In what follows we set, for $f \in \mathcal{H}_{\mathbf{d}}^{\mathbb{R}}$ and $x \in \mathbb{S}^n$,

$$\overline{B}_f(x) := \left\{ y \in \mathbb{S}^n \mid d_{\mathbb{S}}(x,y) \le 2\overline{\beta}(f,x) \right\}.$$

The following result is a version of Smale's α-theorem.

Theorem 19.9 *There exists a universal constant $\alpha_0 := 0.02$ such that for all $f \in \mathcal{H}_{\mathbf{d}}^{\mathbb{R}}$ and $x \in \mathbb{S}^n$, if $\overline{\alpha}(f,x) < \alpha_0$, then:*

(a) *x is an approximate zero of f.*

(b) *If ζ denotes its associated zero, then $\zeta \in \overline{B}_f(x)$.*

(c) *Furthermore, for each point y in $\overline{B}_f(x)$, y is an approximate zero of f with associated zero ζ. In particular, the Newton sequence starting at y converges to ζ.*

The proof of Theorem 19.9 requires some preliminary steps. The first such step is the kth-derivative version of Lemma 15.6.

Lemma 19.10 *For $-1 < u < 1$ and a positive integer k we have*

$$\sum_{\ell=0}^{\infty} \frac{(k+\ell)!}{k!\ell!} u^\ell = \frac{1}{(1-u)^{k+1}}.$$

Proof We compute the kth derivative on both sides of the geometric series $\sum_{i=0}^{\infty} u^i = \frac{1}{1-u}$. By induction it is easy to see that

$$\left(\sum_{i=1}^{\infty} u^i \right)^{(k)} = \sum_{\ell=0}^{\infty} \frac{(k+\ell)!u^\ell}{\ell!} \quad \text{and that} \quad \left(\frac{1}{1-u} \right)^{(k)} = \frac{k!}{(1-u)^{k+1}}. \quad □$$

Recall from Sect. 16.6 the family of functions ψ_δ as well as the quantities $\delta(r)$ and $u(r)$ defined for $r \in [\frac{\pi}{2}, 1]$.

Lemma 19.11 *Fix $\frac{2}{\pi} \leq r \leq 1$ such that $u(r) \leq \frac{1}{8}$. Let $x, y \in \mathbb{S}^n$, $\delta := d_{\mathbb{S}}(x, y)$, and $u := \delta\overline{\gamma}(f, x)$, and assume that $\delta \leq \delta(r)$ and $u \leq u(r)$. Then $\psi_{\delta}(u) > 0$, the map $Df(y)|_{T_y}$ is invertible, and*

$$\left\| Df(x)|_{T_x}^{-1} f(y) \right\| \leq \beta_{\text{proj}}(f, x) + \| y - x \| \left(\frac{1}{1-u} + \sqrt{D}\,\overline{\beta}(f, x) \right).$$

Proof First note that $Df(x)|_{T_x}$ is invertible since we assume $\gamma_{\text{proj}}(f, x)$ to be finite. Lemma 16.37 gives $\psi_{\delta}(u) > 0$. Lemma 16.41 implies that $Df(y)|_{T_y}$ is invertible.

Using the Taylor expansion of f at y around x, we obtain

$$\left\| Df(x)|_{T_x}^{-1} f(y) \right\| \leq \left\| Df(x)|_{T_x}^{-1} f(x) \right\| + \left\| Df(x)|_{T_x}^{-1} Df(x)(y - x) \right\|$$

$$+ \left\| \sum_{k=2}^{\infty} \frac{Df(x)|_{T_x}^{-1} D^k f(x)}{k!} (y - x)^k \right\|. \tag{19.3}$$

The first term on the right-hand side equals $\beta_{\text{proj}}(f, x)$. In order to estimate the second contribution, we decompose $y - x = p + \lambda x$ with $p \in T_x$ (similarly as in Fig. 19.1). Then $\langle x, y - x \rangle = \lambda$ and hence $|\lambda| \leq \| y - x \|$. Further, $\| p \| \leq \| y - x \|$. Using this decomposition we get

$$Df(x)|_{T_x}^{-1} Df(x)(y - x) = p + \lambda Df(x)|_{T_x}^{-1} Df(x)(x)$$

and hence, using Lemma 19.8(c),

$$\left\| Df(x)|_{T_x}^{-1} Df(x)(y - x) \right\| \leq \| p \| + |\lambda| \left\| Df(x)|_{T_x}^{-1} Df(x)(x) \right\|$$

$$\leq \| y - x \| \left(1 + \sqrt{D}\,\overline{\beta}(f, x) \right).$$

We can estimate the third term in (19.3) by

$$\| y - x \| \left(\sum_{k=2}^{\infty} \gamma_{\text{proj}}(f, x)^{k-1} \| y - x \|^{k-1} \right) \leq \| y - x \| \left(\frac{1}{1-u} - 1 \right),$$

taking into account that $\gamma_{\text{proj}}(f, x)\| y - x \| \leq \overline{\gamma}(f, x)\| y - x \| \leq u \leq u(r) < 1$ due to Lemma 19.8(a) and using $\| y - x \| \leq \delta$. Putting these estimates together, the assertion follows. $\qquad\square$

The next proposition estimates $\widetilde{\alpha}$, β_{proj}, and $\overline{\gamma}$ for a point y near x in terms of the values of these quantities at x.

Proposition 19.12 *Under the hypotheses of Lemma 19.11 we have:*

(a) $\beta_{\text{proj}}(f, y) \leq \frac{(1-u)}{\psi_{\delta}(u)} \left((1 - u)\beta_{\text{proj}}(f, x) + (1 + \sqrt{D}\,\overline{\beta}(f, x)) \| y - x \| \right),$

(b) $\overline{\gamma}(f, y) \leq \frac{\overline{\gamma}(f, x)}{1 - 4u},$

(c) $\widetilde{\alpha}(f, y) \leq \frac{1-u}{(1-4u)\psi_{\delta}(u)} \left((1 - u)\widetilde{\alpha}(f, x) + u + \sqrt{D}\,\overline{\alpha}(f, x) \| y - x \| \right).$

Proof (a) We have

$$\beta_{\text{proj}}(f, y) = \left\| Df(y)|_{T_y}^{-1} f(y) \right\| \le \left\| Df(y)|_{T_y}^{-1} Df(x)|_{T_x} \right\| \left\| Df(x)|_{T_x}^{-1} f(y) \right\|$$

$$\le \frac{(1-u)^2}{\psi_\delta(u)} \cdot \left(\beta_{\text{proj}}(f, x) + \|y - x\| \left(\frac{1}{1-u} + \sqrt{D}\, \overline{\beta}(f, x) \right) \right),$$

where we used Lemmas 16.41(b) and 19.11 for the last inequality.

(b) Let $\varepsilon = 2u$. Then $\varepsilon \le \frac{1}{4}$ and $D^{3/2}\mu_{\text{norm}}(f, x)\delta = 2\overline{\gamma}(f, x)\delta = 2u = \varepsilon$. Hence the hypotheses of Corollary 16.54 are satisfied, from which it follows that $\mu_{\text{norm}}(f, y) \le \frac{1}{1-2\varepsilon}\mu_{\text{norm}}(f, x)$. Multiplying both sides by $\frac{1}{2}D^{3/2}$ yields the desired inequality.

(c) Multiplying the inequalities in (a) and (b) and noting that $\|y - x\| \le \delta$ proves (c).   □

A zero x of f is the same as a fixed point of N_f (provided $Df(x)|_{T_x \mathbb{S}^n}$ has full rank). For studying the latter, the Banach fixed-point theorem is a standard tool.

Definition 19.13 Let (X, d) be a metric space and $0 \le c < 1$. A map $F: X \to X$ satisfying $d(F(x), F(y)) \le c\, d(x, y)$ for all x, y in X is called a *contraction map* with *contraction constant* c.

Theorem 19.14 (Banach fixed-point theorem) *Suppose that (X, d) is a complete metric space and let $F : X \to X$ be a contraction map with contraction constant c. Then F has a unique fixed point $p \in X$. For any start value $x \in X$, the nth iteration $F^n(x)$ of F converges to p as $n \to \infty$. Moreover,*

$$\frac{1}{1+c} d(x, F(x)) \le d(x, p) \le \frac{1}{1-c} d(x, F(x)).$$

Proof Note first that F can have at most one fixed point: namely, if $F(p) = p$ and $F(q) = q$, then $d(p, q) = d(F(p), F(q)) \le cd(p, q)$ implies that $p = q$, since $c < 1$.

Let $x \in X$. By induction it follows that $d(F^n(x), F^{n+1}(x)) \le c^n d(x, F(x))$ for $n \ge 1$. By summing the geometric series, we have for all $m \ge n \ge 1$,

$$d\left(F^n(x), F^m(x)\right) \le \sum_{i=m}^{n-1} d\left(F^i(x), F^{i+1}(x)\right) \le \sum_{i=m}^{n-1} c^i d\left(x, F(x)\right)$$

$$\le \frac{c^m}{1-c} d\left(x, F(x)\right).$$

Hence $(F^n(x))_{n \ge 1}$ is a Cauchy sequence that converges to a point p in X, since X is complete. The sequence $\{F^{n+1}(x)\}_{n \ge 1}$ also converges to p, so by continuity of F we have $F(p) = p$. Thus p is the unique fixed point of F. Since $d(x, p) \le d(x, F(x)) + d(F(x), F^2(x)) + \cdots \le \sum_{i=0}^{\infty} c^i d(x, F(x))$, by summing the geometric sequence once again, it follows that $d(x, p) \leq \frac{1}{1-c} d(x, F(x))$. Finally, by the triangle inequality,

$$d(x, F(x)) \leq d(x, p) + d(p, F(x))$$
$$= d(x, p) + d(F(p), F(x)) \leq (1+c)d(x, p). \qquad \square$$

In our case X will be a closed spherical cap in $\mathbb{S}^n$ and d will be $d_\mathbb{S}$. To follow standard notation we will write $B_\mathbb{S}(x, \delta)$ instead of $\mathsf{cap}(x, \delta)$, since this set is the closed ball of radius δ around x in $\mathbb{S}$ with respect to its Riemannian distance.

Lemma 19.15 *Suppose* $g \colon B_\mathbb{S}(x, \delta) \to B_\mathbb{S}(x, \delta)$ *is continuously differentiable with* $\|Dg(y)\| \leq c$ *for all* $y \in B_\mathbb{S}(x, \delta)$. *Then* $d_\mathbb{S}(g(y), g(z)) \leq c \, d_\mathbb{S}(y, z)$ *for all* $y, z \in B_\mathbb{S}(x, \delta)$.

Proof Let $\gamma \colon [0, 1] \to \mathbb{S}^n$ be a parameterization of the segment of the great circle connecting y and z. Then $\tilde{\gamma} := g \circ \gamma$ is a parameterization of a curve connecting $g(y)$ with $g(z)$. We have $\frac{d}{dt}\tilde{\gamma}(t) = Dg(\gamma(t))\frac{d}{dt}\gamma(t)$ and hence

$$\left\| \frac{d}{dt}\tilde{\gamma}(t) \right\| \leq c \left\| \frac{d}{dt}\gamma(t) \right\|.$$

Hence the length of $\tilde{\gamma}$ can be bounded as

$$L(\tilde{\gamma}) = \int_0^1 \left\| \frac{d}{dt}\tilde{\gamma}(t) \right\| dt \leq c \int_0^1 \left\| \frac{d}{dt}\gamma(t) \right\| dt = cL(\gamma) = c \, d_\mathbb{S}(y, z).$$

This implies $d_\mathbb{S}(g(y), g(z)) \leq c \, d_\mathbb{S}(y, z)$. $\qquad \square$

As we pointed out above, we will take N_f as the function F. We next bound the derivative of this map in terms of $\overline{\alpha}$.

Proposition 19.16 *We have* $\|DN_f(x)\| \leq 3.71 \, \widetilde{\alpha}(f, x)$ *for all* $x \in \mathbb{S}^n$.

Proof We may assume $\|f\| = 1$ without loss of generality. Consider the map

$$\varphi \colon \mathbb{R}^{n+1} \setminus \Lambda_f^\mathbb{R} \to \mathbb{R}^{n+1}, \quad x \mapsto x - Df(x)|_{T_x}^{-1} f(x),$$

where $\Lambda_f^\mathbb{R}$ denotes the set of $x \in \mathbb{R}^{n+1}$ such that $Df(x)|_{T_x}$ is not invertible; cf. Fig. 19.1. Note that $\|\varphi(x)\| \geq \|x\| = 1$.

Moreover, consider the map $\pi \colon (\mathbb{R}^{n+1})_* \to \mathbb{S}^n$, $y \mapsto \frac{y}{\|y\|}$. We can factor the Newton operator as $N_f = \pi \circ \varphi$ and hence $DN_f(x) = D\pi(\varphi(x))D\varphi(x)$.

It is easy to check that $D\pi(y)$ is given by the orthogonal projection onto T_y, followed by multiplication by the scalar $\|y\|^{-1}$. This implies $\|D\pi(y)\| \leq \|y\|^{-1}$ and hence $\|D\pi(\varphi(x))\| \leq 1$, since $\|\varphi(x)\|^{-1} \leq 1$. Therefore, it is sufficient to prove that $\|D\varphi(x)\| \leq 4\widetilde{\alpha}(f, x)$.

Take a smooth curve $x(t)$ in $\mathbb{S}^n$ and consider the corresponding curves $z(t) :=$ $Df(x(t))|_{T_{x(t)}} f(x(t))$ and $y(t) := \varphi(x(t)) = x(t) - z(t)$ in $\mathbb{R}^{n+1}$. By differentiating $Df(x(t))(z(t)) = f(x(t))$ with respect to t (and omitting the argument t for notational simplicity), we obtain

$$Df(x)(\dot{z}) + D^2 f(x)(z, \dot{x}) = Df(x)(\dot{x}). \tag{19.4}$$

We also have $\langle \dot{z}, x \rangle + \langle z, \dot{x} \rangle = 0$, since $\langle z, x \rangle = 0$.

Let $p \cdot \mathbb{R}^{n+1} \to T_x$ denote the orthogonal projection onto T_x. We decompose

$$\dot{z} = p(\dot{z}) + \lambda x,$$

where $\lambda = \langle \dot{z}, x \rangle = -\langle z, \dot{x} \rangle$. Since $\beta := \beta_{\text{proj}}(f, x) = \|z\|$, we have

$$|\lambda| \leq \beta \|\dot{x}\|. \tag{19.5}$$

Inserting $\dot{z} = p(\dot{z}) + \lambda x$ into (19.4) and taking the inverse $Df(x)|_{T_x}^{-1}$, we obtain

$$p(\dot{z}) + \lambda Df(x)|_{T_x}^{-1} Df(x)(x) + Df(x)|_{T_x}^{-1} D^2 f(x)(z, \dot{x}) = \dot{x}.$$

Therefore,

$$\left\| \dot{x} - p(\dot{z}) \right\| \leq |\lambda| \left\| Df(x)|_{T_x}^{-1} Df(x)(x) \right\| + \left\| Df(x)|_{T_x}^{-1} D^2 f(x) \right\| \|z\| \|\dot{x}\|. \tag{19.6}$$

To simplify notation, in the rest of this proof we write $\gamma := \gamma_{\text{proj}}(f, x)$ and similarly for $\overline{\gamma}, \beta, \overline{\beta}, \alpha, \widetilde{\alpha}, \overline{\alpha}$, and μ_{norm}.

Using this convention, Lemma 19.8, and the definition of γ_{proj}, we bound

$$\left\| Df(x)|_{T_x}^{-1} Df(x)(x) \right\| \leq \overline{\beta} \sqrt{D}, \qquad \left\| Df(x)|_{T_x}^{-1} D^2 f(x) \right\| \leq 2\gamma.$$

Combining these two bounds with (19.5) and (19.6), we get

$$\left\| \dot{x} - p(\dot{z}) \right\| \leq \beta \|\dot{x}\| \overline{\beta} \sqrt{D} + 2\gamma \beta \|\dot{x}\|$$

and hence, using (19.5) again,

$$\|\dot{y}\| = \|\dot{x} - \dot{z}\| \leq \left\| \dot{x} - p(\dot{z}) \right\| + |\lambda| \leq \beta \|\dot{x}\| \overline{\beta} \sqrt{D} + 2\gamma \beta \|\dot{x}\| + \beta \|\dot{x}\|.$$

Since $\dot{y} = D\varphi(x)(\dot{x})$, we have shown that

$$\left\| D\varphi(x) \right\| \leq 2\alpha + (\overline{\beta} \sqrt{D} + 1)\beta \leq 2\widetilde{\alpha} + (\overline{\beta} \sqrt{D} + 1)\beta.$$

Since $\widetilde{\alpha} = \beta \overline{\gamma}$, the right-hand side equals

$$2\widetilde{\alpha} + (\overline{\beta} \sqrt{D} + 1) \frac{\widetilde{\alpha}}{\overline{\gamma}} = 2\widetilde{\alpha} + (\overline{\beta} \sqrt{D} + 1) \frac{2\widetilde{\alpha}}{D^{3/2} \mu_{\text{norm}}},$$

which equals

$$2\tilde{\alpha}\left(1 + \frac{\overline{\beta}\sqrt{D}}{D^{3/2}\mu_{\text{norm}}} + \frac{1}{D^{3/2}\mu_{\text{norm}}}\right) \leq 2\tilde{\alpha}\left(1 + \frac{1}{D} + \frac{1}{D^{3/2}}\right)$$

$$\leq 2\tilde{\alpha}\left(1 + \frac{1}{2} + \frac{1}{2^{3/2}}\right) \leq 3.71\tilde{\alpha},$$

where we have used $\overline{\beta} \leq \mu_{\text{norm}}$ for the first inequality; cf. Lemma 19.8(b). Hence $\|D\varphi(x)\| \leq 4\tilde{\alpha}$, as claimed. □

Theorem 19.17 *Fix $\frac{2}{\pi} \leq r \leq 1$ such that $u(r) \leq \frac{1}{8}$. Further, let $\delta \leq \delta(r)$ and $x \in \mathbb{S}^n$ be given such that $u := \delta\overline{\gamma}(f, x) \leq u(r)$. Put*

$$c := \frac{3.71(1-u)}{(1-4u)\psi_\delta(u)}(\overline{\alpha}(f, x) + u).$$

Then we have

(a) $\|DN_f(y)\| \leq c$ *for all y with $d_{\mathbb{S}}(y, x) \leq \delta$,*
(b) $N_f(B_{\mathbb{S}}(x, \delta)) \subseteq B_{\mathbb{S}}(N_f(x), c\delta)$.

Proof By Proposition 19.16 we have $\|DN_f(y)\| \leq 3.71\tilde{\alpha}(f, y)$. We can estimate the latter with Proposition 19.12(c). Using $u \geq \overline{\gamma}(f, x)\|y - x\| \geq \frac{1}{2}D^{3/2}\|y - x\|$, we can bound as follows:

$$(1-u)\tilde{\alpha}(f, x) + \sqrt{D}\,\overline{\alpha}(f, x)\|y - x\| + u$$

$$\leq (1-u)\overline{\alpha}(f, x) + \sqrt{D}\,\overline{\alpha}(f, x)\|y - x\| + u$$

$$\leq (1-u)\overline{\alpha}(f, x) + \sqrt{D}\,\overline{\alpha}(f, x)\frac{2}{D^{3/2}}u + u$$

$$\leq (1-u)\overline{\alpha}(f, x) + \frac{2}{D}u\,\overline{\alpha}(f, x) + u \leq \overline{\alpha}(f, x) + u,$$

and part (a) follows.

For part (b) we note that by part (a) and Lemma 19.15,

$$d_{\mathbb{S}}\big(N_f(y), N_f(x)\big) \leq cd_{\mathbb{S}}(y, x) \leq c\delta$$

for all y in $B_{\mathbb{S}}(x, \delta)$. □

Corollary 19.18 *Under the hypotheses of Theorem 19.17, we assume that $c < 1$ and $\overline{\alpha}(f, x) \leq (1-c)u$. Then N_f is a contraction map of the ball $B_{\mathbb{S}}(x, \frac{u}{\overline{\gamma}(f,x)})$ into itself with contraction constant c.*

Proof Write $\overline{\gamma} = \overline{\gamma}(f, x)$ and $\overline{\alpha} = \overline{\alpha}(f, x)$. For all $y \in B_{\mathbb{S}}(N_f(x), c\delta)$ we have

$$d_{\mathbb{S}}(y, x) \leq d_{\mathbb{S}}\big(y, N_f(x)\big) + d_{\mathbb{S}}\big(N_f(x), x\big) \leq c\delta + \beta \leq \delta,$$

the last by dividing the hypothesis $\bar{\alpha} + cu \leq u$ by $\bar{\gamma}$. It follows that $B_{\mathbb{S}}(N_f(x), c\delta) \subseteq B_{\mathbb{S}}(x, \delta)$. Hence, by Theorem 19.17(b),

$$N_f\big(B_{\mathbb{S}}(x, \delta)\big) \subseteq B_{\mathbb{S}}\big(N_f(x), c\delta\big) \subseteq B_{\mathbb{S}}(x, \delta),$$

and we deduce that N_f maps $B_{\mathbb{S}}(x, \delta)$ into itself. Furthermore, c is a contraction constant for this map by Lemma 19.15, Theorem 19.17(a), and the hypothesis $c < 1$. $\qquad\square$

We can finally prove the main result of this section.

Proof of Theorem 19.9 Let $r_* := 0.888$. Then (recall Table 16.1) $\delta_* := \delta(r_*) = 0.834\ldots$ and $u_* := u(r_*) = 0.1246\cdots < \frac{1}{8}$ satisfy $u_* < \sqrt{2}\delta_*$.

Now we take $\alpha_0 := 0.02$ and let $u_0 := 2\alpha_0 = 0.04$. These constants satisfy

$$c_0 := \frac{3.71(1 - u_0)}{(1 - 4u_0)\psi_{\delta_*}(u_0)}\left(\frac{u_0}{2} + u_0\right) \leq \frac{1}{2}. \tag{19.7}$$

Furthermore, $\psi_{\delta_*}(u_0) = 0.54\ldots$ and $u_0 \leq u_*$.

The numbers α_0, u_0, and c_0 are universal constants. They depend neither on $f \in \mathcal{H}_{\mathbf{d}}^{\mathbb{R}}$ nor on $x \in \mathbb{S}^n$. Now consider such a pair (f, x) and assume that $\bar{\alpha}(f, x) \leq \alpha_0$. Then the bound $\bar{\gamma}(f, x) \geq \frac{D^{3/2}}{2} \geq \sqrt{2}$ (cf. Lemma 19.8) together with $u_0 \leq u_*$ implies that

$$\delta(x) := \frac{u_0}{\bar{\gamma}(f, x)} \leq \frac{u_0}{\sqrt{2}} \leq \frac{u_*}{\sqrt{2}} \leq \delta_*.$$

Also, let

$$c := \frac{3.71(1 - u_0)}{(1 - 4u_0)\psi_{\delta(x)}(u_0)}\big(\bar{\alpha}(f, x) + u_0\big).$$

Then, $\bar{\alpha}(f, x) \leq \alpha_0 = \frac{u_0}{2}$, $\psi_{\delta(x)}(u_0) \geq \psi_{\delta_*}(u_0)$, together with (19.7) imply $c \leq c_0 \leq \frac{1}{2}$ and therefore

$$\bar{\alpha}(f, x) \leq \alpha_0 = \frac{u_0}{2} \leq (1 - c)u_0.$$

We see that the hypotheses of Corollary 19.18 hold for $r = r_*$, and $\delta = \delta(x)$. Hence N_f is a contraction map on $B_{\mathbb{S}}(x, \delta(x))$ with contraction constant c_0. The Banach fixed point theorem then implies that there exists a zero $\zeta \in B_{\mathbb{S}}(x, \delta(x))$ of f, and for all points $y \in B_{\mathbb{S}}(x, \delta(x))$ we have $d_{\mathbb{S}}(N_f(y), \zeta) \leq \frac{1}{2}d_{\mathbb{S}}(y, \zeta)$. Hence by induction, $d_{\mathbb{S}}(N_f^i(y), \zeta) \leq (\frac{1}{2})^{2^i - 1}d_{\mathbb{S}}(y, \zeta)$, which means that y is an approximate zero of f with associated zero ζ.

It remains to show that $B_{\mathbb{S}}(x, \delta(x)) \subseteq \overline{B}_f(x)$. This follows from the fact that

$$\delta(x) = \frac{u_0}{\bar{\gamma}(f, x)} = \frac{2\alpha_0}{\bar{\gamma}(f, x)} \geq 2\bar{\beta}(f, x). \qquad\square$$

Remark 19.19

(a) Note that the proof above gives a ball of approximate zeros with a radius $\delta(x) = \frac{2\alpha_0}{\overline{\gamma}(f,x)}$, inversely proportional to $\overline{\gamma}(f,x)$. This is reminiscent of Theorems 15.5 and 16.38, but with the basin of attraction now centered at the point x at hand.

(b) Using the Moore–Penrose inverse $Df(x)^\dagger$ instead of $Df(x)|_{T_x}^{-1}$ in the definition of the Newton operator leads to the so-called Moore–Penrose Newton's iteration. The algebraic properties of the Moore-Penrose inverse, close to those of the common inverse for matrices, would lead to versions of Lemma 19.11 and Propositions 19.12 and 19.16 with simpler proofs. We will briefly return to the Moore–Penrose Newton's iteration in Sect. 19.6 below.

19.4 An Algorithm for Real Zero-Counting

In this section we will describe an algorithm for zero-counting, Algorithm 19.1 below, and show that it satisfies the statements claimed in Theorem 19.1.

19.4.1 Grids and Graphs

Our algorithm works on a grid on $\mathbb{S}^n$, which we construct by projecting onto $\mathbb{S}^n$ a grid on the cube $C^n := \{y \mid \|y\|_\infty = 1\}$. We make use of the (easy to compute) bijections $\phi : C^n \to \mathbb{S}^n$ and $\phi^{-1} : \mathbb{S}^n \to C^n$ given by $\phi(y) = \frac{y}{\|y\|}$ and $\phi^{-1}(x) = \frac{x}{\|x\|_\infty}$.

Given $\eta := 2^{-k}$ for some $k \geq 1$, we consider the uniform grid $\mathcal{U}_\eta$ of mesh η on C^n. This is the set of points in C^n whose coordinates are of the form $i2^{-k}$ for $i \in \{-2^k, -2^k + 1, \ldots, 2^k\}$, with at least one coordinate equal to 1 or -1. We denote by $\mathcal{G}_\eta$ its image by ϕ in $\mathbb{S}^n$. An argument in elementary geometry shows that for $y_1, y_2 \in C^n$,

$$d_{\mathbb{S}}(\phi(y_1), \phi(y_2)) \leq \frac{\pi}{2}\|y_1 - y_2\| \leq \frac{\pi}{2}\sqrt{n+1}\,\|y_1 - y_2\|_\infty. \qquad (19.8)$$

Given η as above, we associate to it a graph G_η as follows. We set $A(f) := \{x \in \mathbb{S}^n \mid \overline{\alpha}(f,x) < \alpha_0\}$. The vertices of the graph are the points in $\mathcal{G}_\eta \cap A(f)$. Two vertices $x, y \in \mathcal{G}_\eta$ are joined by an edge if and only if $\overline{B}(x) \cap \overline{B}(y) \neq \emptyset$. We have here (and we will in the rest of this section) dropped the index f in the balls $\overline{B}_f(x)$.

Note that as a simple consequence of Theorem 19.9, we obtain the following lemma.

Lemma 19.20

(a) *For each $x \in A(f)$ there exists $\zeta_x \in Z_{\mathbb{S}}(f)$ such that $\zeta_x \in \overline{B}(x)$. Moreover, for each point z in $\overline{B}(x)$, the Newton sequence starting at z converges to ζ_x.*

(b) *Let $x, y \in A(f)$. Then $\zeta_x = \zeta_y \iff \overline{B}(x) \cap \overline{B}(y) \neq \emptyset$.* □

We define $W(G_\eta) := \bigcup_{x \in G_\eta} \overline{B}(x) \subset \mathbb{S}^n$, where $x \in G_\eta$ has to be understood as x running over all the vertices of G_η. Similarly, for a connected component U of G_η, we define

$$W(U) := \bigcup_{x \in U} \overline{B}(x).$$

The following lemma implies that the connected components of the graph G_η are of a very special nature: they are cliques. It also implies that

$$|Z_\mathbb{S}(f)| \geq \# \text{ connected components of } G_\eta. \tag{19.9}$$

Lemma 19.21

(a) *For each component U of G_η, there is a unique zero $\zeta_U \in Z_\mathbb{S}(f)$ such that $\zeta_U \in W(U)$. Moreover, $\zeta_U \in \bigcap_{x \in U} \overline{B}(x)$.*

(b) *If U and V are different components of G_η, then $\zeta_U \neq \zeta_V$.*

Proof (a) Let $x \in U$. Since $x \in A(f)$, by Lemma 19.20(a) there exists a zero ζ_x of f in $\overline{B}(x) \subseteq W(U)$. This shows the existence. For the uniqueness and the second assertion, assume that there exist zeros ζ and ξ of f in $W(U)$. Let $x, y \in U$ be such that $\zeta \in \overline{B}(x)$, and $\xi \in \overline{B}(y)$. Since U is connected, there exist $x_0 = x, x_1, \ldots, x_{k-1}, x_k := y$ in $A(f)$ such that (x_i, x_{i+1}) is an edge of G_η for $i = 0, \ldots, k-1$, that is, $\overline{B}(x_i) \cap \overline{B}(x_{i+1}) \neq \emptyset$. If ζ_i and ζ_{i+1} are the associated zeros of x_i and x_{i+1} in $Z_\mathbb{S}(f)$ respectively, then by Lemma 19.20(b) we have $\zeta_i = \zeta_{i+1}$, and thus $\zeta = \xi \in \overline{B}(x) \cap \overline{B}(y)$.

(b) Let $\zeta_U \in \overline{B}(x)$ and $\zeta_V \in \overline{B}(y)$ for $x \in U$ and $y \in V$. If $\zeta_U = \zeta_V$, then $\overline{B}(x) \cap \overline{B}(y) \neq \emptyset$ and x and y are joined by an edge; hence $U = V$. $\square$

If equality holds in (19.9), we can compute $|Z_\mathbb{S}(f)|$ by computing the number of connected components of G_η. The reverse inequality in (19.9) amounts to the fact that there are no zeros of f in $\mathbb{S}^n$ that are not in $W(G_\eta)$. To verify that this is the case, we want to find, for each point $x \in G_n \setminus A(f)$, a ball centered at x such that $f \neq 0$ on this ball. In addition, we want the union of these balls to cover $\mathbb{S}^n \setminus W(G_\eta)$. The next result is the key ingredient towards this goal, since it provides radii for these balls.

Lemma 19.22 (Exclusion lemma) *Let $f \in \mathcal{H}_\mathbf{d}^\mathbb{R}$ and $x, y \in \mathbb{S}^n$ be such that $0 < d_\mathbb{S}(x, y) \leq \sqrt{2}$. Then,*

$$\|f(x) - f(y)\| < \|f\|\sqrt{D}\, d_\mathbb{S}(x, y).$$

In particular, if $f(x) \neq 0$, there is no zero of f in the ball $B_\mathbb{S}(x, \frac{\|f(x)\|}{\|f\|\sqrt{D}})$.

Proof Because of (16.1), for all $f_i \in \mathcal{H}_{d_i}^{\mathbb{R}}$ and $x \in \mathbb{R}^{n+1}$,

$$f_i(x) = \langle f_i(X), \langle x, X \rangle^{d_i} \rangle. \tag{19.10}$$

Because of orthogonal invariance, we can assume that $x = e_0$ and $y = e_0 \cos\theta + e_1 \sin\theta$, where $\theta = d_{\mathbb{S}}(x, y) > 0$. Equation (19.10) implies that

$$
\begin{aligned}
f_i(x) - f_i(y) &= \langle f_i(X), \langle x, X \rangle^{d_i} \rangle - \langle f_i(X), \langle y, X \rangle^{d_i} \rangle \\
&= \langle f_i(X), \langle x, X \rangle^{d_i} - \langle y, X \rangle^{d_i} \rangle \\
&= \langle f_i(X), X_0^{d_i} - (X_0 \cos\theta + X_1 \sin\theta)^{d_i} \rangle.
\end{aligned}
$$

Hence, by Cauchy–Schwarz,

$$\left| f_i(x) - f_i(y) \right| \leq \| f_i \| \, \left\| X_0^{d_i} - (X_0 \cos\theta + X_1 \sin\theta)^{d_i} \right\|.$$

Since

$$X_0^{d_i} - (X_0 \cos\theta + X_1 \sin\theta)^{d_i}$$

$$= X_0^{d_i} \left(1 - (\cos\theta)^{d_i} \right) - \sum_{k=1}^{d_i} \binom{d_i}{k} (\cos\theta)^{d_i - k} (\sin\theta)^k X_0^{d_i - k} X_1^k,$$

we have

$$\left\| X_0^{d_i} - (X_0 \cos\theta + X_1 \sin\theta)^{d_i} \right\|^2$$

$$\leq \left(1 - (\cos\theta)^{d_i} \right)^2 + \sum_{k=1}^{d_i} \binom{d_i}{k} (\cos\theta)^{2(d_i - k)} (\sin\theta)^{2k}$$

$$= \left(1 - (\cos\theta)^{d_i} \right)^2 + 1 - (\cos\theta)^{2d_i} = 2 \left(1 - (\cos\theta)^{d_i} \right)$$

$$< 2 \left(1 - \left(1 - \frac{\theta^2}{2} \right)^{d_i} \right) \leq 2 \left(1 - \left(1 - d_i \frac{\theta^2}{2} \right) \right)$$

$$\leq d_i \theta^2.$$

Here the first inequality is due to the fact that for $g = \sum g_k X_0^{d-k} X_1^k$, we have $\| g \|^2 = \sum \binom{d}{k}^{-1} g_k^2$. Also, the second inequality follows from the bound $\cos\theta > 1 - \frac{\theta^2}{2}$, which is true for all $0 < \theta \leq \sqrt{2}$, and the third from the bound $(1 - a)^d \geq 1 - da$, for $a \leq 1$. We conclude that

$$\left| f_i(x) - f_i(y) \right| < \| f_i \| \, \theta \sqrt{d_i}$$

and hence

$$\| f(x) - f(y) \| < \| f \| \, \theta \sqrt{\max_i d_i}.$$

For the second assertion, we have, for all $y \in B(x, \frac{\|f(x)\|}{\|f\|\sqrt{D}})$,

$$\|f(y)\| \geq \|f(x)\| - \|f(x) - f(y)\|$$
$$> \|f(x)\| - \|f\|\sqrt{D}\, d_{\mathbb{S}}(x, y)$$
$$\geq \|f(x)\| - \|f\|\sqrt{D}\, \frac{\|f(x)\|}{\|f\|\sqrt{D}} = 0.$$ $\square$

19.4.2 Proof of Theorem 19.1

We begin by describing our zero-counting algorithm (see Algorithm 19.1 below).

Remark 19.23 Algorithm 19.1 uses a routine for computing the connected components of a graph from the description of this graph. This is a standard task in discrete algorithmics. We will not enter into a discussion of this aspect of Algorithm 19.1 (but see the Notes for pointers to appropriate references).

Algorithm 19.1 Zero_Counting

Input: $f \in \mathcal{H}_{\mathbf{d}}^{\mathbb{R}}$
Preconditions: $f \neq 0$

```
let  η := ½
repeat
      let  U₁,...,Uᵣ be the connected components of  Gη
      if
          (a)  for  1 ≤ i < j ≤ r
                  for all  xᵢ ∈ Uᵢ  and all  xⱼ ∈ Uⱼ
                  d_S(xᵢ,xⱼ) > πη√(n+1)
      and
          (b)  for all  x ∈ Gη \ A(f)
                  ‖f(x)‖ > (π/4)η√((n+1)D)‖f‖
      then return r/2 and halt
      else η := η/2
```

Output: $r \in \mathbb{N}$
Postconditions: The algorithm halts if $f \notin \Sigma_{\mathbb{R}}$. In this case f has exactly r zeros in $\mathbb{P}(\mathbb{R}^{n+1})$.

We will now show that Algorithm 19.1 satisfies the claims (a)–(c) of the statement of Theorem 19.1.

(a) This part claims the correctness of Algorithm 19.1. To prove it, we will use the notions of spherical convexity introduced in Sect. 13.2.

Let H^n be an open hemisphere in $\mathbb{S}^n$ and $x_1, \ldots, x_q \in H^n$. Recall that the spherical convex hull of $\{x_1, \ldots, x_q\}$ is defined by

$$\mathrm{sconv}(x_1, \ldots, x_q) := \mathrm{cone}(x_1, \ldots, x_q) \cap \mathbb{S}^n,$$

where $\mathrm{cone}(x_1, \ldots, x_q)$ is the smallest convex cone with vertex at the origin and containing the points $x_1, \ldots, x_q$.

Lemma 19.24 *Let* $x_1, \ldots, x_q \in H^n \subset \mathbb{R}^{n+1}$. *If* $\bigcap_{i=1}^{q} B_{\mathbb{S}}(x_i, r_i) \neq \emptyset$, *then* $\mathrm{sconv}(x_1, \ldots, x_q) \subseteq \bigcup_{i=1}^{q} B_{\mathbb{S}}(x_i, r_i)$.

Proof Let $x \in \mathrm{sconv}(x_1, \ldots, x_q)$ and $y \in \bigcap_{i=1}^{q} B_{\mathbb{S}}(x_i, r_i)$. We will prove that $x \in B_{\mathbb{S}}(x_i, r_i)$ for some i. Without loss of generality we assume $x \neq y$. Let H be the open half-space

$$H := \left\{ z \in \mathbb{R}^{n+1} : \langle z, y - x \rangle < 0 \right\}.$$

We have

$$
\begin{aligned}
z \in H \quad &\Longleftrightarrow \quad \langle z, y - x \rangle < 0 \quad \Longleftrightarrow \quad -\langle z, x \rangle < -\langle z, y \rangle \\
&\Longleftrightarrow \quad \|z\|^2 + \|x\|^2 - 2\langle z, x \rangle < \|z\|^2 + \|y\|^2 - 2\langle z, y \rangle \\
&\Longleftrightarrow \quad \|z - x\|^2 < \|z - y\|^2,
\end{aligned}
$$

the second line following from $\|x\| = \|y\| = 1$. Therefore the half-space H is the set of points z in $\mathbb{R}^{n+1}$ such that the Euclidean distance $\|z - x\|$ is less than $\|z - y\|$.

On the other hand, H must contain at least one point of the set $\{x_1, \ldots, x_q\}$, since if this were not the case, the convex set $\mathrm{cone}(x_1, \ldots, x_q)$ would be contained in $\{z : \langle z, y - x \rangle \geq 0\}$, contradicting $x \in \mathrm{sconv}(x_1, \ldots, x_q)$. Therefore, there exists i such that $x_i \in H$. It follows that

$$\|x - x_i\| < \|y - x_i\|.$$

Since the function $z \mapsto 2 \arcsin(\frac{z}{2})$ giving the length of an arc as a function of its chord is nondecreasing, we obtain

$$d_{\mathbb{S}}(x, x_i) < d_{\mathbb{S}}(y, x_i) \leq r_i. \qquad \square$$

We can now proceed. Assume that Algorithm 19.1 halts. We want to show that if r equals the number of connected components of G_η, then $\#_{\mathbb{R}}(f) = \#Z_{\mathbb{S}}(f)/2 = r/2$. We already know by Lemma 19.21 that each connected component U of G_η determines uniquely a zero $\zeta_U \in Z_{\mathbb{S}}(f)$. Thus it is enough to prove that $Z_{\mathbb{S}}(f) \subseteq W(G_\eta)$. This would prove the reverse inequality in (19.9).

Assume, by way of contradiction, that there is a zero ζ of f in $\mathbb{S}^n$ such that ζ is not in $W(G_\eta)$. Let $B_\infty(\phi^{-1}(\zeta), \eta) := \{y \in \mathcal{U}_\eta \mid \|y - \phi^{-1}(\zeta)\|_\infty \leq$

$\eta\} = \{y_1, \ldots, y_q\}$, the set of all neighbors of $\phi^{-1}(\zeta)$ in $\mathcal{U}_\eta$, and let $x_i = \phi(y_i)$, $i = 1, \ldots, q$. Clearly, $\phi^{-1}(\zeta)$ is in the cone spanned by $\{y_1, \ldots, y_q\}$, and hence $\zeta \in \mathsf{sconv}(x_1, \ldots, x_q)$.

We claim that there exists $j \leq q$ such that $x_j \notin A(f)$. Indeed, assume this is not the case. We consider two cases.

(i) All the x_i belong to the same connected component U of G_η. In this case Lemma 19.21 ensures that there exists a unique zero $\zeta_U \in \mathbb{S}^n$ of f in $W(U)$ and $\zeta_U \in \bigcap_i \overline{B}(x_i)$. Since $x_1, \ldots, x_q$ lie in an open half-space of $\mathbb{R}^{n+1}$, we may apply Lemma 19.24 to deduce that

$$\mathsf{sconv}(x_1, \ldots, x_q) \subseteq \bigcup \overline{B}(x_i).$$

It follows that for some $i \in \{1, \ldots, q\}$, $\zeta \in \overline{B}(x_i) \subseteq W(U)$, contradicting that $\zeta \notin W(G_\eta)$.

(ii) There exist $\ell \neq s$ and $1 \leq j < k \leq r$ such that $x_\ell \in U_j$ and $x_s \in U_k$. Since condition (a) in the algorithm is satisfied, $d_\mathbb{S}(x_\ell, x_s) > \pi\eta\sqrt{n+1}$. But by the bounds (19.8),

$$d_\mathbb{S}(x_\ell, x_s) \leq \frac{\pi}{2}\sqrt{n+1}\|y_\ell - y_s\|_\infty$$

$$\leq \frac{\pi}{2}\sqrt{n+1}\big(\|y_\ell - \phi^{-1}(\zeta)\|_\infty + \|\phi^{-1}(\zeta) - y_s\|_\infty\big) \leq \pi\eta\sqrt{n+1},$$

a contradiction.

We have thus proved the claim. Let then $1 \leq j \leq q$ be such that $x_j \notin A(f)$. Then, using Lemma 19.22,

$$\|f(x_j)\| = \|f(x_j) - f(\zeta)\| \leq \|f\|\sqrt{D}\,d_\mathbb{S}(x_j, \zeta) \leq \frac{\pi}{2}\eta\sqrt{(n+1)D}\|f\|.$$

This is in contradiction with condition (b) in the algorithm being satisfied.

(b) We next prove the bound for the cost claimed in part (b) of Theorem 19.1. The idea is to show that when η becomes small enough, as a function of $\kappa(f)$, n, N and D, then conditions (a) and (b) in Algorithm 19.1 are satisfied. We spread this task over a few lemmas, the first two of them being extensions of the bounds for separation of zeros we saw in Sect. 16.6.

Lemma 19.25 *For all $\frac{2}{\pi} \leq r \leq 1$, if $\zeta_1 \neq \zeta_2 \in Z_\mathbb{S}(f)$, then*

$$d_\mathbb{S}(\zeta_1, \zeta_2) \geq \min\left\{\delta(r), \frac{2u(r)}{D^{3/2}\kappa(f)}\right\}.$$

Proof The statement follows from Corollary 16.42, the estimate $\gamma_{\mathrm{proj}}(f, z) \leq \frac{D^{3/2}}{2}\mu_{\mathrm{norm}}(f, z)$, and the fact that $\max\{\mu_{\mathrm{norm}}(f, \zeta), \mu_{\mathrm{norm}}(f, \xi)\} \leq \nu(f) = \kappa(f)$. □

Lemma 19.26 *Let $x_1, x_2 \in G_\eta$ with associated zeros $\zeta_1 \neq \zeta_2$. Let r_*, δ_*, and u_* be as in the proof of Theorem 19.9. If*

$$\eta \leq \frac{2u_*}{3D^{3/2}\pi\kappa(f)\sqrt{n+1}},$$

then $d_\mathbb{S}(x_1, x_2) > \pi\eta\sqrt{n+1}$.

Proof Assume $d_\mathbb{S}(x_1, x_2) \leq \pi\eta\sqrt{n+1}$. Since $x_2 \notin \overline{B}(x_1)$, $d_\mathbb{S}(x_1, x_2) > 2\beta(f, x_1)$. Consequently,

$$d_\mathbb{S}(x_1, \zeta_1) \leq 2\beta(f, x_1) < d_\mathbb{S}(x_1, x_2) \leq \pi\eta\sqrt{n+1},$$

and similarly, $d_\mathbb{S}(x_2, \zeta_2) < \pi\eta\sqrt{n+1}$. But then,

$$d_\mathbb{S}(\zeta_1, \zeta_2) \leq d_\mathbb{S}(\zeta_1, x_1) + d_\mathbb{S}(x_1, x_2) + d_\mathbb{S}(x_2, \zeta_2) < 3\pi\eta\sqrt{n+1} \leq \frac{2u_*}{D^{3/2}\kappa(f)}.$$

In particular, $d_\mathbb{S}(\zeta_1, \zeta_2) < \frac{u_*}{\sqrt{2}} \leq \delta_*$, since $\kappa(f) \geq 1$. These two inequalities are in contradiction with Lemma 19.25 for $r = r_*$. $\qquad\square$

Lemma 19.27 *Let $x \in \mathbb{S}^n$ be such that $x \notin A(f)$. Suppose $\eta \leq \frac{\alpha_0}{(n+1)D^2\kappa(f)^2}$. Then $\|f(x)\| > \frac{\pi}{4}\eta\sqrt{(n+1)D}\|f\|$.*

Proof Since $x \notin A(f)$, we have $\overline{\alpha}(f, x) \geq \alpha_0$. Also, $\kappa(f) = v(f) \geq v(f, x)$. This implies, by (19.2),

$$\kappa(f)^{-2} \leq 2\max\left\{\mu_{\text{norm}}(f, x)^{-2}, \frac{\|f(x)\|^2}{\|f\|^2}\right\}.$$

We accordingly divide the proof into two cases.

Assume firstly that $\max\{\mu_{\text{norm}}(f, x)^{-2}, \frac{\|f(x)\|^2}{\|f\|^2}\} = \frac{\|f(x)\|^2}{\|f\|^2}$.
In this case

$$\eta \leq \frac{\alpha_0}{(n+1)D^2\kappa(f)^2} \leq \frac{2\alpha_0\|f(x)\|^2}{(n+1)D^2\|f\|^2},$$

which implies

$$\|f(x)\| \geq \frac{\sqrt{\eta}\sqrt{n+1}D\|f\|}{\sqrt{2\alpha_0}} > \frac{\pi}{4}\eta\sqrt{(n+1)D}\|f\|,$$

the second inequality since $\eta \leq \frac{1}{2} < \frac{8D}{\pi^2\alpha_0}$.

Now assume instead that $\max\{\mu_{\text{norm}}(f, x)^{-2}, \frac{\|f(x)\|^2}{\|f\|^2}\} = \mu_{\text{norm}}(f, x)^{-2}$.
In this case

$$\eta \leq \frac{\alpha_0}{(n+1)D^2\kappa(f)^2} \leq \frac{2\alpha_0}{(n+1)D^2\mu_{\text{norm}}(f, x)^2},$$

which implies $\alpha_0 \geq \frac{1}{2}\eta(n+1)D^2\mu_{\text{norm}}(f,x)^2$. Also,

$$\alpha_0 \leq \overline{\alpha}(f,x) = \frac{1}{2}\beta(f,x)\mu_{\text{norm}}(f,x)D^{3/2} \leq \frac{1}{2\|f\|}\mu_{\text{norm}}(f,x)^2 D^{3/2}\|f(x)\|.$$

Putting both inequalities together, we obtain

$$\frac{1}{2}\eta(n+1)D^2\mu_{\text{norm}}(f,x)^2 \leq \frac{1}{2\|f\|}\mu_{\text{norm}}(f,x)^2 D^{3/2}\|f(x)\|,$$

which implies

$$\|f(x)\| \geq \eta(n+1)D^{1/2}\|f\| > \frac{\pi}{4}\eta\sqrt{(n+1)D}\|f\|. \qquad \square$$

We can now conclude the proof of part (b) of Theorem 19.1. Assume

$$\eta \leq \eta_0 := \min\left\{\frac{2u_*}{3\pi D^{3/2}\sqrt{n+1}\kappa(f)}, \frac{\alpha_0}{(n+1)D^2\kappa(f)^2}\right\}.$$

Then the hypotheses of Lemmas 19.26 and 19.27 hold. The first of these lemmas ensures that condition (a) in Algorithm 19.1 is satisfied, the second, that condition (b) is satisfied as well. Therefore, the algorithm halts as soon as $\eta \leq \eta_0$. This gives a bound of $\mathcal{O}(\log_2(nD\kappa(f)))$ for the number of iterations.

At each iteration there are $K := 2(n+1)(\frac{2}{\eta})^n$ points in the grid. For each such point x we evaluate $\mu_{\text{norm}}(f,x)$ and $\|f(x)\|$, both with cost $\mathcal{O}(N)$, by Proposition 16.45 and Lemma 16.31, respectively. We can therefore decide with cost $\mathcal{O}(KN)$ which of these points are vertices of G_η and for those points x compute the radius $2\overline{\beta}(f,x)$ of the ball $\overline{B}_f(x)$. Therefore, with cost $\mathcal{O}(K^2N)$ we can compute the edges of G_η. The number of connected components of G_η is then computed with $\mathcal{O}(K^2N)$ operations as well by standard algorithms in graph theory (see the Notes for references).

Since $d_{\mathbb{S}}$ is computed with $\mathcal{O}(n)$ operations, the total cost of verifying condition (a) is at most $\mathcal{O}(K^2n)$, and the additional cost of verifying (b) is $\mathcal{O}(K)$. It follows that the cost of each iteration is $\mathcal{O}(K^2N)$. Furthermore, since at these iterations $\eta \geq \eta_0$, we have $K \leq (\mathbf{C}(n+1)D^2\kappa(f)^2)^{n+1}$. Using this estimate in the $\mathcal{O}(K^2N)$ cost of each iteration and multiplying by the bound $\mathcal{O}(\log_2(nD\kappa(f)))$ for the number of iterations, the claimed bound for the total cost follows.

(c) To prove part (c) of Theorem 19.1 just note that for $i = 1, \ldots, r$, any vertex x_i of U_i is an approximate zero of the only zero of f in $W(U_i)$. $\qquad \square$

Remark 19.28 A finite-precision version of Algorithm 19.1 can be implemented as well. The running time remains the same (with α_0 replaced by a smaller universal constant α_*), and the returned value is $\#_{\mathbb{R}}(f)$ as long as the round-off unit satisfies

$$\epsilon_{\text{mach}} \leq \frac{1}{\mathcal{O}(D^2n^{5/2}\kappa(f)^3(\log_2 N + n^{3/2}D^2\kappa(f)^2))}.$$

19.5 On the Average Number of Real Zeros

The real solution variety $\hat{V}_{\mathbb{R}} \subseteq \mathcal{H}_{\mathbf{d}}^{\mathbb{R}} \times \mathbb{S}^n$ is defined in the obvious way, and so is $\hat{W}_{\mathbb{R}} \subseteq \mathcal{M}_{\mathbb{R}} \times \mathbb{S}^n$, where $\mathcal{M}_{\mathbb{R}} = \mathbb{R}^{n \times (n+1)}$. Let $\#_{\mathbb{R}}(q)$ denote the number of real zeros in $\mathbb{P}^n(\mathbb{R})$ of $q \in \mathcal{H}_{\mathbf{d}}^{\mathbb{R}}$. Thus the number of real zeros in the sphere $\mathbb{S}^n = \mathbb{S}(\mathbb{R}^{n+1})$ equals $2\#_{\mathbb{R}}(q)$. In what follows we denote the density of the standard Gaussian distribution on $\mathcal{H}_{\mathbf{d}}^{\mathbb{R}}$ by $\varphi_{\mathcal{H}_{\mathbf{d}}^{\mathbb{R}}}$.

Theorem 19.2 states that the expectation of $\#_{\mathbb{R}}$ equals the square root of the Bézout number $\mathcal{D}$. We now provide the proof.

Proof of Theorem 19.2 Applying the coarea formula (Theorem 17.8) to the projection $\pi_1 \colon \hat{V}_{\mathbb{R}} \to \mathcal{H}_{\mathbf{d}}^{\mathbb{R}}$ yields

$$\int_{\mathcal{H}_{\mathbf{d}}^{\mathbb{R}}} \#_{\mathbb{R}}\, \varphi_{\mathcal{H}_{\mathbf{d}}^{\mathbb{R}}}\, d\mathcal{H}_{\mathbf{d}}^{\mathbb{R}} = \int_{q \in \mathcal{H}_{\mathbf{d}}^{\mathbb{R}}} \varphi_{\mathcal{H}_{\mathbf{d}}^{\mathbb{R}}}(q) \frac{1}{2} \int_{\pi_1^{-1}(q)} d\pi_1^{-1}(q)\, d\mathcal{H}_{\mathbf{d}}^{\mathbb{R}}$$

$$= \int_{\hat{V}_{\mathbb{R}}} \frac{1}{2} \varphi_{\mathcal{H}_{\mathbf{d}}^{\mathbb{R}}} \mathrm{NJ}\pi_1\, d\hat{V}_{\mathbb{R}}.$$

We can factor the standard Gaussian $\varphi_{\mathcal{H}_{\mathbf{d}}}^{\mathbb{R}}$ into standard Gaussian densities φ_{C_ζ} and φ_{L_ζ} on C_ζ and L_ζ, respectively, as was done in (18.2) over $\mathbb{C}$ (denoting them by the same symbol will not cause any confusion). We also have an isometry $W_\zeta \to L_\zeta$ as in (18.3), and φ_{L_ζ} induces the standard Gaussian density φ_{W_ζ} on W_ζ. The fiber of $\Phi_{\mathbb{R}} \colon \hat{V}_{\mathbb{R}} \to \hat{W}_{\mathbb{R}}$, $(q, \zeta) \mapsto (N, \zeta)$, over (N, ζ) has the form $\Phi_{\mathbb{R}}^{-1}(N, \zeta) = \{(g_{M,\zeta} + h, \zeta) \mid h \in R_\zeta\}$, where $M = \Delta^{-1}N$; cf. (18.18). We therefore have $\varphi_{\mathcal{H}_{\mathbf{d}}^{\mathbb{R}}}(g_{M,\zeta} + h) = \varphi_{C_\zeta}(0)\, \varphi_{W_\zeta}(M)\, \varphi_{R_\zeta}(h)$.

Remark 18.7 states that the normal Jacobian of the map

$$\Phi_{\mathbb{R}} \colon \hat{V}_{\mathbb{R}} \to \hat{W}_{\mathbb{R}}, \qquad (q, \zeta) \mapsto \big(Dq(\zeta), \zeta\big),$$

has the constant value $\mathcal{D}^{n/2}$. The coarea formula applied to $\Phi_{\mathbb{R}}$, using Lemma 17.13, yields

$$\int_{\hat{V}_{\mathbb{R}}} \frac{1}{2} \varphi_{\mathcal{H}_{\mathbf{d}}^{\mathbb{R}}} \mathrm{NJ}\pi_1\, d\hat{V}_{\mathbb{R}}$$

$$= \frac{1}{2\,\mathrm{NJ}\Phi_{\mathbb{R}}} \int_{(N,\zeta) \in \hat{W}_{\mathbb{R}}} \varphi_{C_\zeta}(0)\, \varphi_{W_\zeta}(M)\, \mathrm{NJ}p_1(N, \zeta) \int_{h \in R_\zeta} \varphi_{R_\zeta}(h)\, dR_\zeta\, d\hat{W}_{\mathbb{R}}$$

$$= \frac{1}{2\,\mathrm{NJ}\Phi_{\mathbb{R}}} \int_{(N,\zeta) \in \hat{W}_{\mathbb{R}}} \varphi_{C_\zeta}(0)\, \varphi_{W_\zeta}(M)\, \mathrm{NJ}p_1(N, \zeta)\, d\hat{W}_{\mathbb{R}}.$$

Applying the coarea formula to the projection $p_1 \colon \hat{W}_{\mathbb{R}} \to \mathcal{M}_{\mathbb{R}}$, we can simplify the above to

$$\frac{1}{\mathrm{NJ}\Phi_{\mathbb{R}}} \int_{N \in \mathcal{M}_{\mathbb{R}}} \varphi_{C_\zeta}(0)\, \varphi_{W_\zeta}(M) \frac{1}{2} \int_{\zeta \in p_1^{-1}(N)} dp_1^{-1}(N)\, d\mathcal{M}_{\mathbb{R}}$$

$$= \frac{1}{\mathrm{NJ}\Phi_{\mathbb{R}}} \int_{N \in \mathscr{M}_{\mathbb{R}}} \varphi_{C_\zeta}(0)\, \varphi_{W_\zeta}(M)\, d\mathscr{M}_{\mathbb{R}}$$

$$= \frac{\mathcal{D}^{\frac{n+1}{2}}}{\mathrm{NJ}\Phi_{\mathbb{R}}} \int_{M \in \mathscr{M}_{\mathbb{R}}} \varphi_{C_\zeta}(0)\, \rho_{W_\zeta}(M)\, d\mathscr{M}_{\mathbb{R}},$$

where the last equality is due to the change of variables $\mathscr{M}_{\mathbb{R}} \to \mathscr{M}_{\mathbb{R}}$, $N \mapsto M$, which has Jacobian determinant $\mathcal{D}^{-\frac{n+1}{2}}$. Now we note that

$$\rho_{C_\zeta}(0) \cdot \rho_{W_\zeta}(M) = (2\pi)^{-n/2}\, (2\pi)^{-n^2/2}\, \exp\left(-\frac{1}{2}\|M\|_F^2\right)$$

is the density of the standard Gaussian distribution on $\mathscr{M}_{\mathbb{R}} \simeq \mathbb{R}^{n \times (n+1)}$, so that the last integral (over $M \in \mathscr{M}_{\mathbb{R}}$) equals one. Altogether, we obtain, using $\mathrm{NJ}\Phi_{\mathbb{R}} = \mathcal{D}^{n/2}$,

$$\int_{\mathcal{H}_{\mathbf{d}}^{\mathbb{R}}} \#_{\mathbb{R}}\, \varphi_{\mathcal{H}_{\mathbf{d}}^{\mathbb{R}}}\, d\mathcal{H}_{\mathbf{d}}^{\mathbb{R}} = \frac{\mathcal{D}^{\frac{n+1}{2}}}{\mathrm{NJ}\Phi_{\mathbb{R}}} = \sqrt{\mathcal{D}}. \tag{19.11}$$

This finishes the proof.   □

19.6 Feasibility of Underdetermined and Semialgebraic Systems

The grid method used in Algorithm 19.1 can be put to use as well to decide feasibility of underdetermined systems. For $m \leq n$ we denote by $\mathcal{H}_{\mathbf{d}}^{\mathbb{R}}[m]$ the linear space of systems $f = (f_1, \ldots, f_m)$ of m homogeneous polynomials in $n + 1$ variables.

We want to decide whether a system $f \in \mathcal{H}_{\mathbf{d}}^{\mathbb{R}}[m]$ is feasible, that is, whether there exists $x \in \mathbb{P}^n$ (or equivalently, $x \in \mathbb{S}^n$) such that $f(x) = 0$. In the complex setting this would always be the case. Over the reals, it does not need to be so; for instance, the polynomial $X_0^2 + X_1^2 + X_2^2$ has no zeros on $\mathbb{S}^n$ and this is also true for any small perturbation of it.

A first observation on our way towards an algorithm for this problem is that the projective Newton's method cannot be used in this context. But it turns out that a slightly different form of this method works. For $f \in \mathcal{H}_{\mathbf{d}}^{\mathbb{R}}[m]$ and $x \in \mathbb{R}^{n+1}$ such that $Df(x)$ is surjective define

$$MP_f(x) := x - Df(x)^\dagger f(x).$$

This *Moore–Penrose Newton's iteration* satisfies the basic property of Newton's method, namely, that if we start at a point x close enough to a simple zero ζ of f, the sequence of iterates converges to ζ immediately, quadratically fast. In particular, we can define approximate zeros as in Definition 16.34. Furthermore, versions $\gamma_\dagger$, $\beta_\dagger$, and $\alpha_\dagger$ of γ_{proj}, β_{proj}, and α_{proj}, respectively, are defined in the obvious manner, as well as the natural extension

$$\mu_\dagger(f, x) := \|f\| \cdot \left\| Df(x)^\dagger \mathrm{diag}\left(\sqrt{d_i}\,\|x\|^{d_i - 1}\right) \right\|$$

of μ_{norm} to this context. The main results we proved for the projective versions of these quantities in Chapter 16 can be extended to their Moore–Penrose counterparts. In particular, the following Moore–Penrose α-theorem holds.

Theorem 19.29 *There exists a universal positive constant α_* such that if $\alpha_\dagger(f, x) \leq \alpha_*$, then x is an approximate zero of f.* ☐

Furthermore, if we define $\overline{\alpha_\dagger}(f, x) := \frac{D^{3/2}}{2}\mu_\dagger^2(f, x)\frac{\|f(x)\|}{\|f\|}$, the bound $\alpha_\dagger(f, x) \leq \overline{\alpha_\dagger}(f, x)$ holds as well, so that the computation of the bound $\overline{\alpha_\dagger}(f, x)$ for $\alpha_\dagger(f, x)$ reduces to that of $\mu_\dagger(f, x)$.

We also have the following counterpart of Corollary 16.54.

Proposition 19.30 *There exist constants $C, \overline{\varepsilon} > 0$ such that the following is true. For all $\varepsilon \in [0, \overline{\varepsilon}]$, all $f \in \mathcal{H}_{\mathbf{d}}^{\mathbb{R}}[m]$, and all $x, y \in \mathbb{S}^n$, if $D^{3/2}\mu_\dagger(f, y)d_{\mathbb{S}}(x, y) \leq C\varepsilon$, then*

$$\frac{1}{1+\varepsilon}\mu_\dagger(f, x) \leq \mu_\dagger(f, y) \leq (1+\varepsilon)\mu_\dagger(f, x).$$ ☐

The constants $\alpha_*, \overline{\varepsilon}$, and C in Theorem 19.29 and Proposition 19.30 may be different from those occurring in Theorem 19.9 and Corollary 16.54, but the methods of proof are the same (and some proofs may become simpler; cf. Remark 19.19(b)). We therefore omit these proofs.

The algorithm deciding feasibility is the following (recall Algorithm 19.1 for the notation; see below for the meaning of $\kappa_{\text{feas}}(f)$).

Algorithm 19.2 Underdetermined_Feasibility

Input: $f \in \mathcal{H}_{\mathbf{d}}^{\mathbb{R}}[m]$
Preconditions: $f_1, \ldots, f_m \neq 0$

```
let η := ½
repeat
      if α̅₊(f, x) ≤ α₀ for some x ∈ 𝒰η
         then return "feasible" and halt
      if ‖f(x)‖ > π/2 η √((n + 1)D)‖f‖ for all x ∈ 𝒰η
         then return "infeasible" and halt
      η := η/2
```

Output: a tag in {feasible, infeasible}
Postconditions: The algorithm halts if $\kappa_{\text{feas}}(f) < \infty$. In this case the tag is feasible iff f has a zero in $\mathbb{P}(\mathbb{R}^{n+1})$.

To analyze this algorithm we need a notion of condition for the input system. For $f \in \mathcal{H}_{\mathbf{d}}^{\mathbb{R}}[m]$ we define

$$
\kappa_{\text{feas}}(f) = \begin{cases} \min_{\zeta \in Z_{\mathbb{S}}(f)} \mu_{\dag}(f, \zeta) & \text{if } Z_{\mathbb{S}}(f) \neq \emptyset, \\ \max_{x \in \mathbb{S}^n} \frac{\|f\|}{\|f(x)\|} & \text{otherwise.} \end{cases}
$$

We call f *well-posed* when $\kappa_{\text{feas}}(f) < \infty$. Note that $\kappa_{\text{feas}}(f) = \infty$ if and only if f is feasible and all its zeros are multiple.

For feasible systems f the condition number κ_{feas} is reminiscent of the GCC condition number $\mathscr{C}$. In both cases, condition is defined in terms of the best-conditioned solution (recall the discussion in Sect. 6.8). The absence of a "dual" for the feasibility problem of real polynomial systems forces a different approach for the condition in the infeasible case.

Theorem 19.31 *Algorithm 19.2 works correctly: with input a well-posed system it returns "feasible" (resp. "infeasible") if and only if the system is so. The number of iterations is bounded by $\mathcal{O}(\log_2(Dn\kappa_{\text{feas}}(f)))$.*

Proof The correctness in the feasible case is a trivial consequence of Theorem 19.29 and the inequality $\alpha_{\dag}(f, x) \leq \overline{\alpha_{\dag}}(f, x)$. The correctness in the infeasible case follows from Lemma 19.22 along with the inequalities (19.8).

To see the complexity bound, assume first that f is feasible and let ζ in the cube C^n, $\zeta \in Z(f)$, be such that $\kappa_{\text{feas}}(f) = \mu_{\dag}(f, \zeta)$. Let k be such that

$$
\eta = 2^{-k} \leq \frac{\min\{4\alpha_*, 2C\overline{\varepsilon}\}}{\pi D^2 \sqrt{n+1}\, \kappa_{\text{feas}}^2(f)}.
$$

Here C and $\overline{\varepsilon}$ are the constants in Proposition 19.30. Let $x \in \mathcal{U}_\eta$ be such that $\|x - \zeta\|_\infty \leq \eta$. Then, by (19.8),

$$
d_{\mathbb{S}}(x, \zeta) \leq \frac{\min\{2\alpha_*, C\overline{\varepsilon}\}}{D^2 \kappa_{\text{feas}}^2(f)}.
$$

Proposition 19.30 applies, and we have

$$
\mu_{\dag}(f, x) \leq (1 + \overline{\varepsilon})\mu_{\dag}(f, \zeta) = (1 + \overline{\varepsilon})\kappa_{\text{feas}}(f). \tag{19.12}
$$

Also, by Lemma 19.22,

$$
\|f(x)\| \leq \|f\| \sqrt{D}\, d_{\mathbb{S}}(x, \zeta) \leq \|f\| \frac{2\alpha_*}{D^{3/2} \kappa_{\text{feas}}^2(f)}.
$$

We then have

$$
\overline{\alpha_{\dag}}(f, x) = \frac{D^{3/2}}{2} \mu_{\dag}^2(f, x) \frac{\|f(x)\|}{\|f\|} \leq \frac{D^{3/2}}{2} \kappa_{\text{feas}}^2(f) \frac{2\alpha_*}{D^{3/2} \kappa_{\text{feas}}^2(f)} = \alpha_*.
$$

It follows that Algorithm 19.2 halts at this point, and therefore the number k of iterations performed is at most $\mathcal{O}(\log_2(Dn\kappa_{\text{feas}}(f)))$.

Assume finally that f is infeasible and let k be such that

$$\eta = 2^{-k} < \frac{2}{\pi \sqrt{(n+1)D}\,\kappa_{\text{feas}}(f)}.$$

Then, at any point $y \in \mathcal{U}_\eta$ we have

$$\|f(x)\| \geq \frac{\|f\|}{\kappa_{\text{feas}}(f)} > \frac{\pi}{2}\eta\sqrt{(n+1)D}\|f\|.$$

Again, Algorithm 19.2 halts for this value of η, and the number k of iterations performed is also bounded by $\mathcal{O}(\log_2(Dn\kappa_{\text{feas}}(f)))$. □

Remark 19.32 We finish this section by noting that the ideas above can be used to further decide feasibility of *semialgebraic systems*. These are systems of the form

$$\begin{aligned} f_i(x) &= 0, & i &= 1, \ldots, s, \\ g_i(x) &\geq 0, & i &= s+1, \ldots, t, \\ h_i(x) &> 0, & i &= t+1, \ldots, m, \end{aligned}$$

with $f_i, g_i, h_i \in \mathbb{R}[X_1, \ldots, X_n]$. A solution for such a system is a point $x \in \mathbb{R}^n$ satisfying the equalities and inequalities above, and we say that the system is feasible when solutions for it exist. Details of an algorithm deciding feasibility of semialgebraic systems and its analysis in terms of a condition number close to κ_{feas} are in [70].

Chapter 20
Probabilistic Analysis of Conic Condition Numbers: I. The Complex Case

The smoothed analysis of condition numbers in the preceding chapters was done on a case-by-case basis. For each considered condition number we proved a result giving bounds on either expectation or probability tails or both. In this chapter and the next we proceed differently—the theme of both chapters is the same, but the focus of this is on problems over $\mathbb{C}$, while the focus on the next is on problems over $\mathbb{R}$. We will consider a reasonably large class of condition numbers and obtain smoothed analysis estimates for elements in this class depending only on geometric invariants of the corresponding sets of ill-posed inputs.

This class is a subclass of the condition numbers à la Renegar introduced in Intermezzo II. To be precise, assume that $\Sigma \neq \{0\}$ is an *algebraic cone* included in the data space $\mathbb{C}^{p+1}$, i.e., a Zariski closed subset that is closed under multiplication by complex scalars. We call a function $\mathscr{C} \colon \mathbb{C}^{p+1} \setminus \{0\} \to \mathbb{R}$ a *conic condition number* when it has the form

$$\mathscr{C}(a) = \frac{\|a\|}{d(a, \Sigma)},$$

where the norm and distance d in the quotient above are those induced by the standard Hermitian product on $\mathbb{C}^{p+1}$. We call Σ the set of *ill-posed* inputs for $\mathscr{C}$.

The fact that Σ is a cone implies that for all $a \in \mathbb{C}^{p+1}$ and all $\lambda \in \mathbb{C}_*$, we have $\mathscr{C}(a) = \mathscr{C}(\lambda a)$. Hence, we may restrict attention to data $a \in \mathbb{P}^p := \mathbb{P}(\mathbb{C}^{p+1})$ in complex projective space for which the condition number takes the form

$$\mathscr{C}(a) = \frac{1}{d_{\mathsf{sin}}(a, \Sigma)}, \tag{20.1}$$

where abusing notation, Σ is interpreted now as a subset of $\mathbb{P}^p$ and $d_{\mathsf{sin}} = \sin d_{\mathbb{P}}$ denotes the sine distance in $\mathbb{P}^p$ (cf. Fig. 20.1).

Since $\mathbb{P}^p$ is a Riemannian manifold (cf. Sect. 14.2), we have a well-defined volume measure on it. The total volume of $\mathbb{P}^p$ for this measure is finite (recall Example 17.9). Hence, it makes sense to talk about the uniform probability distribution on the closed ball $B(\bar{a}, \sigma)$ of radius σ around $\bar{a} \in \mathbb{P}^p$ with respect to d_{sin}. So it is

P. Bürgisser, F. Cucker, *Condition*,
Grundlehren der mathematischen Wissenschaften 349,
DOI 10.1007/978-3-642-38896-5_20, © Springer-Verlag Berlin Heidelberg 2013

Fig. 20.1 Three distances

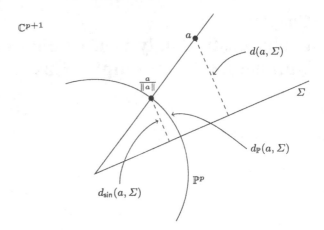

natural to ask for a smoothed analysis of $\mathscr{C}$ whereby a random perturbation a of $\bar{a} \in \mathbb{P}^p$ is modeled by $a \in B(\bar{a}, \sigma)$ chosen from the uniform distribution on $B(\bar{a}, \sigma)$.

Because Σ is a projective variety, it has associated with it a number of geometric invariants, notably a (complex) dimension $m = \dim_{\mathbb{C}} \Sigma$ and a *degree* $d = \deg \Sigma$ (see Sect. A.3.4 for the definition). It is remarkable that a smoothed analysis of $\mathscr{C}$ can be done in terms of these invariants only.

Our main result holds for any conic condition number for which the set of ill-posed inputs Σ is purely dimensional, that is, all of its irreducible components have the same dimension.

Theorem 20.1 *Let $\mathscr{C}$ be a conic condition number with set of ill-posed inputs $\Sigma \subseteq \mathbb{P}^p$, of pure dimension m, $0 < m < p$. Let $K(p, m) := 2 \frac{p^{3p}}{m^{3m}(p-m)^{3(p-m)}}$. Then, for all $\bar{a} \in \mathbb{P}^p$, all $\sigma \in (0, 1]$, and all $t \geq \frac{p\sqrt{2m}}{p-m}$, we have*

$$\operatorname*{Prob}_{a \in B(\bar{a},\sigma)} \{\mathscr{C}(a) \geq t\} \leq K(p, m) \deg \Sigma \left(\frac{1}{t\sigma}\right)^{2(p-m)} \left(1 + \frac{p}{p-m}\frac{1}{t\sigma}\right)^{2m},$$

as well as

$$\operatorname*{\mathbb{E}}_{a \in B(\bar{a},\sigma)} (\mathscr{C}(a)) \leq 2e K(p, m)^{\frac{1}{2(p-m)}} (\deg \Sigma)^{\frac{1}{2(p-m)}} \frac{pm}{p-m}\frac{1}{\sigma}$$

and

$$\operatorname*{\mathbb{E}}_{a \in B(\bar{a},\sigma)} \left(\log_\beta \mathscr{C}(a)\right)$$

$$\leq \frac{1}{2(p-m)} \left(\log_\beta K(p, m) + \log_\beta \deg \Sigma + 3\log_\beta e\right) + \log_\beta \frac{pm}{p-m} + \log_\beta \frac{1}{\sigma}.$$

Taking $\sigma = 1$, one obtains an average-case analysis.

We will devote Sect. 20.6 to deriving applications of Theorem 20.1 to a few condition numbers, some of which we have already encountered in the preceding chapters.

In most of our applications, the set of ill-posed inputs Σ is a hypersurface. That is, Σ is the zero set $Z_{\mathbb{P}}(f)$ of a nonzero homogeneous polynomial f, and thus $\deg \Sigma$ is at most the degree of f. In this case, we have the following corollary.

Corollary 20.2 *Let $\mathscr{C}$ be a conic condition number with set of ill-posed inputs $\Sigma \subseteq \mathbb{P}^p$. Assume $\Sigma \subseteq Z_{\mathbb{P}}(f)$ with $f \in \mathbb{C}[X_0, \ldots, X_p]$ homogeneous of degree d. Then, for all $\bar{a} \in \mathbb{P}^p$, all $\sigma \in (0, 1]$, and all $t \geq \sqrt{2}\, p^{3/2}$,*

$$\Prob_{a \in B(\bar{a},\sigma)} \{\mathscr{C}(a) \geq t\} \leq 2e^3 p^3 d \left(\frac{1}{t\sigma}\right)^2 \left(1 + p\frac{1}{t\sigma}\right)^{2(p-1)}$$

as well as

$$\E_{a \in B(\bar{a},\sigma)} (\mathscr{C}(a)) \leq 4(2e^5)^{\frac{1}{2}} p^{\frac{3}{2}} d^{\frac{1}{2}} \frac{1}{\sigma}$$

and

$$\E_{a \in B(\bar{a},\sigma)} (\log_\beta \mathscr{C}(a)) \leq \frac{3}{2} \log_\beta p + \frac{1}{2} \log_\beta d + \log_\beta \frac{1}{\sigma} + 3 \log_\beta e + \frac{3}{2}.$$

Taking $\sigma = 1$, one obtains an average case analysis.

Remark 20.3 The results above have the beauty of generality. We pay for this beauty with a loss of sharpness. A comparison of the bounds obtained as a consequence of these results with bounds obtained with ad hoc arguments reveals this loss (compare, e.g., the bounds in Sect. 20.6.1 for the condition number $\kappa_F(A)$ with those obtained in Sect. 4.4 for $\kappa(A)$; see also Remark 20.20 at the end of Sect. 20.6.3).

20.1   The Basic Idea

The basic idea towards the proof of Theorem 20.1 is not new to us. We can trace it back to the proof of Theorem 2.39—where we performed a smoothed analysis for the simple example of Sect. O.4—and we find it again at several points in the previous chapters. It consists in reformulating the probability distribution of a conic condition number in terms of a quotient of volumes.

In Sect. 2.2.6 we already introduced caps and tubes in spheres in very special situations and obtained both exact formulas and estimates for the volumes of these sets. We now need to extend these results and to replace the sphere as ambient space by a complex projective space. We start with the obvious definitions.

The volume of a measurable subset $A \subseteq \mathbb{P}^p$ is given by $\operatorname{vol} A = \int_A d\mathbb{P}^p$, where $d\mathbb{P}^p$ denotes the volume form induced by the Riemannian metric on $\mathbb{P}^p$. For $\bar{a} \in$

$\mathbb{P}^p$ and $\sigma \in [0, 1]$ we denote by $B(\overline{a}, \sigma) := \{a \in \mathbb{P}^p \mid d_{\sin}(a, \overline{a}) \leq \sigma\}$ the closed ball of radius σ around $\overline{a}$ in $\mathbb{P}^p$ with respect to the metric $d_{\sin} = \sin d_{\mathbb{P}}$ introduced in Sect. 14.2.2. For a nonempty subset $V \subseteq \mathbb{P}^p$ and $0 \leq \varepsilon \leq 1$ we define the ε-*neighborhood* around U in $\mathbb{P}^p$ as

$$T(U, \varepsilon) := \{x \in \mathbb{P}^p \mid d_{\sin}(x, U) \leq \varepsilon\},$$

where as usual, $d_{\sin}(x, U) := \inf\{d_{\sin}(x, y) \mid y \in U\}$. With this notation, we have

$$\Prob_{a \in B(\overline{a}, \sigma)} \{\mathscr{C}(a) \geq \varepsilon^{-1}\} = \Prob_{a \in B(\overline{a}, \sigma)} \{d_{\sin}(a, \Sigma) \leq \varepsilon\} = \frac{\text{vol}(T(\Sigma, \varepsilon) \cap B(a, \sigma))}{\text{vol}(B(a, \sigma))}.$$

The first claim in Theorem 20.1 will thus follow from the following purely geometric statement.

Theorem 20.4 *Let V be a projective variety in $\mathbb{P}^p$ of pure dimension m, $0 < m < p$. Moreover, let $\overline{a} \in \mathbb{P}^p$, $\sigma \in (0, 1]$, and $0 < \varepsilon \leq \frac{1}{\sqrt{2m}} \frac{p-m}{p}$. Then we have*

$$\frac{\text{vol}(T(V, \varepsilon) \cap B(\overline{a}, \sigma))}{\text{vol}\, B(\overline{a}, \sigma)} \leq K(p, m) \deg V \left(\frac{\varepsilon}{\sigma}\right)^{2(p-m)} \left(1 + \frac{p}{p-m} \frac{\varepsilon}{\sigma}\right)^{2m},$$

where $K(p, m)$ is defined as in Theorem 20.1.

As a first goal towards the proof of Theorem 20.4 we shall first study the case that $V = \mathbb{P}^m$ is a projective linear subspace of $\mathbb{P}^p$.

20.2 Volume of Tubes Around Linear Subspaces

We first study the corresponding situation in a sphere $\mathbb{S}^p$. In Lemma 2.36 we determined the volume of the ε-neighborhood of a subsphere of $\mathbb{S}^p$ codimension one. We now generalize this result to subspheres of higher codimension. Thus we determine the volume $\mathcal{O}_{p,k}(\varepsilon)$ of the ε-neighborhood

$$T(\mathbb{S}^k, \varepsilon) := \{x \in \mathbb{S}^p \mid d_{\sin}(x, \mathbb{S}^k) \leq \varepsilon\}$$

of $\mathbb{S}^k := \{x \in \mathbb{S}^p \mid x_{k+1} = \cdots = x_p = 0\}$ embedded in $\mathbb{S}^p$. Recall that $\mathcal{O}_p$ denotes the p-dimensional volume of $\mathbb{S}^p$.

Lemma 20.5 *For $0 \leq k \leq p - 1$ and $0 < \varepsilon \leq 1$ we have*

$$\mathcal{O}_{p,k}(\varepsilon) := \text{vol}\, T(\mathbb{S}^k, \varepsilon) = \mathcal{O}_k \mathcal{O}_{p-1-k} \int_0^{\arcsin \varepsilon} (\cos \rho)^k (\sin \rho)^{p-1-k} \, d\rho.$$

Proof Let $U \subseteq \mathbb{R}^{p+1}$ denote the subspace of dimension $k+1$ given by $x_{k+1} = \cdots = x_p = 0$ and let $U^{\perp}$ be its orthogonal complement. The unit spheres of these spaces satisfy $\mathbb{S}(U) = \mathbb{S}^k$ and $\mathbb{S}(U^{\perp}) \simeq \mathbb{S}^{p-1-k}$. Consider the open subset $T_{\varepsilon} := \{x \in \mathbb{S}^p \mid 0 < d_{\sin}(x, \mathbb{S}^k) < \varepsilon\}$ of $\mathbb{S}^p$, which has the same volume as $T(\mathbb{S}^k, \varepsilon)$. Moreover, set $\alpha := \arcsin \varepsilon$. We claim that the map

$$\psi : \mathbb{S}(U) \times \mathbb{S}(U^{\perp}) \times (0, \alpha) \to T_{\varepsilon}, \quad (p, q, \rho) \mapsto x = p \cos \rho + q \sin \rho,$$

is a bijection. In order to see this, let $(p, q, \rho) \in \mathbb{S}(U) \times \mathbb{S}(U^{\perp}) \times (0, \alpha)$. Then $x - q \sin \rho = p \cos \rho$ equals the orthogonal projection of x onto U; hence $d_{\mathbb{S}}(x, \mathbb{S}(U)) = \rho$, and so $x \in T_{\varepsilon}$. Conversely, for given $x \in T_{\varepsilon}$, let x' be its orthogonal projection onto U. Then $x' \neq 0$ and $x' \neq x$, since x is not contained in $U \cup U^{\perp}$. Hence we can define $p := \frac{x'}{\|x'\|} \in \mathbb{S}(U)$ and $q := \frac{x-x'}{\|x-x'\|} \in \mathbb{S}(U^{\perp})$. Then $x = p \cos \rho + q \sin \rho$, where ρ is the angle between x and p.

In order to compute the derivative of ψ at the point (p, q, ρ), take a smooth curve $t \mapsto (p(t), q(t), \rho(t))$ in $\mathbb{S}(U) \times \mathbb{S}(U^{\perp}) \times (0, \alpha)$ passing through this point for $t = 0$ and differentiate $\gamma(t) := \psi(p(t), q(t), \rho(t))$ at $t = 0$. This yields

$$\dot{\gamma} = \dot{p} \cos \rho + \dot{q} \sin \rho + (-p \sin \rho + q \cos \rho) \dot{\rho}.$$

Moreover, note that we have the orthogonal decomposition

$$T_x \mathbb{S}^p = T_p \mathbb{S}(U) \oplus T_q \mathbb{S}(U^{\perp}) \oplus \mathbb{R}(-p \sin \rho + q \cos \rho).$$

It follows that the Jacobian of ψ is given by

$$J\psi(p, q, \rho) = \left| \det \begin{pmatrix} (\cos \rho)\mathbf{I}_k & & \\ & (\sin \rho)\mathbf{I}_{p-1-k} & \\ & & 1 \end{pmatrix} \right| = (\cos \rho)^k (\sin \rho)^{p-1-k}.$$

Hence, using the transformation formula (Theorem 2.1), we obtain

$$\text{vol}\, T_{\varepsilon} = \int_{T_{\varepsilon}} d\mathbb{S}^p = \int_{\mathbb{S}(U) \times \mathbb{S}(U^{\perp}) \times (0, \alpha)} (\cos \rho)^k (\sin \rho)^{p-1-k} \, d\rho$$

$$= \text{vol}\, \mathbb{S}(U)\, \text{vol}\, \mathbb{S}(U^{\perp}) \int_0^{\alpha} (\cos \rho)^k (\sin \rho)^{p-1-k} \, d\rho,$$

which completes the proof, as $\text{vol}\, \mathbb{S}(U) = \mathcal{O}_k$ and $\text{vol}\, \mathbb{S}(U^{\perp}) = \mathcal{O}_{p-1-k}$. $\qquad\square$

As a consequence we retrieve a formula for the volume of a spherical cap, a result we had already obtained in Lemma 2.31:

$$\text{vol}\, B(\overline{a}, \varepsilon) = \frac{1}{2} \text{vol}\, T(S^0, \varepsilon) = \frac{1}{2} \mathcal{O}_{p,0}(\varepsilon) = \mathcal{O}_{p-1} \int_0^{\alpha} (\sin \rho)^{p-1} \, d\rho.$$

Recall that Lemma 2.34 states the following bounds on $\text{vol}\, B(\overline{a}, \varepsilon)$:

$$\frac{1}{\sqrt{2\pi(p+1)}} \mathcal{O}_p \varepsilon^p \leq \text{vol}\, B(\overline{a}, \varepsilon) \leq \frac{1}{2} \mathcal{O}_p \varepsilon^p. \tag{20.2}$$

The next result provides upper bounds on $\mathcal{O}_{p,k}$ if $k > 0$.

Lemma 20.6 *For $0 < k \leq p - 1$ and $0 \leq \varepsilon \leq 1$ we have*

$$\mathcal{O}_{p,k}(\varepsilon) \leq \frac{1}{p-k}\, \mathcal{O}_k \mathcal{O}_{p-1-k}\, \varepsilon^{p-k}.$$

Moreover, equality holds if $k = 1$.

Proof Putting $u := \arcsin \rho$, we have

$$\int_0^\alpha (\cos \rho)^k (\sin \rho)^{p-1-k}\, d\rho \leq \int_0^\alpha (\cos \rho)(\sin \rho)^{p-1-k}\, d\rho$$

$$= \int_0^\varepsilon u^{p-1-k}\, du = \frac{\varepsilon^{p-k}}{p-k}.$$

In the case $k = 1$ the inequality is actually an equality. $\square$

Remark 20.7 Since $T(\mathbb{S}^k, 1) = \mathbb{S}^p$, we get from Lemma 20.5 the following formula:

$$\int_0^{\pi/2} (\cos \rho)^k (\sin \rho)^{p-1-k}\, d\rho = \frac{\mathcal{O}_p}{\mathcal{O}_k \mathcal{O}_{p-1-k}}. \tag{20.3}$$

We now extend the estimates above to complex projective space. Let us consider $\mathbb{P}^m \subseteq \mathbb{P}^p$ as the subset given by the equations $z_{m+1} = \cdots = z_p = 0$.

Lemma 20.8 *For $\mathbb{P}^m$ embedded in $\mathbb{P}^p$ and $0 < \varepsilon \leq 1$ we have*

$$\operatorname{vol} T\left(\mathbb{P}^m, \varepsilon\right) \leq \operatorname{vol} \mathbb{P}^m \operatorname{vol} \mathbb{P}^{p-m}\, \varepsilon^{2(p-m)}.$$

For the volume of a ball of radius ε around $\overline{a} \in \mathbb{P}^p$ we have

$$\operatorname{vol} B(\overline{a}, \varepsilon) = \operatorname{vol} \mathbb{P}^p\, \varepsilon^{2p}.$$

Proof By definition, $\mathbb{S}^{2m+1}$ equals the inverse image of the linear subspace $\mathbb{P}^m$ under the natural projection $\pi_{\mathbb{S}} \colon \mathbb{S}^{2p+1} \to \mathbb{P}^p$. Moreover, by Proposition 14.12 we have

$$\pi_{\mathbb{S}}^{-1}\left(T\left(\mathbb{P}^m, \varepsilon\right)\right) = T\left(\mathbb{S}^{2m+1}, \varepsilon\right).$$

Therefore, Eq. (17.8) implies

$$\operatorname{vol} T\left(\mathbb{P}^m, \varepsilon\right) = \frac{1}{2\pi} \operatorname{vol} T\left(\mathbb{S}^{2m+1}, \varepsilon\right) = \frac{1}{2\pi} \mathcal{O}_{2p+1.2m+1}(\varepsilon).$$

Now note that using (17.9),

$$\operatorname{vol} \mathbb{P}^{p-m} = \frac{\pi}{p-m} \operatorname{vol} \mathbb{P}^{p-m-1} = \frac{\pi}{p-m} \frac{1}{2\pi} \mathcal{O}_{2p-2m-1} = \frac{\mathcal{O}_{2p-2m-1}}{2p-2m}.$$

Using Lemma 20.6 and the above identity, we obtain

$$\frac{1}{2\pi}\mathcal{O}_{2p+1,2m+1}(\varepsilon) \le \frac{1}{2\pi}\mathcal{O}_{2m+1}\frac{\mathcal{O}_{2p-2m-1}}{2p-2m}\varepsilon^{2p-2m}$$

$$= \operatorname{vol}\mathbb{P}^m \operatorname{vol}\mathbb{P}^{p-m}\varepsilon^{2p-2m}.$$

This proves the first assertion. In the case $m = 0$, Lemma 20.5 actually gives an equality, so that

$$\operatorname{vol}T\left(\mathbb{P}^0,\varepsilon\right) = \operatorname{vol}\mathbb{P}^0 \operatorname{vol}\mathbb{P}^p\,\varepsilon^{2p}.$$

But $\operatorname{vol}\mathbb{P}^0 = 1$ and $\operatorname{vol}T(\mathbb{P}^0,\varepsilon) = \operatorname{vol}B(\overline{a},\varepsilon)$. $\qquad\qquad\square$

20.3   Volume of Algebraic Varieties

Let $V \subseteq \mathbb{P}^p$ be an irreducible m-dimensional subvariety. To goal of this section is to define a volume measure on V.

Assume first that V does not have singular points. Then V is a smooth submanifold of $\mathbb{P}^p$ of dimension $2m$. Moreover, V inherits a Riemannian metric from the ambient space $\mathbb{P}^p$. In particular, there is an associated volume element dV on V, which allows us to define the $2m$-dimensional volume $\operatorname{vol}_{2m} A := \int_A dV$ of a measurable subset $A \subseteq V$. Clearly, if V equals the projective linear subspace $\mathbb{P}^m$ of $\mathbb{P}^p$, then vol_{2m} coincides with the usual volume on $\mathbb{P}^m$.

Suppose now that V is singular. Then the set $\operatorname{Reg}(V)$ of regular points of V is a smooth submanifold of V with real dimension $2m$ (cf. Theorem A.33). Hence we have a well-defined volume measure vol_{2m} on $\operatorname{Reg}(V)$, which we extend to V by setting $\operatorname{vol}_{2m}(A) := \operatorname{vol}_{2m}(A \cap \operatorname{Reg}(V))$ for all measurable $A \subseteq V$. In particular, the set $\operatorname{Sing}(V) := V \setminus \operatorname{Reg}(V)$ of singular points satisfies $\operatorname{vol}_{2m}(\operatorname{Sing}(V)) = 0$ and can be neglected. This definition is motivated by the fact that $\operatorname{Sing}(V)$ is a projective subvariety of (complex) dimension strictly less than m.

We shall see shortly that $\operatorname{vol}_{2m} V$ is closely related to the degree $\deg V$ of the projective variety V. Recall from Sect. A.3.4 that for almost all projective linear subspaces $L \subseteq \mathbb{P}^p$ of (complex) dimension $p-m$, the intersection $V \cap L$ is finite and contains exactly $\deg V$ points. If we replace V by a Euclidean open subset U, then this assertion is not true anymore. Still, a quantitative statement in a probabilistic sense can be made. For this, we need to put a probability measure on the set of m-dimensional linear subspaces L of $\mathbb{P}^p$. Since any such L is obtained as the image $u\mathbb{P}^m$ of the fixed subspace $\mathbb{P}^m$ under some element u of the unitary group $\mathscr{U}(p+1)$, it suffices to define a probability measure on the latter.

For the following compare Sect. A.2.6. The group $\mathscr{U}(p + 1)$ is a compact Lie group. Indeed, it is a smooth submanifold of $\mathbb{C}^{(p+1)\times(p+1)}$, and hence it inherits from the ambient space a Riemannian metric with a corresponding volume element (cf. Sect. A.2.5). Normalizing the corresponding volume measure, we obtain the uniform probability measure on $\mathscr{U}(p + 1)$, which is referred to as the *normalized*

Haar measure. We shall denote it by vol. It is important that vol is invariant under the action of $\mathscr{U}(p+1)$ on itself: we have $\operatorname{vol} uB = \operatorname{vol} B$ for all measurable subsets $B \subseteq \mathscr{U}(p+1)$ and $u \in \mathscr{U}(p+1)$.

20.4   A Crash Course on Probability: V

Suppose that U is a measurable subset of an m-dimensional irreducible projective variety $V \subseteq \mathbb{P}^p$. From the definition of degree we know that for almost all $u \in \mathscr{U}(p+1)$, the intersection $V \cap u\mathbb{P}^{p-m}$ has exactly $\deg V$ points. In particular, $U \cap u\mathbb{P}^{p-m}$ is finite for almost all $u \in \mathscr{U}(p+1)$ and it makes sense to ask about the expectation of the random variable $\mathbb{E}\#(U \cap u\mathbb{P}^{p-m})$. A fundamental result in integral geometry (or geometric probability) of "Crofton type" provides a close link of this expectation to the m-dimensional volume of U. The proof of this result will be provided in Sect. A.4.1.

Theorem 20.9 *Let $V \subseteq \mathbb{P}^p$ be an m-dimensional irreducible projective variety and $U \subseteq V$ an open subset in the classical topology. Then we have*

$$\mathop{\mathbb{E}}_{u \in \mathscr{U}(p+1)} \#(U \cap u\mathbb{P}^{p-m}) = \frac{\operatorname{vol}_{2m} U}{\operatorname{vol}_{2m} \mathbb{P}^m}. \qquad \square$$

The following beautiful result is an immediate consequence of Theorem 20.9 (with $U = V$) and the characterization of degree.

Corollary 20.10 *For an m-dimensional irreducible projective variety $V \subseteq \mathbb{P}^p$ we have*

$$\operatorname{vol}_{2m} V = \deg V \operatorname{vol} \mathbb{P}^m = \deg V \frac{\pi^m}{m!}. \qquad \square$$

In essentially the same way we can find an upper bound on $\operatorname{vol}_{2m}(V \cap B(\bar{a}, \varepsilon))$. But first we need to prove a lemma.

Lemma 20.11 *Let $\bar{a} \in \mathbb{P}^p$ and consider the orbit map $\varphi \colon \mathscr{U}(p+1) \to \mathbb{P}^p$, $u \mapsto u\bar{a}$. Then the pushforward v of the normalized Haar measure on $\mathscr{U}(p+1)$ with respect to φ equals the uniform measure on $\mathbb{P}^p$.*

Proof First note that φ is equivariant under the action of the unitary group, that is, $\varphi(uv) = u\varphi(e)$ for all $u, v \in \mathscr{U}(p+1)$. Hence for a measurable subset $A \subseteq \mathbb{P}^p$ and $u \in \mathscr{U}(p+1)$ we have $\varphi^{-1}(uA) = u\varphi^{-1}(A)$. Since the normalized Haar measure on $\mathscr{U}(p+1)$ is invariant, we get

$$v(uA) = \operatorname{vol}(\varphi^{-1}(uA)) = \operatorname{vol}(u\varphi^{-1}(A)) = \operatorname{vol}(\varphi^{-1}(A)) = v(A).$$

On the other hand, φ is a surjective smooth map, so that the pushforward measure ν has a continuous density ρ; see Sect. 17.3. The invariance of ν implies that

$$\int_A \rho \, d\mathbb{P}^p = \nu(A) = \nu(uA) = \int_{uA} \rho \, d\mathbb{P}^p = \int_A \rho \circ u^{-1} \, d\mathbb{P}^p,$$

where the last equality is due to the fact that $u^{-1} \colon \mathbb{P}^p \to \mathbb{P}^p$, $a \mapsto u^{-1}a$, preserves the volume. Since the above equality holds for arbitrary A, we get $\rho = \rho \circ g^{-1}$. Hence ρ must be constant, and hence ν is the uniform measure on $\mathbb{P}^p$. $\square$

The following lemma is in the spirit of Corollary 20.10.

Lemma 20.12 *Let* $V \subseteq \mathbb{P}^p$ *be an irreducible* m*-dimensional variety,* $\bar{a} \in \mathbb{P}^p$, *and* $0 < \varepsilon \le 1$. *Then*

$$\frac{\mathrm{vol}_{2m}(V \cap B(\bar{a}, \varepsilon))}{\mathrm{vol}_{2m} \mathbb{P}^m} \le \binom{p}{m} \deg V \, \varepsilon^{2m}.$$

Proof Let U denote the interior of $V \cap B(\bar{a}, \varepsilon)$. According to Theorem 20.9, it is sufficient to show that $\mathbb{E}_{u \in \mathscr{U}(p+1)} \#(U \cap u\mathbb{P}^{p-m}) \le \binom{p}{m} \deg V \, \varepsilon^{2m}$, since we have $\mathrm{vol}_{2m} U = \mathrm{vol}_{2m}(V \cap B(\bar{a}, \varepsilon))$.

To estimate this expectation, note that

$$\underset{u \in \mathscr{U}(p+1)}{\mathbb{E}} \#\bigl(U \cap u\mathbb{P}^{p-m}\bigr) \le \deg V \underset{u \in \mathscr{U}(p+1)}{\mathrm{Prob}} \bigl\{U \cap u\mathbb{P}^{p-m} \ne \emptyset\bigr\},$$

since $\#(U \cap u\mathbb{P}^{p-m}) \le \#(V \cap u\mathbb{P}^{p-m}) \le \deg V$ for almost all $u \in \mathscr{U}(p+1)$. Moreover, since $U \subseteq B(\bar{a}, \varepsilon)$, we have

$$\underset{u \in \mathscr{U}(p+1)}{\mathrm{Prob}} \bigl\{U \cap u\mathbb{P}^{p-m} \ne \emptyset\bigr\} \le \underset{u \in \mathscr{U}(p+1)}{\mathrm{Prob}} \bigl\{B(\bar{a}, \varepsilon) \cap u\mathbb{P}^{p-m} \ne \emptyset\bigr\}$$

$$= \underset{u \in \mathscr{U}(p+1)}{\mathrm{Prob}} \bigl\{B(u^{-1}\bar{a}, \varepsilon) \cap \mathbb{P}^{p-m} \ne \emptyset\bigr\}$$

$$= \underset{x \in \mathbb{P}^p}{\mathrm{Prob}} \bigl\{B(x, \varepsilon) \cap \mathbb{P}^{p-m} \ne \emptyset\bigr\},$$

where the last equality is due to Lemma 20.11. By definition, $B(x, \varepsilon) \cap \mathbb{P}^{p-m} \ne \emptyset$ iff $x \in T(\mathbb{P}^{p-m}, \varepsilon)$, whence

$$\underset{x \in \mathbb{P}^p}{\mathrm{Prob}} \bigl\{B(x, \varepsilon) \cap \mathbb{P}^{p-m} \ne \emptyset\bigr\} = \frac{\mathrm{vol}\, T(\mathbb{P}^{p-m}, \varepsilon)}{\mathrm{vol}\, \mathbb{P}^p}.$$

We can bound the latter by Lemma 20.8, which yields

$$\frac{\mathrm{vol}\, T(\mathbb{P}^{p-m}, \varepsilon)}{\mathrm{vol}\, \mathbb{P}^p} \le \frac{\mathrm{vol}\, \mathbb{P}^{p-m}\, \mathrm{vol}\, \mathbb{P}^m}{\mathrm{vol}\, \mathbb{P}^p}\, \varepsilon^{2m} = \binom{p}{m} \varepsilon^{2m},$$

where we have used (17.9) for the last equality. $\square$

In Theorem 20.9 we intersected a subset U of an irreducible variety V with a random linear subspace of complementary dimension. We also need a variant of this theorem in which we intersect U with a ball $B(a, \varepsilon)$ of fixed radius ε and random center a. Both these results are special cases of Poincaré's formula (Theorem A.55), which is a fundamental result of spherical integral geometry that will be proven in Sect. A.4.1.

Theorem 20.13 Let $V \subseteq \mathbb{P}^p$ be an m-dimensional irreducible projective variety and $U \subseteq V$ an open subset in the classical topology. Then we have for $0 < \varepsilon \le 1$,

$$\operatorname*{\mathbb{E}}_{a \in \mathbb{P}^p} \left(\operatorname{vol}_{2m} \left(U \cap B(a, \varepsilon) \right) \right) = \varepsilon^{2p} \operatorname{vol}_{2m} U. \qquad \square$$

We also need a lower bound on $\operatorname{vol}_{2m}(V \cap B(\overline{a}, \varepsilon))$, which is provided in the following result. Its proof, which is a consequence of Wirtinger's inequality, is outlined in Sect. A.3.6.

Theorem 20.14 Let $V \subseteq \mathbb{P}^p$ be an m-dimensional irreducible projective variety and $\overline{a} \in V$, $0 < \varepsilon \le \frac{1}{\sqrt{m}}$. Then we have

$$\operatorname{vol}_{2m} \left(V \cap B(\overline{a}, \varepsilon) \right) \ge \varepsilon^{2m} \left(1 - m\varepsilon^2 \right) \operatorname{vol}_{2m} \mathbb{P}^m. \qquad \square$$

Remark 20.15 The assertions of Theorems 20.9 and 20.13 are not confined to $\mathbb{C}$ and hold in much more generality. However, the above Theorem 20.14 fails to be true over $\mathbb{R}$.

20.5 Proof of Theorem 20.1

Proof of Theorem 20.4 We first show that it suffices to prove the assertion for an irreducible V. Indeed, suppose that $V = V_1 \cup \cdots \cup V_s$ is the decomposition of V into its irreducible components V_i. We have $T(V, \varepsilon) = T(V_1, \varepsilon) \cup \cdots \cup T(V_s, \varepsilon)$. Moreover, the degree of V is defined as $\deg V = \sum_i \deg V_i$, since we assume that $\dim V_i = m$ for all i. Hence it is clear that the bounds (in the statement) for V_i imply the bound for V.

We therefore assume that V is irreducible and fix $\overline{a} \in \mathbb{P}^p$. We shall see that we can bound the $2p$-dimensional volume of $T(V, \varepsilon) \cap B(\overline{a}, \sigma)$ in terms of the $2m$-dimensional volume of the intersection $U := V \cap B(\overline{a}, \sigma + \varepsilon_1)$ of V with a ball of slightly larger radius $\sigma + \varepsilon_1$. Here $\varepsilon_1 \in (0, 1]$ is assumed to satisfy $0 < \varepsilon_1 - \varepsilon \le \frac{1}{\sqrt{2m}}$; its actual value will be specified later on.

We claim that

$$\inf_{z \in T(V, \varepsilon) \cap B(\overline{a}, \sigma)} \frac{\operatorname{vol}_{2m}(U \cap B(z, \varepsilon_1))}{\operatorname{vol}_{2m} \mathbb{P}^m} \ge \frac{1}{2}(\varepsilon_1 - \varepsilon)^{2m}. \qquad (20.4)$$

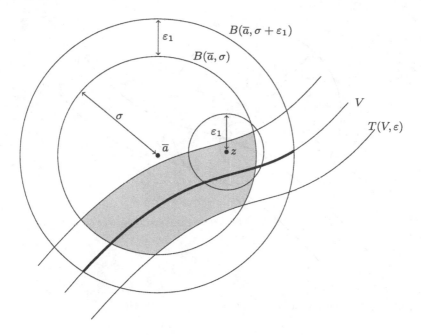

Fig. 20.2 The *thick curve* segment is $U := V \cap B(\overline{a}, \sigma + \varepsilon_1)$, and the *shaded region* is $T(V, \varepsilon) \cap B(\overline{a}, \sigma)$

In order to see this, let $z \in T(V, \varepsilon) \cap B(\overline{a}, \sigma)$. For $x \in V \cap B(z, \varepsilon_1)$ we have

$$d_{\sin}(x, \overline{a}) \leq d_{\sin}(x, z) + d_{\sin}(z, \overline{a}) \leq \varepsilon_1 + \sigma.$$

Hence $V \cap B(z, \varepsilon_1) \subseteq U$; see also Fig. 20.2.

Further, there exists $y \in V$ such that $d_{\sin}(z, y) \leq \varepsilon$, and hence for $x' \in B(y, \varepsilon_1 - \varepsilon)$ we have

$$d_{\sin}(x', z) \leq d_{\sin}(x', y) + d_{\sin}(y, z) \leq \varepsilon_1 - \varepsilon + \varepsilon = \varepsilon_1.$$

Therefore, $B(y, \varepsilon_1 - \varepsilon) \subseteq B(z, \varepsilon_1)$; see also Fig. 20.3. So we have the inclusions

$$V \cap B(y, \varepsilon_1 - \varepsilon) \subseteq V \cap B(z, \varepsilon_1) \subseteq U \cap B(z, \varepsilon_1).$$

Theorem 20.14 implies

$$\mathrm{vol}_{2m}\left(U \cap B(z, \varepsilon_1)\right) \geq \mathrm{vol}_{2m}\left(V \cap B(y, \varepsilon_1 - \varepsilon)\right) \geq \frac{1}{2}(\varepsilon_1 - \varepsilon)^{2m}\,\mathrm{vol}_{2m}\,\mathbb{P}^m,$$

since $1 - m(\varepsilon_1 - \varepsilon)^2 \geq \frac{1}{2}$, which proves the claim (20.4).

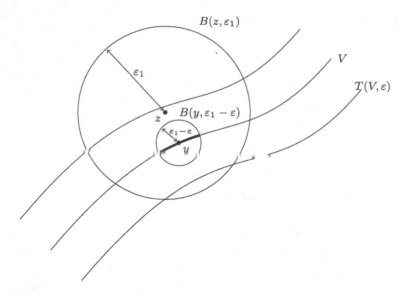

Fig. 20.3 The *thick curve* segment is $V \cap B(y, \varepsilon_1 - \varepsilon)$

Using (20.4) and Theorem 20.13 we now argue as follows:

$$\frac{1}{2}(\varepsilon_1 - \varepsilon)^{2m} \frac{\mathrm{vol}_{2p}(T(V, \varepsilon) \cap B(\bar{a}, \sigma))}{\mathrm{vol}_{2p} \mathbb{P}^p}$$

$$= \quad \frac{1}{\mathrm{vol}_{2p} \mathbb{P}^p} \int_{z \in T(V, \varepsilon) \cap B(\bar{a}, \sigma)} \frac{1}{2}(\varepsilon_1 - \varepsilon)^{2m} \, d\mathbb{P}^p(z)$$

$$\overset{(20.4)}{\leq} \quad \frac{1}{\mathrm{vol}_{2p} \mathbb{P}^p} \int_{z \in T(V, \varepsilon) \cap B(\bar{a}, \sigma)} \frac{\mathrm{vol}_{2m}(U \cap B(z, \varepsilon_1))}{\mathrm{vol}_{2m} \mathbb{P}^m} \, d\mathbb{P}^p(z)$$

$$\leq \quad \frac{1}{\mathrm{vol}_{2m} \mathbb{P}^m} \frac{1}{\mathrm{vol}_{2p} \mathbb{P}^p} \int_{z \in \mathbb{P}^p} \mathrm{vol}_{2m}\big(U \cap B(z, \varepsilon_1)\big) \, d\mathbb{P}^p(z)$$

$$\overset{\text{Theorem 20.13}}{=} \quad \frac{1}{\mathrm{vol}_{2m} \mathbb{P}^m} \varepsilon_1^{2m} \, \mathrm{vol}_{2m} U.$$

It follows that

$$\frac{\mathrm{vol}_{2p}(T(V, \varepsilon) \cap B(\bar{a}, \sigma))}{\mathrm{vol}_{2p} \mathbb{P}^p} \leq \frac{2\varepsilon_1^{2p}}{(\varepsilon_1 - \varepsilon)^{2m}} \cdot \frac{\mathrm{vol}_{2m} U}{\mathrm{vol}_{2m} \mathbb{P}^m}.$$

Lemma 20.12 tells us that

$$\frac{\mathrm{vol}_{2m} U}{\mathrm{vol}_{2m} \mathbb{P}^m} \leq \binom{p}{m} \deg V \, (\sigma + \varepsilon_1)^{2m}.$$

So we obtain

$$\frac{\mathrm{vol}_{2p}(T(V,\varepsilon)\cap B(\overline{a},\sigma))}{\mathrm{vol}_{2p}\,\mathbb{P}^p} \leq \frac{2\varepsilon_1^{2p}}{(\varepsilon_1-\varepsilon)^{2m}}\binom{p}{m}\deg V\,(\sigma+\varepsilon_1)^{2m}.$$

Using $\mathrm{vol}(B(\overline{a},\sigma))=\mathrm{vol}(\mathbb{P}^p)\sigma^{2p}$ (Lemma 20.8) it follows that

$$\frac{\mathrm{vol}_{2p}(T(V,\varepsilon)\cap B(\overline{a},\sigma))}{\mathrm{vol}_{2p}(B(\overline{a},\sigma))} \leq \frac{2}{(\varepsilon_1-\varepsilon)^{2m}}\left(\frac{\varepsilon_1}{\sigma}\right)^{2p}\binom{p}{m}\deg(V)\,(\sigma+\varepsilon_1)^{2m}.$$

We finally choose $\varepsilon_1:=\frac{p}{p-m}\varepsilon$. Then the required inequality

$$\varepsilon_1-\varepsilon=\frac{m}{p-m}\varepsilon\leq\frac{1}{\sqrt{2m}}$$

holds due to the assumption $\varepsilon\leq\frac{1}{\sqrt{2m}}\frac{p-m}{m}$. We obtain now

$$\frac{\mathrm{vol}_{2p}(T(V,\varepsilon)\cap B(\overline{a},\sigma))}{\mathrm{vol}_{2p}(B(\overline{a},\sigma))}$$

$$\leq 2\left(\frac{p-m}{m}\right)^{2m}\frac{1}{\varepsilon^{2m}}\left(\frac{p}{p-m}\right)^{2p}\left(\frac{\varepsilon}{\sigma}\right)^{2p}\binom{p}{m}\deg(V)\sigma^{2m}\left(1+\frac{p}{p-m}\frac{\varepsilon}{\sigma}\right)^{2m}$$

$$=\frac{2p^{2p}}{m^{2m}(p-m)^{2(p-m)}}\binom{p}{m}\deg(V)\left(\frac{\varepsilon}{\sigma}\right)^{2(p-m)}\left(1+\frac{p}{p-m}\frac{\varepsilon}{\sigma}\right)^{2m}.$$

Taking into account the estimate $\binom{p}{m}\leq\frac{p^p}{m^m(p-m)^{p-m}}$, which readily follows from the binomial expansion of $p^p=(m+(p-m))^p$, the assertion follows.   □

We remark for future use the following estimate from the end of the proof:

$$\frac{1}{2}K(p,m)\geq\binom{p}{m}^3. \tag{20.5}$$

Proof of Theorem 20.1 The inequality for the tail follows directly from Theorem 20.4. For the expectation estimates, let $\varepsilon_*:=\frac{p-m}{pm}\sigma$ and $t_*:=\varepsilon_*^{-1}$. Note that for $\varepsilon\leq\varepsilon_*$,

$$\left(1+\frac{p}{p-m}\frac{\varepsilon}{\sigma}\right)^{2m}\leq\left(1+\frac{1}{m}\right)^{2m}\leq e^2,$$

and thus

$$\frac{\mathrm{vol}(T(V,\varepsilon)\cap B(\overline{a},\sigma))}{\mathrm{vol}(B(\overline{a},\sigma))}\leq K(p,m)\deg\Sigma\left(\frac{\varepsilon}{\sigma}\right)^{2(p-m)}e^2.$$

Therefore, for all $t \geq t_*$, writing $\varepsilon = 1/t$,

$$\operatorname*{Prob}_{a \in B(\bar{a}, \sigma)} \{\mathscr{C}(a) \geq t\} \leq K(p, m) \deg \Sigma \, \frac{1}{\sigma^{2(p-m)}} e^2 \cdot t^{-2(p-m)}.$$

Now write $\alpha := 2(p - m) \geq 2$ and put

$$K := K(p, m) \deg \Sigma \, \frac{1}{\sigma^\alpha} e^2, \quad t_0 := K^{\frac{1}{\alpha}} \frac{pm}{p - m}.$$

Note that $t_0 \geq t_*$ (use that $K(p, m) \geq 1$ due to (20.5)). Then we have $\operatorname{Prob}_{a \in B(\bar{a}, \sigma)} \{\mathscr{C}(a) \geq t\} \leq K t^{-\alpha}$ for all $t \geq t_0$. Since in addition, $t_0^\alpha \geq K$, we can apply Proposition 2.26 to bound $\mathbb{E}_{z \in B(\bar{a}, \sigma)} (\log_\beta \mathscr{C}(z))$ from above by

$$\frac{1}{2(p - m)} \left(\log_\beta K(p, m) + \log_\beta \deg \Sigma + 3 \log_\beta e \right) + \log_\beta \frac{pm}{p - m} + \log_\beta \frac{1}{\sigma}.$$

For the expectation of $\mathscr{C}$ take again $\alpha = 2(p - m)$ and define the new value

$$K := K(p, m) \deg(\Sigma) e^2 \frac{1}{\sigma^\alpha} \left(\frac{pm}{p - m} \right)^\alpha$$

and set $B := t_*$. Then, $\operatorname{Prob}_{z \in B(\bar{a}, \sigma)} \{\mathscr{C}(z) \geq t\} \leq K t^{-\alpha}$, for all $t \geq B$, and $B \leq K^{\frac{1}{\alpha}}$. We can therefore use Proposition 2.27 with $k = 1$ to deduce

$$\operatorname*{\mathbb{E}}_{z \in B(\bar{a}, \sigma)} (\mathscr{C}(z)) \leq \frac{\alpha}{\alpha - 1} K^{\frac{1}{\alpha}}.$$

The claimed bound now follows from the fact that $\frac{2(p-m)}{2(p-m)-1} \leq 2$ and $e^{\frac{2}{\alpha}} \leq e$.   □

Proof of Corollary 20.2 Put $\Sigma' = Z_{\mathbb{P}}(f)$ and note that since $\Sigma \subseteq \Sigma'$, we have $\mathscr{C}(a) = \frac{1}{d_{\sin}(a, \Sigma)} \leq \frac{1}{d_{\sin}(a, \Sigma')}$. Moreover, $\deg \Sigma' \leq d$. The assertion then follows from Theorem 20.1 applied to Σ' and the inequality

$$K(p, p - 1) = 2 \frac{p^{3p}}{(p-1)^{3p-3}} = 2 \left[\left(1 + \frac{1}{p-1} \right)^{p-1} \right]^3 p^3 \leq 2 e^3 p^3.$$

The bounds on the expectations follow from the corresponding bounds in Theorem 20.1, using $p/(p - 1) \leq 2$ and $\log_\beta 8 \leq \log_2 8 = 3$.   □

20.6 Applications

20.6.1 Linear Equation-Solving

The first natural application of our results should be for the classical condition number $\kappa(A)$. We note, however, that $\kappa(A)$ is not conic, since in Corollary 1.8 both the norm $\|A\|_{rs}$ and the distance $d_{rs}(A, \Sigma)$ are induced by an operator norm, and this norm is not induced by a Hermitian product on $\mathbb{C}^{n \times n}$.

We can nevertheless consider, for $A \in \mathbb{C}^{n \times n}$, the *Frobenius condition number*

$$\kappa_F(A) := \|A\|_F \|A^{-1}\| = \frac{\|A\|_F}{d_F(A, \Sigma)},$$

the last equality following from Corollary 1.19. This condition number does not arise from perturbation considerations, but it satisfies $\kappa(A) \leq \kappa_F(A) \leq \sqrt{n}\,\kappa(A)$ and is amenable to our analysis, since $\| \ \|_F$ is induced by the Hermitian product $(A, B) \mapsto \mathsf{trace}(AB^*)$. In other words, κ_F is conic.

Proposition 20.16 *For all $n \geq 1$, $0 < \sigma \leq 1$, and $\overline{A} \in \mathbb{C}^{n \times n}$ we have*

$$\mathop{\mathbb{E}}_{A \in B(\overline{A}, \sigma)} \left(\kappa_F(A) \right) \leq 4\left(2e^5\right)^{\frac{1}{2}} n^{\frac{7}{2}} \frac{1}{\sigma}$$

and

$$\mathop{\mathbb{E}}_{A \in B(\overline{A}, \sigma)} \left(\log_\beta \kappa_F(A) \right) \leq \frac{7}{2} \log_\beta n + \log_\beta \frac{1}{\sigma} + 3 \log_\beta e + \frac{3}{2},$$

where the expectations are over A uniformly distributed in the disk of radius σ centered at $\overline{A}$ in projective space $\mathbb{P}^{n^2-1}$ with respect to d_{sin}. •

Proof The variety Σ of singular matrices is the zero set of the determinant and hence a hypersurface in $\mathbb{P}^{n^2-1}$ of degree n. We now apply Corollary 20.2. □

Note that the bound in Proposition 20.16 for the logarithm of $\kappa_F(A)$ is of the same order of magnitude as the one obtained (for real matrices, but this is of no relevance) in Theorem 2.46 for the logarithm of $\kappa_{2\infty}(A)$.

20.6.2 Eigenvalue Computations

Let $A \in \mathbb{C}^{n \times n}$ and let $\lambda \in \mathbb{C}$ be a simple eigenvalue of $\mathbb{C}$. Let $v, u \in \mathbb{C}^n$ be the right and left eigenvectors associated to λ, respectively, that is, $Av = \lambda v$ and $u^*A = \lambda u^*$, cf. Sect. 14.3.1. We saw in Proposition 14.15 (with $r = s = 2$) that the absolute condition number for computing the eigenvalue λ satisfies

$$\mathsf{acond}^{G_2}(A, \lambda) = \left\| DG_2(A) \right\| = \frac{\|u\| \|v\|}{|\langle u, v \rangle|}.$$

Following the lines of thought in Sect. 6.8, we can then define the (absolute) condition number of A for eigenvalue computations by taking

$$\kappa_{\mathsf{eigen}}(A) := \max_\lambda \mathsf{acond}^{G_2}(A, \lambda),$$

where the maximum is over all the eigenvalues λ of A and is taken to be ∞ if one of them is not simple. Note that $\kappa_{\text{eigen}}(A)$ is homogeneous of degree 0 in A. Also, the set Σ where κ_{eigen} is infinite is the set of matrices having multiple eigenvalues.

Proposition 20.17 *For all nonzero $A \in \mathbb{C}^{n \times n}$ we have*

$$\kappa_{\text{eigen}}(A) \le \frac{\sqrt{2}\|A\|_F}{d_F(A, \Sigma)}.$$

Proof The statement is true if $\kappa_{\text{eigen}}(A) = \infty$. We can therefore assume that all the eigenvalues of A are simple.

We consider two cases. If $\kappa_{\text{eigen}}(A) \le \sqrt{2}$, then the statement is true, since $\frac{\|A\|_F}{d_F(A, \Sigma)} \ge 1$. If instead $\kappa_{\text{eigen}}(A) > \sqrt{2}$, then there exists an eigenvalue λ such that $\kappa_{\text{eigen}}(A) = \text{acond}^{G_2}(A, \lambda) > \sqrt{2}$.

Let v and u be right and left eigenvectors of λ, respectively. Lemma 14.17(a) states that $\langle u, v \rangle \ne 0$. Without loss of generality we may assume $\|v\| = \|u\| = 1$ and $\varepsilon := \langle u, v \rangle \in (0, 1]$. Since $\text{acond}^{G_2}(A, \lambda) = \varepsilon^{-1}$, we actually have $\varepsilon < \frac{1}{\sqrt{2}}$.

Our argument is based essentially on the unitary invariance of Σ: if $A' \in \Sigma$ and $M \in \mathcal{U}(n)$, then $MA'M^* \in \Sigma$.

Let $M \in \mathcal{U}(n)$ be such that $Mv = e_1$ and put $z := Mu$. Then,

$$\langle z, e_1 \rangle = \langle u, v \rangle = \varepsilon \quad \text{and} \quad z = (\varepsilon, w) \tag{20.6}$$

for some $w \in \mathbb{C}^{n-1}$ such that $\|w\|^2 = (1 - \varepsilon^2)$. Let now $B := MAM^*$. Then the relations $Av = \lambda v$ and $u^*A = \lambda u^*$ (along with the equality $MM^* = M^*M = I$) imply

$$MAM^*Mv = \lambda Mv \quad \text{and} \quad u^*M^*MAM^* = \lambda u^*M^*,$$

that is,

$$Be_1 = \lambda e_1 \quad \text{and} \quad z^*B = \lambda z^*. \tag{20.7}$$

From the first of these equalities it follows that B has the form

$$B = \left[\begin{array}{c|c} \lambda & b^* \\ \hline 0 & B_1 \end{array}\right]$$

for some $b \in \mathbb{C}^{n-1}$ and $B_1 \in \mathbb{C}^{(n-1) \times (n-1)}$. This form, together with the right-hand equalities in (20.6) and (20.7), yields

$$\left[\varepsilon, w^*\right]\left[\begin{array}{c|c} \lambda & b^* \\ \hline 0 & B_1 \end{array}\right] = \lambda\left[\varepsilon, w^*\right],$$

which implies $\varepsilon b^* + w^*B_1 = \lambda w^*$. We can rewrite this equality as

$$w^*\left[B_1 + \frac{\varepsilon}{w^*w}wb^*\right] = \lambda w^*.$$

This shows that λ is an eigenvalue of $B_1 + E_1$ for the rank-one perturbation $E_1 := \frac{\varepsilon}{w^*w} wb^*$ and hence that λ is a double eigenvalue of $B + E$ with

$$E = \begin{bmatrix} 0 & 0 \\ \hline 0 & E_1 \end{bmatrix}.$$

From the unitary invariance of Σ it follows that

$$d_F(A, \Sigma) = d_F(B, \Sigma) \le \|E\|_F = \|E_1\|_F = \frac{\varepsilon \|w\| \|b\|}{\|w\|^2}.$$

Using $\|b\| \le \|B\|_F = \|A\|_F$ as well as $\|w\| = (1 - \varepsilon^2)^{1/2}$, we get

$$d_F(A, \Sigma) \le \frac{\varepsilon \|A\|_F}{(1 - \varepsilon^2)^{1/2}},$$

and finally, using $\varepsilon < \frac{1}{\sqrt{2}}$,

$$\kappa_{\text{eigen}}(A) = \text{acond}^{G_2}(A, \lambda) = \frac{1}{\varepsilon} \le \frac{\|A\|_F}{d_F(A, \Sigma)} \frac{1}{(1-\varepsilon^2)^{1/2}} < \frac{\sqrt{2}\|A\|_F}{d_F(A, \Sigma)}. \qquad \square$$

The right-hand side in the bound of Proposition 20.17 is conic. We can therefore obtain smoothed analysis estimates for this quantity.

Proposition 20.18 *For all $n \ge 1, 0 < \sigma \le 1$, and $\overline{A} \in \mathbb{C}^{n \times n}$ we have*

$$\underset{A \in B(\overline{A}, \sigma)}{\mathbb{E}} \left(\log_\beta \kappa_{\text{eigen}}(A) \right) \le 4 \log_\beta n + \log_\beta \frac{1}{\sigma} + 3 \log_\beta e + \frac{3}{2}.$$

Proof Let $\chi_A(z) = \det(zI - A)$ be the characteristic polynomial of A. Clearly, A has multiple eigenvalues if and only if χ_A has multiple roots. This happens if and only if the discriminant $\text{disc}(\chi_A)$ of A is zero. Therefore, the set of ill-posed matrices A equals the zero set of the discriminant. It remains to show that $\text{disc}(\chi_A)$ is a homogeneous polynomial in the entries of A and to determine its degree.

The discriminant $\text{disc}(\chi_A)$ is a polynomial in the entries of A, which can be expressed in terms of the eigenvalues $\lambda_1, \ldots, \lambda_n$ of A as follows:

$$\text{disc}(\chi_A) = \prod_{i<j} (\lambda_i - \lambda_j)^2.$$

Note that $\alpha\lambda_1, \ldots, \alpha\lambda_n$ are the eigenvalues of αA, for $\alpha \in \mathbb{C}$. Hence

$$\text{disc}(\chi_{\alpha A}) = \prod_{i<j} (\alpha\lambda_i - \alpha\lambda_j)^2 = \alpha^{n^2-n} \prod_{i<j} (\lambda_i - \lambda_j)^2.$$

We conclude that $\text{disc}(\chi_A)$ is homogeneous of degree $n^2 - n$ in the entries of A.

We now apply Corollary 20.2 with $p = n^2 - 1$ and $d = n^2 - n$ to get the assertion. $\qquad \square$

20.6.3 Complex Polynomial Systems

Fix a degree pattern $\mathbf{d} = (d_1, \ldots, d_n)$ and recall the setting of systems $f \in \mathcal{H}_\mathbf{d}$ of multivariate polynomials treated in Chap. 16. Recall also the condition number $\mu_{\max}(f)$ introduced in Eq. (18.1), which was shown in Theorem 18.2 to control the randomized cost of Algorithm LV for finding an approximate zero of f. In Proposition 17.31 we stated bounds for the tail of $\mu_{\max}(f)$ for $f \in \mathcal{H}_\mathbf{d}$ chosen according to the standard Gaussian distribution. We can now easily obtain a smoothed analysis of $\mu_{\max}(f)$. To do so, we first bound $\mu_{\max}(f)$ by a conic condition number.

Recall the discriminant variety $\Sigma \subseteq \mathbb{P}(\mathcal{H}_\mathbf{d})$, which consists of the systems $f \in \mathbb{P}(\mathcal{H}_\mathbf{d})$ having multiple zeros; see Remark 16.26. This variety is the zero set of the discriminant polynomial $\mathsf{disc}_\mathbf{d}$, which is an irreducible polynomial with integer coefficients. Moreover,

$$\deg \mathsf{disc}_\mathbf{d} = \mathcal{D}\left(1 + \left(\sum_{i=1}^n d_i + 1 - n\right) \sum_{i=1}^n \frac{1}{d_i}\right) \le \mathcal{D}n^2 D; \qquad (20.8)$$

see Corollary A.48 and Remark A.49.

Theorem 16.19 states that for $(f, \zeta) \in V$ we have

$$\mu_{\mathsf{norm}}(f, \zeta) = \frac{1}{d_{\sin}(f, \Sigma_\zeta)},$$

where Σ_ζ consists of the $\tilde{f} \in \mathcal{H}_\mathbf{d}$ for which ζ is a multiple zero (we identify f with $[f] \in \mathbb{P}(\mathcal{H}_\mathbf{d})$). In particular, $\Sigma_\zeta \subseteq \Sigma$ and hence $d_{\sin}(f, \Sigma_\zeta) \ge d_{\sin}(f, \Sigma)$. We can now estimate $\mu_{\max}$ by the conic condition number associated with Σ:

$$\mu_{\max}(f) = \max_{\zeta \mid f(\zeta) = 0} \mu_{\mathsf{norm}}(f, \zeta_i) = \max_{\zeta \mid f(\zeta) = 0} \frac{1}{d_{\sin}(f, \Sigma_\zeta)} \le \frac{1}{d_{\sin}(f, \Sigma)}. \quad (20.9)$$

Proposition 20.19 *For all $\overline{f} \in \mathcal{H}_\mathbf{d}$, $\overline{f} \ne 0$, all $\sigma \in (0, 1]$, and all $t \ge N\sqrt{2}$,*

$$\Prob_{f \in B(\overline{f}, \sigma)} \{\mu_{\max}(f) \ge t\} \le 2e^3 N^3 \mathcal{D}n^2 D \left(\frac{1}{t\sigma}\right)^2 \left(1 + N\frac{1}{t\sigma}\right)^{2(N-2)}$$

and

$$\mathbb{E}_{f \in B(\overline{f}, \sigma)} \left(\log_\beta \mu_{\max}(f)\right) \le \frac{3}{2} \log_\beta N + \frac{1}{2} \log_\beta \mathcal{D} + \log_\beta n + \frac{1}{2} \log_\beta D$$

$$+ \log_\beta \frac{1}{\sigma} + 3 \log_\beta e + \frac{3}{2}.$$

Proof We apply Corollary 20.2 to the zero set Σ in $\mathbb{P}(\mathcal{H}_\mathbf{d})$ of the discriminant $\mathsf{disc}_\mathbf{d}$ and the associated conic condition number. Recall that $N := \dim_\mathbb{C} \mathcal{H}_\mathbf{d}$ so that $p = N - 1$. Further, $\deg \mathsf{disc}_\mathbf{d} \le n(D - 1) + 1$ by (20.8). $\qquad \square$

Remark 20.20 There is a noticeable difference between the tail estimate in Proposition 17.31 and that in Proposition 20.19. The former decays as t^{-4}, whereas the latter decays as t^{-2}. This difference arises from the inequality in (20.9) which bounds $\mu_{\max}$ by the relativized inverse of the distance to a complex hypersurface. It appears that this bound is too generous.

Chapter 21
Probabilistic Analysis of Conic Condition Numbers: II. The Real Case

Our final goal can be succinctly stated: we want to extend the main result of the preceding chapter to problems with real data. Again this boils down to providing bounds on the volume of tubes. However, the technical development will be quite different. One of the key constituents of the proof of Theorem 20.1, Lemma 20.14, is false over the reals. For this reason, a more sophisticated line of argument involving integrals of curvature and the kinematic formula will be required.

We assume that our data space is $\mathbb{R}^{p+1}$ and fix a subset $\Sigma \neq \{0\}$ of "ill-posed inputs" that we assume to be closed under multiplication with real scalars. In other words, Σ is a cone satisfying the symmetry condition $-\Sigma = \Sigma$. (Note that we don't assume Σ to be convex.) Again, we call a function $\mathscr{C} \colon \mathbb{R}^{p+1} \setminus \{0\} \to \mathbb{R}$ a *conic condition number* when it has the form

$$\mathscr{C}(a) = \frac{\|a\|}{d(a, \Sigma)},$$

where $\| \ \|$ and d are the norm and distance induced by the standard inner product $\langle \ , \ \rangle$. Similarly as in the previous chapter, we may restrict to data a lying in the sphere $\mathbb{S}^p = \{x \in \mathbb{R}^{p+1} \mid \|x\| = 1\}$ and express the conic condition number $\mathscr{C}$ as

$$\mathscr{C}(a) = \frac{1}{d_{\sin}(a, \Sigma \cap \mathbb{S}^p)}.$$

Note that $\mathscr{C}(-a) = \mathscr{C}(a)$ due to $-\Sigma = \Sigma$.

Let $B_{\sin}(\bar{a}, \sigma) := \{x \in \mathbb{S}^p \mid d_{\sin}(x, \bar{a}) \leq \sigma\}$ denote the closed ball of radius σ, with respect to $d_{\sin}$, around $\bar{a}$ in $\mathbb{S}^p$. We have $B_{\sin}(\bar{a}, \sigma) = B(\bar{a}, \sigma) \cup B(-\bar{a}, \sigma)$, where $B(\bar{a}, \sigma)$ denotes the spherical cap around $\bar{a}$ with angular radius $\arcsin \sigma$; compare Sect. 2.2.6.

We will endow $B_{\sin}(\bar{a}, \sigma)$ with the uniform probability measure. The main result in this chapter is the following.

Theorem 21.1 *Let $\mathscr{C}$ be a conic condition number with set of ill-posed inputs Σ. Assume that Σ is contained in a real algebraic hypersurface, given as the zero set*

P. Bürgisser, F. Cucker, *Condition*,
Grundlehren der mathematischen Wissenschaften 349,
DOI 10.1007/978-3-642-38896-5_21, © Springer-Verlag Berlin Heidelberg 2013

of a homogeneous polynomial of degree d. Then, for all $0 < \sigma \leq 1$ and all $t \geq (2d+1)\frac{p}{\sigma}$, we have

$$\sup_{\overline{a} \in \mathbb{S}^p} \Prob_{a \in B_{\sin}(\overline{a}, \sigma)} \{ \mathscr{C}(a) \geq t \} \leq 4e\,dp\,\frac{1}{\sigma t}$$

and

$$\sup_{\overline{a} \in \mathbb{S}^p} \mathbb{E}_{a \in B_{\sin}(\overline{a}, \sigma)} \log_\beta \mathscr{C}(a) \leq \log_\beta p + \log_\beta d + \log_\beta \frac{1}{\sigma} + \log_\beta(4e^2).$$

In particular (take $\sigma = 1$), for all $t \geq (2d+1)p$,

$$\Prob_{a \in \mathbb{S}^p} \{ \mathscr{C}(a) \geq t \} \leq 4e\,dp\,\frac{1}{t}$$

and

$$\mathbb{E}_{a \in \mathbb{S}^p} \left(\log_\beta \mathscr{C}(a) \right) \leq \log_\beta p + \log_\beta d + \log_\beta(4e^2).$$

Remark 21.2 (a) In Theorem 21.1 we could replace $B_{\sin}(\overline{a}, \sigma)$ by $B(\overline{a}, \sigma)$. This is because $\mathscr{C}(-a) = \mathscr{C}(a)$ and hence

$$\Prob_{a \in B_{\sin}(\overline{a}, \sigma)} \{ \mathscr{C}(a) \geq t \} = \Prob_{a \in B(\overline{a}, \sigma)} \{ \mathscr{C}(a) \geq t \}.$$

In fact, Theorem 21.1 could be stated for real projective space $\mathbb{P}^p$ with the same bounds. While such a statement is the most natural over the complex numbers, it does not follow the tradition over the reals.

(b) The extension of Theorem 21.1 from hypersurfaces to higher codimension is an open problem.

The techniques developed in this chapter will also provide a proof of Theorem 13.18 bounding the volume of tubes around spherical convex sets.

Let us point out that this chapter is mathematically more demanding than the previous ones, since the proofs rely on various techniques from differential and integral geometry. For this reason, in a few places, we just give a sketch of the arguments and refer to the Appendix or the Notes for more details.

21.1 On the Volume of Tubes

Recall from Definition 13.16 the notion $T(U, \varepsilon)$ of the *ε-neighborhood* around a subset U of $\mathbb{S}^p$, for $0 < \varepsilon \leq 1$.

We have, for $\overline{a} \in \mathbb{S}^p$ and $\sigma \in (0, 1]$,

$$\Prob_{a \in B_{\sin}(\overline{a}, \sigma)} \{ \mathscr{C}(a) \geq \varepsilon^{-1} \} = \frac{\operatorname{vol}(T(\Sigma \cap \mathbb{S}^p, \varepsilon) \cap B_{\sin}(\overline{a}, \sigma))}{\operatorname{vol} B_{\sin}(a, \sigma)}.$$

Clearly, our task is to provide bounds on the volume of $T(\Sigma \cap \mathbb{S}^p, \varepsilon) \cap B_{\sin}(\overline{a}, \sigma)$.

As a first step towards this goal we shall derive in this section a general formula for the volume of the ε-neighborhood around a smooth hypersurface in $\mathbb{S}^p$.

In Sect. 2.2.6 we introduced the notion of an ε-tube $T^\perp(U, \varepsilon)$ around a closed subset of a subsphere $\mathbb{S}^{p-1}$ of $\mathbb{S}^p$ (see Fig. 2.1). We now extend this notion. Let M be a compact smooth hypersurface in $\mathbb{S}^p$, $U \subseteq M$ a subset, and $0 < \varepsilon \leq 1$. We define the ε-tube $T^\perp(U, \varepsilon)$ around U by

$$T^\perp(U, \varepsilon) := \{x \in \mathbb{S}^p \mid \text{there is an arc of a great circle in } \mathbb{S}^p \text{ from } x \text{ to a point}$$
$$\text{in } U \text{ of length } \leq \arcsin \varepsilon \text{ that intersects } M \text{ orthogonally}\}.$$

Lemma 21.3 *We have* $T^\perp(U, \varepsilon) \subseteq T(U, \varepsilon)$ *with equality holding in the case* $U = M$.

Proof Let $q \in T(M, \varepsilon)$ and assume that $x_0 \in M$ minimizes $d_{\mathbb{S}}(q, x)$ for $x \in M$. Then, for any smooth curve $x(t)$ in M with $x(0) = x_0$, the function $t \mapsto \theta(t) = d_{\mathbb{S}}(q, x(t))$, defined in a neighborhood of 0, has a minimum at $t = 0$, and hence $\dot{\theta}(0) = 0$. On the other hand, $\langle q, x(t) \rangle = \cos\theta(t)$, which implies $\langle q, \dot{x} \rangle = -\dot{\theta}\sin\theta = 0$ by differentiating at 0. Hence $\dot{x}$ is orthogonal to $q - x_0$. This implies that $q \in T^\perp(M, \varepsilon)$. $\qquad\square$

Remark 21.4 If U is a submanifold of M with a smooth boundary, then $T(U, \varepsilon)$ equals the union of $T^\perp(U, \varepsilon)$ and a "half-tube" around the boundary of U (cf. Fig. 2.1).

21.1.1 Curvature Integrals

For a short review of the few elementary concepts from differential geometry needed for this section we refer to Sect. A.2.7.

Let M be a compact smooth oriented hypersurface of $\mathbb{S}^p$. Assume that a unit normal vector field $\nu \colon M \to \mathbb{R}^{p+1}$ has been chosen on M that corresponds to the choice of the orientation of M. Consider the *Weingarten map* (see (A.4)) $L_M(x) \colon T_x M \to T_x M$, $L_M(x) := -D\nu(x)$, which is a well-defined self-adjoint linear map (cf. Lemma A.20). The eigenvalues $\kappa_1(x), \ldots, \kappa_{p-1}(x)$ of the Weingarten map $L_M(x)$ are called the *principal curvatures* of the hypersurface M at x. For $1 \leq i \leq p - 1$ one defines the *ith curvature* $K_{M,i}(x)$ of M at x as the ith elementary symmetric polynomial in $\kappa_1(x), \ldots, \kappa_{p-1}(x)$, and one puts $K_{M,0}(x) := 1$. We remark that the ith curvatures are the coefficients of the characteristic polynomial of the Weingarten map:

$$\det\left(I_{p-1} + t L_M(x)\right) = \sum_{i=0}^{p-1} K_{M,i}(x) t^i. \tag{21.1}$$

Note that for $i = p - 1$ one gets

$$K_{M,p-1}(x) = \kappa_1(x) \cdots \kappa_{p-1}(x) = \det L_M(x) = (-1)^{p-1} \det Dv(x), \qquad (21.2)$$

which is called the *Gaussian curvature* of M at x.

Example 21.5 Consider the case of $M = \mathbb{S}^{p-1}$, the subsphere of $\mathbb{S}^p$ given by the equation $x_p = 0$. Then $L_M(x) = 0$ for all $x \in M$; see Example A.22. Hence all the principal curvatures of M are zero. This example makes clear that the principal curvatures are relative to the ambient space $\mathbb{S}^p$. (Of course, $\mathbb{S}^{p-1}$ is curved; however, its "curvature relative to the ambient sphere" is zero.)

Recall that a hypersurface M in $\mathbb{S}^p$ has a Riemannian metric inherited from the ambient space $\mathbb{S}^p$ and that dM denotes the corresponding volume element, cf. Sect. A.2.5. We continue with a fundamental definition.

Definition 21.6 Let U be a measurable subset of the given oriented compact smooth hypersurface M of $\mathbb{S}^p$. The *normalized integral $\mu_i(U)$ of the ith curvature* of U in M, for $0 \le i \le p - 1$, is defined as

$$\mu_i(U) := \frac{2}{\mathcal{O}_i \mathcal{O}_{p-i-1}} \int_U K_{M,i}\, dM.$$

Moreover, we define the *normalized integral $|\mu_i|(U)$ of the ith absolute curvature* of U by

$$|\mu_i|(U) := \frac{2}{\mathcal{O}_i \mathcal{O}_{p-i-1}} \int_U |K_{M,i}|\, dM. \qquad (21.3)$$

The reason for the choice of the scaling factors will become clear soon. We note thats for $i = 0$,

$$\mu_0(U) = |\mu_0|(U) = \frac{\mathrm{vol}_{p-1}\, U}{\mathcal{O}_{p-1}},$$

equals the ratio of the $(p-1)$-dimensional volume of U to the volume of $\mathbb{S}^{p-1}$. We also note that the normalized integral of the top absolute curvature ($i = p - 1$),

$$|\mu_{p-1}|(U) = \frac{1}{\mathcal{O}_{p-1}} \int_U |\det Dv|\, dM, \qquad (21.4)$$

equals the integral of the absolute value of the Gaussian curvature over U, normalized by the factor $\frac{1}{\mathcal{O}_{p-1}}$.

Let us summarize some basic properties of these notions. Their obvious proofs are left to the reader.

Lemma 21.7

(a) $|\mu_i(U)| \le |\mu_i|(U)$.

(b) $|\mu_i|(U_1) \le |\mu_i|(U_2)$ for $U_1 \subseteq U_2$.

(c) For $g \in \mathcal{O}(p+1)$ we have $\mu_i(U) = \mu_i(gU)$ and $|\mu_i|(U) = |\mu_i|(gU)$, where gU is interpreted as a subset of the smooth hypersurface gM.

(d) $|\mu_i|(\mathbb{S}^{p-1}) = 0$ for $i > 0$; see Example 21.5.   □

Example 21.8 In Example A.22 we will show that the boundary M_α of a spherical cap $\mathsf{cap}(\bar{a}, \alpha)$ in $\mathbb{S}^p$ of radius $0 < \alpha \le \pi/2$ has an isotropic Weingarten map given by $L_{M_\alpha} = (\cot \alpha) I_{T_x M}$. Therefore the ith curvature of M_α satisfies $K_{M_\alpha, i}(x) = \binom{p-1}{i} (\cot \alpha)^i$, a quantity independent of $x \in M_\alpha$. Hence we obtain for the normalized integral of the ith curvature

$$\mu_i(M_\alpha) = \frac{2 K_{M_\alpha, i} \, \mathsf{vol}_{p-1} M_\alpha}{\mathcal{O}_i \mathcal{O}_{p-1-i}} = \frac{2\mathcal{O}_{p-1}}{\mathcal{O}_i \mathcal{O}_{p-1-i}} \binom{p-1}{i} (\sin \alpha)^{p-i-1} (\cos \alpha)^i,$$

(21.5)

using that $\mathsf{vol}_{p-1} M_\alpha = \mathcal{O}_{p-1}(\sin \alpha)^{p-1}$. We note that $\mu_i(U) = |\mu_i|(U)$ for all measurable subsets U of M_α, since all the principal curvatures are nonnegative.

21.1.2 Weyl's Tube Formula

We show now that the volume of the tube $T^\perp(U, \varepsilon)$ around a measurable subset U of the smooth hypersurface M in $\mathbb{S}^p$ can be bounded in terms of the integrals of absolute curvature $|\mu_i|(U)$. Recall the definition of $\mathcal{O}_{p,k}(\varepsilon)$ in Lemma 20.5.

Theorem 21.9 *Let M be a compact smooth oriented hypersurface of $\mathbb{S}^p$ and let U be a measurable subset of M. Then we have for all $0 < \varepsilon \le 1$,*

$$\mathsf{vol}\, T^\perp(U, \varepsilon) \le \sum_{i=0}^{p-1} |\mu_i|(U) \, \mathcal{O}_{p, p-1-i}(\varepsilon).$$

Proof Let $v \colon M \to \mathbb{S}^p$ be the unit normal vector field on M corresponding to its orientation. For $x \in M$ we consider the parameterization

$$\varphi_x \colon \mathbb{R} \to \mathbb{S}^p, \quad \varphi_x(t) = \frac{x + t v(x)}{\|x + t v(x)\|} = \frac{x + t v(x)}{(1 + t^2)^{\frac{1}{2}}},$$

of the half great circle intersecting M at x orthogonally (cf. Fig. 21.1). Note that if we set $\rho := d_{\mathbb{S}}(x, \varphi_x(t))$, then $t = \tan \rho$.

Consider the following smooth map:

$$\varphi \colon M \times \mathbb{R} \to \mathbb{S}^p, \quad (x, t) \mapsto \varphi_x(t).$$

Fig. 21.1 The point $\varphi_x(t)$ and the quantities t and α

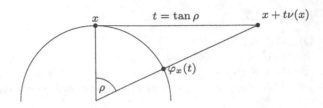

Let $\alpha := \arcsin \varepsilon$ and put $\tau = \tan \alpha$. We denote by $T^+(U, \varepsilon)$ and $T^-(U, \varepsilon)$ the images of $U \times (0, \tau)$ and $U \times (-\tau, 0)$ under the map φ, respectively. Clearly, $T^\perp(U, \varepsilon) = U \cup T^+(U, \varepsilon) \cup T^-(U, \varepsilon)$.

We apply the transformation formula of Corollary 17.10 to the surjective differentiable map $\varphi \colon U \times (0, \tau) \to T^+(U, \varepsilon)$ of Riemannian manifolds. This yields

$$\int_{U \times (0,\tau)} \left|\det D\varphi\right| d(M \times \mathbb{R}) = \int_{y \in T^+(U,\alpha)} \left|\varphi^{-1}(y)\right| d\mathbb{S}^p \geq \mathsf{vol}\, T^+(U, \alpha).$$

By Fubini's theorem, $\int_{U \times (0,\tau)} |\det D\varphi|\, d(M \times \mathbb{R}) = \int_0^\tau g(t)\, dt$, where

$$g(t) := \int_{x \in U} |\det D\varphi|(x, t)\, dM(x). \tag{21.6}$$

Claim A. The determinant of the derivative $D\varphi(x, t)$ of φ at $(x, t) \in M \times \mathbb{R}$ satisfies

$$\left|\det D\varphi(x, t)\right| = \frac{1}{(1 + t^2)^{(p+1)/2}} \left|\det\!\left(I_{T_x M} - t L_M(x)\right)\right|, \tag{21.7}$$

where, we recall from (A.4), $L_M(x)$ is the Weingarten map.

Using this claim, whose proof is postponed, we obtain from (21.6)

$$g(t) = \int_{x \in U} \frac{1}{(1 + t^2)^{(p+1)/2}} \left|\det\!\left(I_{T_x M} - t L_M(x)\right)\right| dM(x) \quad \text{(by Claim A)}$$

$$\leq \sum_{i=0}^{p-1} \frac{|t|^i}{(1 + t^2)^{(p+1)/2}} \int_U |K_{M,i}|\, dM \quad \text{(by (21.1))}$$

$$= \sum_{i=0}^{p-1} \frac{|t|^i}{(1 + t^2)^{(p+1)/2}} \frac{1}{2} \mathcal{O}_i \mathcal{O}_{p-1-i} |\mu_i|(U) \quad \text{(by (21.3))}.$$

By making the substitution $t = \tan \rho$ (recall $\tau = \tan \alpha$) we get

$$\int_0^\tau \frac{t^i}{(1 + t^2)^{(p+1)/2}}\, dt = \int_0^\alpha (\cos \rho)^{p-1-i} (\sin \rho)^i\, d\rho$$

$$= \frac{1}{\mathcal{O}_i \mathcal{O}_{p-1-i}}\, \mathcal{O}_{p,p-1-i}(\varepsilon).$$

Altogether we conclude that

$$\operatorname{vol} T^+(U,\alpha) \le \int_0^\tau g(t)\,dt \le \frac{1}{2}\sum_{i=0}^{p-1}|\mu_i|(U)\,\mathcal{O}_{p,p-1-i}(\varepsilon).$$

The same estimate can be proved for $\operatorname{vol}_p T^-(U,\alpha)$, which implies the desired estimate of $\operatorname{vol} T^\perp(U,\alpha)$.

It remains to prove Claim A. Fix $x \in M$ and choose a local parameterization of M around x. More specifically, let $O \subseteq \mathbb{R}^{p-1}$ and $O' \subseteq M$ be open subsets such that $0 \in O$ and $x \in O'$ and let

$$O \to O', \quad v = (v_1,\dots,v_{p-1}) \mapsto x(e)$$

be a diffeomorphism with $x(0) = x$. Then the partial derivatives $\partial_{v_1}x,\dots,\partial_{v_{p-1}}x$ form a basis of T_xM. We extend the above map to the diffeomorphism

$$\gamma: O \times \mathbb{R} \to O' \times \mathbb{R}, \quad (v,t) \mapsto (x(e),t).$$

Note that the $(p-1)$-dimensional volume $\operatorname{vol}_{p-1}(\partial_{v_1}x,\dots,\partial_{v_{p-1}}x)$ of the parallelepiped spanned by the vectors $\partial_{v_1}x,\dots,\partial_{v_{p-1}}x$ equals the absolute value of the determinant of $D\gamma(v,t)$, that is, $|\det D\gamma(v,t)| = \operatorname{vol}_{p-1}(\partial_{v_1}x,\dots,\partial_{v_{p-1}}x)$. Put $R(v,t) := x(e) + tv(e) \in \mathbb{R}^{p+1}$. Then the map

$$\psi: O \times \mathbb{R} \to \mathbb{S}^p, \quad (v,t) \mapsto \psi(v,t) := \varphi(x(e),t) = \frac{R(v,t)}{(1+t^2)^{\frac{1}{2}}},$$

is a local parameterization of $\mathbb{S}^p$. Since $\psi = \varphi \circ \gamma$, we can characterize $|\det D\varphi(x,t)|$ by

$$\left|\det D\varphi(x(e),t)\right| = \frac{|\det D\psi(v,t)|}{|\det D\gamma(v,t)|} = \frac{|\det D\psi(v,t)|}{\operatorname{vol}_{p-1}(\partial_{v_1}x,\dots,\partial_{v_{p-1}}x)}. \tag{21.8}$$

We compute now the determinant $|\det D\psi(v,t)|$. It equals the p-dimensional volume of the parallelepiped spanned by the vectors $\partial_t\psi, \partial_{v_1}\psi,\dots,\partial_{v_{p-1}}\psi$. Since ψ is orthogonal to T_xM and $\|\psi\| = 1$, we have

$$\left|\det D\psi(v,t)\right| = \left|\det(\psi, \partial_t\psi, \partial_{v_1}\psi,\dots,\partial_{v_{p-1}}\psi)\right|,$$

where $(\psi, \partial_t\psi, \partial_{v_1}\psi,\dots,\partial_{v_{p-1}}\psi)$ denotes the square matrix of size $p+1$ whose rows are ψ and the partial derivatives of ψ. Using

$$\partial_{v_i}\psi = (1+t^2)^{-1/2}\partial_{v_i}R, \qquad \partial_t\psi = (1+t^2)^{-1/2}\partial_t R - t(1+t^2)^{-3/2}R,$$

and the multilinearity of the determinant, we obtain

$$\left|\det(\psi, \partial_t \psi, \partial_{v_1}\psi, \ldots, \partial_{v_{p-1}}\psi)\right|$$

$$= \frac{1}{(1+t^2)^{(p+1)/2}}\left|\det\left(R, \partial_t R - t(1+t^2)^{-1}R, \partial_{v_1}R, \ldots, \partial_{v_{p-1}}R\right)\right|$$

$$= \frac{1}{(1+t^2)^{(p+1)/2}}\left|\det(R, \partial_t R, \partial_{v_1}R, \ldots, \partial_{v_{p-1}}R)\right|.$$

By the definition (A.4) of the Weingarten map, the equation

$$-\partial_{v_i}v = -Dv(x)\partial_{v_i}x = \sum_j \lambda_{ij}(x)\partial_{v_j}x$$

defines the matrix $(\lambda_{ij}(x))$ of $L_M(x)$ with respect to the basis $(\partial_{v_j}x)$ of T_xM. Using this, we get

$$\partial_{v_i}R = \partial_{v_i}x + t\partial_{v_i}v = \sum_j \left(\delta_{ij} - t\lambda_{ij}(x)\right)\partial_{v_j}x.$$

Hence we obtain now, using $\partial_t R = v$,

$$\det(R, \partial_t R, \partial_{v_1}R, \ldots, \partial_{v_{p-1}}R) = \det(x + tv, v, \partial_{v_1}R, \ldots, \partial_{v_{p-1}}R)$$

$$= \det(x, v, \partial_{v_1}R, \ldots, \partial_{v_{p-1}}R)$$

$$= \det\left(\delta_{ij} - t\lambda_{ij}(x)\right)\det(x, v, \partial_{v_1}x, \ldots, \partial_{v_{p-1}}x)$$

$$= \det\left(\delta_{ij} - t\lambda_{ij}(x)\right)\operatorname*{vol}_{p-1}(\partial_{v_1}x, \ldots, \partial_{v_{p-1}}x),$$

where the second-to-last equality follows from the multiplicativity of the determinant and the last equality is due to the fact that x and v have norm 1 and are orthogonal to T_xM.

Summarizing, we have shown that

$$\left|\det D\psi(v, t)\right| = \frac{1}{(1+t^2)^{(p+1)/2}}\left|\det\left(\delta_{ij} - t\lambda_{ij}(x)\right)\right|\operatorname*{vol}_{p-1}(\partial_{v_1}x, \ldots, \partial_{v_{p-1}}x).$$

The claim follows now by comparing this with (21.8). $\qquad\square$

21.2 A Crash Course on Probability: VI

This is the last installment of our crash course. We recall in it a basic formula of integral geometry; cf. Sect. A.4 for more details.

The discussion of the Lie group $\mathcal{U}(p+1)$ in Sect. 20.3 carries over to the orthogonal group $\mathcal{O}(p+1)$. In fact, $\mathcal{O}(p+1)$ is a smooth submanifold of $\mathbb{R}^{(p+1)\times(p+1)}$; hence it inherits from the ambient space the structure of a compact Riemannian manifold and thus has a volume form that defines a volume measure vol on $\mathcal{O}(p+1)$ (cf.

Sect. A.2.6). This volume measure is invariant under the action of $\mathcal{O}(p+1)$ on itself: we have $\mathrm{vol}\, gB = \mathrm{vol}\, B$ for all measurable subsets $B \subseteq \mathcal{O}(p+1)$ and $g \in \mathcal{O}(p+1)$. Since $\mathcal{O}(p+1)$ has finite volume, we can normalize it such that $\mathrm{vol}\,\mathcal{O}(p+1) = 1$ to obtain a uniform probability measure on $\mathcal{O}(p+1)$, which is called the *normalized Haar measure* of $\mathcal{O}(p+1)$.

Note that if M is a smooth hypersurface of $\mathbb{S}^p$ and $g \in \mathcal{O}(p+1)$ is random, then $gM \cap \mathbb{S}^1$ is almost surely finite by Proposition A.18. In this setting, the following Crofton-type formula holds.

Proposition 21.10 *Let U be an open subset of a smooth hypersurface M of $\mathbb{S}^p$ and let $0 \leq i < p$. Then we have*

$$\frac{\mathrm{vol}_{p-1} U}{\mathcal{O}_{p-1}} = \mathop{\mathbb{E}}_{g \in \mathcal{O}(p+1)} \left(\frac{\#(gU \cap \mathbb{S}^1)}{2} \right).$$

$\qquad\square$

Proposition 21.10 is an immediate consequence of Poincaré's formula (Theorem A.55) proved in Sect. A.4.1. The reader should note the similarity of this result with Theorem 20.9.

We now fix a compact smooth oriented hypersurface M of $\mathbb{S}^p$. For $i < p$ we will interpret $\mathbb{S}^{i+1}$ as the submanifold of $\mathbb{S}^p$ given by the equations $x_{i+1} = \cdots = x_p = 0$. We now take a uniformly random $g \in \mathcal{O}(p+1)$ and intersect gM with $\mathbb{S}^{i+1}$. By Proposition A.18, for almost all $g \in \mathcal{O}(p+1)$, the intersection $gM \cap \mathbb{S}^{i+1}$ is either empty or a smooth hypersurface of $\mathbb{S}^{i+1}$.

Suppose that $gM \cap \mathbb{S}^{i+1}$ is a smooth hypersurface. We fix an orientation on $gM \cap \mathbb{S}^{i+1}$ as follows. Let ν be the distinguished unit normal vector field of M; we require that the distinguished unit normal vector of the hypersurface $gM \cap \mathbb{S}^{i+1}$ in $\mathbb{S}^{i+1}$ at x lie in the positive half-space of $T_x M$ determined by ν. Let now U be a measurable subset of M. Then $gU \cap \mathbb{S}^{i+1}$ is a measurable subset of the hypersurface $gM \cap \mathbb{S}^{i+1}$ of $\mathbb{S}^{i+1}$, and hence the integral of the ith curvature $\mu_i(gU \cap \mathbb{S}^{i+1})$ is well defined.

After setting $\mu_i(\emptyset) := 0$, we see that $\mu_i(gU \cap \mathbb{S}^{i+1})$ is well defined for almost all $g \in \mathcal{O}(p+1)$.

The following result is a special case of the *principal kinematic formula* of integral geometry for spheres. We refer to Sect. A.4.2 for a discussion of this result and its context.

Theorem 21.11 *Let U be an open subset of a compact smooth oriented hypersurface M of $\mathbb{S}^p$ and $0 \leq i < p$. Then we have*

$$\mu_i(U) = \mathop{\mathbb{E}}_{g \in \mathcal{O}(p+1)} \left(\mu_i(gU \cap \mathbb{S}^{i+1}) \right).$$

$\qquad\square$

We remark that Proposition 21.10 is obtained for $i = 0$ as a special case.

21.3 Bounding Integrals of Curvature

In this section let $f \in \mathbb{R}[X_0, \ldots, X_p]$ be homogeneous of degree $d \geq 1$ with nonempty zero set $V \subseteq \mathbb{S}^p$ such that the derivative of the restriction of f to $\mathbb{S}^p$ does not vanish on V. Then V is a compact smooth hypersurface of $\mathbb{S}^p$; cf. Theorem A.9. We orient V by the following unit normal vector field, which is called the *Gauss map*:

$$v : V \to \mathbb{S}^p, \quad v(x) = \|\operatorname{grad} f(x)\|^{-1} \operatorname{grad} f(x). \tag{21.9}$$

We next show that the integral of the top absolute curvature of V can be bounded in terms of the dimension p and the degree d only.

Lemma 21.12 *We have* $|\mu_{p-1}|(V) \leq d(d-1)^{p-1}$.

Sketch of proof. For simplicity, we assume that the image N of the Gauss map $v : V \to \mathbb{S}^p$ is a smooth hypersurface of $\mathbb{S}^p$ (this can be achieved by removing lower-dimensional parts). Theorem A.9 combined with Theorem A.12 implies that for almost all $y \in N$, the fibers $v^{-1}(y)$ are zero-dimensional.

Using (21.4) and applying Corollary 17.10 to the Gauss map v yields

$$\mathcal{O}_{p-1} |\mu_{p-1}|(V) = \int_M |\det Dv| \, dV = \int_{y \in N} \#v^{-1}(y) \, dN. \tag{21.10}$$

We decompose N according to the cardinality of the fibers, obtaining $N = \cup_{\ell \in \mathbb{N}} F_\ell$, where $F_\ell := \{y \in N \mid \#v^{-1}(y) = \ell\}$. If F_ℓ° denotes the interior of F_ℓ in N, we have

$$\int_{y \in N} \#v^{-1}(y) \, dN = \sum_{\ell \in \mathbb{N}} \int_{y \in F_\ell} \#v^{-1}(y) \, dN = \sum_{\ell \in \mathbb{N}} \ell \operatorname*{vol}_{p-1} F_\ell^\circ. \tag{21.11}$$

Proposition 21.10 applied to the open subset F_ℓ° of N implies

$$\operatorname*{vol}_{p-1} F_\ell^\circ = \frac{\mathcal{O}_{p-1}}{2} \operatorname*{\mathbb{E}}_{g \in \mathcal{O}(p+1)} \#\left(F_\ell^\circ \cap g\mathbb{S}^1\right) = \frac{\mathcal{O}_{p-1}}{2} \operatorname*{\mathbb{E}}_{g \in \mathcal{O}(p+1)} \#\left(F_\ell \cap g\mathbb{S}^1\right).$$

The last equality is a consequence of the fact that $F_\ell \setminus F_\ell^\circ$ is a finite union of submanifolds of dimension strictly less than $\dim N$ (which is true because F_ℓ is semialgebraic; cf. [39, Chap. 9]). Therefore,

$$\sum_{\ell \in \mathbb{N}} \ell \operatorname*{vol}_{p-1} F_\ell^\circ = \frac{\mathcal{O}_{p-1}}{2} \operatorname*{\mathbb{E}}_{g \in \mathcal{O}(p+1)} \left(\sum_{\ell \in \mathbb{N}} \ell \#\left(F_\ell \cap g\mathbb{S}^1\right) \right). \tag{21.12}$$

But we have

$$v^{-1}\left(g\mathbb{S}^1\right) = v^{-1}\left(N \cap g\mathbb{S}^1\right) = \bigcup_{\ell \in \mathbb{N}} v^{-1}\left(F_\ell \cap g\mathbb{S}^1\right),$$

where the union is disjoint and hence $\#v^{-1}(g\mathbb{S}^1) = \sum_\ell \ell\#(F_\ell \cap g\mathbb{S}^1)$. Combined with (21.12) this gives

$$\frac{\mathcal{O}_{p-1}}{2} \mathop{\mathbb{E}}_{g\in\mathcal{O}(p+1)} \#v^{-1}(g\mathbb{S}^1) = \frac{\mathcal{O}_{p-1}}{2} \sum_{\ell\in\mathbb{N}} \ell \operatorname{vol}_{p-1} F_\ell.$$

From this we conclude with (21.10) and (21.11) that

$$|\mu_{p-1}|(V) = \frac{1}{2} \mathop{\mathbb{E}}_{g\in\mathcal{O}(p+1)} \left(\#v^{-1}(g\mathbb{S}^1)\right). \tag{21.13}$$

A point $x \in \mathbb{R}^{n+1}$ lies in $v^{-1}(\mathbb{S}^1)$ iff it satisfies the following system of equations:

$$\sum_{i=0}^n x_i^2 - 1 = 0, \ \ f(x) = 0, \ \ \ \partial_2 f(x) = \cdots = \partial_n f(x) = 0.$$

By a variant of Bézout's theorem, stated as Corollary A.46 in the Appendix, the number of complex simple solutions to this system of equations is bounded by $2d(d-1)^{n-1}$.

Proposition A.18 states that $g\mathbb{S}^1$ intersects N transversally for almost all $g \in \mathcal{O}(p+1)$. In this case, all the zeros of this system in $\mathbb{S}^p$ are simple. Combined with the above reasoning, we conclude that

$$\#v^{-1}(g\mathbb{S}^1) \le 2d(d-1)^{n-1}$$

for almost all $g \in \mathcal{O}(p+1)$. The assertion follows now from (21.13). $\square$

Example 21.13 Let V_ε denote the zero set of $f = \sum_{i=1}^p X_i^2 - \varepsilon^2 X_0^2$ in $\mathbb{S}^p$. Then $V_\varepsilon = M_\alpha^+ \cup M_\alpha^-$, where M_α^+ denotes the boundary of the spherical cap of angular radius $\alpha = \arctan\varepsilon$ centered at $(1, 0, \ldots, 0)$, and M_α^- denotes the cap of radius α centered at $(-1, 0, \ldots, 0)$. In Example 21.8 we have seen that $\mu_{p-1}(M_\alpha^\pm) = |\mu_{p-1}|(M_\alpha^\pm) = (\cos\alpha)^{p-1}$. Hence $\mu_{p-1}(V_\varepsilon) = 2(\cos\alpha)^{p-1}$, which converges to 2, as $\varepsilon \to 0$. This shows that the bound in Lemma 21.12 is sharp for $d = 2$.

We now extend Lemma 21.12 and give bounds for the integrals of absolute curvature of V intersected with a ball.

Proposition 21.14 *For $\bar{a} \in \mathbb{S}^p, 0 < \sigma \le 1$, and $0 \le i < p-1$ we have*

$$|\mu_i|\left(V \cap B_{\sin}(\bar{a}, \sigma)\right) \le 2d(d-1)^i \frac{\mathcal{O}_{p,i+1}(\sigma)}{\mathcal{O}_p}.$$

Proof Put $U := V \cap B_{\sin}(\bar{a}, \sigma)$ and let U_+ be the set of points of U where $K_{V,i}$ is positive and let U_- be the set of points of U where $K_{V,i}$ is negative. Then $|\mu_i|(U) = |\mu_i(U_+)| + |\mu_i(U_-)|$.

Put $G := \mathcal{O}(p+1)$ and let $g \in G$ be such that gV intersects $\mathbb{S}^{i+1}$ transversally. Proposition A.18 states that this is the case for almost all $g \in G$. We apply Lemma 21.12 to the hypersurface $gV \cap \mathbb{S}^{i+1}$ of the sphere $g\mathbb{S}^{i+1}$, which yields $|\mu_i|(V \cap g\mathbb{S}^{i+1}) \leq d(d-1)^i$. By monotonicity, we obtain

$$\left| \mu_i\left(U_+ \cap g\mathbb{S}^{i+1}\right) \right| \leq |\mu_i|\left(U_+ \cap g\mathbb{S}^{i+1}\right) \leq |\mu_i|\left(V \cap g\mathbb{S}^{i+1}\right) \leq d(d-1)^i.$$

The kinematic formula, Theorem 21.11, applied to the interior U_+° of U_+ in V, implies that

$$\mu_i(U_+) = \mu_i\left(U_+^\circ\right) = \mathop{\mathbb{E}}_{g \in G}\left(\mu_i\left(gU_+^\circ \cap \mathbb{S}^{i+1}\right)\right) = \mathop{\mathbb{E}}_{g \in G}\left(\mu_i\left(gU_+ \cap \mathbb{S}^{i+1}\right)\right).$$

For the right-hand equality, note that the boundary ∂U_+ of U_+ is a finite union of submanifolds of dimension strictly less than $\dim N$, since U_+ is semialgebraic (cf. [39, Chap. 9]). Therefore, $g\partial U_+ \cap \mathbb{S}^{i+1}$ is empty for almost all $g \in G$.

We estimate now as follows:

$$\left| \mu_i(U_+) \right| \leq \mathop{\mathbb{E}}_{g \in G}\left(\left| \mu_i\left(gU_+ \cap \mathbb{S}^{i+1}\right) \right|\right)$$

$$\leq d(d-1)^i \mathop{\mathrm{Prob}}_{g \in G}\left\{ gU_+ \cap \mathbb{S}^{i+1} \neq \emptyset \right\}$$

$$\leq d(d-1)^i \mathop{\mathrm{Prob}}_{g \in G}\left\{ B_{\sin}(g\overline{a}, \sigma) \cap \mathbb{S}^{i+1} \neq \emptyset \right\}.$$

The pushforward distribution of the normalized Haar measure on G under the orbit map $G \to \mathbb{S}^p$, $g \mapsto g\overline{a}$, is the uniform distribution on $\mathbb{S}^p$; see Lemma 20.11. Hence,

$$\mathop{\mathrm{Prob}}_{g \in G}\left\{ B_{\sin}(g\overline{a}, \sigma) \cap \mathbb{S}^{i+1} \neq \emptyset \right\} = \mathop{\mathrm{Prob}}_{a \in \mathbb{S}^p}\left\{ B_{\sin}(\overline{a}, \sigma) \cap \mathbb{S}^{i+1} \neq \emptyset \right\}$$

$$= \frac{\mathrm{vol}\, T(\mathbb{S}^{i+1}, \sigma)}{\mathcal{O}_p} = \frac{\mathcal{O}_{p,i+1}(\sigma)}{\mathcal{O}_p}.$$

We have shown that

$$\left| \mu_i(U_+) \right| \leq d(d-1)^i \frac{\mathcal{O}_{p,i+1}(\sigma)}{\mathcal{O}_p}.$$

The same upper bound holds for $|\mu_i(U_-)|$, and hence the assertion follows. □

21.4 Proof of Theorem 21.1

21.4.1 The Smooth Case

As in the previous section, let $f \in \mathbb{R}[X_0, \ldots, X_p]$ be homogeneous of degree $d \geq 1$ and assume that the derivative of the restriction of f to $\mathbb{S}^p$ does not vanish on the zero set $V = Z_\mathbb{S}(f)$. Then V is a smooth hypersurface in $\mathbb{S}^p$ that we orient as before with the Gauss map (21.9).

By combining Theorem 21.9 with the bounds on the integrals of absolute curvature in Proposition 21.14 we can now derive bounds on the volume of the ε-tube over $V \cap B_{\sin}(\bar{a}, \sigma)$.

But first we need to verify an identity involving the $\mathcal{O}_k$.

Lemma 21.15 *For $1 \le k \le p - 1$ we have*

$$\mathcal{O}_{k-1}\mathcal{O}_k\mathcal{O}_{p-1-k}\mathcal{O}_{p-k} = 2(p-k)\binom{p-1}{k-1}\mathcal{O}_{p-1}\mathcal{O}_p.$$

Proof Using Proposition 2.19, we see that the claim is equivalent to the following identity:

$$2(p-k)\binom{p-1}{k-1}\Gamma\left(\frac{k}{2}\right)\Gamma\left(\frac{k+1}{2}\right)\Gamma\left(\frac{p-k}{2}\right)\Gamma\left(\frac{p-k+1}{2}\right)$$
$$= 4\sqrt{\pi}\,\Gamma\left(\frac{p}{2}\right)\Gamma\left(\frac{p+1}{2}\right). \tag{21.14}$$

We define the *double factorials* $k!!$ by $k!! := k(k-2)\cdots 2$ if $k \ge 2$ is even and $k!! := k(k-2)\cdots 3 \cdot 1$ if $k \ge 1$ is odd. Moreover, we set $0!! := 1$. Note that $k!! \cdot (k-1)!! = k!$.

From the functional equation $\Gamma(x+1) = x\Gamma(x)$ it is straightforward to derive the following formula:

$$\Gamma\left(\frac{r+1}{2}\right) = \begin{cases} \sqrt{\frac{\pi}{2}}(k-1)!!2^{\frac{1-k}{2}} & \text{if } k \text{ is even,} \\ (k-1)!!\,2^{\frac{1-k}{2}} & \text{if } k \text{ is odd.} \end{cases}$$

This implies

$$\Gamma\left(\frac{k}{2}\right)\cdot\Gamma\left(\frac{k+1}{2}\right) = \sqrt{\pi}\,\frac{(k-1)!}{2^{k-1}}.$$

Using this identity, (21.14) is easily verified. $\qquad\square$

Proposition 21.16 *Let $\bar{a} \in \mathbb{S}^p$ and $0 < \varepsilon, \sigma \le 1$. Then*

$$\operatorname{vol} T^\perp\big(V \cap B_{\sin}(\bar{a},\sigma),\varepsilon\big) \le \frac{4\mathcal{O}_{p-1}}{p}\sum_{k=1}^{p-1}\binom{p}{k}d^k\varepsilon^k\sigma^{p-k} + \mathcal{O}_p d^p\varepsilon^p.$$

Proof Put $U := V \cap B_{\sin}(\bar{a}, \sigma)$. Theorem 21.9 implies

$$\operatorname{vol} T^\perp(U,\varepsilon) \le \sum_{i=0}^{p-2}|\mu_i|(U)\,\mathcal{O}_{p,p-1-i}(\varepsilon) + |\mu_{p-1}|(U)\,\mathcal{O}_{p,0}(\varepsilon).$$

Estimating this with Proposition 21.14 and Lemma 21.12, which gives $|\mu_{p-1}(U)| \leq |\mu_{p-1}(V)| \leq d^p$, we get

$$\operatorname{vol} T^\perp(U, \varepsilon) \leq \frac{2}{\mathcal{O}_p} \sum_{i=0}^{p-2} d^{i+1} \mathcal{O}_{p,i+1}(\sigma) \mathcal{O}_{p,p-1-i}(\varepsilon) + d^p \mathcal{O}_{p,0}(\varepsilon).$$

Setting $k = i + 1$ and using the estimates of Lemma 20.6, as well as the bound $\mathcal{O}_{p,0}(\varepsilon) \leq \mathcal{O}_p \varepsilon^p$ (cf. Lemma 2.34), we obtain

$$\operatorname{vol} T^\perp(U, \varepsilon) \leq \frac{2}{\mathcal{O}_p} \sum_{k=1}^{p-1} d^k \frac{\mathcal{O}_k \mathcal{O}_{p-1-k} \mathcal{O}_{p-k} \mathcal{O}_{k-1}}{(p-k)k} \sigma^{p-k} \varepsilon^k + d^p \mathcal{O}_p \varepsilon^p.$$

Simplifying this last expression with Lemma 21.15, we get

$$\operatorname{vol} T^\perp(U, \varepsilon) \leq \frac{2}{\mathcal{O}_p} \sum_{k=1}^{p-1} d^k \binom{p-1}{k-1} \frac{2}{k} \mathcal{O}_{p-1} \mathcal{O}_p \sigma^{p-k} \varepsilon^k + d^p \mathcal{O}_p \varepsilon^p.$$

Finally, using $\binom{p-1}{k-1} = \frac{k}{p} \binom{p}{k}$, the assertion follows.   □

21.4.2 The General Case

We now extend Proposition 21.16 to the case that the real algebraic variety may have singularities, and we also replace tubes by neighborhoods.

Theorem 21.17 *Let $W \subseteq \mathbb{S}^p$ be a real algebraic variety defined by homogeneous polynomials of degree at most $d \geq 1$ such that $W \neq \mathbb{S}^p$. Then we have for $\overline{a} \in \mathbb{S}^p$ and $0 < \varepsilon, \sigma \leq 1$,*

$$\frac{\operatorname{vol}(T(W, \varepsilon) \cap B_{\sin}(\overline{a}, \sigma))}{\operatorname{vol} B_{\sin}(\overline{a}, \sigma)}$$

$$\leq 2 \sum_{k=1}^{p-1} \binom{p}{k} (2d)^k \left(1 + \frac{\varepsilon}{\sigma}\right)^{p-k} \left(\frac{\varepsilon}{\sigma}\right)^k + 2\sqrt{p} (2d)^p \left(\frac{\varepsilon}{\sigma}\right)^p.$$

Proof Assume $W = Z_{\mathbb{S}}(f_1, \ldots, f_r)$ with homogeneous polynomials f_i of degree d_i. Then W is the zero set in $\mathbb{S}^p$ of the polynomial

$$f(X) := \sum_{i=1}^{r} f_i(X)^2 \|X\|^{2d - 2d_i},$$

which is homogeneous of degree $2d$. Our assumption $W \neq \mathbb{S}^p$ implies that $\dim W < p$.

Let $\delta > 0$ be smaller than any positive critical value (cf. Definition A.7) of the restriction of f to $\mathbb{S}^p$. Then $D_\delta := \{\xi \in \mathbb{S}^p \mid f(\xi) \leq \delta\}$ is a compact domain in $\mathbb{S}^p$ with smooth boundary

$$\partial D_\delta = \{\xi \in \mathbb{S}^p \mid f(\xi) = \delta\}.$$

Indeed, by Euler's relation $\sum_i x_i \partial_i f(x) = 2df(x)$, the derivative of f does not vanish on ∂D_δ. Moreover, note that $W = \cap_{\delta>0} D_\delta$ and $D_\delta \subseteq D_{\delta'}$ for $\delta \leq \delta'$; hence $\lim_{\delta \to 0} \mathrm{vol}_p D_\delta = \mathrm{vol}_p(W)$. Moreover, $\mathrm{vol}_p(W) = 0$, since $\dim W < p$.

Claim A. We have $T(W, \varepsilon) \subseteq D_\delta \cup T(\partial D_\delta, \varepsilon)$ for $0 < \varepsilon \leq 1$.

In order to see this, let $x \in T(W, \varepsilon) \setminus D_\delta$ and $\gamma \colon [0, 1] \to \mathbb{S}^p$ be a segment of Riemannian length less than $\arcsin \varepsilon$ such that $\gamma(a) = x$ and $\gamma(0) \in W$. Consider $F \colon [0, 1] \to \mathbb{R}$, $F(t) := f(\gamma(t))$. By assumption, $F(a) = f(x) > \delta$ and $F(0) = 0$. Hence there exists $\tau \in (0, 1)$ such that $F(\tau) = \delta$. Thus $\gamma(\tau) \in \partial D_\delta$ and $d_{\sin}(x, \partial D_\delta) \leq d_{\sin}(x, \gamma(\tau)) \leq \varepsilon$, which proves the claim.

Claim B. We have $T(\partial D_\delta, \varepsilon) \cap B_{\sin}(\overline{a}, \sigma) \subseteq T^{\perp}(\partial D_\delta \cap B_{\sin}(\overline{a}, \sigma + \varepsilon), \varepsilon)$.

Indeed, assume $x \in T(\partial D_\delta, \varepsilon) \cap B_{\sin}(\overline{a}, \sigma)$. We have $T(\partial D_\delta, \varepsilon) = T^{\perp}(\partial D_\delta, \varepsilon)$ by Lemma 21.3. Hence there exists $y \in \partial D_\delta$ with $d_{\sin}(x, y) \leq \varepsilon$ such that the great circle segment $[x, y]$ intersects ∂D_δ orthogonally at y. By the triangle inequality for the projective distance, $d_{\sin}(y, \overline{a}) \leq d_{\sin}(y, x) + d_{\sin}(x, \overline{a}) \leq \varepsilon + \sigma$. Hence $y \in \partial D_\delta \cap B_{\sin}(\overline{a}, \sigma + \varepsilon)$, and it follows that $x \in T(\partial D_\delta \cap B_{\sin}(\overline{a}, \sigma + \varepsilon), \varepsilon)$, which establishes the claim. (Compare Fig. 20.2 for a related situation.)

Combining Claims A and B, we arrive at

$$T(W, \varepsilon) \cap B_{\sin}(\overline{a}, \sigma) \subseteq D_\delta \cup T^{\perp}(\partial D_\delta \cap B_{\sin}(\overline{a}, \sigma + \varepsilon), \varepsilon). \tag{21.15}$$

We now apply Proposition 21.16 to the smooth hypersurface $V = \partial D_\delta = Z_{\mathbb{S}}(f - \delta \|x\|^{2d})$ intersected with the ball $B_{\sin}(\overline{a}, \sigma + \varepsilon)$. This implies

$$\mathrm{vol}\, T^{\perp}(\partial D_\delta \cap B_{\sin}(\overline{a}, \sigma + \varepsilon), \varepsilon)$$

$$\leq \frac{4\mathcal{O}_{p-1}}{p} \sum_{k=1}^{p-1} \binom{p}{k} (2d)^k \varepsilon^k (\sigma + \varepsilon)^{p-k} + \mathcal{O}_p (2d)^p \varepsilon^p.$$

Lemma 2.34 states that

$$\mathrm{vol}\, B_{\sin}(\overline{a}, \sigma) \geq \frac{2\mathcal{O}_{p-1}}{p} \sigma^p \geq \frac{1}{2\sqrt{p}} \mathcal{O}_p \sigma^p.$$

Using this, we obtain from (21.15),

$$\frac{\mathrm{vol}(T(W,\varepsilon) \cap B_{\sin}(\overline{a},\sigma))}{\mathrm{vol}\, B_{\sin}(\overline{a},\sigma)}$$

$$\leq \frac{\mathrm{vol}\, D_\delta}{\mathrm{vol}\, B_{\sin}(\overline{a},\sigma)} + \frac{\mathrm{vol}\, T^\perp(\partial D_\delta \cap B_{\sin}(\overline{a},\sigma+\varepsilon),\varepsilon)}{\mathrm{vol}\, B_{\sin}(\overline{a},\sigma)}$$

$$\leq \frac{\mathrm{vol}\, D_\delta}{\mathrm{vol}\, B_{\sin}(\overline{a},\sigma)} + 2 \sum_{k=1}^{p-1} \binom{p}{k}(2d)^k \left(1+\frac{\varepsilon}{\sigma}\right)^{p-k}\left(\frac{\varepsilon}{\sigma}\right)^k + 2\sqrt{p}(2d)^p \left(\frac{\varepsilon}{\sigma}\right)^p.$$

Taking the limit for $\delta \to 0$, the first term vanishes and the assertion follows. $\qquad\square$

21.4.3 Proof of Theorem 21.1

Assume that we are in the situation of Theorem 21.1. By Theorem 21.17, the probability tail $\mathrm{Prob}_{a \in B_{\sin}(\overline{a},\sigma)}\{\mathscr{C}(a) \geq \varepsilon^{-1}\}$ is bounded by

$$2\left[\sum_{k=1}^{p-1}\binom{p}{k}(2d)^k\left(1+\frac{\varepsilon}{\sigma}\right)^{p-k}\left(\frac{\varepsilon}{\sigma}\right)^k + \sqrt{p}\,(2d)^p\left(\frac{\varepsilon}{\sigma}\right)^p\right]$$

$$\leq 4d\frac{\varepsilon}{\sigma}\left[\sum_{k=1}^{p-1}\binom{p}{k}(2d)^{k-1}\left(1+\frac{\varepsilon}{\sigma}\right)^{p-k}\left(\frac{\varepsilon}{\sigma}\right)^{k-1} + p\,(2d)^{p-1}\left(\frac{\varepsilon}{\sigma}\right)^{p-1}\right]$$

$$\leq \frac{4dp\varepsilon}{\sigma}\left[\sum_{i=0}^{p-2}\binom{p-1}{i}(2d)^i\left(1+\frac{\varepsilon}{\sigma}\right)^{p-1-i}\left(\frac{\varepsilon}{\sigma}\right)^i + (2d)^{p-1}\left(\frac{\varepsilon}{\sigma}\right)^{p-1}\right]$$

$$= \frac{4dp\varepsilon}{\sigma}\left(\frac{2d\varepsilon}{\sigma}+\left(1+\frac{\varepsilon}{\sigma}\right)\right)^{p-1} = \frac{4dp\varepsilon}{\sigma}\left(1+\frac{(2d+1)\varepsilon}{\sigma}\right)^{p-1}.$$

Here we used in the third line $\binom{p}{k} = \frac{p}{k}\binom{p-1}{k-1} \leq p\binom{p-1}{k-1}$, and we have set $i = k-1$. We assume now that $\varepsilon \leq \frac{\sigma}{(1+2d)p}$. Then the above can be bounded by

$$\mathrm{Prob}\{\mathscr{C}(a) \geq \varepsilon^{-1}\} \leq \frac{4dp\varepsilon}{\sigma}\left(1+\frac{1}{p-1}\right)^{p-1} \leq 4e\frac{dp\varepsilon}{\sigma},$$

which proves the asserted tail bound.

For the bound on the expectation we put $K := 4e\frac{dp}{\sigma}$ and $t_0 := K$. We have shown above that $\mathrm{Prob}\{\mathscr{C}(a) \geq \varepsilon^{-1}\} \leq Kt^{-1}$ for all $t \geq t_0 \geq (2d+1)\frac{p}{\sigma}$. Proposition 2.26 implies $\mathbb{E}\log_\beta \mathscr{C} \leq \log_\beta K + \log_\beta e$, which proves the asserted tail bound. $\qquad\square$

21.5   An Application

In Sect. 20.6 we discussed applications to the Frobenius condition number, to eigenvalue computations, and to complex polynomial equation-solving, all for complex data. With the help of Theorem 21.1, all these applications extend now to real data with minimal modifications. We won't be repetitious. Instead, we will limit this section to a single application: probabilistic estimates for the condition number $\kappa(f)$ introduced in Chap. 19 for the problem of counting real zeros of systems of real polynomials. This application implies a smoothed analysis for the number of iterations of the algorithm in Theorem 19.1 as well as for the largest number of bits, or digits, required in the finite-precision version of this algorithm (recall, by Remark 19.28, that this number is bounded as $\mathcal{O}(\log \kappa(f)) + \mathcal{O}(\log(Dn \log N))$).

Proposition 21.18 *For all $d_1, \ldots, d_n \in \mathbb{N} \setminus \{0\}$ and all $\sigma \in (0, 1]$ we have*

$$\sup_{\overline{f} \in \mathbb{S}(\mathcal{H}_{\mathbf{d}})} \ \mathbb{E}_{f \in B_{\sin}(\overline{f}, \sigma)} \left(\log_\beta \kappa(f) \right) \le \log_\beta N + \log_\beta \mathcal{D} + \log_\beta \left(Dn^2 \right) + \log_\beta \frac{1}{\sigma} + 5,$$

where $N = \dim_\mathbb{R} \mathcal{H}_{\mathbf{d}}$ and $\mathcal{D} = d_1 \cdots d_n$ is the Bézout number.

Proof Let $\Sigma \subseteq \mathcal{H}_{\mathbf{d}}$ denote the set of complex polynomial systems having a multiple zero. From Sect. 20.6.3 we know that Σ is the zero set of the discriminant $\mathsf{disc}_{\mathbf{d}}$, which is a homogeneous polynomial with integer coefficients. Moreover, by (20.8), we have $\deg \mathsf{disc}_{\mathbf{d}} \le Dn^2\mathcal{D}$.

On the other hand, by Definition 19.1, $\kappa(f) = \frac{\|f\|}{d(f, \Sigma_\mathbb{R})}$, where $\Sigma_\mathbb{R}$ denotes the set of systems in $\mathcal{H}_{\mathbf{d}}^\mathbb{R}$ having a multiple real zero. Since $\Sigma_\mathbb{R} \subseteq \Sigma_\mathbb{C}$, we can apply Theorem 21.1 to the real zero set W of $\mathsf{disc}_{\mathbf{d}}$. The assertion follows immediately, using $\log_\beta(4e^2) \le \log_2(4e^2) < 5$. $\square$

21.6   Tubes Around Convex Sets

We fill here a gap by providing a proof of Theorem 13.18, stated in Sect. 13.3. In fact, this proof will be similar to that for Theorem 21.17, but considerably simpler. The overall idea is to replace the degree argument involving Bézout's theorem by the simple geometric fact that the boundary of a spherically convex subset of $\mathbb{S}^1$ consists of at most two points.

21.6.1   Integrals of Curvature for Boundaries of Convex Sets

By a *convex body* K in $\mathbb{S}^p$ we will understand a closed convex set K such that both K and its dual $\check{K}$ have nonempty interior, i.e., both are properly convex. In

Sect. 13.2 we have also seen that the map $K \mapsto \check{K}$ is an involution on the set of convex bodies in $\mathbb{S}^p$.

By a *smooth convex body* K in $\mathbb{S}^p$ we will understand a convex body such that its boundary $V := \partial K$ is a smooth hypersurface in $\mathbb{S}^p$ and its Gaussian curvature does not vanish at any point of V. We denote by $\nu \colon V \to \mathbb{S}^p$ the unit normal vector field of the hypersurface V that points towards the interior of K. Then $T_x \mathbb{S}^p = T_x V \oplus \mathbb{R}\nu(x)$ for $x \in V$, and we have $-\nu(x) \in \check{K}$. That is, $\langle y, \nu_0 \rangle \geq 0$ for all $y \in K$.

Lemma 21.19 *Let K be a smooth convex body in $\mathbb{S}^p$ with boundary $V := \partial K$ and unit normal vector field ν. Then, for all $x \in V$, the Weingarten map $L_M(x)$ is positive definite. In particular, the principal curvatures of V at x are positive.*

Proof Let $x_0 \in V$ and $\nu_0 := \nu(x_0)$. For all $x \in K$ we have $\langle x, \nu_0 \rangle \geq 0$. Let $x(t)$ denote a smooth curve in K such that $x(0) = x_0$ and put $f(t) := \langle x(t), \nu_0 \rangle$. Then $f(0) = 0$ and $f'(0) = \langle \dot{x}, \nu_0 \rangle = 0$. Since $f(t) \geq 0$ for all t, we must have $f''(0) \geq 0$. On the other hand, since $\langle \dot{x}(t), \nu(x(t)) \rangle = 0$, we have $f''(0) = \langle \ddot{x}, \nu_0 \rangle = -\langle \dot{x}, \dot{\nu} \rangle$. It follows that $\langle \dot{x}, -\dot{\nu} \rangle \geq 0$. Since $-\dot{\nu} = L_M(x)(\dot{x})$, this implies that $L_M(x)$ is positive semidefinite. $\qquad \square$

This lemma implies that the ith curvatures of V at x are positive, and hence $\mu_i(U) = |\mu_i|(U)$ for any measurable subset U of V.

The following observation is obvious, but essential.

Lemma 21.20 *If $K \subseteq \mathbb{S}^1$ is spherically convex, nonempty, and $K \neq \mathbb{S}^1$, then we have $\#(\partial K) = 2$.* $\qquad \square$

Using some integral geometry, we can quickly derive from this observation the following bound. Considering spherical caps with angular radius almost $\pi/2$ shows that the bound is in fact sharp.

Corollary 21.21 *Any smooth convex body K in $\mathbb{S}^p$ satisfies $\mathrm{vol}\,\partial K \leq \mathcal{O}_{p-1}$.*

Proof By Proposition A.18, for almost all $g \in \mathcal{O}(p+1)$, the intersection $\partial K \cap g\mathbb{S}^1$ is zero-dimensional. Then, by Lemma 21.20, it consists of at most two points. Proposition 21.10 now implies the assertion. $\qquad \square$

Lemma 21.22 *We have $-\nu(\partial K) = \partial \check{K}$ for a smooth convex body K.*

Proof It is clear that $-\nu(\partial K) \subseteq \partial \check{K}$ from the definition of ν. For the other inclusion, let $-\nu \in \partial \check{K}$. From (13.3) we get $d_{\mathbb{S}}(-\nu, K) = \pi/2$. Hence there exists $x \in K$ such that $\langle \nu, x \rangle = 0$ and hence $\nu \in T_x \mathbb{S}^p$. From the assumption $-\nu \in \partial \check{K}$ it follows that $-\nu = \nu(x)$. $\qquad \square$

The following bound will be crucial. Again, considering spherical caps with radius almost $\pi/2$ shows the optimality of the bound.

Proposition 21.23 *If K is a smooth convex body in $\mathbb{S}^p$, then $\mu_{p-1}(\partial K) \leq 1$.*

Proof Again put $V := \partial K$. The map $v \colon V \to \partial \check{K}$ is surjective by Lemma 21.22. By (21.2) we have $K_{V,p-1}(x) = \det(-Dv(x))$ for $x \in V$. Since we assume that the Gaussian curvature does not vanish, the map v has no singular values.

We claim that v is injective. Otherwise, we would have $v(x) = v(y)$ for distinct $x, y \in V$. Since $\langle v(x), x \rangle = 0$ and $\langle v(y), y \rangle = 0$, we would have $\langle v(x), z \rangle = 0$ for all $z \in [x, y]$. Hence v would be constant along this segment. Therefore, the derivative $Dv(x)$ would vanish in the direction towards y. This implies $\det Dv(x) = 0$, contradicting our assumption.

We conclude that $-v \colon V \to v(V)$ is a diffeomorphism onto the smooth hypersurface $\partial \check{K}$. The transformation theorem yields

$$\int_V K_{V,p-1} \, dV = \int_V \det(-Dv) \, dV = \text{vol } \partial \check{K}.$$

Corollary 21.21 now implies the assertion. $\square$

Here is an analogue of Proposition 21.14, with a similar proof.

Lemma 21.24 *Let K be a smooth convex body in $\mathbb{S}^p$. For $\bar{a} \in \mathbb{S}^p$, $0 < \sigma \leq 1$, and $0 \leq i < p - 1$ we have*

$$\mu_i\big(\partial K \cap B_{\sin}(\bar{a}, \sigma)\big) \leq \frac{\mathcal{O}_{p,i+1}(\sigma)}{\mathcal{O}_p}.$$

Proof Put $U := V \cap B_{\sin}(\bar{a}, \sigma)$ and $G := \mathcal{O}(p+1)$. Let $g \in G$ be such that gV intersects $\mathbb{S}^{i+1}$ transversally. By Proposition A.18, this is the case for almost all $g \in G$. We apply Lemma 21.23 to the smooth convex body $gK \cap \mathbb{S}^{i+1}$ of the sphere $g\mathbb{S}^{i+1}$, which yields $|\mu_i|(V \cap g\mathbb{S}^{i+1}) \leq 1$. Theorem 21.11 applied to U implies that

$$\mu_i(U) = \underset{g \in G}{\mathbb{E}} \big(\mu_i\big(gU \cap \mathbb{S}^{i+1}\big)\big)$$

$$\leq \underset{g \in G}{\text{Prob}}\big\{gU \cap \mathbb{S}^{i+1} \neq \emptyset\big\}$$

$$\leq \underset{g \in G}{\text{Prob}}\big\{B_{\sin}(g\bar{a}, \sigma) \cap \mathbb{S}^{i+1} \neq \emptyset\big\}.$$

As in the proof of Proposition 21.14, we have

$$\underset{g \in G}{\text{Prob}}\big\{B_{\sin}(g\bar{a}, \sigma) \cap \mathbb{S}^{i+1} \neq \emptyset\big\} = \frac{\text{vol } T(\mathbb{S}^{i+1}, \sigma)}{\mathcal{O}_p} = \frac{\mathcal{O}_{p,i+1}(\sigma)}{\mathcal{O}_p}$$

and the assertion follows. $\square$

21.6.2 Proof of Theorem 13.18

Let K be a smooth convex body in $\mathbb{S}^p$ and let U be a measurable subset of ∂K. For $0 < \varepsilon \leq 1$ we define the *outer ε-tube* $T_o^{\perp}(U, \varepsilon)$ and the *inner ε-tube* $T_i^{\perp}(U, \varepsilon)$ of U as follows:

$$T_o^{\perp}(U, \varepsilon) := T^{\perp}(U, \varepsilon) \setminus K \quad \text{and} \quad T_i^{\perp}(U, \varepsilon) := T^{\perp}(U, \varepsilon) \cap K.$$

The following is similar to Proposition 21.16.

Lemma 21.25 *Let K be a smooth convex body with boundary $V = \partial K$. Further, let $\bar{a} \in \mathbb{S}^p$ and $0 < \sigma, \varepsilon \leq 1$. Then,*

$$\mathsf{vol}\, T_o^{\perp}\big(V \cap B_{\sin}(\bar{a}, \sigma), \varepsilon\big) \leq \frac{\mathcal{O}_{p-1}}{p} \sum_{k=1}^{p-1} \binom{p}{k} \varepsilon^k \sigma^{p-k} + \frac{1}{2} \mathcal{O}_p \varepsilon^p.$$

The same upper bound holds for the volume of $T_i^{\perp}(V \cap B_{\sin}(\bar{a}, \alpha), \varepsilon)$.

Proof Put $U := V \cap B_{\sin}(\bar{a}, \sigma)$. The proof of Theorem 21.9 actually yields separate bounds for the inner and outer tubes, having the same magnitude. Thus,

$$\mathsf{vol}\, T_o^{\perp}(U, \varepsilon) \leq \frac{1}{2} \sum_{i=0}^{p-2} \mu_i(U)\, \mathcal{O}_{p,p-1-i}(\varepsilon) + \frac{1}{2} \mu_{p-1}(U)\, \mathcal{O}_{p,0}(\varepsilon).$$

We insert the bound $\mu_i(U) \leq \mathcal{O}_{p,i+1}(\sigma)/\mathcal{O}_p$ from Lemma 21.24 as well as $\mu_{p-1}(U) \leq \mu_{p-1}(\partial K) \leq 1$ from Lemma 21.23. The assertion follows by arguing as in the proof of Proposition 21.16. $\square$

We define the *Hausdorff distance* $d_H(K, K')$ of two convex sets K and K' in $\mathbb{S}^p$ as the infimum of the real numbers $\delta \geq 0$ satisfying $K \subseteq T(K', \delta)$ and $K' \subseteq T(K, \delta)$. This defines a metric and allows us to speak about convergence of sequences of convex sets.

We state the following result without proof.

Lemma 21.26 *Any properly convex set K in $\mathbb{S}^p$ is the limit of a sequence of smooth convex bodies.* $\square$

Proof of Theorem 13.18 We assume first that K is a smooth convex body in $\mathbb{S}^p$. Let $\bar{a} \in \mathbb{S}^p$ and $0 < \sigma, \varepsilon \leq 1$. We claim that

$$T_o(\partial K, \varepsilon) \cap B_{\sin}(\bar{a}, \alpha) \subseteq T_o^{\perp}\big(\partial K \cap B_{\sin}(\bar{a}, \sigma + \varepsilon), \varepsilon\big). \tag{21.16}$$

In order to see this, note that $T_o(\partial K, \varepsilon) = T_o^{\perp}(\partial K, \varepsilon)$ by Lemma 21.3. Suppose now that $x \in T_o(\partial K, \varepsilon) \cap B_{\sin}(\bar{a}, \alpha)$. Then there exists $y \in \partial K$ with $d_{\sin}(x, y) \leq \varepsilon$

such that the arc $[x, y]$ of a great circle intersects ∂K orthogonally at y. The triangle inequality for projective distance implies $d_{\sin}(a, y) < d_{\sin}(a, x) + d_{\sin}(x, y) \leq \sigma + \varepsilon$. Hence $y \in \partial K \cap B_{\sin}(\overline{a}, \sigma + \varepsilon)$, which proves the claim.

By combining (21.16) with the bound of Lemma 21.25 we get

$$\operatorname{vol}\left(T_o(\partial K, \varepsilon) \cap B_{\sin}(\overline{a}, \sigma)\right) \leq \frac{\mathcal{O}_{p-1}}{p} \sum_{k=1}^{p-1} \binom{p}{k} \varepsilon^k (\sigma + \varepsilon)^{p-k} + \frac{1}{2} \mathcal{O}_p \varepsilon^p.$$

Lemma 2.34 states that

$$\operatorname{vol} B_{\sin}(\overline{a}, \sigma) \geq \frac{2\mathcal{O}_{p-1}}{p} \sigma^p \geq \frac{1}{2\sqrt{p}} \mathcal{O}_p \sigma^p.$$

Using this, we obtain

$$\frac{\operatorname{vol}(T_o(\partial K, \varepsilon) \cap B_{\sin}(\overline{a}, \sigma))}{\operatorname{vol} B_{\sin}(\overline{a}, \sigma)} \leq \frac{1}{2} \sum_{k=1}^{p-1} \binom{p}{k} \left(1 + \frac{\varepsilon}{\sigma}\right)^{p-k} \left(\frac{\varepsilon}{\sigma}\right)^k + \sqrt{p} \left(\frac{\varepsilon}{\sigma}\right)^p.$$

Bounding this as in Sect. 21.4.3 we obtain

$$\frac{\operatorname{vol}(T_o(\partial K, \varepsilon) \cap B_{\sin}(\overline{a}, \sigma))}{\operatorname{vol} B_{\sin}(\overline{a}, \sigma)} \leq \frac{p\varepsilon}{\sigma} \left(1 + \frac{2\varepsilon}{\sigma}\right)^{p-1} \leq e \frac{p\varepsilon}{\sigma},$$

the second inequality holding when $\varepsilon \leq \frac{\sigma}{2p}$.

This proves the assertion of Theorem 13.18 for the outer neighborhood in the case that K is a smooth convex body. The bound for the inner neighborhood is proved similarly.

The general case, in which K is any properly convex set in $\mathbb{S}^p$, follows now by a perturbation argument using Lemma 21.26. $\square$

21.7   Conic Condition Numbers and Structured Data

We may now return to a discussion we pursued in our first Intermezzo. We described there two ways of approaching condition for structured data: either by considering structured perturbations or by taking a relativized inverse of the distance to structured ill-posedness. In the latter case, letting Triang be the class of lower triangular real $n \times n$ matrices, definition (I.2) becomes

$$\mathscr{D}_{\mathsf{Triang}}(L) = \frac{\|L\|}{d(L, \Sigma \cap \mathsf{Triang})},$$

where Σ denotes the set of singular matrices. We can apply Theorem 21.1 to this condition number: using that $\mathsf{Triang} \simeq \mathbb{R}^{\frac{n^2+n}{2}}$ and the fact that $\Sigma \cap \mathsf{Triang}$ is the zero set of the polynomial $\prod_{i=1}^{n} x_{ii}$ having degree n, we obtain

$$\underset{L \in \mathbb{S}(\text{Triang})}{\mathbb{E}} \left(\log_\beta \mathscr{Q}_{\text{Triang}}(L) \right) = \mathcal{O}(\log_\beta n).$$

Note that the restriction to structured ill-posedness is crucial here. If we take $\mathscr{Q}(L) = \frac{\|L\|}{d(L, \Sigma)}$, we obtain instead

$$\underset{L \in \mathbb{S}(\text{Triang})}{\mathbb{E}} \left(\log_\beta \mathscr{Q}(L) \right) = \underset{L \in \mathbb{S}(\text{Triang})}{\mathbb{E}} \left(\log_\beta \|L\| \|L^{-1}\| \right)$$

$$= \underset{L \in \mathbb{S}(\text{Triang})}{\mathbb{E}} \left(\log_\beta \kappa(L) \right) = \Omega(n),$$

the first equality by the condition number theorem (Theorem 1.7) and the inequality by Theorem 3.1. We conclude that, broadly speaking, triangular matrices are generally close to singular matrices but not to triangular singular matrices.

Regarding the first (and most basic) definition of structured condition number for triangular matrices we can say little. For, say, the problem of linear-equation solving, definition (I.1) becomes

$$\text{cond}_{\text{Triang}}(L, b) = \lim_{\delta \to 0} \underset{\substack{\text{RelError}(L, b) \le \delta \\ \tilde{a} \in \text{Triang}}}{\sup} \frac{\text{RelError}(L^{-1} b)}{\text{RelError}(L, b)}.$$

We did not need to estimate this quantity, because the backward error analysis in Proposition 3.5 revealed a dependence on the componentwise condition number, which, by its definition, is structured for triangular matrices (and, more generally, for all structures given by sparsity patterns).

Remark 21.27 We finally observe that probability bounds for conic condition numbers $\mathscr{Q}$, as obtained in this and in the previous chapter, readily imply probability bounds for their structured counterparts $\mathscr{Q}_S$, when the subclass S of input data is defined by a sparsity pattern. This is due to the fact that the degree of the set of structured ill-posed data $\Sigma \cap S$ is in this case bounded above by the degree of Σ.

Triangular matrices are but one instance of this phenomenon.

21.8 Smoothed Analysis for Adversarial Distributions

We close this chapter, and with it the third and last part of this book by returning to a statement we made at the end of Sect. 2.2.7. We mentioned there that "there is an emerging impression that smoothed analysis is robust in the sense that its dependence on the chosen family of measures is low." We may now give substance to this statement.

We will do so while remaining in the context of this chapter, that is, a conic condition number

$$\mathscr{C}(a) = \frac{\|a\|}{\mathrm{dist}(a, \Sigma)},$$

induced by a cone of ill-posed inputs $\Sigma \subseteq \mathbb{R}^{p+1}$ satisfying the symmetry condition $\Sigma = -\Sigma$. The main result in this chapter, Theorem 21.1, provided a smoothed analysis for the uniform measures on the balls $B(\bar{a}, \sigma)$ (or equivalently, $B_{\sin}(\bar{a}, \sigma)$), for $\bar{a} \in \mathbb{S}^p$ and $\sigma \in [0, 1]$. The goal in this section is to show similar bounds when we replace this uniform measure by one denser around $\bar{a}$. Indeed, we will consider an *adversarial distribution*, that is, one given by a continuous probability density f, radially symmetric, and having a pole of order $-\gamma$ at $\bar{a}$ for some $\gamma > 0$. To formally define this class of distributions it will be helpful to introduce some notation. For $s \in \mathbb{R}$ with $s > 0$ and $0 \le \sigma \le 1$ we define

$$J_s(\sigma) := \int_0^\sigma \frac{r^{s-1}}{\sqrt{1 - r^2}} \, dr.$$

This is a well-defined quantity in the sense that the integral is finite. Moreover, when $s \ge 1$ is an integer, we have $J_s(\sigma) = \frac{\mathcal{O}_{s,0}(\sigma)}{2\mathcal{O}_{s-1}}$ (cf. Lemma 20.5). Furthermore, by Lemma 2.31, and writing $\alpha := \arcsin \sigma$, $\alpha \in [0, \frac{\pi}{2}]$,

$$\mathrm{vol}\, B(\bar{a}, \sigma) = \mathrm{vol}\, \mathrm{cap}(\bar{a}, \alpha) = \mathcal{O}_{p-1} \cdot \int_0^\alpha (\sin\theta)^{p-1} \, d\theta = \mathcal{O}_{p-1} \cdot J_p(\sigma), \quad (21.17)$$

the last equality following from the change of variables $r = \sin\theta$.

Definition 21.28 Fix $\bar{a} \in \mathbb{S}^p$ and $\sigma \in [0, 1]$. An *adversarial distribution* on $B(\bar{a}, \sigma)$ of *order* γ, for $0 \le \gamma < p$, is a probability measure given by a density $f : B(\bar{a}, \sigma) \to [0, \infty)$ of the form $f(x) = g(d_{\sin}(x, \bar{a}))$, with a monotonically decreasing function $g : [0, \sigma] \to [0, \infty]$ of the form

$$g(r) = r^{-\gamma} \cdot h(r).$$

Here $h : [0, \sigma] \to \mathbb{R}_+$ is a continuous function satisfying $h(0) \ne 0$. We require that $\int_{B(\bar{a}, \sigma)} f(x) \, dx = 1$, the integral being with respect to the volume measure on $\mathbb{S}^p$.

The simplest choice of h is a constant function $h = C$. In this case, we can use polar coordinates on $\mathbb{S}^p$ to deduce (cf. Theorem 2.3)

$$\int_{B(\bar{a}, \sigma)} f(x) \, dx = \int_{u \in \mathbb{S}^{p-1}} du \int_0^{\arcsin \sigma} g(\sin\theta)(\sin\theta)^{p-1} \, d\theta$$

$$= \mathcal{O}_{p-1} C \int_0^\sigma \frac{r^{p-\gamma-1}}{\sqrt{1-r^2}} \, dr = \mathcal{O}_{p-1} C J_{p-\gamma}(\sigma) = 1.$$

So we must have $C = C_{\gamma,\sigma} := (\mathcal{O}_{p-1} J_{p-\gamma}(\sigma))^{-1}$.

In what follows, we fix an adversarial distribution with density f and denote by μ the corresponding probability measure on $B(\overline{a}, \sigma)$. We associate to this distribution the quantity $H := \sup_{0 \le r \le \sigma} h(r)/C_{\gamma,\sigma}$. Note that $H \ge 1$, since otherwise, using Theorem 2.3 as above,

$$\mu\big(B(\overline{a}, \sigma)\big) = \int_{B(\overline{a},\sigma)} f(x)\,dx < \mathcal{O}_{p-1} C_{\gamma,\sigma} \int_0^\sigma \frac{r^{p-\gamma-1}}{\sqrt{1-r^2}}\,dr = 1,$$

and f would not be a density. This also shows that $H = 1$ implies $h = C_{\gamma,\sigma}$.

The main result of this section is the following.

Theorem 21.29 *Let $\mathscr{C}$ be a conic condition number with set of ill-posed inputs $\Sigma \subseteq \mathbb{S}^p$. Assume that Σ is contained in a real algebraic hypersurface, given as the zero set of a homogeneous polynomial of degree d. Then, for all $\overline{a} \in \mathbb{S}^p$, all $0 < \sigma \le 1$, and all adversarial distributions μ on $B(\overline{a}, \sigma)$,*

$$\mathop{\mathbb{E}}_{a\sim\mu} \log \mathscr{C}(a) \le 2\log p + \log d + |\log\sigma| + \log(6\pi)$$

$$+ \frac{2}{1 - \frac{\gamma}{p}} \log\left(eH\sqrt{\frac{2p}{\ln(\pi p/2)}}\right).$$

Here $\log$ stands for $\log_\beta$, γ is the order of μ, and H is as above.

The proof of Theorem 21.29 relies on Proposition 21.30 below, which allows us to bound the μ-measure of "small sets" $B \subseteq B(\overline{a}, \sigma)$ in terms of their uniform measure on $B(\overline{a}, \sigma)$. To distinguish between the two measures, we will denote the latter by ν, i.e.,

$$\nu(B) := \frac{\mathrm{vol}(B \cap B(\overline{a}, \sigma))}{\mathrm{vol}\, B(\overline{a}, \sigma)}.$$

Proposition 21.30 *For $0 < \varepsilon < 1 - \frac{\gamma}{p}$ there exists $0 < \delta_\varepsilon \le \sigma$ such that for all $B \subseteq \mathbb{S}^p$ with $\nu(B) \le \delta_\varepsilon$ we have $\mu(B) \le \nu(B)^{1-\frac{\gamma}{p}-\varepsilon}$. Moreover, we may take $\delta_\varepsilon := J_p(\rho_\varepsilon)/J_p(\sigma) \le 1$, where*

$$\rho_\varepsilon := \sigma \cdot \left(\frac{1}{H} \cdot \sqrt{1 - \left(\frac{2}{\pi p}\right)^{(1-\frac{\gamma}{p}-\varepsilon)/(p\varepsilon)}}\right)^{\frac{1}{\varepsilon p}} \left(\sqrt{\frac{2}{\pi p}}\right)^{(1-\frac{\gamma}{p}-\varepsilon)\frac{1}{\varepsilon p}}.$$

The following result is similar to Lemma 2.34. We omit the proof.

Lemma 21.31 *We have for $0 \le \sigma < 1$,*

$$\frac{\sigma^p}{p} \le J_p(\sigma) \le \min\left\{\frac{1}{\sqrt{1-\sigma^2}}, \sqrt{\frac{\pi p}{2}}\right\} \cdot \frac{\sigma^p}{p}. \qquad \square$$

Using Proposition 21.30, it is not hard to give a proof of the main result in this section.

Proof of Theorem 21.29 From Theorem 21.1 it follows that for all $0 < \sigma \leq 1$, all $\bar{a} \in \mathbb{S}^p$, and all $t \geq t_* := \frac{12dp}{\sigma}$, we have

$$\operatorname*{Prob}_{a \sim v}\{\mathscr{C}(a) \geq t\} \leq 4e\,dp\,\frac{1}{\sigma t}.$$

Set $\varepsilon := \frac{1}{2}(1 - \frac{\gamma}{p})$ and $t_\varepsilon := \frac{t_*}{\delta_\varepsilon}$. Then we have $1 - \frac{\gamma}{p} - \varepsilon = \varepsilon$ and hence

$$\rho_\varepsilon = \sigma \cdot \left(\frac{1}{H} \cdot \sqrt{1 - \left(\frac{2}{\pi p}\right)^{\frac{1}{p}}}\right)^{\frac{1}{\varepsilon p}} \left(\sqrt{\frac{2}{\pi p}}\right)^{\frac{1}{p}}. \tag{21.18}$$

Since $\delta_\varepsilon \leq 1$, we have $t_\varepsilon \geq t_*$. Moreover,

$$\frac{4edp}{\sigma t_\varepsilon} = 4edp\,\frac{\delta_\varepsilon}{\sigma t_*} = \frac{4edp}{12dp}\delta_\varepsilon \leq \delta_\varepsilon.$$

Therefore, we may apply Proposition 21.30 to deduce, for all $t \geq t_\varepsilon$,

$$\operatorname*{Prob}_{a \sim \mu}\{\mathscr{C}(a) \geq t\} \leq \left(\operatorname*{Prob}_{a \sim v}\{\mathscr{C}(a) \geq t\}\right)^\varepsilon \leq \left(\frac{4edp}{\sigma t}\right)^\varepsilon.$$

Thus the hypotheses of Proposition 2.26 are satisfied with $\alpha = \varepsilon$, $K = (\frac{4edp}{\sigma})^\varepsilon$, and $t_0 = t_\varepsilon$. Clearly, $t_0^\alpha \geq K$. Therefore, this proposition implies that

$$\operatorname*{\mathbb{E}}_{a \sim \mu} \log \mathscr{C}(a) \leq \log t_0 + \frac{1}{\varepsilon}\log e, \tag{21.19}$$

where $\log$ denotes $\log_\beta$. Furthermore,

$$\log t_0 = \log p + \log d + |\log \sigma| + \log(12) + \log \frac{1}{\delta_\varepsilon}, \tag{21.20}$$

so we only need to bound the last term on the right-hand side. But Lemma 21.31 implies that

$$\delta_\varepsilon = \frac{J_p(\rho_\varepsilon)}{J_p(\sigma)} \geq \sqrt{\frac{2}{\pi p}}\left(\frac{\rho_\varepsilon}{\sigma}\right)^p.$$

Hence, using (21.18), we get

$$\delta_\varepsilon \geq \frac{2}{\pi p}\left(\frac{1}{H} \cdot \sqrt{1 - \left(\frac{2}{\pi p}\right)^{\frac{1}{p}}}\right)^{\frac{1}{\varepsilon}}.$$

A small calculation shows that $(1 - (\frac{2}{\pi p})^{\frac{1}{p}})^{-1/2} \le \sqrt{\frac{2p}{\ln(\pi p/2)}}$. Consequently,

$$\log \frac{1}{\delta_\varepsilon} \le \log p + \log \frac{\pi}{2} + \frac{1}{\varepsilon} \log\left(H\sqrt{\frac{2p}{\ln(\pi p/2)}}\right). \qquad (21.21)$$

We conclude from (21.19), (21.20), and (21.21) that

$$\mathop{\mathbb{E}}_{a\sim\mu} \log \mathscr{C}(a) \le 2\log p + \log d + |\log \sigma| + \log(6\pi) + \frac{1}{\varepsilon} \log\left(eH\sqrt{\frac{2p}{\ln(\pi p/2)}}\right).$$

$\square$

It remains to prove Proposition 21.30. The next lemma shows that we may restrict attention to the case that B is a ball centered at $\bar{a}$.

Lemma 21.32 *Let $0 < \delta < 1$. Then among all measurable sets $B \subseteq B(\bar{a}, \sigma)$ with $0 < v(B) \le \delta$, the quantity $\mu(B)$ is maximized by $B(\bar{a}, \rho)$, where $\rho \in (0, \sigma)$ is chosen so that $v(B(\bar{a}, \rho)) = \delta$.*

Proof It clearly suffices to show that

$$\int_B f(x)\,dx \le \int_{B(\bar{a},\rho)} f(x)\,dx$$

for all $B \subseteq B(\bar{a}, \sigma)$ such that $v(B) = \delta$. To prove this inequality, first note that

$$v\big(B \setminus B(\bar{a}, \rho)\big) = v(B) - v\big(B \cap B(\bar{a}, \rho)\big) = \delta - v\big(B \cap B(\bar{a}, \rho)\big)$$
$$= v\big(B(\bar{a}, \rho)\big) - v\big(B \cap B(\bar{a}, \rho)\big) = v\big(B(\bar{a}, \rho) \setminus B\big). \quad (21.22)$$

Then,

$$\begin{aligned}
\int_B f(x)\,dx &= \int_{B\cap B(\bar{a},\rho)} f(x)\,dx + \int_{B\setminus B(\bar{a},\rho)} f(x)\,dx \\
&\le \int_{B\cap B(\bar{a},\rho)} f(x)\,dx + g(\rho)v\big(B \setminus B(\bar{a},\rho)\big) \\
&\stackrel{(21.22)}{=} \int_{B\cap B(\bar{a},\rho)} f(x)\,dx + g(\rho)\,v\big(B(\bar{a},\rho) \setminus B\big) \\
&\le \int_{B\cap B(\bar{a},\rho)} f(x)\,dx + \int_{B(\bar{a},\rho)\setminus B} f(x)\,dx \\
&= \int_{B(\bar{a},\rho)} f(x)\,dx,
\end{aligned}$$

where we have used the monotonicity of g in the two inequalities. This proves our claim. $\square$

Proof of Proposition 21.30 According to Lemma 21.32 we may take $B = B(\bar{a}, \rho)$. The uniform measure of B is given by (cf. (21.17))

$$\nu\big(B(\bar{a}, \rho)\big) = \frac{J_p(\rho)}{J_p(\sigma)}. \tag{21.23}$$

To estimate the μ-measure of B we again use spherical coordinates on $\mathbb{S}^p$. Recalling the definition of the parameters $C_{\gamma,\sigma}$ and H, we obtain

$$\mu\big(B(\bar{a}, \rho)\big) = \int_{B(\bar{a},\rho)} f(x)\,dx = \mathcal{O}_{p-1} \int_0^\rho r^{-\gamma} h(r) \frac{r^{p-1}}{\sqrt{1-r^2}}\,dr$$

$$= \frac{1}{J_{p-\gamma}(\sigma)} \int_0^\rho \frac{h(r)}{C_{\gamma,\sigma}} \frac{r^{p-\gamma-1}}{\sqrt{1-r^2}}\,dr \leq H \cdot \frac{J_{p-\gamma}(\rho)}{J_{p-\gamma}(\sigma)}. \tag{21.24}$$

By (21.23) and (21.24) our task amounts to showing that

$$H \cdot \frac{J_{p-\gamma}(\rho)}{J_{p-\gamma}(\sigma)} \leq \left(\frac{J_p(\rho)}{J_p(\sigma)}\right)^{1-\frac{\gamma}{p}-\varepsilon}$$

for $\rho \leq \rho_\varepsilon$. And indeed, using Lemma 21.31, we get

$$H \cdot \frac{J_{p-\gamma}(\rho)}{J_{p-\gamma}(\sigma)} \leq H \frac{1}{\sqrt{1-\rho^2}} \cdot \left(\frac{\rho}{\sigma}\right)^{p-\gamma}$$

$$\leq H \frac{1}{\sqrt{1-\rho^2}} \cdot \left(\left(\frac{\rho}{\sigma}\right)^p\right)^{1-\frac{\gamma}{p}-\varepsilon} \left(\frac{\rho_\varepsilon}{\sigma}\right)^{\varepsilon p}$$

$$\leq \frac{\sqrt{1-(\frac{2}{\pi p})^{(1-\frac{\gamma}{p}-\varepsilon)/(p\varepsilon)}}}{\sqrt{1-\rho^2}} \cdot \left(\sqrt{\frac{2}{\pi p}}\left(\frac{\rho}{\sigma}\right)^p\right)^{1-\frac{\gamma}{p}-\varepsilon}$$

$$\leq \frac{\sqrt{1-(\frac{2}{\pi p})^{(1-\frac{\gamma}{p}-\varepsilon)/(p\varepsilon)}}}{\sqrt{1-\rho^2}} \cdot \left(\frac{J_p(\rho)}{J_p(\sigma)}\right)^{1-\frac{\gamma}{p}-\varepsilon},$$

where for the last inequality we used Lemma 21.31 again, and for the one before the last, the definition of ρ_ε. Moreover, we have

$$\rho \leq \rho_\varepsilon \leq \left(\sqrt{\frac{2}{\pi p}}\right)^{(1-\frac{\gamma}{p}-\varepsilon)\frac{1}{\varepsilon p}}.$$

Therefore, $\sqrt{1-(\frac{2}{\pi p})^{(1-\frac{\gamma}{p}-\varepsilon)\frac{1}{\varepsilon p}}} \leq \sqrt{1-\rho^2}$, completing the proof. $\qquad\square$

Remark 21.33 Theorem 21.29 admits a complex version extending Theorem 20.1. We will not spell out the details of this result.

Appendix

A.1 Big Oh, Little Oh, and Other Comparisons

The possibility of having more than one algorithm available for solving a given problem raises the matter of a comparison between these algorithms. Such a comparison may be difficult to do, due to the conflicting nature of some of the criteria one wants to optimize, but simplifying to an extreme, we may assume that we are interested here in comparing speed, that is, the number of arithmetic operations performed by these algorithms. Suppose therefore that we have two algorithms $\mathscr{F}$ and $\mathscr{G}$ and let us denote by $f(n)$ and $g(n)$ the cost of these algorithms over inputs of size n (these costs can be worst-case or average-case: this is irrelevant to our present discussion). Ideally, we would like to compare f and g, but as soon as we try to do so we face two obstacles:

(a) Both f and g may be hard to determine exactly, so that the best we can do is to approximate them.
(b) Even if we had exact expressions for f and g, the sequences of values $(f(n))_{n \in \mathbb{N}}$ and $(g(n))_{n \in \mathbb{N}}$ may be difficult to compare because neither of these sequences dominates the other.

A way out of both obstacles is to compare the behaviors of f and g "near infinity." For this, one first replaces f and g by approximations that are simple to manipulate and, hopefully, accurate enough for large values of n. In what follows we provide the definitions and notation commonly used to carry out this procedure.

Given functions $f, h : \mathbb{N} \to \mathbb{R}$ such that $h(n) > 0$ for all sufficiently large values of n, we say that f is big oh of h—and we write $f = \mathcal{O}(h)$—when

$$\exists n_0, C > 0 \quad \text{s.t.} \quad \forall n \geq n_0 \quad |f(n)| \leq Ch(n). \tag{A.1}$$

In fact, the condition $f = \mathcal{O}(h)$ just means that $|f(n)|/h(n)$ is bounded. But sometimes we may want to speak about the implicit constant C in (A.1): note that the infimum of the possible constants $C > 0$ equals $\limsup_{n \to \infty} \frac{|f(n)|}{h(n)}$.

P. Bürgisser, F. Cucker, *Condition*,
Grundlehren der mathematischen Wissenschaften 349,
DOI 10.1007/978-3-642-38896-5, © Springer-Verlag Berlin Heidelberg 2013

Similarly, we say that f *is big omega of* h—and we write $f = \Omega(h)$— when

$$\liminf_{n \to \infty} \frac{|f(n)|}{h(n)} > 0.$$

In our initial discussion, if, for instance, $f = \mathcal{O}(n^2)$ and $g = \Omega(n^3)$, then we should choose algorithm $\mathscr{F}$. This does not mean that $\mathscr{F}$ will necessarily be faster in practice than $\mathscr{G}$. The constants n_0 and C in (A.1) could be both so large as to make the comparison $n^2 < n^3$ irrelevant. But while it is important to keep this warning in mind, it is also true that much more often than not, asymptotic estimates are useful in practice.

There are other notations that are worth introducing. We say that f *is theta of* h—and we write $f = \Theta(h)$—when $f = \mathcal{O}(h)$ and $f = \Omega(h)$. Finally, we say that f *is little oh of* h—and we write $f = o(h)$—when

$$\lim_{n \to \infty} \frac{f(n)}{h(n)} = 0.$$

In particular, a function f is $o(1)$ when $\lim_{n \to \infty} f(n) = 0$.

These definitions allow one to concisely express the growth of some functions such as

$$f(n) = 6n^3 + \mathcal{O}(n \log n).$$

This means that there exists a function $g \colon \mathbb{N} \to \mathbb{R}$ such that $f(n) = 6n^3 + g(n)$ and $g(n) = \mathcal{O}(n \log n)$. Roughly, the error incurred in approximating $f(n)$ by $6n^3$ grows at most as a multiple of $n \log n$. Similarly, one defines

$$f(n) = 6n^3 + o(n^2),$$

which asserts that this error is (asymptotically) negligible when compared with n^2.

In the discussion above there are two issues that deserve to be pointed out. Firstly, there is no need for the argument of the function at hand to be a natural number. It can perfectly well be a real argument, and the definitions above apply with only the obvious modifications. Secondly, there is no need to consider asymptotics for the argument approaching infinity. An often occurring case is that of the argument approaching 0 (from the right). Again, the definitions above apply mutatis mutandis. It is a must, however, to specify, in using asymptotic notation, which argument we are considering and which limit this argument is approaching.

We won't elaborate more on this topic. The interested reader can find a more detailed exposition in [110, Chap. 9].

A.2 Differential Geometry

We briefly outline the concepts from differential geometry that were used in Part III of this book. The reader should be familiar with basic notions from calculus, in particular with the derivative $Df(x)\colon \mathbb{R}^m \to \mathbb{R}^n$ of a multivariate map $f\colon \mathbb{R}^m \to \mathbb{R}^n$.

A.2.1 Submanifolds of $\mathbb{R}^n$

By a *smooth map* $O \to \mathbb{R}^n$ defined on an open subset O of $\mathbb{R}^m$ we shall understand a map that has continuous partial derivatives of every order. A *diffeomorphism* is a smooth bijective map such that its inverse is smooth (i.e., C^∞) as well.

The implicit function theorem is a fundamental result in analysis; see, for instance, [209].

Theorem A.1 (Implicit function theorem) *Let $F\colon O \to \mathbb{R}^{n_2}$ be a smooth map defined on an open subset $O \subseteq \mathbb{R}^{n_1} \times \mathbb{R}^{n_2}$ and let $(x_0, y_0) \in O$ be such that $F(x_0, y_0) = 0$. Further, assume that the matrix*

$$\frac{\partial F}{\partial y}(x_0, y_0) := \left[\frac{\partial F_i}{\partial y_j}(x_0, y_0) \right]_{1 \le i, j \le n_2}$$

is invertible. Then there exist open subsets $O_1 \subseteq \mathbb{R}^{n_1}$ and $O_2 \subseteq \mathbb{R}^{n_2}$ such that $(x_0, y_0) \in O_1 \times O_2 \subseteq O$ and with the property that for all $x \in O_1$ there exists exactly one $y \in O_2$ such that $F(x, y) = 0$. Moreover, the function $G\colon O_1 \to O_2$ mapping x to y is smooth.   □

A k-dimensional submanifold M of the Euclidean space $\mathbb{R}^n$ is a subset that, locally around any point of M, looks like $\mathbb{R}^k$ embedded in $\mathbb{R}^n$. Here is the formal definition.

Definition A.2 A nonempty subset $M \subseteq \mathbb{R}^n$ is called a *k-dimensional submanifold* if for all $x \in M$, there exists a diffeomorphism φ from an open neighborhood $U \subseteq \mathbb{R}^n$ of x to an open neighborhood $V \subseteq \mathbb{R}^n$ of 0 such that $\varphi(M \cap U) = (\mathbb{R}^k \times \{0\}) \cap V$.

Let $M \subseteq \mathbb{R}^n$ be a submanifold and $p \in M$. A smooth map $\gamma\colon \mathbb{R} \to M$ such that $\gamma(0) = p$ parameterizes a curve on M passing through p. Its derivative $\dot{\gamma} := \frac{d}{dt}\gamma(0) \in \mathbb{R}^n$ shall be called a *tangent vector* of M at p. We define the *tangent space* T_pM of M at p as the set of all tangent vectors of M at p. In order to see that T_pM is a k-dimensional linear subspace of M, suppose that $\varphi(M \cap U) = (\mathbb{R}^k \times \{0\}) \cap V$ as in Definition A.2 and $\varphi(p) = 0$. Let ψ be the restriction of φ^{-1} to $\mathbb{R}^k \times \{0\}$. Then it is easy to check that the derivative $D\psi(0)$ is nonsingular and that T_pM equals the image of $D\psi(0)$.

Corollary A.3 *Let $F\colon O \to \mathbb{R}^{n_2}$ be a smooth map defined on the open subset $O \subseteq \mathbb{R}^n$ and assume $M := F^{-1}(0)$ to be nonempty. Further, assume that 0 is a regular value of F, that is, the derivative $DF(p)$ is surjective for all $p \in M$. Then M is a submanifold of $\mathbb{R}^n$ of dimension $n - n_2$. Moreover, $T_pM = \ker D\varphi(x)$.*

Proof Let $(x_0, y_0) \in F^{-1}(0)$. After a permutation of the coordinates, we may assume without loss of generality that $\frac{\partial F}{\partial y}(x_0, y_0)$ is invertible. In the setting of Theorem A.1, we have the diffeomorphism $O_1 \times \mathbb{R}^{n_2} \to O_1 \times \mathbb{R}^{n_2}$, $(x, y) \mapsto (x, G(x) + z)$, which maps $O_1 \times \{0\}$ to $M \cap (O_1 \times O_2)$. This shows that M is an n_1-dimensional submanifold of $\mathbb{R}^n$.

Suppose that $\gamma \colon \mathbb{R} \to M$ is smooth and $\gamma(0) = p$. Then $F \circ \gamma = 0$ and hence, by the chain rule, $DF(\dot{\gamma}) = 0$. This implies that $T_p M \subseteq \ker DF(p)$. Comparing dimensions we see that equality holds. $\qquad\square$

For a first application of this corollary, consider $F \colon \mathbb{R}^n \to \mathbb{R}$, $x \mapsto \|x\|^2 - 1$. Since $DF(x) \neq 0$ for all $x \neq 0$, the sphere $\mathbb{S}^{n-1} = F^{-1}(0)$ is a submanifold of $\mathbb{R}^n$ with dimension $n - 1$.

We analyze two further important examples of submanifolds using Corollary A.3

Proposition A.4 *Let $1 \leq m \leq n$. The set $\mathrm{St}_{n,m}$ of $n \times m$ matrices A satisfying $A^T A = I_m$ is a compact submanifold of $\mathbb{R}^{n \times m}$ of codimension $m(m+1)/2$. It is called a* Stiefel manifold. *The tangent space of $\mathrm{St}_{n,m}$ at $A = (I_m, 0)^T$ is the set of matrices $(\dot{B}, \dot{C})^T$ where $\dot{B} + \dot{B}^T = 0$ and $\dot{C} \in \mathbb{R}^{m \times (n-m)}$ is arbitrary.*

Proof Let S denote the vector space of symmetric $m \times m$ matrices and consider the map $F \colon \mathbb{R}^{m \times n} \to S$, $A \mapsto A^T A - I_m$. Note that $\mathrm{St}_{n,m} = F^{-1}(0)$, which is compact, since the columns of A have norm 1. The derivative of F at A is given by $\mathbb{R}^{m \times n} \to S$, $\dot{A} \mapsto \dot{A}^T A + A^T \dot{A}$. We claim that this derivative is surjective if A has full rank. In order to see this, write $A = (B, C)^T$ and assume without loss of generality that B is invertible. Further, put $\dot{A} = (\dot{B}, 0)^T$. Now it is easy to see that $\dot{B} \mapsto \dot{B}^T B + B^T \dot{B}$ surjectively maps $\mathbb{R}^{m \times m}$ onto S. Hence I_m is a regular value of F. Corollary A.3 implies the assertion. $\qquad\square$

An important special case is the orthogonal group $\mathcal{O}(n) := \{A \in \mathbb{R}^{n \times n} \mid A^T A = I_n\} = \mathrm{St}_{n,n}$, which is, by Proposition A.4, a compact submanifold of $\mathbb{R}^{n \times n}$ having dimension $n(n-1)/2$.

Proposition A.5 *The set M_r of $m \times n$ real matrices of rank r is a submanifold of $\mathbb{R}^{m \times n}$ of codimension $(m - r)(n - r)$.*

Proof Let U denote the open subset of M_r given by the matrices A having the block form

$$A = \begin{pmatrix} B & C \\ D & E \end{pmatrix},$$

where $B \in \mathbb{C}^{r \times r}$ is invertible and $C \in \mathbb{R}^{r \times (n-r)}$, $D \in \mathbb{R}^{(m-r) \times r}$, $E \in \mathbb{R}^{(m-r) \times (n-r)}$ are arbitrary. By multiplying A by the nonsingular matrix

$$\begin{pmatrix} I_r & -B^{-1}C \\ 0 & I_{n-r} \end{pmatrix},$$

we see that $\operatorname{rank} A = r$ iff $E - DB^{-1}C = 0$. Hence $M_r \cap U$ is obtained as the fiber over zero of the smooth map $\mathbb{R}^{m \times n} \to \mathbb{R}^{m-r(n-r)}$, $A \mapsto E - DB^{-1}C$. It is easy to check that 0 is a regular value of this map. Corollary A.3 implies that $M_r \cap U$ is a submanifold of $\mathbb{R}^{m \times n}$ with codimension $(m - r)(n - r)$. Since M_r is the union of the sets U' obtained by requiring the nonvanishing of other $r \times r$ minors, the assertion follows.   □

A.2.2 Abstract Smooth Manifolds

Complex projective space and its relatives play an important role in Part III of this book. They are not naturally embedded as submanifolds of Euclidean spaces. For this reason, we briefly introduce the abstract concept of smooth manifolds. The emphasis here is on the definition of concepts—the proofs of the stated facts are all straightforward consequences of the corresponding facts for Euclidean spaces. A more detailed treatment can be found, for instance, in [133] or [40].

Let M be a topological space. By a n-dimensional *chart* (U, φ) of M we understand a homeomorphism $\varphi \colon U \to V$ of a nonempty open subset $U \subseteq M$ to an open subset $V \subseteq \mathbb{R}^n$. Note that φ allows us to represent points in $p \in U$ by their *coordinates* $x(p) = (x_1(p), \ldots, x_n(p))$ in n-dimensional Euclidean space $\mathbb{R}^n$. Two charts (U_1, φ_1) and (U_2, φ_2) are called *compatible* if the transition map

$$\varphi_2 \circ \varphi_1^{-1} \colon \varphi_1(U_1 \cap U_2) \to \varphi_2(U_1 \cap U_2)$$

is a diffeomorphism. An n-dimensional *atlas* is a family (U_i, φ_i), $i \in I$, of n-dimensional charts that are pairwise compatible and such that the U_i cover M, i.e., $\bigcup_{i \in I} U_i = M$. Two atlases of M are called *equivalent* if each chart of one atlas is compatible to each chart of the other atlas.

Definition A.6 A (smooth) n-dimensional manifold M is a topological space that is Hausdorff and has a countable basis, together with an equivalence class of n-dimensional atlases on it. One writes $\dim M = n$ for the dimension of M.

The assumptions on the topology on M are required to exclude bizarre situations and need not bother us.

Here are two obvious examples of this general concept. A nonempty open subset U of $\mathbb{R}^n$ naturally becomes an n-dimensional manifold: just take the atlas consisting of the identity map on U. Further, a submanifold M of $\mathbb{R}^n$ is a manifold. Indeed, let $M \subseteq \mathbb{R}^n$ be a k-dimensional submanifold. By definition, it comes with a family of diffeomorphisms $\varphi \colon U \to V$ such that $\varphi(M \cap U) = (\mathbb{R}^k \times \{0\}) \cap V$. Restricting those φ to $M \cap U$ yields an atlas for M.

The complex projective spaces $\mathbb{P}(\mathbb{C}^{n+1})$ discussed in Sect. 14.2 provide an interesting family of manifolds. We note that the charts exhibited there for $\mathbb{P}(\mathbb{C}^{n+1})$ have an additional structure: they map to an open subset of $\mathbb{C}^n \simeq \mathbb{R}^{2n}$, and the transition maps are even complex differentiable. In this case, one speaks about a *holomorphic atlas*, and manifolds M endowed with it are called *complex manifolds*. One calls $\dim_{\mathbb{C}} M := n$ the complex dimension of M. Clearly, complex manifolds are smooth manifolds and $\dim M = 2 \dim_{\mathbb{C}} M$. Another family of natural examples of manifolds are the real projective spaces $\mathbb{P}(\mathbb{R}^{n+1})$, which are constructed similarly to the complex ones.

Oriented manifolds are obtained as in Definition A.6 but requiring a stronger compatibility condition between charts (U_1, φ_1) and (U_2, φ_2), namely, that the transition maps $\varphi_2 \circ \varphi_1^{-1}$ be orientation-preserving diffeomorphisms, i.e., that their Jacobians be positive. Complex manifolds are always naturally oriented. The reason is that if we interpret $A \in GL(\mathbb{C}^n)$ as a linear isomorphism of $\mathbb{R}^{2n}$, then its determinant is given by $|\det A|^2$, which is positive.

Let $f \colon M \to N$ be a map between two manifolds. We define f to be a *smooth map* if the map

$$f_{U,V} := \psi \circ f \circ \varphi^{-1} \colon \varphi(U) \to \varphi(V)$$

is smooth for all charts (U, φ) of an atlas of M and all charts (V, ψ) of an atlas of N. We call f a *diffeomorphism* if it is bijective and f and its inverse are both smooth.

The concept of a *submanifold* M of a manifold Ω is now an immediate extension of Definition A.2, replacing $\mathbb{R}^n$ by Ω. We call $\mathrm{codim}_\Omega M := \dim \Omega - \dim M$ the *codimension* of M in Ω.

It is important to define the concept of the tangent space $T_p M$ of a manifold M at a point $p \in M$ without reference to any possible embedding. There are different, but equivalent, ways of doing so and we just outline one.

Let (U_1, φ_1) be a chart of M such that $p \in U_1$. Just as an element in $\varphi_1(U_1) \subseteq \mathbb{R}^n$ represents a point in U_1 by coordinates, we can also let a vector $v_1 \in \mathbb{R}^n$ represent a tangent vector as follows. Let (U_2, φ_2) be another chart of M such that $p \in U_2$ and $v_2 \in \mathbb{R}^n$. We say that (U_1, φ_1, v_1) and (U_2, φ_2, v_2) are *equivalent* if the derivative $D(\varphi_2 \circ \varphi_1^{-1})(\varphi_1(p))$ maps v_1 to v_2. An equivalence class is called a *tangent vector* of M at p. The set of such tangent vectors is called the *tangent space* of M at p and denoted by $T_p M$. Note that each chart (U_1, φ_1) determines a bijection of $T_p M$ with $\mathbb{R}^n$ via $(U_1, \varphi_1, v) \mapsto v$. The resulting vector space structure on $T_p M$ is easily seen to be independent of the choice of the chart.

Now, if $f \colon M \to N$ is a smooth map and $p \in M$, we can define the *derivative* $Df(p) \colon T_p M \to T_{f(p)} N$, which maps the equivalence class of (U, φ, v) to the equivalence class of (V, ψ, w), where $w := Df_{U,V}(\varphi(p))(e)$. Of course, (V, ψ) stands here for a chart of N with $f(p) \in V$. It is immediate to check that $Df(p)$ is a well-defined linear map. The functorial property

$$D(g \circ f)(p) = Dg\bigl(f(p)\bigr) \circ Df(p)$$

for smooth maps $f \colon M \to N$ and $g \colon N \to P$ is an immediate consequence of the chain rule.

It goes without saying that the tangent spaces of a complex manifold M are complex vector spaces and the corresponding derivatives are $\mathbb{C}$-linear.

Definition A.7 Let $f: M \to N$ be a smooth map between smooth manifolds and $x \in M$, $y \in N$. We call x a *regular point* of f if $\operatorname{rank} Df(x) = \dim N$. Further, we call y a *regular value* of f if all $x \in f^{-1}(y)$ are regular points of f. Finally, y is called a *critical value* of f if it is not a regular value of f.

We observe that if $\dim M < \dim N$, then f has no regular points (and hence, no regular values).

Example A.8 Let $\zeta \in \mathbb{C}^{n+1}$ be a zero of $f \in \mathcal{H}_d$. By (16.5), ζ is a simple zero of f iff $\operatorname{rank} Df(\zeta) = n$. This just means that ζ is a regular value for the map $f: \mathbb{C}^{n+1} \to \mathbb{C}^n$.

The following result follows immediately from Corollary A.3 (applied to the maps $f_{U,V}$).

Theorem A.9 *Let M, N be smooth manifolds with $m = \dim M \geq n = \dim N$, and let $f: M \to N$ be a smooth map. Suppose that $y \in f(M)$ is a regular value of f. Then the fiber $f^{-1}(y)$ over y is a smooth submanifold of M of dimension $m - n$. Further, the tangent space of $f^{-1}(y)$ at x equals the kernel of $Df(x)$.*   □

Remark A.10 Any submanifold of a manifold M can be obtained locally as the inverse image of a regular value as in Theorem A.9. (This is almost immediate from the definition of submanifolds.)

Finally, we note that the Cartesian product $M \times N$ of manifolds M, N is a manifold and the tangent space $T_{(x,y)} M \times N$ can be identified with $T_x M \times T_y N$.

A.2.3 Integration on Manifolds

Let V be an n-dimensional real vector space. It is a well-known fact from linear algebra that the vector space $\Omega^n(V)$ of alternating multilinear forms $V^n \to \mathbb{R}$ is one-dimensional. Moreover, $\Omega^n(\mathbb{R}^n)$ is generated by the determinant $\det$, interpreted as the multilinear map $(v_1, \ldots, v_n) \mapsto \det[v_1, \ldots, v_n]$. A linear map $f: V \to W$ of n-dimensional vector spaces induces a linear map $f^*: \Omega^n(W) \to \Omega^n(V)$, called the *pullback* of f, which is defined by

$$f^*(\omega)(v_1, \ldots, v_n) := \omega(f(v_1), \ldots, f(v_n)).$$

Clearly, the functorial property $(f \circ g)^* = g^* \circ f^*$ holds.

Let now M be an n-dimensional manifold and ω a function associating to any $p \in M$ an n-form $\omega(p) \in \Omega^n(T_p M)$. Let (U, φ) be a chart of M and $\psi: V \to U$ the inverse of φ. Then the pullback of $\omega(\psi(x))$ under the linear map $D\psi(x): T_x \mathbb{R}^n \to T_{\psi(x)} M$ defines the n-form $D\psi(x)^* \omega(\psi(x)) \in \Omega^n(T_x \mathbb{R}^n)$. Since we can identify

$T_x \mathbb{R}^n$ with $\mathbb{R}^n$, there is a function $\rho \colon V \to \mathbb{R}$ such that $D\psi(x)^*\omega(\psi(x)) = \rho(x)\det$.
If for all charts (U, φ), the resulting function ρ is smooth, then we say that ω is an
n-form of M. The vector space of n-forms on M is denoted by $\Omega^n(M)$.

It is obvious that by the same procedure as above, a smooth map $f \colon M \to N$ of
n-dimensional manifolds induces a linear map $f^* \colon \Omega^n(N) \to \Omega^n(M)$, called the
pullback of f. In the special case that M and N are open subsets of $\mathbb{R}^n$, it is easy to
check that $f^*(\det)(x) = \det Df(x) \cdot \det$.

We next define the *integral* of a continuous function $a \colon M \to \mathbb{R}$ with compact
support supp(a) with respect to the absolute value $|\omega|$ of an n-form $\omega \in \Omega^n(M)$.

Suppose first that supp$(a) \subseteq U$ for a chart (U, φ). Let $\psi := \varphi^{-1}$ and write
$\psi^*(\omega) = \rho \det$ with ρ as above. Then we define

$$\int_M a|\omega| := \int_{\mathbb{R}^n} a\big(\psi(x)\big)\big|\rho(x)\big| \, dx,$$

where the right-hand integral is the usual (Lebesgue) integral. The crucial observation is that this value does not change when we replace the chart $\varphi \colon U \to V$
by another chart $\tilde{\varphi} \colon U \to \tilde{V}$. Indeed, let $\tilde{\psi} := \tilde{\varphi}^{-1}$, $\tilde{\psi}^*(\omega) = \tilde{\rho} \det$, and set $\Phi :=
\tilde{\varphi} \circ \psi \colon V \to \tilde{V}$. Then we have $\psi = \tilde{\psi} \circ \Phi$, and hence by functoriality,

$$\psi^*(\omega) = \Phi^*\big(\tilde{\psi}^*(\omega)\big) = \Phi^*(\tilde{\rho}\det) = (\tilde{\rho} \circ \Phi)\,\Phi^*(\det) = (\tilde{\rho} \circ \Phi)\det D\Phi\det,$$

which implies $|\rho| = |\tilde{\rho} \circ \Phi|J\Phi$ with the Jacobian $J\Phi := |\det D\Phi|$. Hence

$$\int_V a\big(\psi(x)\big)\big|\rho(x)\big| \, dx = \int_{\tilde{V}} a\big(\tilde{\psi}(\tilde{x})\big)\big|\tilde{\rho}(\tilde{x})\big| \, d\tilde{x}$$

by the transformation formula (Theorem 2.1).

If the manifold M is oriented, then one can omit absolute values and define the
integral $\int_M a\omega := \int_{\mathbb{R}^n} a(\psi(x)\rho(x)) \, dx$. This is well defined by the same reasoning
as above.

Suppose now that supp(a) is not contained in a chart. If M is compact, then one
can show that there is a finite collection of smooth functions $\chi_1, \ldots, \chi_r$ on M with
values in $[0, 1]$ such that $\sum_i \chi_i = 1$, and such that each supp(χ_i) is contained in
some chart of M. (This collection is called a *partition of unity*; see [209].) Then we
can define

$$\int_M a|\omega| := \sum_i \int_M (a\chi_i)|\omega|,$$

which is easily seen to be independent of the choice of the partition of unity. In the
case that M is not compact, one can proceed by a slightly more general notion of
partition of unity [209].

Actually, to define $\int_M a|\omega|$, it is sufficient to require that a be measurable and
that $\int_M a_+|\omega|$ and $\int_M a_-|\omega|$ both be finite ($a_+ := \max\{a, 0\}$, $a_- := \max\{-a, 0\}$), in
which case we say that a is *integrable* with respect to $|\omega|$.

Again, all these definitions and facts extend to integrals $\int_M a\omega$ when M is oriented.

A.2.4 Sard's Theorem and Transversality

A subset A of $\mathbb{R}^n$ is said to have *measure zero* if for every $\varepsilon > 0$, there exists a countable collection $R_1, R_2, \ldots$ of rectangles such that $A \subseteq \bigcup_i R_i$ and $\sum_{i=1}^{\infty} \text{vol } R_i < \varepsilon$. By a rectangle R we understand here a Cartesian product of intervals, and $\text{vol } R$ is defined as the product of its lengths.

It is not difficult to show that if $f : U \to \mathbb{R}^n$ is a smooth map defined on an open subset $U \subseteq \mathbb{R}^n$ and A has measure zero, then $f(A)$ has measure zero as well; cf. [209]. We define a subset A of a manifold M to be of *measure zero in M* if for all charts (U, φ), $\varphi(A \cap U)$ has measure zero. This is well defined by the above observation.

Proposition A.11 *Let M be a k-dimensional submanifold of a manifold Ω of dimension n and assume $k < n$. Then M has measure zero in Ω.*

Proof Since a manifold can be covered by countably many charts, it suffices to prove the assertion in a chart. By the definition, it is enough to show that $(\mathbb{R}^k \times \{0\}) \cap V$ has measure zero in V for an open subset V of $\mathbb{R}^n$. But this is obvious. $\square$

The following is a deep and important result; see [145] for a proof.

Theorem A.12 (Sard's theorem) *Let $\varphi : M \to N$ be a smooth map between manifolds. Then the set of singular values of f has measure zero in N.* $\square$

We note that in the case $\dim M < \dim N$, the theorem just states that $f(M)$ has measure zero in N. Here is a first application of this observation.

Proposition A.13 *Let $M \subseteq \mathbb{R}^n$ be a submanifold with $\dim M \leq n - 2$. Further, assume $0 \notin M$. Then the set $B := \{v \in \mathbb{R}^n \mid \mathbb{R}v \cap M \neq \emptyset\}$ has measure zero in $\mathbb{R}^n$.*

Proof We obtain B as the image of the smooth map $\mathbb{R}_* \times M \to \mathbb{R}^n$, $(\lambda, x) \mapsto \lambda x$. Since $\dim(\mathbb{R}_* \times M) = 1 + \dim M < n$, the image B has measure zero in $\mathbb{R}^n$ by Theorem A.12. $\square$

Theorem A.9 states that the inverse image of a regular value under a smooth map $f : M \to \Omega$ is a submanifold of M. But when is the inverse image $f^{-1}(N)$ of a submanifold N of Ω a submanifold of M? For analyzing this question, the concept of transversality is useful.

Definition A.14 Let $f : M \to \Omega$ be a smooth map between manifolds and N a submanifold of Ω. We call f *transversal* to N if for all $x \in f^{-1}(N)$,

$$\text{Im } Df(x) + T_{f(x)}N = T_{f(x)}\Omega.$$

Theorem A.15 *Let $f: M \to \Omega$ be a smooth map between manifolds that is transversal to the submanifold N of Ω. Then $f^{-1}(N)$ is a submanifold of M and $\operatorname{codim}_M f^{-1}(N) = \operatorname{codim}_\Omega N$, unless $f^{-1}(N) = \emptyset$. Moreover, $T_x f^{-1}(N) = Df(x)^{-1}(T_{f(x)}N)$ for $x \in f^{-1}(N)$.*

Sketch of proof. We first note that the assertion is local: it suffices to prove it for the restrictions $f^{-1}(U_i) \to U_i$ and the submanifolds $N \cap U_i$, for a covering of Ω by open sets U_i. So we are allowed to replace Ω by a small open subset. By Remark A.10 we may further assume that $N = g^{-1}(0) \neq \emptyset$ for a smooth map $g: \Omega \to \mathbb{R}^\ell$ such that 0 is a regular value for g. Moreover, $T_y N = \ker Dg(y)$ for all $y \in N$ and $\ell = \operatorname{codim}_\Omega N$; cf. Theorem A.9.

Setting $h := g \circ f$, we have $f^{-1}(N) = h^{-1}(0)$. Now note that for $x \in f^{-1}(N)$, by the transversality assumption, we have $\operatorname{Im} Df(x) + \ker Dg(f(x)) = T_{f(x)}\Omega$, and hence $Dh(x): T_x M \to \mathbb{R}^\ell$ is surjective. Hence 0 is a regular value of h and Theorem A.9 implies that $h^{-1}(0)$ is a submanifold of codimension ℓ in M. Moreover, $T_x f^{-1}(N) = \ker Dh(0) = Df(x)^{-1}(T_{f(x)}N)$. $\qquad\qquad\square$

The most important special case concerns the transversality of the inclusion map $i: M \to \Omega$ to another submanifold N of Ω, in which case we call the submanifolds M and N *transversal*. This means that $T_x M + T_x N = T_x \Omega$ for all $x \in M \cap N$. This implies $\dim M + \dim N \geq \dim \Omega$ if $M \cap N \neq \emptyset$. Note that M and N are considered transversal if they don't meet.

Theorem A.15 immediately implies the following.

Corollary A.16 *Let M and N be transversal submanifolds of the manifold Ω such that $M \cap N \neq \emptyset$. Then their intersection $M \cap N$ is a submanifold and $T_x(M \cap N) = T_x M \cap T_x N$ for all $x \in M \cap N$. Moreover, $\operatorname{codim}_\Omega(M \cap N) = \operatorname{codim}_\Omega M + \operatorname{codim}_\Omega N$.* $\qquad\qquad\square$

We next derive a result that will be crucial for integral geometry. We already noted in Sect. A.2.1 that the orthogonal group $G := \mathcal{O}(n + 1)$ is a compact submanifold of $\mathbb{R}^{(n+1)\times(n+1)}$. It is clear that G acts transitively on the sphere $\mathbb{S}^n$. The *stabilizer* of $x \in \mathbb{S}^n$ is defined as $G_x := \{g \in G \mid gx = x\}$, which is a subgroup of G isomorphic to $\mathcal{O}(n)$. Clearly, G_x acts on $T_x \mathbb{S}^n$.

Lemma A.17

(a) *Let $x_0 \in \mathbb{S}^n$. Then the orbit map $G \to \mathbb{S}^n$, $g \mapsto gx_0$, is a submersion, that is, all of its derivatives are surjective.*

(b) *The derivative of the map $\mu: G \times \mathbb{S}^n \to \mathbb{S}^n$, $(g, x) \mapsto g^{-1}x$, at $(g, x) \in G \times \mathbb{S}^n$ is given by*

$$D\mu(g, x): T_g G \times T_x \mathbb{S}^n \to T_x \mathbb{S}^n, \quad (\dot{g}, \dot{x}) \mapsto g^{-1}\dot{x} - g^{-1}\dot{g}g^{-1}x.$$

Proof (a) Let $v \in T_{x_0}\mathbb{S}^n$ and let $D(t) \in G$ denote the rotation with angle t in the plane spanned by x_0 and v. Then $x(t) := D(t)x_0 = x_0 \cos t + v \sin t$ and $\frac{dx}{dt}(0) = v$.

Hence the derivative of the orbit map is surjective at the identity in G. By homogeneity, all the derivatives must be surjective as well.

(b) In Example 14.2 we showed that $G \to G$, $g \mapsto g^{-1}$, has derivative $T_g G \to T_g G$, $\dot{g} \mapsto -g^{-1}\dot{g}g^{-1}$, at $g \in G$. From this, the assertion about the derivative of μ easily follows. $\square$

A property is said to hold for *almost all* points of a manifold when the set of points for which it fails to hold has measure zero.

Proposition A.18 *Let M and N be submanifolds of $\mathbb{S}^n$. Then M and gN intersect transversally, for almost all $g \in G$. In particular, for almost all $g \in G$, the intersection $M \cap gN$ is either empty or a smooth submanifold of $\mathbb{S}^n$ with dimension $\dim M + \dim N - n$.*

Proof By Lemma A.17, the map $f: G \times M \to \mathbb{S}^n$, $(g, x) \mapsto g^{-1}x$, has surjective derivatives. In particular, f is transversal to N. Hence Theorem A.15 implies that $R := f^{-1}(N) = \{(g, x) \in G \times M \mid g^{-1}x \in N\}$ is a submanifold of $G \times M$. Moreover, setting $y = g^{-1}x$, we have

$$T_{(g,x)}R = Df(g,x)^{-1}(T_y N) = (T_g G \times T_x M) \cap D\mu(g,x)^{-1}(T_y N), \qquad \text{(A.2)}$$

where $\mu: G \times \mathbb{S}^n \to \mathbb{S}^n$, $(g, x) \mapsto g^{-1}x$.

Consider the projection $p_1: R \to G$, $(g, x) \mapsto g$, and note that $p_1^{-1}(g) = \{g\} \times (M \cap gN)$. Suppose that $Dp_1(g, x)$ is surjective. Then, using (A.2), we see that for all $\dot{g} \in T_g G$ there exist $\dot{x} \in T_x M$, $\dot{y} \in T_y N$ such that $D\mu(g, x)(\dot{g}, \dot{x}) = \dot{y}$. By Lemma A.17, this means $g^{-1}\dot{x} - g^{-1}\dot{g}y = \dot{y}$. Hence $-\dot{g}y = -\dot{x} + g\dot{y}$. Since $T_g G \to T_x \mathbb{S}^n$, $\dot{g} \mapsto \dot{g}y$, is surjective, we conclude that $T_x \mathbb{S}^n = T_x M + T_x gN$. (Note that this argument is reversible.) Hence, Theorem A.9 implies that M and gN are transversal if g is a regular value of p_1. Sard's theorem completes the proof. $\square$

A.2.5 Riemannian Metrics

In $\mathbb{R}^n$ we have the standard inner product $\langle x, y \rangle_{\text{st}} = \sum_i x_i y_i$ that allows us to define the length of vectors, the angle between vectors, the length of curves, etc. These concepts can be extended to abstract manifolds as follows. Recall that an *inner product* $\langle, \rangle$ on $\mathbb{R}^n$ is given by a positive definite matrix (g_{ij}) by taking $\langle x, y \rangle = \sum_{i,j} g_{ij} x_i y_j$.

Let M be an n-dimensional manifold and suppose that $\langle \, , \, \rangle_p$ is an inner product on $T_p M$ for each $p \in M$. If (U, φ) is a chart of M, then this induces a family of inner products $\langle \, , \, \rangle_x$ on $\mathbb{R}^n$ for $x \in \varphi(U)$ by setting $\psi := \varphi^{-1}$ and

$$\langle v_1, v_2 \rangle_x := \big\langle D\psi(x)(v_1), D\psi(x)(v_2) \big\rangle_{\psi(x)}.$$

We require now that the symmetric matrix corresponding to $\langle\, ,\, \rangle_x$ be a smooth function of $x \in \varphi(U)$. If this is the case for all charts of M, then we say that the inner product $\langle\, ,\, \rangle_p$ *varies smoothly* with $p \in M$.

Definition A.19 A *Riemannian manifold* is a manifold together with a family of inner products $\langle\, ,\, \rangle_p$ on T_pM that varies smoothly with $p \in M$. This family of inner products is called a *Riemannian metric* on M. Thus a tangent vector $v \in T_pM$ has a well-defined norm $\|v\|_p := \sqrt{\langle v, v \rangle_p}$.

The most natural examples are provided by the submanifolds M of $\mathbb{R}^n$. For $\langle\, ,\, \rangle_p$ we just take the restriction of the standard inner product to T_pM.

A more interesting example is provided by the real projective space $\mathbb{P}(\mathbb{R}^{n+1})$, which is obtained from $\mathbb{S}^n$ by identifying antipodal points via the canonical map $\pi : \mathbb{S}^n \to \mathbb{P}(\mathbb{R}^{n+1})$, $p \mapsto [p] := \{p, -p\}$. Since π is a local diffeomorphism, $D\pi(p)$ provides an isomorphism of $T_p\mathbb{S}^n$ with $T_{[p]}\mathbb{P}(\mathbb{R}^{n+1})$. We define the Riemannian metric on $\mathbb{P}(\mathbb{R}^{n+1})$ by requiring $D\pi(p)$ to be isometric. An example of great importance in Part III of this book is the complex projective space $\mathbb{P}(\mathbb{C}^{n+1})$, which is turned into a Riemannian manifold by the Fubini–Study metric; see Sect. 14.2.

We note that the product $M \times N$ of two Riemannian manifolds has the Riemannian metric defined by $\langle (v, w), (v', w') \rangle_{(x,y)} := \langle v, v' \rangle_x + \langle w, w' \rangle_y$ for $v, v' \in T_xM$ and $w, w' \in T_yN$.

In any Riemannian manifold M, we have a well-defined notion of length of curves. Let $\gamma : [0, 1] \to M$ be a continuous map that is piecewise smooth. We define the *length* of γ as

$$L(\gamma) := \int_0^1 \left\| \frac{d}{dt} \gamma(t) \right\|_{\gamma(t)} dt. \tag{A.3}$$

The *Riemannian distance* $d_M(x, y)$ between points $x, y \in M$ is defined as $d_M(x, y) := \inf_\gamma L(\gamma)$, where the infimum is over all piecewise smooth curves γ connecting x and y. Clearly, d_M turns M into a metric space.

It is a well-known fact that for the sphere $\mathbb{S}^n$, the Riemannian distance $d_\mathbb{S}(x, y)$ between $x, y \in \mathbb{S}^n$ equals the angle between x and y, that is, $d_\mathbb{S}(v, w) = \arccos\langle v, w \rangle$. The Riemannian distance for $\mathbb{P}(\mathbb{C}^{n+1})$ is described in Proposition 14.12.

Before discussing volumes on Riemannian manifolds, we proceed with a general observation from linear algebra. Let V be an n-dimensional real vector space with an inner product $\langle\, ,\, \rangle$. Fix an orthonormal basis $(v_1, \ldots, v_n)$, so that $\langle v_i, v_j \rangle = \delta_{ij}$. This basis determines an orientation of V in the following sense. If $(v'_1, \ldots, v'_n)$ is another orthonormal basis, then we have $v'_k = \sum_j a_{kj} v_j$ with a transformation matrix $A = (a_{kj})$ that is easily seen to be orthogonal. By the multiplicativity of the determinant,

$$\det[v'_1, \ldots, v'_n] = \det(A) \det[v_1, \ldots, v_n].$$

We say that (v'_j) is *positively oriented* if the corresponding transformation matrix A satisfies $\det A = 1$. Since $\Omega^n(V)$ is one-dimensional, there is a unique n-form

$\Omega \in \Omega^n(V)$ satisfying $\Omega(v_1, \ldots, v_n) = 1$ for positively oriented orthonormal bases. Thus we have assigned to the inner product $\langle \, , \, \rangle$ an n-form Ω.

Let now M be an n-dimensional Riemannian manifold M. We further assume that M is *oriented*; hence it is possible to orient each of the tangent spaces $T_p M$ such that the induced orientations on $\mathbb{R}^n$ in the charts are positive. Then, as above, the Riemannian metric defines a distinguished n-form $\Omega_M(x)$ on each tangent space $T_x M$. The resulting differential form Ω_M on M is called the *volume form* of M. Its absolute value $dM := |\Omega_M|$ is called the *volume element* of M. In Sect. A.2.3 we defined the integral $\int_M a \, dM$ with respect to dM. We note that the volume element is still defined when M is not orientable and the integral $\int_M a \, dM$ can still be defined. The *volume B* of a measurable subset $B \subseteq M$ is defined as $\mathrm{vol}\, B := \int_M \mathbb{1}_B \, dM$ with $\mathbb{1}_B$ denoting the indicator function of B. If M is oriented, we have $\int_M \mathbb{1}_B \Omega_M = \int_M \mathbb{1}_B |\Omega_M|$.

A.2.6 Orthogonal and Unitary Groups

We already observed in Sect. A.2.1 that the orthogonal group $\mathcal{O}(n)$ is a compact submanifold of $\mathbb{R}^{n \times n}$. Hence it inherits a Riemannian metric from the ambient space. Consider the multiplication $\phi \colon \mathcal{O}(n) \to \mathcal{O}(n)$, $h \mapsto gh$, with a fixed group element g. Since ϕ is the restriction of an isometric linear map, the derivative $D\phi(h)$ is isometric as well. Hence $J\phi(h) = |\det D\phi(h)| = 1$. The coarea formula (along with Remark 17.7) implies that ϕ preserves the volume on $\mathcal{O}(n)$ induced by the Riemannian metric. Since $\mathcal{O}(n+1)$ is compact, it has a finite volume, and we can introduce the normalized volume $\mathrm{rvol}\, B := \mathrm{vol}\, B / \mathrm{vol}\, \mathcal{O}(n+1)$, which defines a probability measure on $\mathcal{O}(n+1)$. This is called the *normalized Haar measure* on $\mathcal{O}(n+1)$.

One calls $\mathcal{O}(n)$ a *Lie group*, since the inverse $\mathcal{O}(n) \to \mathcal{O}(n)$, $g \mapsto g^{-1}$, and the group multiplication $\mathcal{O}(n) \times \mathcal{O}(n) \to \mathcal{O}(n)$, $(g, h) \mapsto gh$, are smooth maps.

Similar observations apply to the *unitary group*, which is defined as $\mathcal{U}(n) := \{A \in \mathbb{C}^{n \times n} \mid AA^* = I_n\}$, where A^* denotes the complex transpose of A. As in Proposition A.4 one can prove that $\mathcal{U}(n)$ is a compact submanifold of $\mathbb{C}^{n \times n}$ with dimension n^2. It is a Lie group, and its tangent space at I_n consists of the matrices $\dot{B} \in \mathbb{C}^{n \times n}$ such that $\dot{B} + \dot{B}^* = 0$.

A.2.7 Curvature of Hypersurfaces

Let M be a *hypersurface* of $\mathbb{S}^n$, that is, a submanifold of codimension 1. Hence the orthogonal complement of $T_x M$ in $T_x \mathbb{S}^n$ is one-dimensional. We assume that it is possible to select one of the two unit normal vectors in this complement such that it depends continuously on $x \in M$. (This assumption is easily seen to be equivalent to the orientability of M.) Let ν denote the resulting unit normal vector field on M. This defines the smooth map $\nu \colon M \to \mathbb{R}^{n+1}$, from which we can take the derivative $D\nu(x) \colon T_x M \to \mathbb{R}^{n+1}$.

Lemma A.20 *We have* $\operatorname{Im} Dv(x) \subseteq T_x M$. *The resulting linear map* $T_x M \to T_x M$ *induced by* $Dv(x)$ *is self-adjoint.*

Proof Let $x(t)$ be a parameterization of a smooth curve in M passing through $x = x(0)$. From $\langle v(x(t)), v(x(t)) \rangle = 1$ we obtain by taking the derivative that $\langle v(x(t)), \frac{d}{dt} v(x(t)) \rangle = 0$. Hence $Dv(x(t))(\dot{x}) = \frac{d}{dt} v(x(t))$ is indeed contained in $T_{x(t)} M$, which proves the first claim.

For the second claim let (U, φ) be a chart around x, denote the resulting coordinates by $(v_1, \ldots, v_{n-1})$, and write $x = \psi(e)$ for the inverse of φ. If we fix all but the jth coordinate, then v_j parameterizes a curve in M via ψ. Its derivative $\frac{\partial x}{\partial v_j}$ is a tangent vector, and hence

$$\left\langle v, \frac{\partial x}{\partial v_j} \right\rangle = 0.$$

Taking the derivative on both sides of this equality, now with respect to v_i, we obtain

$$\left\langle \frac{\partial v}{\partial v_i}, \frac{\partial x}{\partial v_j} \right\rangle + \left\langle v, \frac{\partial^2 x}{\partial v_i \partial v_j} \right\rangle = 0.$$

Since $Dv(x)(\frac{\partial x}{\partial v_i}) = \frac{\partial v}{\partial v_i}$, we get

$$\left\langle Dv(x)\left(\frac{\partial x}{\partial v_i}\right), \frac{\partial x}{\partial v_j} \right\rangle = -\left\langle v, \frac{\partial^2 x}{\partial v_i \partial v_j} \right\rangle = 0 = \left\langle Dv(x)\left(\frac{\partial v}{\partial v_j}\right), \frac{\partial x}{\partial v_i} \right\rangle.$$

But $\frac{\partial x}{\partial v_1}, \ldots, \frac{\partial x}{\partial v_{n-1}}$ form a basis of $T_x M$. So we conclude that $\langle Dv(x)(e), w \rangle = \langle v, Dv(x)(w) \rangle$ for $v, w \in T_x M$.   □

The *Weingarten map* of M at x is the self-adjoint map defined as

$$L_M(x) \colon T_x M \to T_x M, \quad L_M(x) := -Dv(x). \tag{A.4}$$

Definition A.21 Let M be a compact smooth oriented hypersurface of $\mathbb{S}^n$. The eigenvalues $\kappa_1(x), \ldots, \kappa_{n-1}(x)$ of the Weingarten map $L_M(x)$ are called the *principal curvatures* of the hypersurface M at x. For $1 \leq i \leq n-1$ one defines the ith curvature $K_{M,i}(x)$ of M at x as the ith elementary symmetric polynomial in $\kappa_1(x), \ldots, \kappa_{n-1}(x)$, and one puts $K_{M,0}(x) := 1$.

Example A.22 Let $\overline{a} = (1, 0, \ldots, 0) \in \mathbb{S}^n$. Consider the boundary

$$M_\alpha := \left\{ \cos\alpha \overline{a} + \sin\alpha (0, y) \mid y \in \mathbb{S}^{n-1} \right\}$$

of the spherical cap $\mathsf{cap}(\overline{a}, \alpha)$ in $\mathbb{S}^n$ of radius $0 < \alpha \leq \pi/2$ centered at $\overline{a}$. We orient M_α by the unit normal vector field on $\mathbb{S}^n$ pointing towards $\overline{a}$, namely

$$v(x) = \sin\alpha \overline{a} - \cos\alpha (0, y), \quad \text{where } x = \cos\alpha \overline{a} + \sin\alpha (0, y).$$

Take a smooth curve in M_α given by a smooth curve $y(t)$ in $\mathbb{S}^{n-1}$. Then,

$$\dot{v} = -\cos\alpha(0, \dot{y}) = -\cot\alpha \, \sin\alpha(0, \dot{y}) = -\cot\alpha \dot{x}.$$

Hence $L_{M_\alpha} = (\cot\alpha) I_{T_x M}$, and all the principal curvatures of M_α at x are equal to $\cot\alpha$. Therefore the ith curvature of M_α satisfies $K_{M_\alpha,i}(x) = \binom{n-1}{i}(\cot\alpha)^i$, a quantity independent of $x \in M_\alpha$.

For more information on this we refer to the textbooks [218] and [88, p. 129].

A.3 Algebraic Geometry

Here we outline the basic concepts from complex algebraic geometry needed in Part III of the book. We have to be brief, and so we omit most of the proofs. An excellent reference for the material introduced here is Mumford's classic textbook [148]. Another good and appropriate reference is Shafarevich [186].

A.3.1 Varieties

The basic objects of study in algebraic geometry are the sets of solutions of systems of polynomial equations.

Definition A.23 An (affine) *algebraic variety* Z in $\mathbb{C}^n$ is defined as the set of zeros of finitely many polynomials $f_1, \ldots, f_s \in \mathbb{C}[X_1, \ldots, X_n]$, that is,

$$Z = Z(f_1, \ldots, f_s) := \left\{ x \in \mathbb{C}^n \mid f_1(x) = 0, \ldots, f_s(x) = 0 \right\}.$$

More generally, one writes $Z(I) := \{x \in \mathbb{C}^n \mid \forall f \in I \;\; f(x) = 0\}$ for the *zero set* of a subset I of $\mathbb{C}[X] := \mathbb{C}[X_1, \ldots, X_n]$. It is clear that $Z(f_1, \ldots, f_s) = Z(I)$, where $I = \{\sum_{i=1}^s g_i f_i \mid g_i \in \mathbb{C}[X]\}$ denotes the *ideal* in the ring $\mathbb{C}[X]$ generated by $f_1, \ldots, f_s$. The *vanishing ideal* $I(Z)$ of Z is defined as $I(Z) := \{f \in \mathbb{C}[X] \mid \forall x \in Z \; f(x) = 0\}$. It is not hard to check that $Z = Z(I(Z))$.

A fundamental result providing the first link between algebra and geometry is the following.

Theorem A.24 (Hilbert's Nullstellensatz)

(Weak form) *For an ideal $I \subseteq \mathbb{C}[X]$ we have*

$$Z(I) = \emptyset \quad \Longleftrightarrow \quad 1 \in I.$$

(Strong form) *If a polynomial f vanishes on the zero set $Z(I)$ of some ideal I, then $f^e \in I$ for some $e \in \mathbb{N}$.* □

Another fundamental result in algebraic geometry is *Hilbert's basis theorem,* which states that any ideal in $\mathbb{C}[X]$ is finitely generated. Hence, using infinitely many f_i's in Definition A.23 does not lead to a different notion of algebraic variety.

The following properties are easy to check:

$$Z(I_1) \cup Z(I_2) = Z(I_1 \cap I_2), \qquad \bigcap_{\alpha \in A} Z(I_\alpha) = Z\left(\bigcup_{\alpha \in A} I_\alpha\right).$$

As a consequence, the sets $Z(I)$ satisfy the axioms for the closed sets of a topology on $\mathbb{C}^n$, called *Zariski topology.* So Zariski closed subsets of $\mathbb{C}^n$, by definition, are the same as affine algebraic varieties in $\mathbb{C}^n$. For instance, the nonempty Zariski open subsets of $\mathbb{C}^1$ are the complements of finite subsets of $\mathbb{C}^1$. So the Zariski topology violates the Hausdorff separation axiom. It is clear that Zariski closed subsets are also closed in the *classical topology,* which is the one defined by the Euclidean distance metric.

Definition A.25 A Zariski closed subset Z is called *irreducible* if it is nonempty and cannot be written as the union $Z = Z_1 \cup Z_2$ of two Zariski closed proper subsets Z_i.

We note that the above definition of irreducibility could be given for any topological space. However, this concept is not interesting for a Hausdorff space, since there, the only irreducible sets are those consisting of a point only.

An ideal I in $\mathbb{C}[X]$ is called *prime* if $I \neq \mathbb{C}[X]$ and $f_1 f_2 \in I$ implies either $f_1 \in I$ or $f_2 \in I$. It is easy to check that $Z(I)$ is irreducible iff I is a prime ideal. This implies that $\mathbb{C}^n = Z(0)$ is irreducible. More generally, one concludes that the complex linear subspaces of $\mathbb{C}^n$ are irreducible. It is a nontrivial fact that irreducible varieties are connected in the classical topology; see Theorem A.28 below for a more general statement. The converse is false, as shown by the example $Z = Z(X_1 X_2) = Z(X_1) \cup Z(X_2)$ of two intersecting lines in $\mathbb{C}^2$, which is connected, but not irreducible.

The Hilbert basis theorem implies that there are no infinite strictly descending chains $Z_1 \supset Z_2 \supset Z_2 \supset \cdots$ of Zariski closed sets in $\mathbb{C}^n$. The following result is a straightforward consequence of this fact.

Proposition A.26 *Any Zariski closed subset Z can be written as a finite union $Z = Z_1 \cup \cdots \cup Z_r$ of irreducible Zariski closed sets. Moreover, if we require that $Z_i \nsubseteq Z_j$ for $i \neq j$, then the Z_i are uniquely determined. They are called the* irreducible *components of Z.*   □

Example A.27 Let $f \in \mathbb{C}[X_1, \ldots, X_n] \setminus \mathbb{C}$. Then $Z(f)$ is irreducible iff f is irreducible, i.e., $f = f_1 f_2$ implies $f_1 \in \mathbb{C}$ or $f_2 \in \mathbb{C}$. Moreover, if $f = f_1 \cdots f_r$ is the factorization of f into irreducible polynomials f_i, then $Z(f) = Z(f_1) \cup \cdots \cup Z(f_s)$ and the $Z(f_i)$ are the irreducible components of $Z(f)$.

If we assume in Definition A.23 that the polynomials f_i are homogeneous, then the resulting zero set Z is a *cone*, i.e., it satisfies $\lambda x \in Z$ for all $\lambda \in \mathbb{C}$ and $x \in Z$. We call the corresponding subset

$$Z_{\mathbb{P}}(f_1, \ldots, f_s) := \{[x] \in \mathbb{P}^{n-1} \mid f_1(x) = 0, \ldots, f_s(x) = 0\}$$

of the complex projective space $\mathbb{P}^{n-1}$ a *projective variety* and say that Z is its *affine cone* (cf. Sect. 14.2). One defines the Zariski topology on $\mathbb{P}^{n-1}$ as the topology whose closed sets are the projective varieties. Then Proposition A.26 extends from $\mathbb{C}^n$ to $\mathbb{P}^{n-1}$. Also, there is a version of Example A.27 for homogeneous polynomials.

The classical topology on $\mathbb{P}^{n-1}$ is the one induced from the classical topology on $\mathbb{C}^n_*$ via the canonical map $\mathbb{C}^n_* \to \mathbb{P}^{n-1}$. A proof of the following result can be found in ([148, Cor. 4.16] or [187, VII.2]).

Theorem A.28 *A Zariski open subset of an irreducible projective algebraic variety is connected in the classical topology.* ☐

A.3.2 Dimension and Regular Points

In general, varieties are considerably more complicated objects than just submanifolds of $\mathbb{C}^n$ or $\mathbb{P}^n$. Here we investigate this difference. We start with a topological definition of the fundamental notion of dimension.

Definition A.29 The *dimension* dim Z of a Zariski closed set Z is defined as the maximum length n of a chain $Z_0 \subset Z_1 \subset \cdots \subset Z_n$ of distinct irreducible Zariski closed subsets contained in Z.

Looking at the chain $\mathbb{C}^1 \subset \mathbb{C}^2 \subset \cdots \subset \mathbb{C}^n$, we see that $\dim \mathbb{C}^n \geq n$, and one can show that equality holds. Similarly, $\dim \mathbb{P}^n = n$. More generally, if $Z \subseteq \mathbb{P}^{n-1}$ is a projective variety and $\hat{Z} \subseteq \mathbb{C}^n$ denotes the corresponding affine cone, then one can prove that $\dim Z = \dim \hat{Z} - 1$.

The above definition of dimension implies the following important observation: suppose that Z is an irreducible variety and $Y \subseteq Z$ is a Zariski closed subset. Then $\dim Y = \dim Z$ implies $Y = Z$.

Definition A.29 implies that $\dim Z$ equals the maximum of the dimensions of the irreducible components of Z. A variety Z is called *pure dimensional* if all of its irreducible components have the same dimension.

We discuss now the notion of a regular point of a variety. Let $Z \subseteq \mathbb{C}^n$ be a Zariski closed subset with vanishing ideal $I(Z)$. Then we have $Z = \bigcap_{f \in I(Z)} Z(f)$. Now we fix $p \in Z$ and replace any $Z(f)$ in this intersection by the zero set of its linearization $Df(p) := \sum_{i=1}^n \partial_{X_i} f(p)(X - p_i)$ at p. The vector space

$$T_p Z := \bigcap_{f \in I(Z)} Z(Df(p)) \tag{A.5}$$

is called the *Zariski tangent space* of Z at p. If $Z \subseteq \mathbb{P}^{n-1}$ is a Zariski closed subset and $\hat{Z}$ its affine cone, $p \in \hat{Z}$, then we call the projective linear space corresponding to $T_p \hat{Z}$ the *projective tangent space* of Z at $[p]$ and denote it by $T_{[p]}Z$. The following result is well known.

Theorem A.30 *We have* $\dim Z \leq \dim T_p Z$ *for an irreducible variety* Z *and any* $p \in Z$. *Moreover, equality holds for at least one point* p. $\qquad\qquad\Box$

We can now proceed with the definition of regular points.

Definition A.31 A point p of an irreducible variety Z is called *regular* if $\dim Z = \dim T_p Z$. Otherwise, p is called a *singular point* of Z. One denotes by $\mathsf{Reg}(Z)$ the set of regular points and by $\mathsf{Sing}(Z)$ the set of singular points of Z.

The next result is a useful criterion for showing that a point of a variety is regular, based on linear independence. It will provide a link to the concepts introduced in Sect. A.2 on differential geometry. For a proof, see [148, Thm. 1.16].

Lemma A.32 *Let* $f_1, \dots, f_s \in \mathbb{C}[X_1, \dots, X_n]$ *and* $p \in Z(f_1, \dots, f_s)$ *be such that the derivatives* $Df_1(p), \dots, Df_s(p)$ *are linearly independent. Then*

$$I := \left\{ f \in \mathbb{C}[X] \mid \exists g_1, \dots, g_r, h \in \mathbb{C}[X] \ s.t. \ h(p) \neq 0, \ hf = \sum_i g_i f_i \right\}$$

is a prime ideal and $W = Z(I)$ *is an irreducible variety* W *of dimension* $n - s$ *containing* p *as a regular point. Moreover, there is a Zariski closed set* Y *not containing* p *such that* $Z(f_1, \dots, f_s) = W \cup Y$. $\qquad\qquad\Box$

The next result clarifies the relation of varieties to complex manifolds.

Theorem A.33 *Let* Z *be an irreducible variety. Then* $\mathsf{Sing}(Z)$ *is a Zariski closed subset of dimension strictly less than* $\dim Z$. *Furthermore,* $\mathsf{Reg}(Z)$ *is a complex manifold of dimension* $\dim Z$ *and hence a smooth manifold of dimension* $2 \dim Z$.

Proof Let $Z \subseteq \mathbb{C}^n$ be an irreducible affine variety of dimension d and let $f_1, \dots, f_s$ be generators of its vanishing ideal $I(Z)$. Then the Zariski tangent space $T_p Z$ defined in (A.5) is the kernel of the Jacobian matrix $[\partial_{X_j} f_i(p)]$ at p. It follows that $\{p \in Z \mid \dim T_p Z \geq n - k\}$ is a Zariski closed set. Indeed, $\dim T_p Z \geq n - k$ means that $\mathsf{rank}[\partial_{X_j} f_i(p)] \leq k$, and the latter can be expressed by the vanishing of all of the $k \times k$ minors of the Jacobian matrix of $(f_1, \dots, f_s)$ at p. It follows that $\mathsf{Sing}(Z) = \{p \in Z \mid \dim T_p Z \geq d + 1\}$ is a Zariski closed subset. Since $\mathsf{Reg}(Z) \neq \emptyset$ by Theorem A.30, $\mathsf{Sing}(Z)$ is strictly contained in Z and hence $\dim \mathsf{Sing}(Z) < \dim Z$.

It remains to analyze $\mathsf{Reg}(Z)$. Let p be a regular point. So $d = \dim T_p Z$, and we may assume without loss of generality that $T_p Z$ is the zero set of

$Df_1(p), \ldots, Df_{n-d}(p)$. By Lemma A.32 the zero set $Z' := Z(f_1, \ldots, f_{n-d})$ de-composes as $Z' = W \cup Y$ for Zariski closed sets W, Y, where W is irreducible, $\dim W = d$, and $p \notin Y$. Since $Z \subseteq Z'$ and Z is irreducible, we must have $Z \subseteq W$, since $Z \subseteq Y$ is impossible. Since $\dim Z = \dim W$, we get $Z = W$. So we obtain $Z' \cap U = Z \cap U$ for the Zariski open neighborhood $U := \mathbb{C}^n \setminus Y$ of p. After shrinking U, we may assume that $Df_1(x), \ldots, Df_{n-d}(x)$ are linearly independent for all $x \in U$. Hence 0 is a regular value of the polynomial map $U \to \mathbb{C}^{n-d}, x \mapsto (f_1(x), \ldots, f_{n-d}(x))$. Its fiber over 0 equals $Z \cap U$. The complex version of Corollary A.3 implies that $Z \cap U$ is a complex manifold of complex dimension d.

If Z is a projective variety, one can argue similarly.   □

Corollary A.34 *Any affine or projective variety Z is a disjoint union of finitely many complex manifolds. The largest complex dimension of the manifolds occurring in this decomposition equals the dimension of Z as a variety.*

Proof Let $Z = Z_1 \cup \cdots \cup Z_r$ be the decomposition of Z into irreducible components and further decompose $Z_i = \text{Reg}(Z_i) \cup \text{Sing}(Z_i)$. Proposition A.33 states that $\text{Reg}(Z_i)$ is a complex manifold of dimension $\dim Z_i$, and we note that $\dim Z = \max_i \dim Z_i$. We apply the same procedure to the varieties $\text{Sing}(Z_i)$, which satisfy $\dim \text{Sing}(Z_i) < \dim Z_i$ by the same proposition, and iterate. The procedure stops after finitely many steps.   □

Corollary A.34 combined with Proposition A.11 implies the following.

Corollary A.35 *Any Zariski closed set Z properly contained in $\mathbb{C}^n$ has measure zero in $\mathbb{C}^n$. Similarly for a Zariski closed set in Z in $\mathbb{P}^n$.*   □

This is a good juncture to introduce a common terminology: a property of points in $\mathbb{C}^n$ (or $\mathbb{P}^n$) is said to hold for *Zariski almost all* points if the property holds for all points outside a Zariski closed subset Z of $\mathbb{C}^n$ (or $\mathbb{P}^n$). By Corollary A.35, this implies that the property holds for all points outside a subset of measure zero.

At some moment in this book we also have to deal with *real algebraic varieties*. They are defined as in Definition A.23, with $\mathbb{C}$ replaced by $\mathbb{R}$. Many of the concepts defined over $\mathbb{C}$ extend to $\mathbb{R}$, for instance the notion of dimension. Again it is true that $\dim Z < n$ for an algebraic variety Z properly contained in $\mathbb{R}^n$. Also, we state without proof the following fact: any real algebraic variety $Z \subseteq \mathbb{R}^n$ is a disjoint union of smooth submanifolds of $\mathbb{R}^n$ having dimension at most $\dim Z$. Proofs of these facts can be found in [39].

Corollary A.36

(a) A real algebraic variety $Z \subseteq \mathbb{R}^n$ such that $Z \neq \mathbb{R}^n$ has measure zero in $\mathbb{R}^n$.
(b) Let $Z \subseteq \mathbb{R}^n$ be a real algebraic variety of dimension at most $n - 2$. Further, assume $0 \notin Z$. Then the set $\{v \in \mathbb{R}^n \setminus \{0\} \mid \mathbb{R}v \cap Z \neq \emptyset\}$ has measure zero in $\mathbb{R}^n$.

Proof The first assertion follows from Proposition A.11, using the stratification of Z into a union of submanifolds.

The second assertion is an immediate consequence of Proposition A.13, using the stratification of Z into submanifolds. $\qquad\square$

Remark A.37 Real algebraic varieties are wilder than their complex counterparts. For instance, Theorem A.28 fails to hold over the reals. The plane curve given as the zero set of $Y^2 - X^3 + X^2$ is irreducible but has two connected components. To aggravate things, one of these components is an isolated point.

A.3.3 Elimination Theory

We begin with a homogeneous version of Hilbert's Nullstellensatz. Recall that $\mathcal{H}_d$ denotes the complex vector space of homogeneous polynomials of degree d in $X_0, \ldots, X_n$.

Proposition A.38 *Let I denote the ideal generated by the homogeneous polynomials $f_1, \ldots, f_s \in \mathbb{C}[X_0, \ldots, X_n]$. Then $Z_{\mathbb{P}}(I) = \emptyset$ iff there exists $d \in \mathbb{N}$ such that $\mathcal{H}_d \subseteq I$.*

Proof Consider the dehomogenizations $\tilde{f}_i := f_i(1, X_1, \ldots, X_n)$ and note that $f_i = X_0^{d_i} \tilde{f}_i(X_1/X_0, \ldots, X_n/X_0)$, where $d_i = \deg f_i$. Since the zero set of $\tilde{f}_1, \ldots, \tilde{f}_s$ in $\mathbb{C}^n$ is empty, the weak form of Hilbert's Nullstellensatz (Theorem A.24) implies that there are polynomials $\tilde{g}_i$ such that $1 = \sum_i \tilde{g}_i \tilde{f}_i$. Substituting X_i by X_i/X_0 and multiplying by a sufficiently high power $X_0^{d_0}$ we obtain that $X_0^{d_0} = \sum_i g_i f_i$, where g_i denotes the homogenization of g_i. Hence $X_0^{d_0} \in I$. The same argument shows that $X_i^{d_i} \in I$. Now put $d := (n+1) \max_i d_i$. It follows that I contains all monomials of degree d. $\qquad\square$

One defines the Zariski topology on the product $\mathbb{P}^m \times \mathbb{P}^n$ of complex projective spaces by taking for the closed sets the zero sets of polynomials that are homogeneous in both groups of variables $X_0, \ldots, X_m$ and $Y_0, \ldots, Y_n$.

The following result is sometimes called the main theorem of elimination theory. It is the algebraic counterpart of the compactness of $\mathbb{P}^n$ in the classical topology. Let us point out that this result was essential in our proof of Bézout's theorem; compare Proposition 16.25.

Theorem A.39 *The projection $\pi_2 \colon \mathbb{P}^m \times \mathbb{P}^n \to \mathbb{P}^n$ maps Zariski closed subsets of $\mathbb{P}^m \times \mathbb{P}^n$ to Zariski closed subsets of $\mathbb{P}^n$.*

Proof Consider the zero set $Z \subseteq \mathbb{P}^m \times \mathbb{P}^n$ of polynomials $f_1, \ldots, f_s$ that are homogeneous in both the X and the Y variables. For all $y \in \mathbb{C}_*^{n+1}$, we have

$$y \notin \pi_2(Z) \quad \Longleftrightarrow \quad f_1(X, y), \ldots, f_s(X, y) \text{ have no common zero in } \mathbb{P}^m$$

$$\Longleftrightarrow \quad \exists d \in \mathbb{N} \text{ such that } \mathcal{H}_d \subseteq (f_1(X, y), \ldots, f_s(X, y)),$$

where the last equivalence is a consequence of Proposition A.38. It therefore suffices to prove that for each $d \in \mathbb{N}$, the set

$$A_d := \left\{ y \in \mathbb{C}^{n+1} \mid \mathcal{H}_d \subseteq (f_1(X, y), \ldots, f_s(X, y)) \right\}$$

is an open subset of $\mathbb{C}^{n+1}$ in the Zariski topology.

Fix $y \in \mathbb{C}^{m+1}$. We have $\mathcal{H}_d \subseteq (f_1(X, y), \ldots, f_s(X, y))$ iff the linear map

$$T^y : \mathcal{H}_{d-d_1} \times \cdots \times \mathcal{H}_{d-d_s} \to \mathcal{H}_d, \quad (g_1, \ldots, g_s) \mapsto \sum_{i=1}^{s} g_i f_i(X, y),$$

is surjective, or $\operatorname{rank} T^y \geq \dim \mathcal{H}_d =: N$. The matrix M^y of T^y with respect to the monomial bases has entries that are homogeneous polynomials in y. Moreover, $\operatorname{rank} T^y \geq N$ if there is an $N \times N$ submatrix with nonvanishing determinant. This shows that A_d is Zariski open. □

The Zariski topology on a projective variety $Z \subseteq \mathbb{P}^n$ is defined as the one induced by the Zariski topology on $\mathbb{P}^n$. Similarly, one defines the Zariski topology on a Zariski open subset $U \subseteq Z$. The following is an immediate consequence of Theorem A.39.

Corollary A.40 *Let V_1 and V_2 be projective varieties and U a Zariski open subset of V_2. Then the projection $\pi_2 : V_1 \times U \to U$, $(x, y) \mapsto y$, maps closed subsets to closed subsets (with respect to the Zariski topologies).* □

A.3.4 Degree

In Chap. 20 we encountered the notion of the degree of an algebraic variety. We now give a very brief introduction to this concept. For more details we refer to [148, Chap. 5].

Let $\mathbb{G}(m, n)$ denote the set of m-dimensional projective linear subspaces of $\mathbb{P}^n$ (known as a *Grassmann manifold* or *Grassmannian*). Alternatively, this may be seen as the set of complex linear subspaces of $\mathbb{C}^{n+1}$ having dimension $m + 1$. Note that $\mathbb{G}(0, n) = \mathbb{P}^n$. An extension of the construction in Sect. 14.2 shows that $\mathbb{G}(m, n)$ is a complex manifold of dimension $(m + 1)(n - m)$; compare also Sect. 14.3.2. It is possible to view $\mathbb{G}(m, n)$ as a projective variety; cf. [186].

Recall that in Sect. A.2.4 we introduced the general notion of transversality for submanifolds of a manifold. It therefore makes sense to talk about the transversal intersection of a complex linear subspace of $\mathbb{P}^n$ with a submanifold of $\mathbb{P}^n$, such as the set $\mathsf{Reg}(Z)$ of regular points of an irreducible projective variety Z.

The *degree* $\deg Z$ of the irreducible projective variety Z is defined as the natural number d characterized in the theorem below. While this theorem usually is proved by algebraic methods, it is possible to give a differential-topological proof, much as we did for Bézout's theorem. We shall indicate this proof below, leaving out some details.

Theorem A.41 *Let* $Z \subseteq \mathbb{P}^n$ *be an irreducible projective variety and assume* $m +$ $\dim Z = n$. *There is a uniquely determined* $d \in \mathbb{N}$ *such that for all* $L \in \mathbb{G}(m,n)$, *if* L *is transversal to* $\mathsf{Reg}(Z)$ *and* $L \cap \mathsf{Sing}(Z) = \emptyset$, *then* $\#(L \cap Z) = d$.

Sketch of proof. The set $R := \{(L,x) \in \mathbb{G}(m,n) \times Z \mid z \in L\}$ is a Zariski closed subset of $\mathbb{G}(m,n) \times Z$. Moreover, $R' := R \cap (\mathbb{G}(m,n) \times \mathsf{Reg}(Z))$ is a complex manifold of the same dimension as $\mathbb{G}(m,n)$.

Consider the projection $\varphi\colon R \to \mathbb{G}(m,n)$, $(L,x) \mapsto L$, and its restriction $\varphi'\colon R' \to \mathbb{G}(m,n)$, which is a smooth map between manifolds. Let S' be the set of singular points of φ'. Then $S' \cup \mathsf{Sing}(Z)$ is a Zariski closed subset of $\mathbb{G}(m,n) \times Z$. Corollary A.40 implies that $S := \varphi(S \cup \mathsf{Sing}(Z))$ is a Zariski closed subset of $\mathbb{G}(m,n)$. Theorem A.28 implies that $U := \mathbb{G}(m,n) \setminus S$ is connected in the classical topology.

As in the proof of Theorem 16.23, we can argue that for $L \in U$, the fibers $\varphi^{-1}(L)$ are finite, and moreover, by the inverse function theorem, the function $U \to \mathbb{N}$, $L \mapsto \#(\varphi^{-1}(L))$, is locally constant (with respect to the classical topology). Hence, since U is connected, this function must be constant.

Finally, we note that for $x \in L \cap \mathsf{Reg}(Z)$, L is transversal to $\mathsf{Reg}(Z)$ iff (L,x) is a regular point of φ'. So for $L \in U$ we have $\#(\varphi^{-1}(L)) = \#(L \cap Z)$. $\qquad\square$

It is clear that $\deg \mathbb{P}^m = 1$ for a projective linear subspace $\mathbb{P}^m$ of $\mathbb{P}^n$. One can also show that $\deg Z \geq 1$ for every projective algebraic variety Z.

The unitary group $\mathcal{U}(n+1)$ acts transitively on $\mathbb{G}(m,n)$. Thus, if $\mathbb{P}^m$ denotes a fixed linear subspace of $\mathbb{P}^n$, then $u\mathbb{P}^m$ runs through all of $\mathbb{G}(m,n)$ when $u \in \mathcal{U}(n+1)$.

Corollary A.42 *Let* $Z \subseteq \mathbb{P}^n$ *be an irreducible projective variety and assume* $\dim Z + m = n$. *Then, for almost all* $u \in \mathcal{U}(n+1)$, *the intersection* $Z \cap u\mathbb{P}^m$ *has exactly* $\deg Z$ *points.*

Proof The proof of Proposition A.18 immediately extends from spheres to complex projective space with the transitive action of $\mathcal{U}(n+1)$ on $\mathbb{P}^n$. Therefore, for almost all $u \in \mathcal{U}(n+1)$, $u\mathbb{P}^m$ is transversal to $\mathsf{Reg}(Z)$.

Let $\mathsf{Sing}(Z) = M_1 \cup \cdots \cup M_r$ be a stratification into complex manifolds as in Corollary A.34. Then $\dim_{\mathbb{C}} M_i \leq \dim \mathsf{Sing}(Z) < \dim Z$. By the same reasoning as above, for almost all $u \in \mathcal{U}(n+1)$, $u\mathbb{P}^m$ is transversal to each of the M_i. But since $\dim_{\mathbb{C}} M_i + m < n$, this means that $u\mathbb{P}^m$ does not meet M_i. Hence $u\mathbb{P}^m \cap \mathrm{Sing}(Z) = \emptyset$ for almost all $u \in \mathcal{U}(n+1)$. Theorem A.41 now completes the proof. $\qquad\square$

The degree of a hypersurface is what we would expect it to be.

Proposition A.43 *Let $f \in \mathbb{C}[X_0, \ldots, X_n]$ be an irreducible, homogeneous polynomial of degree $d \geq 1$. Then $\deg Z_{\mathbb{P}}(f) = d$.*

Sketch of proof Let $L \in \mathbb{G}(1, n)$ satisfy the assumptions of Theorem A.41. To simplify notation, assume without loss of generality that $L = Z_{\mathbb{P}}(X_0, X_1)$ and $[(0, 1, 0, \ldots, 0)] \notin L$. Then the univariate polynomial $g(X_1) := f(1, X_1, 0, \ldots, 0)$ has degree d and $\#(L \cap Z_{\mathbb{P}}(f))$ equals the number of complex zeros of g. One can check that by assumption, all the zeros of g are simple. Thus $\#(L \cap Z_{\mathbb{P}}(f)) = d$ by the fundamental theorem of algebra. $\qquad\square$

In Sect. 16.5 we proved a version of Bézout's theorem, stating that $\#(Z_{\mathbb{P}}(f_1) \cap \cdots \cap Z_{\mathbb{P}}(f_n)) = d_1 \cdots d_n$ if $f \in \mathcal{H}_{\mathbf{d}} \setminus \Sigma$. The latter condition means that the hypersurfaces $Z_{\mathbb{P}}(f_i)$ intersect transversally (cf. Sect. A.2.4, where this notion was defined for the intersection of two submanifolds).

For the sake of completeness let us mention a more general version of Bézout's theorem. The *degree* of a projective variety of pure dimension is defined as the sum of the degrees of its irreducible components.

Theorem A.44 *Suppose that Z and W are irreducible projective varieties in $\mathbb{P}^n$ such that $\dim Z + \dim W \geq n$. If $Z \cap W = \mathrm{Reg}(Z) \cap \mathrm{Reg}(W)$ and $\mathrm{Reg}(Z)$ and $\mathrm{Reg}(W)$ intersect transversally, then $Z \cap W$ is of pure dimension $\dim Z + \dim W - n$ and $\deg(Z \cap W) = \deg Z \cdot \deg W$.* $\qquad\square$

When the assumptions on Z and W are violated, subtle phenomena may appear. Not only may intersections of higher multiplicities arise: it may also be the case that $Z \cap W$ contains irreducible components of different dimensions. For the purpose of estimation, the so-called *Bézout's inequality* has proven to be of great value in algebraic complexity theory; cf. [47, Sect. 8.2]. Let us state it in full generality for the sake of completeness. We define the *(cumulative) degree* of a Zariski closed subset in $\mathbb{P}^n$ as the sum of the degrees of its irreducible components. A subset V of $\mathbb{P}^n$ is called *locally closed* if it is the intersection of an open with a closed subset in the Zariski topology. We define the degree of V as the degree of its closure in the Zariski topology.

Theorem A.45 (Bézout's inequality) *Suppose that Z and W are locally closed subsets of $\mathbb{P}^n$. Then $\deg(Z \cap W) \leq \deg Z \cdot \deg W$.* $\qquad\square$

In Sect. 21.3 we shall need a corollary of Bézout's inequality.

Corollary A.46 *Let $f \in \mathcal{H}_\mathbf{d}$ and recall that $d_i = \deg f_i$. The number of simple zeros in $\mathbb{P}^n$ of the system $f_1(\zeta) = 0, \ldots, f_n(\zeta) = 0$ is bounded by $d_1 \cdots d_n$.*

Proof Proposition A.43 implies that $\deg Z_\mathbb{P}(f_i) \leq d_i$ (use the factorization of f_i into irreducible polynomials). Theorem A.45 implies that $\deg Z_\mathbb{P}(f) \leq \prod_i \deg Z_\mathbb{P}(f_i) \leq d_1 \cdots d_n$. Now note that by the implicit function theorem (Theorem A.1), each simple zero of f is isolated and thus constitutes an irreducible component of $Z_\mathbb{P}(f)$. $\square$

A.3.5 Resultant and Discriminant

We study now the solvability of overdetermined systems of polynomial equations, where we have $n + 1$ equations in $n + 1$ homogeneous variables.

For a degree pattern $\mathbf{d} = (d_0, \ldots, d_n)$ consider the set $S_\mathbf{d} := \{f \in \mathcal{H}_\mathbf{d} \mid \exists \zeta \in \mathbb{P}^n \ f(\zeta) = 0\}$ of feasible polynomial systems. Theorem A.39 implies that $S_\mathbf{d}$ is a Zariski closed subset of $\mathcal{H}_\mathbf{d}$, since it is obtained as the projection over $\mathcal{H}_\mathbf{d}$ of the Zariski closed set $\{(f, \zeta) \in \mathcal{H}_\mathbf{d} \times \mathbb{P}^n \mid f(\zeta) = 0\}$. But much more can be said. (For a proof see [222, Chap. XI] or [134, Chap. IX §3].)

Theorem A.47 *For any fixed degree pattern $\mathbf{d} = (d_0, \ldots, d_n)$, the set $S_\mathbf{d}$ is a hypersurface. It is the zero set of an irreducible polynomial $\mathrm{res}_\mathbf{d}(f)$ in the coefficients of $f \in \mathcal{H}_\mathbf{d}$. Moreover, for all i, $\mathrm{res}_\mathbf{d}$ is homogeneous of degree $\prod_{j \neq i} d_j$ in the coefficients of f_i.* $\square$

The polynomial $\mathrm{res}_\mathbf{d}$ is uniquely determined up to a scalar, and it is called the *multivariate resultant* corresponding to the degree pattern $\mathbf{d}$.

We return to systems of n homogeneous polynomial equations in $n + 1$ variables and ask for a criterion to determine whether f has a multiple zero. In other words, we seek a more explicit characterization of the discriminant variety Σ introduced in Proposition 16.25. The following corollary was needed for the application in Sect. 20.6.3.

Corollary A.48 *For any fixed degree pattern $\mathbf{d} = (d_1, \ldots, d_n)$, the discriminant variety Σ is a hypersurface in $\mathcal{H}_\mathbf{d}$, given as the zero set of a polynomial $\mathrm{disc}_\mathbf{d}$ of degree*

$$\deg \mathrm{disc}_\mathbf{d} = \mathcal{D}\left(1 + \left(\sum_{i=1}^{n} d_i + 1 - n\right)\sum_{i=1}^{n}\frac{1}{d_i}\right).$$

So for all $f = (f_1, \ldots, f_n) \in \mathcal{H}_\mathbf{d}$, the system $f = 0$ has a multiple zero in $\mathbb{P}^n$ iff $\mathrm{disc}_\mathbf{d}(f) = 0$.

Proof Consider the $(n + 1) \times (n + 1)$ matrix M obtained from the Jacobian matrix $[\partial_{X_j} f_i]_{1 \leq i \leq n, 0 \leq j \leq n}$ by appending the vector $[X_0, \ldots, X_n]$ as the last row. We put

$g := \det M$ and note that g is a homogeneous polynomial of degree

$$\deg g = 1 + \sum_{i=1}^{n}(d_i - 1) = \sum_{i=1}^{n} d_i + 1 - n.$$

Now we define

$$\mathrm{disc}_{\mathbf{d}}(f_1, \ldots, f_n) := \mathrm{res}(g, f_1, \ldots, f_n).$$

A solution ζ to the system $f = 0$ is degenerate if and only if the first n rows $[\partial_{x_j} f_i]_{0 \le j \le n}(\zeta)$, for $1 \le i \le n$, are linearly dependent, which is the case if and only if $g(\zeta) = 0$ (here we used Euler's identity (16.3)). It follows that the zero set of $\mathrm{disc}_{\mathbf{d}}$ equals the discriminant variety Σ. We thus obtain

$$\deg \mathrm{disc}_{\mathbf{d}}(f_1, \ldots, f_n) = \mathcal{D} + \deg g \sum_{i=1}^{n} \frac{\mathcal{D}}{d_i}. \qquad \square$$

Remark A.49 One can show that $\mathrm{disc}_{\mathbf{d}}$ is irreducible and uniquely determined up to scaling. It is called the *discriminant* corresponding to the degree pattern $\mathbf{d}$.

A.3.6 Volumes of Complex Projective Varieties

The goal of this subsection is to outline a proof of Theorem 20.14. We achieve this by adapting the proof in Stolzenberg [215] for $\mathbb{C}^n$ to the situation of $\mathbb{P}^n$.

We shall assume here a basic familiarity with differential forms and Stokes's Theorem; see [209] for more information.

We begin with a result from multilinear algebra. Let V be a complex vector space of dimension n with a Hermitian inner product H on it. Then the real part $q := \Re H$ of H defines an inner product of the real vector space V, and the imaginary part $\omega := \Im H$ defines a 2-form on V, i.e., an alternating real bilinear form. The volume form Ω associated with q (cf. Sect. A.2.5) can be expressed by the n-fold wedge product of ω with itself as follows (cf. [187, chap. VIII, §4.1]):

$$\Omega = \frac{1}{n!} \omega \wedge \cdots \wedge \omega = \frac{1}{n!} \omega^{\wedge n}. \tag{A.6}$$

We can now state a fundamental inequality.

Lemma A.50 (Wirtinger's inequality) *Let V be a complex vector space of dimension n with a Hermitian inner product H on it. Put $q := \Re H$ and $\omega := \Im H$. Further, let $W \subseteq V$ be a real $2k$-dimensional subspace and let Ω_W denote the volume form corresponding to the restriction of q to W. Then we have for any $w_1, \ldots, w_{2k} \in W$,*

$$\frac{1}{k!} |\omega^{\wedge k}(w_1, \ldots, w_{2k})| \le |\Omega_W(w_1, \ldots, w_{2k})|.$$

Proof First we note that it is sufficient to verify the stated inequality for a basis $w_1, \ldots, w_{2k}$.

A standard result on the normal forms of skew-symmetric linear maps (cf. [134]) implies that there exist an orthonormal basis $e_1, \ldots, e_k, f_1, \ldots, f_k$ of W and $\alpha_i \in \mathbb{R}$, such that for all $1 \leq i, j \leq k$,

$$\omega(e_i, e_j) = 0, \qquad \omega(f_i, f_j) = 0, \qquad \omega(e_i, f_j) = \alpha_i \delta_{ij}.$$

We can therefore decompose ω as a sum $\omega = \omega_1 + \cdots + \omega_k$, where ω_j is obtained as the pullback of a 2-form on $V_j := \mathbb{R}e_j \oplus \mathbb{R}f_j$ via the orthogonal projection. In particular, $\omega_j \wedge \omega_j = 0$. Therefore (note that $\omega_j \wedge \omega_\ell = \omega_\ell \wedge \omega_j$, since we deal with 2-forms),

$$\omega^{\wedge k} = (\omega_1 + \cdots + \omega_k)^{\wedge k} = k! \omega_1 \wedge \cdots \wedge \omega_k.$$

It follows that

$$\frac{1}{k!} \omega^{\wedge k}(e_1, f_1, \ldots, e_k, f_k) = \omega_1(e_1, f_1) \cdots \omega_k(e_k, f_k) = \alpha_1 \cdots a_k.$$

The restriction of H on V_j has the matrix

$$M := \begin{pmatrix} 1 & i\alpha_j \\ -i\alpha_j & 1 \end{pmatrix}$$

with respect to the basis (e_j, f_j). Since M must be positive semidefinite, we have $\det M = 1 - \alpha_j^2 \geq 0$. This implies $|\alpha_j| \leq 1$.

We obtain $\frac{1}{k!} |\omega^{\wedge k}(e_1, f_1, \ldots, e_k, f_k)| \leq |\alpha_1 \cdots \alpha_k| \leq 1 = |\Omega(e_1, \ldots, f_k)|$, which proves the lemma. $\square$

We can define a Hermitian inner product on the tangent spaces $T_{[x]}\mathbb{P}^n$ of the projective space $\mathbb{P}^n$ by setting, for $a, b \in T_x$,

$$H_x(a, b) := \frac{\langle a, b \rangle}{\|x\|^2},$$

where $\langle \, , \, \rangle$ is the standard Hermitian inner product on $\mathbb{C}^{n+1}$. Note that the Riemannian metric defined in (14.13) is just the real part of this inner product. We now define $\omega_x(a, b) := \Im H_x(a, b)$ and thus obtain a 2-form ω on $\mathbb{P}^n$. It can be shown that ω is closed, that is, its exterior derivative $d\omega$ vanishes (see [148, Lemma (5.20)] for an elegant proof). This is commonly expressed by saying that $\mathbb{P}^n$ is a *Kähler manifold*.

We proceed with a brief discussion of the exponential maps of $\mathbb{P}^n$. Fix a representative $\overline{a} \in \mathbb{S}(\mathbb{C}^{n+1})$ of a point in $\mathbb{P}^n$ (denoted by the same symbol) and recall that $T_{\overline{a}} := \{z \in \mathbb{C}^{n+1} \mid \langle z, \overline{a} \rangle = 0\}$ is a model for the tangent space of $\mathbb{P}^n$ at $\overline{a}$ (cf. Sect. 14.2). Consider the map

$$\psi : \mathbb{S}(T_{\overline{a}}) \times \mathbb{R} \to \mathbb{P}^n, \qquad (w, \varphi) \mapsto \exp_{\overline{a}}(\varphi w) := [\overline{a} \cos \varphi + w \sin \varphi].$$

It is clear that $B(\overline{a}, \varepsilon) \setminus \{\overline{a}\}$ is obtained as the diffeomorphic image of $\mathbb{S}(T_{\overline{a}}) \times (0, \alpha]$ under ψ, where $\varepsilon = \sin \alpha$. Further, $S(\overline{a}, \varepsilon) := \{x \in \mathbb{P}^n \mid d_{\mathbb{P}}(x, \overline{a}) = \varepsilon\}$ is obtained as the image of $\mathbb{S}(T_{\overline{a}})$. We can thus define a projection map by

$$B(\overline{a}, \varepsilon) \setminus \{\overline{a}\} \to S(\overline{a}, \varepsilon), \qquad \psi(w, \varphi) \mapsto \psi(w, \alpha).$$

The *cone* over a subset $A \subseteq S(\overline{a}, \varepsilon)$, denoted by $\mathsf{cone}(A)$, is defined as the inverse of A under this projection map.

Lemma A.51 *Let* $A \subseteq S(\overline{a}, \varepsilon)$ *be a submanifold of dimension* $m - 1$. *Then* $\mathsf{cone}(A)$ *is a submanifold of dimension* m *and*

$$\underset{m}{\mathsf{vol}} \, \mathsf{cone}(A) \leq \frac{\varepsilon}{m} \frac{1}{1 - \varepsilon^2} \underset{m-1}{\mathsf{vol}} \, A.$$

Proof We shall apply the coarea formula to ψ. First we calculate the derivative of ψ (compare the proof of Lemma 20.5). Put $q := \overline{a} \cos \varphi + w \sin \varphi$ and $v := -\overline{a} \sin \varphi + w \cos \varphi$. If $T_{w,\overline{a}}$ denotes the orthogonal complement of $\mathbb{C}w + \mathbb{C}\overline{a}$ in $\mathbb{C}^{n+1}$, we have the following orthogonal decompositions of the tangent spaces:

$$T_w \mathbb{S}(T_{\overline{a}}) = T_{w,\overline{a}} \oplus \mathbb{R}iw, \quad T_q = T_{q,v} \oplus \mathbb{C}v = T_{w,\overline{a}} \oplus \mathbb{R}iv \oplus \mathbb{R}v$$

(for the first decomposition see Lemma 14.9). We claim that $D\psi(w, \varphi)$ splits according to the above decompositions as follows: for $\dot{w}_1 \in T_{w,\overline{a}}$ and $\lambda_1, \lambda_2 \in \mathbb{R}$,

$$D\psi(w, \varphi)(\dot{w}_1 \oplus \lambda_1 iw, \lambda_2) = \dot{w}_1 \sin \varphi \oplus iv\lambda_1 \sin \varphi \cos \varphi \oplus v\lambda_2. \tag{A.7}$$

In order to see this, take curves $w(t), \varphi(t)$ and differentiate

$$q(t) := \psi\big(w(t), \varphi(t)\big) = \overline{a} \cos \varphi(t) + w(t) \sin \varphi(t)$$

with respect to t. This gives

$$\dot{q} = -\overline{a}\dot{\varphi} \sin \varphi + w\dot{\varphi} \cos \varphi + \dot{w} \sin \varphi = \dot{w} \sin \varphi + v\dot{\varphi}.$$

To complete the proof of the claim, recall from Lemma 14.8 that $\frac{d}{dt}[q(t)] = \pi(\dot{q})$, where $\pi \colon \mathbb{C}^{n+1} \to T_q$ denotes the orthogonal projection. Further, it is immediate to check that $\pi(iw) = iv \cos \varphi$.

Let A' denote the inverse image of the submanifold $A \subseteq S(\overline{a}, \varepsilon)$ under the map $w \mapsto \psi(w, \alpha)$. Then $A' \times [0, \alpha] = \psi^{-1}(\mathsf{cone}(A))$ by the definition of $\mathsf{cone}(A)$. Let ψ_{res} denote the restriction of ψ to $A' \times [0, \alpha]$ and recall that $\dim A = m - 1$. The Jacobian $\mathrm{J}\psi_{\mathrm{res}}$ of ψ_{res} can, due to Eq. (A.7), be bounded as follows:

$$(\sin \varphi)^{m-1} \cos \varphi \leq \mathrm{J}\psi_{\mathrm{res}}(w, \varphi) \leq (\sin \varphi)^{m-1}.$$

Thus, using the coarea formula, we get

$$\underset{m}{\mathsf{vol}} \, \mathsf{cone}(A) = \int_{A' \times [0, \alpha]} \mathrm{J}\psi_{\mathrm{res}} \, dA' d\varphi \leq \underset{m-1}{\mathsf{vol}} \, A' \int_0^\alpha (\sin \varphi)^{m-1} d\varphi$$

and

$$\operatorname*{vol}_{m-1} A = \int_{w \in A} \mathrm{J}\psi_{\mathrm{res}}(w, \alpha)\, dA' \geq \operatorname*{vol}_{m-1} A'(\sin\alpha)^{m-1} \cos\alpha.$$

This implies

$$\frac{\operatorname{vol}_m \operatorname{cone}(A)}{\operatorname{vol}_{m-1} A} \leq \frac{1}{(\sin\alpha)^{m-1}\cos\alpha} \int_0^\alpha (\sin\varphi)^{m-1}\, d\varphi.$$

Further,

$$\int_0^\alpha (\sin\varphi)^{m-1}\, d\varphi \leq \int_0^\alpha (\sin\varphi)^{m-1} \frac{\cos\varphi}{\cos\alpha}\, d\varphi = \frac{1}{\cos\alpha} \frac{(\sin\alpha)^m}{m}.$$

We conclude that

$$\frac{\operatorname{vol}_m \operatorname{cone}(A)}{\operatorname{vol}_{m-1} A} \leq \frac{\sin\alpha}{m} \frac{1}{(\cos\alpha)^2} = \frac{\varepsilon}{m} \frac{1}{1-\varepsilon^2}. \qquad \square$$

Let $M \subseteq \mathbb{P}^n$ be an m-dimensional submanifold of complex projective space $\mathbb{P}^n$. We fix $\overline{a} \in M$ and define for $0 < \varepsilon \leq 1$ the *level sets*

$$M_{\leq\varepsilon} := \{x \in M \mid d_{\mathbb{P}}(x, \overline{a}) \leq \varepsilon\}, \qquad M_\varepsilon := \{x \in M \mid d_{\mathbb{P}}(x, \overline{a}) = \varepsilon\}.$$

If ε is a regular value of the map $M \to \mathbb{R}$, $x \mapsto d_{\mathbb{P}}(x, \overline{a})$, then, by a variant of Theorem A.9, $M_{\leq\varepsilon}$ is a smooth manifold with boundary M_ε (cf. [209] for a definition of this notion). By Sard's theorem, this is the case for almost all $\varepsilon > 0$. Moreover, an orientation of M induces an orientation of $M_{\leq\varepsilon}$.

One can check that the normal Jacobian of $M \to \mathbb{R}$, $x \mapsto d_{\mathbb{P}}(x, \overline{a})$, equals 1. Hence $F(\varepsilon) := \operatorname{vol}_m M_{\leq\varepsilon} = \int_0^\varepsilon \operatorname{vol}_{m-1} M_\rho\, d\rho$ and $F'(\varepsilon) = \operatorname{vol}_{m-1} M_\varepsilon$.

Proposition A.52 *Let $V \subseteq \mathbb{P}^n$ be a complex submanifold of complex dimension k and $\overline{a} \in V$. Then, for almost all $0 < \varepsilon < 1$,*

$$\operatorname*{vol}_{2k} V_{\leq\varepsilon} \leq \frac{\varepsilon}{2k} \frac{1}{1-\varepsilon^2} \operatorname*{vol}_{2k-1} V_\varepsilon.$$

Proof Let ω denote the 2-form on $\mathbb{P}^n$ defined as the imaginary part of the Hermitian metric on $\mathbb{P}^n$. Since $d\omega = 0$, we have $d\omega^{\wedge k} = 0$ by the product rule for the exterior differentiation of differential forms. Since $B(\overline{a}, \varepsilon)$ is contractible to $\overline{a}$, there is a $(2k+1)$-form Φ on $B(\overline{a}, \varepsilon)$ such that $d\Phi = \omega^{\wedge k}$, due to Poincaré's lemma; cf. [209].

We can express the volume form Ω_V of V as $\frac{1}{k!}\omega^{\wedge k}$; cf. (A.6). We thus obtain

$$\operatorname*{vol}_{2k} V_{\leq\varepsilon} = \int_{V_{\leq\varepsilon}} \Omega_V = \frac{1}{k!} \int_{V_{\leq\varepsilon}} \omega^{\wedge k} = \frac{1}{k!} \int_{V_\varepsilon} \Phi,$$

where we used Stokes's theorem for the last equality, noting that V_ε is the boundary of the manifold with boundary $V_{\leq\varepsilon}$. One can show that the singularity of the apex $\overline{a}$

of cone(V_ε) does not harm, so that we can apply Stokes's theorem again to obtain

$$\frac{1}{k!} \int_{V_\varepsilon} \Phi = \frac{1}{k!} \int_{\text{cone}(V_\varepsilon)} \omega^{\wedge k}.$$

Wirtinger's inequality (Lemma A.50) applied to the real subspaces $T_x \text{cone}(V_\varepsilon)$ of $T_x \mathbb{P}^n$ implies that $\frac{1}{k!} |\omega^{\wedge k}| \le |\Omega_{\text{cone}(V_\varepsilon)}|$, where $\Omega_{\text{cone}(V_\varepsilon)}$ denotes the volume form of cone(V_ε). Therefore,

$$\left| \frac{1}{k!} \int_{\text{cone}(V_\varepsilon)} \omega^{\wedge k} \right| \le \frac{1}{k!} \int_{\text{cone}(V_\varepsilon)} |\omega^{\wedge k}| \le \int_{\text{cone}(V_\varepsilon)} |\Omega_{\text{cone}(V_\varepsilon)}| = \underset{2k}{\text{vol}} \, \text{cone}(V_\varepsilon).$$

Bounding the latter with Lemma A.51, the assertion follows. $\qquad\square$

The proof of the next lemma is straightforward and therefore omitted.

Lemma A.53 *Let $V \subseteq \mathbb{P}^n$ be a complex submanifold of complex dimension k and $\bar{a} \in V$. Then, for any k-dimensional projective linear subspace $\mathbb{P}^k$ of $\mathbb{P}^n$ containing $\bar{a}$, we have*

$$\lim_{\varepsilon \to 0} \frac{\text{vol}_{2k} \, V_\varepsilon}{\text{vol}_{2k}(B(\bar{a}, \varepsilon) \cap \mathbb{P}^k)} = 1. \qquad\square$$

Proof of Theorem 20.14 Suppose first that $\bar{a}$ is a regular point of V. Then there exists $\varepsilon_0 > 0$ such that $F(\varepsilon) := \text{vol}_{2k} \, V_{\le \varepsilon}$ is well defined for almost all $0 < \varepsilon \le \varepsilon_0$. We already noted that $F(\varepsilon) = \int_0^\varepsilon \text{vol}_{2k-1} V_\rho \, d\rho$ and hence $F'(\varepsilon) = \text{vol}_{2k-1} V_\varepsilon$.

Put $G(\varepsilon) := \frac{F(\varepsilon)}{\varepsilon^{2k}(1-k\varepsilon^2)}$. Lemma A.53 combined with $\text{vol}_{2k}(B(\bar{a}, \varepsilon) \cap \mathbb{P}^k) = \varepsilon^{2k} \, \text{vol}_{2k} \, \mathbb{P}^k$ (cf. Lemma 20.8) implies that $\lim_{\varepsilon \to 0} G(\varepsilon)\varepsilon^{-2k} = \text{vol}_{2k} \, \mathbb{P}^k$. It is therefore sufficient to prove that G is monotonically increasing. By calculating its derivative we get

$$G'(\varepsilon) = \frac{1}{\varepsilon^{2k}(1-k\varepsilon^2)} \left(F'(\varepsilon) - \frac{2k}{\varepsilon} \frac{(1-(k+1)\varepsilon^2)}{(1-k\varepsilon^2)} F(\varepsilon) \right)$$

$$\ge \frac{1}{\varepsilon^{2k}(1-k\varepsilon^2)} \left(F'(\varepsilon) - \frac{2k}{\varepsilon}(1-\varepsilon^2) F(\varepsilon) \right) \ge 0,$$

where we used Proposition A.52 for the last inequality.

The case in which $\bar{a}$ is a singular point of V can be reduced to the above case by a continuity argument, whose details are harmless, but shall be omitted. $\qquad\square$

Remark A.54 In a similar way one can prove the bound

$$\underset{2k}{\text{vol}}\big(V \cap B(\bar{a}, r)\big) \ge r^{2k} \underset{2k}{\text{vol}}\big(\mathbb{C}^k \cap B(\bar{a}, r)\big)$$

for a k-dimensional irreducible affine variety $V \subseteq \mathbb{C}^n$, $\overline{a} \in V$, and $r > 0$. Here $\mathbb{C}^k$ stands for any k-dimensional linear subspace of $\mathbb{C}^n$ containing $\overline{a}$, and $B(\overline{a}, r)$ denotes the Euclidean ball of radius r and center $\overline{a}$ in $\mathbb{C}^n$ or $\mathbb{C}^k$, respectively. See Stolzenberg [215, Thm. B], who attributes the result to Federer.

A.4 Integral Geometry

For the reader's convenience, we collect here the results from integral geometry that are relevant in the last two chapters of this book. For those used in Chap. 20, we shall be able to provide complete proofs.

A.4.1 Poincaré's Formula

Suppose that $M, N \subseteq \mathbb{S}^p$ are smooth submanifolds of dimension m and n, respectively, such that $m + n \geq p$. Pick a uniform random $g \in \mathcal{O}(p+1)$. Proposition A.18 states that the intersection of M with the random translate gN of N is almost surely a submanifold of dimension $m + p - n$, or empty. In particular, the volume $\mathrm{vol}_{m+p-n}(M \cap gN)$ is almost surely well defined. Poincaré's formula gives a beautifully simple expression for the expectation of this volume in terms of the volumes of M and N, respectively.

Theorem A.55 (Poincaré's formula) *Suppose that $M, N \subseteq \mathbb{S}^p$ are smooth submanifolds of dimension m and n, respectively, such that $m + n \geq p$. Then, for a uniform random $g \in G := \mathcal{O}(p+1)$, we have*

$$\mathop{\mathbb{E}}_{g \in G} \left(\frac{\mathrm{vol}_{m+n-p}(M \cap gN)}{\mathcal{O}_{m+n-p}} \right) = \frac{\mathrm{vol}_m M}{\mathcal{O}_m} \cdot \frac{\mathrm{vol}_n N}{\mathcal{O}_n}.$$

The proof relies on the coarea formula (Theorem 17.8) and the following obvious transitivity property of the action of the orthogonal group $\mathcal{O}(p+1)$ on the sphere $\mathbb{S}^p$ and its tangent spaces.

Lemma A.56 *Let $x_0, y_0 \in \mathbb{S}^p$ and $U_0 \subseteq T_{x_0}\mathbb{S}^p$, $V_0 \subseteq T_{y_0}\mathbb{S}^p$ be n-dimensional linear subspaces. Then there exists $g \in \mathcal{O}(p+1)$ such that $gx_0 = y_0$ and $gU_0 = V_0$.* $\square$

Proof of Theorem A.55 Consider the smooth map $\mu \colon G \times \mathbb{S}^p \to \mathbb{S}^p$, $(g, x) \mapsto g^{-1}x$, and its restriction $f \colon G \times M \to \mathbb{S}^p$. We define

$$R := f^{-1}(N) = \left\{ (g, x) \in G \times M \mid g^{-1}x \in N \right\}.$$

In the proof of Proposition A.18 it was shown that R is a submanifold, and we determined its tangent spaces $T_{(g,x)}R$; see (A.2). We will consider the surjective projections

$$p_1: R \to G, \quad (g, x) \mapsto g, \quad \text{and} \quad p_2: R \to M, \quad (g, x) \mapsto x.$$

Note that the fibers of p_1 are given by $p_1^{-1}(g) = \{g\} \times (M \cap gN)$. Moreover, the fibers of p_2,

$$p_2^{-1}(x) = \{g \in G \mid g^{-1}x \in N\} \times \{x\},$$

are submanifolds (this follows by taking $M = \{x\}$ in the above argument). Note also that by (A.2), the normal Jacobians $\mathrm{NJ}p_1(g, x)$ and $\mathrm{NJ}p_2(g, x)$ depend on the submanifolds M and N only through their tangent spaces $T_x M$ and $T_y N$, where $y = g^{-1}x$. (See Sect. 17.3 for the definition of normal Jacobians.)

The orthogonal group G acts isometrically on $G \times \mathbb{S}^p$ via $h(g, x) := (hg, hx)$. Also, the submanifold R is invariant under the action of G, since $(hg)^{-1}hx = g^{-1}h^{-1}hx = g^{-1}x \in N$ for $(g, x) \in R$ and $h \in G$. It is clear that both p_1 and p_2 are G-equivariant. This easily implies that their normal Jacobians $\mathrm{NJ}p_1$ and $\mathrm{NJ}p_2$ are G-invariant.

The coarea formula applied to the smooth map p_1 implies

$$\int_R \mathrm{NJ}p_1 \, dR = \int_{g \in G} \underset{m+n-p}{\mathrm{vol}} \, (M \cap gN) \, dG. \tag{A.8}$$

Moreover, the coarea formula applied to the smooth map p_2 yields

$$\int_R \mathrm{NJ}p_1 \, dR = \int_{x \in M} \int_{p_2^{-1}(x)} \frac{\mathrm{NJ}p_1}{\mathrm{NJ}p_2} \, dp_2^{-1}(x) \, dM. \tag{A.9}$$

The function $F := \frac{\mathrm{NJ}p_1}{\mathrm{NJ}p_2}$ is G-invariant, since the normal Jacobians of p_1 and p_2 are G-invariant.

Fix $x_0 \in M$. Any $x \in M$ is of the form $x = hx_0$ for some $h \in G$. Moreover, we have the isometric bijection $p_2^{-1}(x_0) \to p_2^{-1}(x)$, $(g, x_0) \mapsto h(g, x_0)$, which implies that

$$\int_{p_2^{-1}(x_0)} F \, dp_2^{-1}(x_0) = \int_{p_2^{-1}(x)} F \, dp_2^{-1}(x),$$

using that F is G-equivariant. Hence Eq. (A.9) translates to

$$\int_R \mathrm{NJ}p_1 \, dR = \underset{m}{\mathrm{vol}} \, M \cdot \int_{p_2^{-1}(x_0)} F \, dp_2^{-1}(x_0). \tag{A.10}$$

We now consider the smooth map

$$\psi: p_2^{-1}(x_0) \to N, \quad (g, x_0) \mapsto g^{-1}x_0.$$

Note that $\psi(hg, x_0) = \psi(g, x_0)$ for h lying in the stabilizer $G_{x_0} := \{h \in G \mid hx_0 = x_0\}$ of x_0. It follows that $\mathrm{NJ}\psi(hg, x_0) = \mathrm{NJ}\psi(g, x_0)$. The coarea formula applied to

ψ yields

$$\int_{p_2^{-1}(x_0)} F \, dp_2^{-1}(x_0) = \int_{y \in N} \int_{\psi^{-1}(y)} \frac{F}{\mathrm{NJ}\psi} \, d\psi^{-1}(y) \, dN.$$

Fix $y_0 \in N$. We have the isometry $\psi^{-1}(y_0) \to \psi^{-1}(h^{-1}y_0)$, $(g, x_0) \mapsto (gh, x_0)$, for any $h \in G$, which implies that

$$\int_{p_2^{-1}(x_0)} F \, dp_2^{-1}(x_0) = \underset{n}{\mathrm{vol}} \, N \cdot \int_{\psi^{-1}(y_0)} \frac{F}{\mathrm{NJ}\psi} \, d\psi^{-1}(y_0). \qquad (A.11)$$

Fix $g_0 \in G$ such that $y_0 = g_0^{-1} x_0$. Then

$$\psi^{-1}(y_0) = \{ g \in G \mid g^{-1} x_0 = y_0 \} \times \{ x_0 \} = \{ (hg_0, x_0) \mid h \in G_{x_0} \}.$$

By the G_{x_0}-invariance of the normal Jacobians of p_1, p_2 and of ψ, we obtain

$$C := \int_{\psi^{-1}(y_0)} \frac{F}{\mathrm{NJ}\psi} \, d\psi^{-1}(y_0) = \frac{F}{\mathrm{NJ}\psi}(g_0, x_0) \cdot \int_{G_{x_0}} dG_{x_0}.$$

Combining this with (A.8), (A.10), and (A.11), we obtain

$$\int_{g \in G} \underset{m+n-p}{\mathrm{vol}} \, (M \cap gN) \, dG = C \cdot \underset{m}{\mathrm{vol}} \, M \cdot \underset{n}{\mathrm{vol}} \, N. \qquad (A.12)$$

It remains to investigate the dependence of the value C on the manifolds M and N. We already noted that $\mathrm{NJ}p_1(g_0, x_0)$ and $\mathrm{NJ}p_2(g_0, x_0)$ depend only on the tangent spaces $T_{x_0} M$ and $T_{y_0} N$. Similarly, $\mathrm{NJ}\psi(g_0, x_0)$ depends only on $T_{y_0} N$. In order to determine C we may therefore realize M as an m-dimensional sphere through x_0 having the prescribed tangent space at x_0, see Lemma A.56. Similarly, we choose N as an n-dimensional sphere through y_0 with the prescribed tangent space at y_0. Then we have $M \cap gN \simeq \mathbb{S}^{m+n-p}$ for almost all g. Equation (A.12) for this particular choice of M and N implies that

$$\mathcal{O}_{m+n-p} = \mathcal{O}_m \cdot \mathcal{O}_n \cdot C,$$

yielding $C = \frac{\mathcal{O}_{m+n-p}}{\mathcal{O}_m \cdot \mathcal{O}_n}$, and the same equation, now with arbitrary M and N, completes the proof. $\qquad \square$

By essentially the same proof one obtains the following version of Poincaré's formula for submanifolds of the complex projective space $\mathbb{P}^p$. Recall that the unitary group $\mathscr{U}(p+1)$ acts on $\mathbb{P}^p$.

Theorem A.57 *Suppose that $M, N \subseteq \mathbb{P}^p$ are smooth submanifolds of real dimension $2m$ and $2n$, respectively, such that $m + n \geq p$. Then, for a uniform random*

$u \in G := \mathcal{U}(p+1)$, we have

$$\mathop{\mathbb{E}}_{u \in G} \left(\frac{\mathrm{vol}_{2m+2n-2p}(M \cap uN)}{\mathrm{vol}_{2m+2n-2p} \mathbb{P}^{2m+2n-2p}} \right) = \frac{\mathrm{vol}_{2m} M}{\mathrm{vol}_{2m} \mathbb{P}^m} \cdot \frac{\mathrm{vol}_{2n} N}{\mathrm{vol}_{2n} \mathbb{P}^n}. \qquad \square$$

We can now provide the missing proofs from Chap. 20.

Proof of Theorem 20.9 Let $V \subseteq \mathbb{P}^p$ be an m-dimensional irreducible projective variety and $U \subseteq V$ an open subset in the Euclidean topology. By Theorem A.33, $\mathrm{Reg}(V)$ is a manifold of real dimension $2m$ and $\mathrm{Sing}(V)$ is a manifold of dimension less than $2m$. We put $U_0 := U \cap \mathrm{Reg}(V)$ and $U_1 := U \cap \mathrm{Sing}(V)$. Then U_0 is a submanifold of dimension $2m$ (or empty). Theorem A.57 applied to $M := U_0$ and $N := \mathbb{P}^{p-m}$ implies

$$\mathop{\mathbb{E}}_{u \in G} \#\left(U_0 \cap u\mathbb{P}^{p-m}\right) = \frac{\mathrm{vol}_{2m} U_0}{\mathrm{vol}_{2m} \mathbb{P}^m} = \frac{\mathrm{vol}_{2m} U}{\mathrm{vol}_{2m} \mathbb{P}^m}.$$

On the other hand, by Proposition A.18, $\mathrm{Sing}(V) \cap u\mathbb{P}^{p-m}$ is empty for almost all $u \in G$. We conclude that

$$\mathop{\mathbb{E}}_{u \in G} \#\left(U \cap u\mathbb{P}^{p-m}\right) = \mathop{\mathbb{E}}_{u \in G} \#\left(U_0 \cap u\mathbb{P}^{p-m}\right) = \frac{\mathrm{vol}_{2m} U}{\mathrm{vol}_{2m} \mathbb{P}^m}. \qquad \square$$

Proof of Theorem 20.13 Let $V \subseteq \mathbb{P}^p$ be an m-dimensional irreducible projective variety and $U \subseteq V$ an open subset in the Euclidean topology. We put $U_0 := U \cap \mathrm{Reg}(V)$ and $U_1 := U \cap \mathrm{Sing}(V)$ as before. Then U_0 is a submanifold of dimension $2m$ (or empty). Fix $\overline{a} \in \mathbb{P}^p$ and let N be the open ball around $\overline{a}$ of radius ε (with respect to d_{sin}). Lemma 20.8 tells us that $\mathrm{vol}_{2p} N = \mathrm{vol}_{2p} B(\overline{a}, \varepsilon) = \varepsilon^{2p} \mathrm{vol}_{2p} \mathbb{P}^p$. Theorem A.57 applied to $M := U_0$ and N implies

$$\mathop{\mathbb{E}}_{u \in G} \frac{\mathrm{vol}_{2m}(U_0 \cap uN)}{\mathrm{vol}_{2m} \mathbb{P}^m} = \frac{\mathrm{vol}_{2m} U_0}{\mathrm{vol}_{2m} \mathbb{P}^m} \cdot \frac{\mathrm{vol}_{2p} N}{\mathrm{vol}_{2p} \mathbb{P}^p} = \frac{\mathrm{vol}_{2m} U}{\mathrm{vol}_{2m} \mathbb{P}^m} \cdot \varepsilon^p.$$

Let ∂N denote the boundary of the ball $B(\overline{a}, \varepsilon)$. By Proposition A.18, $U_0 \cap u\partial N$ is a manifold of dimension strictly less than $2m$, for almost all $u \in G$. Hence $\mathrm{vol}_{2m}(U_0 \cap u\partial N) = 0$ for almost all u; cf. Proposition A.11. We thus obtain

$$\mathop{\mathbb{E}}_{u \in G} \mathrm{vol}_{2m}\left(U_0 \cap u B(\overline{a}, \varepsilon)\right) = \mathop{\mathbb{E}}_{u \in G} \mathrm{vol}_{2m}(U_0 \cap uN) = \varepsilon^p \mathrm{vol}_{2m} U.$$

In the same way we see that $\mathrm{vol}_{2m}(U_1 \cap u B(\overline{a}, \varepsilon)) = 0$ for almost all $u \in G$. So we obtain, using Lemma 20.11,

$$\mathop{\mathbb{E}}_{a \in \mathbb{P}^p} \mathrm{vol}_{2m}\left(U \cap B(a, \varepsilon)\right) = \mathop{\mathbb{E}}_{u \in G} \mathrm{vol}_{2m}\left(U \cap u B(\overline{a}, \varepsilon)\right) = \varepsilon^p \mathrm{vol}_{2m} U. \qquad \square$$

A.4.2 The Principal Kinematic Formula

The integral geometric result stated in Theorem 21.11 was essential in Chap. 21. We cannot provide its proof for lack of space, but we would like to indicate briefly how this result is related with the so-called principal kinematic formula of spherical integral geometry.

For the following compare Sect. 21.1. Let $M \subseteq \mathbb{S}^p$ be a smooth submanifold of dimension m. For $x \in M$ let $S_x := \mathbb{S}(T_x M^\perp)$ denote the sphere of unit normal vectors v in $T_x \mathbb{S}^n$ that are orthogonal to $T_x M$. Let us denote by $K_{M,i}(x, v)$ the ith elementary symmetric polynomial in the eigenvalues of the second fundamental form of the embedding $M \hookrightarrow \mathbb{S}^p$ at x in the direction v; see [88, p. 128]. Definition 21.6 dealt with the case of an oriented hypersurface. There we had a well-defined unit normal direction v, and up to scaling, we defined the normalized integral $\mu_i(M)$ of the ith curvature of M by integrating $K_{M,i}(x, v)$ over M. In general, we don't have a distinguished direction v, but we can eliminate this deficiency by averaging over all normal directions in S_x. We thus define the *(modified) normalized integral* $\tilde{\mu}_i(M)$ *of the ith curvature* of M ($0 \leq i \leq m$) as follows:

$$\tilde{\mu}_i(M) := \frac{1}{\mathcal{O}_{m-i} \mathcal{O}_{p-m+i-1}} \int_{x \in M} \int_{v \in S_x} K_{M,i}(x, v) \, dS_x(e) \, dM(x). \qquad (A.13)$$

Note that $\tilde{\mu}_0(M) = \frac{\mathrm{vol}_m M}{\mathcal{O}_m}$. Since $K_{M,i}(x, -v) = (-1)^i K_{M,i}(x, v)$, we have $\tilde{\mu}_i(M) = 0$ if i is odd. So the quantities $\tilde{\mu}_i(M)$ are of interest for even i only. Note also that if M is a hypersurface ($m = p - 1$), then we retrieve the quantities from Definition 21.6: we have $\tilde{\mu}_i(M) = \mu_i(M)$, provided i is even. (However, the values $\mu_i(M)$ for odd i are not captured by the $\tilde{\mu}_j(M)$.)

Remark A.58 One can show that $\tilde{\mu}_i(M)$ does not change when we embed M in a sphere $\mathbb{S}^{p'}$ of larger dimension via $\mathbb{S}^p \hookrightarrow \mathbb{S}^{p'}$. This is a main reason for the choice of the normalizing factors.

The extension of Weyl's tube formula (N.6) from hypersurfaces to submanifolds of higher codimension states that for sufficiently small ε, we have

$$\mathrm{vol}\, T^\perp(M, \varepsilon) = \sum_{\substack{0 \leq i \leq m \\ i \text{ even}}} \tilde{\mu}_i(M) \mathcal{O}_{p,m-i}(\varepsilon). \qquad (A.14)$$

We define the *curvature polynomial* $\tilde{\mu}(M; X)$ of M by

$$\tilde{\mu}(M; X) := \sum_{i=0}^{m} \tilde{\mu}_i(M) X^i,$$

where X denotes a formal variable. Note that the degree of $\tilde{\mu}(M; X)$ is at most the dimension m of M. For example, we have $\tilde{\mu}(\mathbb{S}^m; X) = 1$.

The *principal kinematic formula for spheres* is the following result. It is considered the most important result of spherical integral geometry.

Theorem A.59 *Let M and N be submanifolds of $\mathbb{S}^p$ having dimension m and n, respectively, and assume $m + n \geq p$. Then we have*

$$\mathop{\mathbb{E}}_{g \in G} \left(\tilde{\mu}(M \cap gN; X) \right) \equiv \tilde{\mu}(M; X) \cdot \tilde{\mu}(N; X) \mod X^{m+n-p+1},$$

where the expectation on the left-hand side is defined coefficientwise, and on the right-hand side we have polynomial multiplication modulo $X^{m+n-p+1}$.   □

This result contains Poincaré's formula (Theorem A.55) as a special case (consider the constant coefficients of the curvature polynomials). Moreover, choosing $N = \mathbb{S}^n$ in Theorem A.59, we obtain $\mathbb{E}_{g \in G}(\tilde{\mu}(M \cap g\mathbb{S}^n; X)) \equiv \tilde{\mu}(M; X) \mod X^{m+n-p+1}$. This means that $\mathbb{E}_{g \in G}(\tilde{\mu}_i(M \cap g\mathbb{S}^n)) = \tilde{\mu}_i(M; X)$ for $0 \leq i \leq m + n - p$. In particular, if $m = p - 1$, so that M is a hypersurface, we obtain $\mathbb{E}_{g \in G}(\tilde{\mu}_i(M \cap g\mathbb{S}^n)) = \tilde{\mu}_i(M; X)$ for $0 \leq i \leq n - 1$. These equalities recover Theorem 21.11 for even indices i.

Notes

Overture   Although the loss of accuracy due to an accumulation of round-off errors in a computation had been mentioned before (the initial quotation from Gauss is an example), the systematic analysis of this subject begins with two papers published independently by Herman Goldstine and John von Neumann [226] and by Alan Turing [221]. Both these papers dealt with the solution of linear systems of equations. The latter introduced most of the subject's terminology such as the term "condition number" and the adjective "ill-conditioned." However, it appears that the notion of "ill-posedness" had been in use long before in the context of partial differential equations; see Courant and Hilbert [67].

Backward-error analysis is also present in these two papers, but its place in contemporary numerical analysis is due to the strong advocacy of it made by James Wilkinson in the 1960s and 70s. A concise exposition of Wilkinson's views appears in his 1970 SIAM John von Neumann lecture [237]. A detailed treatment of these views is found in his books [235, 236].

The themes we have collected under the heading *The Many Faces of Condition* in Sect. 0.5 arose in the last 60 years in a somehow unordered manner. A goal of this book is to attempt a unified presentation. Some of these themes—e.g., the computation of condition numbers—obey an immediate need demanded by applications. Others grew up out of a need of understanding. An example of the latter is the relation of condition to distance to ill-posedness. Probably the first instance of this phenomenon is the fact that for an invertible square matrix A one has $\|A^{-1}\|^{-1} = d(A, \Sigma)$. While this result is usually attributed to Carl Eckart and Gale Young [91], it actually dates back to much earlier work by Erhard Schmidt [182] and Hermann Weyl [230], as pointed out by Stewart [213]. The systematic search for relations between condition and distance to ill-posedness was promoted by Jim Demmel in [84]. A further twist on these relations was pioneered by Jim Renegar, who proposed to *define* condition as the relativized inverse to the distance to ill-posedness for those problems in which the usual definition is meaningless (e.g., decision problems; see the notes to Chap. 6 below).

An early attempt at a general theory of condition appears in the paper [170] by John Rice.

P. Bürgisser, F. Cucker, *Condition*,
Grundlehren der mathematischen Wissenschaften 349,
DOI 10.1007/978-3-642-38896-5, © Springer-Verlag Berlin Heidelberg 2013

The idea of randomizing the data and looking for the expected condition (or the tail of the condition number) was, as we mentioned in Sect. O.5.3, introduced by Goldstine and von Neumann in the sequel [108] to their paper [226] and subsequently strongly advocated by Steve Smale [201].

As for the relations between complexity and conditioning, one can identify specific instances as early as in the 1950s (see Notes of Chap. 5 below). The suggestion of a complexity theory for numerical algorithms parameterized by a condition number $\mathscr{C}(a)$ for the input data (in addition to input size) was first made, to the best of our knowledge, by Lenore Blum in [35]. It was subsequently supported by Smale [201, Sect. 1], who extended it, as pointed out above, by proposing to obtain estimates on the probability distribution of $\mathscr{C}(a)$. By combining both ideas, he argued, one can give probabilistic bounds on the complexity of numerical algorithms.

Chapter 1 Linear algebra is doubtless the most highly cultivated part of numerical analysis. This is unsurprising, since ultimately, most of the problems for which a numerical solution is available are so because they reduce to a linear algebra problem. Due to this prominence, there is no short supply of books on the subject. A classic reference is by Gene Golub and Charles van Loan [109]. Three excellent modern books are those by Jim Demmel [86], by Nick Higham [121], and by Lloyd Trefethen and David Bau [219]. A book with a focus on perturbation theory is that by Pete Stewart and Ji-Guang Sun [214].

Theorem 1.1 is a particular case of Theorem 19.3 of [121] and is due to Nick Higham, who first published it in a technical report [118].

The characterization of the normwise condition number for linear equation solving goes back to the work of Turing, von Neumann, and Goldstine that we mentioned above. Early results in componentwise analysis were obtained by Oettli and Prager [153]. The first mixed perturbation analysis appears in the paper [198] by Robert Skeel, where a mixed error analysis of Gaussian elimination is performed. In this work Skeel defined a condition number of mixed type: it uses componentwise perturbations on the input data and the infinity norm in the solution. In [172], Jiří Rohn introduced a new relative condition number measuring both perturbation in the input data and error in the output componentwise. It was Gohberg and Koltracht [107] who named Skeel's condition number *mixed* to distinguish it from componentwise condition numbers such as those in [172]. They also gave explicit expressions for both mixed and componentwise condition numbers.

The paragraph above refers to square systems of linear equations. Perturbation theory for rectangular matrices and linear least squares problems has existed quite a while for the normwise case (cf. [211, 227]) and has been further studied in [103, 112, 143]. In particular, the bounds in (1.13) follow from a result of Per-Åke Wedin in [227] (see also [121, Theorem 19.1]). For the mixed and componentwise settings for the problem of linear least squares, bounds for both condition numbers (or first-order perturbation bounds) and unrestricted perturbation bounds appear in [9, 34, 119]. A characterization of these condition numbers is given in [72].

Theorem 1.7 for spectral norms is usually attributed to Eckart and Young, but as we pointed out before, it actually dates back much earlier. See Stewart's survey [213] for this and on the history of the fundamental singular value decomposition. The more general version we presented in Theorem 1.7 was proved by Kahan [124], who attributes it to Gastinel (cf. Higham [121, Thm. 6.5]).

The inclusion in this chapter of the characterization of condition in Sect. 1.4 was suggested to us by Javier Peña. It follows a line of thought that has proved to be useful in conic programming [137, 157, 167].

A notion we have not mentioned in this book is that of *stochastic condition number*. Condition numbers as defined in the Overture measure the worst-case magnification of the output error with respect to a small input perturbation. An idea advanced by Fletcher [97] is to replace "worst-case" by "average" in this measure. This idea was further pursued in [10, 212, 228]. The bottom line of the results in these works, however, is somehow disappointing: stochastic condition numbers are smaller than their worst-case counterparts but not substantially so.

The vast amount of work in numerical linear algebra in general, and of conditioning in this context in particular, makes it infeasible for us to do justice to its authors. Readers interested in history and references for numerical linear algebra will find a carefully wrought account in the set of "Notes and References" closing each of the chapters in [121].

Chapter 2   We relied on many sources to write the crash courses in this chapter, taking just the minimum we needed to proceed with the probabilistic analyses of condition numbers. In particular, we tailored the notion of *data space*, since this notion was well fitted to cover these minima.

A detailed exposition of integration theory, including proofs of the theorems by Fubini and Tonelli, can be found, e.g., in [17]. For a proof of the transformation formula we refer to [209].

There are many books on probability, but few seem to be at the same time elementary (avoiding measure theory) and yet containing sufficient information about continuous distributions. In this respect we found [96] helpful.

We remark that Proposition 2.22 is a nontrivial result from [65, Corollary 6].

Smoothed analysis was proposed by Daniel Spielman and Shang-Hua Teng [206, 207] and initially used to give an explanation of the superb performance of the simplex algorithm in practice [208]. This kind of explanation has gained currency since its introduction, as witnessed by the fact that Spielman and Teng were awarded the Gödel 2008 and Fulkerson 2009 prizes for it (the former by the theoretical computer science community and the latter by the optimization community). Also, in 2010, Spielman was awarded the Nevanlinna prize, and smoothed analysis appears in the laudatio of his work.

A smoothed analysis of Turing's condition number was first performed by Sankar et al. [179] and later improved by Mario Wschebor [244] to Theorem 2.50. Its optimality follows from [14]. These results rely on the assumption of (isotropic) Gaussian perturbations, and the proofs make essential use of orthogonal invariance (see also the notes of Chap. 4 for more information). For random matrices with entries from discrete distributions (e.g., independent Bernoulli ± 1), the situation is considerably more complicated. Recently, a general "average-case" result in this direction

was obtained by Terence Tao and Van Vu [216], which was subsequently extended
to a smoothed analysis [217] by the same authors.

The average and smoothed analysis presented in Sect. 2.3 is, to the best of our
knowledge, one of the simplest instances of such analyses. It goes back to discussions with Martin Lotz.

Proposition 2.44 is often referred to as "Renegar's trick." It was communicated
in a personal letter to Shub and Smale in 1985. The letter mentions that "the bound
could be made better with more sophisticated arguments," clearly pointing out that
the goal was simplicity.

Chapter 3   The fact that random triangular matrices are poorly conditioned (with
respect to normwise condition) was given a precise statement by Viswanath and
Trefethen in [225]: if L_n denotes a random triangular $n \times n$ matrix (whose entries
are independent standard Gaussians) and $\kappa_n = \|L_n\| \|L_n^{-1}\|$ is its condition number,
then

$$\sqrt[n]{\kappa_n} \to 2 \quad \text{almost surely}$$

as $n \to \infty$. A straightforward consequence of this result is that the expected value
of $\log \kappa_n$ satisfies $\mathbb{E}(\log \kappa_n) = \Omega(n)$.

Theorem 3.1 is a less ambitious version of this result with an equally devastating
lower bound and a much simpler proof.

The probabilistic analysis of sparse matrices that occupies most of the rest of the
chapter is taken from [59]. An extension with a smoothed analysis can be found
in [53].

Chapter 4   Clearly, the probabilistic analysis of condition numbers is linked to
understanding the eigenvalues (or singular values) of random matrices. For Gaussian distributions, this is a thoroughly studied topic that originated from multivariate
statistics (John Wishart [239]) and later was taken up in physics by Eugene Wigner;
see [233]. Recall that AA^T is called Wishart distributed if $A \in \mathbb{R}^{m \times n}$ is standard
Gaussian, and that the singular values of A are just the square roots of the eigenvalues of AA^T. An excellent treatment of random matrices in multivariate statistics
can be found in Muirhead [147].

Even though the joint distribution of the eigenvalues of Wishart distributed random matrices is known in closed form, deriving from this the distribution of the
largest eigenvalue $\sigma_{\max}^2$ or the smallest one $\sigma_{\min}^2$ is a nontrivial task. Early probabilistic analyses of $\sigma_{\max}$ and $\sigma_{\min}$ for rectangular random matrices appear in the
work of Geman [102] and Silverstein [197], respectively. Their results imply that for
a sequence (m_n) of integers such that $\lim_{n \to \infty} m_n/n = \lambda \in (0, 1)$ and a sequence of
standard Gaussian random matrices $A_n \in \mathbb{R}^{m_n \times n}$, we have

$$\kappa(A_n) \longrightarrow \frac{1 + \sqrt{\lambda}}{1 - \sqrt{\lambda}} \quad \text{almost surely.} \tag{N.1}$$

Alan Edelman [92] made a thorough study of the distribution of the smallest
eigenvalue of a Wishart matrix AA^T. He gave closed formulas for its density in the

cases $n = m$ and $n = m + 1$, a recurrence for computing the density for $n > m$, and also derived asymptotic limit distributions. As a consequence of this, Edelman obtained that for both real and complex standard Gaussian $n \times n$ matrices,

$$\mathbb{E}(\log \kappa(A)) = \log n + C + o(1), \quad \text{as } n \to \infty, \tag{N.2}$$

where $C = 1.537$ in the real case and $C = 0.982$ in the complex case.

For rectangular matrices, explicit nonasymptotic tail estimates for the condition number $\kappa(A)$ were derived by Zizhong Chen and Jack Dongarra [51], who showed that for $A \in \mathbb{R}^{m \times n}$ with $n \geq m$ and $x \geq n - m + 1$ we have

$$\frac{1}{\sqrt{2\pi}} \left(\frac{1}{5x}\right)^{n-m+1} \leq \Prob_{A \sim N(0,\mathrm{I})} \left\{\kappa(A) \geq \frac{x}{1-\lambda}\right\} \leq \frac{1}{\sqrt{2\pi}} \left(\frac{7}{x}\right)^{n-m+1}. \tag{N.3}$$

Here $\lambda = \frac{m-1}{n}$ is the elongation of A.

We note that Mark Rudelson and Roman Vershynin [174] have a recent result on the distribution of the smallest singular value of a random rectangular matrix for very general distributions.

All the results mentioned above are average-case analyses. As for smoothed analysis, we already reviewed in the notes of Chap. 2 what is known about the condition number of square matrices.

Our Chap. 4, which is taken from [45], provides a smoothed analysis of the condition number of rectangular matrices. Theorem 4.16 can be seen as an extension to smoothed analysis of the upper bound in (N.3). We note that the decay in z in this tail bound is the same as in (N.3) up to the logarithmic factor $\sqrt{\ln z}$. We believe that the latter is an artefact of the proof that could be omitted. In fact, the exponent $n - m + 1$ is just the codimension of the set $\Sigma := \{A \in \mathbb{R}^{m \times n} \mid \mathrm{rk} A < m\}$ of rank-deficient matrices; cf. [115]. From the interpretation of $\Prob\{\kappa(A) \geq t\}$ as the volume of a tube around Σ, as discussed in Chaps. 20 and 21, one would therefore expect a decay of order $1/z^{n-m+1}$. (Compare Theorem 20.1, which, however, is over $\mathbb{C}$.)

The proof techniques employed for the proof of Theorem 4.16 are an extension of methods by Sankar et al. [179]. In particular, the proof of Proposition 4.19 is based on an idea in [179]. The proof of Theorem 4.4 is taken from [136].

We remark that the bounds in Sect. 4.1.3 can be slightly improved: Let $\sigma_{\max}(X)$ and $\sigma_{\min}(X)$ denote the maximal and minimal singular values of $X \in \mathbb{R}^{m \times n}$, $m \leq n$. For standard Gaussian X it is known that $\sqrt{n} - \sqrt{m} \leq \mathbb{E}\sigma_{\min}(X) \leq \mathbb{E}\sigma_{\max}(X) \leq \sqrt{n} + \sqrt{m}$; cf. [78]. This implies $Q(m, n) \leq 1 + \sqrt{m/n} \leq 2$, which improves Lemma 4.14.

Chapter 5 Complexity theory aims at proving lower bounds on the cost of all algorithms belonging to a certain class that solve a particular problem. This requires a formal development of models of computation that we do not address in this book, since we have not dealt here with the issue of lower complexity bounds. Instead we have limited ourselves to estimating costs of algorithms (which provide upper complexity bounds for the underlying problem). The primer in Sect. 5.1 succinctly sets up the context for these estimates. In particular, it implicitly fixes a model of computation whose associated cost is the algebraic, that is, one that performs arithmetic operations and comparisons of real numbers with unit cost. For more information on these models and on the complexity theories built upon them we refer to the books [38, 47]. The monographs [12, 156] are excellent expositions of complexity theory for discrete computations.

The method of conjugate gradients is an important algorithm for solving large sparse linear systems. It is due to Hestenes and Stiefel [117]. Our treatment follows [141].

The cost analyses for the steepest descent and conjugate gradient algorithms may have been the first examples of condition-based complexity analysis. We could not, however, confirm (or refute) this fact in the literature.

Intermezzo I Literature on the issue of structured condition is scattered. A central reference regarding linear algebra are the papers by Siegfried Rump [175, 176], which show that for a significant number of matrix structures, the condition numbers obtained by restricting perturbations to those respecting the structure coincide with their unrestricted versions. The two papers deal with normwise and componentwise perturbations, respectively. Another example is the paper [59] cited above, where the emphasis is on probabilistic analysis and the structures considered are given by sparsity patterns.

Other instances of condition for structured data occur, for instance, in [127, 158].

Chapter 6 Carathéodory's theorem, the separating hyperplane theorem, and Helly's theorem are classic results in convex analysis. One can find proofs for them in [171].

The idea of defining condition numbers for feasibility problems in terms of distance to ill-posedness goes back to Jim Renegar [165–167]. The condition number $\mathscr{C}(A)$ in Sect. 6.4 can be seen as a variant of this idea. It had been introduced before Renegar's series of papers by Jean-Louis Goffin [106] for dual feasible problems only. Goffin's definition was in terms of the quantities $\xi(A, y)$ as in Sect. 6.7. The extension of $\mathscr{C}(A)$ to infeasible data as well, along with the characterization as the inverse of the distance to ill-posedness, was part of the PhD thesis of Dennis Cheung and appeared in [54]. Goffin's use of $\mathscr{C}(A)$ was for the analysis of relaxation methods such as the perceptron algorithm presented in Sect. 6.9. This explains the fact that he considered only feasible data.

The characterization of $\mathscr{C}(A)$ in terms of spherical caps making the substance of Sect. 6.5 is taken from [63]. That in Sect. 6.6 was suggested to us by Javier Peña.

The perceptron algorithm was introduced in [173]. It is a relaxation method in the sense of [2, 146]. In Sect. 6.9 we showed that its complexity is quadratically bounded in $\mathscr{C}(A)$. A more efficient version, known as *rescaled perceptron*, has recently been devised by Alexandre Belloni, Bob Freund, and Santosh Vempala [21]. Its complexity's dependence on condition is $\mathcal{O}(\log \mathscr{C}(A))$.

Besides $\mathscr{C}(A)$, several condition numbers have been proposed (and used) to analyze algorithms for polyhedral feasibility problems. Renegar's $C(A)$ is paramount among them and features in many condition-based analyses in the literature. Other condition measures are Steve Vavasis and Yinyu Ye's $\bar{\chi}_A$ [223, 224], Ye's $\sigma(A)$ [245], and Marina Epelman and Freund's μ_A [94]. A comparison between many of these measures can be found in [62].

Chapter 7 The possible polynomial cost of linear programming problems (with integer data) had been an open question for several years when Leonid Khachiyan [129] gave a positive answer in 1979. The idea his proof relied on, the ellipsoid method, had been used in the Soviet Union by Naum Shor, Arkady Nemirovsky, and David Yudin since early in that decade for other purposes, but it was Khachiyan's result that brought the ellipsoid method into the limelight.

Our treatment in Sect. 7.1 is brief. A more detailed exposition on ellipsoids can be found in Grötschel, Lovász, and Schrijver [114]. The exposition in Sect. 7.2 was inspired by Bob Freund and Jorge Vera [98]. The idea of analyzing a real data algorithm in terms of its condition number and then proving a "gap result" (such as Proposition 7.9) when the data is restricted to integer coefficients goes back to Renegar [167].

Chapter 8 Books on linear programming are legion. Most of them, moved by the understandable goal of providing an elementary exposition, are based on the simplex method, which allows for short, clear developments. There is a price, however, in both conciseness and clarity, because, on the one hand, the complexity analysis of simplex is not easy. Its worst-case complexity is exponential, and the analyses of its average-case complexity, which in general are not condition-based, inherit the complication of many ad hoc arguments. On the other hand, simplex does not generalize to nonpolyhedral contexts.

A goal of Part II of this book is to provide an exposition of linear programming with a condition-based approach to algorithmic analysis (and amenable to more general contexts). Because of this, our account in this chapter does not follow any existing exposition. In particular, some of the terminology we used has been introduced by us. For additional material on linear programming the reader might find useful the textbooks [33, 144, 178].

The origins of linear programming can be traced back to the work of Joseph Fourier, who in 1827 published a method for solving systems of linear inequalities (see [111] for a history of this contribution). The first algorithm for solving linear programs, the simplex method, was announced in 1947 by George Dantzig. In that same year, John von Neumann is credited with the development of duality theory. An authoritative account of these early years of linear programming is in [77].

Chapter 9 Modern interior-point methods were developed by Narendra Karmarkar [126]. The first goal, as the title of Karmarkar's paper reveals, was to give another algorithm solving linear programming problems in polynomial time. As with Khachiyan's work, one of the key ideas in Karmarkar's algorithm was much older—Karmarkar's choice of projection's direction was given by a steepest descent after making a projective transformation, but this turned out to be equivalent to a Newton direction for an earlier barrier function, introduced in the barrier methods of the 1950s and 60s (see [243]). Shortly after Karmarkar's paper it was realized that interior-point methods were both faster and more stable than the ellipsoid method. Furthermore, fundamental work of Yury Nesterov and Arkady Nemirovsky [149] extended the use of this method to general convex programming problems. These discoveries created a substantial interest in interior-point methods whose effects have lasted till today.

An overview of the history of interior-point methods is given in [243]. Two books devoted to the subject are [168, 242].

Regarding our exposition, the relaxation scheme in Sect. 9.4 has its origins in work by Peña and Renegar [159], and a variant of it was also used in [69]. The simple form presented in Sect. 9.4 was suggested to us by Javier Peña. The primal–dual perspective is partly motivated by Vavasis and Ye's formulation in [223]. The proof in Sect. 9.2 follows [242].

As in many other parts in this book, in proving Lemma 9.6 we aimed for simplicity and not for optimality. We remark that with a little more work, the factor 2^{-1} in that lemma can be improved to $2^{-3/2}$; see [151, Lemma 14.1].

Chapter 10   The condition number $C(S)$ was introduced by Renegar [166, 167], and it has been extensively used in relation to several aspects of interior-point methods (see, e.g., [99, 152, 159]). Its history is interwoven with that of the GCC condition number, and they are closely related when the data has been adequately normalized. Proposition 10.3 is an instance of this relationship. We have taken it from [58].

Chapter 11   The condition number $\mathscr{K}(d)$ was introduced in [56]. Most of the results shown in this chapter are taken either from this paper or from its sequel [58]. Actually, it is in this sequel where Theorem 11.21 is proved. Also, Algorithm 11.2 is an infinite-precision (and hence simplified) version of the main algorithm in this paper.

Problems in linear programming are related in the sense that they often reduce to each other. Different problems, however, have associated different measures of condition. The relations between these measures are studied in [61], where a single problem is stated whose condition number yields a host of condition measures as particular cases.

In the acronym RCC, the two last letters refer to the authors of [56, 58] and the initial letter to Jim Renegar, who was the first to suggest using relativized inverses to the distance to ill-posedness as a condition number for finite-valued problems.

Chapter 12   This short chapter is taken from [60], a paper that one may say was crafted with the idea of completing the second part of this book. An issue conspicuously left open in [60] is the smoothed analysis of $\mathscr{K}(d)$. Such an analysis is yet to be done.

Chapter 13   In recent years there has been a stream of results around the probability analysis of the GCC condition number $\mathscr{C}(A)$. A bound for $\mathbb{E}(\ln \mathscr{C}(A))$ of the form $\mathcal{O}(\min\{n, m \ln n\})$ was shown in [55]. This bound was improved in [71] to $\max\{\ln m, \ln\ln n\} + \mathcal{O}(1)$, assuming that n is moderately larger than m. Still, in [63], the asymptotic behavior of both $\mathscr{C}(A)$ and $\ln\mathscr{C}(A)$ was exhaustively studied, and these results were extended in [116] to matrices $A \in (\mathbb{S}^m)^n$ drawn from distributions more general than the uniform. Independently of this stream of results, in [89], a smoothed analysis for Renegar's condition number $C(A)$ was performed from which it follows that $\mathbb{E}(\ln C(A)) = \mathcal{O}(\ln n)$ (and hence, a similar bound holds for $\mathbb{E}(\ln\mathscr{C}(A))$). The finest bound for the average analysis of $\ln\mathscr{C}(A)$ was recently given in [50], namely, $\mathbb{E}(\ln\mathscr{C}(A)) \leq 2\ln(m+1) + 3.31$.

Theorem 13.1 shows a result of this kind (a bound for $\mathbb{E}(\ln\mathscr{C}(A))$ linear in $\log m$ and independent of n). The constant in the $\mathcal{O}$ notation is not made explicit, and we can expect it to be greater than the 2 of the bound in [50], but the result extends to a smoothed analysis (cf. Corollary 13.4).

Dunagan et al. [89] were the first to provide a smoothed analysis of Renegar's condition number. They proved that for $A \in \mathbb{R}^{m \times n}$,

$$\sup_{\|\overline{A}\|=1} \quad \mathop{\mathbb{E}}_{A \sim N(\overline{A}, \sigma^2 I)} \left(\ln C(A) \right) = \mathcal{O}\left(\ln \frac{mn}{\sigma} \right). \tag{N.4}$$

In [7], Amelunxen and Bürgisser derived a robust variant of this result for the GCC condition number in the model of radially symmetric probability distributions supported on a spherical disk. Their proof combined ideas from [49] on the volume of tubes (cf. Chap. 21) with techniques in [89].

The exposition of Chap. 13 is based on [7]. Theorem 13.3 and Corollary 13.4 are special cases of the main result obtained there. Theorem 13.6 is due to Wendel [229]. The reductions in Sect. 13.4 are inspired by [89].

Intermezzo II The level-2 condition number was introduced by Jim Demmel in [84], where he proved, for some specific problems, that their level-2 condition numbers coincide with their original condition numbers up to a multiplicative constant. Subsequently, Des Higham [120] improved this result by sharpening the bounds for the problems of matrix inversion and linear systems solving. He actually proved Theorem II.1 for these two problems. The full version of this result was shown in [57].

We called *à la Renegar* the class of condition numbers dealt with in the intermezzo to acknowledge the role of Jim Renegar in the definition of condition as relativized distance to ill-posedness. The relation between these two notions had been noticed previously for a number of problems. It was Jim Demmel, in [84], who first suggested that this was a general phenomenon and considered a class of condition numbers given as inverses to distance to ill-posedness, doubtless with the general probabilistic analysis of his subsequent paper [85] in view (more on this in the notes to Chaps. 20 and 21). A few years after that, as mentioned in the notes to Chap. 6, Jim Renegar gave additional strength to this view by proposing to define condition as relativized distance to ill-posedness for problems in which the usual definition is meaningless.

Chapter 14 The paper by John Rice [170] defines absolute and relative condition numbers in the general abstract framework of a smooth map between smooth manifolds. Condition numbers for computing zeros of polynomials were studied by Wilkinson [235], Woźniakowski [241], Demmel [84], and in the so-called *Bézout series* authored by Mike Shub and Steve Smale [192–196] in the early 1990s. The presentation of this chapter mainly follows [196].

For some basic information on the geometry of complex projective spaces, the reader may consult Mumford [148].

The condition of computing eigenvalues and eigenvectors was studied in detail by Wilkinson [235, 236]. For a comprehensive treatment of the perturbation theory for eigenvalues we refer to Stewart and Sun [214]. The material in Sect. 14.3.2 on the computation of the kernel is based on a paper by Diego Armentano [10].

Chapter 15 The idea of homotopy has been fundamental in the development of algebraic topology. As a construction for equation solving, it goes back at least to E. Lahaye [132] in 1934, and it got impetus in the 1970s and 1980s. The practical success of homotopy methods triggered the appearance of a number of books on the subject such as [101, 128, 169]. A more recent account on these methods is in [205] and a comprehensive survey article in [138].

Newton's method is due, no surprises here, to Sir Isaac Newton. He applied it to polynomial equations in 1669 (*De analysi per æquationes numero terminorum infinitas*) and to nonpolynomial analytic functions in 1687 (*Philosophiæ Naturalis Principia Mathematica*). A careful historical description of the development of Newton's method (and of the work of other actors playing a role in this development) is in [246].

The modern approach to Newton's method, in which estimates on the size of the basin of quadratic attraction are sought, was initiated by Leonid Kantorovich [125]. The state of the art on this theme is Smale's γ-theory, which first appeared in [200]. Theorem 15.5 is taken from there. This theory was accompanied in [200] by another body of results, known as Smale's α-theory, providing conditions for a point z to be an approximate zero that can be measured at the point z itself (instead of at its associated zero ζ). This theory occupies us in Chap. 19, and we will return to it in the notes to that chapter. Up-to-date expositions of Newton's method include [8, 81, 87].

Chapter 16 A fundamental reference for our exposition of the material in this chapter is the Bézout series [192–196] by Shub and Smale mentioned above. All the basic ingredients in our exposition—the space $\mathcal{H}_\mathbf{d}$ endowed with a unitarily invariant Hermitian product, homotopy continuation, projective Newton's method, the condition numbers $\mu_{\mathrm{norm}}(f, \zeta)$ and $\mu_{\mathrm{max}}(f)$, etc.—are present in these papers and play a role equally basic. Some of these ingredients predate the Bézout series; see Shub and Smale's earlier works [190, 191]. We point out that our account omits an interesting relationship, established in [194], between condition and dense packings on the sphere $\mathbb{S}^2$. The latter is closely related to Smale's 7th Problem [202].

Weyl's Hermitian product was introduced by Weyl; see e.g., [231]. The orthogonal decomposition in Sect. 16.3 was introduced by Beltrán and Pardo [28].

Bézout's theorem is a classic result in algebraic geometry. The proof for it in Sect. 16.5 is taken from [38, Chap. 10]. We remark that the underlying idea to prove the fundamental theorem of algebra via a homotopy goes back to Gauss (see Smale [199] for a detailed account). There exist more sophisticated versions of Bézout's theorem, e.g., dealing with multiplicities [148, 186]. We did not need this version for our purposes.

Projective Newton was proposed by Shub in [188], and its γ- and α-theories were developed within the Bézout series. The sustained need for results from these theories for the development of the algorithmic content of the series gave rise to an exposition of them that is fragmented, notationally inconsistent, and at times repetitious. A cleaner exposition is given in [38, Chap. 14] but is limited to the γ-theorem, which is stated in terms of the function d_T defined by $d_T(x, y) = \tan d_{\mathbb{P}}(x, y)$. This approach yields a result that corresponds, word by word, with the Euclidean γ-theorem (Theorem 15.5) but pays for this by relying on a function that is not a distance: d_T does not satisfy the triangle inequality. In particular, proceeding to an extension to an α-theorem does not appear to be easy.

A different, and more general, approach to these issues was proposed by Jean-Pierre Dedieu, Pierre Priouret, and Gregorio Malajovich [82]. They tailored a version of Newton's method that works in Riemannian manifolds. The underlying idea is to measure the Newton step not on the tangent space but on the manifold itself. In Fig. 19.1 this corresponds to moving the distance $\|x - \bar{x}\|$ directly on the circle. The resulting point is therefore more distant from x than the point $N_f(x)$ in that figure. That is, Newton's steps in this approach are longer than in the projective Newton's method. An obvious benefit of this Newton's method on manifolds is its generality. Another advantage is the elegance of the statements corresponding to its γ- and α-theories. A possible drawback is its technical difficulty.

Yet one more approach consists in replacing the inverse $Df(z)|_{T_z}^{-1}$ in the definition of the projective Newton operator by the Moore–Penrose inverse $Df(x)^{\dagger}$. This is sometimes referred to as the Moore–Penrose Newton's method, and an early exposition of it is in [3]. Subsequent accounts of this variant are in [81, 196]. We use Moore–Penrose Newton in Sect. 19.6.

The contents of Sect. 16.6 are based on the projective Newton's method as proposed by Shub. Its γ-theory (and its α-theory as described in Sect. 19.3) is based on the distance $d_{\mathbb{P}}$ together with a parameter $r \in [\frac{2}{\pi}, 1]$ and some functions of it that are used to bound how much the resulting γ-theorem departs from that of the Euclidean case. This compromise allows for a short development whose proofs follow the same lines as those in the Euclidean setting.

The Lipschitz estimate in Proposition 16.2 first appeared in [189, Theorem 1], without attention to the constants.

The $\mathcal{O}(N)$ algorithm for computing the Jacobi matrix mentioned in the proof of Proposition 16.32 was first found by Linnainmaa [139] and then rediscovered by Baur and Strassen [20].

Chapter 17 There is a vast body of work on polynomial equation-solving by exact symbolic algorithms that we haven't discussed at all in this monograph: we cannot even attempt to survey this multifaceted theory here. But let us point out some common features of these symbolic algorithms as well as the differences from the numerical approaches.

Suppose that the polynomials $f_1, \ldots, f_s \in \mathbb{C}[X_1, \ldots, X_n]$ of degree at most D are given by their lists of coefficients. The *Hilbert Nullstellensatz problem* is to decide whether these polynomials have a common zero in $\mathbb{C}^n$. This problem is known to be NP-complete over $\mathbb{C}$ in the sense of Blum, Shub, and Smale [37]. Algorithms solving this feasibility problem, with a very high cost, have long been known. The first one having only exponential complexity was exhibited in [64] by Alexander Chistov and Dimitri Grigoriev. Marc Giusti and Joos Heintz [104] designed a randomized (Monte Carlo) algorithm for deciding the Hilbert Nullstellensatz problem with $(sD^n)^{\mathcal{O}(1)}$ arithmetic operations and tests for equality (if the polynomials f_i have integer coefficients of bit-size at most ℓ, then $(\ell s D^n)^{\mathcal{O}(1)}$ bit operations suffice). It is also possible to decide finiteness of the set of solutions and to compute exact representations of the zeros (described as zeros of univariate polynomials) within these bounds. We refer to the surveys [90, 177] for more information and references. These algorithms have running time exponential in n. In the case $D \geq n^{1+\epsilon}$ (for fixed $\epsilon > 0$) the running time is polynomially bounded in the input size N; cf. Lemma 18.25. However, in the more interesting case of bounded D, the running time is exponential in the input size N. In view of the above mentioned NP-completeness result, this seems unavoidable. There are similar results for solving systems of polynomial equations over the reals, going back to Grigoriev and Vorobjov [113]; see Renegar [163] and the monograph by Basu, Pollack, and Roy [18] for detailed information and references.

In contrast with the numerical algorithms discussed in this monograph, the running times of these symbolic algorithms do not depend on a notion of condition that measures the well-posedness of the given instance, and they have a worst-case exponential complexity. But a closer look at them reveals that in the presence of finite precision, they are likely to be unstable (due to their work with exponentially large matrices). This is also in contrast with numerical algorithms, most of which have been designed with accuracy as a goal.

We now turn attention to numerical algorithms.

The fifth paper in the Bézout series concludes with a nonuniform algorithm that computes approximate zeros of systems in $\mathcal{H}_\mathbf{d}$ within average polynomial time. The qualifier "nonuniform," unfortunately, makes the algorithm inapplicable, and the series stops short of producing an implementable algorithm. The main obstruction to doing so is identified to be the difficulty in constructing a well-conditioned initial pair for the homotopy (that is, a pair (g, ζ) with $\mu_{\max}(g)$ small).

A measure of the importance that Smale attached to this problem is given by the fact that when requested to propose a list of problems for the mathematicians of the 21st century, Smale included two problems in the list whose possible solution would lead to the desired algorithm: the 17th, which plainly asks for the algorithm, and the 7th, which asks for an algorithm producing sets of points well separated on $\mathbb{S}^2$ (a set that then could be used to produce the desired initial pair (g, ζ)). The list, with technical descriptions of the problems and the state of the art for them in the year 1998, appeared in [202]. A regular update of the state of the art for these problems appears at the page "Smale's problems" in Wikipedia.

The first breakthrough in the 17th problem was announced at the *Foundations of Computational Mathematics* conference in 2005 [24] and expanded in [26]. This article was subsequently improved (both in terms of quality of results and readability) in [27] and then in [28].

One of the advances making possible the improvements in [28] was the publication of [189], a new installment in the Bézout series authored by Mike Shub. Besides the Lipschitz estimate for μ_{norm} mentioned in the notes to the previous chapter, the most important result of this paper was an upper bound on the necessary steps in the homotopy with origin at (g, ζ) and end at f in terms of the integral of $\mu_{\text{norm}}^2(q_\tau, \zeta_\tau)$ along the lifting on V of the segment $E_{g,f}$. This result was not constructive, but constructive versions of it could be easily derived. Such versions occur for instance in [22] and [46]. Our exposition of Sect. 17.1 follows the second of these references, which is less general but considerably shorter.

The papers of Beltrán and Pardo introduced an idea that is absent in the formulation of Smale's 17th problem: randomizing the initial pair of the homotopy. While this absence prevents their main result from being a positive answer to the problem as stated, it removes none of the applicability of a deterministic uniform algorithm. Furthermore, it brought a wave of fresh air to the problem, triggering a new stream of research on the subject.

Together with the idea of randomizing the initial pair, Beltrán and Pardo proposed the randomization procedure we described in Sect. 17.6 with only one essential difference (compare with [28]). Their routine for drawing the system h proceeds in a different manner from that of random_h. They observe that given ζ, one can easily construct a unitary matrix H_ζ such that $H_\zeta(e_0) = \zeta$, also, that it is trivial to draw a system in

$$R_{e_0} = \{q \in \mathcal{H}_\mathbf{d} \mid q_{i,\alpha} = 0 \text{ if } \alpha_0 < 2\}$$

from $N(0, I)$. They further note that if $q \in R_{e_0}$, then $h := H_\zeta(q) \in R_\zeta$, and that since $H_\zeta : R_{e_0} \to R_\zeta$ preserves standard Gaussianity, the system h follows the standard Gaussian distribution in R_ζ.

This procedure possesses the elegance that most invariance arguments have. Unfortunately, though, its cost is comparable to the average cost of the homotopy continuation. Indeed, the computation of $H_\zeta(q)$ requires the evaluation of q not at a point in $\mathbb{C}^{n+1}$ but on an $(n+1)$-tuple of linear forms, and this appears to have cost $\Omega(N^2)$. This cost would become prohibitive if the attempts to reduce the cost of the homotopy continuation to $\mathcal{O}(DnN \log N)$ were successful. The ideas behind these attempts are laid down in the latest (as of today) issue of the Bézout series [29].

The routine random_h presented in Sect. 17.6 is taken from [46].

The complexity analysis in Sect. 17.7 is taken mostly from [46]. The proof of Proposition 17.27 is, however, simpler and relies on the ideas of [193]. It was suggested to us by Diego Armentano.

Theorem 17.28 is a variation of the results obtained in [193] (in that paper the distribution of μ_{norm} was studied with respect to the uniform distribution of V embedded in $\mathbb{P}(\mathcal{H}_\mathbf{d}) \times \mathbb{P}^n$). A proof of Proposition 17.31 can be found in [193].

Algorithm LV has been implemented and a number of experiments run over this implementation. Details are in [23].

The coarea formula (Theorem 17.8) is due to Federer [95, Thm. 3.1], who stated this result in more generality. A proof of Theorem 17.8, avoiding the measure-theoretic machinery needed in the case of Lipschitz maps can be found in [123, Appendix]. The results by Solovay and Strassen and Adleman and Huang that we mentioned in Sect. 17.2.1 appeared in [1, 203, 204]. Another early instance of the use of randomization for primality testing, by Michael Rabin, is [160].

Our treatment focuses on numerical polynomial equation-solving in the case of many variables. For the case of one variable, considerably more can be said; see Schönhage [185] and Pan [155].

Chapter 18 Smoothed analysis was not an issue at the time Smale published his list of problems, in 1998. It would make its first public appearance in 2001 [206]. By the end of that decade, however, the consideration of a smoothed analysis version of Smale's 17th problem was all but natural.

The extension of the results in [28] from average analysis to smoothed analysis was carried out in [46]. All the results in this chapter are taken from this paper. A remarkable difference with the previous work in the subject is the systematic use of Gaussians. This choice of distribution simplified the exposition of many results (by eliminating the use of integral geometry) and allowed for the unified treatment in Sects. 17.7, 18.3, and 18.4.

The important Proposition 18.6 first appears in [28].

Renegar's Algorithm Ren, which is based on the factorization of the u-resultant, is described and analyzed in [162]. The algorithm even finds the multiplicities of the zeros ζ; see [162] for the precise statement.

Chapter 19 The bounds in Theorem 19.1 are not precisely of the kind that encourages implementation. Yet this order of magnitude was somehow to be expected. The reasons for this have a complexity-theoretic nature, and therefore, the following explanation barely hints at them.

The most elementary way of comparing the complexity of two problems is to compare the costs of algorithms solving these problems. A drawback of this approach is that these algorithms may not be (and in general aren't) optimal. Hence, a dramatic improvement in the algorithmics for one of the problems can completely change the outcome of this comparison. Complexity theorists use a more intrinsic way to compare problems—the so-called *reductions*—which allows one to group computational problems in classes and to identify, within these classes, the most difficult problems: those for which a dramatic improvement in the algorithmics means the same improvement for all problems in the class. Such problems are said to be *complete* in their classes. The reader can find textbook expositions of these ideas—in the context of discrete computations—in [12, 156] or—in the context of numerical computations—in [38].

A class that appears to be hopeless with regard to efficient computations is the class #P$_\mathbb{R}$ (this is the real counterpart of the discrete counting class #P). And the discouraging news is that the problem of counting the real zeros of a polynomial system is complete in #$P_{\mathbb{R}}$ (see [44]).

Our exposition in this chapter is based on [74, 75] for Sects. 19.2 to 19.4, on [46] for Sect. 19.5, and on [70] for Sect. 19.6.

Smale's α-theory, just like his γ-theorem, was first published in [200]. The development in Sect. 19.3 adapts the general lines in this paper to our context.

The condition number $\kappa_{\text{feas}}(f)$ was introduced in [70] to analyze a finite-precision version of Algorithm 19.2. The condition number $\nu(f)$ was introduced shortly thereafter, in [68], where it was used to strengthen this algorithm to make it return an approximate zero of the input f when it is feasible. The condition number theorem for $\nu(f)$ (Theorem 19.3) is proved in [75].

Theorem 19.2 is due to Shub and Smale [193]. Our proof, taken from [46], is perhaps less involved. A different proof, based on the Rice formula, was given by Jean-Marc Azaïs and Mario Wschebor [13]. An early proof for the case of all degrees equal ($d_i = D$ for $i = 1, \ldots, n$) was given by Eric Kostlan [131].

The exposition in Sect. 19.6 relies on the extension of Newton's method (both the iteration and its corresponding α-theory) to surjective mappings and the derived use of Moore–Penrose inversion. As we mentioned earlier, an exposition of Moore–Penrose Newton's iteration is in [3].

A standard reference for algorithms that deal, among others, with graph problems is [66].

Chapter 20 The idea of reformulating probability distributions as quotients of volumes to estimate condition measures goes back at least to Smale's influential paper [199]. There we can already find the core idea of estimating the volume of tubes by means of Poincaré's formula from integral geometry combined with Bézout's theorem.

Blum and Shub [36] took up these ideas in [199] for establishing bounds on the average loss of precision in evaluating rational functions. Renegar [161] further extended the ideas in [199]. In particular, he proved bounds on the probability distribution of a certain random variable in the average-case analysis of the complexity of Newton's method. Central to his argument is the fact that this random variable can be bounded by a conic condition number. The set of ill-posed inputs in [161] is a hypersurface. An affine version of Theorem 20.14 (stated as Remark A.54) was already used by Renegar [161].

An extension of these results to the case of codimension greater than one was done by Demmel in [85], where in addition, an average-case analysis of several conic condition numbers is performed. Most of these results are for problems over the complex numbers. An extension in another direction, namely, to possibly singular ambient spaces, was done by Beltrán and Pardo [25]. Another extension of Demmel's result, now to smoothed analysis for complex problems, was achieved in [48], an article that has been the main source for the exposition of this chapter.

A version of Theorem 20.14 with the slightly better correction factor $1/2$ was deduced in [25] from the corresponding bound for $\mathbb{C}^n$, which can be found in Stolzenberg [215, Thm. B] (see Remark A.54 for its statement). Stolzenberg attributes the

idea of the proof to Federer. We have included in the Appendix a direct derivation
of Theorem 20.14 based on similar ideas.

Luís Santaló's monograph [181], which is the standard reference on integral geometry, refers to Theorem A.55 as *Poincaré's formula* (cf. §7.1 in [181]). Apparently, Henri Poincaré stated this result for the case of $\mathbb{S}^2$, and in such form it was also
known to Joseph-Émile Barbier [15]. This result is close in spirit to Buffon's needle
problem [135], which is the earliest appearance of a problem in geometric probability. Even though Theorem A.55 is stated in Sect. 18.6 of Santaló's book, a proof is
given there in Sect. 15.2 only for an analogous statement for Euclidean space. John
Brothers [42] has proved versions of Poincaré's formula for homogeneous spaces
in great generality. We refer to the book by Ralph Howard [123] for a more ac-
cessible treatment. In particular, Poincaré's formula holds for complex projective
space (Theorem A.57), as was first shown in [180]. We used Poincaré's formula for
complex projective space in this chapter to derive Theorems 20.9 and 20.13.

The Frobenius condition number $\kappa_F(A) = \|A\|_F \|A^{-1}\|$ was first studied by
Demmel [85]. Edelman [93] obtained the following beautiful exact formula for its
tail probability:

$$\mathrm{Prob}\{\kappa_F(A) \geq t\} = 1 - \left(1 - \frac{n}{t^2}\right)^{n^2-1}, \quad \text{for } t > \sqrt{n}, \tag{N.5}$$

where A is a standard Gaussian complex $n \times n$ matrix.

Proposition 20.17 is due to James Wilkinson [238].

Chapter 21 Demmel's paper [85] dealt with both complex and real problems.
For complex problems he provided complete proofs. For real problems, Demmel's
bounds relied on an unpublished (and apparently unavailable) result by Adrian Ocneanu on the volumes of tubes around real algebraic varieties. In [240], Richard
Wongkew gave a bound for the volume of these tubes. A number of constants in his
bounds, however, are not explicit and only shown to be independent of the variety.
The first explicit bounds were given, to the best of our knowledge, in [49], from
where Theorem 21.1, along with most of this chapter, is taken.

Theorem 21.9 is from [49], but the proof closely follows the lines of a seminal
paper of Hermann Weyl [232], in which he determined the volume of small tubes
around submanifolds in Euclidean spaces and in spheres. Weyl showed that in the
setting of Theorem 21.9 there exists $\varepsilon_M > 0$ such that for all $0 < \varepsilon \leq \varepsilon_M$ and all
measurable subsets $U \subseteq M$ we have the *equality*

$$\mathrm{vol}\, T^\perp(U, \varepsilon) = \sum_i \mu_i(U) \mathcal{O}_{p,p-1-i}(\varepsilon), \tag{N.6}$$

where the sum runs over all even indices i such that $0 \leq i \leq p - 1$. (There is a
cancellation effect between the contributions of "outer" and "inner" neighborhoods
that results in the sum being only over even indices i.) In fact, Weyl's result is more
general, since it applies also to submanifolds of higher codimension; see (A.14) in
the Appendix. We remark that Jakob Steiner [210] had already discovered a related

formula for the volume of the ε-neighborhood around a convex body in Euclidean space. There one gets a polynomial in ε and the arising coefficients (up to scaling) are called the *inner volumes* of the convex body. These quantities were investigated in detail by Hermann Minkowski, who called them *Quermass integrals*; see the book by Rolf Schneider [183] for detailed information. When M is the boundary of a smooth convex body K in $\mathbb{S}^p$, one also calls the $\mu_i(M)$ the *spherical inner volumes* of K; cf. [100, 130].

In this chapter, besides Poincaré's formula (Theorem A.55), we also relied on another result from integral geometry, Theorem 21.11, whose relation to the principal kinematic formula for spheres (Theorem A.59) is explained in Theorem A.59.

To the best of our knowledge, the principal kinematic formula, along with the corresponding degree-based estimations of (absolute) integrals of curvature, was applied in [49] for the first time in the context of the probabilistic analysis of condition numbers.

In spite of its importance, it is not at all easy to locate the kinematic formula for spheres in the literature. Santaló in his book attributes the principal kinematic formula in the plane to Wilhelm Blaschke, and in Euclidean spaces to Shiing-Shen Chern [52] and Herbert Federer [95]. The normalization of integrals of curvatures leading to the simple formula of reduced polynomial multiplication was discovered by Albert Nijenhuis [150], again for Euclidean space. Santaló [181] derives the principal kinematic formula for the special case of intersections of domains in spheres, but he does not care about the scaling coefficients. In fact, the principal kinematic formulas for submanifolds of spheres and Euclidean spaces take exactly the same form. An indication of this at first glance astonishing fact can be found, somewhat hidden, in Santaló's book on page 320. The situation was clarified by Howard [123], who gave a unified treatment of kinematic formulas in homogeneous spaces. But Howard does not care about the scaling constants either. For the purpose of explicitly bounding the volumes of tubes, a good understanding of the scaling factors is relevant. The thesis of Stefan Glasauer [105] contains a detailed treatment of the principal kinematic formulas for spheres, however only in the special case of spherically convex sets. The recent book by Rolf Schneider and Wolfgang Weil [184, Sect. 6.5] contains an account of this work. To the best of our knowledge, the kinematic formula for spheres was first stated in the form of Theorem A.59 in [43]. An elementary and unconventional introduction to geometric probability and the kinematic formula for Euclidean spaces can be found in the textbook by Daniel Klain and Gian-Carlo Rota [130].

The application in Sect. 21.5 is taken from [75]. Improved bounds for the average analysis case appear in [76].

The PhD thesis by Dennis Amelunxen [4], cf. [5, 6], provides a detailed analysis of the notion of condition for the feasibility problem for any regular closed convex cone $C \subseteq \mathbb{R}^n$. It rephrases the feasibility problem in a coordinate-free way as deciding the alternative

$$(P) \quad W \cap C \neq 0 \quad \text{or} \quad (D) \quad W^{\perp} \cap \check{C} \neq 0,$$

for an input W in the *Grassmann manifold* $\mathbb{G}(m, n)$ of m-dimensional linear subspaces of $\mathbb{R}^n$. The set $\Sigma_{\mathbb{G}}$ of ill-posed instances consists of the subspaces W touching the cone C. Amelunxen defines the *Grassmann condition number* of $W \in \mathbb{G}$ as

$$\mathscr{C}_{\mathbb{G}}(W) := \frac{1}{\sin d(W, \Sigma_{\mathbb{G}})}, \tag{N.7}$$

where d denotes the geodesic distance in $\mathbb{G}$. By extending the framework of this chapter, replacing the ambient space $\mathbb{S}^p$ by the Grassmann manifold $\mathbb{G}$ (and considering the special hypersurface $\Sigma_{\mathbb{G}}$ therein), Amelunxen manages to provide a satisfactory average analysis of the Grassmann condition showing that for $W \in \mathbb{G}$ chosen uniformly at random,

$$\operatorname*{Prob}_{W \in \mathbb{G}}\{\mathscr{C}_{\mathbb{G}}(W) \geq t\} \leq 6\sqrt{m(n-m)}\frac{1}{t} \quad \text{if } t > n^{\frac{1}{2}} \tag{N.8}$$

and

$$\operatorname*{\mathbb{E}}_{W \in \mathbb{G}} \left(\ln \mathscr{C}_{\mathbb{G}}(W)\right) \leq 1.5 \ln n + 2. \tag{N.9}$$

The result on adversarial distributions, Theorem 21.29, is taken from [73] but has its roots in a paper [116] by Raphael Hauser and Tobias Müller, where a more general result is shown.

Coda: Open Problems

We do not want to bring this book to a close leaving an impression of a finished edifice. Whereas we believe that the general lines of this edifice are elegant, we have no doubts that it has, as of now, an uncompleted character. Indeed, the understanding of the role played by condition in the performance of numerical algorithms has had a phenomenal advance in the last decades, but our understanding of the performance of these algorithms—and of other possible algorithms for the same problems—has still more shadow than light. In what follows we point to some of these shadows by drawing a list with some open problems. All these problems are related to themes exposed in the previous chapters, but otherwise, there is no homogeneity in their (perceived) level of difficulty nor in the role played by condition in a possible solution.

P.1. Probabilistic Analysis of Growth Factors

"The problem of stability of Gaussian elimination is an embarrassing theoretical gap at the heart of numerical analysis." Thus wrote Trefethen and Schreiber in 1990 [220]. The origins of the embarrassment go back to an early perception, back in the 1940s, of Gaussian elimination as a numerically unstable method [16, 122]. Computational experience provided evidence to the contrary, and in 1961, Wilkinson [234] proved a bound on the relative error of a solution $\bar{x}$ of $Ax = b$ computed with Gaussian elimination, namely

$$\frac{\|\bar{x} - x\|_\infty}{\|x\|_\infty} \leq 4n^2 \kappa_\infty(A) \, \rho \, \epsilon_{\mathsf{mach}}. \tag{P.1}$$

Here $x = A^{-1}b$, and ρ is the *growth factor* of A, which is defined as

$$\rho := \frac{\|U\|_{\max} \|L\|_{\max}}{\|A\|_{\max}}, \tag{P.2}$$

P. Bürgisser, F. Cucker, *Condition*,
Grundlehren der mathematischen Wissenschaften 349,
DOI 10.1007/978-3-642-38896-5, © Springer-Verlag Berlin Heidelberg 2013

where L is lower triangular, U upper triangular, and $A = LU$ is the result of Gaussian elimination on A (without pivoting). We used $\| \ \|_{\max}$ to denote the $\| \ \|_{1\infty}$ norm (recall Corollary 1.3). In practice, Gaussian elimination is performed with *partial pivoting* (that is, the pivot is chosen to be the entry of largest absolute value in the column considered at each step). In this case, at the end of Gaussian elimination we obtain $PA = LU$ with L and U as before and P a permutation matrix. Furthermore, Wilkinson's bound (P.1) is still valid with the growth factor ρ also given by (P.2).

Wilkinson's result set a framework for understanding stability of Gaussian elimination. It also showed a hopeless worst-case scenario, since there are examples of $n \times n$ matrices A for which (with partial pivoting) $\rho = 2^{n-1}$. The obvious way out was to prove a stable behavior of Gaussian elimination in a probabilistic sense. The paper by Trefethen and Schreiber mentioned above provides ample experimental evidence of such behavior for Gaussian elimination with partial pivoting in the average case. More recently, Sankar, Spielman, and Teng [179] performed a smoothed analysis of ρ that shows stability for Gaussian elimination without pivoting. This is a puzzling result, since on the one hand, Gaussian elimination with partial pivoting is widely accepted to be more stable than without pivoting at all, whereas on the other hand, examples of matrices A for which the growth factor without pivoting is smaller than with partial pivoting are known to exist. Hence, a smoothed analysis for Gaussian elimination with partial pivoting cannot be deduced from [179]. Since this is the algorithm used in practice, the following problem is posed:

Prove average and/or smoothed polynomial bounds for the growth factor ρ for Gaussian elimination with partial pivoting.

P.2. Eigenvalue Problem

The computation of eigenvalues and eigenvectors is, in practice, amazingly successful. Yet we know of no analysis showing both low cost and numerical stability, not even in an average-case setting. In a naive approach, given a matrix $A \in \mathbb{C}^{n \times n}$, one computes the characteristic polynomial $\chi_A(z) := \det(z\mathbf{I} - A)$ and then approximates its zeros $\lambda_1, \ldots, \lambda_n$ up to a predetermined precision δ. These approximations can be obtained with low complexity. (For instance, we can apply Algorithm Ren for one polynomial in one variable and the bound (18.20), together with the fact that $|\lambda_i| \leq \|A\|$ for all $i \leq n$, which yields a cost bounded by $\mathcal{O}(n^4 \log(n) \log \log(\|A\|/\delta) + n^8)$. Algorithms tailored to the one-variable situation yield bounds that have lower degree in n; see [154, Thm. 1.1(d)].) This approach, however, is hardly ever used. The reason is that the map $A \mapsto \chi_A$ may destroy stability. To see this, consider the matrix

$$A = \begin{bmatrix} 1 & 0 & \ldots & 0 \\ 0 & 2 & \ldots & 0 \\ & & \ddots & \\ 0 & 0 & & 0 \\ 0 & 0 & \ldots & 20 \end{bmatrix}.$$

Its characteristic polynomial $\chi_A(z) = \prod_{i=1}^{20}(z - i)$ is referred to as the *Wilkinson polynomial* and is well known to behave badly under small perturbations (see, e.g., [235, Ch.2, §9]). Yet Example 14.16 shows that all the eigenvalues of A are well-conditioned. Trefethen and Bau [219, p. 190] summarize this in their textbook:

> Perhaps the first method one might think of would be to compute the coefficients of the characteristic polynomial and use a rootfinder to extract its roots. Unfortunately [...] this strategy is a bad one, because polynomial rootfinding is an ill-conditioned problem in general, even when the underlying eigenvalue problem is well-conditioned.

The results in Chap. 17 (see Proposition 17.31 and Remark 17.32) give one reasons to doubt that, at least over the complex numbers, "polynomial rootfinding is an ill-conditioned problem in general." Nonetheless, it appears that in practice, the polynomials arising as characteristic polynomials of complex matrices are indeed ill-conditioned. One may guess that for the pushforward measure on $\mathbb{C}[t]$ induced by the standard Gaussian on $\mathbb{C}^{n \times n}$ and the map $A \mapsto \chi_A(t)$, the expectation of μ_{max} is large (in contrast with the results just mentioned, which hold for the standard Gaussian in $\mathbb{C}[t]$ with respect to the Weyl Hermitian product, and in spite of the small expected condition for Gaussian matrices A with respect to the eigenvalue problem (Proposition 20.18)). In simpler words, this map amplifies, in general, condition.

In any case, one is left with the need to use a different approach, and many have been proposed that work efficiently in practice. But a theoretical explanation of their success is still due, as pointed out by Demmel [86, p. 139] when describing the performance of one of the most commonly used (the QR iteration with implicit shifts):

> It is interesting that after more than 30 years of dependable service, convergence failures of this algorithm have quite recently been observed, analyzed, and patched [...]. But there is still no global convergence proof, even though the current algorithm is considered quite reliable. So the problem of devising an algorithm that is numerically stable and globally (and quickly!) convergent remains open.

We now note that on top of this lack of results for eigenvalue computations, there is the issue of computing eigenvectors from approximations of the eigenvalues. We thus state the following open problem.

Provide a rigorous analysis of the eigenpair problem. That is, exhibit an algorithm computing one eigenpair (or all eigenpairs) (λ, v) of a matrix A along with an analysis showing, on average, low cost and numerical stability.

The word "average" here refers to a Gaussian distribution. Low cost may refer to a bound polynomial in n and $\log\log(\|A\|/\delta)$, but other forms of approximation (e.g., à la Smale) will entail different forms for the cost bounds. A similar remark applies to numerical stability.

A step towards the solution of this problem was recently made by Diego Armentano, who described and analyzed a homotopy method for the computation of eigenpairs [11]. His analysis provides an estimate of the cost of following a path with extremities the data A and an initial triple (B, λ, v) that is strongly reminiscent of Theorem 17.3. As in the case of complex polynomial systems that occupied us in Chaps. 17 and 18, the problem remains to find good initial triples, with the difference that as of today, we don't even know whether there is an efficient randomized procedure to do so. One can also expect that a stability analysis for this homotopy method could be done along the lines of that done for the algorithm MD in [41].

The eigenpair problem in the real case presents, needless to say, additional difficulties.

P.3. Smale's 9th Problem

Provide a solution to Smale's 9th problem. That is, give an answer to the following question:

Is there a polynomial-time algorithm over the real numbers that decides, on input $A \in \mathbb{R}^{m \times n}, b \in \mathbb{R}^m$, the feasibility of the linear system of inequalities $Ax = b, x \geq 0$?

Here the model of computation is the real Turing machine (BSS model), and the running time counts the number of arithmetic operations with real numbers (including comparisons). For rational input data, polynomial time is a well-known result [129] in the model of Turing machines, measuring the number of bit operations (see also Remark 10.5). This problem was posed by Steve Smale [202].

P.4. Smoothed Analysis of RCC Condition Number

In Chap. 11 we introduced the RCC condition number $\mathscr{K}(d)$ for a data triple $d = (A, b, c) \in \mathbb{R}^{m \times n} \times \mathbb{R}^m \times \mathbb{R}^n$ specifying a pair of primal and dual linear programming problems in standard form $(n \geq m)$, and in Chap. 12 an average analysis of $\mathscr{K}(d)$ was shown, conditioned to $d \in \mathcal{W}$, where, we recall, $\mathcal{W}$ denotes the set of feasible well-posed data triples for this problem. This average result, Theorem 12.1, suggests the following question:

Is it true that

$$\sup_{\|\bar{d}\| \leq 1} \mathbb{E}_{d \sim N(\bar{d}, \sigma^2 I)} \left(\ln \mathscr{K}(d) \mid d \in \mathcal{W} \right) = \mathcal{O}\left(\log \frac{n}{\sigma} \right)?$$

Note that the bound in Theorem 12.1 is consistent with this equality when $\bar{d} = 0$ and $\sigma = 1$.

P.5. Improved Average Analysis of Grassmann Condition

Let $\mathbb{G}(m, n)$ denote the Grassmann manifold of m-dimensional linear subspaces of $\mathbb{R}^n$. The Grassmann condition number $\mathscr{C}_\mathbb{G}(W)$ of $W \in \mathbb{G}(m, n)$ naturally arises in analyzing the feasibility problem for a regular closed convex cone $C \subseteq \mathbb{R}^n$; see (N.7) in the Notes for its definition. In [5] the average analysis stated in (N.8)–(N.9) was achieved. We conjecture the following:

For the uniform distribution on $\mathbb{G}(m, n)$ and all $t > m$,

$$\operatorname*{Prob}_{W \in \mathbb{G}(m,n)} \left\{ \mathscr{C}_\mathbb{G}(W) \geq t \right\} \leq \mathcal{O}\left(v(C)\sqrt{m}\, \frac{1}{t} \right)$$

with a quantity $v(C)$ depending only on the cone C.

This conjecture is due to Amelunxen and Bürgisser; compare [5, Theorem 1.4].

P.6. Smoothed Analysis of Grassmann Condition

For $\overline{W} \in \mathbb{G}(m, n)$ and $0 < \sigma \leq 1$ let $B(\overline{W}, \sigma)$ denote the ball of radius σ in the Grassmann manifold $\mathbb{G}(m, n)$ with respect to the sine of the geodesic distance in $\mathbb{G}(m, n)$.

Is it true that

$$\sup_{\overline{W} \in \mathbb{G}(m,n)} \ \operatorname*{\mathbb{E}}_{W \sim B(\overline{W},\sigma)} \left(\ln \mathscr{C}_\mathbb{G}(W) \right) = \mathcal{O}\left(\log \frac{n}{\sigma} \right)?$$

This question was posed by Amelunxen and Bürgisser [5]. The average analysis in [5] proves that this is the case for $\sigma = 1$.

P.7. Robustness of Condition Numbers

The probabilistic analyses of condition numbers done in this book were based on the assumption of either Gaussian distributions or uniform distributions in spherical disks, the only exception being the discussion of adversarial distributions in Sect. 21.8. It would be valuable to have corresponding results for more general distributions, e.g., for input matrices whose entries are chosen independently from a fixed (say discrete) distribution.

Provide probabilistic analyses for condition numbers of linear optimization (like Renegar's, GCC, or RCC) for such general distributions.

For Turing's condition number, results for such general distributions have been obtained by Terence Tao and Van Vu [216, 217]; compare the notes of Chap. 2.

P.8. Average Complexity of IPMs for Linear Programming

The bound for the number of interior-point iterations in Theorem 9.10 is proportional to $\sqrt{n}$. This is due to the choice of the centering parameter $\sigma = 1 - \frac{\xi}{\sqrt{n}}$ (with $0 < \xi \leq \frac{1}{4}$) and the fact that the duality gap μ is decreased by a factor of σ at each iteration of Algorithm 9.1. Our average (or smoothed) complexity bounds for the different applications of this algorithm invariably relied on the following two steps. Firstly, to bound, in terms of a relevant condition number, how small the duality gap should be to guarantee a correct output. Secondly, to make an average (or smoothed) analysis for this condition number. On these complexity bounds, therefore, the worst-case $\sqrt{n}$ factor for the number of iterations is intrinsic and cannot be removed.

A number of different IPM schemes to follow the central path (known as "long-step" as opposed to the one described in Chap. 9 referred to as "short-step") have been proposed, which, even though they do not remove the worst-case $\sqrt{n}$ factor in their complexity bounds, certainly behave much better in practice, with a number of iterations that appears to be bounded as $\mathcal{O}(\log n)$. This observation suggests the following problem:

> *Show average, and/or smoothed, bounds for the number of iterations of "long-step" interior-point methods of order* $\log n$.

For reasons similar to those behind the proof of Theorem 17.3, it is conjectured that the number of iterations in long-step methods is a function of the total curvature of the central path. Estimates for the average value of this total curvature have been consequently sought and can be found in [79, 83]. These estimates do not solve the problem above but may provide a step towards its solution.

P.9. Smale's 17th Problem

Provide a complete solution to Smale's 17th problem. That is, give a solution to the following:

> *Describe a deterministic algorithm that finds an approximate zero of a given system of n complex polynomial equations in n unknowns (or n homogeneous equations in $n + 1$ unknowns) in average polynomial time.*

This problem was posed by Steve Smale [202]; see Chaps. 17 and 18 for partial solutions.

P.10. The Shub–Smale Starting System

For a degree pattern $\mathbf{d} = (d_1, \ldots, d_n)$ consider the system $\overline{g} = (g_1, \ldots, g_n) \in \mathcal{H}_\mathbf{d}$, where $g_i := X_0^{d_i - 1} X_i$. Remark 16.18 tells us that the system $\overline{g}$ is the only one, up to scaling and unitary invariance, having a zero that is best possibly conditioned, namely $e_0 := (1, 0, \ldots, 0)$.

> Run the algorithm ALH with the starting system $(\overline{g}, e_0)$. Does this algorithm run in average polynomial time on input $f \in \mathcal{H}_{\mathbf{d}}$?

If the answer is yes, then this would solve Smale's 17th problem in the affirmative, indeed in a beautifully explicit way. Computational experiments suggest that this is in fact the case [23].

We note that in [195] it was conjectured that the starting system $(\overline{g}, e_0)$ leads to an average polynomial-time algorithm for finding an approximate zero for given $f \in \mathcal{H}_{\mathbf{d}}$, although using an algorithm different from ALH.

P.11. Equivariant Morse Function

For $(g, \zeta) \in V$ and $f \in \mathcal{H}_{\mathbf{d}}$ we connect f and g by the straight-line segment $q_t := (1 - t)g + tf$, $0 \le t \le 1$. If none of the q_t has a multiple zero, we can uniquely extend the zero ζ of g to a zero ζ_t of q_t. Consider the function

$$I(f, g, \zeta) := \int_0^1 \mu_{\mathrm{norm}}(q_t, \zeta_t) \left\| \frac{d}{dt}(q_t, \zeta_t) \right\| dt. \tag{P.3}$$

In [189] it is shown that $\mathcal{O}(D^{3/2} I(f, g, \zeta))$ Newton steps are sufficient to continue the zero ζ from g to f; compare Sect. 17.1. Put $A(g, \zeta) := \mathbb{E}_f I(f, g, \zeta)$, where the expectation is taken with respect to the standard Gaussian on $\mathcal{H}_{\mathbf{d}}$. Recall from Sect. 16.2 the solution manifold $V := \{(f, \zeta) \in \mathcal{H}_{\mathbf{d}} \times \mathbb{P}^n \mid f(\zeta) = 0\}$ and its subset $\Sigma' \subseteq V$ of ill-posed solutions.

The following conjecture is due to Beltrán and Shub [30]:

> The map $A \colon V \setminus \Sigma' \to \mathbb{R}$ is a Morse function that is equivariant with respect to the action of the unitary group $\mathcal{U}(n + 1)$. Further, A has exactly one orbit of nondegenerate minima and no other critical points. The latter is the orbit of $(\overline{g}, e_0)$.

This conjecture would answer the previous problem: it would imply that ALH runs in average polynomial time on the starting system $(\overline{g}, e_0)$. In particular, it would solve Smale's 17th problem. The conjecture is consistent with the topology of $V \setminus \Sigma'$, as analyzed in [30].

P.12. Good Starting Pairs in One Variable

In Remark 17.32 we noted that most univariate polynomials q of degree d satisfy $\mu_{\max}(q) = \mathcal{O}(d)$.

Is there an algorithm computing on input $d \in \mathbb{N}$, in time polynomial in d, a univariate complex polynomial g_d, and $\xi \in \mathbb{C}$, such that $\mu_{\max}(g_d) = d^{\mathcal{O}(1)}$ and ξ is an approximate zero of g_d?

This question was raised in [194]. It is related to Smale's 7th problem on computing well-distributed points on the 2-dimensional sphere [202].

P.13. Approximating Condition Geodesics

Recall the (normalized) condition number $\mu_{\mathrm{norm}} \colon V \setminus \Sigma' \to [1, \infty)$ from (16.11). The *condition metric* on the Riemannian manifold $V \setminus \Sigma'$ is obtained by multiplying its Riemannian metric by the square of the condition number μ_{norm}. However, the condition metric is not a metric in the usual sense, since μ_{norm}^2 is not a smooth function. Still, we can define the *condition length* $L_C(\gamma)$ of an absolutely continuous curve $\gamma \colon [0, 1] \to V \setminus \Sigma'$ connecting two pairs (g, ζ) and (f, ξ) by $L_C(\gamma) := \int_0^1 \|\dot{\gamma}(t)\| \mu_{\mathrm{norm}}(\gamma(t)) \, dt$. In [189] it is shown that $\mathcal{O}(D^{3/2} L_C(\gamma))$ Newton steps are sufficient to continue the zero ζ from g to f along γ. In fact, the quantity $I(f, g, \zeta)$ defined in (P.3) is nothing but the condition length of the solution curve $(q_t, \zeta_t)_{0 \le t \le 1}$ obtained using a linear homotopy.

We call a curve γ in $V \setminus \Sigma'$ a *condition geodesic* if it minimizes the condition length between any two of its points.

Beltrán and Shub [29] constructed for any two pairs (f, ξ) and (g, ζ) in $V \setminus \Sigma'$ a curve γ in $V \setminus \Sigma'$ connecting those pairs with a condition length bounded by

$$L_C(\gamma) = \mathcal{O}\left(nD^{3/2} + \sqrt{n} \ln \frac{\mu_{\mathrm{norm}}(f, \xi) \mu_{\mathrm{norm}}(g, \zeta)}{n} \right).$$

Note that in contrast with Theorem 18.2, the dependence on the condition numbers is only logarithmic here. But unfortunately, the construction of the curve in [29] is not efficient.

Find efficient numerical algorithms to approximately follow condition geodesics.

P.14. Self-Convexity of μ_{norm} in Higher Degrees

A condition geodesic joining two pairs (g, ζ) and (f, ξ) in $V \setminus \Sigma'$ has the property that it strikes a balance between moving efficiently in V from one pair to the other and keeping the condition number small. As pointed out by Shub in [189], understanding the properties of condition geodesics should help in understanding and designing efficient homotopy algorithms. Beltrán et al. [31, 32] raised the following question:

Let γ be an arbitrary condition geodesic of $V \setminus \Sigma'$. Is $t \mapsto \log \mu_{\mathrm{norm}}(\gamma(t))$ a convex function?

An affirmative answer to this question would imply, for any condition geodesic γ, that $L_C(\gamma) \leq L \max\{\mu_{\mathrm{norm}}(g, \zeta), \mu_{\mathrm{norm}}(f, \xi)\}$, where $L_C(\gamma)$ is the condition length and L denotes the length of γ in the usual Riemannian metric of V.

In the linear case $d_1 = \cdots = d_n = 1$ the self-convexity property stated above was confirmed in [31, 32]. As an additional indication of why to expect a positive answer, we note the following observation from [31]. Let $M \subset \mathbb{R}^n$ be a smooth submanifold and $U \subseteq \mathbb{R}^n \setminus M$ the largest open set such that every point in U has a unique closest point in M with respect to the Euclidean distance d. Then the function $\mu(x) := d(x, M)^{-1}$ has the property that $t \mapsto \log \mu(\gamma(t))$ is convex for any geodesic in U with respect to the Riemannian metric $\mu^2\langle\,,\,\rangle$ of U.

P.15. Structured Systems of Polynomial Equations

Systems of polynomial equations arising in practice often have a special structure. For instance, they lie in a linear subspace of $\mathcal{H}_{\mathbf{d}}$ that depends on a few parameters. An important case is provided by "sparse polynomials" having only a few monomial terms. Dedieu [80] has defined condition numbers for structured systems of polynomials and analyzed the cost of homotopy methods in terms of this condition. However, there are very few probabilistic analyses of such condition numbers; see Malajovich and Rojas [142] for a result in this direction.

Provide probabilistic analyses of condition numbers for structured polynomial systems.

Such results would help to explain the success of numerical practice, as in the software package "Bertini"; see [19].

P.16. Systems with Singularities

The homotopy methods described in this book focus on systems of polynomial equations with simple zeros only. However, as pointed out by Andrew Sommese, polynomial systems in practice often have singularities, and algorithms can benefit from this information. Insofar as it is misleading to consider all polynomial systems in the discriminant variety Σ as ill-posed, one may consider the following problem:

Extend the Shub–Smale theory from Chaps. 15–17 to systems with multiple zeros.

P.17. Conic Condition Numbers of Real Problems with High Codimension of Ill-posedness

The main result in Chap. 20, Theorem 20.1, exhibits a bound for the probability tail $\mathrm{Prob}\{\mathscr{C}(a) \geq t\}$ that decays as $t^{2(m-p)}$, where $p - m$ is the (complex) codimension of the set Σ of ill-posed data. This decay ensures the finiteness of $\mathbb{E}(\mathscr{C}(a)^k)$ for all $k < 2(p - m)$.

In contrast with the above, the main result in Chap. 21, Theorem 21.1, exhibits a bound for $\mathrm{Prob}\{\mathscr{C}(a) \geq t\}$ that decays as t^{-1}, independently of the (now real) codimension of the set Σ. In particular, no matter what this codimension is, we cannot deduce a finite bound for $\mathbb{E}(\mathscr{C}(a))$.

> *For a real conic condition number $\mathscr{C}$ associated to a set of ill-posed data Σ, prove tail bounds (both average-case and smoothed analysis) that decay as t^{-s}, where s is the codimension of Σ.*

The article [85] we cited in the notes to Chap. 21 states a result of this kind in the case that Σ is a complete intersection. As we mentioned in these notes, that result relied on an unpublished and apparently unavailable result by Adrian Ocneanu. A proof can now be found in [140]. Since most Σ of interest for applications are not complete intersections, the challenge remains to derive good bounds on $\mathscr{C}$ for these cases.

P.18. Feasibility of Real Polynomial Systems

In Sect. 19.6 we briefly described a numerical algorithm to detect feasibility of real polynomial systems. The cost analysis of this algorithm featured the condition number κ_{feas}, which, for a system $f \in \mathcal{H}_{\mathbf{d}}^{\mathbb{R}}[m]$, is defined by

$$
\kappa_{\mathsf{feas}}(f) = \begin{cases} \min_{\zeta \in Z_{\mathbb{S}}(f)} \mu_\dagger(f, \zeta) & \text{if } Z_{\mathbb{S}}(f) \neq \emptyset, \\[2mm] \max_{x \in \mathbb{S}^n} \frac{\|f\|}{\|f(x)\|} & \text{otherwise.} \end{cases}
$$

Recall that $\kappa_{\mathsf{feas}}(f) = \infty$ if and only if f is feasible and all its zeros are multiple.

As of today, there are no known bounds for either the probability tail $\mathrm{Prob}\{\kappa_{\mathsf{feas}}(f) \geq t\}$ or the expectations $\mathbb{E}(\kappa_{\mathsf{feas}}(f))$ and $\mathbb{E}(\log \kappa_{\mathsf{feas}}(f))$. An obstacle to obtaining such bounds is the fact that $\kappa_{\mathsf{feas}}(f)$ is defined in two different ways according to whether f is feasible or not. The set Σ of ill-posed data for the feasibility problem is, however, an algebraic cone, and the problem has therefore a conic condition number $\mathscr{C}(f)$ naturally associated to it. This fact suggest the following problem.

> *Can $\kappa_{\mathsf{feas}}(f)$ be bounded by a polynomial function in $\mathscr{C}(f)$?*

Note that a positive answer to this question would immediately yield (via Theorem 21.1) bounds for both $\mathrm{Prob}\{\kappa_{\mathsf{feas}}(f) \geq t\}$ and $\mathbb{E}(\log \kappa_{\mathsf{feas}}(f))$. Furthermore, should Problem P.17 above be solved as well, one could deduce bounds for $\mathbb{E}(\kappa_{\mathsf{feas}}(f)^k)$ for a wide range of values of k. This is so because the set Σ consists of the systems *all* of whose zeros are multiple, and this is a set having high codimension in $\mathcal{H}_{\mathbf{d}}^{\mathbb{R}}[m]$.

Bibliography

1. L.M. Adleman and M.-D. Huang. *Primality Testing and Abelian Varieties over Finite Fields*, volume 1512 of *Lecture Notes in Mathematics*. Springer, Berlin, 1992.
2. S. Agmon. The relaxation method for linear inequalities. *Canadian Journal of Mathematics*, 6:382–392, 1954.
3. E.L. Allgower and K. Georg. *Numerical Continuation Methods*. Springer, Berlin, 1990.
4. D. Amelunxen. Geometric analysis of the condition of the convex feasibility problem. PhD thesis, University of Paderborn, 2011.
5. D. Amelunxen and P. Bürgisser. Probabilistic analysis of the Grassmann condition number. arXiv:1112.2603v1, 2011.
6. D. Amelunxen and P. Bürgisser. A coordinate-free condition number for convex programming. *SIAM Journal on Optimization*, 22(3):1029–1041, 2012.
7. D. Amelunxen and P. Bürgisser. Robust smoothed analysis of a condition number for linear programming. *Mathematical Programming Series A*, 131(1):221–251, 2012.
8. I.K. Argyros. *Convergence and Applications of Newton-Type Iterations*. Springer, New York, 2008.
9. M. Arioli, I.S. Duff and P.P.M. de Rijk. On the augmented system approach to sparse least-squares problems. *Numerische Mathematik*, 55(6):667–684, 1989.
10. D. Armentano. Stochastic perturbations and smooth condition numbers. *Journal of Complexity*, 26(2):161–171, 2010.
11. D. Armentano. Complexity of path-following methods for the eigenvalue problem. To appear at *Foundations of Computational Mathematics*, 2013.
12. S. Arora and B. Barak. A modern approach. In *Computational Complexity*. Cambridge University Press, Cambridge, 2009.
13. J.-M. Azaïs and M. Wschebor. On the roots of a random system of equations. The theorem of Shub and Smale and some extensions. *Foundations of Computational Mathematics*, 5(2):125–144, 2005.
14. J.M. Azaïs and M. Wschebor. Upper and lower bounds for the tails of the distribution of the condition number of a Gaussian matrix. *SIAM Journal on Matrix Analysis and Applications*, 26(2):426–440, 2004/05.
15. E. Barbier. Note sur le problème de l'aguille et le jeu du joint couvert. *Journal de Mathématiques Pures et Appliquées*, 5(2):273–286, 1860.
16. V. Bargmann, D. Montgomery, and J. von Neumann. Solution of linear systems of high order (Princeton, 1946). In A.H. Taub, editor, *John von Neumann Collected Works*, volume 5. Pergamon, Elmsford, 1963.
17. R.G. Bartle. *The Elements of Integration and Lebesgue Measure*, Wiley Classics Library. Wiley, New York, 1995.

P. Bürgisser, F. Cucker, *Condition*,
Grundlehren der mathematischen Wissenschaften 349,
DOI 10.1007/978-3-642-38896-5, © Springer-Verlag Berlin Heidelberg 2013

18. S. Basu, R. Pollack, and M.-F. Roy. *Algorithms in Real Algebraic Geometry*, volume 10 of *Algorithms and Computation in Mathematics*. Springer, Berlin, 2003.

19. D.J. Bates, J.D. Hauenstein, A.J. Sommese, and C.W. Wampler. Software for numerical algebraic geometry: a paradigm and progress towards its implementation. In *Software for Algebraic Geometry*, volume 148 of *IMA Vol. Math. Appl.*, pages 1–14. Springer, New York, 2008.

20. W. Baur and V. Strassen. The complexity of partial derivatives. *Theoretical Computer Science*, 22(3):317–330, 1983.

21. A. Belloni, R.M. Freund, and S. Vempala. An efficient rescaled perceptron algorithm for conic systems. *Mathematics of Operations Research*, 34:621–641, 2009.

22. C. Beltrán. A continuation method to solve polynomial systems and its complexity. *Numerische Mathematik*, 117(1):89–113, 2011.

23. C. Beltrán and A. Leykin. Certified numerical homotopy tracking. *Experimental Mathematics*, 21(1):69–83, 2012.

24. C. Beltrán and L.M. Pardo. On the complexity of non universal polynomial equation solving: old and new results. In *Foundations of Computational Mathematics*, Santander 2005, volume 331 of *London Math. Soc. Lecture Note Ser.*, pages 1–35. Cambridge Univ. Press, Cambridge, 2006.

25. C. Beltrán and L.M. Pardo. Estimates on the distribution of the condition number of singular matrices. *Foundations of Computational Mathematics*, 7(1):87–134, 2007.

26. C. Beltrán and L.M. Pardo. On Smale's 17th problem: a probabilistic positive solution. *Foundations of Computational Mathematics*, 8:1–43, 2008.

27. C. Beltrán and L.M. Pardo. Smale's 17th problem: average polynomial time to compute affine and projective solutions. *Journal of the American Mathematical Society*, 22(2):363–385, 2009.

28. C. Beltrán and L.M. Pardo. Fast linear homotopy to find approximate zeros of polynomial systems. *Foundations of Computational Mathematics*, 11(1):95–129, 2011.

29. C. Beltrán and M. Shub. Complexity of Bézout's Theorem VII: distance estimates in the condition metric. *Foundations of Computational Mathematics*, 9:179–195, 2009.

30. C. Beltrán and M. Shub. On the geometry and topology of the solution variety for polynomial system solving. *Foundations of Computational Mathematics*, 12:719–763, 2012.

31. C. Beltrán, J.-P. Dedieu, G. Malajovich, and M. Shub. Convexity properties of the condition number. *SIAM Journal on Matrix Analysis and Applications*, 31(3):1491–1506, 2009.

32. C. Beltrán, J.-P. Dedieu, G. Malajovich, and M. Shub. Convexity properties of the condition number II. arXiv:0910.5936v3, 7 May 2012.

33. D. Bertsimas and J. Tsitsiklis. *Introduction to Linear Optimization*. Athena Scientific, Nashua, 1997.

34. Å. Björck. Component-wise perturbation analysis and error bounds for linear least squares solutions. *BIT*, 31(2):238–244, 1991.

35. L. Blum. Lectures on a theory of computation and complexity over the reals (or an arbitrary ring). In E. Jen, editor, *Lectures in the Sciences of Complexity II*, pages 1–47. Addison-Wesley, Reading, 1990.

36. L. Blum and M. Shub. Evaluating rational functions: infinite precision is finite cost and tractable on average. *SIAM Journal on Computing*, 15(2):384–398, 1986.

37. L. Blum, M. Shub, and S. Smale. On a theory of computation and complexity over the real numbers: NP-completeness, recursive functions and universal machines. *Bulletin of the American Mathematical Society*, 21:1–46, 1989.

38. L. Blum, F. Cucker, M. Shub, and S. Smale. *Complexity and Real Computation*. Springer, New York, 1998. With a foreword by R.M. Karp.

39. J. Bochnak, M. Coste, and M.-F. Roy. *Real Algebraic Geometry*, volume 36 of *Ergebnisse der Mathematik und ihrer Grenzgebiete (3) [Results in Mathematics and Related Areas (3)]*. Springer, Berlin, 1998. Translated from the 1987 French original. Revised by the authors.

40. G.E. Bredon. *Topology and Geometry*, volume 139 of *Graduate Texts in Mathematics*. Springer, New York, 1993.

41. I. Briquel, F. Cucker, J. Peña, and V. Roshchina. Fast computation of zeros of polynomial systems with bounded degree under variable-precision. To appear at *Mathematics of Computation*, 2013.

42. J.E. Brothers. Integral geometry in homogeneous spaces. *Transactions of the American Mathematical Society*, 124:480–517, 1966.

43. P. Bürgisser. Average Euler characteristic of random real algebraic varieties. *Comptes Rendus Mathematique. Academie Des Sciences. Paris*, 345(9):507–512, 2007.

44. P. Bürgisser and F. Cucker. Counting complexity classes for numeric computations II: algebraic and semialgebraic sets. *Journal of Complexity*, 22:147–191, 2006.

45. P. Bürgisser and F. Cucker. Smoothed analysis of Moore-Penrose inversion. *SIAM Journal on Matrix Analysis and Applications*, 31(5):2769–2783, 2010.

46. P. Bürgisser and F. Cucker. On a problem posed by Steve Smale. *Annals of Mathematics*, 174:1785–1836, 2011.

47. P. Bürgisser, M. Clausen, and M.A. Shokrollahi. *Algebraic Complexity Theory*, volume 315 of *Grundlehren der Mathematischen Wissenschaften [Fundamental Principles of Mathematical Sciences]*. Springer, Berlin, 1997.

48. P. Bürgisser, F. Cucker, and M. Lotz. Smoothed analysis of complex conic condition numbers. *Journal de Mathématiques Pures et Appliquées*, 86(4):293–309, 2006.

49. P. Bürgisser, F. Cucker, and M. Lotz. The probability that a slightly perturbed numerical analysis problem is difficult. *Mathematics of Computation*, 77:1559–1583, 2008.

50. P. Bürgisser, F. Cucker, and M. Lotz. Coverage processes on spheres and condition numbers for linear programming. *Annals of Probability*, 38:570–604, 2010.

51. Z.-Z. Cheng and J.J. Dongarra. Condition numbers of Gaussian random matrices. *SIAM Journal on Matrix Analysis and Applications*, 27:603–620, 2005.

52. S. Chern. On the kinematic formula in integral geometry. *Journal of Mathematics and Mechanics*, 16:101–118, 1966.

53. D. Cheung and F. Cucker. Smoothed analysis of componentwise condition numbers for sparse matrices. Available at arXiv:1302.6004.

54. D. Cheung and F. Cucker. A new condition number for linear programming. *Mathematical Programming Series A*, 91(1):163–174, 2001.

55. D. Cheung and F. Cucker. Probabilistic analysis of condition numbers for linear programming. *Journal of Optimization Theory and Applications*, 114:55–67, 2002.

56. D. Cheung and F. Cucker. Solving linear programs with finite precision: I. Condition numbers and random programs. *Mathematical Programming*, 99:175–196, 2004.

57. D. Cheung and F. Cucker. A note on level-2 condition numbers. *Journal of Complexity*, 21:314–319, 2005.

58. D. Cheung and F. Cucker. Solving linear programs with finite precision: II. Algorithms. *Journal of Complexity*, 22:305–335, 2006.

59. D. Cheung and F. Cucker. Componentwise condition numbers of random sparse matrices. *SIAM Journal on Matrix Analysis and Applications*, 31:721–731, 2009.

60. D. Cheung and F. Cucker. On the average condition of random linear programs. *SIAM Journal on Optimization*, 23(2):799–810, 2013.

61. D. Cheung, F. Cucker, and J. Peña. Unifying condition numbers for linear programming. *Mathematics of Operations Research*, 28(4):609–624, 2003.

62. D. Cheung, F. Cucker, and Y. Ye. Linear programming and condition numbers under the real number computation model. In Ph. Ciarlet and F. Cucker, editors, *Handbook of Numerical Analysis*, volume XI, pages 141–207. North-Holland, Amsterdam, 2003.

63. D. Cheung, F. Cucker, and R. Hauser. Tail decay and moment estimates of a condition number for random linear conic systems. *SIAM Journal on Optimization*, 15(4):1237–1261, 2005.

64. A.L. Chistov and D.Yu. Grigor'ev. Complexity of quantifier elimination in the theory of algebraically closed fields. In *Mathematical Foundations of Computer Science*, Prague, 1984, volume 176 of *Lecture Notes in Comput. Sci.*, pages 17–31. Springer, Berlin, 1984.

65. K.P. Choi. On the medians of gamma distributions and an equation of Ramanujan. *Proceedings of the American Mathematical Society*, 121(1):245–251, 1994.

66. T.H. Cormen, C.E. Leiserson, R.L. Rivest, and C. Stein. *Introduction to Algorithms*. 3rd edition. MIT Press, Cambridge, 2009.

67. R. Courant and D. Hilbert. Partial differential equations. In *Methods of Mathematical Physics. Vol. II, Wiley Classics Library*. Wiley, New York, 1989. Reprint of the 1962 original, A Wiley-Interscience Publication.

68. F. Cucker. Approximate zeros and condition numbers. *Journal of Complexity*, 15:214–226, 1999.

69. F. Cucker and J. Peña. A primal-dual algorithm for solving polyhedral conic systems with a finite-precision machine. *SIAM Journal on Optimization*, 12(2):522–554, 2001/02.

70. F. Cucker and S. Smale. Complexity estimates depending on condition and round-off error. *Journal of the ACM*, 46:113–184, 1999.

71. F. Cucker and M. Wschebor. On the expected condition number of linear programming problems. *Numerische Mathematik*, 94:419–478, 2003.

72. F. Cucker, H. Diao, and Y. Wei. On mixed and componentwise condition numbers for Moore-Penrose inverse and linear least squares problems. *Mathematics of Computation*, 76:947–963, 2007.

73. F. Cucker, R. Hauser, and M. Lotz. Adversarial smoothed analysis. *Journal of Complexity*, 26:255–262, 2010.

74. F. Cucker, T. Krick, G. Malajovich, and M. Wschebor. A numerical algorithm for zero counting. I: Complexity and accuracy. *Journal of Complexity*, 24:582–605, 2008.

75. F. Cucker, T. Krick, G. Malajovich, and M. Wschebor. A numerical algorithm for zero counting. II: Distance to ill-posedness and smoothed analysis. *Journal of Fixed Point Theory and Applications*, 6:285–294, 2009.

76. F. Cucker, T. Krick, G. Malajovich, and M. Wschebor. A numerical algorithm for zero counting. III: Randomization and condition. *Advances in Applied Mathematics*, 48:215–248, 2012.

77. G.B. Dantzig. Reminiscences about the origins of linear programming. In *Mathematical Programming: The State of the Art*, Bonn, 1982, pages 78–86. Springer, Berlin, 1983.

78. K.R. Davidson and S.J. Szarek. Local operator theory, random matrices and Banach spaces. In *Handbook of the Geometry of Banach Spaces. Vol. I*, pages 317–366. North-Holland, Amsterdam, 2001.

79. J. De Loera, B. Sturmfels, and C. Vinzant. The central curve of linear programming. *Foundations of Computational Mathematics*, 12:509–540, 2012.

80. J.P. Dedieu. Condition number analysis for sparse polynomial systems. In *Foundations of Computational Mathematics*, Rio de Janeiro, 1997, pages 75–101. Springer, Berlin, 1997.

81. J.-P. Dedieu. *Points Fixes, Zéros et la Méthode de Newton*, volume 54 of *Mathématiques & Applications (Berlin) [Mathematics & Applications]*. Springer, Berlin, 2006. With a preface by Steve Smale.

82. J.-P. Dedieu, P. Priouret, and G. Malajovich. Newton's method on Riemannian manifolds: convariant alpha theory. *IMA Journal of Numerical Analysis*, 23(3):395–419, 2003.

83. J.-P. Dedieu, G. Malajovich, and M. Shub. On the curvature of the central path of linear programming theory. *Foundations of Computational Mathematics*, 5(2):145–171, 2005.

84. J.W. Demmel. On condition numbers and the distance to the nearest ill-posed problem. *Numerische Mathematik*, 51:251–289, 1987.

85. J.W. Demmel. The probability that a numerical analysis problem is difficult. *Mathematics of Computation*, 50:449–480, 1988.

86. J.W. Demmel. *Applied Numerical Linear Algebra*. SIAM, Philadelphia, 1997.

87. P. Deuflhard. Affine invariance and adaptive algorithms. In *Newton Methods for Nonlinear Problems*, volume 35 of *Springer Series in Computational Mathematics*. Springer, Berlin, 2004.

88. M.P. do Carmo. *Riemannian Geometry, Mathematics: Theory & Applications*. Birkhäuser, Boston, 1992. Translated from the second Portuguese edition by Francis Flaherty.

89. J. Dunagan, D.A. Spielman, and S.-H. Teng. Smoothed analysis of condition numbers and complexity implications for linear programming. *Mathematical Programming Series A*, 126(2):315–350, 2011.

90. C. Durvye and G. Lecerf. A concise proof of the Kronecker polynomial system solver from scratch. *Expositiones Mathematicae*, 26(2):101–139, 2008.

91. C. Eckart and G. Young. The approximation of one matrix by another of lower rank. *Psychometrika*, 1(3):211–218, 1936.

92. A. Edelman. Eigenvalues and condition numbers of random matrices. *SIAM Journal on Matrix Analysis and Applications*, 9(4):543–560, 1988.

93. A. Edelman. On the distribution of a scaled condition number. *Mathematics of Computation*, 58(197):185–190, 1992.

94. M. Epelman and R.M. Freund. A new condition measure, preconditioners, and relations between different measures of conditioning for conic linear systems. *SIAM Journal on Optimization*, 12(3):627–655, 2002.

95. H. Federer. Curvature measures. *Transactions of the American Mathematical Society*, 93:418–491, 1959.

96. M. Fisz. *Probability Theory and Mathematical Statistics*. 3rd edition. Wiley, New York, 1963. Authorized translation from the Polish. Translated by R. Bartoszynski.

97. R. Fletcher. Expected conditioning. *IMA Journal of Numerical Analysis*, 5(3):247–273, 1985.

98. R.M. Freund and J.R. Vera. Condition-based complexity of convex optimization in conic linear form via the ellipsoid algorithm. *SIAM Journal on Optimization*, 10(1):155–176, 1999.

99. R.M. Freund and J.R. Vera. Some characterizations and properties of the "distance to ill-posedness" and the condition measure of a conic linear system. *Mathematical Programming*, 86:225–260, 1999.

100. F. Gao, D. Hug, and R. Schneider. Intrinsic volumes and polar sets in spherical space. *Mathematicae Notae*, 41:159–176, 2003. 2001/02, Homage to Luis Santaló. Vol. 1 (Spanish).

101. C.B. García and W.I. Zangwill. *Pathways to Solutions, Fixed Points, and Equilibria*. Prentice-Hall, Englewood Cliffs, 1981.

102. S. Geman. A limit theorem for the norm of random matrices. *Annals of Probability*, 8(2):252–261, 1980.

103. A.J. Geurts. A contribution to the theory of condition. *Numerische Mathematik*, 39:85–96, 1982.

104. M. Giusti and J. Heintz. La détermination des points isolés et de la dimension d'une variété algébrique peut se faire en temps polynomial. In *Computational Algebraic Geometry and Commutative Algebra*, Cortona, 1991, volume XXXIV of *Sympos. Math.*, pages 216–256. Cambridge Univ. Press, Cambridge, 1993.

105. S. Glasauer. Integral geometry of spherically convex bodies. *Dissertation Summaries in Mathematics*, 1(1–2):219–226, 1996.

106. J.-L. Goffin. The relaxation method for solving systems of linear inequalities. *Mathematics of Operations Research*, 5(3):388–414, 1980.

107. I. Gohberg and I. Koltracht. Mixed, componentwise, and structured condition numbers. *SIAM Journal on Matrix Analysis and Applications*, 14:688–704, 1993.

108. H.H. Goldstine and J. von Neumann. Numerical inverting matrices of high order, II. *Proceedings of the American Mathematical Society*, 2:188–202, 1951.

109. G.H. Golub and C.F. Van Loan. *Matrix Computations, Johns Hopkins Studies in the Mathematical Sciences*, 4th edition. Johns Hopkins University Press, Baltimore, 2013.

110. R. Graham, D. Knuth, and O. Patashnik. *Concrete Mathematics*. Addison-Wesley, Reading, 1989.

111. I. Grattan-Guinness. Joseph Fourier's anticipation of linear programming. *Operational Research Quarterly*, 3:361–364, 1970.

112. S. Gratton. On the condition number of linear least squares problems in a weighted Frobenius norm. *BIT*, 36(3):523–530, 1996.

113. D.Yu. Grigor'ev and N.N. Vorobjov Jr. Solving systems of polynomial inequalities in subexponential time. *Journal of Symbolic Computation*, 5(1–2):37–64, 1988.

114. M. Grötschel, L. Lovász, and A. Schrijver. *Geometric Algorithms and Combinatorial Optimization*, volume 2 of *Algorithms and Combinatorics: Study and Research Texts*. Springer, Berlin, 1988.

115. J. Harris. A first course. In *Algebraic Geometry*, volume 133 of *Graduate Texts in Mathematics*. Springer, New York, 1992.

116. R. Hauser and T. Müller. Conditioning of random conic systems under a general family of input distributions. *Foundations of Computational Mathematics*, 9:335–358, 2009.

117. M.R. Hestenes and E. Stiefel. Methods of conjugate gradients for solving linear systems. *Journal of Research of the National Bureau of Standards*, 49:409 436, 1952.

118. N.J. Higham. Iterative refinement enhances the stability of QR factorization methods for solving linear equations. Numerical Analysis Report No. 182, University of Manchester, Manchester, England, 1990.

119. N.J. Higham. A survey of componentwise perturbation theory in numerical linear algebra. In *Mathematics of Computation 1943–1993: A Half-Century of Computational Mathematics*, Vancouver, BC, 1993, volume 48 of *Proc. Sympos. Appl. Math.*, pages 49–77. Am. Math. Soc., Providence, 1994.

120. D. Higham. Condition numbers and their condition numbers. *Linear Algebra and Its Applications*, 214:193–215, 1995.

121. N.J. Higham. *Accuracy and Stability of Numerical Algorithms*, 2nd edition. SIAM, Philadelphia, 2002.

122. H. Hotelling. Some new methods in matrix calculation. *The Annals of Mathematical Statistics*, 14:1–34, 1943.

123. R. Howard. The kinematic formula in Riemannian homogeneous spaces. *Memoirs of the American Mathematical Society*, 106(509):69, 1993.

124. W. Kahan. Numerical linear algebra. *Canadian Mathematical Bulletin*, 9:757–801, 1966.

125. L.V. Kantorovich. *On Newton's Method*, volume 28 of *Trudy Mat. Inst. Steklov.*, pages 104–144. Acad. Sci. USSR, Moscow–Leningrad, 1949. In Russian.

126. N. Karmarkar. A new polynomial time algorithm for linear programming. *Combinatorica*, 4:373–395, 1984.

127. M. Karow, D. Kressner, and F. Tisseur. Structured eigenvalue condition numbers. *SIAM Journal on Matrix Analysis and Applications*, 28(4):1052–1068, 2006.

128. H.B. Keller. *Lectures on Numerical Methods in Bifurcation Problems*, volume 79 of *Tata Institute of Fundamental Research Lectures on Mathematics and Physics*. Tata Institute of Fundamental Research, Bombay, 1987. With notes by A. K. Nandakumaran and Mythily Ramaswamy.

129. L.G. Khachiyan. A polynomial algorithm in linear programming. *Doklady Akademii Nauk SSSR*, 244:1093–1096, 1979. (In Russian, English translation in *Soviet Math. Dokl.*, 20:191–194, 1979.)

130. D.A. Klain and G.-C. Rota. *Introduction to Geometric Probability*, *Lezioni Lincee [Lincei Lectures]*. Cambridge University Press, Cambridge, 1997.

131. E. Kostlan. On the distribution of the roots of random polynomials. In M. Hirsch, J.E. Marsden, and M. Shub, editors, *From Topology to Computation: Proceedings of the Smalefest*, pages 419–431. Springer, Berlin, 1993.

132. E. Lahaye. Une méthode de resolution d'une categorie d'equations transcendantes. *Comptes Rendus Mathematique. Academie Des Sciences. Paris*, 198:1840–1842, 1934.

133. S. Lang. *Real Analysis*, 2nd edition. Addison-Wesley, Reading, 1983.

134. S. Lang. *Algebra*, volume 211 of *Graduate Texts in Mathematics*. 3rd edition. Springer, New York, 2002.

135. G.-L. Leclerc, Comte de Buffon. Essai d'arithmétique morale. In *Supplément à l'Histoire Naturelle, volume 4*, pages 46–148. Imprimerie Royale, Paris, 1777.

136. M. Ledoux and M. Talagrand. Isoperimetry and processes. In *Probability in Banach Spaces*, volume 23 of *Ergebnisse der Mathematik und ihrer Grenzgebiete (3) [Results in Mathematics*

and Related Areas], page 3. Springer, Berlin, 1991.

137. A. Lewis. Ill-conditioned convex processes and linear inequalities. *Mathematics of Operations Research*, 24:829–834, 1999.

138. T.Y. Li. Numerical solution of polynomial systems by homotopy continuation methods. In Ph. Ciarlet and F. Cucker, editors, *Handbook of Numerical Analysis, volume XI*, pages 209–304. North-Holland, Amsterdam, 2003.

139. S. Linnainmaa. Taylor expansion of the accumulated rounding error. *BIT*, 16(2):146–160, 1976.

140. M. Lotz. On the volume of tubular neighborhoods of real algebraic varieties. Preprint arXiv:1210.3742.

141. D.G. Luenberger. *Linear and Nonlinear Programming*, 2nd edition. Kluwer Academic, Boston, 2003.

142. G. Malajovich and J.M. Rojas. High probability analysis of the condition number of sparse polynomial systems. *Theoretical Computer Science*, 315(2–3):524–555, 2004.

143. A.N. Malyshev. A unified theory of conditioning for linear least squares and Tikhonov regularization solutions. *SIAM Journal on Matrix Analysis and Applications*, 24(4):1186–1196, 2003.

144. J. Matousek and B. Gärtner. *Understanding and Using Linear Programming*. Springer, Berlin, 2007.

145. J.W. Milnor. *Topology from the Differentiable Viewpoint, Princeton Landmarks in Mathematics*. Princeton University Press, Princeton, 1997. Based on notes by David W. Weaver, Revised reprint of the 1965 original.

146. T. Motzkin and I.Y. Schönberg. The relaxation method for linear inequalities. *Canadian Journal of Mathematics*, 6:393–404, 1954.

147. R.J. Muirhead. *Aspects of Multivariate Statistical Theory*. Wiley, New York, 1982. Wiley Series in Probability and Mathematical Statistics.

148. D. Mumford. Complex projective varieties. In *Algebraic Geometry. I, Classics in Mathematics*. Springer, Berlin, 1995. Reprint of the 1976 edition.

149. Y. Nesterov and A. Nemirovsky. *Interior-Point Polynomial Algorithms in Convex Programming*. SIAM, Philadelphia, 1994.

150. A. Nijenhuis. On Chern's kinematic formula in integral geometry. *Journal of Differential Geometry*, 9:475–482, 1974.

151. J. Nocedal and S.J. Wright. *Numerical Optimization, Springer Series in Operations Research and Financial Engineering*, 2nd edition. Springer, New York, 2006.

152. M. Nunez and R.M. Freund. Condition measures and properties of the central trajectory of a linear program. *Mathematical Programming*, 83:1–28, 1998.

153. W. Oettli and W. Prager. Compatibility of approximate solution of linear equations with given error bounds for coefficients and right-hand sides. *Numerische Mathematik*, 6:405–409, 1964.

154. V.Y. Pan. Optimal and nearly optimal algorithms for approximating polynomial zeros. *Computer Mathematics and Its Applications*, 31(12):97–138, 1996.

155. V.Y. Pan. Solving a polynomial equation: some history and recent progress. *SIAM Review*, 39(2):187–220, 1997.

156. C.H. Papadimitriou. *Computational Complexity*. Addison-Wesley, Reading, 1994.

157. J. Peña. Understanding the geometry on infeasible perturbations of a conic linear system. *SIAM Journal on Optimization*, 10:534–550, 2000.

158. J. Peña. A characterization of the distance to infeasibility under block-structured perturbations. *Linear Algebra and Its Applications*, 370:193–216, 2003.

159. J. Peña and J. Renegar. Computing approximate solutions for conic systems of constraints. *Mathematical Programming*, 87:351–383, 2000.

160. M.O. Rabin. Probabilistic algorithms. In J. Traub, editor, *Algorithms and Complexity: New Directions and Results*, pages 21–39. Academic Press, San Diego, 1976.

161. J. Renegar. On the efficiency of Newton's method in approximating all zeros of a system of complex polynomials. *Mathematics of Operations Research*, 12(1):121–148, 1987.

162. J. Renegar. On the worst-case arithmetic complexity of approximating zeros of systems of polynomials. *SIAM Journal on Computing*, 18:350–370, 1989.
163. J. Renegar. On the computational complexity and geometry of the first-order theory of the reals. I, II, III. *Journal of Symbolic Computation*, 13(3):255–352, 1992.
164. J. Renegar. Is it possible to know a problem instance is ill-posed? *Journal of Complexity*, 10:1–56, 1994.
165. J. Renegar. Some perturbation theory for linear programming. *Mathematical Programming*, 65:73–91, 1994.
166. J. Renegar. Incorporating condition measures into the complexity theory of linear programming. *SIAM Journal on Optimization*, 5:506–524, 1995.
167. J. Renegar. Linear programming, complexity theory and elementary functional analysis. *Mathematical Programming*, 70:279–351, 1995.
168. J. Renegar. *A Mathematical View of Interior-Point Methods in Convex Optimization*. SIAM, Philadelphia, 2000.
169. W.C. Rheinboldt. *Numerical Analysis of Parametrized Nonlinear Equations*, volume 7 of *University of Arkansas Lecture Notes in the Mathematical Sciences*. Wiley, New York, 1986.
170. J.R. Rice. A theory of condition. *SIAM Journal on Numerical Analysis*, 3:217–232, 1966.
171. R.T. Rockafellar. *Convex Analysis*, Princeton Landmarks in Mathematics. Princeton University Press, Princeton, 1997. Reprint of the 1970 original, Princeton Paperbacks.
172. J. Rohn. Systems of linear interval equations. *Linear Algebra and Its Applications*, 126:39–78, 1989.
173. R. Rosenblatt. *Principles of Neurodynamics: Perceptrons and the Theory of Brain Mechanisms*. Spartan Books, East Lansing, 1962.
174. M. Rudelson and R. Vershynin. Smallest singular value of a random rectangular matrix. *Communications on Pure and Applied Mathematics*, 62(12):1707–1739, 2009.
175. S.M. Rump. Structured perturbations part I: normwise distances. *SIAM Journal on Matrix Analysis and Applications*, 25:1–30, 2003.
176. S.M. Rump. Structured perturbations part II: componentwise distances. *SIAM Journal on Matrix Analysis and Applications*, 25:31–56, 2003.
177. J. Sabia. Algorithms and their complexities. In *Solving Polynomial Equations*, volume 14 of *Algorithms Comput. Math.*, pages 241–268. Springer, Berlin, 2005.
178. R. Saigal. A modern integrated analysis. In *Linear Programming*, volume 1 of *International Series in Operations Research Management Science*. Kluwer Academic, Boston, 1995.
179. A. Sankar, D.A. Spielman, and S.-H. Teng. Smoothed analysis of the condition numbers and growth factors of matrices. *SIAM Journal on Matrix Analysis and Applications*, 28(2):446–476, 2006.
180. L.A. Santaló. Integral geometry in Hermitian spaces. *American Journal of Mathematics*, 74:423–434, 1952.
181. L.A. Santaló. *Integral Geometry and Geometric Probability*, volume 1 of *Encyclopedia of Mathematics and Its Applications*. Addison-Wesley, Reading, 1976. With a foreword by Mark Kac.
182. E. Schmidt. Zur Theorie der linearen und nichtlinearen Integralgleichungen. *Mathematische Annalen*, 63(4):433–476, 1907.
183. R. Schneider. *Convex Bodies: The Brunn-Minkowski Theory*, volume 44 of *Encyclopedia of Mathematics and Its Applications*. Cambridge University Press, Cambridge, 1993.
184. R. Schneider and W. Weil. *Stochastic and Integral Geometry*, Probability and Its Applications (New York). Springer, Berlin, 2008.
185. A. Schönhage. The fundamental theorem of algebra in terms of computational complexity. Technical Report, Institute of Mathematics, University of Tübingen, 1982.
186. I.R. Shafarevich. Varieties in projective space. In *Basic Algebraic Geometry. 1*, 2nd edition. Springer, Berlin, 1994. Translated from the 1988 Russian edition and with notes by Miles Reid.
187. I.R. Shafarevich. Schemes and complex manifolds. In *Basic Algebraic Geometry. 2*, 2nd edition. Springer, Berlin, 1994. Translated from the 1988 Russian edition by Miles Reid.

188. M. Shub. Some remarks on Bézout's theorem and complexity theory. In *From Topology to Computation: Proceedings of the Smalefest*, Berkeley, CA, 1990, pages 443–455. Springer, New York, 1993.

189. M. Shub. Complexity of Bézout's Theorem VI: geodesics in the condition (number) metric. *Foundations of Computational Mathematics*, 9:171–178, 2009.

190. M. Shub and S. Smale. Computational complexity: on the geometry of polynomials and a theory of cost. I. *Annales Scientifiques de L'Ecole Normale Supérieure*, 18(1):107–142, 1985.

191. M. Shub and S. Smale. Computational complexity: on the geometry of polynomials and a theory of cost. II. *SIAM Journal on Computing*, 15(1):145–161, 1986.

192. M. Shub and S. Smale. Complexity of Bézout's Theorem I: geometric aspects. *Journal of the American Mathematical Society*, 6:459–501, 1993.

193. M. Shub and S. Smale. Complexity of Bézout's Theorem II: volumes and probabilities. In F. Eyssette and A. Galligo, editors, *Computational Algebraic Geometry*, volume 109 of *Progress in Mathematics*, pages 267–285. Birkhäuser, Basel, 1993.

194. M. Shub and S. Smale. Complexity of Bézout's Theorem III: condition number and packing. *Journal of Complexity*, 9:4–14, 1993.

195. M. Shub and S. Smale. Complexity of Bézout's Theorem V: polynomial time. *Theoretical Computer Science*, 133:141–164, 1994.

196. M. Shub and S. Smale. Complexity of Bézout's Theorem IV: probability of success; extensions. *SIAM Journal on Numerical Analysis*, 33:128–148, 1996.

197. J.W. Silverstein. The smallest eigenvalue of a large-dimensional Wishart matrix. *Annals of Probability*, 13(4):1364–1368, 1985.

198. R.D. Skeel. Scaling for numerical stability in Gaussian elimination. *Journal of the ACM*, 26:494–526, 1979.

199. S. Smale. The fundamental theorem of algebra and complexity theory. *Bulletin of the American Mathematical Society*, 4:1–36, 1981.

200. S. Smale. Newton's method estimates from data at one point. In, R. Ewing, K. Gross, and C. Martin, editors, *The Merging of Disciplines: New Directions in Pure, Applied, and Computational Mathematics*. Springer, Berlin, 1986.

201. S. Smale. Complexity theory and numerical analysis. In A. Iserles, editor, *Acta Numerica*, pages 523–551. Cambridge University Press, Cambridge, 1997.

202. S. Smale. Mathematical problems for the next century. *The Mathematical Intelligencer*, 20(2):7–15, 1998.

203. R. Solovay and V. Strassen. A fast Monte-Carlo test for primality. *SIAM Journal on Computing*, 6:84–85, 1977.

204. R. Solovay and V. Strassen. Erratum on "A fast Monte-Carlo test for primality". *SIAM Journal on Computing*, 7:118, 1978.

205. A.J. Sommese and C.W. Wampler II. *The Numerical Solution of Systems of Polynomials*. World Scientific, Hackensack, 2005.

206. D.A. Spielman and S.-H. Teng. Smoothed analysis of algorithms: why the simplex algorithm usually takes polynomial time. In *Proceedings of the Thirty-Third Annual ACM Symposium on Theory of Computing*, pages 296–305. ACM, New York, 2001.

207. D.A. Spielman and S.-H. Teng. Smoothed analysis of algorithms. In *Proceedings of the International Congress of Mathematicians*, volume I, pages 597–606, 2002.

208. D.A. Spielman and S.-H. Teng. Smoothed analysis: why the simplex algorithm usually takes polynomial time. *Journal of the ACM*, 51(3):385–463, 2004.

209. M. Spivak. *Calculus on Manifolds. a Modern Approach to Classical Theorems of Advanced Calculus*. W. A. Benjamin, New York, 1965.

210. J. Steiner. Über parallele Flächen. *Monatsber. Preuss. Akad. Wiss.*, 114–118, 1840.

211. G.W. Stewart. On the perturbation of pseudo-inverses, projections and linear least squares problems. *SIAM Review*, 19(4):634–662, 1977.

212. G.W. Stewart. Stochastic perturbation theory. *SIAM Review*, 32(4):579–610, 1990.

213. G.W. Stewart. On the early history of the singular value decomposition. *SIAM Review*, 35(4):551–566, 1993.

214. G.W. Stewart and J.-G. Sun. *Matrix Perturbation Theory, Computer Science and Scientific Computing*. Academic Press, Boston, 1990.

215. G. Stolzenberg. *Volumes, Limits, and Extensions of Analytic Varieties*, volume 19 of *Lecture Notes in Mathematics*. Springer, Berlin, 1966.

216. T. Tao and V. Vu. Inverse Littlewood-Offord theorems and the condition number of random discrete matrices. *Annals of Mathematics. Second Series*, 169(2):595–632, 2009.

217. T. Tao and V. Vu. Smooth analysis of the condition number and the least singular value. *Mathematics of Computation*, 79(272):2333–2352, 2010.

218. J.A. Thorpe. Elementary topics in differential geometry. In *Undergraduate Texts in Mathematics*. Springer, New York, 1994. Corrected reprint of the 1979 original.

219. L.N. Trefethen and D. Bau III. *Numerical Linear Algebra*. SIAM, Philadelphia, 1997.

220. L.N. Trefethen and R.S. Schreiber. Average-case stability of Gaussian elimination. *SIAM Journal on Matrix Analysis and Applications*, 11:335–360, 1990.

221. A.M. Turing. Rounding-off errors in matrix processes. *Quarterly Journal of Mechanics and Applied Mathematics*, 1:287–308, 1948.

222. B.L. van der Waerden. *Modern Algebra. Vol. II*. Frederick Ungar, New York, 1950. Translated from the second revised German edition by Theodore J. Benac.

223. S.A. Vavasis and Y. Ye. Condition numbers for polyhedra with real number data. *Operations Research Letters*, 17:209–214, 1995.

224. S.A. Vavasis and Y. Ye. A primal-dual interior point method whose running time depends only on the constraint matrix. *Mathematical Programming*, 74:79–120, 1996.

225. D. Viswanath and L.N. Trefethen. Condition numbers of random triangular matrices. *SIAM Journal on Matrix Analysis and Applications*, 19:564–581, 1998.

226. J. von Neumann and H.H. Goldstine. Numerical inverting matrices of high order. *Bulletin of the American Mathematical Society*, 53:1021–1099, 1947.

227. P.-Å. Wedin. Perturbation theory for pseudo-inverses. *BIT*, 13:217–232, 1973.

228. N. Weiss, G.W. Wasilkowski, H. Woźniakowski, and M. Shub. Average condition number for solving linear equations. *Linear Algebra and Its Applications*, 83:79–102, 1986.

229. J.G. Wendel. A problem in geometric probability. *Mathematica Scandinavica*, 11:109–111, 1962.

230. H. Weyl. Das asymptotische Verteilungsgesetz der Eigenwerte linearer partieller Differentialgleichungen (mit einer Anwendung auf die Theorie der Hohlraumstrahlung). *Mathematische Annalen*, 71(4):441–479, 1912.

231. H. Weyl. *The Theory of Groups and Quantum Mechanics*. Dover, New York, 1932.

232. H. Weyl. On the volume of tubes. *American Journal of Mathematics*, 61(2):461–472, 1939.

233. E. Wigner. Random matrices in physics. *SIAM Review*, 9:1–23, 1967.

234. J.H. Wilkinson. Error analysis of direct methods of matrix inversion. *Journal of the Association for Computing Machinery*, 8:281–330, 1961.

235. J.H. Wilkinson. *Rounding Errors in Algebraic Processes*. Prentice Hall, New York, 1963.

236. J.H. Wilkinson. *The Algebraic Eigenvalue Problem*. Clarendon Press, Oxford, 1965.

237. J.H. Wilkinson. Modern error analysis. *SIAM Review*, 13:548–568, 1971.

238. J.H. Wilkinson. Note on matrices with a very ill-conditioned eigenproblem. *Numerische Mathematik*, 19:176–178, 1972.

239. J. Wishart. The generalized product moment distribution in samples from a normal multivariate population. *Biometrika*, 20A(272):32–43, 1928.

240. R. Wongkew. Volumes of tubular neighbourhoods of real algebraic varieties. *Pacific Journal of Mathematics*, 159(1):177–184, 1993.

241. H. Woźniakowski. Numerical stability for solving nonlinear equations. *Numerische Mathematik*, 27(4):373–390, 1976/77.

242. S. Wright. *Primal-Dual Interior-Point Methods*. SIAM, Philadelphia, 1997.

243. M.H. Wright. The interior-point revolution in optimization: history, recent developments, and

lasting consequences. *Bulletin, New Series, of the American Mathematical Society*, 42(1):39–56, 2005.

244. M. Wschebor. Smoothed analysis of $\kappa(A)$. *Journal of Complexity*, 20(1):97–107, 2004.

245. Y. Ye. Toward probabilistic analysis of interior-point algorithms for linear programming. *Mathematics of Operations Research*, 19:38–52, 1994.

246. T.J. Ypma. Historical development of the Newton-Raphson method. *SIAM Review*, 37(4):531–551, 1995.

Notation ...

P. Bürgisser, F. Cucker, *Condition*,
Grundlehren der mathematischen Wissenschaften 349,
DOI 10.1007/978-3-642-38896-5, © Springer-Verlag Berlin Heidelberg 2013

...Concepts...

P. Bürgisser, F. Cucker, *Condition*,
Grundlehren der mathematischen Wissenschaften 349,
DOI 10.1007/978-3-642-38896-5, © Springer-Verlag Berlin Heidelberg 2013

...and the People Who Crafted Them

P. Bürgisser, F. Cucker, *Condition*,
Grundlehren der mathematischen Wissenschaften 349,
DOI 10.1007/978-3-642-38896-5, © Springer-Verlag Berlin Heidelberg 2013